Non-Homogeneous Markov Chains and Systems

Non-Homogeneous Markov Chains and Systems: Theory and Applications fulfills two principal goals. It is devoted to the study of non-homogeneous Markov chains in the first part, and to the evolution of the theory and applications of non-homogeneous Markov systems (populations) in the second. The book is self-contained, requiring a moderate background in basic probability theory and linear algebra, common to most undergraduate programs in mathematics, statistics, and applied probability. There are some advanced parts, which need measure theory and other advanced mathematics, but the readers are alerted to these so they may focus on the basic results.

Features

- A broad and accessible overview of non-homogeneous Markov chains and systems
- Fills a significant gap in the current literature
- A good balance of theory and applications, with advanced mathematical details separated from the main results
- Many illustrative examples of potential applications from a variety of fields
- Suitable for use as a course text for postgraduate students of applied probability, or for self-study
- Potential applications included could lead to other quantitative areas

The book is primarily aimed at postgraduate students, researchers, and practitioners in applied probability and statistics, and the presentation has been planned and structured in a way to provide flexibility in topic selection so that the text can be adapted to meet the demands of different course outlines. The text could be used to teach a course to students studying applied probability at a postgraduate level or for self-study. It includes many illustrative examples of potential applications in order to be useful to researchers from a variety of fields.

P.-C.G. Vassiliou is an Honorary Professor in the Department of Statistical Sciences, University College London. He has been affiliated with UCL for almost a decade.

Non-Homogeneous Markov Chains and Systems

Theory and Applications

P.-C.G. Vassiliou
University College London, UK

CRC Press
Taylor & Francis Group
Boca Raton London New York

CRC Press is an imprint of the
Taylor & Francis Group, an **informa** business
A CHAPMAN & HALL BOOK

Front cover image: Photo of pond lilies in the garden of author's house, created by author's wife, Fevronia Vassiliou. See Section 14.2.

First edition published 2023
by CRC Press
6000 Broken Sound Parkway NW, Suite 300, Boca Raton, FL 33487-2742

and by CRC Press
4 Park Square, Milton Park, Abingdon, Oxon, OX14 4RN

Library of Congress Cataloging-in-Publication Data

Names: Vassiliou, P-C. G. (Panos-C. G.), author.
Title: Non-homogeneous Markov chains and systems : theory and applications / P.-C.G. Vassiliou, University College London, UK.
Description: First edition. | Boca Raton : C&H/CRC Press, 2023. | Includes bibliographical references and index.
Identifiers: LCCN 2022023854 (print) | LCCN 2022023855 (ebook) | ISBN 9781138034525 (hardback) | ISBN 9781032378046 (paperback) | ISBN 9781315269986 (ebook)
Subjects: LCSH: Markov processes.
Classification: LCC QA274.7 .V37 2023 (print) | LCC QA274.7 (ebook) | DDC 519.2/33--dc23/eng20221006
LC record available at https://lccn.loc.gov/2022023854
LC ebook record available at https://lccn.loc.gov/2022023855

ISBN: 978-1-138-03452-5 (hbk)
ISBN: 978-1-032-37804-6 (pbk)
ISBN: 978-1-315-26998-6 (ebk)

DOI: 10.1201/b23204

Typeset in Computer Modern Font
by KnowledgeWorks Global Ltd.

Publisher's note: This book has been prepared from camera-ready copy provided by the authors.

*To my daughter **Olympia**, the bravest fighter in life-threatening situations I have ever known. To my daughter **Aglaia**, the greatest supporter and joy to all of us. To my wife **Fevronia**, the most patient and reliable person in my life. To my grandsons **Alexander** and **Philip**, the greatest sources of joy and love.*

Contents

Preface

Are prefaces important? Are there any rules for writing them? Are prefaces idiosyncratic and individual?

It is the belief of many authors and readers that they are important indeed for many various reasons. We will only refer to a few of these reasons. The preface for a reader is like visiting the author's house or mental castle for the first time. The reader is hesitant of what kind of treatment he will receive, what kind of individual he will meet and which idiosyncrasy he is going to unfold. For the author, on the other hand, it is like having someone visiting his mental images for the first time, and he has to be kind while providing explanations of the role and contents of the book; be encouraging and maybe apologetic about some of the mathematical armor and barriers present, which may well discourage many readers; provide some motives by creating some intuition about the potential applicability of the content of the book, and many more. Certainly there are no rules for writing them. It is almost entirely how the author chooses to introduce himself and the context of the book to a first-time visitor. From the above, it is more than apparent that they are idiosyncratic and individual.

This book is devoted to the study of non-homogeneous Markov chains and processes in the first part and to the evolution of the theory of non-homogeneous Markov systems (NHMS) or populations in discrete and continuous time and in discrete and general state spaces. Homogeneous Markov chains have been introduced by Markov in 1906, and his motive was to show that independence is not necessary for the law of large numbers. As an example, he used the alternation of consonants and vowels in Puskin's *Eugene Onegin* that is a two-state Markov chain. Markov models were already around before Markov systematically studied the sequence of random variables having the Markov property. Almost at the same time, Poincare studied Markov chains on finite groups with applications to card shuffling. Later, the Austrian physicists Paul and Tatiana Ehrenfest proposed in 1907 a Markov chain model which clarified the controversial issue of thermodynamical irreversibility. However, the rigorous theory of Markov attracted the attention of the "big" probabilists such as Kolmogorov, Doeblin, Frechet and others. The outcome was a sound and compact theory ready for applications. This path of creation of homogeneous Markov chains was followed up to now by a myriad applications. Actually, the list of applications of Markov chains is virtually countably infinite, and one is perhaps allowed to say that it is not a simple approximation of the course of nature but rather a way randomness in natural phenomena evolves. As it might be apparent, homogeneous Markov chains and processes are very special cases of non-homogeneous Markov chains and processes with which we naturally go up a few steps in the ladder toward reality. As it will be shown later, a non-homogeneous Markov chain is a special case of an NHMS, and certainly a very special case of an NHMS is a homogeneous Markov chain.

While I hope that the book will be of much interest to researchers in the field, the presentation has been planned and structured in a way to provide flexibility in topic selection so that the text can be easily adapted to meet the demands of different course outlines without suffering breaks in continuity. The text accommodates a variety of presentation levels by allowing instructors to select sections, discussions, and examples of appropriate sophistication. I tend to believe that the book would be a primary for students studying applied probability at graduate levels (BSc (Hons) in a four-year degree, MSc and PhD). Such programmes are in applied mathematics, statistics, operations research, engineering science, mathematical economics, computer science and physics. For a general course on applied probability, it would be a supplementary text, but if the lecturers/professors wish to concentrate on the topics of the book, it would be an ideal text. The nature of the illustrative potential applications presented could provide intuition for applications in many fields.

I made an effort to root the results and the text in history and to say things (sometimes very human things that are not always complimentary) about the lives of the people who contributed very greatly to the development of the subject. Omitting to look at the history deprives students and researchers of an important part of their mathematical heritage and deprives them from experience which could be gained without pain. With historical notes, you quickly realize that many sources provided are an interpretation of the facts rather than a statement of the facts. I made an effort not to include too much and I also tried to corroborate only "facts."

The book is organized as follows and from the following description emerge various approaches to its reading.

A serious effort has been made to make the book as self-contained as possible. In this respect, the first chapter, Foundations of Probability Theory, attempts to be a preview of the founding results of probability theory, which will be useful to some extent to what follows. The chapter attempts to introduce the reader in a soft way to measure theory and its strong link with probability theory. Depending on the reader's background and education, this may well be a mountain to climb. However, we suggest that in such a case only a superficial reading is sufficient to provide only the flavor of the results. No deeper understanding or knowledge is necessary, and in the sections that this is needed, the reader or the lecturer will be warned to avoid them if he wishes without damage to the continuity of the basic results. Many of the results included are standards in many books and so no particular credit is provided for some of them, although many standard texts are cited which elaborate more on them.

The second chapter, A Small Review of Matrix Analysis, includes very useful results from the theory of matrices for the present book. However, it is very useful as a founding ground for any course in applied mathematics, applied probability, mathematical statistics and stochastic finance. Again, many of the results included are standards in many books and so no particular credit is provided for some of them although many standard texts are cited which elaborate more on them.

Chapter 3, Non-Homogeneous Markov Chains; Weak Ergodicity, starts the study of weak ergodicity for non-homogeneous Markov chains in discrete time and space. After a short categorization of stochastic processes in terms of state space and time, we proceed to define the Markov property and the class of stochastic processes that processes it, that is, the Markov chains which are in discrete time and the Markov processes which are in continuous time. A more advanced definition of a non-homogeneous Markov chain follows in Section 3.3.1. This could be omitted if the reader/lecturer desires to avoid measure theory. A tribute to the life and work of A.A. Markov follows. In the section that follows, the well-known Chapman-Kolmogorov equations are founded. In Section 3.6.1, a more advanced result on the relation between non-homogeneous Markov chains and martingales is proved. Again, some use of measure theory is used and the readers that find this an obstacle might only read the final results without proofs. The next section exploits various coefficients of ergodicity and studies the relations among them and characterizes various sets of stochastic matrices using various properties of coefficients of ergodicity. The book then evolves by studying the weak ergodicity under various conditions of non-homogeneous Markov chains with general products of stochastic matrices, that is, all possible products from a given sequence of transition probability matrices. The chapter is concluded by studying conditions for weak ergodicity for the case where the transition probability matrices are given in a chronological order. Also, a set of examples are given illustrating weak ergodicity for the various cases.

In Chapter 4, Non-Homogeneous Markov Chains; Strong Ergodicity, the book evolves with founding criteria for strong ergodicity for non-homogeneous Markov chains when the transition probability sequence is given in a chronological order. It then proceeds to study convergence in the Cesaro sense. The chapter proceeds with the study of strong ergodicity with the use of mean visit times. Then the study of strong ergodicity for general products of matrices follows, and criteria are provided with conditions that are easily checked by the practitioner. This is followed by proving criteria for strong ergodicity for sets of transition probability matrices, all non-homogeneous products of which converge. An interesting section follows where a geometric approach to ergodic non-homogeneous Markov chains is presented. Geometric approaches usually have a magic atmosphere, which is rather unique for certain readers. We conclude the study of strong ergodicity by studying the case were there is chronological order in the sequence of transition probabilities but the stochastic matrices are of arbitrary type. Illustrative examples are provided at the end of the chapter.

In Chapter 5, The Non-Homogeneous Markov System, the definition of an NHMS in discrete space and time is provided, firstly for an expanding population and secondly for a fluctuating one but without forced wastage. The study evolves by evaluating the expected and relative expected population structure as functions of what are called the basic parameters of the system. That is, parameters that uniquely determine an NHMS and are estimated from available data on the phenomenon under study. In the section that follows, a synopsis of areas of applications of NHMS is discussed, illustrating the breadth of applications and little of the reasons why the theoretical results that will follow are central for the study of the population at hand. These areas include manpower planning in corporate organizations such as in hierarchical manpower systems, sociology and labor markets, and military establishments. Also, ecological modeling for various ecological systems, healthcare systems specializing in various health problems for patients, fish populations restricted in fisheries and being interested in problems of optimal harvesting and cost. Then non-homogeneous Markov jump systems, which attracted much attention for their practical applications and are closely related with NHMS. There is also immense literature on biological Markov population models, which are closely related, or potentially related, or partially related, to the present NHMS theory. There is also a large amount of interesting literature on open Markov populations, which actually are special cases of the process NHMS and hence the results in the present book are directly applicable. The eighth area of application entitled social physics and statistical mechanics collects open Markov population models in discrete space and continuous time closely related with statistical physics. Social physics, biological physics and physics of particles are related probably through the strong concepts of entropy, stability and energy, which exists in all kind of populations. Finally, we conclude the areas of applications with non-homogeneous semi-Markov systems, which has been a very successful extension of the NHMS theory. Sections 5.5 and 5.6 contain the more general definition of $\mathcal{G}$-non-homogeneous Markov systems and change of measure in an NHMS and uses some measure theory with some insight. For those unfamiliar with measure theory, the reader/lecturer could omit these sections or read them only superficially. Section 5.7 studies the space or random population structures as a Hilbert space. The mathematical level is rather advanced and not usual in applied probability, but I believe it could be accessible. We conclude the chapter with a section on the estimation of the transition probabilities of an NHMS.

In Chapter 6, Asymptotic Behavior of a Non-Homogeneous Markov System, the asymptotic behavior of the expected population structure and the relative expected population structure are studied for an expanding finite NHMS. Conditions for the existence of an asymptotic distribution are founded and it is expressed in a closed analytic form as a function of the basic parameters of the population. Then the case for a contracting NHMS is studied and the same questions are answered. In the next section, a basic theorem is provided for the asymptotic behavior of an NHMS when the input flow into the population is a non-homogeneous Poisson process. In Section 6.6, the concept of asymptotic stability, which is of central importance in theoretical and applied studies of populations, is studied in detail. Section 6.7 studies *$\mathcal{G}-$non-homogeneous Markov systems* with input flow of a non-homogeneous Poisson in the population, and a basic theorem is proved providing conditions when the asymptotically compensated population structure is a multidimensional martingale. Again, this section makes use of measure theory and could be treated as the previous ones like it. The chapter is concluded with the section of research notes useful to researchers or practitioners looking for a more complete picture of the available results.

In Chapter 7, Asymptotic Variability of NHMS, the next logical, important question is studied; that is, if there are conditions under which the variances and covariances of the state sizes of the expected population structures converge. We call this the asymptotic variability problem for an NHMS. A new type of matrices, the **V** matrices, are introduced and their properties are studied. Then a basic theorem is proved, where when the transition probability matrices of the inherent non-homogeneous Markov chain consists of arbitrary stochastic matrices, then under certain conditions the vector of means, variances and covariances of the state sizes converge and the limit is found in closed analytic form as a function of the basic parameters of the system. An illustrative probable application is given in Section 7.3. Then the book evolves into the next section where the rate of convergence in the previous theorem is studied. In particular, a theorem is founded were conditions are provided under which the convergence is at a geometric rate. Geometrical rate of convergence is very important for practitioners and for approximating non-equilibrium analytical solutions that are difficult to be found. The chapter concludes with a section of research notes.

In Chapter 8, Cyclic Behavior of Non-Homogeneous Markov Systems (Cycl-NHMS) are introduced, the basic parameters of which are undergoing a cyclic behavior. A large part of these processes are applied in ecological phenomena where the four epochs naturally create a cycle in the random variables involved. The central problem studied is the asymptotic behavior of the expected population structure, the relative expected population structure of the Cycl-NHMS, and the behavior of their asymptotic variability. A series of basic theorems are proved where under various conditions there exist a stationary distribution for the expected relative population structure and also the variability vector. Also, the important questions, under what conditions the rate of convergence of the expected population structure and the variability vector is geometrical, are answered in the form of theorems. An illustrative example from a manpower system follows. We conclude the chapter with a section of research notes on the central problem.

Chapter 9, Stochastic Control in NHMS, is an important class of problems in various populations concerned with the control of the expected population structure or the control of the relative expected population structure. Both structures are crucially important for the wellbeing and the constructive productivity in whatever way these are defined as part of these systems. In Section 9.2, we study the problem of maintainability of a desired expected relative population structure by controlling the input flow into the population. A series of theorems is provided and it is proved that the maintainable sets of expected relative structures are convex sets with vertices, which are functions of the transition probability matrices among the internal states and the rate of increase of the total memberships in the system. Then the theory evolves by solving the next logical problem, that is, how to attain a desirable expected population structure, which is also maintainable. An algorithm is provided for solving this problem in an optimal way, which is a mixture of dynamic and integer programming techniques. This algorithm works for any cost function that we wish to optimize. In Section 9.4, we formulate and study a new aspect of the classical problem of the asymptotic behavior for an NHMS. Our interest is transferred in finding the set of all possible limiting expected relative population structures provided that we control the limiting vector of input probabilities. A basic theorem is proved in which the set of all possible limiting expected relative population structures is given as the convex set with vertices being functions of the limiting transition probability matrix among the internal states. In Section 9.5, we relax basic assumptions of the two basic theorems of the previous section by allowing the inherent non-homogeneous Markov chain to converge to a periodic homogeneous Markov chain. We provide a series of theorems that address the problems posed in Section 9.4 for the periodic case. Illustrative examples are provided that help significantly to clarify and explain the results and which will be useful to the practitioner. In Section 9.7, we study the problems posed in the previous two sections for the asymptotic variability in NHMS, that is for the limiting vector of means variances and covariances. In Section 9.8, we modify one of the basic assumptions, that is, of assuming that the sequence of transition probabilities are given in chronological order. We generalize by assuming that at each step in time a transition probability matrix is chosen from a finite set of matrices by a compromise non-homogeneous Markov chain, and we call this process non-homogeneous Markov systems in a stochastic environment (S-NHMS). A series of theorems is proved, which evaluates the expected population structure as a function of the basic parameters of the process S-NHMS. Then the chapter evolves by studying the maintainability aspect in a stochastic environment. Next we study strategies for attaining a desirable maintainable structure in an optimal way optimizing a cost function. An algorithm is provided to this respect and an illustrative potential application is described for the practitioner. We conclude the chapter with a section of important research notes.

In Chapter 10, Laws of Large Numbers for Non-Homogeneous Markov Systems, laws of large numbers and central limit theorems for homogeneous Markov chains were known from the time of Markov himself. It is proved that for an NHMS under certain conditions the fraction of time that a membership is in a certain state, asymptotically converges in the mean square to the limit of the expected relative population structure of memberships in that state. Then it is proved under what conditions the convergence is almost sure. In Section 10.5, laws of large numbers are proved for the important class of stochastic processes Cycl-NHMS. It is proved in two theorems under what conditions the expected relative population structure of a Cycl-NHMS asymptotically converges in the strongly Cesaro sense. It is proved that for a Cycl-NHMS under certain conditions the fraction of time that a membership is in a certain state, it asymptotically converges in the mean square to the limit of the expected relative population structure in the strongly Cesaro sense of memberships in that state. We then proceed to establish a founding theorem, which provides the conditions under which the expected relative population structure asymptotically converges in the strongly Cesaro sense with geometrical rate. This theorem is the basic instrument missing to prove under what conditions the law of large numbers for a Cycl-NHMS is almost surely with the mode of convergence. In Section 10.6, we provide applications of the present results in Section 10.3 to geriatric and stroke patients. Also, we provide applications of the results in Sections 10.4 and 10.5 for the movements of students in a university system. In Section 10.7, we study laws of large numbers for NHMS, the inherent non-homogeneous Markov chain of which the sequence of transition probabilities become arbitrary stochastic matrices. We conclude the chapter with an illustrative example useful for practitioners.

In Chapter 11, The Non-Homogeneous Markov System in a Stochastic Environment, we turn to the development of an NHMS in continuous time. Firstly, we present the study of non-homogeneous Markov chains in continuous time and some useful results on their ergodicity. We will then proceed to study S-NHMS in continuous time and solving problems of asymptotic behavior of expected population structures and the set of all asymptotically attainable expected relative population structures when we control the limiting input probabilities. Apparently, the study of NHMS in continuous time is a special case of the above.

An illustrative potential application useful for the practitioner and useful research notes conclude the chapter.

In Chapter 12, The Perturbed Non-Homogeneous Markov System, we relax the basic assumption of the previous chapters when studying the asymptotic behavior of NHMS. That is, we relax the assumption that the inherent non-homogeneous Markov chain is asymptotically homogeneous. We start by studying what happens to the limit of a transition probability matrix of a homogeneous Markov chain if we perturb it. We prove that the new limit is given with the help of the group inverse of a certain matrix. We then proceed to define the perturbed non-homogeneous Markov system (P-NHMS), where the perturbation of all sequences of matrices and vectors is being stochastically done through a selection of perturbations from a finite set of perturbations. The asymptotic behavior of the expected population structure of the P-NHMS is then studied in detail in a series of theorems. Also, the sensitivity of the asymptotic relative expected population structure of an NHMS in changes in the limiting input probabilities is being founded in a series of theorems. The chapter concludes with the study of the asymptotic variability of the P-NHMS and a section of research notes on the subject of the chapter.

In Chapter 13, Non-Homogeneous Markov Set System (NHMSS), a novel idea is introduced for an NHMS, that is, the values of the basic parameters are viewed as being contained in intervals. After the definition of the tight interval and the lemma that a tight interval is a convex polytope, then the chapter proceeds in the formal definition of an NHMSS. In Section 13.3, after a few useful lemmas, we prove that under certain conditions the set of the expected relative population structures of NHMSS is convex. We then prove that at each time step this set is getting smaller and as $t \to \infty$ it converges to a non-empty convex set. In Section 13.4, we study the asymptotic behavior of NHMSS. It is proved that an NHMSS starting with two different sets of initial structures and allocation probabilities under certain conditions easily met in practice and asymptotically converge to the same limiting set geometrically fast. Then it is proved that for an NHMSS if the sets of initial structures and allocation probabilities are compact and convex then under certain easily met conditions the limiting set of the expected relative population structures exist and the convergence is geometrically fast. In Section 13.5, we study important properties of the limiting set of the expected relative population structures. The chapter is concluded with the presentation of an illustrative potential application where all the possible difficulties in applying the present results are being resolved in two novel lemmas and an Algorithm which lead step by step to the desired results.

In Chapter 14, Markov Systems on a General State Space (MSGS), now the state space is a general space. That is, the population lives on space $(X, \mathcal{B}(X))$, where X is an abstract space and $\mathcal{B}(X)$ a countably generated $\sigma-$field on X. When X is topological, then X is equipped with a locally compact, separable, metrizable topology with $\mathcal{B}(X)$ as the Borel $\sigma-$field, but otherwise may be arbitrary. The chapter uses measure theory and advance hard mathematics and could be read superficially in a first reading. In Section 14.3, the foundation of a Markov system on a general state space is presented. In the section that follows, the ergodicity of MSGS is studied, that is, it is founded under what conditions the limiting expected and relative population structure exists and finding them in closed analytic forms. The chapter evolves by studying a coupling theorem for a MSGS. Also, a theorem is proved by studying under what conditions the rate of convergence of expected population structure to its limit is geometrical. In Section 14.7, asymptotic periodicity for a MSGS is studied, that is, the periodicity in the limit of the expected population structure as t goes to infinity. The chapter is concluded with two basic theorems studying the total variability from the invariant measure for a MSGS.

In Chapter 15, The $\mathcal{G}$-Non-Homogeneous Markov System of High Order, the process $\mathcal{G}$-NHMS introduced in Chapter 5 is studied when the order of the inherent Markov chain and other processes are of order $m > 1$. The chapter uses measure theory and could be read superficially in a first reading. In Section 15.3, the expected population structure is studied while in Section 15.4, the inhomogeneous mixture approximation distribution model is introduced to deal with the problem of large numbers of parameters to be estimated which increase geometrically as the number of states increase. In Section 15.5, the asymptotic behavior of the inherent Markov chain is founded in the form of a theorem, and in the next section the existence of the asymptotic behavior of the expected population structure is proved and found in close analytic form. Finally, an illustrative example from manpower systems is given and the chapter concludes with some research notes on its subject.

Acknowledgments

This book evolved over a period of five and a half difficult years, through many tough different phases, and hence there are many people I wish to thank. I would like firstly to thank Prof. Sir David Cox who first insisted that I take on the present project. He also continued to encourage me for a few decades until I started working on it. Unfortunately, he "left" a couple of months before the book was submitted to print. I would also especially like to thank Prof. Valerie Isham for her continuous support during this long endeavor. If it was not for her I probably would not have started the book, let alone finish it. I would also like to thank the Department of Statistical Sciences, University College London, with which I have been affiliated for almost a decade, for supporting my research in many ways.

I owe special debt to the five reviewers of my proposal for the book to Chapman & Hall/CRC Press for their stunning welcoming reviews for the project and their flattering comments for my work. I confess that whenever I was almost ready to quit the project, I read their reviews again and kept going. I only hope that a large portion of their expectations have been met. Also, my sincere thanks go to Mr. Rob Calver, senior publisher at Chapman & Hall, for mentoring my project with real interest.

There are many people from the medical world who I would like to deeply thank for supporting us as a family. However, I would like to especially mention Dr. Alison Jones, Mr. Christian Brown and Prof. Heather Payne, who I feel are friends of our family by now and without their kind interest and support this project would have never come to a conclusion.

Finally, I would like to deeply thank my family: Olympia my daughter, who is the greatest fighter in life-threatening situations that I have ever known; Aglaia, my daughter, a great support for all of us; Alexander and Philip, my grandsons, a great source for joy and love; and, last but not least, my wife Febronia, whose patience and ability to bear with me made the completion of the book possible.

Prof. Panos Vassiliou
March 2022
London

Chapter 1

FOUNDATIONS OF PROBABILITY THEORY

1.1 Introductory notes

Modern Probability has a face with two different sides. One side is the closest possible to be a branch of Pure Mathematics, derived from certain axioms as classical areas of Algebra does. The addition rule dictates the need to measure length, area and volume of general sets, that cannot be dealt by elementary calculus. This fact makes Probability part of a subject called Measure Theory. Measure theory is about *measurable sets.* An *event* in probability is a measurable set. In addition, Probability enriches with the multiplication rule, which describes independence, and links closely to the *product measure* in Measure theory. However the relation is a two sided one, since Probability theory enlivens and enriches things in a stunning way by dealing with lots of σ-algebras, not just the one σ-algebra which is the concern of Measure Theory.

On the other side, Probability is an Applied Mathematics (although the usual distinction between Pure and Applied mathematics, I believe, is more or less unsuccessful) discipline most commonly known as Applied Probability. This side without being less demanding mathematically or stochastically, creates very useful models for everyday phenomenon in the real world. Markov processes, Brownian motions, Levy processes and other processes have very important influence on important decisions in the real world. On this side one would strongly argue that lies the foundation of Probability via Expectation, which is done in the natural way, that mathematics enter in Natural Sciences. An excellent book on the subject of Probability via Expectation, and one it should be in books collection of every stochastically minded scientist is that by Whittle (2000). Moreover, but equally importantly, there seems to be a two way influence. That is, all too often real applications provide strong motives to the foundations of the stochastic world. Such an example is the area of stochastic finance, which provided a strong motive for the evolution of Brownian motion and more generally of stochastic differential equations. Intuitively I believe that we actually live in a much more stochastic world than we really anticipate. Certainly, I am not referring only to the evolution of life of an individual where this is apparent. A little study of nowadays beliefs of Theoretical Physics for the structure of an atom, and of the stochastic evolution in Genetics, will provide easily the reader with strong evidence toward this end.

There is an unavoidable dilemma confronting the authors of a chapter on Foundations of Probability for a book of a certain advanced stochastic level. There are more reasons than merely rigor that the reader should acquire some knowledge of Measure theory. This is a hard subject; unfortunately, as in other branches of Mathematics, it has to come first and not later as the book evolves. This fact engenders the risk, that even readers that have been exposed to a gentle introduction to Probability, will be discouraged to proceed reading the chapters that follow. Be aware though that the results of Measure theory are not difficult to understand. The solution I am suggesting is to read the chapter even superficially at least

DOI: 10.1201/b23204-1

once, and come back from time to time as needs arise, and as the level of your stochastic foundation grows. In this way also your intuition on the subject has something to gain from the various remarks and comments. The present *textbook is meant among other goals to teach and to be as much as possible self-contained.* In this respect, we will always try to sharpen the reader's intuition by intuitive explanations for many entities, although they will sometimes look nebulous. I almost agree with Williams (2004) since I believe that: "Intuition is as much important at least as rigor", and we do need to know how to back up intuition with rigor. It is exactly for that reason that I wish I had his book about forty years ago, when I first discovered my love for stochastic thinking. In addition, I intuitively believe that there is some rewards for our intuition, if we poke at the roots of Probability.

With these thoughts in mind, the present chapter was written to provide the reader with a review, that will be helpful, for those who are coming to live at this level of stochastic thinking for the first time and a rather important refresher for others. For those wishing to expand, further there are excellent references for that, and I only mention a few of my favorites, such as: Doob (1990), Kallenberg (2002), Grimmet and Stirzaker (2001), Durrett (1996), Billingsley (1965,1968), Williams (2000, 2004).

1.2 Some set theory and topology

The reader should not feel uncomfortable, when she/he reads in this chapter and most probably in others about some areas of mathematics, that she/he didn't even hear that existed before. Almost everything is a matter of taking the time to visualize mentally the new definitions and concepts, and learning the properties that the new entities have in the space they live. It also helps to use in many cases as an example the real line. I am not going to hide the fact that you also need a *readable* book.

1.2.1 Set theory useful in probability

We assume in our small review of sets that the concepts of an element and of a set of elements are intuitively clear. A *set* is a collection of elements, considered as whole. For example, the set of integers which will be denoted by $\mathbb{N} = \{1, 2, ...\}$. Another notation useful for sets is for example the following:

$$\{x : x \text{ is a real number and irrational}\},$$

which means the set of all numbers that are real and cannot be written as the quotient of two integers. Two sets A and B are equal if they consist of exactly the same elements; We say that A is a subset of B if each element of A is also an element of B. We denote this relation by $A \subseteq B$, allowing for the possibility, that B does not contain any other elements than those of A, in which case it is equal with A. We denote by $A \subset B$ when A is a *proper subset* of B, that is, B contains more elements than A. The following properties are palpably true:

$$(a)\ A \subseteq A \text{ for any set } A;$$

$$(b)\ \ A \subseteq B \text{ and } B \subseteq A \ \Leftrightarrow\ A = B;$$

$$(c)\ \ A \subseteq B \text{ and } B \subseteq C \ \Rightarrow\ A \subseteq C;$$

where I believe the meaning of the universal in science symbols $\Leftrightarrow$ and $\Rightarrow$ are obvious. A set which contains no elements will be denoted by $\varnothing$.

Let Ω be a set and let $\mathcal{P}(\Omega)$ be the power set of Ω, i.e., the collection of all subsets of Ω including the empty set $\varnothing$. The *union* of A and B in $\mathcal{P}(\Omega)$ is denoted by $A \cup B$, and all its elements belong at least to A or B, and there is no element of A or B that does not belong to their union. If we have a finite number of sets A_i where $i \in I$ with $I = \{1, 2, ..., n\}$, then their union is denoted by $\cup_{i=1}^{n} A_i$ and its elements belong at least to one of the sets A_i with $i \in I$, and there is no element of A_i with $i \in I$ that does not belong to their union. The definition of the union is easily extended in the case, where the set I is the set of integers denoted by $\mathbb{N}$. In this case, we have a countable collection of sets A_i with $i \in \mathbb{N}$ and their union is denoted by $\cup_{i\in\mathbb{N}} A_i$.

The intersection of A and B in $\mathcal{P}(\Omega)$ is denoted by $A \cap B$ or AB and its elements are the common elements of A and B. The definition of intersection is naturally extended for a finite and a countable number of sets A_i respectively and is denoted by $\cap_{i=1}^{n} A_i$ and $\cap_{i\in\mathbb{N}} A_i$. The complement of A in $\mathcal{P}(\Omega)$ is denoted by A^c and its elements are the elements that belong to Ω and are not elements of A. For A and B in $\mathcal{P}(\Omega)$ their *difference* is defined as $A \backslash B = A \cap B^c$ and the *symmetric difference* by

$$A \triangle B = (A \setminus B) \cup (B \setminus A) = (A \cup B) \setminus (A \cap B) .$$

If $A \subset B$, the *strict difference* $B - A$ is the set of elements belonging to B and not belonging to A. It is not difficult to prove following the usual path for sets, that is, proving first that an element on the left side of the equation is an element also on the right side of the equation and vice versa, that

$$A\,(B \cup C) = AB \cup BC,$$
$$(A \triangle B) \triangle C = A \triangle (B \triangle C) = A \triangle B \triangle C,$$
$$A\,(B \triangle C) = AB \triangle AC.$$

The *distributive laws* are

$$A \cap \cup_k B_k = \cup_k\,(A \cap B_k)\,, \quad A \cup \cap_k B_k = \cap_k\,(A \cup B_k)\,,$$

and *De Morgan's laws are*

$$\{\cup_k A_k\}^c = \cap_k A_k^c, \quad \{\cap_k A_k\}^c = \cup_k A_k^c,$$

valid for arbitrary (not necessarily countable) unions and intersections. De Morgan laws allow us to convert any relations involving unions (intersections) into the dual formula for intersections (unions).

Let Ω again being our superset and $\mathcal{P}(\Omega)$ its power set. Then the class of sets in $\mathcal{P}(\Omega)$ is *partially ordered with the relation* $\subseteq$ because it has the following properties:

(a) For every $A \in\mathcal{P}(\Omega)$, we have $A \subseteq A$ (*reflexivity*);

(b) For $A, B \in\mathcal{P}(\Omega)$ for which $A \subseteq B$ and $B \subseteq A \Rightarrow A = B$ (antisymmetry);

(c) For $A, B, C \in\mathcal{P}(\Omega)$ if $A \subseteq B$ and $B \subseteq C \Rightarrow A \subseteq C$ (transitivity).

More generally any set equipped with a relation among its elements, which satisfies the above properties, will be a partially ordered set. For example, the set of real numbers $\mathbb{R}$ equipped with the relation $\leq$ is a partially ordered set.

In $\mathcal{P}(\Omega)$, it is always possible to find two subsets of Ω neither of which is a subset of the other. In this case, we say that these two sets are not *comparable.* This is why we use the word *partially* in the terminology "partially ordered set". However, in general there are sets for which any two elements are comparable. Such sets are called *totally ordered sets, or linearly ordered sets.*

Let A be a set in the partially ordered set $\mathcal{P}(\Omega)$. A set $B \in\mathcal{P}(\Omega)$ is a lower bound for A if $B \subseteq A$; and a lower bound of A is called a *greatest lower bound* of A, if it is greater than or equal to every lower bound of A. Now, an element $C \in\mathcal{P}(\Omega)$ is said to be an upper bound of A if $A \subseteq C$. An upper bound of A is called a *least upper bound* of A, if it is less than or equal to every upper bound of A. Apparently, in general there may be many lower bounds and many upper bounds. Though, it is easy to see that **if** a least upper bound or a greatest lower bound **exists**, then it is **unique**.

We call the greatest lower bound of a set A, if it exists, its *infimum* and denote it by $\inf A$. Respectively, if A has a least upper bound, we call it its *supremum* and denoted by $\sup A$. Note that when A is a finite set, then its infimum and supremum always exist and are called minimum and maximum respectively.

Let $\{A_i, i \in I\} \in \mathcal{P}(\Omega)$ be a class of subsets of Ω. Then the infimum of the class of sets $\{A_i, i \in I\}$ is the intersection of all its sets, that is:

$$\inf_{i \in I} A_i = \cap_{i \in I} A_i.$$

Respectively, the supremum of the class of sets $\{A_i, i \in I\} \in \mathcal{P}(\Omega)$ is the union of all its sets, that is:

$$\sup_{i \in I} A_i = \cup_{i \in I} A_i.$$

Important insight

⁕ As the reader already suspects all concepts discussed above in what follows will be more enliven, when in fact Ω is the sample space of an experiment of Goddess Tyche ($T\upsilon\chi\eta$=luck).[1] That is, Ω will represent the set of all possible outcomes of the experiment we are interested in. A *possible outcome, or sample point,* ω is mathematically a point of the abstract set Ω. The actual outcome of the experiment of Tyche will be an element of the set Ω, which we will denote by $\hat{\omega}$. The sample point $\hat{\omega}$ will also be called the *realization* of the experiment of Tyche. Apparently, the execution once of an experiment of Tyche results in one possible outcome $\omega \in \Omega$, the sample point or realization $\hat{\omega} \in \Omega$. ⁕

An important experiment for reasons that will be revealed later on, is that of choosing at random a point in the interval $[0, 1]$. In this case, the sample space Ω is actually the interval $[0, 1]$, an uncountable set. A possible outcome ω is any point in the interval $[0, 1]$ and the realization $\hat{\omega}$ will be the point actually chosen.

Another example is the experiment to observe for three consecutive days if it will rain. We use the number 1 to denote that it did not rain and the number 0 if rain has fallen at any time during a day. Then the sample space of the experiment is

$$\Omega = \{111, 110, 101, 011, 001, 010, 100, 000\}.$$

In this case, Ω is a finite set. The reader can easily think of many experiments where Ω is a countable set.

Let us visit again our abstract set Ω, that is, with no special structure, and let $\{A_n\}_{n \in \mathbb{N}}$ be a countable sequence in $\mathcal{P}(\Omega)$. The sequence $\{A_n\}_{n \in \mathbb{N}}$ is said to be increasing, if $A_n \subset A_{n+1}$ for every $n \in \mathbb{N}$, and the union $A = \cup_{n \in \mathbb{N}} A_n$ is called the increasing limit of A_n and is usually denoted by $A = \lim \uparrow A_n$.The sequence $I_n = \cap_{m \geq n} A_m = \inf_{m \geq n} A_m$ is easily seen that is increasing and its increasing limit is denoted by

$$\liminf A_n = \cup_{n \in \mathbb{N}} \cap_{m \geq n} A_m = \lim \uparrow \inf_{m \geq n} A_m. \tag{1.1}$$

The sequence $\{A_n\}_{n \in \mathbb{N}}$ is said to be decreasing, if $A_{n+1} \subset A_n$ for every $n \in \mathbb{N}$, and the intersection $A = \cap_{n \in \mathbb{N}} A_n$ is called the decreasing limit of A_n and is usually denoted by

[1] Tyche: Goddess of destiny in Greek mythology, daughter of the Ocean and Tythis (one of the Titans). She is usually illustrated as a nude young woman blindfolded and holding in her right hand the horn of Amaltheia, throwing goods randomly from it, while travelling fast on a wheel with wings on it. Amaltheia was the goat, which milked Zeous the Head of the Department of Gods. Her horn was a symbol of prosperity and richness.

$A = \lim \downarrow A_n$. The sequence $S_n = \cup_{m \geq n} A_m = \sup A_m$ is easily seen that is decreasing and its decreasing limit is denoted by

$$\limsup A_n = \cap_{n \in \mathbb{N}} \cup_{m \geq n} A_m = \lim \downarrow \sup_{m \geq n} A_m.$$

Let us now get a small taste, how the above results become alive, when Ω is a sample space of an experiment of Tyche. We will provide the following propositions:

Proposition 1 *Let Ω a sample space and let $\{A_n\}_{n \in \mathbb{N}}$ a countable sequence of events, that is, $A_n \in \mathcal{P}(\Omega)$ for every $n \in \mathbb{N}$. Consider the event*

$$E_{i.o.} = \{\omega : \omega \in A_n \text{ for infinitely many } n\} = \{\omega : \omega \in A_n, i.o.\}. \tag{1.2}$$

Then

$$E_{i.o.} = \cap_{n=1}^{\infty} \cup_{k=n}^{\infty} A_k.$$

Proof. We have that $\omega \in A_n$ *i.o. iff* for every positive integer n, there exist $k \geq n$ such that $\omega \in A_k$. That is, iff for every positive integer n, we have $\omega \in \cup_{k=n}^{\infty} A_k$. The last statement is equivalent with: iff $\omega \in \cap_{n=1}^{\infty} \cup_{k=n}^{\infty} A_k$. ■

Similarly the following Proposition could be proved.

Proposition 2 *Let Ω a sample space and let $\{A_n\}_{n \in \mathbb{N}}$ a countable sequence of events, that is, $A_n \in \mathcal{P}(\Omega)$ for every $n \in \mathbb{N}$. Consider the event*

$$E_f = \{\omega : \omega \in A_n \text{ for all but a finite number of indices}\}.$$

Then

$$E_f = \cup_{n=1}^{\infty} \cap_{k=n}^{\infty} A_k.$$

Remark 3 *Note that* $\liminf A_n$ *is not always equal with* $\limsup A_n$. *When* $\liminf A_n = \limsup A_n = A$ *we say that* $\{A_n\}_{n \in \mathbb{N}}$ *has a limit, denoted by* $\lim_{n \to \infty} A_n = A$.

1.2.2 Interesting topological spaces

The word topology is a Greek word and is the synthesis of two words: "$T o\pi o s$" which means space and "$\Lambda o \gamma o s$" which has the general meaning of "logic" (as it does in ancient Greek) or "saying". Hence, topology has the literal meaning as the "logic of a certain space". Where the word "logic" should be taken as having the meaning it also has in the word methodology which is again the synthesis of two Greek words.

A lot of research has been done as to which are the basic underlined elements to found a topological space and its properties. It has been found that there are many different ways to do so, but they are all equivalent. After several years, there seems that most mathematicians have been convinced that the *open set* approach is the most appropriate. In the present subsection, we will only provide some definitions and possible insight together with a few results without proofs. In later chapters, whatever we need we will be built on that. A very readable book in topology is that by Simmons (1963).

On any abstract space X a topology $\mathcal{T}$ is a collection of subsets of X called *open sets* for which the following holds:

(i) arbitrary unions of members of $\mathcal{T}$ belong to $\mathcal{T}$.

(ii) finite intersections of members of $\mathcal{T}$ belong to $\mathcal{T}$.

(iii) The space X and the empty set $\varnothing$ belong to $\mathcal{T}$.

We call $(X, \mathcal{T})$ a *topological space.* On any abstract space, what subsets of X we define as open makes a topology distinct and makes the pair $(X, \mathcal{T})$ a distinct topological space. On a certain set X, we may consider several topologies which will lead to several topological spaces. Though, it is important to note that the null set $\varnothing$ and the set X are open sets in any topology. If we let the open sets to be all possible subsets of X, then this topology is called the *discrete topology* on X, and the corresponding topological space $(X, \mathcal{T})$ is called a *discrete space*. The complements of open sets in X are called *close* sets. Any set $A \in \mathcal{T}$ is called a *neighborhood* for any point $x \in X$ if $x \in A$. Topological spaces are generalizations of metric spaces, which we will discuss in what immediately follows:

Let M a non-empty set and consider a function $d : M \longmapsto \mathbb{R}$ which for any ordered pair of elements in $(x, y) \in M$ satisfies the following three conditions.

(1) $d(x, y) \geq 0$ and $d(x, y) = 0$ if and only if $x = y$;
(2) $d(x, y) = d(y, x)$ (symmetry);
(3) $d(x, y) \leq d(x, z) + d(z, y)$ (the triangle inequality).

The function $d(x, y)$ is called a *metric* in M, the pair (M, d) is called a *metric space* and the function $d(x, y)$ represents the *distance* between the points $x, y \in M$. An example of a metric space is the set of real numbers $(\mathbb{R}, d)$, where as a distance between two real numbers x, y is used the absolute function of their difference, i.e., $d(x, y) = |x - y|$. This is called the *usual metric* on $\mathbb{R}$. Another not so obvious example is if we consider as the set M_f, the set of all bounded continuous real functions defined on the interval $[0, 1]$. That is, for each real function f, there is a real number $w \in \mathbb{R}$ such that $|f(x)| \leq w$ for every $x \in [0, 1]$, and f is continuous on that interval. In this set, we define a function which is actually a norm by

$$\|f\| = \int_0^1 |f(x)|\, dx,$$

and as a metric the function

$$d(f, g) = \|f - g\| = \int_0^1 |f(x) - g(x)|\, dx.$$

Then the pair (M_f, d) is a metric space. This example gives us an idea of the diversity of the different metric spaces that we can have.

In a metric space (X, d), we define as the distance of an element $a \in X$ from a set $B \subseteq X$ by

$$d(a, B) = \inf\{d(a, b) : b \in B\}.$$

The diameter of a set is defined by

$$d(B) = \sup\{d(b_1, b_2) : b_1, b_2 \in B\}.$$

Note that $d(\varnothing) = -\infty$.

We will now define in a metric space (M, d) open sets, which are induced by the metric $d(x, y)$. In this way we will define the induced topology by the metric $d(x, y)$ in the metric space (M, d).

We will start by defining an *open sphere* of an element a of the metric space (M, d).

Let $a \in (M, d)$ then we define the *open sphere* $S_r(a)$ with *center* a and *radius* $r \in \mathbb{R}$ the subset of M defined by

$$S_r(a) = \{x : d(x, a) < r\}.$$

Let us consider as an example the metric space $(\mathbb{R}, |.|)$. The open sphere with center a and radius r is the bounded open interval $(a - r, a + r)$ which has total length $2r$. Apparently, every bounded interval in the real line is an open sphere, since a midpoint and a suitable radius always exist.

A subset B of the metric space (M, d) is called an *open set,* if given any point $b \in B$, there exists a positive real number r such that $S_r(b) \subseteq B$, that is, each point $b \in B$ is the centre of some open sphere contained in B. In any metric space (M, d), the empty set $\varnothing$

and the full space M are open sets, in addition each open sphere is an open set and any subset A of M is open if and only if it is a union of open spheres. It is important to note that it could be proved by the reader as an exercise, that in a metric space (M, d) any union of open sets in M is open; and any finite intersection of open sets in M is open.

Let $(X, \mathcal{T})$ a topological space and assume, that there exists at least one metric, whose class of generated open sets is precisely the given topology. Then the topological $(X, \mathcal{T})$ is called *metrizable.* That is, a topological space $(X, \mathcal{T})$ is said to be metrizable, if there is a metric $d : X \times X \mapsto [0, \infty)$ such that the topology induced by d is $\mathcal{T}$. There are many topological spaces which are not metrizable. However, there is another interesting characterization of a metrizable space, which we will discuss now. Let $(X, \mathcal{T}_X)$ and $(Y, \mathcal{T}_Y)$ be topological spaces and f a mapping of X into Y. Then the mapping f is called a *continuous mapping,* if for every open set $B \in Y$ we have $f^{-1}(B) \subseteq A$ an open set in $(X, \mathcal{T}_X)$. The mapping f is called an *open mapping,* if for every open set $A \in X$ then $f(A) \subseteq Y$ an open set in $(Y, \mathcal{T}_Y)$.

Let $(X, \mathcal{T}_X)$ and $(Y, \mathcal{T}_Y)$ be two topological spaces, then a one to one mapping from $X \mapsto Y$ which is simultaneously a continuous mapping and an open mapping is called a *homeomorphism* and the two topological spaces are called *homeomorphic.* It is apparent, that if $(X, \mathcal{T}_X)$ and $(Y, \mathcal{T}_Y)$ are two homeomorphic topological spaces, there is a one to one correspondence between their open sets. A metrizable space is a topological space that is homeomorphic to a metric space.

Let $(X, \mathcal{T})$ a topological space and $A \subset X$. We define as the *closure* of A denoted by $cl(A)$ the smallest closed set containing A. Equivalently, $cl(A)$ is the intersection of all closed sets containing A. It is easy to see that a set A is closed if and only if $A = cl(A)$. For example in $\mathbb{R}$ with the usual topology, we have that $cl((0,1)) = [0,1]$. Also in $\mathbb{R}$ let $\mathbb{Q}$ be the set of rational numbers, then $cl(\mathbb{Q}) = \mathbb{R}$. Now, consider a sphere in the 3 dimensional Euclidean space $\mathbb{R}^3$. The open sphere is the interior of this ball and the closed sphere is the ball with its surface which is in fact the closure of the open sphere. In general the closure of a set depends on the topology chosen for the underlying space and also upon in which space we are taking the closure. In any topological space $(X, \mathcal{T})$, $\varnothing = cl(\varnothing)$ and $X = cl(X)$. Let $A_1, A_2, ..., A_n$ be subsets of X then

$$cl(A_1 \cup A_2 \cup ... \cup A_n) = cl(A_1) \cup cl(A_2) \cup ... \cup cl(A_n),$$

$$cl(A_1 \cap A_2 \cap ... \cap A_n) \subseteq cl(A_1) \cap cl(A_2) \cap ... \cap cl(A_n).$$

If we have a countable number of sets $A_1, A_2, ...$ subsets of X, then we get that

$$cl(\cup_{n=1}^{\infty} A_n) \supseteq \cup_{n=1}^{\infty} cl(A_n).$$

Let $(X, \mathcal{T})$ a topological space and $A \subset X$ then A is *dense* if $cl(A) = X$. For example, as we have seen above $cl(\mathbb{Q}) = \mathbb{R}$ and so the set $\mathbb{Q}$ is a dense subset of $\mathbb{R}$. A topological space $(X, \mathcal{T})$ is called *separable* if it contains a countable dense subset. Hence, $\mathbb{R}$ is separable. In fact the set of all vectors $(q_1, q_2, ..., q_n) \in \mathbb{R}^n$ with q_i rational for all i is a countable dense subset of $\mathbb{R}^n$, and consequently the n-dimensional Euclidean space is separable. Any finite or countably infinite topological space is separable. Any topological space which is the union of a countable number of separable subspaces is separable. Every open subspace of a separable space is separable (Willard 1970). Also every subspace of a separable metric space is separable.

1.3 Important family of sets in probability theory

Let Ω be a set and again $\mathcal{P}(\Omega)$ the power set of Ω. There are some important families of sets, which are subfamilies of $\mathcal{P}(\Omega)$ and which play a significant role in probability theory.

We will start presenting them from the more primitive and moving to those with more rich properties. However, let us firstly discuss some terminology. Let $\mathcal{L}$ be a subfamily of $\mathcal{P}(\Omega)$ then one says that $\mathcal{L}$ is *stable by* or *closed under* **finite (countable, arbitrary)** *intersection (union),* if the intersection (union) of any finite (countable, arbitrary) collection of elements of $\mathcal{L}$ are in $\mathcal{L}$. Stability by or closeness under complementation, strict difference, etc., is defined in an analogous manner. Finally, $\mathcal{L}$ is said to be *stable by or closed under sequential increasing (decreasing) limit* if the limit of any increasing (decreasing) sequence in $\mathcal{L}$ is still in $\mathcal{L}$. If $\mathcal{L}$ is stable by sequential increasing limit and by sequential decreasing limit, it is said to be *stable by or closed under sequential monotone limit.*

π-system. $\mathcal{L} \subset \mathcal{P}(\Omega)$ is called a *π-system,* if and only if, it is stable by finite intersection.

Semialgebra. $\mathcal{L} \subset \mathcal{P}(\Omega)$ is called a *semialgebra,* if and only if:

(*i*) $\Omega, \varnothing \in \mathcal{L}$,

(*ii*) $\mathcal{L}$ is stable by finite intersection,

(*iii*) if $A \in \mathcal{L}$, then the complement A^c is the union of a finite collection of pairwise disjoint sets of $\mathcal{L}$.

d-system. $\mathcal{L} \subset \mathcal{P}(\Omega)$ is called a *d-system,* if and only if:

(*i*) $\Omega, \varnothing \in \mathcal{L}$,

(*ii*) $\mathcal{L}$ is stable by strict difference,

(*iii*) $\mathcal{L}$ is stable by sequential increasing limit,

Algebra. $\mathcal{A} \subset \mathcal{P}(\Omega)$ is called an *Algebra,* if and only if:

(*i*) $\Omega, \varnothing \in \mathcal{L}$,

(*ii*) $\mathcal{A}$ is stable by complementation,

(*iii*) $\mathcal{A}$ is stable by finite intersection or equivalently is stable by finite union.

σ-algebra. $\mathcal{F} \subset \mathcal{P}(\Omega)$ is called a *σ-algebra,* if and only if:

(*i*) $\Omega, \varnothing \in \mathcal{L}$,

(*ii*) $\mathcal{F}$ is stable by complementation,

(*iii*) $\mathcal{F}$ is stable by countable intersection or equivalently is stable by countable union.

For any class of σ-algebras in Ω, the intersection is again a σ-algebra. The same is not always true for the union of σ-algebras. In fact the reader could show as an exercise the following:

If $\mathcal{F}_1 \subset \mathcal{F}_2 \subset ...$, are σ-algebras in Ω, then

$$\bigcup_i \mathcal{F}_i \quad \text{is a } \sigma\text{-algebra.}$$

If $\mathcal{L} \subset \mathcal{P}(\Omega)$, then the intersection of all σ-algebras in Ω that contain $\mathcal{L}$ is the σ-algebra *generated by* $\mathcal{L}$ and is denoted by $\sigma(\mathcal{L})$. Apparently $\sigma(\mathcal{L})$ is the smallest σ-algebra containing $\mathcal{L}$. If the σ-algebra $\mathcal{F}$ is generated by a countable collection of sets $\{A_n\}_{n=1}^{\infty} \in \mathcal{F}$, then we say that the σ-algebra $\mathcal{F}$ is *countably generated.* Let $\mathcal{F}_1$and $\mathcal{F}_2$ be two σ-algebras then the σ-algebra generated by $\mathcal{F}_1$and $\mathcal{F}_2$ is denoted by $\mathcal{F}_1 \vee \mathcal{F}_2 = \sigma(\mathcal{F}_1 \cup \mathcal{F}_2)$ drawing the notation from the fact that the $\sigma(\mathcal{F}_1 \cup \mathcal{F}_2)$ is the minimum σ-algebra containing $\mathcal{F}_1$and $\mathcal{F}_2$. The same is true for d-systems and algebras, that is, the intersection of d-systems (algebras) is a d-system (algebra). In consequence we denote by $d(\mathcal{L})$ $(\mathcal{A}(\mathcal{L}))$ the d-system (algebra) generated by $\mathcal{L}$ as the intersection of all d-systems (algebras) in Ω that contain $\mathcal{L}$. We will now prove the following proposition:

Proposition 4 *Let $\mathcal{L} \subset \mathcal{P}(\Omega)$ be a d-system and a π-system, then it is a σ-algebra.*

Proof. In order to show that $\mathcal{L}$ is a σ-algebra, we should prove that the following three conditions hold.

(*i*) $\Omega, \varnothing \in \mathcal{L}$

(*ii*) $\mathcal{L}$ is stable by complementation,

(*iii*) $\mathcal{L}$ is stable by countable intersection or equivalently is stable by countable union.

Condition (i) holds since $\mathcal{L}$ is a d-system. In order to prove that condition (ii) holds it is sufficient to show that if $A \in \mathcal{L}$ then $A^c \in \mathcal{L}$. Since $A \in \mathcal{L}$ and $\Omega \in \mathcal{L}$ and $\mathcal{L}$ a d-system, we get that $A^c = \Omega - A \in \mathcal{L}$. Moreover, since $\mathcal{L}$ is a π-system and so it is stable under finite intersection we immediately arrive at the conclusion that $\mathcal{L}$ is an algebra. It remains to show that $\mathcal{L}$ is stable under countable intersection.

We will start by showing that the union of two sets $A_1, A_2 \in \mathcal{L}$ belongs again to $\mathcal{L}$. Since $A_1 \in \mathcal{L}$ we have already showed that $A_1^c, A_2^c \in \mathcal{L}$. Now, $\mathcal{L}$ is a π-system, hence $A_1^c \cap A_2^c \in \mathcal{L}$. Moreover, $\mathcal{L}$ is a d-system thus

$$\Omega - A_1^c \cap A_2^c \in \mathcal{L},$$

and consequently

$$A_1 \cup A_2 = \Omega - A_1^c \cap A_2^c \in \mathcal{L},$$

and the process of thought could continue in the same line to show, that for any finite number of sets $A_1, A_2, ..., A_k \in \mathcal{L}$ their union $A_1 \cup A_2 \cup ... \cup A_k \in \mathcal{L}$.

Let $\{A_n\}_{n=1}^{\infty}$ be an arbitrary countable sequence in $\mathcal{L}$. Then consider the union $\cup_{n=1}^{\infty} A_n$. In order to complete the proof we need to show that $\cup_{n=1}^{\infty} A_n \in \mathcal{L}$. We have that

$$\cup_{n=1}^{\infty} A_n = \cup_{n=1}^{\infty} \cap_{m \geq n} A_m,$$

where $\cap_{m \geq n} A_m$ is increasing and the right hand side of the previous equation is its increasing limit which since $\mathcal{L}$ is a d-system it belongs to $\mathcal{L}$. ■

Remark 5 *The converse, that is, if $\mathcal{L}$ is a collection of subsets of Ω and it is a σ-algebra then it is a d-system and a π-system is trivial.*

With similar arguments we could prove the following proposition:

Proposition 6 *Let $\mathcal{L} \subset \mathcal{P}(\Omega)$ then it is true that $d(\mathcal{L}) \subseteq \sigma(\mathcal{L})$.*

Proof. This is so because $\sigma(\mathcal{L})$ contains, in addition, all countable unions of subsets of $\mathcal{L}$. ■

We continue with one of the fundamental tools of probability theory. The proof could be found in Rogers and Williams (2000, p.87).

Theorem 7 *Let $\mathcal{L}$ be a π-system defined on Ω. Then $d(\mathcal{L}) = \sigma(\mathcal{L})$. In other words the d-system generated by a π-system is the σ-algebra generated by that π-system.*

1.4 Measurable spaces

Let Ω be an abstract set and $\mathcal{F} \subset \mathcal{P}(\Omega)$ a σ-algebra on Ω. The pair $(\Omega, \mathcal{F})$ is called a *measurable space.* An element $A \in \mathcal{F}$ is called an $\mathcal{F}$-measurable subset of Ω.

Let $\mathcal{A} \subset \mathcal{P}(\Omega)$ an algebra on Ω. Let ν be a non-negative *set function,* i.e.,

$$\nu : \mathcal{A} \to [0, \infty],$$

then ν is called *additive* if $\nu(\varnothing) = 0$ and for $A, B \in \mathcal{A}$, with $A \cap B = \varnothing$ we have

$$\nu(A \cup B) = \nu(A) + \nu(B).$$

The set function ν is called *countably additive or σ-additive* if $\nu(\varnothing) = 0$, and for every sequence $\{A_n\}_{n=1}^{\infty}$ of disjoint sets in S with the union $\cup_{n=1}^{\infty} A_n \in S$, then

$$\nu\left(\cup_{n=1}^{\infty} A_n\right) = \sum_{n=1}^{\infty} \nu(A_n).$$

Let the measurable space $(\Omega, \mathcal{F})$. A set function $\mu : \mathcal{F} \to [0, \infty]$ is called a measure on $(\Omega, \mathcal{F})$ if μ is countably additive. The triple $(\Omega, \mathcal{F}, \mu)$ is called a *measure space.* The measure space is called *finite* if $\mu(\Omega) < \infty$.

The measure space $(\Omega, \mathcal{F}, \mu)$ is called σ-finite, if there is a sequence $\{A_n\}_{n=1}^{\infty}$ of elements of $\mathcal{F}$ such that

$$\mu(A_n) < \infty \ \forall n \in \mathbb{N} \ \text{ and } \ \cup_{n=1}^{\infty} A_n = \Omega.$$

We say that μ is *positive* if $\mu(A) \geq 0$ for any $A \in \mathcal{F}$. An element $A \in \mathcal{F}$ is called *μ-null* if $\mu(A) = 0$. The measure μ is called a *probability measure* if $\mu(\Omega) = 1$.

Let $(X, \mathcal{T})$ be a topological space, where as we have denoted in section 1.2.2, $\mathcal{T}$ is the collection of open sets. By definition, the σ-algebra generated by $\mathcal{T}$ is called the *Borel* σ-algebra of X, and is denoted by $\mathcal{B}(X)$. When X is the real line we will denote the Borel σ-algebra $\mathcal{B}(\mathbb{R}) = \mathcal{B}$. When the set X is the extended real line $\hat{\mathbb{R}} = [-\infty, +\infty]$ then the Borel σ-algebra $\mathcal{B}\left(\hat{\mathbb{R}}\right)$ will be denoted by $\hat{\mathcal{B}}$. The Borel[2] σ-algebra $\mathcal{B}$ is countably generated by the countable collection of sets $\mathcal{A} = (a, b]$, where a, b range over the rationals $\mathbb{Q}$. We will state slightly more generally and prove this result in the form of a theorem.

Theorem 8 *The Borel σ-algebra of $\hat{\mathbb{R}}$ is generated by the intervals of the form $(-\infty, a]$ were $a \in \mathbb{Q}$.*

Proof. Let $\mathcal{O}$ denote all open intervals in the topological structure of $\hat{\mathbb{R}}$. Since every open set in $\hat{\mathbb{R}}$ is the countable union of open intervals, then apparently $\mathcal{O}$ is a σ-algebra and obviously it is also the smallest σ-algebra generated by $\mathcal{O}$. That is, we have $\sigma(\mathcal{O})$ is the Borel σ-algebra of $\hat{\mathbb{R}}$.

Let $\mathcal{I}$ denote the set of all intervals of the form $(-\infty, a]$ where $a \in \mathbb{Q}$. Now, let $(a, b) \in \mathcal{O}$, and let $\{a_n\}_{n=1}^{\infty}$ be a sequence of rationals decreasing to a and $\{b_n\}_{n=1}^{\infty}$ be a sequence of rationals increasing to b. Then

$$\begin{aligned}(a, b) &\in \mathcal{O} = \bigcup_{n=1}^{\infty} (a_n, b_n] \\ &= \bigcup_{n=1}^{\infty} \left((-\infty, b_n] \cap (-\infty, a_n]^c\right) \in \sigma(\mathcal{I}).\end{aligned}$$

Thus, $\mathcal{O} \subset \sigma(\mathcal{I})$, and therefore $\sigma(\mathcal{O}) \subset \sigma(\mathcal{I})$. Now, each element of $\mathcal{I}$ is a Borel set, and consequently $\sigma(\mathcal{I})$ is contained in the Borel sets $\hat{\mathcal{B}}$. Hence, we have

$$\hat{\mathcal{B}} = \sigma(\mathcal{O}) \subset \sigma(\mathcal{I}) \subset \hat{\mathcal{B}},$$

and therefore $\sigma(\mathcal{I}) = \hat{\mathcal{B}}$. ■

Remark 9 *Cantor[3] very early in his turbulent mental life, while studying infinite sets proved*

---

[2]Felix Eduard Justin Emile Borel (7 January 1871–3 February 1956) was a French mathematician and politician. He was one of the pioneers of measure theory and its applications in probability theory. The concept of a Borel set, among other mathematical entities, is named in his honor. He is also known for the amusing thought experiment under the name *the infinite monkey theorem.* He also published a number of research papers on game theory.

[3]Georg Ferdinand Ludwig Phillip Cantor (March 3 1845–January 6, 1918) was a German mathematician (not German in origin), born in Russia. He is best known as the creator of set theory, which has become a fundamental theory in Mathematics. Cantor's theory of transfinite numbers was originally regarded as so counter intuitive that it encounter resistance from mathematical contemporaries such as Kronecker. David Hilbert defended it from its critics by famously declaring: "No one shall expel us from the Paradise that Cantor has created". More on Cantor and his mental health which tortured him through a large part of his life could be found in the interesting book by Rucker (2005). In addition, there is a very interesting chapter (16) in Penrose (2004), *The Rood of Reality,* where he illustrates how Cantorian thinking intrigues a leading contemporary theoretical physicist.

that the set of all positive rational numbers is countable. Note that the set of all positive rational numbers is a very large set containing the set $\mathbb{N}$ *and many other numbers. Between any two rational numbers, there are infinitely many others, and in addition, it is not possible to list the positive rational numbers in order of size, for there is no identifiable a smallest positive rational number.*

For any subset A of a topological space X, we define the Borel σ-algebra generated by A to be

$$\mathcal{B}(A) = \{C : C = A \cap B, B \in \mathcal{B}(X)\}.$$

When X is the n-dimensional Euclidean space $\mathbb{R}^n$, we will denote the Borel σ-algebra $\mathcal{B}(\mathbb{R}^n) = \mathcal{B}^n$. Let $\mathcal{L}$ be the class of all rectangles (open, semiopen, closed) of $\mathbb{R}^n$. Then it is easy to see that $\mathcal{L}$ is a semialgebra. In addition, it is true that $\sigma(\mathcal{L}) = \mathcal{B}^n$. A measurable space $(\Omega, \mathcal{F})$ is separable if $\mathcal{F}$ is generated by a countable collection of sets, i.e., it is countably generated. In particular, if Ω is a locally compact space with a countable basis, the σ-algebra of its Borel sets is separable.

If $(X_1, \mathcal{B}(X_1))$ is a measurable space and $(X_2, \mathcal{B}(X_2))$ is another measurable space, then a mapping $f : X_1 \to X_2$ is called a *measurable function* if

$$f^{-1}\{A\} := \{x : f(x) \in A\} \in \mathcal{B}(X_1),$$

for all sets $A \in \mathcal{B}(X_2)$.

We will now provide a useful proposition which although it looks apparent it needs a rigorous proof.

Proposition 10 *Let* $(\Omega, \mathcal{F}, \mu)$ *be a measure space. If* $\{A_n\}_{n=1}^{\infty} \in \mathcal{F}$ *with* $A_n \subseteq A_{n+1}$ *for every* $n \in \mathbb{N}$ *and* $A_n \uparrow A$ *then* $\mu(A_n) \uparrow \mu(A)$. *If* $\{A_n\}_{n=1}^{\infty} \in \mathcal{F}$ *with* $A_{n+1} \subseteq A_n$ *for every* $n \in \mathbb{N}$ *and* $A_n \downarrow A$ *then* $\mu(A_n) \downarrow \mu(A)$.

Proof. Since $A_n \subseteq A_{n+1}$ for every $n \in \mathbb{N}$ then the strict difference of the sets $A_{n+1} \backslash A_n = B_n$ defines in fact the following family of disjoint sets.

Define $B_1 = A_1$ and $A_{n+1} \backslash A_n = B_n$ for $n \geq 2, n \in \mathbb{N}$. Then it is easy to see that

$$B_i \cap B_j = \varnothing \text{ for every } i, j \geq 1 \text{ and } i \neq j.$$

Hence, we have that

$$\mu(A_n) = \mu(B_1 \cup B_2 \cup ... \cup B_n) = \sum_{k=1}^{n} \mu(B_k). \tag{1.5}$$

Since $A_n \uparrow A$, we get that

$$\cup_{n=1}^{\infty} A_n = A,$$

and consequently

$$\mu(A_{n+1}) = \sum_{k=1}^{n} \mu(B_k) \uparrow \sum_{k=1}^{\infty} \mu(B_k) = \mu(A). \tag{1.6}$$

The second part of the Proposition is easily proved in an analogous way as that of the first part. ■

We will now proceed to state a Lemma, which plays an important role, whenever measures are involved and is often called the *uniqueness of extension* Lemma. The proof could be found in Rogers and Williams (2000).

Lemma 11 *Let* $(\Omega, \mathcal{F}, \mu)$ *be a measure space. Let* $\mathcal{L}$ *be a* π*-system on* Ω *and let that* $\mathcal{F} = \sigma(\mathcal{L})$. *Assume that* μ_1 *and* μ_2 *are measures on* $(\Omega, \mathcal{F}, \mu)$ *such that*

$$\mu_1(\Omega) = \mu_2(\Omega) < \infty \text{ and } \mu_1 = \mu_2 \text{ on } \mathcal{L}.$$

Then

$$\mu_1 = \mu_2 \text{ on } \mathcal{F}.$$

We will now provide some properties of measures which could be proved as exercises by the reader.

Let $(\Omega, \mathcal{F}, \mu)$ be a measure space. Then the following properties hold.

(1) For any two sets $A, B \in \mathcal{F}$, we have

$$\mu(A \cup B) \leq \mu(A) + \mu(B),$$

which is easily generalized by induction in the following:

(2) For any finite number of sets $A_1, A_2, ..., A_n \in \mathcal{F}$,

$$\mu(\cup_{i \leq n} A_i) \leq \sum_{i \leq n} \mu(A_i).$$

Now, if $\mu(\Omega) < \infty$, then for any $A, B \in \mathcal{F}$,

$$\mu(A \cup B) = \mu(A) + \mu(B) - \mu(A \cap B),$$

which could be generalized by induction into the following:

For any finite number of sets $A_1, A_2, ..., A_n \in \mathcal{F}$, we have that:

$$\mu(\cup_{i \leq n} A_i) = \sum_{i \leq n} \mu(A_i) - \sum\sum\nolimits_{i<j \leq n} \mu(A_i \cap A_j)$$

$$+ \sum\sum\sum\nolimits_{i<j<k \leq n} \mu(A_i \cap A_j \cap A_k) - ... + (-1)^{n-1} \mu(A_1 \cap A_2 \cap ... \cap A_n).$$

We will now state without proof the Caratheodory[4]'s Extension Theorem, which is very useful in constructing measures for measurable spaces.

Theorem 12 ***(Caratheodory's Extension Theorem).*** *Let Ω be a set, let $\mathcal{A}$ be an algebra on Ω, and let*

$$\mathcal{F} := \sigma(\mathcal{A}).$$

If μ_0 is a countably additive map $\mu_0 : \mathcal{A} \to [0, \infty)$, then there exists a measure μ on $(\Omega, \mathcal{F})$ such that

$$\mu = \mu_0 \text{ on } \mathcal{A}.$$

If $\mu_0(\Omega) < \infty$, then, this extension is unique.

Proof. The proof could be found in Williams (2000, p. 197). ■

We will now provide the following theorem concerning various countable additivity properties of a measure. The proof of the theorem could be found in Chow and Teicher (1997).

Theorem 13 *Let μ be a non-negative additive set function on an algebra $\mathcal{A}$ and $\{A_n, n \geq 0\} \subset \mathcal{A}$, then we have (a) If $\{A_n, n \geq 0\}$ is a disjoint class and $\cup_0^\infty A_n \subset A$, then*

$$\mu(A) \geq \sum_0^\infty \mu(A_n).$$

[4]Constantin Carathéodory (or Constantine Karatheodoris) (September 13, 1873–February 2, 1950) was a Greek mathematician. He made significant contributions to the theory of functions of a real variable, the calculus of variations, and measure theory. His work also includes important results in conformal representations and in the theory of boundary correspondence. In 1909, Carathéodory pioneered the Axiomatic Formulation of Thermodynamics along a purely geometrical approach. He is known to the wider audience in mathematics from his correspondence with Albert Einstein on the mathematical foundation of relativity theory.

(b) If $A \subset \cup_0^k A_n$ *for* $k = 0, 1, 2, ...,$ *then*

$$\mu(A) \geq \sum_0^k \mu(A_n).$$

(c) The set function μ *is* σ*-additive iff for every increasing sequence* $\{A_n, n \geq 0\} \subset \mathcal{A}$ *with* $\lim_{n\to\infty} A_n = A \in \mathcal{A}$, *then* $\lim_{n\to\infty} \mu(A_n) = \mu(A)$. *Under these conditions* μ *is subadditive on* $\mathcal{A}$, *that is, if* $\cup_0^\infty A_n \in \mathcal{A}$, *then*

$$\mu(\cup_0^\infty A_n) \leq \sum_0^\infty \mu(A_n).$$

Theorem 14 *Let* μ *be a* σ*-additive set function on an algebra* $\mathcal{A}$, *then for every decreasing sequence* $A_n \in \mathcal{A}$ *with* $\mu(A_1) < \infty$ *and* $lim_{n\to\infty} A_n = A \in \mathcal{A}$, *then we have that*

$$\lim_{n\to\infty} \mu(A_n) = \mu(A).$$

Conversely, if for every decreasing sequence $A_n \in \mathcal{A}$ *with* $\lim_{n\to\infty} A_n = \varnothing$ *and* $\lim_{n\to\infty} \mu(A_n) = 0$, *then we have that* μ *is* σ*-additive.*

We have already discussed that the Borel σ-algebra $\mathcal{B}$ in $\mathbb{R}$ is countably generated by the countable collection of sets $\mathcal{A} = (a, b]$, where a, b range over the rationals $\mathbb{Q}$. In addition, we proved that the Borel σ-algebra $\hat{\mathcal{B}}$ of $\hat{\mathbb{R}}$ is generated by the intervals of the form $(-\infty, a]$ were $a \in \mathbb{Q}$. These results provide in fact the intuition of how with the use of Caratheodoty's extension theorem to construct a measure on the interval $(0, 1]$ and by analogy on $\hat{\mathbb{R}}$. Let $S = (0, 1]$. Consider the algebra $\mathcal{A}$ of subsets of S defined as

$$\mathcal{A} := \{A : A = (a_1, b_1] \cup (a_2, b_2] \cup ... \cup (a_k, b_k]\},$$

where $k \in \mathbb{N}$, $0 \leq a_1 \leq b_1 \leq ... \leq a_k \leq b_k \leq 1$, that is, $\mathcal{A}$ is the algebra of all the subsets of $S = (0, 1]$ that may be written as a finite union of sets of the type $(a, b]$. Now, consider the σ-algebra $\sigma(\mathcal{A})$ generated by $\mathcal{A}$, then apparently we have that

$$\sigma(\mathcal{A}) = \mathcal{B}(0, 1].$$

For any set $A \in \mathcal{A}$ define $\mu_0(\mathcal{A})$ to be

$$\mu_0(\mathcal{A}) = \sum_{i \leq k} (b_i - a_i).$$

Then it is easy to see that μ_0 is additive on $\mathcal{A}$ and in addition it could be proved that μ_0 is *countably additive on* $\mathcal{A}$ (see Williams (2000) p. 200). Thus by Caratheodoty's extension theorem, there exists a unique measure μ on $((0, 1], \mathcal{B}(0, 1])$, which coincides with μ_0 on $\mathcal{A}$. This is called the *Lebesque*[5] *measure on* $((0, 1], \mathcal{B}(0, 1])$. In order to distinguish the Lebesque from other measures, we will often denote it by *Leb*. In a similar way, we can construct *σ-finite Lebesque measure,* which we also denote by *Leb* on $(\mathbb{R}, \mathcal{B})$.

[5]Henri Leon Lebesgue (June 28, 1875–July 26, 1941). He is most famous for Lebesgue's theory of integration, which was somehow a generalization of the by then known Rieman integral. Lebesgue's great Ph.D. thesis *Integrale, longueur, aire,* with the full account of Lebesgue's theory of integration, appeared in the Annali di Mathematica in 1902. Emile Borel was his thesis advisor.

1.5 Probability spaces

In this chapter, we will attempt to raise the level of the insight and precision of a probability space, based on the usual gentle first course in probability theory and the small ladder constructed in the previous sections. Our goal is among others to reach the level, where we will be able to study Probability spaces with uncountable sample spaces.

Let us consider a random experiment or an experiment of Tyche, which is performed and all possible results, or *outcomes,* are grouped into a set Ω, called the *sample space.* We denote by ω the generic element of Ω called a *sample point.* Again let $\wp(\Omega)$ be the power set of Ω. Let $A \in \wp(\Omega)$ then A is called *event,* if it represents a property of interest from the point of view of the modeler. Therefore, in general not all members of $\wp(\Omega)$ are necessarily events. Let $\mathcal{F}$ be the σ-algebra of events. Hence, an event is an $\mathcal{F}$-measurable subset of Ω. Now, let $\mathbb{P}$ be a probability measure on $(\Omega, \mathcal{F})$ mapping $\mathcal{F}$ into the interval $[0,1]$. The choice of $[0,1]$ as the set of values of $\mathbb{P}(.)$ is somewhat arbitrary, but through the decades proved to provide the necessary intuition. The family of events and the probability measure $\mathbb{P}$ satisfy the following, imposed in an axiomatic way properties, which are compatible with the intuitive notion of probability.

A probability measure $\mathbb{P}$ on $(\Omega, \mathcal{F})$ is a function $\mathbb{P} : \mathcal{F} \to [0,1]$ satisfying:

(a) $\mathbb{P}(\varnothing) = 0, \ \ \mathbb{P}(\Omega) = 1.$

(b) If $A_1, A_2, ...$ is a countable collection of disjoint events, that is, $A_i \cap A_j = \varnothing$ for all pairs with $i \neq j$, then

$$\mathbb{P}\left(\bigcup_{i=1}^{\infty} A_i\right) = \sum_{i=1}^{\infty} \mathbb{P}(A_i). \tag{1.7}$$

The relation (1.7) is often called the Full Addition-Rule axiom. The triple $(\Omega, \mathcal{F}, \mathbb{P})$ comprising a set Ω, a σ-algebra of subsets of Ω, and a probability measure $\mathbb{P}$, on $(\Omega, \mathcal{F})$, is called a *probability space.*

Apart from a collection of very simple cases, the Full Addition-Rule axiom does not apply if we demand that all members of $\wp(\Omega)$ are events. An illuminating example in this respect is the known as Banach[6]-Tarski[7] paradox. These Polish magisian-mathematicians (as Williams (2004) calls them) showed assuming the Axiom of Choice, that one can find a subset of the surface S of the familiar unit ball in $\mathbb{R}^3$, that is not measurable, that is, the event that a point chosen at random on S belongs to A is not a true event. For more on this see Williams (2004) and a very detailed account of that in Wagon (1994).

Note that in Probability there is no attempt to provide the physical meaning of a probability in the real world. The closest such an attempt is the definition of probability, as a limit of the relative frequency of occurrence of an event in an infinitely countable number of repetitions of an experiment. A definition that eventually fails, in later stages one climbs the ladder. Probably, the foundation of probability theory axiomatizing on the expected value of a random variable, which is older, is more near to how the human mind functions. People usually think of what to expect in a situation, where luck plays a role and not what the probability of each possible event is.

---

[6]Stefan Banach (1892–1945) was a self-taught mathematics prodigy who is by many a founder of modern functional analysis. Among his most prominent achivements was the mongraph *Theory of Linear Operations.* His name is met in Banach-Tarski paradox, Hahn-Banach theorem, Banach-Mazur game and Banach space. His group of researchers was named the "Lwow School of Mathematics" published its own journal *Studia Mathematica".*

[7]Alfred Tarski (January 14, 1901–October 26, 1983) was a Polish logician and mathematician. He taught and carried out research on model theory, metamathematics, and algebraic logic for which he is famous. He also contributed to abstract algebra, topology, geometry, measure theory, mathematical logic, set theory and analytic phylosophy. He is thought by many that along with Kurt Gödel, they changed the face of logic in the twentieth century. At Berkley since 1942 up to his death acquired a reputation as an awesome and demanding teacher.

Null sets and completion of a probability space.

Let $(\Omega, \mathcal{F}, \mathbb{P})$ be a probability space used to model some experiment of Tyche. Any event $A \in \mathcal{F}$ which has zero probability, that is, $\mathbb{P}(A) = 0$, is called *null.* It may seem reasonable to suppose, that any subset B of a null set A will itself be null, but this may be without meaning since B may not be an event, and thus $\mathbb{P}(B)$ may not be defined. Null events should not be confused with the impossible event $\varnothing$. Null events are happening, even though they have zero probability. That is, the impossible event is null, but null events need not be impossible. For example, if one follows a series of games of basket ball, let say one hundred, it may very easily deduce, that the event that a player shooting from one basket to the opposite basket he will succeed, is a null event, that is, has probability zero. However, extremely rarely though, it could happen and makes you either fiercely applaud or tear your clothes. A probability space $(\Omega, \mathcal{F}, \mathbb{P})$ is called *complete if all subsets of null sets are events.* Therefore, any incomplete space can be completed. Let $\mathcal{N}$ be the collection of all subsets of null sets in $\mathcal{F}$ and let $\mathcal{G} = \sigma(\mathcal{F} \cup \mathcal{N})$. It can be shown that the domain of $\mathbb{P}$ may be extended in an obvious way from $\mathcal{F}$ to $\mathcal{G}$. In this case $(\Omega, \mathcal{G}, \mathbb{P})$ is called the *completion* of $(\Omega, \mathcal{F}, \mathbb{P})$.

Almost surely.

A statement S is said to be *almost surely true* or *to be true with probability one,* if the set A of outcomes ω for which S holds, is an element of $\mathcal{F}$, and $\mathbb{P}(A) = 1$. Hence, the probability of an *almost sure* event is one. If a statement is *certain* the set of outcomes ω for which it is true, is the entire sample space Ω, for which we have that $\mathbb{P}(\Omega) = 1$, that is, it is almost surely true. However, in general there are statements or events, that are almost surely true without being necessarily certain.

Borel-Cantelli Lemma.

The Borel-Cantelli lemma concerns the appearance of infinitely often events, that was first studied in the present chapter in Proposition 1.

Theorem 15 *Let $(\Omega, \mathcal{F}, \mathbb{P})$ be a probability space and $\{A_n\}_{n\in\mathbb{N}}$ a countable sequence of events. Then, (a) If $\sum_n^{\infty} \mathbb{P}(A_n) < \infty$ then the event $E_{i.o} = \{\omega : \omega \in A_n, i.o.\}$ has probability zero, i.e., $\mathbb{P}(E_{i.o}) = 0$. (b) If $\sum_n^{\infty} \mathbb{P}(A_n) = \infty$ and $\{A_n\}_{n\in\mathbb{N}}$ are independent then $\mathbb{P}(E_{i.o}) = 1$.*

Proof. (a) From Proposition 1, we have that

$$E_{i.o} = \cap_{n=1}^{\infty} \cup_{k=n}^{\infty} A_k.$$

Hence, we have that

$$E_{i.o} \subset \cup_{k=n}^{\infty} A_k,$$

from which we get that

$$\mathbb{P}(E_{i.o}) \leq \sum_{k=n}^{\infty} \mathbb{P}(A_k) \to 0 \text{ as } n \to \infty,$$

whenever $\sum_n^{\infty} \mathbb{P}(A_n) < \infty$. From this relation, we get that $\mathbb{P}(E_{i.o}) = 0$.

(*b*) It is a well-known fact that

$$E_{i.o}^c = [\cap_{n=1}^{\infty} \cup_{k=n}^{\infty} A_k]^c = \cup_{n=1}^{\infty} \cap_{k=n}^{\infty} A_k^c.$$

Now, by Proposition 10, we have that

$$\begin{aligned}
\mathbb{P}\left(\cap_{k=n}^{\infty} A_k^c\right) &= \lim_{t\to\infty} \mathbb{P}\left(\cap_{k=n}^{t} A_k^c\right) \\
&= \text{(by the independence assumption)} \\
&= \prod_{k=n}^{\infty} \left[1 - \mathbb{P}\left(A_k\right)\right] \\
&\leq \text{(due to that for } x \geq 0, \text{ we have } 1 - x \leq e^{-x}) \\
&\leq \prod_{k=n}^{\infty} \exp\left[-\mathbb{P}\left(A_k\right)\right] \\
&= \exp\left(-\sum_{k=n}^{\infty} \mathbb{P}\left(A_k\right)\right) \\
&= \left(\text{since } \sum_k \mathbb{P}\left(A_k\right) = \infty\right) \\
&= 0.
\end{aligned}$$

From which we get that

$$\mathbb{P}(E_{i.o}^c) = \lim_{n\to\infty} \mathbb{P}\left(\cap_{k=n}^{\infty} A_k^c\right) = 0,$$

hence $\mathbb{P}(E_{i.o}) = 1$. ∎

Product space.

The probability space $(\Omega, \mathcal{F}, \mathbb{P})$ we discussed so far, has usually been constructed around outcomes of one random experiment. However, think for example the case where two fair dices are thrown together. In this case, we need to combine the outcomes of two independent experiments into one space. It is easy also to think of cases, where several random experiments have to be combined into one probability space. Let, that in the cases of the two fair dices the associated probability spaces are $(\Omega_1, \mathcal{F}_1, \mathbb{P}_1)$ and $(\Omega_2, \mathcal{F}_2, \mathbb{P}_2)$ respectively. The sample space of the simultaneous experiments, considered jointly, is the collection of ordered pairs

$$\Omega_1 \times \Omega_2 = \left\{(\omega_1, \omega_2) : \omega_1 \in \Omega_1, \omega_2 \in \Omega_2\right\},$$

called the Cartesian product of the two sets Ω_1 and Ω_2. The appropriate σ-algebra of events in $\Omega_1 \times \Omega_2$ should contain all subsets of the form

$$A_1 \times A_2 = \left\{(a_1, a_2) : a_1 \in A_1, a_2 \in A_2\right\},$$

where A_1 and A_2 are typical members of $\mathcal{F}_1$ and $\mathcal{F}_2$ respectively. However, the family of all such sets, $\mathcal{F}_1 \times \mathcal{F}_2 = \{A_1 \times A_2 : A_1 \in \mathcal{F}_1, A_2 \in \mathcal{F}_2\}$, is not in general a σ-algebra. However, as we discussed previously, there exists a unique smallest σ-algebra $\mathcal{G} = \sigma\left(\mathcal{F}_1 \times \mathcal{F}_2\right)$, which consists of subsets of $\Omega_1 \times \Omega_2$ which contains $\mathcal{F}_1 \times \mathcal{F}_2$. What is left now is to find a suitable probability measure on $(\Omega_1 \times \Omega_2, \mathcal{G})$. Led by the fact, that the two experiments are independent, we define the probability measure $\mathbb{P}_{12} : \mathcal{F}_1 \times \mathcal{F}_2 \to [0, 1]$ be given by:

$$\mathbb{P}_{12}\left(A_1 \times A_2\right) = \mathbb{P}_1\left(A_1\right) \mathbb{P}_2\left(A_2\right) \quad \text{for } A_1 \in \mathcal{F}_1, A_2 \in \mathcal{F}_2.$$

It is not difficult to see, that the domain of $\mathbb{P}_{12}$ can be extended from $\mathcal{F}_1 \times \mathcal{F}_2$ to the whole of $\mathcal{G} = \sigma\left(\mathcal{F}_1 \times \mathcal{F}_2\right)$. The resulting probability space $(\Omega_1 \times \Omega_2, \mathcal{G}, \mathbb{P}_{12})$ is called the *product space* of $(\Omega_1, \mathcal{F}_1, \mathbb{P}_1)$ and $(\Omega_2, \mathcal{F}_2, \mathbb{P}_2)$. The present construction is easily applied to any two independent experiments, and moreover, it is easily generalized to more than two independent random experiments.

1.5.1 Infinite probability spaces

In this subsection, we will present two fundamental random experiments, one with infinite sample space and the other with infinitely countable sample space.

Choose a real number uniformly between 0 and 1.

$\circledast\circledast\circledast$Important insight

Intuitively is an unexpected possible fact, but every random experiment can be reduced to the simple random experiment of choosing a number between 0 and 1 according to the uniform distribution. $\circledast\circledast\circledast$

So in fact the sample space of this random experiment is $\Omega = [0,1]$. We have seen in the previous section that by Caratheodory's extention theorem, there exist a unique measure μ on $((0,1], \mathcal{B}(0,1])$, which coincides with the measure μ_0 on $\mathcal{A}$, the algebra of subsets of Ω defined as

$$\mathcal{A} := \{A : A = (a_1, b_1] \cup (a_2, b_2] \cup ... \cup (a_k, b_k]\},$$

where $k \in \mathbb{N}$, $0 \leq a_1 \leq b_1 \leq ... \leq a_k \leq b_k \leq 1$. In addition, we defined $\mathcal{B}(0,1] = \sigma(\mathcal{A})$ and the measure μ called the Lebesgue measure for an interval $(a,b]$ to be $Leb(a,b] = b - a$. Apparently this unique measure as it was proved by Borel and Lebesgue, due to Caratheodory's extension theorem, on the interval $(0,1]$ is a probability measure and hence we may write for the probability of choosing a point in the interval $(a,b]$

$$\mathbb{P}(a,b] = b - a. \tag{1.8}$$

Apparently then the probability of choosing a single point is equal to zero, that is, $\mathbb{P}(a,a] = 0$. Hence, since single points have zero probability we have that

$$\mathbb{P}(a,b] = \mathbb{P}[a,b] = \mathbb{P}[a,b) = b - a.$$

It is easy to see that also the probability of choosing any point of a countable sequence $\{a_n\}_{n=1}^{\infty}$ is equal to zero by the countable addition rule of probability measures. Hence, single points and countable sequences on $\mathcal{B}[0,1]$ are null sets.

An interesting set in $\mathcal{B}[0,1]$ is the *Cantor set,* which we now present. From the interval $[0,1]$, remove the middle third, that is, the interval $\left(\frac{1}{3}, \frac{2}{3}\right)$. Denote now by

$$C_1 = \left[0, \frac{1}{3}\right] \bigcup \left[\frac{2}{3}, 1\right]. \tag{1.9}$$

Using the probability measure defined in (1.8), we get that $\mathbb{P}(C_1) = \frac{2}{3}$. From each of the two pieces of C_1, remove the middle third, that is, remove the intervals $\left(\frac{1}{3}, \frac{2}{9}\right)$ and $\left(\frac{7}{9}, \frac{8}{9}\right)$. Denote the remaining set by

$$C_2 = \left[0, \frac{1}{9}\right] \cup \left[\frac{2}{9}, \frac{1}{3}\right] \cup \left[\frac{2}{3}, \frac{7}{9}\right] \cup \left[\frac{8}{9}, 1\right].$$

Using the probability measure defined in (1.8), we get that $\mathbb{P}(C_2) = \frac{4}{9}$. Continue this process so as to stage n we have a set C_n that has 2^n pieces, each with probability $\left(\frac{1}{3}\right)^n$ and the set C_n has probability $\left(\frac{2}{3}\right)^n$. The Cantor set is defined to be $C = \cap_{n=1}^{\infty} C_n$. Since $\mathbb{P}$ is a measure by Proposition 10, we get that

$$\mathbb{P}(C) = \lim_{n\to\infty} \mathbb{P}(C_n) = \lim_{n\to\infty} \left(\frac{2}{3}\right)^n = 0.$$

This certainly is a counter-intuitive result, since the Cantor set has an uncountable many points as we shall now prove. Let us assume that the Cantor set is countably infinite. Then there exists a sequence $\{c_n\}_{n=0}^{\infty}$ of points in the Cantor set, which represents a one to one correspondence of the members of the Cantor set with the natural numbers. Assume without loss of generality that c_1 does not belong to the first part of C_1, that is, $c_1 \notin [0, \frac{1}{3}] = H_1$. Now, let us consider $H_1 \cap C_2$ which consists of the sets $[0, \frac{1}{9}]$ and $[\frac{2}{9}, \frac{1}{3}]$. Again, since we have that $c_1 \notin H_1$, without loss of generality we may assume that c_2 does not belong to one of the two sets $[0, \frac{1}{9}]$ and $[\frac{2}{9}, \frac{1}{3}]$. Let $c_2 \notin [0, \frac{1}{9}] = H_2$. If we continue in this line, we in fact construct a sequence of sets with the following properties

$$H_1 \supset H_2 \supset H_3..., \text{ with } c_1 \notin H_1, c_2 \notin H_2, \ c_3 \notin H_3....$$

Therefore, $\cap_{i=1}^{\infty} C_i$ does not contain any point of the sequence $\{c_n\}_{n=0}^{\infty}$. However, by Proposition 10, $\cap_{i=1}^{\infty} C_i$ is not an empty set and thus it contains at least one element which belongs to the Cantor set C but is not a member of $\{c_n\}_{n=0}^{\infty}$. Hence, the cantor set C is not countably infinite and is uncountable infinite.

The infinite number of Bernoulli trials.

In Vassiliou (2010, p.15), we defined as the Bernoulli trial a random experiment which has only two possible outcomes. One of the outcomes is named "success" and the other "failure". Which outcome is named "success" depends on both the experiment and the problem solver but it has no real influence on the study of the random variables to be introduced. We define the random variable X as

$$X = \left\{ \begin{array}{l} 1 \text{ when the outcome of the experiment is success} \\ 0 \text{ when the outcome of the experiment is failure} \end{array} \right\}.$$

We define p to be the probability of success and $q = 1 - p$ the probability of failure.

Assume now that we execute a Bernoulli trial infinitely many times. It may be very useful intuitively to the reader to refresh Example 1.3 in Vassiliou (2010, p. 4). In there we had only a sequence of three Benoulli trials and the sample space of the experiment was

$$\Omega = \{(1,1,1), (1,1,0), (1,0,1), (1,0,0), (0,1,1), (0,1,0), (0,0,1), (0,0,0)\}.$$

In the case of infinite Bernoulli trials, the sample space Ω_∞ will consist of events $\omega_1, \omega_2, ...$, each one of which will be of the form

$$\omega_n = (1,0,1,1,0,1,...),$$

in fact, each $\omega_n, n \in \mathbb{N}$ will consist of a countably infinite combination of one's and zero's, and moreover, they are countably infinite themselves. The problem we are facing is to construct a probability measure corresponding to this random experiment. Naturally, we start with the fact that for our needed probability measure $\mathbb{P}$, we should first have $\mathbb{P}(\varnothing) = 0$ and $\mathbb{P}(\Omega_\infty) = 1$. These two sets as in Example 1.3 in Vassiliou (2010, p.4) form a trivial σ-algebra, which we call $\mathcal{F}_0 = \{\varnothing, \Omega_\infty\}$. Now, consider the sets

$$A_1 = \text{the set of all events } \omega_n, n \in \mathbb{N} \text{ beginning with } 1, \tag{1.10}$$

$$A_2 = \text{the set of all events } \omega_n, n \in \mathbb{N} \text{ beginning with } 0. \tag{1.11}$$

In Example 1.3 in Vassiliou (2010, p.5), the sets A_1 and A_2 were in essence defined in the same way and they were

$$A_1 = \{(1,1,1), (1,1,0), (1,0,1), (1,0,0)\},$$

$$A_2 = \{(0,1,1), (0,1,0), (0,0,1), (0,0,0)\},$$

and in addition they were the atoms of the σ-algebra $\mathcal{F}_1$, that is, they generated $\mathcal{F}_1$ with all their possible unions and complements. Analogously, we call $\mathcal{F}_1$ the σ-algebra generated by A_1 and A_2 as in (1.10) and (1.11). It is easy to see that in the three Bernoulli trial experiment, we have $\mathbb{P}(A_1) = p$ and $\mathbb{P}(A_2) = q$. Intuitively thinking, we set for the infinite Bernoulli trials also

$$\mathbb{P}(A_1) = p \text{ and } \mathbb{P}(A_2) = q.$$

For the infinite Bernoulli trials experiment, in an analogous way as in the three Bernoulli trials experiment, we define by

$$A_{11} = \text{the set of all sequences beginning with } 1,1,$$

$$A_{10} = \text{the set of all sequences beginning with } 1,0,$$

$$A_{01} = \text{the set of all sequences beginning with } 0,1,$$

$$A_{00} = \text{the set of all sequences beginning with } 0,0.$$

Thinking in an analogous way as in setting the probabilities of the sets in $\mathcal{F}_1$, we set

$$\mathbb{P}(A_{11}) = p^2, \ \mathbb{P}(A_{10}) = pq, \mathbb{P}(A_{01}) = pq, \mathbb{P}(A_{00}) = q^2. \tag{1.12}$$

We denote by $\mathcal{F}_2$ the σ-algebra generated by the sets $A_{11}, A_{10}, A_{01}, A_{00}$. Then as we have already seen in Vassiliou (2010, p.5) the σ-algebra $\mathcal{F}_2$ is given by

$$\mathcal{F}_2 = \left\{ \begin{array}{c} \varnothing, \Omega, A_1, A_2, A_{11}, A_{10}, A_{01}, A_{00}, A_{11}^c, A_{10}^c, A_{01}^c, A_{00}^c, \\ A_{11} \cup A_{01}, A_{11} \cup A_{00}, A_{10} \cup A_{01}, A_{10} \cup A_{00} \end{array} \right\}.$$

In $\mathcal{F}_2$ as is given above one may wonder where is $A_{11} \cup A_{10}$. The reason that is missing is that $A_{11} \cup A_{10} = A_1$. The same reasoning apply to $A_{01} \cup A_{00} = A_0$. From (1.12), we immediately get that

$$\mathbb{P}(A_{11}^c) = 1 - p^2, \mathbb{P}(A_{10}^c) = \mathbb{P}(A_{01}^c) = 1 - pq, \mathbb{P}(A_{00}^c) = 1 - q^2.$$

For the remaining sets in $\mathcal{F}_2$, since $A_{11} \cap A_{01} = \varnothing$, we get that

$$\mathbb{P}(A_{11} \cup A_{01}) = \mathbb{P}(A_{11}) + \mathbb{P}(A_{01}) = p(p+q) = p.$$

With the same reasoning, we may determine the probabilities of the remaining sets in $\mathcal{F}_2$. We may continue this process of defining the probability of every set that can be described in terms of the first finitely many Bernoulli trials. However, not all the sets in Ω_∞ can be described in this way. It is not, in fact, difficult intuitively to see that every sequence $\omega =$ (infinitely countable $1's$ and $0's$) has probability zero, since we have infinitely many independent random experiments, each with probability less than one. Let us define by $\mathcal{F}_\infty$ the σ-algebra, which consists of all the sets that can be described in terms of the first finitely many Bernoulli trials together with all the necessary sets in order to satisfy the requirements of a σ-algebra. Define by $Z_n(\omega)$ the rate of zeros in the first n Bernoulli trials. Denote by A the set of sequences of $1's$ and $0's$ for which the long-run average of zero's is

$$\lim_{n\to\infty} Z_n(\omega) = \frac{1}{2}.$$

Then based on the Strong Law of Large Numbers it could be proved (see Williams (2004)) that

$$\mathbb{P}(A) = 1 \text{ if } p = \frac{1}{2} \text{ and } \mathbb{P}(A) = 0 \text{ if } p \neq \frac{1}{2}.$$

1.6 Filtration

A filtration $\{\mathcal{F}_t, t \geq 0\}$ is a family of σ-algebras $\mathcal{F}_t$ on the same probability space $(\Omega, \mathcal{F}, \mathbb{P})$, which is increasing. That is,

$$\mathcal{F}_s \subset \mathcal{F}_t \text{ for } s < t \, .$$

In other words, a filtration is a family of σ-algebras $\mathcal{F}_t$ in the same probability space $(\Omega, \mathcal{F}, \mathbb{P})$, for which if $t = 0, 1, 2, ...$ then

$$\mathcal{F}_0 \subset \mathcal{F}_1 \subset \mathcal{F}_2 \subset ... \subset \mathcal{F}_t \subset ...$$

We denote by $\mathcal{F}_\infty$ the minimum σ-algebra generated by the filtration $\{\mathcal{F}_t, t \geq 0\}$, that is,

$$\mathcal{F}_\infty = \vee_{t \in \mathbb{R}} \mathcal{F}_t.$$

We assume that the filtration satisfies the so-called *usual hypothesis,* that is,

(a) The filtration is right continuous, i.e., $\mathcal{F}_t = \cap_{s>t} \mathcal{F}_s$,

(b) The σ-algebra $\mathcal{F}_0$ contains the null sets of $\mathcal{F}_\infty$.

A *filtered probability space* $(\Omega, \mathcal{F}, \mathcal{F}_t, \mathbb{P})$ is a probability space endowed with a filtration which satisfies the usual hypotheses.

If in a filtered probability space $(\Omega, \mathcal{F}, \mathcal{F}_t, \mathbb{P})$ the index t expresses the time, then a σ-algebra $\mathcal{F}_t$ contains all the information for the true value of ω from 0 up to time t.

Example 16 *Assume that we execute three times a Bernoulli trial, where we define the two possible outcomes as "success" and "failure". We assign the number* 1 *if the outcome of a Bernoulli trial is success and the number* 0 *if the outcome is failure. Then the sample space of the experiment is the following:*

$$\Omega = \{(1,1,1), (1,1,0), (1,0,1), (1,0,0), (0,1,1), (0,1,0), (0,0,1), (0,0,0)\}.$$

(*a*) *Let us define by*

$$\mathcal{F}_0 = \{\varnothing, \Omega\},$$

then it is very easy to see that $\mathcal{F}_0$ *is a* σ*-algebra. The* σ*-algebra* $\mathcal{F}_0$ *contains no information for the true outcome* ω *for any trial of the experiment. This is so since the set* $\varnothing$ *and the sample space* Ω *are always resolved,even without information. The* σ*-algebra* $\mathcal{F}_0$ *is called the trivial* σ*-algebra.*

(*b*) *Let us define by*

$$A_1 = [(1,1,1), (1,1,0), (1,0,1), (1,0,0)],$$

and

$$A_2 = [(0,1,1), (0,1,0), (0,0,1), (0,0,0)].$$

Then define as

$$\mathcal{F}_1 = \{\varnothing, \Omega, A_1, A_2\},$$

then it is easy to see that $\mathcal{F}_1$*is a* σ*-algebra. We say that the* σ*-algebra* $\mathcal{F}_1$ *contains the information learned by observing the realization of the first trial. Specifically, if* $\omega \in \mathcal{F}_1$ *we know that on the first trial we had as outcome a success. The sets* A_1 *and* A_2 *that generate with all their possible unions and complements the elements of* $\mathcal{F}_1$ *are called the atoms of the* σ*-algebra* $\mathcal{F}_1$*. Note, that*

$$\mathcal{F}_0 \subset \mathcal{F}_1.$$

(*c*) *Now consider the sets*

$$A_{11} = [(1,1,1), (1,0,1)], \qquad A_{10} = [(1,0,1), (1,0,0)],$$

$$A_{01} = [(0,1,1),(0,1,0)], \qquad A_{00} = [(0,0,1),(0,0,0)],$$

and consider the set that these sets generate as atoms, i.e., the set containing all their possible unions and compliments,

$$\mathcal{F}_2 = \left\{ \begin{array}{c} \varnothing, \Omega, A_1, A_2, A_{11}, A_{10}, A_{01}, A_{00}, A_{11}^c, A_{10}^c, A_{01}^c, A_{00}^c, \\ A_{11} \cup A_{01}, A_{11} \cup A_{00}, A_{10} \cup A_{01}, A_{10} \cup A_{00} \end{array} \right\}.$$

In $\mathcal{F}_2$ as is given above one may wonder where is $A_{11} \cup A_{10}$. The reason that is missing is that $A_{11} \cup A_{10} = A_1$. The same reasoning apply to $A_{01} \cup A_{00} = A_0$. Then it is easy to see that $\mathcal{F}_2$ is a σ-algebra. We say that the σ-algebra $\mathcal{F}_2$ contains the information learned by observing the first two trials being realized. More specifically, if we are told for which set in $\mathcal{F}_2$ the realized $\hat{\omega}$ belongs, we know the outcome of the first two trials. Again note that:

$$\mathcal{F}_0 \subset \mathcal{F}_1 \subset \mathcal{F}_2.$$

Now consider the σ-algebra $\mathcal{F}_3$ to be the power set $\mathrm{P}(\Omega)$ of Ω. There are 256 subsets in this set. We observe that

$$\mathcal{F}_0 \subset \mathcal{F}_1 \subset \mathcal{F}_2 \subset \mathcal{F}_3.$$

Remark 17 *The way $\mathcal{F}_2$ was constructed from the four sets $A_{11}, A_{10}, A_{01}, A_{00}$, suggests that the σ-algebras in a filtration can be built by taking unions and complements of certain fundamental sets and not considering all other sets. However, in uncountable sample spaces, there are sets that cannot be constructed as countable unions of atoms, and uncountable unions are forbidden because we cannot add probabilities of such unions.*

1.7 Random variables

In the present section, we consider random variables defined on a probability space $(\Omega, \mathcal{F}, \mathbb{P})$ where Ω is an abstract sample space. At the same time, we wish to generalize the definition of the random variable given in an elementary course in probability by considering more general spaces for the values of the random variables than that of the line of real numbers or its multivariate extensions.

Let $(E, \mathcal{E})$ be a measurable space and consider a function $X : \Omega \to E$. What is of central importance all too often in stochastic mathematics is to assign probabilities on sets of $\mathcal{E}$. Hence, we will want to compute for $A \in \mathcal{E}$

$$\mathbb{P}(\{\omega : X(\omega) \in A\}) = \mathbb{P}(X \in A) = \mathbb{P}\left(X^{-1}(A)\right).$$

From these three equivalent ways of writing the required quantity the third is the more enlighten, since in order to compute $\mathbb{P}\left(X^{-1}(A)\right)$, we need $X^{-1}(A)$ to be an element of $\mathcal{F}$. This motivates the following definition.

Definition 18 (*i*) *Let $(S, \mathcal{S})$ and $(E, \mathcal{E})$ be two measurable spaces. A mapping $X : S \to E$ is called measurable relative or with respect to $\mathcal{S}$ and $\mathcal{E}$ if $X^{-1}(A) \in \mathcal{S}$ for all $A \in \mathcal{E}$. In this case we may write $X^{-1}(\mathcal{E}) \subset \mathcal{S}$.*

(*ii*) *When $(S, \mathcal{S}) = (\Omega, \mathcal{F})$, a measurable map X is called a random variable.*

(*iii*) *When $E = \mathbb{R}$, we usually take $\mathcal{E}$ to be the Borel σ-algebra $\mathcal{B}$ of $\mathbb{R}$.*

It is straightforward to see that the mapping $X^{-1} : \mathcal{E} \to \mathcal{S}$ preserves the set of operations of complementation, union, and intersection, that is to say, for any $A \in \mathcal{E}$ and any family A_i, $i \in I$ of elements of $\mathcal{E}$,

$$X^{-1}(A^c) = \left(X^{-1}(A)\right)^c,$$

$$X^{-1}\left(\bigcup_{i \in I} A_i\right) = \bigcup_{i \in I} X^{-1}(A_i),$$

$$X^{-1}\left(\bigcap_{i\in I} A_i\right) = \bigcap_{i\in I} X^{-1}(A_i).$$

From this it follows immediately, that if $\mathcal{E}$ is a π-system [algebra, σ-algebra] on E, then $X^{-1}(\mathcal{E})$ is a π-system [algebra, σ-algebra] on S.

σ-algebras generated by random variables.

We will now provide the definition of a σ-algebra that is generated by a random variable X.

Definition 19 *Let $(\Omega, \mathcal{F}, \mathbb{P})$ be a probability space, and X a random variable on it, that is a function*

$$X : \Omega \to \mathbb{R}.$$

We call the generated by X σ-algebra and we denote it by $\sigma(X)$, the minimum σ-algebra in Ω that is being generated by the set $X^{-1}(\mathbb{R}) \subseteq \Omega$.

We will now describe how to construct $\sigma(X)$ for the case of a discrete random variable that takes values in $\mathbb{N}$. This method of construction applies to any discrete random variable since we can always have a one-to-one mapping of the values of a discrete random variable and the set $\mathbb{N}$.

We define the sets

$$A_i = \{\omega : X(\omega) = i\} \text{ for } i = 1, 2, ...,$$

then the sets A_i for $i = 1, 2, ...$ are a countable partitioning of Ω, since

$$\Omega = \cup_{i=1}^{\infty} A_i \text{ and } A_i \cap A_j = \varnothing \text{ if } i \neq j.$$

Then the σ-algebra $\sigma(X)$ consists of the sets A_i for $i = 1, 2, ..., \varnothing, \Omega$ and every other set that is the union of any of the sets A_i.

Example 20 *Let us visit Example 16 and define the random variable $X = \{$the number of successes in the first two trials$\}$. Define by*

$$A_0 = \{\omega : X(\omega) = 0\},$$

which in fact means that

$$A_0 = \{(0,0,1), (0,0,0)\}.$$

Similarly we have

$$A_1 = \{\omega : X(\omega) = 1\} \quad , \quad A_2 = \{\omega : X(\omega) = 2\}$$

and

$$A_1 = \{(1,0,1), (0,1,1), (1,0,0), (0,1,0)\} \quad , \quad A_2 = \{(1.1.1), (1,1,0)\}.$$

We have that

$$\Omega = A_0 \cup A_1 \cup A_2, \quad \text{with} \quad A_0 \cap A_1 = \varnothing, \quad A_0 \cap A_2 = \varnothing, \quad A_1 \cap A_2 = \varnothing.$$

Now, the σ-algebra $\sigma(X)$ consists of the sets A_i for $i = 1, 2, 3, \varnothing, \Omega$ and every other set that is the union of any of the sets A_i. As we have seen this is σ-algebra $\mathcal{F}_2$ of Example 16.

We will now define the σ-algebra generated by two or more random variables

Definition 21 *Let $(\Omega, \mathcal{F}, \mathbb{P})$ be a probability space and X, Y two random variable on it, that is,*

$$X : \Omega \to \mathbb{R}, \quad Y : \Omega \to \mathbb{R}.$$

We call the generated by X and Y σ-algebra and we denote it by $\sigma(X, Y)$, the minimum σ-algebra in Ω that the random variables X and Y are $\sigma(X, Y)$-measurable in Ω. For more than two random variables in Ω the σ-algebra generated by them is defined analogously.

We will now describe how to construct $\sigma(X, Y)$ for the case of a discrete random variable that takes values in $\mathbb{N}$. This method of construction applies for any discrete random variable since we can always have a one to one mapping of the values of a discrete random variable and the set $\mathbb{N}$.

We define the sets

$$A_{ij} = \{\omega : X(\omega) = i, Y(\omega) = j\}, \quad \text{for } i, j = 1, 2, ...$$

Then we have that

$$\Omega = \cup_{i,j}^{\infty} A_{ij} \text{ and } A_{ij} \cap A_{mn} = \varnothing \text{ if } (i, j) \neq (m, n).$$

The σ-algebra $\sigma(X, Y)$ is the set containing the sets A_{ij} for $i, j = 1, 2, ...$, $\varnothing$, Ω and all possible unions of the sets A_{ij} for $i, j = 1, 2, ...$

Example 22 *We extend Example 20 by using the random variable X defined there, and by defining as the random variable Y in Ω to be*

$$Y = \{\text{the number of failures in the first two trials}\}.$$

Then we have that

$$A_{02} = \{\omega : X(\omega) = 0, Y(\omega) = 2\} = \{(1,1,1), (1,1,0)\},$$
$$A_{01} = \{\omega : X(\omega) = 0, Y(\omega) = 1\} = \{\varnothing\},$$
$$A_{00} = \{\omega : X(\omega) = 0, Y(\omega) = 0\} = \{\varnothing\},$$
$$A_{10} = \{\omega : X(\omega) = 1, Y(\omega) = 0\} = \{\varnothing\},$$
$$A_{11} = \{\omega : X(\omega) = 1, Y(\omega) = 1\} = \{(1,0,1), (0,1,1), (1,0,0), (0,1,0)\},$$
$$A_{12} = \{\omega : X(\omega) = 1, Y(\omega) = 2\} = \{\varnothing\},$$
$$A_{20} = \{\omega : X(\omega) = 2, Y(\omega) = 1\} = \{(0,0,1), (0,0,0)\},$$
$$A_{21} = \{\omega : X(\omega) = 2, Y(\omega) = 1\} = \{\varnothing\},$$
$$A_{22} = \{\omega : X(\omega) = 2, Y(\omega) = 2\} = \{\varnothing\}.$$

It is easy to see that

$$\Omega = \cup_{i,j=1}^{2} A_{ij}$$
$$= \{\varnothing, (0,0,1), (0,0,0), (1,0,1), (0,1,1), (1,0,0), (0,1,0), (1,1,1), (1,1,0)\},$$

and

$$A_{ij} \cap A_{mn} = \varnothing \text{ for every } (i, j) \neq (m, n).$$

According to Definition 21 and the method of construction of σ(X, Y) that followed, we have that σ(X, Y) consists of the elements

$$\left\{ \begin{array}{c} \varnothing, \Omega, [(1,1,1), (1,1,0)], [(1,0,1), (0,1,1), (1,0,0), (0,1,0)], \\ [(0,0,1), (0,0,0)], \end{array} \right\},$$

and all their possible unions. It is straight forward to check that this is the σ-algebra $\mathcal{F}_2$.

Distribution measures of random variables

Random variables have distribution measures. Two different random variables can have the same distribution measure. A random variable can have two different distribution measures under different probability measures. As examples of these remarks consider the Fundamental experiment of choosing a point uniformly on the interval $[0,1]$ and the Lebesque measure on $\mathbb{R}$. Define the two real random variables $X(\omega) = 1+\omega$ and $Y(\omega) = 2-\omega$ for all $\omega \in [0,1]$. These two random variables are different since if $X = \frac{1}{2}$ then $Y = \frac{5}{2}$. The distribution measure of $X(\omega)$ is uniform on the interval of its values, and equal with

$$\begin{aligned} \mu_X[a,b] &= \mathbb{P}\left(X^{-1}[a,b]\right) = \mathbb{P}\{\omega : a \le 1+\omega \le b\} = \mathbb{P}[a-1, b-1] \\ &= b-1-a+1 = b-a. \end{aligned}$$

It turns out that $Y(\omega) = 2-\omega$ has the same distribution as $X(\omega)$ on the common interval of their values. That is,

$$\begin{aligned} \mu_Y[a,b] &= \mathbb{P}\left(Y^{-1}[a,b]\right) = \mathbb{P}\{\omega : a \le 2-\omega \le b\} = \mathbb{P}[2-b, 2-a] \\ &= 2-a-2+b = b-a. \end{aligned}$$

We will now provide the definition of two equivalent probability measures.

Definition 23 *Consider a probability space $(\Omega, \mathcal{F}, \mathbb{P})$ where Ω is a finite set. Assume that on this probability space two probability measures $\mathbb{P}$ and $\mathbb{Q}$ are defined. We say that the two probability measures are equivalent if for every $\omega \in \Omega$, we have*

$$\mathbb{P}(\omega) > 0 \text{ and } \mathbb{Q}(\omega) > 0.$$

In more depth, by changing probability measures, we assign different probabilities to the same elementary events of the sample space.

Now let us now define a probability measure $\hat{\mathbb{P}}$ equivalent with $\mathbb{P}$ by specifying

$$\hat{\mathbb{P}}[a,b] = \int_a^b 3\omega^2 d\omega = b^3 - a^3.$$

Then it is apparent that $\hat{\mathbb{P}}$ is a measure satisfying the necessary properties and in addition $\hat{\mathbb{P}}[0,1] = 1$. Hence, for any Borel subset A of $\mathbb{R}$ then $\hat{\mathbb{P}}(A)$ is well defined. Under $\hat{\mathbb{P}}$ the random variables $X(\omega)$ and $Y(\omega)$, as we will see, have different distributions. Let us define by $\hat{\mu}_X$ the distribution measure of $X(\omega)$ under the probability measure $\hat{\mathbb{P}}$, then

$$\begin{aligned} \hat{\mu}_X[a,b] &= \hat{\mathbb{P}}\left(X^{-1}[a,b]\right) = \hat{\mathbb{P}}\{\omega : a \le 1+\omega \le b\} = \hat{\mathbb{P}}[a-1, b-1] \\ &= (b-1)^3 - (a-1)^3. \end{aligned}$$

Similarly, let us define by $\hat{\mu}_Y$ the distribution measure of $Y(\omega)$ under the probability measure $\hat{\mathbb{P}}$, then

$$\begin{aligned} \hat{\mu}_Y[a,b] &= \hat{\mathbb{P}}\left(Y^{-1}[a,b]\right) = \hat{\mathbb{P}}\{\omega : a \le 2-\omega \le b\} = \hat{\mathbb{P}}[2-b, 2-a] \\ &= (2-a)^3 - (2-b)^3. \end{aligned}$$

When X is a real valued random variable, its distribution measure is a probability measure on $\mathbb{R}$, which is entirely characterized by its *cumulative distribution function:*

$$F_X(x) = \mu_X((-\infty, x]) = \mathbb{P}(X \le x), \quad x \in \mathbb{R}. \tag{1.13}$$

On the other hand, if we know the cumulative distribution function $F_X(x)$, then from (1.13) we immediately get that

$$\mu_X (x, y] = F_X (y) - F_X (x),$$

and remains to find the distribution measure $\mu_X [x, y]$. In order to do so we proceed as follows

$$\begin{aligned}\mu_X [x, y] &= \mu_X \bigcap_{n \to \infty} \left(x - \frac{1}{n}, y\right] = \lim_{n \to \infty} \left(x - \frac{1}{n}, y\right] \\ &= F_X (y) - \lim_{n \to \infty} F_X \left(x - \frac{1}{n}\right).\end{aligned}$$

Therefore, knowing the cumulative distribution function for a random variable is the same as knowing its distribution measure.

When the cumulative distribution function F_X of the random variable X admits a density f_X, that is,

$$F_X (x) = \int_{-\infty}^{x} f_X (u)\, du \quad \text{for all } x \in \mathbb{R},$$

we say that the function f_X is the probability function of the random variable X. When the random variable is discrete, that is, it takes a countable or finite number of values with probability one, we then define $p_i = \mathbb{P}\{X = x_i\}$ with $\sum_i p_i = 1$. Then the probability mass assigned to a Borel set $A \subset \mathbb{R}$ by the distribution measure of X is

$$\mu_X (A) = \sum_{\{i; x_i \in A\}} p_i = 1,$$

which we call a *probability function.* There are random variables that have no cumulative probability function nor probability mass function.

As an example consider the case where $\Omega = \mathbb{R}$ and $\mathcal{F} = \mathcal{B}$. Consider the random variable $X(\omega) = \omega$ for all $\omega \in \mathbb{R}$ and consider the probability measure $\mathbb{P}$ defined by

$$\mathbb{P}[a, b] = \int_a^b \phi (u)\, du \quad \text{whenever } -\infty < a \leq b < \infty,$$

where

$$\phi (x) = \frac{1}{\sqrt{2\pi}} \exp\left(-\frac{x^2}{2}\right),$$

then X is called the standard normal random variable.

1.8 Integration with respect to a probability measure

Let $(\Omega, \mathcal{F}, \mathbb{P})$ be a probability space and $X(\omega)$ a random variable on it. We want to define the *expectation,* or what is equivalent, the "integral" of random variables when the sample space Ω is a general space, that is, Ω is uncountable infinite. This case is more delicate than the countably infinite case which is taught in a first course in Probability Theory. We assume that it is known how the Riemann[8] integral of a continuous function f is defined, as

[8]Bernhard Riemann was born in September 17, 1826 in Breselenz, Germany and died of tuberculosis in July 20, 1866. His doctoral advisor was Gauss. Riemann's published works opened up research areas combining analysis with geometry. These would subsequently become major parts of the theories of Riemannian geometry, algebraic geometry and complex manifold theory. Riemann made major contributions to real analysis. He also made some famous contributions to modern analytic number theory, where he introduced the Riemann zeta function. In 1853, Gauss asked his student to prepare a *Habitationsschrift* on the foundations of geometry. Reimann delivered at Gottingen in 1854 the legendary lecture with the title "On the hypotheses which lie at the basis of geometry" (Riemann (1998). The subject founded by this work is Riemannian geometry.

the limit of the upper Riemann sum and the lower Riemann sum, while the partition of the integration interval becomes very thin. The problem of applying Riemann's methodology for integration of a continuous function, to define expectation for the random variable $X(\omega)$ is, that Ω is a general set and more than often not a subset of $\mathbb{R}$. There is the problem of how to partition Ω in a natural way. Instead we partition the set of values of the random variable $X(\omega)$ in the following way:

Assume initially that $0 \leq X(\omega) < \infty$ for every $\omega \in \Omega$ and let the points of partition be if we order them

$$0 = y_0 < y_1 < \ldots < \ldots,$$

where for each subinterval $[y_i, y_{i+1}]$, we set

$$A_i = \{\omega \in \Omega;\ y_i \leq X(\omega) \leq y_{i+1}\}.$$

Now, we define the lower Lebesgue sum to be

$$LS^-(X) = \sum_{i=1}^{\infty} y_i \mathbb{P}(A_i).$$

This limit exists as the maximum distance between the partition points $[y_i, y_{i+1}]$ for every i goes to zero, that is, the partition becomes very thin. We define this limit to be the *Lebesque integral*

$$\int_{\Omega} X(\omega)\, d\mathbb{P}(\omega) \quad \text{or} \quad \int_{\Omega} X d\mathbb{P}. \tag{1.14}$$

If $\mathbb{P}\{\omega : X(\omega) \geq 0\} = 1$ but $\mathbb{P}\{\omega : X(\omega) = \infty\} > 0$ then we define

$$\int_{\Omega} X(\omega)\, d\mathbb{P}(\omega) = \infty.$$

It remains to consider random variables that have the possibility to take both positive and negative values. For such a case, we define the positive and negative parts of $X(\omega)$ as follows

$$X^+(\omega) = \max\{X(\omega), 0\} \quad \text{and} \quad X^-(\omega) = \max\{-X(\omega), 0\},$$

then apparently both random variables $X^+(\omega)$ and $X^-(\omega)$ are non-negative and in addition we have

$$X(\omega) = X^+(\omega) - X^-(\omega) \quad \text{and} \quad |X(\omega)| = X^+(\omega) + X^-(\omega).$$

The Lebesgue integrals

$$\int_{\Omega} X^+(\omega)\, d\mathbb{P}(\omega) \quad \text{and} \quad \int_{\Omega} X^-(\omega)\, d\mathbb{P}(\omega),$$

are both defined analogously as the Lebesgue integral (1.14), and provided that they are both finite we have

$$\int_{\Omega} X(\omega)\, d\mathbb{P}(\omega) = \int_{\Omega} X^+(\omega)\, d\mathbb{P}(\omega) - \int_{\Omega} X^-(\omega)\, d\mathbb{P}(\omega).$$

In this case, we say that the random variable $X(\omega)$ is *integrable.* If

$$\int_{\Omega} X^+(\omega)\, d\mathbb{P}(\omega) = \infty \ \text{and} \int_{\Omega} X^-(\omega)\, d\mathbb{P}(\omega) \quad \text{is finite,}$$

then $\int_\Omega X(\omega)\, d\mathbb{P}(\omega) = \infty$. If

$$\int_\Omega X^+(\omega)\, d\mathbb{P}(\omega) \quad \text{is finite and} \int_\Omega X^-(\omega)\, d\mathbb{P}(\omega) = \infty,$$

then $\int_\Omega X(\omega)\, d\mathbb{P}(\omega) = -\infty$. If

$$\int_\Omega X^+(\omega)\, d\mathbb{P}(\omega) = \infty \quad \text{and} \int_\Omega X^-(\omega)\, d\mathbb{P}(\omega) = \infty,$$

then $\int_\Omega X(\omega)\, d\mathbb{P}(\omega)$ is not defined.

We now provide the definition of an integrable random variable

Definition 24 *Let $(\Omega, \mathcal{F}, \mathbb{P})$ be a probability space and $X(\omega)$ a random variable on it. The random variable X is integrable if and only if*

$$\int_\Omega |X(\omega)|\, d\mathbb{P}(\omega) < \infty.$$

We write $\mathcal{L}^1$or $\mathcal{L}^1(\Omega, \mathcal{F}, \mathbb{P})$ to be more precise in case of possible confusion in the presence of more than one probability spaces to be the space of all integrable random variables on $(\Omega, \mathcal{F}, \mathbb{P})$.

We are now in a position to provide the definition of the expectation of a random variable on an abstract probability space.

Definition 25 *Let $(\Omega, \mathcal{F}, \mathbb{P})$ be a probability space and $X(\omega)$ a random variable on it. The expectation (or expected variable) of X is defined to be*

$$\mathbb{E}[X] = \int_\Omega X(\omega)\, d\mathbb{P}(\omega),$$

which makes sense if X is integrable, that is, if

$$\mathbb{E}[|X|] = \int_\Omega |X(\omega)|\, d\mathbb{P}(\omega) < \infty,$$

or if $X \geq 0$ almost surely; in this case $\mathbb{E}[X]$ might be ∞.

The next theorem contains some of the most important properties of the expectation operator on $\mathcal{L}^1(\Omega, \mathcal{F}, \mathbb{P})$.

Theorem 26 *Let $(\Omega, \mathcal{F}, \mathbb{P})$ be a probability space and $X(\omega)$ a random variable on it. Then the following statements hold.*

(i) $\mathcal{L}^1$ is a vector space, basically due to the linearity property of the expectations of two random variables, that is, if X and Y are random variables and a and b are non-negative constants then

$$\mathbb{E}[aX + bY] = a\mathbb{E}[X] + b\mathbb{E}[Y],$$

exhibiting in fact also the linearity property of the Lebesgue integral. Moreover, if $0 \leq X \leq Y$ *and* $Y \in \mathcal{L}^1$, *then* $X \in \mathcal{L}^1$ *and* $\mathbb{E}[X] \leq \mathbb{E}[Y]$.

(ii) $X \in \mathcal{L}^1$ *if and only if* $|X| \in \mathcal{L}^1$ *and in this case* $|\mathbb{E}[X]| \leq \mathbb{E}[|X|]$. *In particular, any bounded random variable is integrable. Moreover, if* $X = Y$ *almost surely, that is,*

$$\mathbb{P}(\{\omega : X(\omega) = Y(\omega)\}) = 1,$$

then

$$\mathbb{E}[X] = \mathbb{E}[Y].$$

Proof. The proof of the present and the next three important theorems are not essential to the reader's insight and we will skip them. However for the interested reader readable proofs could be found in Jacod and Protter (2004). ∎

Theorem 27 *(Monotone convergence theorem). Let* $(\Omega, \mathcal{F}, \mathbb{P})$ *be a probability space and* $\{X_n\}_{n=0}^{\infty}$ *a sequence of random variables. If* $\{X_n\}_{n=0}^{\infty}$ *is positive and increasing almost surely to* X, *then*

$$\lim_{n\to\infty} \mathbb{E}[X_n] = \mathbb{E}[X].$$

Theorem 28 *(Fatou's lemma). Let* $(\Omega, \mathcal{F}, \mathbb{P})$ *be a probability space and* $\{X_n\}_{n=0}^{\infty}$ *a sequence of random variables. If for all* n, $X_n \geq Y$ *almost surely and* $Y \in \mathcal{L}^1$ *then we have that*

$$\mathbb{E}\left\{\lim\inf_{n\to\infty} X_n\right\} \leq \lim\inf_{n\to\infty} \mathbb{E}[X_n].$$

In particular, if $X_n \geq 0$ *almost surely for all* n *then*

$$\mathbb{E}\left\{\lim\inf_{n\to\infty} X_n\right\} \leq \lim\inf_{n\to\infty} \mathbb{E}[X_n].$$

Theorem 29 *(Lebesgue's dominated convergence theorem). Let* $(\Omega, \mathcal{F}, \mathbb{P})$ *be a probability space and* $\{X_n\}_{n=0}^{\infty}$ *a sequence of random variables. If for all* n, $|X_n| \leq Y$ *almost surely,* $Y \in \mathcal{L}^1$ *and* $\{X_n\}_{n=0}^{\infty}$ *converge to* X *almost surely, then* $X \in \mathcal{L}^1$ *and*

$$\lim_{n\to\infty} \mathbb{E}[X_n] = \mathbb{E}[X].$$

Remark 30 *Let the vector space* $\mathcal{L}^1(\Omega, \mathcal{F}, \mathbb{P})$, *then as we have seen in the above theorems, the almost sure equality of two random variables in* $\mathcal{L}^1$ *is an equivalence relation. In this respect a random variable* X *is a representative of all almost surely random variables equals to it and in fact they all have the same expectation. We may thus consider the space* L^1 *of all such representatives of the random variables in* $\mathcal{L}^1$. *Then, since the addition of random variables preserve almost surely equality, the set* L^1 *is also a vector space.*

We may easily generalize this concept by letting a parameter $1 \leq p < \infty$ and define the space $\mathcal{L}^p$ to be the space of all random variables such that $|X|^p \in \mathcal{L}^1$. Then the space L^p is defined analogously to L^1, as the space of representatives random variables that are almost surely equal and belong to $\mathcal{L}^p$.

We continue now with a useful consequence of the Lebesgue Dominated Convergence Theorem, which establishes the fact that we are allowed under certain conditions to interchange infinite summation and expectation. The proof could be found in Jacod and Protter (2004).

Theorem 31 *Let* $(\Omega, \mathcal{F}, \mathbb{P})$ *be a probability space and* $\{X_n\}_{n=0}^{\infty}$ *a sequence of random variables. Then*

(i) If the random variables $X_n > 0$ for all n,

$$\mathbb{E}\left[\sum_{n=1}^{\infty} X_n\right] = \sum_{n=1}^{\infty} \mathbb{E}\left[X_n\right].$$

and their sum could be finite or infinite.

(ii) If $\sum_{n=1}^{\infty} \mathbb{E}\left[|X_n|\right] < \infty$, then $\sum_{n=1}^{\infty} X_n$ converges almost surely and the sum of this series is integrable. Moreover,

$$\mathbb{E}\left[\sum_{n=1}^{\infty} X_n\right] = \sum_{n=1}^{\infty} \mathbb{E}\left[X_n\right] < \infty.$$

For easy reference, we will call the members of $\mathcal{L}^2$, that is, the random variables X of the probability space $(\Omega, \mathcal{F}, \mathbb{P})$, which are such that $\mathbb{E}\left[X^2\right] < \infty$, *square integrable random variables.* Consider now the space L^2 of the representatives of all equivalent classes of almost surely equality of random variables X such that $\mathbb{E}\left[X^2\right] < \infty$. Therefore, $X \in L^2$ represents all random variables $Y \in \mathcal{L}^2$ which are almost surely equal with X. This has a consequence that if $\mathbb{E}\left[X^2\right] = 0$, we can conclude that $X = 0$ and not only that $X = 0$ almost surely. We will now prove the following theorem which apart from the well known inequality of Caushy-Schwarz will provide the first tools to study the properties of the space L^2.

Theorem 32 *Let a probability space $(\Omega, \mathcal{F}, \mathbb{P})$ and the random variables $X, Y \in L^2$. Then $XY \in L^1$. In addition, the Caushy[9]-Schwarz[10] inequality is true, i.e.,*

$$|\mathbb{E}\left[XY\right]| \leq \left\{\mathbb{E}\left[X^2\right]\right\}^{\frac{1}{2}} \left\{\mathbb{E}\left[Y^2\right]\right\}^{\frac{1}{2}}.$$

Finally we have that $L^2 \subset L^1$, and if $X \in L^2$, then $\left\{\mathbb{E}\left[X\right]\right\}^2 \leq \mathbb{E}\left[X^2\right]$.

Proof. *It is easy to see that*

$$|XY| \leq \frac{1}{2} X^2 + \frac{1}{2} Y^2,$$

and therefore since $\mathbb{E}\left[X^2\right] < \infty$ and $\mathbb{E}\left[Y^2\right] < \infty$, we immediately get that $\mathbb{E}\left[|XY|\right] < \infty$ and consequently $XY \in L^1$.

Now let any $x \in \mathbb{R}$ then we have that

$$x^2 \mathbb{E}\left[X^2\right] + 2x \mathbb{E}\left[XY\right] + \mathbb{E}\left[Y^2\right] = \mathbb{E}\left[(xX + Y)^2\right] \geq 0,$$

is a polynomial of the second degree in x which is non-negative for any $x \in \mathbb{R}$ and thus from its discriminant we immediately get that

$$\left\{\mathbb{E}\left[XY\right]\right\}^2 - \mathbb{E}\left[X^2\right] \mathbb{E}\left[Y^2\right] \leq 0,$$

from which we get the Caushy-Swarz inequality. Finally assume that $X \in L^2$. Then since $X = X.1$ and apparently $\mathbb{E}\left[1^2\right] = 1 < \infty$, that is, $1 \in L^2$ therefore $X.1 = X \in L^1$. ■

[9]Augustin-Louis Caushy was born in 21 August 1789 in France and died in 23 May 1857. He founded compex analysis and permutation groups in algebra. His research work has a great diversity in mathematics and mathematical physics. More concepts and theorems have named after Caushy than any other mathematician. Caushy is one of the most prolific writers in Mathematics.

[10]Herman Schwarz was born in 25 January, 1843 in Jerzmanowa, Polland. He died in 30 November 1921. He made great contributions in the fields of complex analysis, differential geometry and calculus of variations. He worked for many years in Gottingen University. He supervised 20 research students many of which became famous mathematicians.

1.9 Indicator functions

Let $(\Omega, \mathcal{F}, \mathbb{P})$ a probability space. For any event $A \in \mathcal{F}$, we define as the *indicator function* $\mathbf{1}_A$ of the event A the following function

$$\mathbf{1}_A(\omega) = \begin{cases} 1 & \text{if } \omega \in A, \\ 0 & \text{if } \omega \notin A. \end{cases}$$

It is easy to see that

$$\mathbf{1}_{A^c} = 1 - \mathbf{1}_A, \qquad \mathbf{1}_A \leq \mathbf{1}_B \text{ iff } A \subset B, \qquad \mathbf{1}_{A \cup B} \leq \mathbf{1}_A + \mathbf{1}_B,$$

$$\mathbf{1}_{A \cap B} = \mathbf{1}_A \mathbf{1}_B, \qquad \mathbf{1}_{A \cup B} = \mathbf{1}_A + \mathbf{1}_B - \mathbf{1}_A \mathbf{1}_B,$$

It is useful to adopt that

$$\cup_{i \in \varnothing} A_i = \varnothing, \qquad \cap_{i \in \varnothing} A_i = \Omega.$$

Let I be an an arbitrary set and let that the family of sets $\{A_i, i \in I\} \subset \mathcal{F}$. Then as an exercise one might be interested to prove that

$$\mathbf{1}_{\cap_{i \in I} A_i} = \inf_{i \in I} A_i, \qquad \mathbf{1}_{\cup_{i \in I} A_i} = \sup_{i \in I} A_i.$$

Let $\{A_n, n \in \mathbb{N}\}$ be a sequence of events in $(\Omega, \mathcal{F}, \mathbb{P})$. Then $\{\mathbf{1}_{A_n}, n \in \mathbb{N}\}$ is a sequence of functions on Ω with values 0 or 1. It is not difficult to see that

$$\mathbf{1}_{\cup_{n=1}^{\infty} A_n} \leq \sum_{n=1}^{\infty} \mathbf{1}_{A_n},$$

with equality holding iff $\mathbf{1}_{A_n} = 1$ for all $n \in \mathbb{N}$.

Remark 33 *(Important). Indicator functions are very useful when we need to integrate a random variable X over an event A of a probability space $(\Omega, \mathcal{F}, \mathbb{P})$, rather than over all of Ω. In such case, we have that*

$$\int_A X(\omega)\, d\mathbb{P}(\omega) = \int_\Omega \mathbf{1}_A(\omega) X(\omega)\, d\mathbb{P}(\omega) \text{ for all } A \in \mathcal{F}.$$

When $A, B \in \mathcal{F}$ with $A \cap B = \varnothing$, we have seen that $\mathbf{1}_{A \cup B} = \mathbf{1}_A + \mathbf{1}_B$, hence

$$\begin{aligned} \int_{A \cup B} X(\omega)\, d\mathbb{P}(\omega) &= \int_\Omega (\mathbf{1}_A(\omega) + \mathbf{1}_B(\omega)) X(\omega)\, d\mathbb{P}(\omega) \\ &= \int_\Omega \mathbf{1}_A(\omega) X(\omega)\, d\mathbb{P}(\omega) + \int_\Omega \mathbf{1}_B(\omega) X(\omega)\, d\mathbb{P}(\omega) \\ &= \int_A X(\omega)\, d\mathbb{P}(\omega) + \int_B X(\omega)\, d\mathbb{P}(\omega). \end{aligned}$$

The following remark is also very useful and important:

Remark 34 *(Important). Let a probability space $(\Omega, \mathcal{F}, \mathbb{P})$ and $A \in \mathcal{F}$ then*

$$\mathbb{E}[\mathbf{1}_A] = \int_\Omega \mathbf{1}_A(\omega)\, d\mathbb{P}(\omega) = \int_A d\mathbb{P}(\omega) = \mathbb{P}(A).$$

Another very interesting use of the indicator function is in the proof of *Markov's* and *Chebyshev*'s inequalities. Let that X is a non-negative random variable on $(\Omega, \mathcal{F}, \mathbb{P})$ with $X \in L^1$. Then it is rather easy to see that

$$\mathbf{1}_{(X \geq a)} \leq X/a, \quad X \geq 0.$$

Taking expectations on both sides of this inequality and using Remark 33, we get

$$\mathbb{P}(X \geq a) \leq \mathbb{E}(X)/a,$$

which Markov's inequality.

Now, let us consider a random variable X on $(\Omega, \mathcal{F}, \mathbb{P})$ with $X \in L^2$ and with no sign restriction. It is rather easy to see that

$$\mathbf{1}(|X-b| \geq a) \leq \left[\frac{X-b}{a}\right]^2,$$

and by taking expectations in this inequality, we obtain that

$$\mathbb{P}(|X-b| \geq a) \leq \mathbb{E}\left[(X-b)^2\right]/a^2.$$

This bound is minimal when $\mathbb{E}[X] = b$. Now setting $\mathbb{E}[X] = b$ in the relation above we obtain

$$\mathbb{P}(|X - \mathbb{E}[X]| \geq a) \leq \mathbb{V}(X)/a^2.$$

The above relation is called the *Chebychev's*[11] inequality and it is a most useful result. Among other uses it expresses the fact that if we can prove that the variance of a random variable X is small, then the probability that X deviates by more than a given amount from its mean value, is of a similar order of smallness.

1.10 The space L^2 is a Hilbert space

We will start by providing the formal definition of a Hilbert[12] space and then proceed to prove step by step that the space L^2 is a Hilbert space.

Definition 35 *A Hilbert space $\mathcal{H}$ is a complete normed linear space with an inner product satisfying $\langle x, x\rangle^{\frac{1}{2}} = \|x\|$ for all $x \in \mathcal{H}$.*

Hence, we have first to prove that L^2 is a linear space, secondly to define an inner product $\langle X, Y\rangle$ of two random variables X, Y in L^2, thirdly to prove that $\langle X, X\rangle^{1/2} = \|X\|$ is a norm in L^2, and finally that the space L^2 is complete. That is, in the final step we should prove that every convergent sequence of random variables in L^2 has its limit in L^2.

We will start by proving that L^2 is a linear space.

Let $X, Y \in L^2$, and $a, b \in \mathbb{R}$ we will show that L^2 is a *linear space.* We have that

$$(aX + bY)^2 \leq 2a^2X^2 + 2b^2Y^2,$$

[11]Pafnuty Chebyshev was born in May 16, 1821 in Okatovo Russia and died in December 8 1894. He is very well known for his work in the fields of Probability, Statistics, Mechanics, Analytical Geometry and Number Theory. He started teaching at St Petersberg University in 1847 and left in 1882 to devote his remaining life in research. Chebyshev is considered to be a founding father of Russian mathematics. Among his well-known students were the mathematicians Dmitry Grave, Aleksandr Korkin, Aleksandr Lyapunov, and Andrei Markov. There is a long list of things in Mathematics and Mechanics named after Chebychev.

[12]David Hilbert (January 23, 1862–February 14, 1943) was a German Mathematician and Philosopher who discovered a broad range of fundamental ideas in many areas. He is very well known for *Hilbert basis theorem, Hilbert axioms, Hilbert problems, Hilbert program, Einstein-Hilbert action* and *Hilbert space.* During the second International Congress of Mathematics held in Paris he put forward a most influential list of 23 unsolved problems known under the title *Hilbert problems.* Hilbert had a very large number of Doctoral students who became famous mathematicians. The largest part of his mathematical life has been spent at the University of Göttingen but unfortunately he lived to see in 1933 the authorities purge many of its prominent faculty members. On his tombstone, at Göttingen, one can read his epitaph, the famous lines he had spoken at the end of his retirement address to the Society of German Scientists and Physisians in the fall of 1930: "We must know. We will know".

and by taking expectations on both sides of the inequality, we get that

$$\mathbb{E}\left[(aX+bY)^2\right] \leq 2a^2\mathbb{E}\left[X^2\right] + 2b^2\mathbb{E}\left[Y^2\right] < \infty,$$

which proves that $aX + bY \in L^2$.

The second step is to define an inner product $\langle X, Y \rangle$ of two random variables X, Y in L^2:

For any two random variables $X, Y \in L^2$, we define their inner product to be

$$\langle X, Y \rangle = \mathbb{E}\left[XY\right].$$

For any $X \in L^2$, we have that

$$\langle X, X \rangle \geq 0, \text{ and } \langle X, X \rangle = 0 \text{ if and only if } X = 0.$$

Note also that for any $a, b \in \mathbb{R}$, we have

$$\langle aX + bY, Z \rangle = a\mathbb{E}\left[XZ\right] + b\mathbb{E}\left[YZ\right] = a\langle X, Z \rangle + b\langle Y, Z \rangle.$$

The third step is to prove that $\langle X, X \rangle^{1/2} = \|X\|$ is a norm in L^2. We have that

$$\| X \| = \langle X, X \rangle^{\frac{1}{2}} = \left(\mathbb{E}\left[X^2\right]\right)^{\frac{1}{2}}. \tag{1.15}$$

The function $\| \, . \, \|: L^2 \to \mathbb{R}$ which we defined in (1.15), we should show that it is indeed a norm. That is, we should prove that for any $X, Y \in L^2$ the following properties are true:

$$\begin{aligned} 1)\ \|X\| &\geq 0. \\ 2)\ \|aX\| &= |a| \, \| X \| \ \text{ for all } a \in \mathbb{R}. \\ 3)\ \|X+Y\| &\leq \| X \| + \| Y \|. \\ 4)\ \|XY\| &\leq \| X \| \| Y \|. \\ 5)\ \|X\| &= 0 \text{ if and only if } X = 0. \end{aligned}$$

Now we have

$$\| X \| = \left(\mathbb{E}\left[X^2\right]\right)^{\frac{1}{2}} \geq 0.$$

For the second property, we have that

$$\| aX \| = \left(\mathbb{E}\left[a^2X^2\right]\right)^{\frac{1}{2}} = |a| \left(\mathbb{E}\left[X^2\right]\right)^{\frac{1}{2}} = |a| \, \| X \| \ \text{ for all } a \in \mathbb{R}.$$

For the third property with the use of the Caushy-Schwarz inequality, we get that

$$\begin{aligned} \|X+Y\|^2 &= \mathbb{E}\left[X^2\right] + 2\mathbb{E}\left[XY\right] + \mathbb{E}\left[Y^2\right] \\ &\leq \|X\|^2 + 2\|X\|\|Y\| + \|Y\|^2 \\ &= \left(\|X\| + \|Y\|\right)^2, \end{aligned}$$

from which we get

$$\|X+Y\| \leq \|X\| + \|Y\|.$$

It remains to prove the forth property for a norm, which is also called the Schwarz inequality. However, this is an immediate consequence of the Caushy-Schwarz inequality for expectations.

So far we have established that the space L^2 is a normed linear space with inner product

$$\langle . \rangle^{\frac{1}{2}} = \|.\|.$$

It remains to prove the final step, that is, every convergent sequence of random variables in L^2 has its limit in L^2. This is provided in the following theorem the proof of which could be found in Jacod and Protter (2004).

Theorem 36 *Let a probability space $(\Omega, \mathcal{F}, \mathbb{P})$ and L^2 the space of representatives of the random variables X for which $\mathbb{E}\left[X^2\right] < \infty$. The space L^2 is a complete space.*

1.11 Independent σ-algebras and random variables

We will now discuss when two σ-algebras are independent, which is closely related with the independence of two random variables. Where two random variables X and Y are concerned, which in general take their values in two different spaces, to be independent we want knowledge of Y to leave unchanged the probabilities that X will take on certain values. We now provide the definition for two independent σ-algebras which generalizes straightforwardly for a finite number of σ-algebras.

Definition 37 *Let $\mathcal{F}$ a σ-algebra and $\mathcal{F}_1$ and $\mathcal{F}_2$ sub σ-algebras of $\mathcal{F}$. Then $\mathcal{F}_1$ and $\mathcal{F}_2$ are independent if for any pair of sets $A_1 \in \mathcal{F}_1$ and $A_2 \in \mathcal{F}_2$, we have*

$$\mathbb{P}(A_1 \cap A_1) = \mathbb{P}(A_1)\,\mathbb{P}(A_2).$$

Again for notational simplicity, we will provide the definition of two independent random variables, however, the results extend without difficulty to finite families of random variables.

Definition 38 *Let X and Y two random variables with values in $(F_1, \mathcal{F}_1)$ and $(F_2, \mathcal{F}_2)$ respectively. Then X and Y are independent if the generated σ-algebras $X^{-1}(\mathcal{F}_1)$ and $X^{-1}(\mathcal{F}_2)$ are independent.*

It is rather apparent that the above definition is not easily implemented. However, we provide the following theorem which provides necessary and sufficient conditions for two random variables to be independent. The proof exist for any of the standard texts on measure theory already mentioned.

Theorem 39 *Let X and Y two random variables with values in $(F_1, \mathcal{F}_1)$ and $(F_2, \mathcal{F}_2)$ respectively. Then in order for X and Y to be independent, it is necessary and sufficient to have any one of the following two conditions holding:*
(1) For all $A \in \mathcal{F}_1$ and $B \in \mathcal{F}_2$, we have

$$\mathbb{P}(X \in A, Y \in B) = \mathbb{P}(X \in A)\,\mathbb{P}(Y \in B).$$

(2) For all $A \in G_1$ and $B \in G_2$, where G_1, G_2 are respectively classes of sets which generate $\mathcal{F}_1$ and $\mathcal{F}_2$ and in addition are stable under finite intersections we have

$$\mathbb{P}(X \in A, Y \in B) = \mathbb{P}(X \in A)\,\mathbb{P}(Y \in B).$$

In general, if $\mathcal{F}_1$ and $\mathcal{F}_2$ are respectively σ-algebras on spaces F_1 and F_2 then the Cartesian product

$$\mathcal{F}_1 \times \mathcal{F}_2 = \{A \subset F_1 \times F_2 : A = B \times \Gamma, B \in \mathcal{F}_1 \text{ and } \Gamma \in \mathcal{F}_2\},$$

is *not* a σ-algebra on $F_1 \times F_2$. For this reason, we write $\sigma(\mathcal{F}_1 \times \mathcal{F}_2)$ to denote the smallest σ-algebra on $F_1 \times F_2$ generated by $\mathcal{F}_1 \times \mathcal{F}_2$. We denote this σ-algebra by

$$\mathcal{F}_1 \otimes \mathcal{F}_2 = \sigma(\mathcal{F}_1 \times \mathcal{F}_2).$$

1.12 Convergence of sequences of random variables

Let a probability space $(\Omega, \mathcal{F}, \mathbb{P})$ and a sequence of random variables $\{X_n\}_{n=0}^{\infty}$ with $X_n : \Omega \to \mathbb{R}$. Apparently, X_n for each $n \in \mathbb{N}$ is in fact a real function and one might think that following the definition of pointwise convergence, for a sequence of functions may define, that the sequence of random variables $\{X_n\}_{n=0}^{\infty}$ converge pointwise to a random variable $X : \Omega \to \mathbb{R}$ if

$$\lim_{n\to\infty} X_n(\omega) = X(\omega) \text{ for all } \omega \in \Omega.$$

However, this is natural in terms of mathematical analysis definition, but pointless in probability theory. In order to deal with this difficulty in defining pointwise convergence for a sequence of random variable, we will be led to define a number of variants of the pointwise convergence which are not equivalent. We will also discuss how they are interrelated and we will discuss weak convergence which is of a completely different nature.

Definition 40 *Let a probability space $(\Omega, \mathcal{F}, \mathbb{P})$ and a sequence of random variables $\{X_n\}_{n=0}^{\infty}$ with $X_n : \Omega \to \mathbb{R}$ and a random variable $X : \Omega \to \mathbb{R}$. We say that the sequence of random variables $\{X_n\}_{n=0}^{\infty}$ converge almost surely to the random variable X if the event*

$$\{\omega \in \Omega : \; X_n(\omega) \to X(\omega) \;\; as\; n \to \infty\},$$

has probability one. We will denote this type of convergence by

$$X_n \overset{a.s.}{\to} X \quad or \quad \lim_{n\to\infty} X_n = X \quad a.s.$$

Alternatively this mode of convergence is being noted as

$$X_n \to X \text{ almost everywhere, or } X_n \overset{a.e..}{\to} X,$$

$$X_n \to X \text{ with probability one, or } X_n \to X \text{ w.p.1.}$$

Hence, we defined almost sure convergence to deal with the case that naturally occurs when convergence for all points of the sample space Ω fails. However, there are cases where almost sure convergence fails or it is difficult to be proved in an immediate way. For these reasons, we define the next types of convergence of a sequence of random variables.

Definition 41 *Let a probability space $(\Omega, \mathcal{F}, \mathbb{P})$ and a sequence of random variables $\{X_n\}_{n=0}^{\infty}$ with $X_n : \Omega \to \mathbb{R}$ and a random variable $X : \Omega \to \mathbb{R}$. If $|X_n|$ and $|X|$ are in L^p where $1 \le p \le \infty$, then we say that the sequence of random variables $\{X_n\}_{n=0}^{\infty}$ converges to X in p-th mean and we denote it by*

$$X_n \overset{L^p}{\to} X,$$

if and only if

$$\lim_{n\to\infty} \mathbb{E}\{|X_n - X|^p\} = 0.$$

In the present book, we shall consider only integer values for p. However, the theory can be extended to deal with any real p. It is not difficult for the reader to prove as an exercise that

$$\text{if } X_n \overset{L^1}{\to} X \quad \text{then} \quad \mathbb{E}[X_n] \to \mathbb{E}[X] \quad \text{and} \quad \mathbb{E}[|X_n|] \to \mathbb{E}[|X|],$$

also

$$\text{if } X_n \overset{L^p}{\to} X \quad \text{for } p \in (0,\infty) \quad \text{then} \quad \mathbb{E}[|X_n|^p] \to \mathbb{E}[|X|^p].$$

One of the most useful modes of convergence is the *mean square, that is, for* $p = 2$ we have

$$X_n \overset{L^2}{\to} X, \text{ or } X_n \to X \text{ in mean square, or } X_n \overset{m.s.}{\to} X.$$

Definition 42 *Let a probability space $(\Omega, \mathcal{F}, \mathbb{P})$ and a sequence of random variables $\{X_n\}_{n=0}^{\infty}$ with $X_n : \Omega \to \mathbb{R}$ and a random variable $X : \Omega \to \mathbb{R}$. We say that the sequence of random variables $\{X_n\}_{n=0}^{\infty}$ converges to X in probability and we denote it by*

$$X_n \overset{P}{\to} X,$$

if and only if for every $\in > 0$

$$\lim_{n\to\infty} \mathbb{P}(|X_n - X| > \in) = 0.$$

Equivalently, for any $\in > 0$ and $\delta > 0$ there exists a $n^*(\delta)$ such that

$$\mathbb{P}(|X_n - X| > \in) < \delta \quad for\ all \quad n > n^*(\delta).$$

Analogously with the convergence of sequences of real numbers, we define the *Caushy convergence.*

Definition 43 *Let a probability space* $(\Omega, \mathcal{F}, \mathbb{P})$ *and a sequence of random variables* $\{X_n\}_{n=0}^{\infty}$ *with* $X_n : \Omega \to \mathbb{R}$ *and a random variable* $X : \Omega \to \mathbb{R}$. *We say that* $\{X_n\}_{n=0}^{\infty}$ *is almost surely Caushy convergent if the set of* ω*'s for which the real sequence* $\{X_n(\omega) : n \geq 1\}$ *is a Caushy convergence, constitute an event with probability one. That is,*

$$\mathbb{P}\left(\{\omega \in \Omega : X_m(\omega) - X_n(\omega) \to 0 \text{ as } m, n \to \infty\}\right) = 1.$$

The fourth type of convergence known as *weak convergence* is quite different from the first three types defined in the present section. The basic difference in weak convergence is that the actual values of the sequence of random variables do not converge, it is the probability distributions of the random variables that will be converging. This mode of convergence is the weakest, since it is a condition only on the distribution functions of the X_n. It is this fact that gave this kind of convergence the name "weak convergence." It is also called *convergence in distribution* or *convergence in law.*

Another difference which is important to be emphasized is that a sequence of random variables $\{X_n\}_{n=0}^{\infty}$ may weakly converge to a random variable X without necessarily $\{X_n\}_{n=0}^{\infty}$ and X be defined on the same probability space. This fact is in contradiction with almost sure convergence, L^p convergence, and convergence in probability where one always need that $\{X_n\}_{n=0}^{\infty}$ and X are all defined on the same probability space. Also, there is no reference to the dependence or independence of X_n. Hence, weak convergence is not always comparable with other types of convergence.

Definition 44 *Let a sequence of random variables* $\{X_n\}_{n=0}^{\infty}$ *with* X_n *having values on* $\mathbb{R}^k$ *and a random variable* $X : \Omega \to \mathbb{R}^k$. *We say that the sequence of random variables* $\{X_n\}_{n=0}^{\infty}$ *converges in distribution to* X *or equivalently converges in Law to* X *and we write* $X_n \xrightarrow{D} X$ *or equivalently* $X_n \xrightarrow{\mathcal{L}} X$ *if*

$$\mathbb{P}(\omega : X_n \in A) = \mathbb{P}(\omega : X \in A) \quad \text{for every } A \in \mathcal{B}\left(\mathbb{R}^k\right),$$

or more practically

$$\lim_{n\to\infty} \mathbb{P}(X_n \leq x) = \mathbb{P}(X \leq x).$$

It is apparent that the above definition is very difficult in many cases to be checked. However, the following theorem, which is an immediate consequence of the definition of convergence of probability measures, is very useful in this respect.

Theorem 45 *Let a sequence of random variables* $\{X_n\}_{n=0}^{\infty}$ *with* X_n *having values on* $\mathbb{R}^k$ *and a random variable* $X : \Omega \to \mathbb{R}^k$. *Then* $X_n \xrightarrow{D} X$ *if and only if for all continuous bounded functions* f *in* $\mathbb{R}^k$

$$\lim_{n\to\infty} \mathbb{E}\left[f(X_n)\right] = \mathbb{E}\left[f(X)\right].$$

The various modes of convergence defined above are related, in the sense that convergence of a sequence of random variables in one mode of convergence implies also convergence in another mode. In the following basic theorem, we provide all such implications concentrated in one theorem. We will not provide the proof, but the interested reader could find an elegant proof in Grimmett and Stirzaker (2001).

Theorem 46 *Let a probability space* $(\Omega, \mathcal{F}, \mathbb{P})$ *and a sequence of random variables* $\{X_n\}_{n=0}^{\infty}$ *with* $X_n : \Omega \to \mathbb{R}$ *and a random variable* $X : \Omega \to \mathbb{R}$. *Then we have the following implications:*

$$a)\ X_n \xrightarrow{a.s.} X \quad \text{implies} \quad X_n \xrightarrow{P} X \quad \text{implies} \quad X_n \xrightarrow{D} X.$$

For $p \geq 1$, we have that

$$b)\ X_n \stackrel{L^p}{\to} X \quad \text{implies} \quad X_n \stackrel{P}{\to} X \quad \text{implies} \quad X_n \stackrel{D}{\to} X.$$

For $p \geq s \geq 1$, we have

$$c)\ X_n \stackrel{L^p}{\to} X \quad \text{implies} \quad X_n \stackrel{L^s}{\to} X.$$

The converse of the above theorem is not true in general. However, with some added conditions we could get the following theorem acting in a way as a converse:

Theorem 47 *Let a probability space $(\Omega, \mathcal{F}, \mathbb{P})$ and a sequence of random variables $\{X_n\}_{n=0}^{\infty}$ with $X_n : \Omega \to \mathbb{R}$ and a random variable $X : \Omega \to \mathbb{R}$. Then we have the following*

$$a)\ \text{If}\ \ X_n \stackrel{D}{\to} c, \text{ where } c \text{ is constant, then } \ X_n \stackrel{P}{\to} c.$$

$$\begin{aligned} b)\ \text{If}\ \ X_n \stackrel{P}{\to} X \text{ and } \mathbb{P}\left(|X_n \leq k|\right) &= 1 \text{ for all } n \text{ and some } k, \\ \text{then } X_n \stackrel{L^p}{\to} X \text{ for all } p &\geq 1. \end{aligned}$$

$$c)\ \text{If}\ \ \mathbb{P}\left(|X_n - X| > \epsilon\right) \text{ and } \sum_n \mathbb{P}\left(|X_n - X| > \epsilon\right) < \infty \text{ for all } \epsilon > 0,$$

$$\text{then } X_n \stackrel{a.s}{\to} X.$$

The proof of the above theorem and also of the next five theorems could be found in Grimmett and Strirzaker (2001). Note that when all the random variables $\{X_n\}_{n=0}^{\infty}$ and X are all defined on the same probability space then there is a strong link between convergence in probability and convergence in distribution, given in the following theorem:

Theorem 48 *Let a probability space $(\Omega, \mathcal{F}, \mathbb{P})$ and a sequence of random variables $\{X_n\}_{n=0}^{\infty}$ with $X_n : \Omega \to \mathbb{R}$ and a random variable $X : \Omega \to \mathbb{R}$. If $X_n \stackrel{P}{\to} X$ then $X_n \stackrel{D}{\to} X$.*

In general, the integrals of a sequence of functions is not converging to the integral of the limiting function. Random variables are in fact functions from $\Omega \to \mathbb{R}$ or $\mathbb{R}^k$ and their expected values are integrals of the random variables. Hence, the expected values of a sequence of random variables, in general, is not expected to converge to the expected value of the limiting random variable. However, as in the case of functions, with one more condition posed on the sequence of random variables, we have convergence as the following theorem shows.

Theorem 49 ***(Monotone convergence).*** *Let a probability space $(\Omega, \mathcal{F}, \mathbb{P})$ and $\{X_n\}_{n=0}^{\infty}$ be a sequence of random variables on it and in $\mathcal{L}^1$, converging almost surely to a random variable $X \in \mathcal{L}^1$. If $\{X_n\}_{n=0}^{\infty}$ is almost surely monotonically increasing then*

$$\lim_{n \to \infty} \mathbb{E}\left[X_n\right] = \mathbb{E}\left[X\right].$$

There exists another variant of the above theorem which is also useful in many cases. It is the following case:

Theorem 50 ***(Dominated convergence).*** *Let a probability space $(\Omega, \mathcal{F}, \mathbb{P})$ and $\{X_n\}_{n=0}^{\infty}$ be a sequence of random variables on it and in $\mathcal{L}^1$, converging almost surely to a random variable $X \in \mathcal{L}^1$. If there is another random variable Y, such that*

$$|X_n| \leq Y \quad \text{almost surely for every } n,\ Y \in \mathcal{L}^1.$$

then

$$\lim_{n \to \infty} \mathbb{E}\left[X_n\right] = \mathbb{E}\left[X\right].$$

There is another dominated convergence theorem, which is useful to transfer from convergence in probability to convergence in L^p.

Theorem 51 *Let a probability space* $(\Omega, \mathcal{F}, \mathbb{P})$ *and* $\{X_n\}_{n=0}^{\infty}$ *be a sequence of random variables on it. If*

$$X_n \xrightarrow{P} X, \ \ |X_n| \leq Y \text{ for all } n, \text{ and } Y \in L^p,$$

then

$$|X| \in L^p \text{ and } X_n \xrightarrow{L^p} X.$$

The next theorem is easy to prove but useful to have it handy.

Theorem 52 *Let a probability space* $(\Omega, \mathcal{F}, \mathbb{P})$ *and* $\{X_n\}_{n=0}^{\infty}$ *be a sequence of random variables on it. Assume also that* f *is a continuous function, then*

$$\text{If } \lim_{n\to\infty} X_n = X \text{ almost surely, then } \lim_{n\to\infty} f(X_n) = f(X) \ \ a.s.$$

and

$$\text{If } X_n \xrightarrow{P} X, \text{ then } f(X_n) \xrightarrow{P} f(X).$$

1.13 The laws of large numbers and the central limit theorem

Let a probability space $(\Omega, \mathcal{F}, \mathbb{P})$ and a sequence of random variables $\{X_n\}_{n=0}^{\infty}$ with $X_n : \Omega \to \mathbb{R}$. Define the random variable

$$S_n = \sum_{i=1}^{n} X_i.$$

We are interested in the long-term behavior of S_n as $n \to \infty$. Intuitively it is apparent that the asymptotic behavior of S_n will crucially depends upon the sequence $\{X_n\}_{n=0}^{\infty}$, which is true. The results on any type of convergence, that is, almost surely, L^p or in probability of the random variable $\frac{1}{n}S_n$ is called a *law of large numbers.* When the convergence is *almost sure,* we call it a *strong law of large numbers,* otherwise, it is called the *weak law of large numbers.* The weak law is implied by the strong law, since convergence in probability is implied by almost sure convergence. However, there exist sequences which satisfy the weak law but not the strong law.

We seek sufficient, and if possible necessary, conditions on the common distribution of the X_i for the weak and strong laws to hold. We will start with a theorem providing both sufficient and necessary conditions for the weak law of large numbers, when the mean square convergence of the partial sum S_n is used. Under the same conditions, we also have almost sure convergence. The proof could be found in Grimmett and Stirzaker (2001), where also could be found the proof of the next two theorem.

Theorem 53 *Let a probability space* $(\Omega, \mathcal{F}, \mathbb{P})$ *and a sequence of random variables* $\{X_n\}_{n=0}^{\infty}$ *with* $X_n : \Omega \to \mathbb{R}$*, which are independent and identically distributed. Then*

$$\frac{1}{n} S_n \xrightarrow{L^2} \mu,$$

that is,

$$\lim_{n\to\infty} \mathbb{E}\left[\left(\frac{1}{n}\sum_{i=1}^{n} X_i - \mu\right)^2\right] = 0,$$

if and only if, $\mathbb{E}\left(X_i^2\right) < \infty$ *and* $\mathbb{E}(X_i) = \mu$*. Also, under the same sufficient and necessary conditions we have*

$$\frac{1}{n} S_n \xrightarrow{a.s.} \mu \text{ as } n \to \infty.$$

As we shall see below, we can get sharper results for almost sure convergence. In fact, for almost sure convergence the weaker condition $\mathbb{E}(|X_i|) < \infty$ is sufficient and necessary.

Theorem 54 *Let a probability space* $(\Omega, \mathcal{F}, \mathbb{P})$ *and a sequence of random variables* $\{X_n\}_{n=0}^{\infty}$ *with* $X_n : \Omega \to \mathbb{R}$, *which are independent and identically distributed. Then*

$$\frac{1}{n} S_n \overset{a.s.}{\to} \mu \quad as\ n \to \infty,$$

for some constant μ, *if and only if,* $\mathbb{E}(|X_i|) < \infty$. *In this case* $\mathbb{E}(X_i) = \mu$.

There is also a celebrated version of the law of large numbers, which uses convergence in distribution as the mode of convergence.

Theorem 55 *Let a probability space* $(\Omega, \mathcal{F}, \mathbb{P})$ *and a sequence of random variables* $\{X_n\}_{n=0}^{\infty}$ *with* $X_n : \Omega \to \mathbb{R}$, *which are independent and identically distributed with* $\mathbb{E}(X_i) = \mu < \infty$. *Then*

$$\frac{1}{n} S_n \overset{D}{\to} \mu \quad as\ n \to \infty.$$

We will now state without proof a basic theorem in probability, for which there exist a number of generalizations. This is called the *Central Limit Theorem* and a first version of it was founded by de Moivre[13] around 1733. In here, we will restrict ourselves in the following version:

Theorem 56 (*Central Limit Theorem*). *Let a probability space* $(\Omega, \mathcal{F}, \mathbb{P})$ *and a sequence of random variables* $\{X_n\}_{n=0}^{\infty}$ *with* $X_n : \Omega \to \mathbb{R}$, *which are independent and identically distributed with* $\mathbb{E}(X_i) = \mu < \infty$ *and non-zero variance* σ^2. *Then*

$$\frac{S_n - n\mu}{(n\sigma^2)^{1/2}} \overset{D}{\to} N(0,1) \quad as\ n \to \infty.$$

1.14 Conditional distributions and conditional expectations

Nowadays with the evolution of electronic communications, we usually have an overflow of information, which we assume, is acquired by everyone with equal probability. The situation has reached such complexity that a large number of people spend their working time, for collecting an amazing amount of information and report it in a concise and compact form. Apart from the development of electronic instruments, probability spaces, filtrations of σ-algebras and conditional expectations under various real or artificial probability measures have been founded in the last century. These are strong arsenals for dealing with the organization and use of information, for predicting and controlling the evolution of phenomenons with a large degree of randomness. In addition, conditional expectations play one of the most important roles in the foundation of Markov chains and systems which are the object of the present book.

We will assume that our reader has been exposed to a gentle course in probability. If this is not the case or even if it is, for reentering into the spirit and essence of conditional expectation, we suggest the reader to go through chapter 3 in Vassiliou (2010). In there the discrete time is studied and many examples clarify many issues to some depth. In the present, we will try to climb a little bit higher in the ladder. Also, there is an excellent account on conditional expectations in Shreve (2004) and Williams (2000).

We will start by defining the joint distribution measure for two random variables defined on the same probability space.

[13]Abraham de Moivre was born in May 26, 1667 France and died in November 27, 1754 London. He pioneered the development of analytic geometry and the theory of Probability. He wrote one of the first books in Probability Theory with the title: The Doctrine of Chances: a method of calculating the probabilities in play. Also, some results on the Poisson distribution were first indroduced by De Moivre. He was severely influenced by Isaac Newton and his Principia Mathematica.

Definition 57 *Let a probability space $(\Omega, \mathcal{F}, \mathbb{P})$ and X, Y two real random variables defined on it, that is, $X : \Omega \to \mathbb{R}$ and $Y : \Omega \to \mathbb{R}$. Then the random pair (X, Y) is defined on $\Omega \times \Omega \to \mathbb{R}^2$. We define the joint probability distribution measure of (X, Y) to be*

$$\mu_{X,Y}(A) = \mathbb{P}\{(X,Y) \in A\} \text{ for all Borel sets } A \subset \mathbb{R}^2.$$

Now, the joint cumulative distribution function of (X, Y) is given by

$$F_{X,Y}(a,b) = \mu_{X,Y}((-\infty, a] \times (-\infty, b]) = \mathbb{P}\{X \le a, Y \le b\}, a, b \in \mathbb{R}.$$

As in the case of one random variable, we may define the joint probability density function for the random pair (X, Y) as follows:

Definition 58 *Let a probability space $(\Omega, \mathcal{F}, \mathbb{P})$ and X, Y two real random variables defined on it, that is, $X : \Omega \to \mathbb{R}$ and $Y : \Omega \to \mathbb{R}$. Then a non-negative function $f_{X,Y}(x,y)$ is a joint density function for the random pair (X, Y) if*

$$\mu_{X,Y}(A) = \int_{-\infty}^{\infty} \int_{-\infty}^{\infty} \mathbf{1}_{\{A\}}(x,y) f_{X,Y}(x,y)\, dydx \text{ for all Borel sets } A \subset \mathbb{R}^2.$$

The above condition is true if and only if

$$F_{X,Y}(a,b) = \int_{-\infty}^{a} \int_{-\infty}^{b} f_{X,Y}(x,y)\, dydx \text{ for all } a, b \in \mathbb{R}.$$

Now, suppose that X and Y have joint density function f. We wish to discuss the conditional distribution of Y, given that X takes a value in the interval $[x, x + dx]$. Then under the condition that $f_X(x) > 0$, we have that

$$\mathbb{P}(Y \le y \mid x \le X \le x + dx) = \frac{\mathbb{P}(Y \le y, x \le X \le x + dx)}{\mathbb{P}(x \le X \le x + dx)}.$$

As $dx \to 0$ the above equation provides the basis for intuitively define the conditional distribution function $F_{Y|X}(y \mid x)$ and the conditional density function $f_{Y|X}(y \mid x)$ of $F_{Y|X}(y \mid x)$.

Definition 59 *Let a probability space $(\Omega, \mathcal{F}, \mathbb{P})$ and X, Y two real random variables defined on it, that is, $X : \Omega \to \mathbb{R}$ and $Y : \Omega \to \mathbb{R}$. Then the conditional distribution function $F_{Y|X}(y \mid x)$ is given by*

$$F_{Y|X}(y \mid x) = \int_{-\infty}^{y} \frac{f_{XU}(x,u)}{f_X(x)} du$$

for any x such that $f_X(x) > 0$.

The conditional density function $f_{Y|X}(y \mid x)$ of $F_{Y|X}(y \mid x)$ is given by

$$f_{Y|X}(y \mid x) = \frac{f_{YX}(x,y)}{f_X(x)},$$

for any x such that $f_X(x) > 0$.

Since, we defined the conditional density function $f_{Y|X}(y \mid x)$ of two random variables, it is straightforward to go ahead and following the steps of the discrete time case (see Vassiliou (2010, Chapter 3)) to define the conditional expectation of two random variables as

$$\mathbb{E}[X \mid Y = y] = \int_{-\infty}^{+\infty} x f_{X|Y}(x \mid y)\, dx.$$

Now, it is interesting to know as in the discrete case, what happens if we consider, in general, the conditional expectation

$$\mathbb{E}[X \mid Y],$$

As in the discrete case all possible values of $\mathbb{E}[X \mid Y]$ depend on the possible values of Y. Therefore, $\mathbb{E}[X \mid Y]$ is a random variable and hence there is a meaning in finding its expected value. Actually, one following the steps of the proof in the discrete time case, see Vassiliou (2010), could prove again that

$$\mathbb{E}[\mathbb{E}[X \mid Y]] = \mathbb{E}[X].$$

These considerations will help us to create the intuition in order to climb a few steps up the ladder. Let a probability space $(\Omega, \mathcal{F}, \mathbb{P})$ and let $\mathcal{G}$ be a sub-σ-algebra of $\mathcal{F}$. Consider a random variable X on $(\Omega, \mathcal{F}, \mathbb{P})$, which is either non-negative or in $\mathcal{L}^1$. Since we were able to define $\mathbb{E}[X \mid Y]$, we in fact took the first step to define $\mathbb{E}[X \mid \sigma(Y)]$. Then one may logically ask "is it possible to define $\mathbb{E}[X \mid \mathcal{G}]$?". The answer is provided in what follows:

Definition 60 *Let a probability space $(\Omega, \mathcal{F}, \mathbb{P})$ and let $\mathcal{G}$ be a sub-σ-algebra of $\mathcal{F}$. Consider a random variable X on $(\Omega, \mathcal{F}, \mathbb{P})$, which is either non-negative or in $\mathcal{L}^1$. The conditional expectation of X given $\mathcal{G}$, that is, $\mathbb{E}[X \mid \mathcal{G}]$ is any random variable that satisfies*

(i) $\mathbb{E}[X \mid \mathcal{G}]$ is $\mathcal{G}$-measurable.

(ii)

$$\int_A \mathbb{E}[X \mid \mathcal{G}](\omega)\, d\mathbb{P}(\omega) = \int_A X(\omega)\, d\mathbb{P}(\omega) \text{ for all } A \in \mathcal{G}.$$

The last property is called partial averaging.

Condition (i) demands that the values of the random variable $\mathbb{E}[X \mid \mathcal{G}]$ can be determined from the information in $\mathcal{G}$. Condition (ii) guarantees that, in fact, $\mathbb{E}[X \mid \mathcal{G}]$ satisfies, in a more general context, a property of the type expressed by relation (1.14).

The first immediate logical question is:

Does there always exist a random variable $\mathbb{E}[X \mid \mathcal{G}]$
satisfying conditions (i) and (ii)?

Before we provide an answer to the above question, we need to revisit the concept of change of measure in a more general setting than in Vassiliou (2010). However, it is strongly suggested to refresh or read section 6.2 in Vassiliou (2010).

1.15 Change of measure

Let a probability space $(\Omega, \mathcal{F}, \mathbb{P})$ and a random value X_t representing the value of some commodity at time t. Assume that, for some specific time t, the random value X_t follows the normal distribution $N(0,1)$. It is very well known that in this case the probability density function of X_t is given by

$$f(x_t) = \frac{1}{\sqrt{2\pi}} e^{-\frac{1}{2}x_t^2}.$$

Let us denote by $dP(x_t)$ the probability

$$\begin{aligned} d\mathbb{P}(x_t) &= \mathbb{P}\left\{x_t - \frac{1}{2}dx_t \le x_t \le x_t + \frac{1}{2}dx_t\right\} \\ &= \int_{x_t - \frac{1}{2}dx_t}^{x_t + \frac{1}{2}dx_t} \frac{1}{\sqrt{2\pi}} e^{-\frac{1}{2}x_t^2}\, dx_t. \end{aligned}$$

Since the interval $\left[x_t - \frac{1}{2}dx_t, x_t + \frac{1}{2}dx_t\right]$ is chosen to be very small we could assume that within this interval the function

$$\frac{1}{\sqrt{2\pi}}e^{-\frac{1}{2}x_t^2},$$

is a stable, i.e., does not change value and thus we may write

$$\begin{aligned} d\mathbb{P}(x_t) &= \frac{1}{\sqrt{2\pi}}e^{-\frac{1}{2}x_t^2} \int\limits_{x_t-\frac{1}{2}dx_t}^{x_t+\frac{1}{2}dx_t} dx_t \qquad (1.16) \\ &= \frac{1}{\sqrt{2\pi}}e^{-\frac{1}{2}x_t^2}\left[x_t + \frac{1}{2}dx_t - x_t + \frac{1}{2}dx_t\right] \\ &= \frac{1}{\sqrt{2\pi}}e^{-\frac{1}{2}x_t^2} dx_t. \end{aligned}$$

It is easy to see that

$$\int\limits_{-\infty}^{+\infty} d\mathbb{P}(x_t) = 1.$$

It is also known that

$$\mathbb{E}[X_t] = \int\limits_{-\infty}^{+\infty} x_t d\mathbb{P}(x_t). \qquad (1.17)$$

From relation (1.16), it becomes apparent that in order to change the expected value of the random variable X_t, there are two possible options. The first is the very well known transformation of the random variable X_t into another random variable Y_t, which will have the desired mean. This is done all too often in statistics, transforming a random variable with distribution $N(\mu, \sigma^2)$, into another random variable Y_t having the standard normal distribution $N(0,1)$. The second choice would be, if this is possible, to change the probability measure $d\mathbb{P}(x_t)$, with another probability measure let say $d\tilde{\mathbb{P}}(x_t)$, in which case the expected value in (1.17) would be different. This action actually changes the probabilities $\mathbb{P}(X_t \in A)$ with $A \in \mathcal{F}$ with the probabilities $\tilde{\mathbb{P}}(X_t \in A)$, in order that the random variable X_t will have the desired expected value.

Note that this is what we have done in section 6.2 in Vassiliou (2010). The basic discriminating difference being the fact that in the present Ω is uncountable infinite. We will now continue our introductory example by showing how to change the expected value of X_t in such a way as to have normal distribution $N(\mu, 1)$.

We first define the random variable $Z(X_t)$ for specific t to be

$$Z(X_t) = \exp\left(-\mu X_t - \frac{1}{2}\mu^2\right). \qquad (1.18)$$

It is easy to check that $Z(X_t) > 0$ for every $\omega \in \Omega$ because $Z(\omega)$ is an exponential. We will now show that $\mathbb{E}[Z(X_t)] = 1$. We have that

$$\begin{aligned} \mathbb{E}[Z(X_t)] &= \int\limits_{-\infty}^{+\infty} Z(x_t)\, d\mathbb{P}(x_t) \qquad (1.19) \\ &= \int\limits_{-\infty}^{+\infty} \exp\left(-\mu x_t - \frac{1}{2}\mu^2\right) \frac{1}{\sqrt{2\pi}}e^{-\frac{1}{2}x_t^2} dx_t \end{aligned}$$

$$= \frac{1}{\sqrt{2\pi}} \int_{-\infty}^{+\infty} \exp\left\{-\frac{1}{2}\left(x_t^2 + 2\mu x_t + \mu^2\right)\right\} dx_t$$

$$= \frac{1}{\sqrt{2\pi}} \int_{-\infty}^{+\infty} \exp\left\{-\frac{1}{2}\left(x_t + \mu\right)^2\right\} dx_t$$

$$= \frac{1}{\sqrt{2\pi}} \int_{-\infty}^{+\infty} \exp\left\{-\frac{1}{2}\left(y_t\right)^2\right\} dy_t = 1.$$

We use the random variable $Z(X_t)$ in order to create a new probability measure $d\tilde{\mathbb{P}}(x_t)$ defined by

$$d\tilde{\mathbb{P}}(X_t) = Z(X_t)\, d\mathbb{P}(X_t). \tag{1.20}$$

It is easy to check that $d\tilde{\mathbb{P}}(X_t)$ is a probability measure, since as we have shown previously

$$\int_{-\infty}^{+\infty} d\tilde{\mathbb{P}}(x_t) = 1.$$

From (1.16) it is apparent that $d\tilde{\mathbb{P}}(X_t)$ is a probability measure whose values are distributed as $N(\mu, 1)$. Hence, we observe that by multiplying $d\mathbb{P}(X_t)$ by $Z(X_t)$, we in fact created a new probability measure with values distributed also as normal but with a different expected value and the same variance. The random variable $Z(X_t)$, as we have shown has another characteristic and one, which will be proved to be of vital importance, that is, $\mathbb{E}[Z(X_t)] = 1$. Note that from (1.18), we get that

$$[Z(X_t)]^{-1} = \exp\left(\mu X_t + \frac{1}{2}\mu^2\right),$$

and then it is easy to see that

$$[Z(X_t)]^{-1}\, d\tilde{\mathbb{P}}(X_t) = d\mathbb{P}(X_t). \tag{1.21}$$

Relations (1.20) and (1.21) allows us to write

$$\frac{d\tilde{\mathbb{P}}(X_t)}{d\mathbb{P}(X_t)} = Z(X_t). \tag{1.22}$$

The above relation is called the *Random-Nikodym derivative,* due to the fact that it looks like a derivative. Naturally, one may ask, how are we going to find each time the appropriate random variable $Z(X_t)$ for our purpose? The answer will be given by *Girsanov's Theorem,* which we will study later on. Equation (1.22) is not a derivative in any possible sense and the only way that equation (1.22) should be treated, is as a quotient of two probability measures. In addition but more importantly equation (1.22), as we will establish in what follows, in terms of sets $A \in \mathcal{F}$ and $\omega \in \Omega$ could be written as follows

$$\tilde{\mathbb{P}}(A) = \int_A Z(\omega)\, d\mathbb{P} \tag{1.23}$$

We will start by proving the following:

Theorem 61 *(Shreve (2004)). Let a probability space $(\Omega, \mathcal{F}, \mathbb{P})$ and let Z be an almost surely non-negative random variable with $\mathbb{E}[Z] = 1$. For $A \in \mathcal{F}$, define*

$$\tilde{\mathbb{P}}(A) = \int_A Z(\omega)\, d\mathbb{P}(\omega). \tag{1.24}$$

Then $\tilde{\mathbb{P}}(A)$ with $A \in \mathcal{F}$ is a probability measure and in addition if X is a non-negative random variable, then

$$\mathbb{E}_{\tilde{\mathbb{P}}}[X] = \mathbb{E}_{\mathbb{P}}[XZ]. \tag{1.25}$$

If Z is almost surely strictly positive, then for any non-negative random variable X

$$\mathbb{E}_{\mathbb{P}}[X] = \mathbb{E}_{\tilde{\mathbb{P}}}\left[\frac{X}{Z}\right].$$

Proof. We will start by proving that $\tilde{\mathbb{P}}$ is a probability measure. Firstly we should prove that $\tilde{\mathbb{P}}(\Omega) = 1$ and secondly that $\tilde{\mathbb{P}}$ is countably additive. By relation (1.23), we have that

$$\tilde{\mathbb{P}}(\Omega) = \int_\Omega Z(\omega)\, d\mathbb{P}(\omega) = \mathbb{E}[Z] = 1.$$

For countable additivity assume that $\{A_n\}_{n=0}^{\infty}$ is a sequence in $\mathcal{F}$, with $A_i \cap A_j = \varnothing$ for all $i \neq j \in \mathbb{N}$. Now define by

$$C_1 = A_1,\ C_2 = A_1 \cup A_2,\ C_3 = A_1 \cup A_2 \cup A_3, ...,$$

then apparently $C_i \subset C_{i+1}$ for every $i \in \mathbb{N}$. Consequently, we get that

$$\mathbf{1}_{\{C_i\}} \leq \mathbf{1}_{\{C_{i+1}\}} \quad \text{for every } i \in \mathbb{N}.$$

Since the random variable $Z \geq 0$ almost surely then

$$\mathbf{1}_{\{C_i\}} Z \leq \mathbf{1}_{\{C_{i+1}\}} Z \quad \text{for every } i \in \mathbb{N}.$$

Now, define by

$$C_\infty = \bigcup_{n=1}^{\infty} C_n,$$

then

$$\lim_{n\to\infty} \mathbf{1}_{\{C_n\}} Z = \mathbf{1}_{\{C_\infty\}} Z.$$

Then according to the monotone convergence (there are two) theorem, we have that

$$\lim_{n\to\infty} \mathbb{E}\left[\mathbf{1}_{\{C_n\}} Z\right] = \mathbb{E}\left[\mathbf{1}_{\{C_\infty\}} Z\right]. \tag{1.26}$$

Now, according to the definition of an expected value, we have that

$$\begin{aligned}
\lim_{n\to\infty} \mathbb{E}\left[\mathbf{1}_{\{C_n\}} Z\right] &= \lim_{n\to\infty} \int_\Omega \mathbf{1}_{\{C_n\}}(\omega)\, Z(\omega)\, d\mathbb{P}(\omega) \qquad (1.27)\\
&= (\text{due to } (1.23))\\
&= \lim_{n\to\infty} \tilde{\mathbb{P}}(C_n)\\
&= \lim_{n\to\infty} \tilde{\mathbb{P}}\left(\bigcup_{k=1}^{n} A_k\right)\\
&= \lim_{n\to\infty} \sum_{k=1}^{n} \tilde{\mathbb{P}}(A_k).
\end{aligned}$$

Now, from (1.25) and (1.26), we get that

$$\begin{aligned}\mathbb{E}\left[\mathbf{1}_{\{C_\infty\}}Z\right] &= \int_\Omega \mathbf{1}_{\{C_\infty\}}(\omega)\,Z(\omega)\,d\mathbb{P}(\omega)\\ &= \tilde{\mathbb{P}}(C_\infty)=\tilde{\mathbb{P}}\left(\bigcup_{k=1}^{\infty}A_k\right)\\ &= \sum_{k=1}^{\infty}\tilde{\mathbb{P}}(A_k).\end{aligned}$$

We will now prove relation (1.24) with a standard method which has four steps.

Step 1. In the first step, we assume that X is an indicator function. Let say $X=\mathbf{1}_A$ with $A\in\mathcal{F}$. Then

$$\begin{aligned}\mathbb{E}_{\tilde{\mathbb{P}}}[X] &= \mathbb{E}_{\tilde{\mathbb{P}}}[\mathbf{1}_A]=\tilde{\mathbb{P}}(A)\\ &= \int_\Omega \mathbf{1}_A(\omega)\,Z(\omega)\,d\mathbb{P}(\omega)\\ &= \mathbb{E}_{\mathbb{P}}(\mathbf{1}_A Z)=\mathbb{E}_{\mathbb{P}}(XZ).\end{aligned}$$

Step 2. Assume now that X is a linear combination of a finite number of indicator functions, that is,

$$X=\sum_{k=1}^{n}c_k\mathbf{1}_{A_k},$$

where c_k, $k=1,2,...,k$ are non-negative constants and $A_1,A_2,...,A_n\in\mathcal{F}$. Then we have that

$$\begin{aligned}\mathbb{E}_{\tilde{\mathbb{P}}}[X] &= \mathbb{E}_{\tilde{\mathbb{P}}}\left[\sum_{k=1}^{n}c_k\mathbf{1}_{A_k}\right]=\sum_{k=1}^{n}c_k\mathbb{E}_{\tilde{\mathbb{P}}}[\mathbf{1}_{A_k}]\\ &= \sum_{k=1}^{n}c_k\tilde{\mathbb{P}}(A_k)=\sum_{k=1}^{n}c_k\int_\Omega \mathbf{1}_{\{A_k\}}(\omega)\,Z(\omega)\,d\mathbb{P}(\omega)\\ &= \sum_{k=1}^{n}c_k\mathbb{E}_{\mathbb{P}}\left[\mathbf{1}_{\{A_k\}}Z\right]=\mathbb{E}_{\mathbb{P}}\left[\sum_{k=1}^{n}c_k\mathbf{1}_{\{A_k\}}Z\right]\\ &= \mathbb{E}_{\mathbb{P}}[XZ].\end{aligned}$$

Step 3. Let X be a non-negative $\mathcal{F}$-measurable random variable with values in $\mathbb{R}$. For $n\in\mathbb{N}$ define the sets

$$A_{n,k}(\omega)=\left\{\omega;\frac{k}{2^n}\le X(\omega)<\frac{k+1}{2^n}\right\}\text{ for }k=0,1,...,4^n-1.$$

Define by

$$X_n=\sum_{k=0}^{4^n-1}\frac{k}{2^n}\mathbf{1}_{A_{n,k}}(\omega).$$

It is useful in the understanding of the proof to analyze the following two simple cases.

Firstly, consider the case $n = 1$ in which case $k = 0, 1, 2, 3$. Then

$$\begin{aligned} A_{1,0}(\omega) &= \left\{\omega : 0 \leq X_1(\omega) < \frac{1}{2}\right\}, \\ A_{1,1}(\omega) &= \left\{\omega : \frac{1}{2} \leq X_1(\omega) < 1\right\}, \\ A_{1,2}(\omega) &= \left\{\omega : 1 \leq X_1(\omega) < \frac{3}{2}\right\}, \\ A_{1,3}(\omega) &= \left\{\omega : \frac{3}{2} \leq X_1(\omega) < 2\right\}. \end{aligned}$$

In this case

$$X_1(\omega) = \frac{1}{2}\mathbf{1}_{\{A_{1,1}\}}(\omega) + \mathbf{1}_{\{A_{2,1}\}}(\omega) + \frac{3}{2}\mathbf{1}_{\{A_{3,1}\}}(\omega).$$

Secondly, consider the case $n = 2$ in which case $k = 0, 1, 2, ..., 7$. Then

$$\begin{aligned} A_{2,0}(\omega) &= \left\{\omega : 0 \leq X_1(\omega) < \frac{1}{4}\right\}, \\ A_{2,1}(\omega) &= \left\{\omega : \frac{1}{4} \leq X_1(\omega) < \frac{1}{2}\right\}, \\ A_{2,2}(\omega) &= \left\{\omega : \frac{1}{2} \leq X_1(\omega) < \frac{3}{4}\right\}, \\ A_{2,3}(\omega) &= \left\{\omega : \frac{3}{4} \leq X_1(\omega) < 1\right\}, \\ A_{2,4}(\omega) &= \left\{\omega : 1 \leq X_1(\omega) < \frac{5}{4}\right\}, \\ A_{2,5}(\omega) &= \left\{\omega : \frac{5}{4} \leq X_1(\omega) < \frac{6}{4}\right\}, \\ A_{2,6}(\omega) &= \left\{\omega : \frac{6}{4} \leq X_1(\omega) < \frac{7}{4}\right\}, \\ A_{2,7}(\omega) &= \left\{\omega : \frac{7}{4} \leq X_1(\omega) < 2\right\}. \end{aligned}$$

In this case

$$\begin{aligned} X_2(\omega) &= \frac{1}{4}\mathbf{1}_{\{A_{2,1}\}}(\omega) + \frac{1}{2}\mathbf{1}_{\{A_{2,2}\}}(\omega) + \frac{3}{4}\mathbf{1}_{\{A_{2,3}\}}(\omega) + \mathbf{1}_{\{A_{2,4}\}}(\omega) \\ &\quad + \frac{5}{4}\mathbf{1}_{\{A_{2,5}\}}(\omega) + \frac{6}{4}\mathbf{1}_{\{A_{2,6}\}}(\omega) + \frac{7}{4}\mathbf{1}_{\{A_{2,7}\}}(\omega). \end{aligned}$$

It is easy to see that $X_1 \leq X_2$. In fact by extending the same steps for every X_n, $n \in N$, it becomes apparent that

$$X_1 \leq X_2 \leq X_3 \leq X_4 \leq ... \leq X_n \leq X,$$

and as n gets larger, then X_n becomes a better approximation for X. Further, we have that

$$\lim_{n \to \infty} X_n = X.$$

Then according to the Monotone Convergence Theorem, we have that

$$\begin{aligned}
\mathbb{E}_{\tilde{\mathbb{P}}}[X] &= \lim_{n\to\infty} \mathbb{E}_{\tilde{\mathbb{P}}}[X_n] \\
&= \lim_{n\to\infty} \mathbb{E}_{\tilde{\mathbb{P}}}\left[\sum_{k=1}^{4^n-1} \frac{k}{2^n}\mathbf{1}_{A_{n,k}}(\omega)\right] \\
&= \lim_{n\to\infty} \sum_{k=1}^{4^n-1} \frac{k}{2^n}\mathbb{E}_{\tilde{\mathbb{P}}}\left[\mathbf{1}_{A_{n,k}}(\omega)\right] \\
&= \lim_{n\to\infty} \sum_{k=1}^{4^n-1} \frac{k}{2^n}\tilde{\mathbb{P}}(A_{n,k}) \\
&= \lim_{n\to\infty} \sum_{k=1}^{4^n-1} \frac{k}{2^n}\int_{\Omega} \mathbf{1}_{A_{n,k}}(\omega)\, Z(\omega)\, d\mathbb{P}(\omega) \\
&= \int_{\Omega} X(\omega)\, Z(\omega)\, d\mathbb{P}(\omega) = \mathbb{E}_{\mathbb{P}}[XZ].
\end{aligned}$$

The inverse relationship among the expectations under different probability measures could easily be proved along the same lines. ■

We are now in a position to extend the definition of two equivalent probability measures given in Definition 6.1., Vassiliou (2010).

Definition 62 *Let a probability space $(\Omega, \mathcal{F}, \mathbb{P})$. Two probability measures $\mathbb{P}$ and $\tilde{\mathbb{P}}$ on $(\Omega, \mathcal{F})$ are said to be equivalent if they agree which sets in $\mathcal{F}$ have probability zero.*

Remark 63 *Under the assumptions of Theorem 61, including the assumption that $Z > 0$ almost surely, $\mathbb{P}$ and $\tilde{\mathbb{P}}$ are equivalent. Actually this is easy to see. Let $A \in \mathcal{F}$ with $\mathbb{P}(A) = 0$. Then from Theorem 61, we have that*

$$\tilde{\mathbb{P}}(A) = \int_{\Omega} \mathbf{1}_A(\omega)\, Z(\omega)\, d\mathbb{P}(\omega) = 0,$$

since $\mathbb{P}(A) = 0$. Inversely, assume that $A \in \mathcal{F}$ with $\tilde{\mathbb{P}}(A) = 0$. Then

$$\mathbb{P}(A) = \mathbb{E}_{\mathbb{P}}[\mathbf{1}_A] = \mathbb{E}_{\tilde{\mathbb{P}}}\left[\mathbf{1}_A \frac{1}{Z}\right] = 0,$$

since $\mathbf{1}_A \frac{1}{Z} = 0$ almost surely under $\tilde{\mathbb{P}}$.

In fact Theorem 61, Definition 62 and Remark 63 are being related with the following very important Theorem of Radon-Nikodým, which we provide below without proof which could be found in Shreve (2004).

Theorem 64 ***(Radon-Nikodým).*** *Let a probability space $(\Omega, \mathcal{F}, \mathbb{P})$ and two equivalent probability measures $\mathbb{P}$ and $\tilde{\mathbb{P}}$ on $(\Omega, \mathcal{F})$. Then there exists an almost surely positive random variable Z such that $\mathbb{E}_{\mathbb{P}}[Z] = 1$ and*

$$\tilde{\mathbb{P}}(A) = \int_{A} Z(\omega)\, d\mathbb{P}(\omega), \text{ for every } A \in \mathcal{F}.$$

We will now proceed to prove an important result called the *abstract version of the Bayes's formula* (see also Musiela and Rutkowski (2005)). Let a probability space $(\Omega, \mathcal{F}, \mathbb{P})$

and two equivalent probability measures $\mathbb{P}$ and $\tilde{\mathbb{P}}$ on $(\Omega, \mathcal{F})$. Assume that the Radon-Nikodým derivative of $\tilde{\mathbb{P}}$ with respect to $\mathbb{P}$ equals with

$$\frac{d\tilde{\mathbb{P}}}{d\mathbb{P}} = Z, \text{ } \mathbb{P}\text{-almost surely.} \tag{1.28}$$

then the random variable Z is strictly positive $\mathbb{P}$-almost surely, moreover, Z is $\mathbb{P}$-integrable with $\mathbb{E}_{\mathbb{P}}(Z) = 1$. In addition, due to Theorem 61 we have that for any $\tilde{\mathbb{P}}$-integrable random variable X we have that

$$\mathbb{E}_{\tilde{\mathbb{P}}}(X) = \mathbb{E}_{\mathbb{P}}(ZX).$$

Theorem 65 ***(Abstract version of the Bayes's formula).*** *Let a probability space $(\Omega, \mathcal{F}, \mathbb{P})$ and two equivalent probability measures $\mathbb{P}$ and $\tilde{\mathbb{P}}$ on $(\Omega, \mathcal{F})$. Let $\mathcal{G}$ be a sub-σ-algebra of the σ-algebra $\mathcal{F}$ and let X a $\tilde{\mathbb{P}}$-integrable random variable. Then the following formula holds*

$$\mathbb{E}_{\tilde{\mathbb{P}}}(X \mid \mathcal{G}) = \frac{\mathbb{E}_{\mathbb{P}}(ZX \mid \mathcal{G})}{\mathbb{E}_{\mathbb{P}}(Z \mid \mathcal{G})} \tag{1.29}$$

Proof. Since, Z is strictly positive $\mathbb{P}$-almost surely, then $\mathbb{E}_{\mathbb{P}}(Z \mid \mathcal{G})$ is strictly positive $\mathbb{P}$-almost surely, and therefore, (1.29) is well defined. In addition, ZX is $\mathbb{P}$-integrable, and thus it is sufficient to show that

$$\mathbb{E}_{\mathbb{P}}(ZX \mid \mathcal{G}) = \mathbb{E}_{\tilde{\mathbb{P}}}(X \mid \mathcal{G})\, \mathbb{E}_{\mathbb{P}}(Z \mid \mathcal{G}) \tag{1.30}$$

Since, in (1.30) the right hand side defines a $\mathcal{G}$-measurable random variable, we need to show that for any set $A \in \mathcal{G}$, we have

$$\int_A ZX d\mathbb{P} = \int_A \mathbb{E}_{\tilde{\mathbb{P}}}(X \mid \mathcal{G})\, \mathbb{E}_{\mathbb{P}}(Z \mid \mathcal{G})\, d\mathbb{P}. \tag{1.31}$$

In this respect for every $A \in \mathcal{G}$, we get

$$\begin{aligned}
\int_A ZX d\mathbb{P} &= \int_A X d\tilde{\mathbb{P}} = (\text{ due to the partial average property}) \\
&= \int_A \mathbb{E}_{\tilde{\mathbb{P}}}(X \mid \mathcal{G})\, d\tilde{\mathbb{P}} = \int_A \mathbb{E}_{\tilde{\mathbb{P}}}(X \mid \mathcal{G})\, Z d\mathbb{P} \\
&= (\text{due to the partial average property}) \\
&= \int_A \mathbb{E}_{\mathbb{P}}\left[\mathbb{E}_{\tilde{\mathbb{P}}}(X \mid \mathcal{G})\, Z \mid \mathcal{G}\right] d\mathbb{P} \\
&= \int_A \mathbb{E}_{\tilde{\mathbb{P}}}(X \mid \mathcal{G})\, \mathbb{E}_{\mathbb{P}}(Z \mid \mathcal{G})\, d\mathbb{P}.
\end{aligned}$$

■

1.16 Existence and uniqueness of conditional expectations

We now turn back into the question posed after Definition 60, that is, "*Does there always exist a random variable $\mathbb{E}(X \mid \mathcal{G})$ satsfying conditions (i) and (ii)?*". We will also extend our question : "*If it exists, is it unique?*". In the theorem that follows, we provide an answer to these questions for X initially an integrable non-negative random variable. We also provide the reader with the proof which is a very elegant use of the Randon-Nikodým theorem (see also Shreve (2004), Williams (2000)).

Theorem 66 *Let a probability space $(\Omega, \mathcal{F}, \mathbb{P})$ and let $\mathcal{G}$ be a sub-σ-algebra of the σ-algebra $\mathcal{F}$ and X an integrable non-negative random variable. Then there exists a $\mathcal{G}$-measurable random variable $\mathbb{E}(X \mid \mathcal{G})$ such that*

$$\int_A \mathbb{E}_{\mathbb{P}}(X \mid \mathcal{G})(\omega)\, d\mathbb{P}(\omega) = \int_A X(\omega)\, d\mathbb{P}(\omega) \quad \text{for every } A \in \mathcal{G}. \tag{1.32}$$

The random variable $\mathbb{E}_{\mathbb{P}}(X \mid \mathcal{G})$ in (1.32) is unique.

Proof. In the measurable space $(\Omega, \mathcal{F})$, we define a new probability measure $\tilde{\mathbb{P}}$, which for every $A \in \mathcal{F}$ is defined by

$$\tilde{\mathbb{P}}(A) = \int_A \frac{X(\omega)+1}{\mathbb{E}_{\mathbb{P}}[X+1]} d\mathbb{P}(\omega). \tag{1.33}$$

From (1.33), it is evident that the integrand has the following two properties

$$\frac{X(\omega)+1}{\mathbb{E}_{\mathbb{P}}[X+1]} > 0 \text{ a.s. and } \mathbb{E}_{\mathbb{P}}\left[\frac{X(\omega)+1}{\mathbb{E}_{\mathbb{P}}[X+1]}\right] = 1. \tag{1.34}$$

From (1.33), it is evident that the probability measures $\mathbb{P}$ and $\tilde{\mathbb{P}}$ are equivalent on the measurable space $(\Omega, \mathcal{F})$. We will now define two equivalent probability measures on the measurable space $(\Omega, \mathcal{G})$. The first probability measure $\mathbb{Q}$ is defined to be, in fact, equal with $\mathbb{P}$ on the restricted space $(\Omega, \mathcal{G})$ and not defined for sets $B \in \mathcal{F}$ but $B \notin \mathcal{G}$. That is,

$$\mathbb{Q}(A) = \mathbb{P}(A) \text{ for every } A \in \mathcal{G}, \text{ undefined for } B \in \mathcal{F} \text{ but } B \notin \mathcal{G}. \tag{1.35}$$

The second probability measure $\tilde{\mathbb{Q}}$ is equivalent to $\mathbb{Q}$ and is defined as

$$\tilde{\mathbb{Q}}(A) = \tilde{\mathbb{P}}(A) \text{ for every } A \in \mathcal{G}, \text{ undefined for } B \in \mathcal{F} \text{ but } B \notin \mathcal{G}. \tag{1.36}$$

Since the probability measures $\mathbb{Q}$ and $\tilde{\mathbb{Q}}$ are equivalent on the measurable space $(\Omega, \mathcal{G})$, according to the Randon-Nikodým theorem (Theorem 63), we have that there exists a random variable Z, which will be $\mathcal{G}$-measurable, such that

$$\tilde{\mathbb{Q}}(A) = \int_A Z(\omega)\, d\mathbb{Q}(\omega) \quad \text{for every } A \in \mathcal{G}. \tag{1.37}$$

From (1.33), (1.35), (1.36), (1.37), we get that

$$\int_A \frac{X(\omega)+1}{\mathbb{E}_{\mathbb{P}}[X+1]} d\mathbb{P}(\omega) = \int_A Z(\omega)\, d\mathbb{P}(\omega) \quad \text{for every } A \in \mathcal{G}. \tag{1.38}$$

We multiply both sides of (1.38) by $\mathbb{E}_{\mathbb{P}}[X+1]$ and we get that

$$\int_A X(\omega)\, d\mathbb{P}(\omega) = \int_A \{\mathbb{E}_{\mathbb{P}}[X+1]\, Z(\omega) - 1\}\, d\mathbb{P}(\omega) \quad \text{for every } A \in \mathcal{G}. \tag{1.39}$$

From (1.38) if we let

$$\mathbb{E}_{\mathbb{P}}(X \mid \mathcal{G}) = \mathbb{E}_{\mathbb{P}}[X+1]\, Z(\omega) - 1, \tag{1.40}$$

then the second part of the equation is $\mathcal{G}$-measurable, since $Z(\omega)$ is $\mathcal{G}$-measurable. Hence, we proved the existence of a random variable $\mathbb{E}_{\mathbb{P}}(X \mid \mathcal{G})$, which satisfies equation (1.32). What remains is to prove the uniqueness of the solution (1.40). In order to do so we assume that it is not unique and let that there exists two random variables Y and W, which are $\mathcal{G}$-measurable and satisfy (1.32). That is,

$$\int_A \mathbb{Y}(\omega)\, d\mathbb{P}(\omega) = \int_A X(\omega)\, d\mathbb{P}(\omega) \quad \text{for every } A \in \mathcal{G}, \tag{1.41}$$

and

$$\int_A \mathbb{W}(\omega)\, d\mathbb{P}(\omega) = \int_A X(\omega)\, d\mathbb{P}(\omega) \quad \text{for every } A \in \mathcal{G}. \tag{1.42}$$

Now from (1.41) and (1.42) we get that

$$\int_A [\mathbb{Y}(\omega) - \mathbb{W}(\omega)]\, d\mathbb{P}(\omega) = 0 \text{ for every } A \in \mathcal{G}, \tag{1.43}$$

where apparently $[\mathbb{Y}(\omega) - \mathbb{W}(\omega)]$ is $\mathcal{G}$-measurable, since $\mathbb{Y}(\omega)$ and $\mathbb{W}(\omega)$ are. Since (1.43) is satisfied for every $A \in \mathcal{G}$ it will be satisfied also for the set $B = \{Y - W > 0\} \in \mathcal{G}$. Hence,

$$\int_B [\mathbb{Y}(\omega) - \mathbb{W}(\omega)]\, d\mathbb{P}(\omega) = 0 \text{ for } B \in \mathcal{G}. \tag{1.44}$$

Equation (1.44) can be satisfied only if $B \in \mathcal{G}$ has probability zero, which is in contradiction with our assumption that $Y - W > 0$. Hence,

$$Y \geq W \quad \text{almost surely.} \tag{1.45}$$

If we define the set $C = \{Y - W < 0\} \in \mathcal{G}$ and follow the same reasoning we easily end up with

$$Y \leq W \quad \text{almost surely.} \tag{1.46}$$

Hence, from (1.45) and (1.46) we conclude that $Y = W$ almost surely. ■

1.17 Properties of conditional expectation

We have established the existence of the conditional expectation of a random variable X, given a sub-σ-algebra of the σ-algebra $\mathcal{F}$ on the probability space $(\Omega, \mathcal{F}, \mathbb{P})$. It is important for the area of stochastic processes in general, to establish the properties of such conditional expectations, since most of the stochastic foundation of the methods used in what follows are based on these properties. We list the basic properties, as a series of independent propositions and it is their proof that helps the reader to get accustomed with the methods and techniques in the area.

Proposition 67 *Let a probability space $(\Omega, \mathcal{F}, \mathbb{P})$ and let $\mathcal{G}$ be a sub-σ-algebra of $\mathcal{F}$. If X and Y are integrable random variables and $c_1,\ c_2 \in \mathbb{R}$ are constants, then*

$$\mathbb{E}_{\mathbb{P}}[(c_1 X + c_2 Y) \mid \mathcal{G}] = c_1 \mathbb{E}_{\mathbb{P}}[X \mid \mathcal{G}] + c_2 \mathbb{E}_{\mathbb{P}}[Y \mid \mathcal{G}]. \tag{1.47}$$

The same equation holds if we assume that both X and Y are nonnegative and c_1 and c_2 are positive.

Proof. We first observe that both parts of (1.47) are $\mathcal{G}$-measurable so the first property of Definition 60 is satisfied. It is known from the partial average property that

$$\mathbb{E}_{\mathbb{P}}[(c_1 X + c_2 Y) \mid \mathcal{G}] = \int_A (c_1 X(\omega) + c_2 Y(\omega))\, d\mathbb{P}(\omega). \tag{1.48}$$

Hence, it remains to show that the second part of the equation (1.47) satisfies the partial averaging property. We have that

$$\begin{aligned}
& \int_A \left(c_1 \mathbb{E}_{\mathbb{P}}\left[X \mid \mathcal{G}\right](\omega) + c_2 \mathbb{E}_{\mathbb{P}}\left[Y \mid \mathcal{G}\right](\omega)\right) d\mathbb{P}(\omega) \\
= \; & c_1 \int_A \mathbb{E}_{\mathbb{P}}\left[X \mid \mathcal{G}\right](\omega)\, d\mathbb{P}(\omega) + c_2 \int_A \mathbb{E}_{\mathbb{P}}\left[Y \mid \mathcal{G}\right](\omega)\, d\mathbb{P}(\omega) \\
= \; & c_1 \int_A X(\omega)\, d\mathbb{P}(\omega) + c_2 \int_A Y(\omega)\, d\mathbb{P}(\omega) \\
= \; & \int_A \left(c_1 X(\omega) + c_2 Y(\omega)\right) d\mathbb{P}(\omega).
\end{aligned}$$

■

We now proceed to find what is called the Tower property of conditional expectations, which we believe is one of the most important properties.

Proposition 68 ***(Tower property).*** *Let a probability space* $(\Omega, \mathcal{F}, \mathbb{P})$ *and let* $\mathcal{G}$ *be a sub-*σ*-algebra of* $\mathcal{F}$*. If* $\mathcal{G}_1$ *is a* σ*-algebra such that* $\mathcal{G}_1 \subset \mathcal{G} \subset \mathcal{F}$*, then if* X *is an integrable random variable we have that*

$$\mathbb{E}_{\mathbb{P}}\left[\mathbb{E}_{\mathbb{P}}\left[X \mid \mathcal{G}\right] \mid \mathcal{G}_1\right] = \mathbb{E}_{\mathbb{P}}\left[X \mid \mathcal{G}_1\right]. \tag{1.49}$$

Equation (1.48) *also holds if we assume that* X *is non-negative.*

Proof. As we discussed before $\mathbb{E}_{\mathbb{P}}\left[X \mid \mathcal{G}\right]$ is a random variable, which is $\mathcal{G}$-measurable and $\mathbb{E}_{\mathbb{P}}\left[\mathbb{E}_{\mathbb{P}}\left[X \mid \mathcal{G}\right] \mid \mathcal{G}_1\right]$ is a $\mathcal{G}_1$-measurable random variable. The second part of (1.49), that is, $\mathbb{E}_{\mathbb{P}}\left[X \mid \mathcal{G}_1\right]$ is also $\mathcal{G}_1$-measurable. What remains to be proved is that we can replace $\mathbb{E}_{\mathbb{P}}\left[\mathbb{E}_{\mathbb{P}}\left[X \mid \mathcal{G}\right] \mid \mathcal{G}_1\right]$ with $\mathbb{E}_{\mathbb{P}}\left[X \mid \mathcal{G}_1\right]$ in the partial average property for the former. The partial average property for $\mathbb{E}_{\mathbb{P}}\left[\mathbb{E}_{\mathbb{P}}\left[X \mid \mathcal{G}\right] \mid \mathcal{G}_1\right]$ is the following

$$\int_A \mathbb{E}_{\mathbb{P}}\left[\mathbb{E}_{\mathbb{P}}\left[X \mid \mathcal{G}\right] \mid \mathcal{G}_1\right](\omega)\, d\mathbb{P}(\omega) = \int_A \mathbb{E}_{\mathbb{P}}\left[X \mid \mathcal{G}\right](\omega)\, d\mathbb{P}(\omega) \text{ for all } A \in \mathcal{G}_1. \tag{1.50}$$

But when $A \in \mathcal{G}_1$, since $\mathcal{G}_1 \subset \mathcal{G}$ we also have $A \in \mathcal{G}$. Therefore, we have that

$$\int_A \mathbb{E}_{\mathbb{P}}\left[X \mid \mathcal{G}_1\right](\omega)\, d\mathbb{P}(\omega) = \int_A X(\omega)\, d\mathbb{P}(\omega) = \int_A \mathbb{E}_{\mathbb{P}}\left[X \mid \mathcal{G}\right](\omega)\, d\mathbb{P}(\omega), \tag{1.51}$$

hence, from (1.50) and (1.51), we have that $\mathbb{E}_{\mathbb{P}}\left[X \mid \mathcal{G}_1\right]$ satisfies the partial-averaging property for $\mathbb{E}_{\mathbb{P}}\left[\mathbb{E}_{\mathbb{P}}\left[X \mid \mathcal{G}\right] \mid \mathcal{G}_1\right]$, which proves (1.49). ■

Proposition 69 ***(Conditional Jensen's inequality).*** *Let a probability space* $(\Omega, \mathcal{F}, \mathbb{P})$ *and* X *an integrable random variable in it. If* $\phi(X)$ *is a convex function, then*

$$\mathbb{E}_{\mathbb{P}}\left[\phi(X) \mid \mathcal{G}\right] \geq \phi\left(\mathbb{E}_{\mathbb{P}}\left[X \mid \mathcal{G}\right]\right). \tag{1.52}$$

Proof. The proof emerges easily by following the steps of the proof for the unconditional Jensen inequality. ■

Proposition 70 *Let a probability space* $(\Omega, \mathcal{F}, \mathbb{P})$ *and* X, Y *are integrable random variables in it. Let in addition that* XY *is integrable, and* X *is* $\mathcal{G}$*-measurable, then*

$$\mathbb{E}_{\mathbb{P}}\left[XY \mid \mathcal{G}\right] = X \mathbb{E}_{\mathbb{P}}\left[Y \mid \mathcal{G}\right]. \tag{1.53}$$

Proof. The proof could be found in Shreve (2004). ■

Proposition 71 *Let a probability space $(\Omega, \mathcal{F}, \mathbb{P})$ and X an integrable random variable in it. Let $\mathcal{G}$ a sub-σ-algebra of $\mathcal{F}$ and let that X is independent of $\mathcal{G}$. Then, we get that*

$$\mathbb{E}_{\mathbb{P}}[X \mid \mathcal{G}] = \mathbb{E}_{\mathbb{P}}[X]. \tag{1.54}$$

Proof. Firstly, since $\mathbb{E}_{\mathbb{P}}[X]$ is a constant it is measurable under any σ-algebra and so it is measurable under $\mathcal{G}$. Secondly, it remains to verify that $\mathbb{E}_{\mathbb{P}}[X]$ satisfies the partial averaging property that characterizes $\mathbb{E}_{\mathbb{P}}[X \mid \mathcal{G}]$, that is,

$$\int_A \mathbb{E}_{\mathbb{P}}[X]\, d\mathbb{P}(\omega) = \int_A X(\omega)\, d\mathbb{P}(\omega). \tag{1.55}$$

We will use again the standard method in order to prove property (1.55). Let us again start by assuming that X is an indicator random variable, let us say, $\mathbf{1}_{\{C\}}$ where the set C is independent of $\mathcal{G}$. Now, for all $A \in \mathcal{G}$, we have that

$$\begin{aligned}
\int_A X(\omega)\, d\mathbb{P}(\omega) &= \int_A \mathbf{1}_{\{C\}}(\omega)\, d\mathbb{P}(\omega) \\
&= \mathbb{P}(A \cap C) = \mathbb{P}(A)\,\mathbb{P}(C) \\
&= \mathbb{P}(A)\,\mathbb{E}_{\mathbb{P}}[X] = \int_A \mathbb{E}_{\mathbb{P}}[X]\, d\mathbb{P}(\omega),
\end{aligned}$$

which in fact proves (1.56). Following the logic of the proof for X an indicator function, we could prove all remaining three steps of the standard method. ■

In a similar way one could prove the following proposition:

Proposition 72 *Let a probability space $(\Omega, \mathcal{F}, \mathbb{P})$ and X an integrable random variable in it. If $X \geq 0$, then $\mathbb{E}_{\mathbb{P}}[X \mid \mathcal{G}] \geq 0$ almost surely.*

The following three useful propositions could be proved following the corresponding unconditional versions.

Proposition 73 *Let a probability space $(\Omega, \mathcal{F}, \mathbb{P})$ and $\{X_n\}_{n=0}^{\infty}$ an increasing sequence of non-negative random variables in it such that $X_n \uparrow X$, then $\mathbb{E}_{\mathbb{P}}[X_n \mid \mathcal{G}] \uparrow \mathbb{E}_{\mathbb{P}}[X \mid \mathcal{G}]$, almost surely.*

Proposition 74 *Let a probability space $(\Omega, \mathcal{F}, \mathbb{P})$ and $\{X_n\}_{n=0}^{\infty}$ a sequence of non-negative random variables in it such that $X_n \to X$ almost surely. Let also that Y is an integrable random variable on $(\Omega, \mathcal{F}, \mathbb{P})$ such that $|X_n(\omega)| \leq Y(\omega)$ for every $n \in \mathbb{N}$, then*

$$\mathbb{E}_{\mathbb{P}}[X_n \mid \mathcal{G}] \to \mathbb{E}_{\mathbb{P}}[X \mid \mathcal{G}], \text{ almost surely.}$$

Proposition 75 *Let a probability space $(\Omega, \mathcal{F}, \mathbb{P})$ and $\{X_n\}_{n=0}^{\infty}$ a sequence of non-negative random variables in it, then*

$$\mathbb{E}_{\mathbb{P}}[\liminf X_n \mid \mathcal{G}] \leq \liminf \mathbb{E}_{\mathbb{P}}[X_n \mid \mathcal{G}], \text{ almost surely.}$$

Finally, we conclude the properties of the conditional expectations with a proposition the proof of which is given as an exercise.

Proposition 76 *Let a probability space $(\Omega, \mathcal{F}, \mathbb{P})$ and X an integrable random variable in it. If $\mathcal{G}$ and $\mathcal{G}_1$ are sub-σ-algebras of $\mathcal{F}$ and $\mathcal{G}_1$ is independent of $\sigma(\sigma(X), \mathcal{G})$, then*

$$\mathbb{E}_{\mathbb{P}}[X \mid (\mathcal{G}, \mathcal{G}_1)] = \mathbb{E}_{\mathbb{P}}[X \mid \mathcal{G}], \text{ almost surely.}$$

Chapter 2

A SMALL REVIEW OF MATRIX ANALYSIS

2.1 Introductory notes

The purpose of this chapter is to discuss briefly, without many proofs, a number of useful concepts and facts on matrix analysis. These aspects of matrix analysis constitute a set of very useful tools, used in the development of many aspects of stochastic thinking. Much of the material would be included, in some form, in an introductory course in Linear Algebra. However, we will include a number of rather advance but useful items that are not commonly found elsewhere. Although the text will somehow be felt dense, it will be more or less self-contained, it will unify notation and terminology and it will be helpful to the reader to feel the gaps of his knowledge, which arises due to the optic horizon of the present material. Texts that are useful for more details and further reading are Barnett (1979, 1994), Berman and Plemmons (1979), Gantmacher (1959a,b,c), Rao (1973), Searle (1982), Graybill (1983), Horn and Johnson (1985, 1991), Lancaster (1969), Lancaster and Timenetsky (1985), Minc (1988), Pease (1965), Seneta (1981), Meyer (2000), Vassiliou and Tsaklidis (2001), and Schott (2005).

In general, we consider that the basic stage of Matrix Analysis, is a vector space and the field from which we get the coefficients for the act of multiplication. For our purposes as fields we will use the field of real numbers $\mathbb{R}$, and the field of complex numbers $\mathbb{C}$ under the usual addition and multiplication.

A *vector space* V over a field $\mathbf{F}$ is a set V of objects (called vectors), which is closed under a binary operation (addition) and is associative and commutative and has an identity and additive inverses in the set. The set is also closed under an operation of left multiplication of the vectors by elements of the scalar field $\mathbf{F}$. For all $x, y \in V : a(x+y) = ax + ay$, $(a+b)x = ax + bx$, $a(bx) = (ab)x$, and $ex = x$ for the multiplicative identity $e \in \mathbf{F}$.

We denote by $M_{m,n}(\mathbf{F})$ the set of all $m \times n$ matrices with elements from the field $\mathbf{F}$. The set $M_{n,n}(\mathbf{F})$ we simply write it as $M_n(\mathbf{F})$. For the needs of the present we will simply write $M_n = M_n(\mathbb{R})$. The set M_n is a vector space. A subset S of a vector space V is said to *span* V if every element of V may be represented as a linear combination (with coefficients from the underlying scalar field) of elements of S. The vectors $\mathbf{a}_1, \mathbf{a}_2, ..., \mathbf{a}_n$ of a vector space V over the field $\mathbf{F}$ are called *linearly independent* if and only if the relation

$$c_1\mathbf{a}_1 + c_2\mathbf{a}_2 + ... + c_n\mathbf{a}_n = 0 \text{ with } c_1, c_2, ..., c_n \in \mathbf{F},$$

is true when $c_1 = c_2 = ... = c_n = 0$. Conversely they are called *linearly dependent* if the above relation has a solution for $c_1, c_2, ..., c_n$ where not all the $c_i\ (i = 1, 2, ..., n)$ are zero. A linearly independent set which spans a vector space V is called a *basis* for V. Bases are highly non unique, but are very efficient in that each element of V can be represented in terms of the basis in one and only one way. Every vector space has a basis. If some basis of the vector space V consists of a finite number of elements, then all bases have the same

DOI: 10.1201/b23204-2

number of elements; this common number is called the *dimension* of the vector space V, and is denoted by $\dim V$. Any set of linearly independent elements of V may be extended to a basis in V; that is, given a linearly independent set$\{\mathbf{x}_1, \mathbf{x}_2, ..., \mathbf{x}_k\} \in V$, there exists $\{\mathbf{x}_{k+1}, \mathbf{x}_{k+2}, ..., \mathbf{x}_n\} \in V$, where $n = \dim V$, such that $\{\mathbf{x}_1, \mathbf{x}_2, ..., \mathbf{x}_n\}$ is a basis in V.

2.2 Matrices

We assume that addition and multiplication among matrices are well known. Let $\mathbf{A} \in M_{m,n}(\mathbf{F})$, we call *rank* of the matrix $\mathbf{A}$ and we denote it by $rank[\mathbf{A}]$ the larger number of linearly independent rows (columns) of $\mathbf{A}$. Equivalently there exists a $k \times k$ submatrix of $\mathbf{A}$ with $\det[\mathbf{A}] \neq 0$ and all the submatrices of $\mathbf{A}$ with dimension greater than k have determinant zero. We associate a n-dimensional vector space over $\mathbf{F}$ with $\mathbf{F}^n$, and we may think of $\mathbf{A} \in M_{m,n}(\mathbf{F})$ as a linear transformation from $\mathbf{F}^n$ to $\mathbf{F}^m$. The *domain* of such a linear transformation is $\mathbf{F}^n$; its *range* is $\{\mathbf{y} \in \mathbf{F}^m : \mathbf{y} = \mathbf{A}\mathbf{x} \text{ for some } \mathbf{x} \in \mathbf{F}^n\}$. The *null space* of $\mathbf{A}$ is given by $\{\mathbf{x} \in \mathbf{F}^n : \mathbf{A}\mathbf{x} = \mathbf{0}\}$. The range of $\mathbf{A}$ is a subspace of $\mathbf{F}^m$, and the null space of $\mathbf{A}$ is a subspace of $\mathbf{F}^n$. We have the relation

$$n = \text{dimension of null space of } \mathbf{A} + \text{ dimension of the range of } \mathbf{A}$$

between these two subspaces.

A matrix $\mathbf{A} \in M_n(\mathbf{F})$ is *nonsingular if* $\det[\mathbf{A}] \neq 0$ and *invertible* if there is a matrix $\mathbf{A}^{-1}$ called the *inverse* of $\mathbf{A}$ such that $\mathbf{A}^{-1}\mathbf{A} = \mathbf{I}$. In this case, we also have $\mathbf{A}\mathbf{A}^{-1} = \mathbf{I}$ and the inverse of a matrix is unique if it exists. The following are equivalent for a matrix $\mathbf{A} \in M_n(\mathbf{F})$:

(*i*) $\mathbf{A}$ is nonsingular;
(*ii*) $\mathbf{A}^{-1}$ exists;
(*iii*) $rank[\mathbf{A}] = n$;
(*iv*) The rows (columns) of $\mathbf{A}$ are linearly independent;
(*v*) $\det[\mathbf{A}] \neq 0$;
(*vi*) The dimension of the range of $\mathbf{A}$ is n;
(*vii*) The dimension of the null space of $\mathbf{A}$ is 0;
(*viii*) The system $\mathbf{A}\mathbf{x} = \mathbf{b}$ has a unique solution for each $\mathbf{b} \in \mathbf{F}^n$.

Perhaps the single most important concept in matrix theory is the set of n numbers called the *spectrum* of $\mathbf{A}$, denoted by $\sigma(\mathbf{A})$, its eigenvalues.

Definition 77 *Let $\mathbf{A} \in M_n(\mathbb{C})$ and $\mathbf{x} \in \mathbb{C}^n$, we consider the equation*

$$\mathbf{A}\mathbf{x} = \lambda\mathbf{x}, \quad \lambda \in \mathbb{C}, \mathbf{x} \neq \mathbf{0}, \tag{2.1}$$

then λ is called an eigenvalue of $\mathbf{A}$ and $\mathbf{x}$ is called a right eigenvector of $\mathbf{A}$ associated with λ. The spectral radius of $\mathbf{A}$ is the non-negative real number $\rho(\mathbf{A}) = \max\{|\lambda| : \lambda \in \sigma(\mathbf{A})\}$. This is just the radius of the smallest disc centered at the origin in the complex plane that includes all the eigenvalues of $\mathbf{A}$.

It is easy to show as an exercise that

Theorem 78 *Let $p(.)$ be a given polynomial. If λ is an eigenvalue of $\mathbf{A} \in M_n(\mathbb{C})$, while $\mathbf{x}$ is an associated right eigenvector, then $p(\lambda)$ is an eigenvalue of the matrix $p(\mathbf{A})$ and $\mathbf{x}$ is an right eigenvector of $p(\mathbf{A})$ associated with $p(\lambda)$.*

Remark 79 *It is not difficult to show that if $\mathbf{A} = diag(a_1, a_2, ..., a_n)$ is a diagonal matrix then the set $\{a_1, a_2, ..., a_n\} = \sigma(\mathbf{A})$ is the spectrum of $\mathbf{A}$. Also a matrix $\mathbf{A} \in M_n(\mathbb{C})$ is singular if and only if $\mathbf{0} \in \sigma(\mathbf{A})$.*

The eigenvalue-eigenvector equation (2.1) may be rewritten equivalently as

$$(\lambda\mathbf{I}-\mathbf{A})\,\mathbf{x}=\mathbf{0},\ \ \mathbf{x}\neq\mathbf{0},$$

hence, $\lambda\in\sigma(\mathbf{A})$ if and only if $\lambda\mathbf{I}-\mathbf{A}$ is a singular matrix, that is,

$$\det[\lambda\mathbf{I}-\mathbf{A}]=0,$$

which is a formal polynomial in λ called the *characteristic polynomial* of $\mathbf{A}$, the roots of which apparently provide the spectrum of $\mathbf{A}$. It is not difficult to show that for a triangular matrix

$$\mathbf{A}=\begin{pmatrix} a_{11} & a_{12} & ... & a_{1n}\\ 0 & a_{22} & ... & a_{2n}\\ ... & ... & ... & ...\\ 0 & 0 & ... & a_{nn}\end{pmatrix},$$

$\sigma(\mathbf{A})=\{a_{11},a_{22},...,a_{nn}\}$.

A fundamental and nontrivial fact, called the *fundamental theorem of algebra* is, that a polynomial of degree n with complex coefficients has exactly n zeroes, counting multiplicities, among the complex numbers. In view of this each matrix $\mathbf{A}\in M_n(\mathbb{C})$ has, among the complex numbers, exactly n eigenvalues counting multiplicities. The "multiplicity" of an eigenvalue λ of $\mathbf{A}\in M_n(\mathbb{C})$, is the number of times λ occurs as a root of the characteristic polynomial $p_A(.)$.

It is not difficult to show the following useful relationships

$$\det[\mathbf{A}]=\prod_{i=1}^{n}\lambda_i,$$

and

$$tr[\mathbf{A}]=\sum_{i=1}^{n}a_{ii}=\sum_{i=1}^{n}\lambda_i.$$

Also

$$tr\left[\mathbf{A}^k\right]=\sum_{i=1}^{n}\lambda_i^k,$$

the right-hand side of the previous equation is called the k-th moment of the eigenvalues of $\mathbf{A}$. In addition let $\mathbf{A},\mathbf{B}\in M_n(\mathbb{C})$ be given, then $\mathbf{A}$ and $\mathbf{B}$ have the same eigenvalues if and only if $tr\left[\mathbf{A}^k\right]=tr\left[\mathbf{B}^k\right]$ for $k=1,2,...,n$.

Let $\mathbf{A}\in M_n(\mathbb{C})$ with $\mathbf{A}=\{a_{ij}\}_{i,j}$, then the *transpose* of $\mathbf{A}$ denoted by $\mathbf{A}^\top$ is the matrix $\mathbf{A}^\top=\{a_{ji}\}_{i,j}\in M_n(\mathbb{C})$. Also the *Hermitian adjoint* is the conjugate transpose of $\mathbf{A}$, denoted by $\mathbf{A}^*=\{\bar{a}_{ji}\}_{i,j}\in M_n(\mathbb{C})$.

If $\mathbf{A}\in M_n(\mathbb{C})$ then the eigenvalues of $\mathbf{A}^\top$ are the same with those of $\mathbf{A}$ and the eigenvalues of $\mathbf{A}^*$ are the complex conjugate of the eigenvalues of $\mathbf{A}$.

Let $\mathbf{A}\in M_n(\mathbb{C})$. For a given $\lambda\in\sigma(\mathbf{A})$, the set of all eigenvectors of λ is called the *eigenspace* of $\mathbf{A}$ corresponding to the eigenvalue of λ. It could be easily proved that the eigenspace of each $\lambda\in\sigma(\mathbf{A})$ is a subspace of $\mathbb{C}^n$. The dimension of the eigenspace of λ is called the *geometric multiplicity* of λ. The geometric multiplicity of the eigenvalue λ of a matrix $\mathbf{A}\in M_n(\mathbb{C})$ is less or equal than its algebraic multiplicity.

Even though $\mathbf{A}$ and $\mathbf{A}^\top$ have the same eigenvalues, their eigenvectors corresponding to a given eigenvalue may be very different.

A nonzero vector $\mathbf{y}\in\mathbb{C}^n$ is called a *left eigenvector* of $\mathbf{A}\in M_n(\mathbb{C})$ corresponding to $\lambda\in\sigma(\mathbf{A})$ if

$$\mathbf{y}^*\mathbf{A}=\lambda\mathbf{y}^*.$$

If necessary for clarity we refer to an eigenvector as a right eigenvector. When the context does not require distinction, we just say eigenvector.

A left eigenvector $\mathbf{y}$ corresponding to the eigenvalue λ of $\mathbf{A} \in M_n(\mathbb{C})$ is a right eigenvector of $\mathbf{A}^* \in M_n(\mathbb{C})$ corresponding to $\bar{\lambda}$, and also that $\bar{\mathbf{y}}$ is a right eigenvector of $\mathbf{A}^\top$ corresponding to λ. Two vectors $\mathbf{x}$,$\mathbf{y} \in \mathbb{C}^n$ are called *orthogonal* if $\mathbf{y}^*\mathbf{x} = \mathbf{0}$. The following theorem is known as the *principle of biorthogonality.*

Theorem 80 *If $\mathbf{A} \in M_n(\mathbb{C})$ and if $\lambda, \mu \in \sigma(\mathbf{A})$, with $\lambda \neq \mu$, then any left eigenvector of $\mathbf{A}$ corresponding to μ is orthogonal to any right eigenvector of $\mathbf{A}$ corresponding to λ.*

Proof. Let $\mathbf{y} \in \mathbb{C}^n$ one of the left eigenvectors of $\mathbf{A}$ which corresponds to the eigenvalue μ.

Then

$$\mathbf{y}^*\mathbf{A} = \mu\mathbf{y}^*.$$

Let also $\mathbf{x} \in \mathbb{C}^n$ one of the right eigenvectors of $\mathbf{A}$ which corresponds to the eigenvalue λ. Then

$$\mathbf{A}\mathbf{x} = \lambda\mathbf{x}.$$

Hence,

$$\mathbf{y}^*\mathbf{A}\mathbf{x} = \mathbf{y}^*(\lambda\mathbf{x}) = \lambda(\mathbf{y}^*\mathbf{x}),$$

and

$$\mathbf{y}^*\mathbf{A}\mathbf{x} = (\mu\mathbf{y}^*)\mathbf{x} = \mu(\mathbf{y}^*\mathbf{x}),$$

and since $\lambda \neq \mu$, we have

$$\mathbf{y}^*\mathbf{x} = \mathbf{0}.$$

■

Let $\mathbf{A} \in M_{m,n}(\mathbf{F})$. We call a submatrix of $\mathbf{A}$ every matrix which is the result of excluding a number of rows and columns of $\mathbf{A}$. If we exclude the same rows and columns the submatrix is called a *principal submatrix* of $\mathbf{A}$. If $S_1 = \{1, 2, ..., k\}$ and $S_2 = \{1, 2, ..., k\}$ are respectively the sets of rows and columns that are excluded from the matrix $\mathbf{A}$, then we denote the resulting submatrix by $\mathbf{A}(S_1, S_2)$. A principal submatrix since we exclude the same rows and columns we will denote it by $\mathbf{A}(S)$. We will now provide without proof the following important theorem:

Theorem 81 *Let $\mathbf{A} \in M_{m,n}(\mathbb{C})$ and $\lambda \in \mathbb{C}$ an eigenvalue of $\mathbf{A}$ with geometric multiplicity $k \geq 1$. If the matrix $\hat{\mathbf{A}} \in M_m(\mathbb{C})$ is a principal submatrix of $\mathbf{A}$ and $m > n - k$ then λ is an eigenvalue of $\hat{\mathbf{A}}$ with algebraic multiplicity at least k.*

2.2.1 Similarity

It is well known that two matrices $\mathbf{A}, \mathbf{B} \in M_n(\mathbb{C})$ are equivalent, if there exists nonsingular matrices $\mathbf{P}$ and $\mathbf{Q} \in M_n(\mathbb{C})$ such that

$$\mathbf{PAQ} = \mathbf{B}.$$

We will now proceed to a more strong relation among two matrices $\mathbf{A}, \mathbf{B} \in M_n(\mathbb{C})$.

A matrix $\mathbf{A} \in M_n(\mathbb{C})$ is said to be *similar* to a matrix $\mathbf{B} \in M_n(\mathbb{C})$, if there exists a nonsingular matrix $\mathbf{S} \in M_n(\mathbb{C})$ such that

$$\mathbf{A} = \mathbf{S}^{-1}\mathbf{B}\,\mathbf{S}.$$

The transformation $\mathbf{B} \to \mathbf{S}^{-1}\mathbf{B}\,\mathbf{S}$ is called a *similarity transformation* by the *similarity matrix* $\mathbf{S} \in M_n$. The relation "$\mathbf{B}$ is similar to $\mathbf{A}$" is often abbreviated $\mathbf{B} \sim \mathbf{A}$.

The relation $\mathbf{A} \sim \mathbf{B}$ is an equivalent relation since it is easy to see that

$$(i)\ \mathbf{A}\sim \mathbf{A};\ (ii)\ \mathbf{B} \sim \mathbf{A} \Longrightarrow \mathbf{A} \sim \mathbf{B} \text{ and}$$

$$(iii)\quad \mathbf{A} \sim \mathbf{B} \text{ and } \mathbf{B} \sim \mathbf{C} \Longrightarrow \mathbf{A} \sim \mathbf{C}.$$

Matrices in an equivalence class share many important properties and we will mention some of them in here. It is of interest though to observe that a similarity transformation of a matrix $M_n(\mathbb{C})$, corresponds to representation of a linear transformation on $\mathbb{C}^n$ in another basis. We start with the following theorem:

Theorem 82 *Let $\mathbf{A}, \mathbf{B} \in M_n(\mathbb{C})$ and that the matrices are similar, that is, there exists a nonsingular matrix $\mathbf{S} \in M_n(\mathbb{C})$, such that $\mathbf{A} = \mathbf{S}^{-1}\mathbf{B}\ \mathbf{S}$. Then they have the same characteristic polynomials and as a consequence the same eigenvalues, counting multiplicities.*

Proof. For every $\lambda \in \mathbb{C}$, we have that

$$\begin{aligned} p_{\mathbf{A}}(\lambda) &= \det[\lambda\mathbf{I} - \mathbf{A}] = \det\left[\lambda\mathbf{S}^{-1}\mathbf{S} - \mathbf{S}^{-1}\mathbf{B}\ \mathbf{S}\right] \\ &= \det\left[\mathbf{S}^{-1}(\lambda\mathbf{I} - \mathbf{B})\,\mathbf{S}\right] = \det\left[\mathbf{S}^{-1}\right]\det[(\lambda\mathbf{I} - \mathbf{B})]\det[\mathbf{S}] \\ &= \det\left[\mathbf{S}^{-1}\right]\det[\mathbf{S}]\det[(\lambda\mathbf{I} - \mathbf{B})] = p_{\mathbf{B}}(\lambda). \end{aligned}$$

■

If the matrix $\mathbf{A} \in M_n(\mathbb{C})$ is similar to a diagonal matrix then $\mathbf{A}$ is *diagonizable.* The following theorem provides a characterization of when a matrix is diagonizable. We will now provide in the form of a theorem a more usable way of recognizing if a matrix is diagonizable.

Theorem 83 *If the eigenvalues $\lambda_1, \lambda_2, ..., \lambda_n$ of a matrix $\mathbf{A} \in M_n(\mathbb{C})$ are distinct, then $\mathbf{A}$ is diagonizable.*

Proof. We will start by showing that the eigenvectors $\mathbf{x}_1, \mathbf{x}_2, ..., \mathbf{x}_n$ of the matrix $\mathbf{A} \in M_n(\mathbb{C})$, which are associated with the eigenvalues $\lambda_1, \lambda_2, ..., \lambda_n$ are linearly independent. We assume that this is not true. Then let r be the minimum number of nonzero coefficients such that

$$a_1\mathbf{x}_1 + a_2\mathbf{x}_2 + ... + a_r\mathbf{x}_r = 0, \quad 1 \le r \le n, \quad \text{with } \mathbf{x}_i \neq 0. \tag{2.2}$$

We have that

$$\begin{aligned} \mathbf{A}(a_1\mathbf{x}_1 + a_2\mathbf{x}_2 + ... + a_r\mathbf{x}_r) &= a_1\mathbf{A}\mathbf{x}_1 + a_2\mathbf{A}\mathbf{x}_2 + ... + a_r\mathbf{A}\mathbf{x}_r \\ &= a_1\lambda_1\mathbf{x}_1 + a_2\lambda_2\mathbf{x}_2 + ... + a_r\lambda_r\mathbf{x}_r = 0. \end{aligned} \tag{2.3}$$

Multiplying (2.2) by λ_r and subtracting from (2.3), we get that

$$a_1(\lambda_1 - \lambda_r)\mathbf{x}_1 + a_2(\lambda_1 - \lambda_r)\mathbf{x}_2 + ... + a_{r-1}(\lambda_{r-1} - \lambda_r)\mathbf{x}_{r-1} = 0,$$

which is a contradiction with our hypothesis, since we have found $r-1$ coefficients $a_i(\lambda_i - \lambda_r)$ for $i = 1, 2, ..., r-1$, different than zero since $\lambda_i \neq \lambda_r$ for $i = 1, 2, ..., r-1$. Consequently, the eigenvectors $\mathbf{x}_1, \mathbf{x}_2, ..., \mathbf{x}_n$ are linearly independent and therefore the matrix $\mathbf{A} \in M_n(\mathbb{C})$ is diagonizable. ■

A matrix $\mathbf{A} \in M_n(\mathbb{C})$ of the form

$$\mathbf{A} = \begin{pmatrix} \mathbf{A}_{11} & \mathbf{0} & ... & \mathbf{0} \\ \mathbf{0} & \mathbf{A}_{22} & ... & \mathbf{0} \\ ... & ... & ... & ... \\ \mathbf{0} & 0 & ... & \mathbf{A}_{kk} \end{pmatrix}$$

in which $\mathbf{A}_{ii} \in M_{n_i}(\mathbb{C}), i = 1, 2, ..., k$ and $\sum_{i=1}^{k} n_i = n$, is called *block diagonal.* We often denote such a matrix as

$$\mathbf{A} = \mathbf{A}_{11} \oplus \mathbf{A}_{22} \oplus ... \oplus \mathbf{A}_{kk} = \oplus \sum \mathbf{A}_{ii},$$

and we call it the *direct sum* of the matrices $\mathbf{A}_{ii}, i = 1, 2, ..., k$. It is not difficult to see that

$$\det \left[\oplus \sum \mathbf{A}_{ii} \right] = \prod_{i=1}^{k} \det \mathbf{A}_{ii},$$

so that $\mathbf{A} = \mathbf{A}_{11} \oplus \mathbf{A}_{22} \oplus ... \oplus \mathbf{A}_{kk}$, is not singular if and only if each $\mathbf{A}_{ii}, i = 1, 2, ..., k$, is nonsingular. Also,

$$rank \left[\oplus \sum \mathbf{A}_{ii} \right] = \sum_{i=1}^{k} rank \mathbf{A}_{ii}.$$

2.3 The minimal polynomial of A

A polynomial $p_A(\mathbf{A})$ is called an *annihilating polynomial* if $p_A(\mathbf{A}) = 0$. According to the Caley-Hamilton theorem every matrix $\mathbf{A}$ satisfies its characteristic equation. Hence, if we denote by $c(\lambda)$ the characteristic polynomial of $\mathbf{A}$, that is,

$$c_{\mathbf{A}}(\lambda) = \det [\lambda \mathbf{I} - \mathbf{A}],$$

then $c(\mathbf{A}) = 0$. Hence, the characteristic polynomial of $\mathbf{A}$ is an *annihilating polynomial* for $\mathbf{A}$. For a matrix $\mathbf{A}$ there may be many annihilating polynomials, one of which will have the minimum degree. Since if $p_A(\mathbf{A}) = 0$, then also $ap_A(\mathbf{A}) = 0$ for any $a \in \mathbf{C}$, it is apparent, that such a polynomial is possible to be normalized in such a way that the coefficient of the highest degree will be 1. Such a polynomial will be called the *minimum annihilating polynomial* for $\mathbf{A}$ and will be denoted by $q_{\mathbf{A}}(\mathbf{A})$.

The following three results could be proved by the reader as exercises.

Similar matrices have the same minimal annihilating polynomial.

For every matrix $\mathbf{A} \in M_n(\mathbb{C})$ the minimal polynomial $q_{\mathbf{A}}(\mathbf{A})$ divides the characteristic polynomial $c_{\mathbf{A}}(\lambda)$. Moreover, $q_{\mathbf{A}}(\lambda) = 0$, if and only if, λ is an eigenvalue of $\mathbf{A}$, so every root of $c_{\mathbf{A}}(t)$ is a root of $q_{\mathbf{A}}(t)$.

The minimal annihilating polynomial of a matrix $\mathbf{A}$ is unique.

These results are useful in creating an algorithm, for finding the minimal annihilating polynomial of a matrix, given that it is not a matrix of big order. Assume, that for the matrix $\mathbf{A} \in M_n(\mathbb{C})$ the characteristic polynomial $c_{\mathbf{A}}(t)$ has been completely factored as

$$c_{\mathbf{A}}(t) = \prod_{i=1}^{s} (t - \lambda_i)^{m_i}, \quad 1 \leq m_i \leq n, \quad \text{with} \quad m_1 + m_2 + ... + m_s = n, \tag{2.4}$$

with $\lambda_1, \lambda_2, ..., \lambda_s$ distinct, then the minimal annihilating polynomial $q_{\mathbf{A}}(t)$ must have the form

$$q_{\mathbf{A}}(t) = \prod_{i=1}^{s} (t - \lambda_i)^{r_i}, \quad 1 \leq r_i \leq m_i. \tag{2.5}$$

Hence, there are finitely many polynomials of the form of the product in (2.5) that are candidates to be the minimal annihilating polynomial of $\mathbf{A} \in M_n(\mathbb{C})$. Starting with the product in which all $r_i = 1$, we determine by explicit calculation the one of minimal degree that annihilates $\mathbf{A}$.

Example 84 *We will find the minimal annihilating polynomial $q_{\mathbf{A}}(t)$ of the matrix*

$$\mathbf{A} = \begin{pmatrix} 1 & 1 & 1 & 1 \\ 1 & i & -1 & -1 \\ 1 & -1 & 1 & -1 \\ 1 & -i & -1 & i \end{pmatrix}.$$

We have that

$$\lambda\mathbf{I} - \mathbf{A} = \begin{pmatrix} \lambda-1 & 1 & 1 & 1 \\ 1 & \lambda-i & -1 & -1 \\ 1 & -1 & \lambda-1 & -1 \\ 1 & -i & -1 & \lambda-i \end{pmatrix},$$

and

$$c_{\mathbf{A}}(\lambda) = \det[\lambda\mathbf{I} - \mathbf{A}] = (\lambda+2)(\lambda-2)^2(\lambda-2i).$$

Hence, the minimal polynomial is either

$$q_{\mathbf{A}}^{(1)}(\lambda) = (\lambda+2)(\lambda-2)(\lambda-2i),$$

or

$$q_{\mathbf{A}}^{(2)}(\lambda) = (\lambda+2)(\lambda-2)^2(\lambda-2i),$$

but we have that

$$q_{\mathbf{A}}^{(1)}(\mathbf{A}) = (\mathbf{A}+2\mathbf{I})(\mathbf{A}-2\mathbf{I})(\mathbf{A}-2i\mathbf{I}) = 0,$$

hence, the minimal polynomial is

$$q_{\mathbf{A}}(\lambda) = (\lambda+2)(\lambda-2)(\lambda-2i).$$

2.3.1 Invariant polynomials and elementary divisors

Let a matrix $\mathbf{A} \in M_n(\mathbb{C})$ and consider the characteristic matrix $(\lambda\mathbf{I} - \mathbf{A})$. Consider the set of all possible determinants of order k of the matrix $(\lambda\mathbf{I} - \mathbf{A})$. This gives us a set of polynomials in λ. Denote by $d_k(\lambda)$, the greatest common divisor of this set of determinants normalized in such a way that the coefficient of the highest power of λ is unity. Without loss of generality assume that the rank of $\mathbf{A}$ is n. Obtain the set

$$d_n(\lambda), d_{n-1}(\lambda), ..., d_1(\lambda), \text{ and define } d_0(\lambda) = 1.$$

Apparently, $d_n(\lambda)$ is the determinant of $(\lambda\mathbf{I} - \mathbf{A})$. If the rank of $\mathbf{A}$ is $r < n$, then apparently $d_i(\lambda) = 0$ for $i = r+1, ..., n$. In such a case, we do not consider the zero polynomials. It is not difficult to see that $d_k(\lambda)$ is always divisible by $d_{k-1}(\lambda)$. Therefore, $d_{k-1}(\lambda)$ is a common divisor of all determinants of order k and hence is a factor at least in $d_k(\lambda)$. We form (assuming that $rank(\mathbf{A}) = n$) the quotients

$$ip_n(\lambda) = \frac{d_n(\lambda)}{d_{n-1}(\lambda)},$$

$$ip_{n-1}(\lambda) = \frac{d_{n-1}(\lambda)}{d_{n-2}(\lambda)},$$

$$\dots\dots\dots\dots\dots\dots$$

$$ip_1(\lambda) = \frac{d_1(\lambda)}{d_0(\lambda)} = d_1(\lambda),$$

where all $ip_k(\lambda)$ are polynomials in λ. The polynomials $ip_k(\lambda)$ for all k are called the invariant polynomials of $\mathbf{A}$. Note that $ip_n(\lambda)$ is the characteristic polynomial of $\mathbf{A}$. It is evident that

$$\prod_{k=1}^{n} ip_k(\lambda) = d_n(\lambda) = \det[\lambda\mathbf{I} - \mathbf{A}].$$

It can be shown, that the invariant polynomials are invariant under similarity transformations.

If we factor each of the invariant polynomials of the matrix $\mathbf{A}$ into powers of irreducible polynomials, then the set of all non constant factors appearing in the set are called the *elementary divisors of* $\mathbf{A}$. The elementary divisors are again invariant under similarity transformations and the product of all elementary divisors is, again, the characteristic equation.

Example 85 *Let the matrix*

$$\mathbf{A} = \begin{pmatrix} 4 & 1 & 1 & 1 \\ -1 & 2 & -1 & -1 \\ 6 & 1 & -1 & 1 \\ -6 & -1 & 4 & 2 \end{pmatrix},$$

then

$$\lambda\mathbf{I} - \mathbf{A} = \begin{pmatrix} \lambda-4 & -1 & -1 & -1 \\ 1 & \lambda-2 & 1 & 1 \\ -6 & -1 & \lambda+1 & -1 \\ 6 & 1 & -4 & \lambda-2 \end{pmatrix}.$$

We have that

$$d_4(\lambda) = \det[\lambda\mathbf{I} - \mathbf{A}] = (\lambda-3)^3(\lambda+2).$$

We now find all possible subdeterminants of order 3 *of* $\det[\lambda\mathbf{I} - \mathbf{A}]$, *we factor them and find their greatest common divisor* $d_3(\lambda)$, *that is, there are 16 possible determinants of order 3 and*

$$d_3(\lambda) = \lambda - 3.$$

Now, we have to find all possible subdeterminants of order 2. *However, there is one, which is independent of* λ *and consequently*

$$d_2(\lambda) = 1.$$

Similarly, we have

$$d_1(\lambda) = 1 \quad \textit{and by definition } d_0(\lambda) = 1.$$

Hence, the invariant polynomials are the following

$$ip_4(\lambda) = \frac{d_4(\lambda)}{d_3(\lambda)} = (\lambda-3)^2(\lambda+2),$$

$$ip_3(\lambda) = \frac{d_3(\lambda)}{d_2(\lambda)} = \lambda - 3,$$

$$ip_2(\lambda) = \frac{d_2(\lambda)}{d_1(\lambda)} = 1,$$

$$ip_1(\lambda) = \frac{d_1(\lambda)}{d_0(\lambda)} = 1.$$

From the invariant polynomial $ip_4(\lambda)$ *we get the elementary divisors* $(\lambda-3)^2$ *and* $(\lambda+2)$, *and from* $ip_3(\lambda)$ *the elementary divisor* $(\lambda-3)$.

We now provide without proof, three useful theorems, which intuitively seem to be true. The third one could be proved as an exercise, taking for granted that the second one is true.

Theorem 86 *The set of elementary divisors of the matrix* $\mathbf{C}(\lambda)$

$$\mathbf{C}(\lambda) = \begin{pmatrix} \mathbf{A}(\lambda) & \mathbf{0} \\ \mathbf{0} & \mathbf{B}(\lambda) \end{pmatrix},$$

is the union of the sets of elementary divisors of the matrices $\mathbf{A}(\lambda)$ *and* $\mathbf{B}(\lambda)$.

Theorem 87 *Let the matrices* $\mathbf{A}, \mathbf{B} \in M_n(\mathbb{C})$. *The matrices* $\mathbf{A}$ *and* $\mathbf{B}$ *are similar, if and only if, the polynomial matrices* $\lambda\mathbf{I} - \mathbf{A}$ *and* $\lambda\mathbf{I} - \mathbf{B}$ *are equivalent.*

Theorem 88 *Two matrices* $\mathbf{A}, \mathbf{B} \in M_n(\mathbb{C})$ *are similar, if and only if, they have the same invariant polynomials or equivalently if they have the same elementary divisors.*

2.3.2 The Jordan canonical form

We studied similarity via a general non-singular matrix $\mathbf{S} \in M_n(\mathbb{C})$. There are a very special class of nonsingular matrices, acting as similarity matrices, called *unitary matrices* defined as follows:

Definition 89 *A matrix* $\mathbf{U} \in M_n(\mathbb{C})$ *is said to be unitary if* $\mathbf{U}^*\mathbf{U} = \mathbf{I}$. *If, in addition,* $\mathbf{U} \in M_n(\mathbb{R})$, $\mathbf{U}$ *is said to be real orthogonal.*

The main properties of unitary matrices are included in the following theorem, which is not difficult for the reader to prove.

Theorem 90 *Let* $\mathbf{U} \in M_n(\mathbb{C})$ *be a unitary matrix, then it has the following properties:*

(1) $\mathbf{U}$ *is nonsingular and* $\mathbf{U}^* = \mathbf{U}^{-1}$;

(2) $\mathbf{U}\mathbf{U}^* = \mathbf{I}$.

(3) $\mathbf{U}^*$ *is unitary.*

(4) The columns of $\mathbf{U}$ *form an orthonormal set. That is, if* $\mathbf{u}_1, \mathbf{u}_2, ..., \mathbf{u}_n$ *are the columns of* $\mathbf{U}$, *then* $\mathbf{u}_i^*\mathbf{u}_j = 0$ *and* $\mathbf{u}_i^*\mathbf{u}_i = 1$ *for* $i, j = 1, 2, ..., n$.

(5) The rows of $\mathbf{U}$ *form an orthonormal set.*

Perhaps the most fundamentally useful fact of elementary matrix theory is that any matrix $A \in M_n(\mathbb{C})$, is unitary equivalent to an upper triangular matrix T, and also a lower triangular matrix. This is formally called the Schur theorem.

Theorem 91 ***(Schur).*** *Let* $\mathbf{A} \in M_n(\mathbb{C})$ *with eigenvalues* $\lambda_1, \lambda_2, ..., \lambda_n$ *in any prescribed order, then there is a unitary matrix* $\mathbf{U} \in M_n(\mathbb{C})$ *such that*

$$\mathbf{U}^*\mathbf{A}\mathbf{U} = \mathbf{T},$$

is upper triangular, with diagonal entries $t_{ii} = \lambda_i$, $i = 1, 2, ..., n$. *Furthermore, if* $\mathbf{A} \in M_n(\mathbb{R})$ *and if all the eigenvalues of* $\mathbf{A}$ *are real, then* $\mathbf{U}$ *may be chosen to be real and orthogonal.*

Then it can be shown, that if $\mathbf{A} \in M_n(\mathbb{C})$ is an upper triangular matrix with eigenvalues $\lambda_1, \lambda_2, ..., \lambda_k$ distinct, with multiplicities n_i for $i = 1, 2, ..., k$, then it can be transformed by similarity into a block diagonal matrix of the form

$$\begin{pmatrix} \mathbf{T}_1 & \mathbf{0} & \dots & \mathbf{0} \\ \mathbf{0} & \mathbf{T}_2 & \dots & \mathbf{0} \\ \dots & \dots & \dots & \dots \\ \mathbf{0} & \mathbf{0} & \dots & \mathbf{T}_k \end{pmatrix},$$

where $\mathbf{T}_i \in M_{n_i}(\mathbb{C})$ is upper triangular with all diagonal entries equal to λ_i, $i = 1, 2, ..., k$. If $A \in M_n(\mathbb{R})$ and if all the eigenvalues of $\mathbf{A}$ are real, then the same result holds, and the similarity matrix may be taken to be real.

Finally, in a third step it can be shown that an upper triangular matrix, whose main diagonal entries are all equal, is similar to a direct sum of Jordan blocks as defined below. The above results provide a three step reduction of an arbitrary complex matrix, to what in what follows we will call *Jordan Canonical Form.*

All similar square complex matrices constitute an equivalence class, which contains at least one Jordan matrix. Any two Jordan matrices in the same under similarity equivalence class are essentially the same in a trivial way. A Jordan matrix that is similar to a given matrix is called the *Jordan Canonical Form or the Jordan Normal Form* of the matrix. The Jordan canonical form is a set of "almost diagonal" matrices, called Jordan matrices, which include the diagonal matrices. A Jordan matrix is in general a block diagonal matrix, in which each diagonal block is a so called Jordan diagonal block. A *Jordan block* $\mathbf{J}_k(\lambda)$ is a $k \times k$ upper triangular matrix of the form

$$\mathbf{J}_k(\lambda) = \begin{pmatrix} \lambda & 1 & 0 & ... & 0 \\ 0 & \lambda & 1 & ... & 0 \\ ... & ... & ... & ... & ... \\ 0 & 0 & 0 & \lambda & 1 \\ 0 & 0 & 0 & 0 & \lambda \end{pmatrix}.$$

For $k = 1$, we have $\mathbf{J}_1(\lambda) = \lambda$. Then a *Jordan matrix* $\mathbf{J} \in M_n(\mathbb{C})$ is a direct sum of Jordan blocks

$$\mathbf{J} = \begin{pmatrix} \mathbf{J}_{n_1}(\lambda_1) & 0 & ... & 0 \\ 0 & \mathbf{J}_{n_2}(\lambda_2) & ... & 0 \\ ... & ... & ... & ... \\ 0 & 0 & ... & \mathbf{J}_{n_k}(\lambda_k) \end{pmatrix},$$

where $n_1 + n_2 + ... + n_k = n$ in which the orders n_i may not be distinct and the values λ_i need not be distinct.

If each Jordan block $\mathbf{J}_{n_i}(\lambda_i)$ is one dimensional, that is, $n_i = 1$ and $k = n$, then the Jordan matrix $\mathbf{J}$ is diagonal. If any of the Jordan blocks is of order greater than one, then $\mathbf{J}$ is not even diagonizable.

Let $\mathbf{A} \in M_n(\mathbb{C})$ and let that the characteristic polynomial of $\mathbf{A}$ is of the form

$$c_{\mathbf{A}}(\lambda) = \prod_{i=1}^{s} (\lambda - \lambda_i)^{m_i}, \quad 1 \le m_i \le n, \quad \text{with} \quad m_1 + m_2 + ... + m_s = n.$$

Assume, that the *elementary divisors* of the matrix $\mathbf{A}$ are

$$(\lambda - \lambda_i)^{r_i} \quad \text{for} \quad i = 1, 2, ..., k. \tag{2.6}$$

In every elementary divisor $(\lambda - \lambda_i)^{r_i}$ we correspond the $r_i \times r_i$ Jordan block

$$\mathbf{J}_{r_i}(\lambda_i) = \begin{pmatrix} \lambda_i & 1 & 0 & ... & 0 \\ 0 & \lambda_i & 1 & ... & 0 \\ ... & ... & ... & ... & ... \\ 0 & 0 & 0 & \lambda_i & 1 \\ 0 & 0 & 0 & 0 & \lambda_i \end{pmatrix}, \tag{2.7}$$

it could be easily seen that $J_{r_i}(\lambda_i)$ has $(\lambda - \lambda_i)^{r_i}$ as the only elementary divisor. Then, the blog-diagonal matrix

$$\mathbf{J}_{\mathbf{A}} = diag\{\mathbf{J}_{r_1}(\lambda_1), \mathbf{J}_{r_2}(\lambda_2), ..., \mathbf{J}_{r_k}(\lambda_k)\}, \tag{2.8}$$

has due to Theorem 86, as the only elementary divisors the elementary divisors (2.6). Now, since the matrices $\mathbf{A}$ and $\mathbf{J_A}$ have the same elementary divisors, according to Theorem 88 they are similar. The Jordan matrix $\mathbf{J_A}$ is called the *Jordan cannonical form of* $\mathbf{A}$ or simply the Jordan form of $\mathbf{A}$. Equivalent results we could get if instead of the uppertriangular Jordan block we use the downtriangular Jordan form

$$\mathbf{J}_{r_i}(\lambda_i) = \begin{pmatrix} \lambda_i & 0 & 0 & \dots & 0 \\ 1 & \lambda_i & 0 & \dots & 0 \\ \dots & \dots & \dots & \dots & \dots \\ 0 & 0 & 1 & \lambda_i & 0 \\ 0 & 0 & 0 & 1 & \lambda_i \end{pmatrix}.$$

Example 92 *We consider again the matrix of the previous example, that is,*

$$\mathbf{A} = \begin{pmatrix} 4 & 1 & 1 & 1 \\ -1 & 2 & -1 & -1 \\ 6 & 1 & -1 & 1 \\ -6 & -1 & 4 & 2 \end{pmatrix},$$

as we have found in the example 85 the elementary divisors of $\mathbf{A}$ *are*

$$(\lambda-3)^2, \quad (\lambda+2) \quad and \quad (\lambda-3).$$

Consequently, the Jordan cannonical form $\mathbf{J_A}$ *of the matrix* $\mathbf{A}$ *has the Jordan blocks* $\mathbf{J}_2(3)$, $\mathbf{J}_1(-2)$ *and* $\mathbf{J}_1(3)$, *that is,*

$$\mathbf{J_A} = diag\{\mathbf{J}_2(3), \mathbf{J}_1(3), \mathbf{J}_1(-2)\} = \begin{pmatrix} 3 & 1 & 0 & 0 \\ 0 & 3 & 0 & 0 \\ 0 & 0 & 3 & 0 \\ 0 & 0 & 0 & -2 \end{pmatrix}.$$

2.3.3 Remarks on the Jordan canonical form of a matrix

Let a matrix $\mathbf{A} \in M_n(\mathbb{C})$, then the Jordan canonical form of the matrix

$$\mathbf{J_A}(\lambda) = \begin{pmatrix} \mathbf{J}_{r_1}(\lambda_1) & 0 & \dots & 0 \\ 0 & \mathbf{J}_{r_2}(\lambda_2) & \dots & 0 \\ \dots & \dots & \dots & \dots \\ 0 & 0 & \dots & \mathbf{J}_{r_k}(\lambda_k) \end{pmatrix},$$

has a definite structure, that makes apparent certain properties of the matrix and of any matrix that is similar to it.

(1) The number k of Jordan blocks, counting multiple occurrences of the same block, is the number of linearly independent eigenvectors of $\mathbf{J_A}(\lambda)$.

(2) A matrix $\mathbf{A} \in M_n(\mathbb{C})$ is diagonizable, if and only if, its elementary divisors are of the first degree, in which case we have $n = k$.

(3) Let $\mathbf{J}_r(\lambda_i)$ be the Jordan block with the largest order that corresponds to the eigenvalue λ_i in the Jordan form $\mathbf{J_A}(\lambda)$ of a matrix $\mathbf{A} \in M_n(\mathbb{C})$. Then r is the power of $(\lambda - \lambda_i)$ in the minimal annihilating polynomial of $\mathbf{A}$.

(4) The number of Jordan blocks corresponding to a given eigenvalue is the geometric multiplicity of the eigenvalue, which is the dimension of the associate eigenspace.

(5) Let a Jordan block $\mathbf{J}_r(\lambda)$ with $r \geq 1$, $\lambda \in \mathbb{C}$.

(a) Also let that $\lambda \neq 0$. We observe that

$$(\mathbf{J}_r(\lambda))^2 = \begin{pmatrix} \lambda^2 & 2\lambda & 1 & 0 & 0 & \dots & 0 & 0 \\ 0 & \lambda^2 & 2\lambda & 1 & 0 & \dots & 0 & 0 \\ 0 & 0 & \lambda^2 & 2\lambda & 1 & \dots & 0 & 0 \\ 0 & 0 & 0 & \lambda^2 & 2\lambda & \dots & 0 & 0 \\ \dots & \dots & \dots & \dots & \dots & \dots & \dots & \dots \\ 0 & 0 & 0 & 0 & \dots & \lambda^2 & 2\lambda & 1 \\ 0 & 0 & 0 & 0 & \dots & 0 & \lambda^2 & 2\lambda \\ 0 & 0 & 0 & 0 & \dots & 0 & 0 & \lambda^2 \end{pmatrix}.$$

Raising the Jordan block $\mathbf{J}_r(\lambda)$ into the third, fourth power and so on, then from the pattern that emerges we conclude that

$$(\mathbf{J}_r(\lambda))^m = \begin{pmatrix} \lambda^m & \binom{m}{1}\lambda^{m-1} & \binom{m}{2}\lambda^{m-2} & \dots & \binom{m}{r-1}\lambda^{m-r+1} \\ 0 & \lambda^m & \binom{m}{1}\lambda^{m-1} & \dots & \binom{m}{r-2}\lambda^{m-r+2} \\ \dots & \dots & \dots & \dots & \dots \\ 0 & 0 & 0 & \dots & \lambda^m \end{pmatrix},$$

with $m = 1, 2, ...$ Hence, we immediately get that

$$rank\left[(\mathbf{J}_r(\lambda))^m\right] = r, \quad \text{for } \lambda \neq 0 \text{ and } m = 1, 2, ...$$

(b) If $\lambda = 0$, then

$$\mathbf{J}_r(0) = \begin{pmatrix} 0 & 1 & 0 & \dots & 0 \\ 0 & 0 & 1 & \dots & 0 \\ \dots & \dots & \dots & \dots & \dots \\ 0 & 0 & 0 & \dots & 1 \\ 0 & 0 & 0 & \dots & 0 \end{pmatrix},$$

therefore,

$$rank\left[\mathbf{J}_r(0)\right] = r - 1.$$

Also by raising $\mathbf{J}_r(0)$ into the second, third power and so on it is easy to see that

$$rank\left[(\mathbf{J}_r(0))^m\right] = \begin{cases} r-m & \text{for } m=1,2,\dots,r. \\ 0 & \text{for } m>r. \end{cases}$$

(6) Let again a Jordan block $\mathbf{J}_r(\lambda)$ with $r \geq 1$, $\lambda \in \mathbb{C}$ and $\lambda \neq 0$. Observe that the matrix

$$\mathbf{J}_r(\lambda) - \lambda\mathbf{I} = \begin{pmatrix} 0 & 1 & 0 & \dots & 0 \\ 0 & 0 & 1 & \dots & 0 \\ \dots & \dots & \dots & \dots & \dots \\ 0 & 0 & 0 & \dots & 1 \\ 0 & 0 & 0 & \dots & 0 \end{pmatrix} = \mathbf{J}_r(0).$$

Hence, according to the previous paragraph

$$rank\left[(\mathbf{J}_r(\lambda) - \lambda\mathbf{I})^m\right] = \begin{cases} r-m & \text{for } m=1,2,\dots,r. \\ 0 & \text{for } m>r. \end{cases}$$

(7) Since $\mathbf{A} \in M_n(\mathbb{C})$ and $\mathbf{J_A}$ are similar there exist a matrix $\mathbf{U} \in M_n(\mathbb{C})$ such that

$$\mathbf{A} = \mathbf{U}\mathbf{J_A}\mathbf{U}^{-1},$$

and for every $\lambda \in \mathbb{C}$ we get that

$$(\mathbf{A} - \lambda\mathbf{I})^m = \mathbf{U}(\mathbf{J_A} - \lambda\mathbf{I})^m \mathbf{U}^{-1}, \quad \text{for } m = 1, 2, ...$$

Hence, the matrices $(\mathbf{A}-\lambda\mathbf{I})^m$ and $(\mathbf{J_A}-\lambda\mathbf{I})^m$ are similar for every $m \in \mathbb{N}$ and $\lambda \in \mathbb{C}$, and therefore

$$rank\left[(\mathbf{A}-\lambda\mathbf{I})^m\right] = rank\left[(\mathbf{J_A}-\lambda\mathbf{I})^m\right] \quad \text{for every } m \in \mathbb{N} \text{ and } \lambda \in \mathbb{C}.$$

This property is the basis for a second method for finding the Jordan cannonical form of a matrix, see Horn and Jonson (1985).

2.3.4 Construction of the transformation matrix of **A** *to its Jordan form*

Let the matrix $\mathbf{A} \in M_n(\mathbb{C})$ and its elementary divisors

$$(\lambda-\lambda_1)^{r_1}, (\lambda-\lambda_2)^{r_2}, ..., (\lambda-\lambda_k)^{r_k},$$

where $\lambda_1, \lambda_2, ..., \lambda_k$ are eigenvalues, not necessary distinct, of **A**. Now, let that $\mathbf{U} \in M_n(\mathbb{C})$ is such that

$$\mathbf{A} = \mathbf{U}\mathbf{J_A}\mathbf{U}^{-1},$$

or

$$\mathbf{A}\mathbf{U} = \mathbf{U}\mathbf{J_A}, \tag{2.9}$$

where $\mathbf{J_A} = diag\left\{\mathbf{J}_{r_1}(\lambda_1), \mathbf{J}_{r_2}(\lambda_2), ..., \mathbf{J}_{r_k}(\lambda_k)\right\}$ and $\mathbf{J}_{r_i}(\lambda_i)$ for $i = 1, 2, ..., k$, are the Jordan blocks of order $r_i \times r_i$

$$\mathbf{J}_{r_i}(\lambda_i) = \begin{pmatrix} \lambda_i & 1 & 0 & ... & 0 \\ 0 & \lambda_i & 1 & ... & 0 \\ ... & ... & ... & ... & ... \\ 0 & 0 & 0 & \lambda_i & 1 \\ 0 & 0 & 0 & 0 & \lambda_i \end{pmatrix} \quad \text{for } i = 1, 2, ..., k.$$

Let us denote by $\mathbf{u}_i$ the *i-th* column of the matrix **U**. Then from the relation (2.9) we get that

$$\begin{gathered} \mathbf{A}\mathbf{u}_1 = \lambda_1\mathbf{u}_1, \ \mathbf{A}\mathbf{u}_2 = \lambda_1\mathbf{u}_2 + \mathbf{u}_1, ..., \ \mathbf{A}\mathbf{u}_{r_1} = \lambda_1\mathbf{u}_{r_1} + \mathbf{u}_{r_1-1} \\ \mathbf{A}\mathbf{u}_{r_1+1} = \lambda_2\mathbf{u}_{r_1+1}, \ \mathbf{A}\mathbf{u}_{r_1+2} = \lambda_2\mathbf{u}_{r_1+2} + \mathbf{u}_{r_1+1}, ..., \\ \mathbf{A}\mathbf{u}_{r_1+r_2} = \lambda_2\mathbf{u}_{r_1+r_2} + \mathbf{u}_{r_1+r_2-1}. \\ \cdots\cdots\cdots\cdots\cdots\cdots\cdots\cdots\cdots \end{gathered} \tag{2.10}$$

$$\begin{aligned} \mathbf{A}\mathbf{u}_{r_1+...+r_{k-1}+1} &= \lambda_k\mathbf{u}_{r_1+...+r_{k-1}+1}, \\ \mathbf{A}\mathbf{u}_{r_1+...+r_{k-1}+2} &= \lambda_k\mathbf{u}_{r_1+...+r_{k-1}+2} + \mathbf{u}_{r_1+...+r_{k-1}+1}, ..., \\ \mathbf{A}\mathbf{u}_n &= \lambda_k\mathbf{u}_n + \mathbf{u}_{n-1}. \end{aligned}$$

From the system (2.10) we get the equivalent system

$$(\mathbf{A}-\lambda_1\mathbf{I})\mathbf{u}_1 = \mathbf{0}, \ (\mathbf{A}-\lambda_1\mathbf{I})\mathbf{u}_2 = \mathbf{u}_1, ..., \ (\mathbf{A}-\lambda_1\mathbf{I})\mathbf{u}_{r_1} = \mathbf{u}_{r_1-1},$$

$$\begin{aligned} (\mathbf{A}-\lambda_2\mathbf{I})\mathbf{u}_{r_1+1} &= \mathbf{0}, \ (\mathbf{A}-\lambda_2\mathbf{I})\mathbf{u}_{r_1+2} = \mathbf{u}_{r_1+1}, ..., \\ (\mathbf{A}-\lambda_1\mathbf{I})\mathbf{u}_{r_1+r_2} &= \mathbf{u}_{r_1+r_2-1} \end{aligned} \tag{2.11}$$

$$\cdots\cdots\cdots\cdots\cdots\cdots\cdots\cdots\cdots$$

From the first line of (2.11) and the first two relations we get that

$$(\mathbf{A}-\lambda_1\mathbf{I})^2\mathbf{u}_2 = \mathbf{0} \quad \text{while} \quad (\mathbf{A}-\lambda_1\mathbf{I})\mathbf{u}_2 = \mathbf{u}_1 \neq \mathbf{0}.$$

From the first line of of (2.11) and the first three relations we get that

$$(\mathbf{A}-\lambda_1\mathbf{I})^3\,\mathbf{u}_3=\mathbf{0}\quad\text{while}\quad(\mathbf{A}-\lambda_1\mathbf{I})^2\,\mathbf{u}_3=\mathbf{u}_1\neq\mathbf{0}.$$

Continuing with similar arguments we arrive at the relations

$$(\mathbf{A}-\lambda_1\mathbf{I})^i\,\mathbf{u}_i=\mathbf{0}\quad\text{while}\quad(\mathbf{A}-\lambda_1\mathbf{I})^{i-1}\,\mathbf{u}_i\neq\mathbf{0}\ \text{ for }\ i=1,2,...,r_1. \tag{2.12}$$

$$(\mathbf{A}-\lambda_2\mathbf{I})^j\,\mathbf{u}_{r_1+j}=\mathbf{0}\quad\text{while}\quad(\mathbf{A}-\lambda_2\mathbf{I})^{j-1}\,\mathbf{u}_{r_1+j}\neq\mathbf{0}\ \text{ for }\ j=1,2,...,r_2.$$

....

Due to relations (2.12) it is apparent that there is a one to one relation of the Jordan block $\mathbf{J}_{r_1}(\lambda_1)$ with the first r_1 columns $\{\mathbf{u}_1,\mathbf{u}_2,...,\mathbf{u}_{r_1}\}$ of $\mathbf{U}$. Also, there is a one to one relation of the Jordan block $\mathbf{J}_{r_2}(\lambda_2)$ with the columns $\{\mathbf{u}_{r_1+1},\mathbf{u}_{r_1+2},...,\mathbf{u}_{r_1+r_2}\}$ of $\mathbf{U}$, etc.

The set of columns $\{\mathbf{u}_{r_i+1},\mathbf{u}_{r_i+2},...,\mathbf{u}_{r_i+r_{i+1}}\}$ of $\mathbf{U}$ that relates with the Jordan block $\mathbf{J}_{r_i+1}(\lambda_{i+1})$ is called the *Jordan chain* that relates with the Jordan block $\mathbf{J}_{r_i+1}(\lambda_{i+1})$ or equivalently the elementary divisor $(\lambda-\lambda_{i+1})^{r_i+1}$ for $i=1,2,...,k$. The number of vectors in a Jordan chain, such as for example $\{\mathbf{u}_{r_i+1},\mathbf{u}_{r_i+2},...,\mathbf{u}_{r_i+r_{i+1}}\}$, is called the *length of the chain,* which apparently in this example is r_{i+1}. The column $\mathbf{u}_{r_i+j}$ of $\mathbf{U}$ is called a *generalized eigenvector* of rank j of the Jordan chain $\{\mathbf{u}_{r_i+1},\mathbf{u}_{r_i+2},...,\mathbf{u}_{r_i+r_{i+1}}\}$ for $j=1,2,...,r_{i+1}$ and $i=1,2,...,k$. The last generalized eigenvector in a Jordan chain is called the *generator of the Jordan chain.* Taking into account relations (2.12) we could equivalently define generalized eigenvectors as follows:

Definition 93 *A column vector $\mathbf{u}_m$ of the matrix $\mathbf{U}\in M_n(\mathbb{C})$ is called a generalized eigenvector of rank m for the eigenvalue λ of $\mathbf{A}\in M_n(\mathbb{C})$, $n\in\mathbb{N}$, if we have that*

$$(\mathbf{A}-\lambda\mathbf{I})^m\,\mathbf{u}_m=\mathbf{0}\quad\textit{while}\quad(\mathbf{A}-\lambda\mathbf{I})^{m-1}\,\mathbf{u}_m\neq\mathbf{0}\,. \tag{2.13}$$

It is apparent that, if we evaluate the generator eigenvector of a Jordan chain, related with a Jordan block of a matrix $\mathbf{A}\in M_n(\mathbb{C})$, then it is possible from relations (2.11) to evaluate the entire chain. The generator eigenvectors should satisfy both conditions of the type (2.13) .It is not difficult to prove that, the generalized eigenvectors of a Jordan chain are independent.

2.4 The norm of a vector

Let a vector space V with elements from the field F, which for our purposes will be $\mathbb{R}$ or $\mathbb{C}$. We will start with the definition of the norm of a vector.

Definition 94 *A function $\|.\|:V\to\mathbb{R}$ is a vector norm (or norm of a vector) if for all* $\mathbf{x},\mathbf{y}$ *we have*

(1) $\|\mathbf{x}\|\geq 0$
(2) $\|a\mathbf{x}\|=|a|\,\|\mathbf{x}\|$ *for all* $a\in F$.
(3) $\|\mathbf{x}+\mathbf{y}\|\leq\|\mathbf{x}\|+\|\mathbf{y}\|$.
(4) $\|\mathbf{x}\|=\mathbf{0}$, *if and only if* $\mathbf{x}=\mathbf{0}$.

If such a function satisfies the three first properties of the previous definition, but not necessarily the forth, then we call it a vector semi-norm. A semi-norm allows also other vectors to have norms equal with zero apart from the zero vector.

A real or complex vector space V, together with a given norm $\|.\|$, is called a *normed linear space (normed vector space).* We now provide the definition of an inner product of vectors y and $x\in V$.

Definition 95 *Let V be a vector space over the field $\mathbf{F}$. A function $\langle .,.\rangle:V\times V\to\mathbf{F}$ is an inner product if for all $x,y,z\in V$ and all $a\in\mathbf{F}$,*

(1) $\langle x, x\rangle \geq 0$; (2) $\langle x, x\rangle = 0$ if and only if $x = 0$; (3) $\langle x+y, z\rangle = \langle x, z\rangle + \langle y, z\rangle$; (4) $\langle ax, y\rangle = a\langle x, y\rangle$; (5) $\langle x, y\rangle = \overline{\langle y, x\rangle}$.

An important property for inner products is the Cauchy-Schwarz inequality.

Theorem 96 *(Cauchy-Schwarz inequality). Let $\langle .,.\rangle : V \times V \to \mathbf{F}$ be an inner product on V a vector space over the field $\mathbf{F}$. Then*

$$|\langle x, y\rangle|^2 \leq \langle x, x\rangle\langle y, y\rangle \text{ for all } x, y \in V,$$

with equality if and only if x and y are linearly dependent.

The following is an important Corollary that links norms and inner products in a vector space.

Corollary 97 *Let $\langle .,.\rangle : V \times V \to \mathbf{F}$ be an inner product on the vector space V over the field $\mathbf{F}$. Then the function $\|.\| : V \to [0, \infty)$ defined by $\|x\| = \langle x, x\rangle^{1/2}$ is a norm on V.*

We will now present some often used vector norms, which have been proved useful tools in many cases. What we should have in mind is that each norm that we will present, could be proved a useful tool for a different purpose or problem.

(1) **The L_2 norm in C^n (Euclidean).**

If $\mathbf{x} \in \mathbb{C}^n$ with $\mathbf{x} = (x_1, x_2, ..., x_n)^\top$, the L_2 norm is defined by

$$\|\mathbf{x}\|_2 = \left(|x_1|^2, |x_2|^2, ..., |x_n|^2\right)^{1/2}.$$

The L^2 norm has been proved useful and is probably the right one for optimization problems.

(2) **The L_1 norm in C^n (additive).**

If $\mathbf{x} \in \mathbb{C}^n$ with $\mathbf{x} = (x_1, x_2, ..., x_n)^\top$, the L_1 norm is defined by

$$\|\mathbf{x}\|_1 = |x_1| + |x_2| + ... + |x_n|.$$

The L_1 norm is useful in estimation problems in Statistics, because it provides estimators with desired properties (robustness). It has also been proved useful in approximation theory.

(3) **The L_∞ norm in C^n (max norm).**

If $\mathbf{x} \in \mathbb{C}^n$ with $\mathbf{x} = (x_1, x_2, ..., x_n)^\top$, the L_∞ norm is defined by

$$\|\mathbf{x}\|_\infty = \max\left(|x_1|, |x_2|, ..., |x_n|\right).$$

The L_∞ norm is mainly useful in the proofs in problems of convergence and it has been proved an important tool in the theory of stochastic process and the theory of convergence of deterministic and stochastic algorithms.

(4) **The L_p norm in C^n**

If $\mathbf{x} \in \mathbb{C}^n$ with $\mathbf{x} = (x_1, x_2, ..., x_n)^\top$, the L_p norm is defined by

$$\|\mathbf{x}\|_p = \left(\sum_{i=1}^{n} |x_i|^p\right)^{1/p}, \quad \text{for } p \geq 1.$$

In fact, from any vector norm or from many vector norms, it is possible to construct new vector norms. This fact is established in the following theorem.

Theorem 98 *Let $\|.\| : V \to \mathbb{R}$ be a vector norm in $\mathbb{C}^n$ and $\mathbf{A} \in M_n(\mathbb{C})$ a nonsingular matrix, then the function $\|.\|_\mathbf{A} : \mathbb{C}^n \to \mathbb{R}$ which is such that $\|\mathbf{x}\|_\mathbf{A} = \|\mathbf{A}\mathbf{x}\|$, $\|\mathbf{x}\| \in \mathbb{C}^n$, is also a vector norm in $\mathbb{C}^n$.*

Proof. We have that

$$\|\mathbf{x}\|_{\mathbf{A}} = \|\mathbf{Ax}\| \geq 0 \text{ since apparently } \mathbf{Ax} \in \mathbb{C}^n.$$

From property 2 of a vector norm we get that

$$\|a\mathbf{x}\|_{\mathbf{A}} = \|a\mathbf{Ax}\| = |a| \, \|\mathbf{Ax}\| = |a| \, \|\mathbf{x}\|_{\mathbf{A}}, \quad \text{for all } a \in M_n(\mathbb{C}).$$

From property 3 of a vector norm we get that

$$\|\mathbf{x}+\mathbf{y}\|_{\mathbf{A}} = \|\mathbf{A}(\mathbf{x}+\mathbf{y})\| = \|\mathbf{Ax}+\mathbf{Ay}\| \leq \|\mathbf{Ax}\| + \|\mathbf{Ay}\| = \|\mathbf{x}\|_{\mathbf{A}} + \|\mathbf{y}\|_{\mathbf{A}}.$$

The property $\|\mathbf{x}\|_{\mathbf{A}} = 0$ if and only if $\mathbf{x} = \mathbf{0}$ is proved analogously. ■

We will now prove a property of a vector norm which is analogous with a property of the absolute value.

Proposition 99 *Let V a vector space and $\|.\| : V \to \mathbb{R}$ a vector semi-norm, then*

$$|\|\mathbf{x}\| - \|\mathbf{y}\|| \leq \|\mathbf{x}-\mathbf{y}\|, \quad \textit{for all } \mathbf{x}, \mathbf{y} \in V.$$

Proof. We have that $\mathbf{y} = \mathbf{x} + (\mathbf{y}-\mathbf{x})$, from which we get that

$$\|\mathbf{y}\| \leq \|\mathbf{x}\| + \|\mathbf{y}-\mathbf{x}\| = \|\mathbf{x}\| + |-1| \, \|\mathbf{x}-\mathbf{y}\|,$$

and finally

$$\|\mathbf{y}\| - \|\mathbf{x}\| \leq \|\mathbf{x}-\mathbf{y}\|.$$

However, we also have $\mathbf{x} = \mathbf{y} + (\mathbf{x}-\mathbf{y})$ from which we get

$$\|\mathbf{x}\| - \|\mathbf{y}\| \leq \|\mathbf{x}-\mathbf{y}\|,$$

and consequently from the last two equations we get that

$$|\|\mathbf{x}\| - \|\mathbf{y}\|| \leq \|\mathbf{x}-\mathbf{y}\|, \quad \text{for all } \mathbf{x}, \mathbf{y} \in V.$$

■

Vector norms could be used for the convergence of vectors.

Definition 100 *Let V be a vector space with elements from $\mathbb{R}$ or $\mathbb{C}$, and let $\|.\|$ a norm on V. We say that a sequence of vectors $\{\mathbf{x}_n\}$ of V converges to vector $\mathbf{x}$ in norm, if and only if,*

$$\lim_{n\to\infty} \|\mathbf{x}_n - \mathbf{x}\| = 0 \textit{ when } n \to \infty.$$

In the previous definitions, in each particular case it is important to choose the right norm to use in proving convergence. Another delicate question is if the chosen norm influence the limit. In a vector space of infinite dimension, it is possible a sequence of vectors to converge in a norm and not to converge in another. When the vector space is of finite dimension it is not possible for such a thing to happen. Two norms are called *equivalent* if any sequence of vectors in the vector space V that converges with one of the two, converges also with the other. A consequence of that is, that all norms in a finite dimensional vector space are equivalent. Another consequence is that convergence in norm of a sequence of vectors in a finite dimensional vector space with elements from $\mathbb{R}$ or $\mathbb{C}$ is equivalent with convergence elementwise of the respective coordinates of the sequence of vectors.

In order to have a criterion if a sequence of vectors in a vector space with elements from $\mathbb{R}$ or $\mathbb{C}$ converges, when we do not know the possible limit, we provide the definition of a *Caushy sequence* in a vector space.

Definition 101 *A sequence of vectors* $\{\mathbf{x}_n\}$ *in a vector space* V *with elements from* $\mathbb{R}$ *or* $\mathbb{C}$ *is a Caushy sequence in norm* $\|.\|$, *if there exist a positive number* $N(\varepsilon)$ *such that*

$$\|\mathbf{x}_{n_1} - \mathbf{x}_{n_2}\| \leq \varepsilon,$$

for every $n_1, n_2 \geq N(\varepsilon)$.

Now, it is easy to show that a sequence of vectors $\{\mathbf{x}_n\}$ in a finite dimensional vector space V, with elements from $\mathbb{R}$ or $\mathbb{C}$ converges to a vector in V, if it is a Caushy sequence in norm $\|.\|$ on V for every $n_1, n_2 \geq N(\varepsilon)$.

We will now discuss some of the geometric properties of vector norms. One of the most important geometric features of a norm is its unit ball, through which considerable insight into properties of the norm may be gained.

Definition 102 *Let* V *be a vector space with elements from* $\mathbb{R}$ *or* $\mathbb{C}$, *and let* $\|.\|$ *a norm on* V *and* $x \in V, r > 0$. *The ball of radius* r *around* x *is the set*

$$B_{\|.\|}(r; x) = \{y \in V : \|y - x\| \leq r\}.$$

Definition 103 *The unit ball of* $\|.\|$ *is the set*

$$B_{\|.\|}(1; 0) = \{y \in V : \|y\| \leq 1\}.$$

The unit ball is a geometric summary of a norm and characterizes it. For example, if $\|.\|_a$ and $\|.\|_b$ are norms on V then

$$\|x\|_a \leq \|x\|_b \text{ if and only if } B_{\|.\|_b}(1; 0) \subset B_{\|.\|_a}(1; 0)$$

The unit ball of a norm is not necessarily always a sphere. In order to understand this deeper it would be useful for the reader to sketch the unit balls of the norms $\|.\|_1$, $\|.\|_2$, and $\|.\|_\infty$. A norm is called *polyhedral if its unit ball is a polyedron.*

Remark 104 *The unit ball of a norm is equilibrated, that is, for every* $x \in B_{\|.\|}(1; 0)$ *then* $ax \in B_{\|.\|}(1; 0)$ *for all scalars* a *such that* $|a| = 1$.

We now provide the following definition:

Definition 105 *Let* V *be a vector space with elements from* $\mathbb{R}$ *or* $\mathbb{C}$, *and let* $\|.\|$ *a norm on* V *and let* S *be a subset of* V. *The set* S *is bounded if there exists some* $r > 0$ *such that* $S \subset B_{\|.\|}(r; 0)$. *A point* $x \in S$ *is an interior point of* S *if there is an* $\varepsilon > 0$ *such that* $B(\varepsilon, x) \subset S$. *The set* S *is open if every point of* S *is an interior point; the set* S *is compact if from every covering* $\cup_i S_i \supset S$ *by open sets* S_i *there exist a finite number of them let say* $S_1, S_2, ..., S_k$ *such that* $\cup_{i=1}^k S_i \supset S$. *The set* S *is closed if its complement is open.*

Remark 106 *Let* V *a finite normed vector space then the unit ball of any norm is compact.*

Remark 107 *Let* V *a normed vector space then the unit ball of any norm is compact.*

Remark 108 *A set in a normed linear space is closed if and only if it contains all its limit points. That is, a point* $x \in V$ *is a limit point if there exists some sequence* $\{x_k\} \subset S$ *such that* $\lim_{k \to \infty} x_k = x$ *with respect to some norm.*

2.5 The matrix norm

In the present section, we will extend the concept of the vector norm to the norm of a matrix in the vector space $M_n(\mathbb{C})$ which is of dimension n^2.

Definition 109 *A function* $\|.\| : M_n(\mathbb{C}) \to \mathbb{R}$ *is called a matrix norm if for all matrices* $\mathbf{A}, \mathbf{B} \in M_n$ *the following properties hold.*

(1) $\|\mathbf{A}\| \geq 0$.
(2) $\|a\mathbf{A}\| = |a| \|\mathbf{A}\|$ *for all* $a \in \mathbb{C}$.
(3) $\|\mathbf{A} + \mathbf{B}\| \leq \|\mathbf{A}\| + \|\mathbf{B}\|$.
(4) $\|\mathbf{AB}\| \leq \|\mathbf{A}\| \|\mathbf{B}\|$.
(5) $\|\mathbf{A}\| = 0$ *if and only if* $\mathbf{A} = \mathbf{0}$.

If a function satisfies the properties (1), (2), (3) and (5), then it is called the generalized norm of a matrix. Now, based on the above properties the reader could prove as exercises the following two propositions.

Proposition 110 *If* $\|.\| : M_n(\mathbb{C}) \to \mathbb{R}$ *is a generalized norm and* $\mathbf{A}, \mathbf{B} \in M_n(\mathbb{C})$ *then*

$$|\|\mathbf{A}\| - \|\mathbf{B}\|| \leq \|\mathbf{A} - \mathbf{B}\|.$$

Proposition 111 *If* $\|.\| : M_n(\mathbb{C}) \to \mathbb{R}$ *is any norm, then*

(*i*) $\|\mathbf{I}\| \leq 1$.
(*ii*) *If* $\mathbf{A} \in M_n(\mathbb{C})$ *is non singular, then*

$$\|\mathbf{A}^{-1}\| \geq \frac{\|\mathbf{I}\|}{\|\mathbf{A}\|}.$$

(*iii*) *If* $\mathbf{A} \in M_n(\mathbb{C})$, *then* $\|\mathbf{A}^n\| \leq \|\mathbf{A}\|^n$.

We will present now some commonly used norms, which have been proved useful tools in different cases.

(1) The L_1 norm in $M_n(\mathbb{C})$.
If $\mathbf{A} \in M_n(\mathbb{C})$, then the L_1 norm is defined as

$$\|\mathbf{A}\|_1 = \sum_{i,j=1}^{n} |a_{ij}|.$$

(2) The L_2 norm in $M_n(\mathbb{C})$.
If $\mathbf{A} \in M_n(\mathbb{C})$, then the L_2 norm is defined as

$$\|\mathbf{A}\|_2 = \left(\sum_{i,j=1}^{n} |a_{ij}|^2 \right)^{1/2}.$$

(3) The L_∞ norm in $M_n(\mathbb{C})$.
If $\mathbf{A} \in M_n(\mathbb{C})$, then the L_∞ norm is defined as

$$\|\mathbf{A}\|_\infty = \max_{1 \leq i,j \leq n} |a_{ij}|.$$

It is easy to see that L_∞ is a generalized norm.

(4) If $\mathbf{A} \in M_n(\mathbb{C})$, then we define

$$\|\mathbf{A}\|_{m\infty} = n \|\mathbf{A}\|_\infty.$$

It is easy to see that the above is a norm.

(5) If $\mathbf{A} \in M_n(\mathbb{C})$, then we define as the **Frobenious norm**

$$\|\mathbf{A}\|_F^2 = trace(\mathbf{A}^* \mathbf{A}).$$

It is easy to be proved that

$$\|\mathbf{A}\|_F^2 = \sum_{i,j} |a_{ij}|^2 .$$

(6) If $\mathbf{A} \in M_n(\mathbb{C})$, then we define the norm

$$\|\mathbf{A}\|_r = \max_i \sum_{j=1}^{n} |a_{ij}| .$$

The above norm is very useful as a tool for the convergence of infinite products and powers of stochastic matrices. If $\mathbf{A} \in M_\infty(\mathbb{R})$, then the norm $\|.\|_r$ is defined as

$$\|\mathbf{A}\|_r = \sup_i \sum_{j=1}^{\infty} |a_{ij}| .$$

(7) If $\mathbf{A} \in M_n(\mathbb{C})$, then we define the norm

$$\|\mathbf{A}\|_c = \max_j \sum_{i=1}^{n} |a_{ij}| .$$

This is an analogous norm as the previous one except that we have a column sum instead of a row sum.

(8) The spectral norm in $M_n(\mathbb{C})$.

If $\mathbf{A} \in M_n(\mathbb{C})$, then we define the spectral norm

$$\|\mathbf{A}\|_\phi = \max\left\{\sqrt{\lambda} : \lambda \text{ an eigenvalue of } \mathbf{A}^*\mathbf{A}\right\}.$$

(9) Induced Matrix norms.

Because $\mathbb{C}^{mn}$ and $\mathbb{R}^{mn}$ are vector spaces of dimension mn, magnitude of a matrix $\mathbf{A}$ in $M_{m,n}(\mathbb{C})$ or $M_{m,n}(\mathbb{R})$ can be "measured" by employing any vector norm on $\mathbb{C}^{mn}$ and $\mathbb{R}^{mn}$ respectively.

For example if

$$\mathbf{A} = \begin{pmatrix} a_{11} & a_{12} \\ a_{21} & a_{22} \end{pmatrix},$$

by stringing out its entries into a vector in $\mathbb{R}^4$, that is, $[a_{11}, a_{12}, a_{21}, a_{22}]$, the Euclidean norm on $\mathbb{R}^4$ can be applied to provide

$$\|\mathbf{A}\| = \left[a_{11}^2 + a_{12}^2 + a_{21}^2 + a_{22}^2\right]^{1/2} .$$

Which in fact is the *Frobenius norm* $\|\mathbf{A}\|_F$ as defined previously. In fact, every vector norm induces a matrix norm as described below:

A vector norm that is defined on $\mathbb{C}^{mn}$ and $\mathbb{R}^{mn}$ respectively **induces** *a matrix norm on $M_{m,n}(\mathbb{C})$ and $M_{m,n}(\mathbb{R})$ respectively by*

$$\|\mathbf{A}\| = \max_{\|\mathbf{x}\|=1} \|\mathbf{A}\mathbf{x}\| \text{ for } \mathbf{A} \in M_{m,n}(\mathbb{C}) \text{ or } M_{m,n}(\mathbb{R})$$

$$\text{and } \mathbf{x} \in M_{n,1}(\mathbb{C}) \text{ or } M_{n,1}(\mathbb{R}) .$$

Note that the maximum (minimum) above is guaranteed to exist due to the fact that unit spheres in finite dimensional spaces are closed and bounded, and every norm on a vector space is continuous.

The induced matrix norm is compatible with its underlying vector norm in the sense that

$$\|\mathbf{Ax}\| \leq \|\mathbf{A}\| \, \|\mathbf{x}\| .$$

When $\mathbf{A}$ *is nonsingular,*

$$\min_{\|\mathbf{x}\|=1} \|\mathbf{Ax}\| = \frac{1}{\|\mathbf{A}^{-1}\|}.$$

That is, an induced norm $\|\mathbf{A}\|$ represents the maximum extent to which a vector on the unit sphere can be stretched by $\mathbf{A}$, and $\frac{1}{\|\mathbf{A}^{-1}\|}$ measures the extent to which a nonsingular matrix $\mathbf{A}$ can shrink vectors on the unit sphere.

Applications of Induced matrix norms

(a) The induced matrix norm by the Euclidean vector norm is

$$\|\mathbf{A}\|_{2i} = \max_{\|\mathbf{x}\|_2=1} \|\mathbf{Ax}\|_2 = (\lambda_{\max})^{1/2},$$

where $\lambda_{\max}$ *is the largest eigenvalue of* $\mathbf{A}^*\mathbf{A}$.

When $\mathbf{A}$ *is nonsingular,*

$$\|\mathbf{A}^{-1}\|_{2i} = \frac{1}{\max_{\|\mathbf{x}\|_2=1} \|\mathbf{Ax}\|_2} = (\lambda_{\min})^{-1/2},$$

where $\lambda_{\min}$ *is the smallest eigenvalue of* $\mathbf{A}^*\mathbf{A}$.

(b) The induced matrix norm by the vector 1-norm is as follows:

$$\|\mathbf{A}\|_{1i} = \max_{\|\mathbf{x}\|_1=1} \|\mathbf{Ax}\|_1 = \max_j \sum_i |a_{ij}| = \|\mathbf{A}\|_c .$$

That is, the induced matrix norm by the vector 1-norm the largest absolute column sum, that is, the already defined norm $\|\mathbf{A}\|_c$.

(c) The induced matrix norm by the vector ∞*-norm is as follows:*

$$\|\mathbf{A}\|_{\infty i} = \max_{\|\mathbf{x}\|_\infty=1} \|\mathbf{Ax}\|_\infty = \max_i \sum_j |a_{ij}| = \|\mathbf{A}\|_r .$$

That is, the induced matrix norm by the vector ∞*-norm the largest absolute row sum, that is, the already defined norm* $\|\mathbf{A}\|_r$.

Example 112 *Let the stochastic matrix*

$$\mathbf{A} = \begin{pmatrix} 0.7 & 0.2 & 0.1 \\ 0.2 & 0.8 & 0 \\ 0.1 & 0.4 & 0.5 \end{pmatrix},$$

we will calculate all the defined above norms and comment on our findings.

$$\|\mathbf{A}\|_1 = 3, \quad \|\mathbf{A}\|_2 = 1.8, \quad \|\mathbf{A}\|_\infty = 0.8,$$

$$\|\mathbf{A}\|_{m_\infty} = 2.4, \quad \|\mathbf{A}\|_F = 1.28, \quad \|\mathbf{A}\|_r = 1,$$

we have that

$$\mathbf{A}^*\mathbf{A} = \begin{pmatrix} 0.54 & 0.34 & 0.12 \\ 0.34 & 0.84 & 0.22 \\ 0.12 & 0.22 & 0.26 \end{pmatrix},$$

which has eigenvalues: $\{1.1331, 0.32103, 0.18583\}$. *Hence, we have that*

$$\|\mathbf{A}\|_{2i} = 1.06 \quad \text{and} \quad \left\|\mathbf{A}^{-1}\right\|_{2i} = 0.43.$$

We observe that all norms are within the small interval $[0.8, 3]$. *This is not a property of this example but a more general result (see Meyer (2000, p.284)).*

We will now prove that it is possible to transform a norm into another norm with a similarity transformation.

Theorem 113 *Let* $\|.\| : M_n(\mathbb{C}) \to \mathbb{R}$ *be a matrix norm. Then if* $\mathbf{S} \in M_n(\mathbb{C})$ *is non-singular, then for every* $\mathbf{A} \in M_n(\mathbb{C})$ *the function*

$$\|\mathbf{A}\|_{\mathbf{S}} = \left\|\mathbf{S}^{-1}\mathbf{A}\mathbf{S}\right\|,$$

is a matrix norm.

Proof. The only property that presents some difficulty to prove is that

$$\|\mathbf{AB}\|_{\mathbf{S}} \leq \|\mathbf{A}\|_{\mathbf{S}} \|\mathbf{B}\|_{\mathbf{S}}.$$

We have that

$$\begin{aligned} \|\mathbf{AB}\|_{\mathbf{S}} &= \left\|\mathbf{S}^{-1}\mathbf{ABS}\right\| = \left\|\left(\mathbf{S}^{-1}\mathbf{AS}\right)\left(\mathbf{S}^{-1}\mathbf{BS}\right)\right\| \\ &\leq \left\|\mathbf{S}^{-1}\mathbf{AS}\right\| \left\|\mathbf{S}^{-1}\mathbf{BS}\right\| = \|\mathbf{A}\|_{\mathbf{S}} \|\mathbf{B}\|_{\mathbf{S}}. \end{aligned}$$

■

In an analogous way with the vector norms, we could use the matrix norms for the convergence of sequences of matrices.

Definition 114 *Let* $\{\mathbf{A}_n\}_{n=0}^{\infty}$ *be a sequence of matrices in* $M_n(\mathbb{C})$ *and a norm* $\|.\| : M_n(\mathbb{C}) \to \mathbb{R}$. *We say that the sequence* $\{\mathbf{A}_n\}_{n=0}^{\infty}$ *converges in norm to a matrix* $\mathbf{A} \in M_n(\mathbb{C})$, *if and only if,*

$$\lim_{n\to\infty} \|\mathbf{A}_n - \mathbf{A}\| = 0.$$

The convergence in norm of a sequence $\{\mathbf{A}_n\}_{n=0}^{\infty} \in M_n(\mathbb{C})$ to a matrix $\mathbf{A}$ is equivalent with the elementwise convergence of $\{\mathbf{A}_n\}_{n=0}^{\infty}$ to $\mathbf{A}$ when the matrices are of finite order.

The *spectral radious* of a matrix is one of the most useful tools in matrix analysis, which among other uses it could be used for the convergence of power of matrices, infinite products of matrices and generally of sequence of matrices. In what follows we will relate the norm of a matrix with its spectral radius. We start with the formal definition of the spectral radius of a matrix.

Definition 115 *The spectral radius* $\rho(\mathbf{A})$ *of a matrix* $\mathbf{A} \in M_n(\mathbb{C})$ *is defined by*

$$\rho(\mathbf{A}) = \max\{|\lambda| : \lambda \text{ is an eigenvalue of } \mathbf{A}\}.$$

The next two theorems provide relations between the norm and the spectral radius of a matrix.

Theorem 116 *Let a matrix* $\mathbf{A} \in M_n(\mathbb{C})$ *and a matrix norm* $\|.\| : M_n(\mathbb{C}) \to \mathbb{R}$ *then* $\rho(\mathbf{A}) \leq \|\mathbf{A}\|$.

Proof. From the definition of the spectral radious, we know that there exist at least one eigenvalue λ of the matrix $\mathbf{A}$ such that $|\lambda| = \rho(\mathbf{A})$. Then

$$\mathbf{Ax} = \lambda\mathbf{x} \quad \text{with} \quad \mathbf{x} \neq \mathbf{0}.$$

We construct the matrix $\mathbf{S} \in M_n(\mathbb{C})$ with all its columns equal with $\mathbf{x}$, that is,

$$\mathbf{S} = [\mathbf{x}, \mathbf{x}, ..., \mathbf{x}],$$

and consequently

$$\mathbf{AS} = \lambda\mathbf{S}.$$

Let any matrix norm $\|.\| : M_n(\mathbb{C}) \to \mathbb{R}$. Then,

$$|\lambda| \|\mathbf{S}\| = \|\lambda\mathbf{S}\| = \|\mathbf{AS}\| \leq \|\mathbf{A}\| \|\mathbf{S}\|,$$

hence,

$$|\lambda| = \rho(\mathbf{A}) \leq \|\mathbf{A}\|.$$

∎

Theorem 117 *Let $\mathbf{A} \in M_n(\mathbb{C})$ and $\varepsilon > 0$. Then there exists a matrix norm $\|.\| : M_n(\mathbb{C}) \to \mathbb{R}$ such that*

$$\rho(\mathbf{A}) \leq \|\mathbf{A}\| \leq \rho(\mathbf{A}) + \varepsilon.$$

Proof. According to Schur's theorem, we know that for every matrix $\mathbf{A} \in M_n(\mathbb{C})$ with eigenvalues $\lambda_1, \lambda_2, ..., \lambda_n$ there exist a unitary matrix $\mathbf{U}$ such that

$$\mathbf{A} = \mathbf{U}^*\mathbf{DU},$$

where $\mathbf{D}$ is a matrix of the form

$$\mathbf{D} = \begin{pmatrix} \lambda_1 & d_{12} & d_{13} & ... & d_{1n} \\ 0 & \lambda_2 & d_{23} & ... & d_{2n} \\ ... & ... & ... & ... & ... \\ 0 & 0 & 0 & ... & d_{n-1,n} \\ 0 & 0 & 0 & ... & \lambda_n \end{pmatrix}.$$

Consider the matrix $\mathbf{L} \in M_n(\mathbb{C})$, which is of the form

$$\mathbf{L} = \begin{pmatrix} l & 0 & ... & 0 \\ 0 & l^2 & ... & 0 \\ ... & ... & ... & ... \\ 0 & 0 & ... & l^n \end{pmatrix},$$

then

$$\mathbf{L}^{-1} = \begin{pmatrix} l^{-1} & 0 & ... & 0 \\ 0 & l^{-2} & ... & 0 \\ ... & ... & ... & ... \\ 0 & 0 & ... & l^{-n} \end{pmatrix}.$$

Then the matrix $\mathbf{LDL}^{-1}$ is of the form

$$\mathbf{LDL}^{-1} = \begin{pmatrix} \lambda_1 & l^{-1}d_{12} & l^{-2}d_{13} & ... & l^{-n+1}d_{1n} \\ 0 & \lambda_2 & l^{-1}d_{23} & ... & l^{-n+2}d_{2n} \\ ... & ... & ... & ... & ... \\ 0 & 0 & 0 & ... & l^{-1}d_{n-1,n} \\ 0 & 0 & 0 & ... & \lambda_n \end{pmatrix}.$$

Observe that we may always choose the element l of the matrix $\mathbf{L}$ such that

$$\max_i \sum_j \left| l^{-j+i} d_{ij} \right| < \varepsilon.$$

Then we have that

$$\left\|\mathbf{LDL}^{-1}\right\|_r = \max_i \left\{ |\lambda_i| + \sum_j \left| l^{-j+i} d_{ij} \right| \right\} \leq \rho(\mathbf{A}) + \varepsilon.$$

Combining the above result with the previous theorem, we arrive at what we want to prove. ■

We will now prove two theorems which link the spectral radius of a matrix with the asymptotic behavior of powers of a matrix through any matrix norm.

Theorem 118 *Let* $\mathbf{A} \in M_n(\mathbb{C})$ *then for every norm* $\|.\| : M_n(\mathbb{C}) \to \mathbb{R}$, *we have* $\lim \left\|\mathbf{A}^k\right\| = 0$, *when* $k \to \infty$, *if and only if,* $\rho(\mathbf{A}) < 1$.

Proof. Since in a vector space with finite dimension all norms are equivalent in terms of convergence, it is sufficient to show that

$$\lim_{k\to\infty} \left\|\mathbf{A}^k\right\|_r = 0 \text{ when } k \to \infty \text{ if and only if } \rho(\mathbf{A}) < 1.$$

Let any eigenvalue λ of $\mathbf{A}$ and let $\mathbf{x}$ its eigenvector, that is,

$$\mathbf{Ax} = \lambda\mathbf{x}, \text{ with } \mathbf{x} \neq \mathbf{0}.$$

Then

$$\mathbf{A}^k\mathbf{x} = \lambda^k\mathbf{x}, \text{ with } \mathbf{x} \neq \mathbf{0},$$

and

$$\lim_{k\to\infty} \left\|\mathbf{A}^k\mathbf{x}\right\|_r = \lim_{k\to\infty} \left\|\lambda^k\mathbf{x}\right\|_r = \lim_{k\to\infty} \left|\lambda^k\right| \|\mathbf{x}\|_r.$$

From the previous relation we get that, if $\lim_{k\to\infty} \left\|\mathbf{A}^k\mathbf{x}\right\|_r = 0$ then

$$\lim_{k\to\infty} \left|\lambda^k\right| \|\mathbf{x}\|_r = 0, \text{ with } \mathbf{x} \neq \mathbf{0},$$

hence, we have that $|\lambda| < 1$ and since we choose λ randomly to be an eigenvalue of $\mathbf{A}$, we get that $\rho(\mathbf{A}) < 1$.

Inversely, if $\rho(\mathbf{A}) < 1$, then according to the previous theorem for any $\varepsilon > 0$, and consequently for an $\varepsilon > 0$ such that $\rho(\mathbf{A}) + \varepsilon < 1$, there exist a matrix norm $\|.\|$, such that

$$\|\mathbf{A}\| < \rho(\mathbf{A}) + \varepsilon < 1.$$

Hence, we have that $\|\mathbf{A}\| < 1$ and since $\left\|\mathbf{A}^k\right\| \leq \|\mathbf{A}\|^k$ we get that

$$\lim_{k\to\infty} \left\|\mathbf{A}^k\right\| \leq \lim_{k\to\infty} \|\mathbf{A}\|^k = 0.$$

■

Theorem 119 *Let* $\mathbf{A} \in M_n(\mathbb{C})$ *and* $\|.\| : M_n(\mathbb{C}) \to \mathbb{R}$ *a matrix norm. Then*

$$\rho(\mathbf{A}) = \lim_{k\to\infty} \left\|\mathbf{A}^k\right\|^{1/k}.$$

Proof. It is known that if $\lambda_1, \lambda_2, ..., \lambda_n$ are the eigenvalues of the matrix $\mathbf{A}$, then eigenvalues of $\mathbf{A}^k$ are $\lambda_1^k, \lambda_2^k, ..., \lambda_n^k$. Hence,

$$\rho(\mathbf{A})^k = \rho\left(\mathbf{A}^k\right) \text{ for } k = 1, 2, ...$$

From the result that for every $\mathbf{A} \in M_n(\mathbb{C})$, we have $\rho(\mathbf{A}) \leq \|\mathbf{A}\|$, we have that

$$\rho(\mathbf{A})^k \leq \left\|\mathbf{A}^k\right\| \quad \text{for } k = 1, 2, ...,$$

from where we get

$$\rho(\mathbf{A}) \leq \left\|\mathbf{A}^k\right\|^{1/k} \quad \text{for } k = 1, 2, ... \tag{2.14}$$

For any $\varepsilon > 0$ we define the matrix

$$\hat{\mathbf{A}} = [\rho(\mathbf{A}) + \varepsilon]^{-1} \mathbf{A}.$$

Since $\rho\left(\hat{\mathbf{A}}\right) < 1$ we have that

$$\lim_{k \to \infty} \left\|\hat{\mathbf{A}}^k\right\| = 0. \tag{2.15}$$

From (2.15) we get that there exist $\nu = \nu(\varepsilon, \mathbf{A})$ such that

$$\left\|\hat{\mathbf{A}}^k\right\| < 1, \quad \text{for all } k \geq \nu.$$

Then,

$$\left\|\hat{\mathbf{A}}^k\right\| = \left\|[\rho(\mathbf{A}) + \varepsilon]^{-k} \mathbf{A}^k\right\| = [\rho(\mathbf{A}) + \varepsilon]^{-k} \left\|\mathbf{A}^k\right\| < 1 \text{ for all } k \geq \nu,$$

from which we get

$$\left\|\mathbf{A}^k\right\| \leq [\rho(\mathbf{A}) + \varepsilon]^k \quad \text{for all } k \geq \nu. \tag{2.16}$$

Since, (2.14) and (2.16) hold for any $\varepsilon > 0$ we get that

$$\rho(\mathbf{A}) = \lim_{k \to \infty} \left\|\mathbf{A}^k\right\|^{1/k}.$$

■

The following two theorems are useful as criterions, to see in an immediate simple way, if a matrix is invertible. Their proof could be done by the reader as exercises.

Theorem 120 *Let* $\mathbf{A} \in M_n(\mathbb{C})$ *and* $\|.\| : M_n(\mathbb{C}) \to \mathbb{R}$ *a matrix norm. If* $\|\mathbf{A}\| < 1$, *then the matrix* $\mathbf{I} - \mathbf{A}$ *is invertible and*

$$(\mathbf{I} - \mathbf{A})^{-1} = \sum_{k=0}^{\infty} \mathbf{A}^k.$$

Proof. *Hint: Use the identity*

$$(\mathbf{I} - \mathbf{A})\left(\mathbf{I} + \mathbf{A} + \mathbf{A}^2 + ... + \mathbf{A}^{n-1}\right) = \mathbf{I} - \mathbf{A}^n.$$

■

Theorem 121 *Let* $\mathbf{A} \in M_n(\mathbb{C})$. *If*

$$|a_{ii}| > \sum_{j=1, j \neq i}^{n} |a_{ij}|, \quad \textit{for all } \ i = 1, 2, ..., n,$$

then $\mathbf{A}$ *is invertible.*

Matrices with the above property are called diagonally dominant. The proof of the above theorem could be found in Horn and Johnson (1985). Diagonally dominant matrices are often met in applications of Markov chains when the transition probabilities of remaining in all states are larger than the sum of probabilities of moving in another state. Diagonically dominant matrices also are not sensitive when we use subroutines on a computer to find their inverse. In contrast, for matrices that do have an inverse but are not diagonally dominant the stability of the existing subroutines to find their inverse is all too often a problem.

2.6 The Kronecker product of two matrices

Let us start by defining the *Kronecker product* of two matrices:

Definition 122 *Let* $\mathbf{A} \in M_{m,n}(\mathbb{C})$, $\mathbf{B} \in M_{r,k}(\mathbb{C})$ *and* $\mathbf{A} = \{a_{ij}\}, \mathbf{B} = \{b_{ij}\}$. *We define as the Kronecker product of the matrices* $\mathbf{A}$ *and* $\mathbf{B}$, *and we denote it by* $\mathbf{A} \otimes \mathbf{B}$ *the matrix*

$$\mathbf{A} \otimes \mathbf{B} = \begin{pmatrix} a_{11}\mathbf{B} & a_{12}\mathbf{B} & \dots & a_{1n}\mathbf{B} \\ a_{21}\mathbf{B} & a_{22}\mathbf{B} & \dots & a_{2n}\mathbf{B} \\ \dots & \dots & \dots & \dots \\ a_{m1}\mathbf{B} & a_{m2}\mathbf{B} & \dots & a_{mn}\mathbf{B} \end{pmatrix}.$$

Example 123 *Let*

$$\mathbf{A} = \begin{pmatrix} 3 & 2 \\ -2 & 4 \end{pmatrix} \quad \textit{and} \quad \mathbf{B} = \begin{pmatrix} 1 & 3 \\ 2 & 1 \end{pmatrix}$$

According to the definition we have

$$\mathbf{A} \otimes \mathbf{B} = \begin{pmatrix} 3\begin{pmatrix} 1 & 3 \\ 2 & 1 \end{pmatrix} & 2\begin{pmatrix} 1 & 3 \\ 2 & 1 \end{pmatrix} \\ -2\begin{pmatrix} 1 & 3 \\ 2 & 1 \end{pmatrix} & 4\begin{pmatrix} 1 & 3 \\ 2 & 1 \end{pmatrix} \end{pmatrix} = \begin{pmatrix} 3 & 9 & 2 & 6 \\ 6 & 3 & 4 & 2 \\ -2 & -6 & 4 & 12 \\ -4 & -2 & 8 & 4 \end{pmatrix}.$$

We will now find $\mathbf{B} \otimes \mathbf{A}$ *and compare the two Kronecker products*

$$\mathbf{B} \otimes \mathbf{A} = \begin{pmatrix} 1\begin{pmatrix} 3 & 2 \\ -2 & 4 \end{pmatrix} & 3\begin{pmatrix} 3 & 2 \\ -2 & 4 \end{pmatrix} \\ 2\begin{pmatrix} 3 & 2 \\ -2 & 4 \end{pmatrix} & 1\begin{pmatrix} 3 & 2 \\ -2 & 4 \end{pmatrix} \end{pmatrix} = \begin{pmatrix} 3 & 2 & 9 & 6 \\ -2 & 4 & -6 & 12 \\ 6 & 4 & 3 & 2 \\ -4 & 8 & -2 & 4 \end{pmatrix}.$$

We observe that in general $\mathbf{A} \otimes \mathbf{B} \neq \mathbf{B} \otimes \mathbf{A}$.

It is easy to verify that if $\mathbf{A} = \{a_{ij}\} \in M_m$, $\mathbf{B} = \{b_{ij}\} \in M_m$ and we denote as

$$\mathbf{C} = \mathbf{A} \otimes \mathbf{B},$$

with $\mathbf{C} = \{c_{ij}\}$, then

$$c_{ij} = a_{rk} b_{d\nu}, \text{ with } 1 \leq r, k \leq m,\ 1 \leq d, \nu \leq n$$

where

$$i = (r-1)\,n + d \quad \text{and} \quad j = (k-1)\,n + \nu.$$

We will now provide a series of Propositions that in essence are properties of the Kronecker product of two matrices.

It is easy to see that if $c \in \mathbb{C}$ and $\mathbf{A} \in M_{m,n}$, then

$$c \otimes \mathbf{A} = \mathbf{A} \otimes c = c\mathbf{A}.$$

From the definition of the Kronecker product, we have that if

$$\mathbf{a} = \begin{pmatrix} a_1 \\ a_2 \\ \dots \\ a_m \end{pmatrix} \text{ and } \mathbf{b} = \begin{pmatrix} b_1 \\ b_2 \\ \dots \\ b_n \end{pmatrix},$$

then

$$\mathbf{a} \otimes \mathbf{b} = \begin{pmatrix} a_1\mathbf{b} \\ a_2\mathbf{b} \\ \dots \\ a_m\mathbf{b} \end{pmatrix},$$

and it is easy to prove that

$$\mathbf{a}\otimes\mathbf{b}^{\top}=\mathbf{b}^{\top}\otimes\mathbf{a}=\mathbf{a}\mathbf{b}^{\top}.$$

It is easy to prove the following:

Proposition 124 *If* $\mathbf{A}\in M_m$, $\mathbf{B}\in M_n$ *and* $c\in\mathbb{C}$, *then* $c(\mathbf{A}\otimes\mathbf{B})=(c\mathbf{A})\otimes\mathbf{B}=\mathbf{A}\otimes(c\mathbf{B})$.

Proposition 125 *If* $\mathbf{A}\in M_m$, $\mathbf{B}\in M_n$ *and* $\mathbf{C}\in M_n$, *then*

$$(\mathbf{A}+\mathbf{B})\otimes\mathbf{C}=\mathbf{A}\otimes\mathbf{C}+\mathbf{B}\otimes\mathbf{C}.$$

Proof. Let γ_{ij} be the (i,j) element of the matrix $(\mathbf{A}+\mathbf{B})\otimes\mathbf{C}$ and let $\mathbf{A}=\{a_{ij}\}$ and $\mathbf{B}=\{b_{ij}\}$, then

$$\gamma_{ij}=(a_{rk}+b_{rk})\,c_{d\nu}=a_{rk}c_{d\nu}+b_{rk}c_{d\nu},$$

where $i=(r-1)\,n+d$, $j=(k-1)\,n+\nu$ with $1\leq r,k\leq m$ and $1\leq d,\nu\leq n$.

However, $a_{rk}c_{d\nu}$ and $b_{rk}c_{d\nu}$ as defined above are the (i,j) elements of the matrices $\mathbf{A}\otimes\mathbf{C}$ and $\mathbf{B}\otimes\mathbf{C}$ respectively. Hence, the (i,j) element of the matrix $(\mathbf{A}+\mathbf{B})\otimes\mathbf{C}$ is equal with the (i,j) element of the matrix $\mathbf{A}\otimes\mathbf{C}+\mathbf{B}\otimes\mathbf{C}$ for $i,j=1,2,...,nm$. ■

In a similar way we could prove the following proposition

Proposition 126 *If* $\mathbf{A}\in M_m$, $\mathbf{B}\in M_n$ *and* $\mathbf{C}\in M_n$, *then*

$$\mathbf{A}\otimes(\mathbf{B}+\mathbf{C})=\mathbf{A}\otimes\mathbf{B}+\mathbf{A}\otimes\mathbf{C}.$$

It could be proved as an exercise that

Proposition 127 *If* $\mathbf{A}\in M_{m,n}(\mathbb{C})$ *and* $\mathbf{B}\in M_{k,r}(\mathbb{C})$ *then*

$$(\mathbf{A}\otimes\mathbf{B})^{\top}=\mathbf{A}^{\top}\otimes\mathbf{B}^{\top},\qquad(\mathbf{A}\otimes\mathbf{B})^{*}=\mathbf{A}^{*}\otimes\mathbf{B}^{*}$$

Definition 128 *Let* $\mathbf{A}\in M_{m,n}$ *then the n-Kronecker power of the matrix* $\mathbf{A}$, *which is denoted by* $\mathbf{A}^{\otimes n}$ *is defined by the relation*

$$\mathbf{A}^{\otimes n}=\mathbf{A}\otimes\mathbf{A}^{\otimes(n-1)}\ \textit{for}\ n=2,3,...\ \textit{with}\ \mathbf{A}^{\otimes 1}=\mathbf{A}.$$

If $\mathbf{I}_n\in M_n$ and $\mathbf{I}_m\in M_m$ are unit matrices respectively then $\mathbf{I}_r=\mathbf{I}_n\otimes\mathbf{I}_m\in M_{mn}$ is the unit matrix for the common matrix product in M_{mn}.

Example 129 *Let*

$$\mathbf{A}=\begin{pmatrix}1&2\\3&1\end{pmatrix},$$

then

$$\mathbf{A}^{\otimes 2}=\begin{pmatrix}1\begin{pmatrix}1&2\\3&1\end{pmatrix}&2\begin{pmatrix}1&2\\3&1\end{pmatrix}\\3\begin{pmatrix}1&2\\3&1\end{pmatrix}&1\begin{pmatrix}1&2\\3&1\end{pmatrix}\end{pmatrix}=\begin{pmatrix}1&2&2&4\\3&1&6&1\\3&6&1&2\\9&3&3&1\end{pmatrix}.$$

We observe that $\mathbf{A}^{\otimes 2}$ *is of order* $2^2\times 2^2$ *and in general for* $\mathbf{A}\in M_n$, *then* $\mathbf{A}^{\otimes k}$ *is of order* $n^k\times n^k$.

It is easy to prove as an exercise the following proposition:

Proposition 130 *If* $\mathbf{A}\in M_{m,n}$, $\mathbf{B}\in M_{r,k}$ *and* $\mathbf{C}\in M_{st}$, *then*

$$(\mathbf{A}\otimes\mathbf{B})\otimes\mathbf{C}=\mathbf{A}\otimes(\mathbf{B}\otimes\mathbf{C}).$$

We will now prove a result in the form of a theorem that links the Kronecker product with the common product of matrices.

Theorem 131 *Let* $\mathbf{A} \in M_{m,n}$, $\mathbf{B} \in M_{r,s}$, $\mathbf{C} \in M_{nk}$ *and* $\mathbf{D} \in M_{s\nu}$. *Then*

$$(\mathbf{A} \otimes \mathbf{B})(\mathbf{C} \otimes \mathbf{D}) = \mathbf{AC} \otimes \mathbf{BD}.$$

Proof. Let $a_{ij}$the (i, j) element of the matrix $\mathbf{A}$ and c_{jt} the (j, t) element of $\mathbf{C}$. Then we have

$$\mathbf{A} \otimes \mathbf{B} = \begin{pmatrix} a_{11}\mathbf{B} & a_{12}\mathbf{B} & \dots & a_{1n}\mathbf{B} \\ a_{21}\mathbf{B} & a_{22}\mathbf{B} & \dots & a_{2n}\mathbf{B} \\ \dots & \dots & \dots & \dots \\ a_{m1}\mathbf{B} & a_{m2}\mathbf{B} & \dots & a_{mn}\mathbf{B} \end{pmatrix},$$

$$\mathbf{C} \otimes \mathbf{D} = \begin{pmatrix} c_{11}\mathbf{D} & c_{12}\mathbf{D} & \dots & c_{1k}\mathbf{D} \\ c_{21}\mathbf{D} & c_{22}\mathbf{D} & \dots & c_{2k}\mathbf{D} \\ \dots & \dots & \dots & \dots \\ c_{n1}\mathbf{D} & c_{n2}\mathbf{D} & \dots & c_{nk}\mathbf{D} \end{pmatrix},$$

from which we get that

$$(\mathbf{A} \otimes \mathbf{B})(\mathbf{C} \otimes \mathbf{D}) =$$
$$\begin{pmatrix} \sum_{l=1}^{n} a_{1l}c_{l1}\mathbf{BD} & \sum_{l=1}^{n} a_{1l}c_{l2}\mathbf{BD} & \dots & \sum_{l=1}^{n} a_{1l}c_{lk}\mathbf{BD} \\ \sum_{l=1}^{n} a_{2l}c_{l1}\mathbf{BD} & \sum_{l=1}^{n} a_{2l}c_{l2}\mathbf{BD} & \dots & \sum_{l=1}^{n} a_{2l}c_{lk}\mathbf{BD} \\ \dots & \dots & \dots & \dots \\ \sum_{l=1}^{n} a_{ml}c_{l1}\mathbf{BD} & \sum_{l=1}^{n} a_{ml}c_{l2}\mathbf{BD} & \dots & \sum_{l=1}^{n} a_{ml}c_{lk}\mathbf{BD} \end{pmatrix}.$$

We have that the (i, j) element of the product $\mathbf{AC}$, which for the needs of the present proof we denote by $(\mathbf{AC})_{ij}$ is of the form

$$(\mathbf{AC})_{ij} = \sum_{l=1}^{n} a_{il}c_{lj},$$

and hence we could write relation (6.4) as

$$(\mathbf{A} \otimes \mathbf{B})(\mathbf{C} \otimes \mathbf{D}) = \begin{pmatrix} (\mathbf{AC})_{11}\mathbf{BD} & (\mathbf{AC})_{12}\mathbf{BD} & \dots & (\mathbf{AC})_{1k}\mathbf{BD} \\ (\mathbf{AC})_{21}\mathbf{BD} & (\mathbf{AC})_{22}\mathbf{BD} & \dots & (\mathbf{AC})_{2k}\mathbf{BD} \\ \dots & \dots & \dots & \dots \\ (\mathbf{AC})_{m1}\mathbf{BD} & (\mathbf{AC})_{m2}\mathbf{BD} & \dots & (\mathbf{AC})_{mk}\mathbf{BD} \end{pmatrix}, \tag{2.17}$$

According to the definition of the Kronecker product the right hand size of (2.17) is the Kronecker product $\mathbf{AC} \otimes \mathbf{BD}$. Hence, we arrive at the desired conclusion. ■

With the use of the above theorem, we easily get that

$$(\mathbf{A} \otimes \mathbf{B})^{-1} = \mathbf{A}^{-1} \otimes \mathbf{B}^{-1}. \tag{2.18}$$

We will now proceed to found the link between the eigenvalues and eigenvectors of a Kronecker product of two matrices with the eigenvalues and eigenvectors of each one of the matrices. We start with the following Proposition:

Proposition 132 *If* $\mathbf{A} \in M_{m,n}, \mathbf{B} \in M_{r,s}$, $\mathbf{a} \in M_{n,1}$ *and* $\mathbf{b} \in M_{r,1}$, *then*

$$\left(\mathbf{A} \otimes \mathbf{b}^{\top}\right)(\mathbf{a} \otimes \mathbf{B}) = \mathbf{Aab}^{\top}\mathbf{B} = \left(\mathbf{b}^{\top} \otimes \mathbf{A}\right)(\mathbf{B} \otimes \mathbf{a}).$$

Proof. From Theorem 131, we have that

$$\left(\mathbf{A} \otimes \mathbf{b}^{\top}\right)(\mathbf{a} \otimes \mathbf{B}) = \mathbf{Aa} \otimes \mathbf{b}^{\top}\mathbf{B}.$$

As we have shown previously, we have

$$\mathbf{a} \otimes \mathbf{b}^\top = \mathbf{a}\mathbf{b}^\top,$$

hence, we get

$$\left(\mathbf{A} \otimes \mathbf{b}^\top\right)(\mathbf{a} \otimes \mathbf{B}) = \mathbf{A}\mathbf{a}\mathbf{b}^\top\mathbf{B}.$$

Similarly we have

$$\left(\mathbf{b}^\top \otimes \mathbf{A}\right)(\mathbf{B} \otimes \mathbf{a}) = \left(\mathbf{b}^\top\mathbf{B}\right) \otimes (\mathbf{A}\mathbf{a}) = \mathbf{A}\mathbf{a}\mathbf{b}^\top\mathbf{B}.$$

■

In the following two theorems, we will prove the desired link

Theorem 133 *Let $\mathbf{A} \in M_m$ and $\mathbf{B} \in M_n$. If λ is an eigenvalue of $\mathbf{A}$ with corresponding eigenvector $\mathbf{x}$ and μ is an eigenvalue of $\mathbf{B}$ with eigenvector $\mathbf{y}$, then $\lambda\mu$ is an eigenvector of $\mathbf{A} \otimes \mathbf{B}$ with eigenvector $\mathbf{x} \otimes \mathbf{y}$.*

Proof. Since, λ is an eigenvalue of $\mathbf{A}$ with eigenvector $\mathbf{x}$ we have that

$$\mathbf{A}\mathbf{x} = \lambda\mathbf{x}, \ \mathbf{x} \neq \mathbf{0}, \ \mathbf{x} \in \mathbb{C}^n.$$

Since, μ is an eigenvalue of $\mathbf{B}$ with eigenvector $\mathbf{y}$ we have that

$$\mathbf{B}\mathbf{y} = \mu\mathbf{y}, \ y \neq \mathbf{0}, \ \mathbf{y} \in \mathbb{C}^n.$$

Then

$$(\mathbf{A} \otimes \mathbf{B})(\mathbf{x} \otimes \mathbf{y}) = (\mathbf{A}\mathbf{x} \otimes \mathbf{B}\mathbf{y}) = \lambda\mathbf{x} \otimes \mu\mathbf{y} = \lambda\mu(\mathbf{x} \otimes \mathbf{y}),$$

with $(\mathbf{x} \otimes \mathbf{y}) \in \mathbb{C}^{mn}$. ■

We will now prove a theorem which links the $\sigma(\mathbf{A} \otimes \mathbf{B})$ with the $\sigma(\mathbf{A})$ and $\sigma(\mathbf{B})$.

Theorem 134 *Let $\mathbf{A} \in M_n$ and $\mathbf{B} \in M_m$. Let also $\sigma(\mathbf{A}) = \{\lambda_1, \lambda_2, ..., \lambda_n\}$ and $\sigma(\mathbf{B}) = \{\mu_1, \mu_2, ..., \mu_m\}$. Then the matrix $\mathbf{A} \otimes \mathbf{B}$ has nm eigenvalues and*

$$\sigma(\mathbf{A} \otimes \mathbf{B}) = \left\{\lambda_i\mu_j : i = 1, 2, ..., n; j = 1, 2, ..., m\right\}.$$

If the distinct eigenvalues of $\mathbf{A}$ are $\lambda_1, \lambda_2, ..., \lambda_r$ with corresponding algebraic multiplicities $n_1, n_2, ..., n_r$ with $n_1 + n_2 + ... + n_r = n$, and the distinct eigenvalues of $\mathbf{B}$ are $\mu_1, \mu_2, ..., \mu_s$ with respective algebraic multiplicities $m_1, m_2, ..., m_s$ with $m_1 + m_2 + ... + m_s = m$, then

$$\sigma(\mathbf{A} \otimes \mathbf{B}) = \left\{\lambda_i\mu_j : i = 1, 2, ..., r; j = 1, 2, ..., s\right\},$$

with respective algebraic multiplicities for the eigenvalues $\lambda_i\mu_j$

$$n_i m_j, \ i = 1, 2, ..., r, j = 1, 2, ..., s,$$

with

$$\sum n_i m_j = nm.$$

Proof. We know from Schur's Theorem that if $\mathbf{A} \in M_n$ with eigenvalues $\lambda_1, \lambda_2, ..., \lambda_n$ in whatever predetermined order, then there is a unit orthogonal matrix $\mathbf{U} \in M_n$, such that

$$\mathbf{U}^*\mathbf{A}\mathbf{U} = \mathbf{I}_\mathbf{A},$$

where $\mathbf{I}_\mathbf{A}$ is an upper diagonal matrix with diagonal elements $\lambda_1, \lambda_2, ..., \lambda_n$ that is of the form

$$\mathbf{I}_\mathbf{A} = \begin{pmatrix} \lambda_1 & t_{12} & ... & t_{1n} \\ 0 & \lambda_2 & ... & t_{2n} \\ ... & ... & ... & ... \\ 0 & 0 & ... & \lambda_n \end{pmatrix}. \tag{2.19}$$

Similarly, for $\mathbf{B} \in M_m$ with eigenvalues $\mu_1, \mu_2, ..., \mu_m$ in whatever order there is a unit orthogonal matrix $\mathbf{V} \in M_n$, such that

$$\mathbf{V}^*\mathbf{A}\mathbf{V} = \mathbf{I_B},$$

where $\mathbf{I_B}$ is an upper diagonal matrix with diagonal elements $\mu_1, \mu_2, ..., \mu_m$, that is, of the form

$$\mathbf{I_B} = \begin{pmatrix} \mu_1 & d_{12} & ... & d_{1m} \\ 0 & \mu_2 & ... & d_{2m} \\ ... & ... & ... & ... \\ 0 & 0 & ... & \mu_m \end{pmatrix}. \tag{2.20}$$

From (2.19) and (2.20), we have that

$$\begin{aligned} (\mathbf{U} \otimes \mathbf{V})^* (\mathbf{A} \otimes \mathbf{B}) (\mathbf{U} \otimes \mathbf{V}) &= (\mathbf{U}^* \otimes \mathbf{V}^*) (\mathbf{A} \otimes \mathbf{B}) (\mathbf{U} \otimes \mathbf{V}) \\ &= (\mathbf{U}^*\mathbf{A} \otimes \mathbf{V}^*\mathbf{B}) (\mathbf{U} \otimes \mathbf{V}) \\ &= (\mathbf{U}^*\mathbf{A}\mathbf{U}) \otimes (\mathbf{V}^*\mathbf{B}\mathbf{V}) = \mathbf{I_A} \otimes \mathbf{I_B}. \end{aligned} \tag{2.21}$$

From (2.21) we get that the matrix $\mathbf{I_A} \otimes \mathbf{I_B}$ is orthogonally equivalent with the matrix $\mathbf{A} \otimes \mathbf{B}$ and if we evaluate it it could be seen that is upper diagonal. Hence, the diagonal elements of $\mathbf{I_A} \otimes \mathbf{I_B}$ are the eigenvalues of $\mathbf{A} \otimes \mathbf{B}$. In this respect, it easy to see that the diagonal elements of $\mathbf{I_A} \otimes \mathbf{I_B}$ are

$$\lambda_1\mu_1, \lambda_1\mu_2, ..., \lambda_1\mu_m, \lambda_2\mu_1, \lambda_2\mu_2, ..., \lambda_2\mu_m, ..., \lambda_n\mu_1, ..., \lambda_n\mu_m,$$

hence

$$\sigma(\mathbf{A} \otimes \mathbf{B}) = \left\{\lambda_i\mu_j : i = 1, 2, ..., r; j = 1, 2, ..., s\right\}.$$

In order to find their algebraic multiplicities, we observe that for any eigenvalue $\lambda_i\mu_j$ the eigenvalue λ_i will appear in $\mathbf{I_A}$, n_i times and the eigenvalue μ_j in $\mathbf{I_B}$, m_j times. Therefore, the algebraic multiplicity of $\lambda_i\mu_j$ is $n_i m_j$. ■

2.7 The Hadamard product of matrices

In the present section, we will study a third product of matrices the so called *Hadamard product* of matrices which in fact is quite simple as the following definition shows:

Definition 135 *Let* $\mathbf{A} \in M_{m,n}$ *and* $\mathbf{B} \in M_{m,n}$. *We define as the Hadamard product of the two matrices* $\mathbf{A}$ *and* $\mathbf{B}$ *and we denote it by* $\mathbf{A} \triangle \mathbf{B}$ *the matrix, which we get if we multiply the corresponding elements of the two matrices* $\mathbf{A}$ *and* $\mathbf{B}$. *That is,*

$$(\mathbf{A} \triangle \mathbf{B})_{ij} = a_{ij}b_{ij}.$$

Consequently, we immediately get that

$$(\mathbf{A} \triangle \mathbf{B}) = (\mathbf{B} \triangle \mathbf{A}).$$

The unit element of the Hadamard product is apparent that it is not the matrix $\mathbf{I}$, but a matrix which has all its elements equal with one, which we denote it by $\mathbf{E}$. That is, we have

$$\mathbf{A} \triangle \mathbf{E} = \mathbf{E} \triangle \mathbf{A} \quad \text{for every } \mathbf{A}, \mathbf{E} \in M_{m,n}.$$

The inverse of the matrix $\mathbf{A} \in M_{m,n}$ under the Hadamard product is denoted by $\mathbf{A}_H^{-1}$ and is defined by the relation

$$\mathbf{A} \triangle \mathbf{A}_H^{-1} = \mathbf{E}.$$

From this relation, we get that the matrix $\mathbf{A}_H^{-1}$ has as elements in the (i,j) position $1/a_{ij}$. Therefore, a necessary and sufficient condition for a matrix to have an inverse under the Hadamard product is that it will not have zero elements.

It is not difficult to see that in general if $\mathbf{A},\mathbf{B} \in M_{m,n}$ and $\mathbf{C} \in M_{m,r}$ then

$$(\mathbf{A} \triangle \mathbf{B})\,\mathbf{C} \neq \mathbf{AC} \triangle (\mathbf{BC}).$$

However, in this respect, we will prove the following Proposition.

Proposition 136 *If* $\mathbf{A},\mathbf{B} \in M_{m,n}$ *and the matrices* $\mathbf{C} \in M_m$ *and* $\mathbf{D} \in M_n$ *are diagonal, then*

$$\mathbf{C}\,(\mathbf{A} \triangle \mathbf{B})\,\mathbf{D} = (\mathbf{CA}) \triangle (\mathbf{BD}).$$

Proof. Let $\mathbf{A} = \{a_{ij}\}$, $\mathbf{B} = \{b_{ij}\}$, $\mathbf{C} = diag\,\{c_1, c_2, ..., c_m\}$ and $\mathbf{D} = diag\,\{d_1, d_2, ..., d_n\}$. Then

$$\mathbf{A} \triangle \mathbf{B} = \{a_{ij}b_{ij}\} \in M_{m,n}.$$

We also have that

$$\mathbf{C}\,(\mathbf{A} \triangle \mathbf{B}) = \{c_i a_{ij} b_{ij}\} \in M_{m,n}.$$

and

$$\mathbf{C}\,(\mathbf{A} \triangle \mathbf{B})\,\mathbf{D} = \{c_i a_{ij} b_{ij} d_j\} \in M_{m,n}.$$

Now, similarly we have

$$\mathbf{CA} = \{c_i a_{ij}\} \text{ and } \mathbf{BD} = \{b_{ij} d_j\} \in M_{m,n}.$$

Therefore,

$$(\mathbf{CA}) \triangle (\mathbf{BD}) = \{c_i a_{ij} b_{ij} d_j\} \in M_{m,n},$$

hence

$$\mathbf{C}\,(\mathbf{A} \triangle \mathbf{B})\,\mathbf{D} = (\mathbf{CA}) \triangle (\mathbf{BD}).$$

■

Of special interest is the product $\mathbf{H}(\mathbf{A}) = \mathbf{A} \triangle \mathbf{A}^{-1}$. We will now prove the following related Propositions.

Proposition 137 *Let* $\mathbf{A} \in M_n$ *and* $\mathbf{A}$ *non-singular, then*

$$\mathbf{H}(\mathbf{A}) = \mathbf{H}\left(\mathbf{A}^{-1}\right).$$

Proof. We have that

$$\mathbf{H}\left(\mathbf{A}^{-1}\right) = \mathbf{A}^{-1} \triangle \left(\mathbf{A}^{-1}\right)^{-1},$$

therefore,

$$\mathbf{H}\left(\mathbf{A}^{-1}\right) = \mathbf{A}^{-1} \triangle \mathbf{A} = \mathbf{A} \triangle \mathbf{A}^{-1} = \mathbf{H}(\mathbf{A}).$$

■

Proposition 138 *Let* $\mathbf{A} \in M_n$ *and* $\mathbf{A}$ *non-singular, then*

$$\mathbf{H}\left(\mathbf{A}^\top\right) = [\mathbf{H}(\mathbf{A})]^\top.$$

Proof. We have that

$$\mathbf{H}\left(\mathbf{A}^\top\right) = \mathbf{A}^\top \triangle \left(\mathbf{A}^\top\right)^{-1}.$$

Moreover,

$$\mathbf{A}^{-1}\mathbf{A} = \mathbf{I} = \mathbf{I}^\top = \left(\mathbf{A}^{-1}\mathbf{A}\right)^\top = \mathbf{A}^\top \left(\mathbf{A}^{-1}\right)^\top.$$

From the above relation we get if we multiply from the left by $\left(\mathbf{A}^{\top}\right)^{-1}$

$$\left(\mathbf{A}^{\top}\right)^{-1} = \left(\mathbf{A}^{\top}\right)^{-1} \mathbf{A}^{\top} \left(\mathbf{A}^{-1}\right)^{\top} = \left(\mathbf{A}^{-1}\right)^{\top}.$$

Hence,

$$\mathbf{H}\left(\mathbf{A}^{\top}\right) = \mathbf{A}^{\top} \triangle \left(\mathbf{A}^{-1}\right)^{\top} = \left(\mathbf{A} \triangle \mathbf{A}^{-1}\right)^{\top} = \left[\mathbf{H}\left(\mathbf{A}\right)\right]^{\top}.$$

■

If $\mathbf{A} \in M_n$ and $\mathbf{A}$ non-singular, then of special interest is the product $\mathbf{H}_T\left(\mathbf{A}\right) = \mathbf{A} \triangle \left(\mathbf{A}^{-1}\right)^{\top}$ which we study in the next proposition:

Proposition 139 *Let $\mathbf{A} \in M_n$ and $\mathbf{A}$ non-singular, then*

$$\mathbf{H}_T\left(\mathbf{A}^{-1}\right) = \mathbf{H}_T\left(\mathbf{A}^{\top}\right).$$

Proof. We have that

$$\mathbf{H}_T\left(\mathbf{A}^{-1}\right) = \mathbf{A}^{-1} \triangle \left[\left(\mathbf{A}^{-1}\right)^{-1}\right]^{\top} = \mathbf{A}^{-1} \triangle \mathbf{A}^{\top} = \mathbf{A}^{\top} \triangle \mathbf{A}^{-1}.$$

Also, we get that

$$\mathbf{H}_T\left(\mathbf{A}^{\top}\right) = \mathbf{A}^{\top} \triangle \left[\left(\mathbf{A}^{\top}\right)^{-1}\right]^{\top} = \mathbf{A}^{\top} \triangle \mathbf{A}^{-1},$$

from which we get what we want to prove. ■

2.8 Canonical forms of a matrix

There are some operations on a matrix which preserve the rank of a matrix and which are equivalent to premultiplying a matrix with an elementary matrix. We list the following such operations on a matrix:

(*a*) Multiplying a row (or a column) of a matrix $\mathbf{A}$ of order $m \times n$ by a scalar c.

Assume that we want to multiply the kth-row of the matrix $\mathbf{A}$ by c. This is equivalent to multiply the matrix $\mathbf{A}$ from the left (premultiplication) by an $m \times m$ matrix of the form

$$\begin{pmatrix} 1 & 0 & 0 & 0 & 0 & 0 & \dots & 0 \\ 0 & 1 & 0 & 0 & 0 & 0 & \dots & 0 \\ \dots & \dots & \dots & \dots & \dots & \dots & \dots & \dots \\ 0 & 0 & 0 & 1 & 0 & 0 & \dots & 0 \\ 0 & 0 & 0 & 0 & c & 0 & \dots & 0 \\ 0 & 0 & 0 & 0 & 0 & 1 & \dots & 0 \\ 0 & 0 & 0 & 0 & 0 & 0 & \dots & 0 \\ 0 & 0 & 0 & 0 & 0 & 0 & \dots & 1 \end{pmatrix}.$$

That is a matrix which has ones in the diagonal except in the kth-row where it has c and zeros elsewhere. This is known as the *Kronecker matrix* for the scalar c and row k. It is denoted by $\mathbf{\Delta}_k\left(c\right)$. Note that the determinant of any such matrix is equal with c, that is, $\left|\mathbf{\Delta}_k\left(c\right)\right| = c$ for any c, k. The transpose of any Kronecker matrix is apparently the same matrix, that is, $\left(\mathbf{\Delta}_k\left(c\right)\right)^{\top} = \mathbf{\Delta}_k\left(c\right)$. Since the determinant of any Kronecker matrix is different than zero therefore it is not singular and has an inverse. In fact, we have that $\left(\mathbf{\Delta}_k\left(c\right)\right)^{-1} = \mathbf{\Delta}_k\left(1/c\right)$.

(*b*) Replacing the kth-row (or column) by the kth-row (or column) plus ν times the sth-row (or column).

This is equivalent to multiply the matrix $\mathbf{A}$ from the left (premultiplication) by an $m \times m$ matrix which has all the diagonal elements equal with one, the (k, s) element equal with ν and the remaining elements equal to zero. We denote such a matrix by $\mathbf{E}_{ks}(\nu)$. Note that the determinant of any such matrix is equal with 1, that is $|\mathbf{E}_{ks}(\nu)| = 1$ for any ν, k, s. The transpose of any elementary matrix $\mathbf{E}_{ks}(\nu)$ is apparently the same matrix, that is, $(\mathbf{E}_{ks}(\nu))^\top = \mathbf{E}_{ks}(\nu)$. Since the determinant of any elementary matrix $\mathbf{E}_{ks}(\nu)$ is different than zero, therefore it is not singular and has an inverse. In fact, we have that $(\mathbf{E}_{ks}(\nu))^{-1} = \mathbf{E}_{ks}(-\nu)$.

(c) Interchanging two rows.

Let us say that we wish to interchange the rows r and s of the matrix $\mathbf{A}$ which is of order $m \times n$. This is equivalent to multiplying from the left the matrix $\mathbf{A}$ successively by the matrices $\mathbf{E}_{sr}(-1)$, $\mathbf{E}_{rs}(1)$, $\mathbf{E}_{sr}(-1)$ and $\mathbf{\Delta}_s(-1)$. Alternatively, we could interchange the rows r and s of the matrix $\mathbf{A}$ by multiplying from the right by a matrix which has $1's$ in the positions (r, s) and (s, r), zeros in the remaining positions of the respective rows, $1's$ in the remaining positions in the diagonals and zeros everywhere else. We will denote such an elementary matrix by $\mathbf{R}_{rs}$. Note that the determinant of any such matrix is equal with -1, that is $|\mathbf{R}_{rs}| = -1$ for any r, s. For example, if we want to interchange the rows 1 and 2 of an 3×3 matrix $\mathbf{A}$, we multiply from the right by the 3×3 matrix $\mathbf{R}_{12}$, that is,

$$\begin{pmatrix} 0 & 1 & 0 \\ 1 & 0 & 0 \\ 0 & 0 & 1 \end{pmatrix} \begin{pmatrix} a_{11} & a_{12} & a_{13} \\ a_{21} & a_{22} & a_{23} \\ a_{31} & a_{32} & a_{33} \end{pmatrix} = \begin{pmatrix} a_{21} & a_{22} & a_{23} \\ a_{11} & a_{12} & a_{13} \\ a_{31} & a_{32} & a_{33} \end{pmatrix}.$$

The transpose of any elementary matrix $\mathbf{R}_{rs}$ is apparently the same matrix, that is, $(\mathbf{R}_{rs})^\top = \mathbf{R}_{rs}$. Since the determinant of any elementary matrix $\mathbf{R}_{rs}$ is different than zero therefore it is not singular and has an inverse. In fact, we have that $(\mathbf{R}_{rs})^{-1} = \mathbf{R}_{rs}$.

(d) Elementary operators on columns of a Matrix.

Let $\mathbf{O}$ be an elementary operator of any type of the above, then we have

$$\mathbf{AO} = \left(\mathbf{O}^\top \mathbf{A}^\top\right)^\top,$$

where apparently $\mathbf{O}^\top$ is still an elementary matrix. Hence, $\mathbf{O}^\top \mathbf{A}^\top$ represents some elementary operations on the rows of $\mathbf{A}^\top$ which in fact are the columns of $\mathbf{A}$. Therefore, multiplying from the right (postmultiplication) the matrix $\mathbf{A}$ by an elementary matrix performs similar manipulations on the columns of $\mathbf{A}$ as are performed on rows of $\mathbf{A}$ by multiplying from the lefts (premultiplication).

The resulting matrix after any number of premultiplications or postmultiplications by elementary matrices of a matrix $\mathbf{A}$ is called an *equivalent form of the matrix* $\mathbf{A}$. All equivalent forms of a matrix as we have seen preserve the rank of a matrix. Some equivalent forms are called *equivalent canonical forms,* and we will introduce them by example before defining them rigorously.

Example 140 *Finding the rank of a matrix by premultiplications with elementary matrices. Let the* 3×4 *matrix*

$$\mathbf{A} = \begin{pmatrix} 1 & 3 & 4 & 2 \\ -2 & 3 & 4 & 1 \\ 4 & -6 & -8 & -2 \end{pmatrix}.$$

We will describe the steps to find the rank of the matrix. First step is to multiply row 1 *by* 2 *and add it to row* 2 *in order to make the position* $(2, 1)$ *equal with zero. This is achieved by premultiplying matrix* $\mathbf{A}$ *with the following elementary matrix:*

$$\mathbf{E}_{21}(2)\ \mathbf{A} = \begin{pmatrix} 1 & 0 & 0 \\ 2 & 1 & 0 \\ 0 & 0 & 1 \end{pmatrix} \begin{pmatrix} 1 & 3 & 4 & 2 \\ -2 & 3 & 4 & 1 \\ 4 & -6 & -8 & -2 \end{pmatrix} = \begin{pmatrix} 1 & 3 & 4 & 2 \\ 0 & 9 & 12 & 5 \\ 4 & -6 & -8 & -2 \end{pmatrix}.$$

The second step is to multiply row 1 *by* -4 *and add it to row* 3 *in order to make the position* $(3,1)$ *equal with zero in the above matrix. This is achieved by premultiplying matrix* $\mathbf{E}_{21}(2)\ \mathbf{A}$ *with the following elementary matrix:*

$$\begin{aligned}\mathbf{E}_{31}(-4)(\mathbf{E}_{21}(2)\ \mathbf{A}) &= \begin{pmatrix}1&0&0\\0&1&0\\-4&0&1\end{pmatrix}\begin{pmatrix}1&3&4&2\\0&9&12&5\\4&-6&-8&-2\end{pmatrix}\\ &= \begin{pmatrix}1&3&4&2\\0&9&12&5\\0&-18&-24&-10\end{pmatrix} = (\mathbf{E}_{31}(-4)\,\mathbf{E}_{21}(2))\,\mathbf{A}.\end{aligned}$$

In a rectangular matrix $\mathbf{A}$, *we define as diagonal elements the elements* a_{ii} *for* $i=1,2,\ldots$. *In order to find the rank of matrix* $\mathbf{A}$, *it is sufficient to find an equivalent form of* $\mathbf{A}$ *where all the elements below the diagonal are zero. Hence, we need to premultiply matrix* $(\mathbf{E}_{31}(-4)\,\mathbf{E}_{21}(2))\,\mathbf{A}$ *with an elementary matrix which will produce an equivalent matrix form of* $\mathbf{A}$ *where there will be a zero in the position* $(3,2)$. *This is achieved by premultiplying matrix* $(\mathbf{E}_{31}(-4)\,\mathbf{E}_{21}(2))\,\mathbf{A}$ *with the following elementary matrix:*

$$\begin{aligned}\mathbf{E}_{32}(2)\,(\mathbf{E}_{31}(-4)\,\mathbf{E}_{21}(2))\,\mathbf{A} &= \begin{pmatrix}1&0&0\\0&1&0\\0&2&1\end{pmatrix}\begin{pmatrix}1&3&4&2\\0&9&12&5\\0&-18&-24&-10\end{pmatrix}\\ &= \begin{pmatrix}1&3&4&2\\0&9&12&5\\0&0&0&0\end{pmatrix}\\ &= (\mathbf{E}_{32}(2)\,\mathbf{E}_{31}(-4)\,\mathbf{E}_{21}(2))\,\mathbf{A}.\end{aligned}$$

Now we have that rank$(\mathbf{A})$ $=$*rank*$[(\mathbf{E}_{32}(2)\,\mathbf{E}_{31}(-4)\,\mathbf{E}_{21}(2))\,\mathbf{A}]=2$ *the number of non-zero rows. We define by*

$$\begin{aligned}\mathbf{P}_r &= \mathbf{E}_{32}(2)\,\mathbf{E}_{31}(-4)\,\mathbf{E}_{21}(2)\\ &= \begin{pmatrix}1&0&0\\2&1&0\\0&2&1\end{pmatrix}.\end{aligned}$$

From the above example, we may easily deduce the general method for finding the rank of a matrix $\mathbf{A}$. We premultiply matrix $\mathbf{A}$ by the appropriate elementary matrices in order to find an equivalent form of matrix $\mathbf{A}$, let say $\mathbf{B}_r$, where all the subdiagonal elements are zero. That is

$$\mathbf{P}_r\mathbf{A} = \mathbf{B}_r.$$

The number of the non-zero rows in $\mathbf{B}$ is the rank of the matrix $\mathbf{A}$. Note that apparently the matrix $\mathbf{P}_r$ always exist.

Example 141 *The Echelon equivalent form of a matrix. Let us continue the previous example. The matrix* $\mathbf{B}_r$ *is*

$$\mathbf{B}_r = \begin{pmatrix}1&3&4&2\\0&9&12&5\\0&0&0&0\end{pmatrix}.$$

We premultiply the matrix $\mathbf{B}_r$ *by the elementary matrix* $\mathbf{\Delta}_2(1/9)$ *and we get*

$$\begin{aligned}\mathbf{\Delta}_2(1/9)\,\mathbf{B}_r &= \begin{pmatrix}1&0&0\\0&1/9&0\\0&0&1\end{pmatrix}\begin{pmatrix}1&3&4&2\\0&9&12&5\\0&0&0&0\end{pmatrix} = \begin{pmatrix}1&3&4&2\\0&1&\frac{4}{3}&\frac{5}{9}\\0&0&0&0\end{pmatrix}\\ &= (\mathbf{\Delta}_2(1/9)\,\mathbf{E}_{32}(2)\,\mathbf{E}_{31}(-4)\,\mathbf{E}_{21}(2))\,\mathbf{A}.\end{aligned}$$

We premultiply the matrix $\mathbf{\Delta}_2(1/9)\,\mathbf{B}_r$ *by the elementary matrix* $\mathbf{E}_{12}(-3)$ *and we get*

$$\begin{aligned}\mathbf{E}_{12}(-3)\left(\mathbf{\Delta}_2(1/9)\,\mathbf{B}_r\right) &= \begin{pmatrix}1 & -3 & 0\\ 0 & 1 & 0\\ 0 & 0 & 1\end{pmatrix}\begin{pmatrix}1 & 3\cdot & 4 & 2\\ 0 & 1 & \frac{4}{3} & \frac{5}{9}\\ 0 & 0 & 0 & 0\end{pmatrix}\\ &= \begin{pmatrix}1 & 0 & 0 & \frac{1}{3}\\ 0 & 1 & \frac{4}{3} & \frac{5}{9}\\ 0 & 0 & 0 & 0\end{pmatrix}.\end{aligned}$$

The above equivalent form of matrix $\mathbf{A}$ *is called the Echelon equivalent canonical form of* $\mathbf{A}$*. We denote it by* $\mathbf{B}_e$*. Apparently we have*

$$\mathbf{E}_{12}(-3)\,\mathbf{\Delta}_2(1/9)\,\mathbf{B}_r = \mathbf{E}_{12}(-3)\,\mathbf{\Delta}_2(1/9)\,\mathbf{P}_r\mathbf{A} = \mathbf{P}_e\mathbf{A},$$

where in our example

$$\mathbf{P}_e = \mathbf{E}_{12}(-3)\,\mathbf{\Delta}_2(1/9)\,\mathbf{P}_r = \begin{pmatrix}\frac{1}{3} & -\frac{1}{3} & 0\\ \frac{2}{9} & \frac{1}{9} & 0\\ 0 & 2 & 1\end{pmatrix} \quad \textit{and} \quad \mathbf{P}_e\mathbf{A} = \mathbf{B}_e.$$

We generalize now what we have done in the above example. We say that a matrix $\mathbf{A}$ of order $m \times n$ is in the *Echelon equivalent canonical form* $\mathbf{B}_e$ iff:

(a) Each row of $\mathbf{B}_e$ consists of all zeros or a unity preceded by zeros and not necessarily succeded by zeros.

(b) Any column containing the first unity of a row has the rest of the elements as zeros.

From the above example could be intuitively deduced and in fact could be easily proved that:

Let a matrix $\mathbf{A}$ *of order* $m \times n$*. There exists an* $m \times n$ *nonsingular matrix* $\mathbf{P}_e$ *which is the product of elementary matrices such that* $\mathbf{P}_e\mathbf{A} = \mathbf{B}_e$ *the Echelon equivalent form of* $\mathbf{A}$*.*

The general method for finding the Echelon equivalent form of a matrix is apparent from the above two last examples.

Example 142 *Continuing the above example. Let the Echelon equivalent form of matrix* $\mathbf{A}$*:*

$$\mathbf{B}_e = \begin{pmatrix}1 & 0 & 0 & \frac{1}{3}\\ 0 & 1 & \frac{4}{3} & \frac{5}{9}\\ 0 & 0 & 0 & 0\end{pmatrix}.$$

Having obtained $\mathbf{B}_e$ *by the row elementary operators* $\mathbf{P}_e\mathbf{A}$ *we can continue to increase the zero elements above the diagonal using column elementary operators. We postmultiply the matrix* $\mathbf{B}_e$ *by the* 4×4 *elementary matrix* $\mathbf{E}_{14}(-1/3)$ *and we get*

$$\mathbf{B}_e\mathbf{E}_{14}(-1/3) = \begin{pmatrix}1 & 0 & 0 & \frac{1}{3}\\ 0 & 1 & \frac{4}{3} & \frac{5}{9}\\ 0 & 0 & 0 & 0\end{pmatrix}\begin{pmatrix}1 & 0 & 0 & -1/3\\ 0 & 1 & 0 & 0\\ 0 & 0 & 1 & 0\\ 0 & 0 & 0 & 1\end{pmatrix} = \begin{pmatrix}1 & 0 & 0 & 0\\ 0 & 1 & \frac{4}{3} & \frac{5}{9}\\ 0 & 0 & 0 & 0\end{pmatrix}.$$

We now postmultiply the matrix $\mathbf{B}_e\mathbf{E}_{14}(-1/3)$ *by the* 4×4 *elementary matrix* $\mathbf{E}_{23}(-4/3)$ *and we get*

$$\begin{aligned}\mathbf{B}_e\mathbf{E}_{14}(-1/3)\,\mathbf{E}_{23}(-4/3) &= \begin{pmatrix}1 & 0 & 0 & 0\\ 0 & 1 & \frac{4}{3} & \frac{5}{9}\\ 0 & 0 & 0 & 0\end{pmatrix}\begin{pmatrix}1 & 0 & 0 & 0\\ 0 & 1 & -4/3 & 0\\ 0 & 0 & 1 & 0\\ 0 & 0 & 0 & 1\end{pmatrix}\\ &= \begin{pmatrix}1 & 0 & 0 & 0\\ 0 & 1 & 0 & \frac{5}{9}\\ 0 & 0 & 0 & 0\end{pmatrix}.\end{aligned}$$

We now postmultiply the matrix $\mathbf{B}_e\mathbf{E}_{14}(-1/3)\,\mathbf{E}_{23}(-4/3)$ *by the* 4×4 *elementary matrix* $\mathbf{E}_{24}(-5/9)$ *and we get*

$$\begin{aligned}\mathbf{B}_e\mathbf{E}_{14}(-1/3)\,\mathbf{E}_{23}(-4/3)\,\mathbf{E}_{24}(-5/9) &= \begin{pmatrix}1&0&0&0\\0&1&0&\frac{5}{9}\\0&0&0&0\end{pmatrix}\begin{pmatrix}1&0&0&0\\0&1&0&-5/9\\0&0&1&0\\0&0&0&1\end{pmatrix}\\ &= \begin{pmatrix}1&0&0&0\\0&1&0&0\\0&0&0&0\end{pmatrix} = \mathbf{P}_e\mathbf{A}\mathbf{Q}_e,\end{aligned}$$

where

$$\mathbf{Q}_e = \begin{pmatrix}1&0&0&-\frac{1}{3}\\0&1&-\frac{4}{3}&-\frac{5}{9}\\0&0&1&0\\0&0&0&1\end{pmatrix}.$$

Hence, we conclude that

$$\mathbf{P}_e\mathbf{A}\mathbf{Q}_e = \begin{pmatrix}\mathbf{I}_2&\mathbf{0}\\\mathbf{0}&\mathbf{0}\end{pmatrix} = \mathbf{C}_2,$$

where $\mathbf{I}_2$ *is a* 2×2 *identity matrix where* 2 *is the rank of matrix* $\mathbf{A}$. *Note that* $\mathbf{P}_e$ *and* $\mathbf{Q}_e$ *are products of elementary matrices and therefore they are non-singular and their inverses exist. Hence,*

$$\mathbf{A} = \mathbf{P}_\mathbf{e}^{-1}\mathbf{C}_2\mathbf{Q}_e^{-1}.$$

From the above example could be intuitively deduced and in fact could be easily proved that:

Theorem 143 *Any non-null matrix* $\mathbf{A} \in M_{mn}(\mathbb{R})$ *of rank* r *is equivalent by elementary operators*

$$\mathbf{P}_e\mathbf{A}\mathbf{Q}_e = \begin{pmatrix}\mathbf{I}_r&\mathbf{0}\\\mathbf{0}&\mathbf{0}\end{pmatrix} = \mathbf{C}_r,$$

where $\mathbf{I}_r$ *is the identity matrix of order* r *and the rest of the elements are zero. The matrices* $\mathbf{P}_e$ *and* $\mathbf{Q}_e$ *are nonsigular matrices with* $\mathbf{P}_e \in M_n$ *and* $\mathbf{Q}_e \in M_m$ *and they are products of elementary matrices. The matrices* $\mathbf{P}_e$ *and* $\mathbf{Q}_e$ *are not unique. Therefore,*

$$\mathbf{A} = \mathbf{P}_e^{-1}\begin{pmatrix}\mathbf{I}_r&\mathbf{0}\\\mathbf{0}&\mathbf{0}\end{pmatrix}\mathbf{Q}_e^{-1},$$

or

$$\mathbf{A} = \mathbf{A}_{\mathbf{P}_e}\mathbf{A}_{\mathbf{Q}_e},$$

where $\mathbf{A}_{\mathbf{P}_e}$ *is an* $m\times r$ *matrix of rank* r *and* $\mathbf{A}_{\mathbf{Q}_e}$ *is an* $r \times n$ *matrix of rank* r.

Hermite Canonical Form

We provide the definition of the equivalent Hermite canonical form of a matrix.

Definition 144 *Let* $\mathbf{A} \in M_m$ *then if* $\mathbf{H_A}$ *is an equivalent form, then we say that it is a Hermite Canonical Form of* $\mathbf{A}$ *if a). In the main diagonal there are only* $0's$ *and* $1's$. *b) If a row has a zero in the main diagonal then every element in this row is zero. c) If a row has a* 1 *in the main diagonal then every other element of this column is zero.*

Example 145 *Hermite Canonical Form. Let the matrix*

$$\mathbf{A} = \begin{pmatrix}1&4&3\\2&1&6\\3&2&9\end{pmatrix}.$$

The first step is to premultiply matrix $\mathbf{A}$ *with the appropriate elementary matrices in order to find an equivalent form of the matrix where the first column below the element* 1*will have zeros. We start by premultiply with the elementary matrix* $\mathbf{E}_{21}(-2)$*:*

$$\mathbf{E}_{21}(-2)\,\mathbf{A} = \begin{pmatrix} 1 & 0 & 0 \\ -2 & 1 & 0 \\ 0 & 0 & 1 \end{pmatrix} \begin{pmatrix} 1 & 4 & 3 \\ 2 & 1 & 6 \\ 3 & 2 & 9 \end{pmatrix} = \begin{pmatrix} 1 & 4 & 3 \\ 0 & -7 & 0 \\ 3 & 2 & 9 \end{pmatrix}.$$

We continue by premultiplying matrix $\mathbf{E}_{21}(-2)\,\mathbf{A}$ *by the elementary matrix* $\mathbf{E}_{31}(-3)$*:*

$$\mathbf{E}_{31}(-3)\,\mathbf{E}_{21}(-2)\,\mathbf{A} = \begin{pmatrix} 1 & 0 & 0 \\ 0 & 1 & 0 \\ -3 & 0 & 1 \end{pmatrix} \begin{pmatrix} 1 & 4 & 3 \\ 0 & -7 & 0 \\ 3 & 2 & 9 \end{pmatrix} = \begin{pmatrix} 1 & 4 & 3 \\ 0 & -7 & 0 \\ 0 & -10 & 0 \end{pmatrix}$$

Our next step is to premultiply the matrix $\mathbf{E}_{31}(-3)\,\mathbf{E}_{21}(-2)\,\mathbf{A}$ *by* $\boldsymbol{\Delta}_2(-1/7)$ *:*

$$\begin{aligned} \boldsymbol{\Delta}_2(-1/7)\,\mathbf{E}_{31}(-3)\,\mathbf{E}_{21}(-2)\,\mathbf{A} &= \begin{pmatrix} 1 & 0 & 0 \\ 0 & -1/7 & 0 \\ 0 & 0 & 1 \end{pmatrix} \begin{pmatrix} 1 & 4 & 3 \\ 0 & -7 & 0 \\ 0 & -10 & 0 \end{pmatrix} \\ &= \begin{pmatrix} 1 & 4 & 3 \\ 0 & 1 & 0 \\ 0 & -10 & 0 \end{pmatrix}. \end{aligned}$$

We now should arrive in an equivalent form where the second column will have a 1 *in the main diagonal and the remaining elements of the column will be zeros. We premultiply the matrix* $\boldsymbol{\Delta}_2(-1/7)\,\mathbf{E}_{31}(-3)\,\mathbf{E}_{21}(-2)\,\mathbf{A}$ *by the matrix* $\mathbf{E}_{32}(10)$ *:*

$$\begin{aligned} \mathbf{E}_{32}(10)\,\boldsymbol{\Delta}_2(-1/7)\,\mathbf{E}_{31}(-3)\,\mathbf{E}_{21}(-2)\,\mathbf{A} &= \begin{pmatrix} 1 & 0 & 0 \\ 0 & 1 & 0 \\ 0 & 10 & 1 \end{pmatrix} \begin{pmatrix} 1 & 4 & 3 \\ 0 & 1 & 0 \\ 0 & -10 & 0 \end{pmatrix} \\ &= \begin{pmatrix} 1 & 4 & 3 \\ 0 & 1 & 0 \\ 0 & 0 & 0 \end{pmatrix}. \end{aligned}$$

Finally, we premultiply $\mathbf{E}_{32}(10)\,\boldsymbol{\Delta}_2(-1/7)\,\mathbf{E}_{31}(-3)\,\mathbf{E}_{21}(-2)\,\mathbf{A}$ *by the matrix* $\mathbf{E}_{12}(-4)$ *:*

$$\mathbf{E}_{12}(-4)\,\mathbf{E}_{32}(10)\,\boldsymbol{\Delta}_2(-1/7)\,\mathbf{E}_{31}(-3)\,\mathbf{E}_{21}(-2)\,\mathbf{A} =$$

$$\begin{pmatrix} 1 & -4 & 0 \\ 0 & 1 & 0 \\ 0 & 0 & 1 \end{pmatrix} \begin{pmatrix} 1 & 4 & 3 \\ 0 & 1 & 0 \\ 0 & 0 & 0 \end{pmatrix} = \begin{pmatrix} 1 & 0 & 3 \\ 0 & 1 & 0 \\ 0 & 0 & 0 \end{pmatrix} = \mathbf{H_A}.$$

We define by

$$\mathbf{P_H} = \mathbf{E}_{12}(-4)\,\mathbf{E}_{32}(10)\,\boldsymbol{\Delta}_2(-1/7)\,\mathbf{E}_{31}(-3)\,\mathbf{E}_{21}(-2) = \begin{pmatrix} -\frac{1}{7} & \frac{4}{7} & 0 \\ \frac{2}{7} & -\frac{1}{7} & 0 \\ -\frac{1}{7} & -\frac{10}{7} & 1 \end{pmatrix}.$$

Then we have

$$\mathbf{P_H A} = \mathbf{H_A}.$$

Note that $\mathbf{H}_\mathbf{A}^2 = \mathbf{H_A}$.

From the above example intuitively we except the truth of the following statement which could be proved easily:

Hermite Canonical Form: Let $\mathbf{A} \in M_m$. Then there exist a non-singular matrix $\mathbf{P_H} \in M_m$ such that $\mathbf{P_H A} = \mathbf{H_A}$ and in addition $\mathbf{H}_\mathbf{A}^2 = \mathbf{H_A}$. The matrix $\mathbf{P_H}$ is not unique.

2.9 Generalized inverses

The inverse of a matrix $\mathbf{A} \in M_n$ is well know that exists, when the matrix is nonsingular, and then we have $\mathbf{AA}^{-1} = \mathbf{I}_n$. A logical question is what happens when $\mathbf{A} \in M_{mn}$ with $m \neq n$ or if $\mathbf{A} \in M_n$ but singular. In such cases we still need a matrix which somehow behaves like an inverse. There are many reasons for defining such a matrix the most common probably being when we need to find a class of solutions for a system of linear equations which in matrix form could be written as $\mathbf{Ax} = \mathbf{b}$. Where $\mathbf{A} \in M_{mn}$ with $m \neq n$ or if $\mathbf{A} \in M_n$ but singular. There are in Statistics many other cases that we need such a matrix which we will call for rather obvious reasons *generalize inverse.* We now provide a definition of the class of matrices which we call generalized inverses.

Definition 146 *Given any matrix $\mathbf{A} \in M_{mn}$ there are many matrices $\mathbf{A}_g \in M_{nm}$ which satisfy the equation*

$$\mathbf{AA_gA} = \mathbf{A}.$$

We call such matrices $\mathbf{A}_g$ generalized inverses of $\mathbf{A}$.

We will refer now to some cases where we may easily derive a generalized inverse of a matrix and also we will provide a general algorithm.

(a) Derivation from elementary row and column operations

As we have seen in Example 140 by elementary row operations, it might be possible to find an equivalent matrix form of a matrix $\mathbf{A}$ with $\text{rank}(\mathbf{A}) = r_{\mathbf{A}}$ as follows

$$\mathbf{A} \approx \mathbf{P_rA} = \begin{pmatrix} \mathbf{A}_{11} & \mathbf{A}_{12} \\ \mathbf{0} & \mathbf{0} \end{pmatrix},$$

where $\mathbf{A}_{11}$ is a square nonsingular matrix of order $r_{\mathbf{A}} \times r_{\mathbf{A}}$ which is upper triangular, and $\mathbf{P}_r$ represents the elementary row operations needed to arrive in the required equivalent form. Then

$$\mathbf{A_g} = \begin{pmatrix} \mathbf{A}_{11}^{-1} & \mathbf{0} \\ \mathbf{0} & \mathbf{0} \end{pmatrix} \mathbf{P}_r.$$

Example 147 *Continuing Example 140, we have that*

$$\begin{aligned} \mathbf{A} \approx \mathbf{P_rA} = \quad & \begin{pmatrix} \mathbf{A}_{11} & \mathbf{A}_{12} \\ \mathbf{0} & \mathbf{0} \end{pmatrix} = \begin{pmatrix} 1 & 0 & 0 \\ 2 & 1 & 0 \\ 0 & 2 & 1 \end{pmatrix} \begin{pmatrix} 1 & 3 & 4 & 2 \\ -2 & 3 & 4 & 1 \\ 4 & -6 & -8 & -2 \end{pmatrix} \\ = \quad & \begin{pmatrix} 1 & 3 & 4 & 2 \\ 0 & 9 & 12 & 5 \\ 0 & 0 & 0 & 0 \end{pmatrix}, \end{aligned}$$

where

$$\mathbf{A}_{11} = \begin{pmatrix} 1 & 3 \\ 0 & 9 \end{pmatrix} \text{ with inverse } \mathbf{A}_{11}^{-1} = \begin{pmatrix} 1 & -\frac{1}{3} \\ 0 & \frac{1}{9} \end{pmatrix},$$

hence,

$$\mathbf{A_g} = \begin{pmatrix} 1 & -\frac{1}{3} & 0 \\ 0 & \frac{1}{9} & 0 \\ 0 & 0 & 0 \\ 0 & 0 & 0 \end{pmatrix} \begin{pmatrix} 1 & 0 & 0 \\ 2 & 1 & 0 \\ 0 & 2 & 1 \end{pmatrix} = \begin{pmatrix} \frac{1}{3} & -\frac{1}{3} & 0 \\ \frac{2}{9} & \frac{1}{9} & 0 \\ 0 & 0 & 0 \\ 0 & 0 & 0 \end{pmatrix}.$$

It is easy for the reader to verify that $\mathbf{A_g}$ is a generalized inverse for $\mathbf{A}$.

Note that when $\mathbf{A}$ is $m \times n$ then $\mathbf{A_g}$ is $n \times m$. In the case that also elementary column operations will be needed in order to find the desired equivalent form of $\mathbf{A}$ let say

$$\mathbf{A} \approx \mathbf{P_rAQ}_r = \begin{pmatrix} \mathbf{A}_{11} & \mathbf{A}_{12} \\ \mathbf{0} & \mathbf{0} \end{pmatrix},$$

then

$$\mathbf{A_g} = \mathbf{Q}_r \begin{pmatrix} \mathbf{A}_{11}^{-1} & \mathbf{0} \\ \mathbf{0} & \mathbf{0} \end{pmatrix} \mathbf{P_r}.$$

(b) Derivation from the diagonal equivalent form

In the present paragraph we will study the following case. Assume that we are able by premultiplying and postmultiplying by elementary matrices to arrive in an equivalent diagonal form as follows

$$\mathbf{A} \approx \mathbf{P_d A Q_d} = \begin{pmatrix} \mathbf{D}_{r_\mathbf{A}} & \mathbf{0} \\ \mathbf{0} & \mathbf{0} \end{pmatrix},$$

where $\mathbf{D}_{r_\mathbf{A}}$ is a diagonal matrix of order $rank\,(\mathbf{A}) \times rank\,(\mathbf{A})$ with all diagonal elements nonzero. Then

$$\mathbf{A_g} = \mathbf{Q_d} \begin{pmatrix} \mathbf{D}_{r_\mathbf{A}}^{-1} & \mathbf{0} \\ \mathbf{0} & \mathbf{0} \end{pmatrix} \mathbf{P_d}.$$

Note also that a class of generalized inverses of $\mathbf{A}$ in this case are given by

$$\mathbf{A_g} = \mathbf{Q_d} \begin{pmatrix} \mathbf{D}_{r_\mathbf{A}}^{-1} & \mathbf{Y} \\ \mathbf{X} & \mathbf{Z} \end{pmatrix} \mathbf{P_d},$$

for any matrices $\mathbf{X}, \mathbf{Y}$ and $\mathbf{Z}$ of appropriate order. This fact is a good example of the non uniqueness of the generalized inverse of a matrix.

(c) A general Algorithm.

There is a quite simple algorithm for finding a generalize inverse of a matrix $\mathbf{A}$ with $rank\,(\mathbf{A}) = r_\mathbf{A}$. Its foundation is based on properties of elementary permutation matrices and could be found in Searle (1982). The algorithm is as follows:

Algorithm.

(a) In $\mathbf{A}$ find any nonsingular submatrix of order $rank\,(\mathbf{A})$. The submatrix which we denote by $\mathbf{B}$ it does not have to consist from $rank\,(\mathbf{A})$ adjacent rows and columns of $\mathbf{A}$.

(b) Find $\left(\mathbf{B}^{-1}\right)^\top$.

(c) In $\mathbf{A}$ replace each element of $\mathbf{B}$ by the corresponding element of $\left(\mathbf{B}^{-1}\right)^\top$.

(d) Replace all other elements of $\mathbf{A}$ by zero.

(e) Transpose the resulting matrix and you get a generalized matrix $\mathbf{A}_g$ of $\mathbf{A}$.

Example 148 *We choose as* $\mathbf{A}$ *the matrix we used in example 129. That is*

$$\mathbf{A} = \begin{pmatrix} 1 & 3 & 4 & 2 \\ -2 & 3 & 4 & 1 \\ 4 & -6 & -8 & -2 \end{pmatrix}.$$

We have $rank\,(\mathbf{A}) = 2$ *hence we must choose* $\mathbf{B}$ *to be a* 2×2 *nonsingular matrix. We get*

$$\mathbf{B} = \begin{pmatrix} 1 & 2 \\ -2 & 1 \end{pmatrix} \quad \text{and} \quad \mathbf{B}^{-1} = \begin{pmatrix} \frac{1}{5} & -\frac{2}{5} \\ \frac{2}{5} & \frac{1}{5} \end{pmatrix}, \quad \left(\mathbf{B}^{-1}\right)^\top = \begin{pmatrix} \frac{1}{5} & \frac{2}{5} \\ -\frac{2}{5} & \frac{1}{5} \end{pmatrix}.$$

Hence

$$\mathbf{A}_g = \begin{pmatrix} \frac{1}{5} & -\frac{2}{5} & 0 \\ 0 & 0 & 0 \\ 0 & 0 & 0 \\ \frac{2}{5} & \frac{1}{5} & 0 \end{pmatrix}.$$

It is easy to verify that $\mathbf{A}\mathbf{A}_g\mathbf{A} = \mathbf{A}$. *We now make a second choice of* $\mathbf{B}$, *that is*

$$\mathbf{B} = \begin{pmatrix} 4 & 2 \\ 4 & 1 \end{pmatrix} \quad \text{and} \quad \mathbf{B}^{-1} = \begin{pmatrix} -\frac{1}{4} & \frac{1}{2} \\ 1 & -1 \end{pmatrix}, \quad \left(\mathbf{B}^{-1}\right)^\top = \begin{pmatrix} -\frac{1}{4} & 1 \\ \frac{1}{2} & -1 \end{pmatrix}.$$

Therefore,

$$\mathbf{A}_g = \begin{pmatrix} 0 & 0 & 0 \\ 0 & 0 & 0 \\ -\frac{1}{4} & \frac{1}{2} & 0 \\ 1 & -1 & 0 \end{pmatrix}.$$

It is easy to verify that $\mathbf{A}\mathbf{A}_g\mathbf{A} = \mathbf{A}$. *This example exhibits very clearly the nonuniquenence of the generalized matrix.*

2.10 The Moore-Penrose generalized matrix

The Moore-Penrose generalized inverse is developed in Penrose (1955), on foundations laid by Moore (1920). The Moore-Penrose generalized inverse of a matrix $\mathbf{A} \in M_{mn}$, denoted by $\mathbf{A}^+$so as defined to possess additional properties that the inverse of a nonsingular matrix has. It always exists and it is unique.

Definition 149 *The Moore-Penrose generalized inverse of a matrix* $\mathbf{A} \in M_{mn}$, *satisfies the conditions*

$$(i)\ \mathbf{A}\mathbf{A}^+\mathbf{A} = \mathbf{A} \quad (ii)\ \mathbf{A}^+\mathbf{A}\mathbf{A}^+ = \mathbf{A}^+$$

$$(iii)\ \left(\mathbf{A}\mathbf{A}^+\right)^\top = \mathbf{A}\mathbf{A}^+ \text{ is symmetric,}$$

$$(iv)\ \left(\mathbf{A}^+\mathbf{A}\right)^\top = \mathbf{A}^+\mathbf{A} \text{ is symmetric.}$$

The Moore-Penrose generalized inverse of a matrix $\mathbf{A} \in M_{mn}$, that is, $\mathbf{A}^+ \in M_{nm}$ has the following properties which we provide in the form of a theorem. These properties could be proved by the reader using the defining four properties of $\mathbf{A}^+$. It will be a good exercise for someone whishing to get accustomed with $\mathbf{A}^+$.

Theorem 150 *Let* $\mathbf{A} \in M_{mn}$. *Then*

(1) $(\mathbf{A}^+)^+ = \mathbf{A}$.
(2) $\left(\mathbf{A}\mathbf{A}^+\right)^+ = \mathbf{A}\mathbf{A}^+$ *and* $(\mathbf{A}^+\mathbf{A})^+ = \mathbf{A}^+\mathbf{A}$.
(3) $\left(\mathbf{A}^\top\mathbf{A}\right)^+ = \mathbf{A}^+(\mathbf{A}^+)^\top$ *and* $\left(\mathbf{A}\mathbf{A}^\top\right)^+ = (\mathbf{A}^+)^\top\mathbf{A}$.
(4) $\mathbf{A}^+ = \left(\mathbf{A}^\top\mathbf{A}\right)^+\mathbf{A}^\top$.
(5) $(a\mathbf{A})^+ = a^{-1}\mathbf{A}^+$, if $a \neq 0$ *a scalar.*
(6) $rank\,(\mathbf{A}) = rank\,(\mathbf{A}^+) = rank\,\left(\mathbf{A}\mathbf{A}^+\right) = rank\,(\mathbf{A}^+\mathbf{A})$.

Note that the Moore-Penrose generalized inverse is defined also when a matrix $\mathbf{A} \in M_{mn}(\mathbb{C})$ as follows:

Definition 151 *Let* $\mathbf{A} \in M_{mn}(\mathbb{C})$. *Then* $\mathbf{A}^+ \in M_{nm}(\mathbb{C})$ *is a generalized inverse of* $\mathbf{A}$ *if and only if:*

$$(i)\ \mathbf{A}\mathbf{A}^+\mathbf{A} = \mathbf{A} \quad (ii)\ \mathbf{A}^+\mathbf{A}\mathbf{A}^+ = \mathbf{A}^+$$

$$(iii)\ \left(\mathbf{A}\mathbf{A}^+\right)^* = \mathbf{A}\mathbf{A}^* \text{ is Hermite,}$$

$$(iv)\ \left(\mathbf{A}^+\mathbf{A}\right)^* = \mathbf{A}^+\mathbf{A} \text{ is Hermite.}$$

In the case $\mathbf{A} \in M_{mn}(\mathbb{C})$ Theorem 150 is analogously written as:

Theorem 152 *Let* $\mathbf{A} \in M_{mn}(\mathbb{C})$. *Then*

(1) $(\mathbf{A}^+)^+ = \mathbf{A}$. *(2)* $\left(\mathbf{A}\mathbf{A}^+\right)^+ = \mathbf{A}\mathbf{A}^+$ *and* $(\mathbf{A}^+\mathbf{A})^+ = \mathbf{A}^+\mathbf{A}$.
(3) $(\mathbf{A}^*\mathbf{A})^+ = \mathbf{A}^+(\mathbf{A}^+)^*$ *and* $(\mathbf{A}\mathbf{A}^*)^+ = (\mathbf{A}^+)^*\mathbf{A}$.
(4) $\mathbf{A}^+ = (\mathbf{A}^*\mathbf{A})^+\mathbf{A}^*$.
(5) $(a\mathbf{A})^+ = a^{-1}\mathbf{A}^+$, if $a \neq 0$ *a scalar.*
(6) $rank\,(\mathbf{A}) = rank\,(\mathbf{A}^+) = rank\,\left(\mathbf{A}\mathbf{A}^+\right) = rank\,(\mathbf{A}^+\mathbf{A})$.

With the use of the properties of the Moore-Penrose generalized inverse it is easy to verify the following proposition

Proposition 153 *Let* $\mathbf{A} \in M_{mn}(\mathbb{C})$ *and* $\mathbf{A} = \mathbf{A}_{\mathbf{P}_e}\mathbf{A}_{\mathbf{Q}_e}$*, as in Theorem 143, then*

$$\mathbf{A}^+ = \mathbf{A}^*_{\mathbf{Q}_e}\left(\mathbf{A}_{\mathbf{Q}_e}\mathbf{A}^*_{\mathbf{Q}_e}\right)^{-1}\left(\mathbf{A}^*_{\mathbf{P}_e}\mathbf{A}_{\mathbf{P}_e}\right)^{-1}\mathbf{A}_{\mathbf{P}_e}.$$

If $\mathbf{A} \in M_{mn}(\mathbb{R})$ *then*

$$\mathbf{A}^+ = \mathbf{A}^\top_{\mathbf{Q}_e}\left(\mathbf{A}_{\mathbf{Q}_e}\mathbf{A}^\top_{\mathbf{Q}_e}\right)^{-1}\left(\mathbf{A}^\top_{\mathbf{P}_e}\mathbf{A}_{\mathbf{P}_e}\right)^{-1}\mathbf{A}_{\mathbf{P}_e}.$$

Example 154 *Continuing Examples 140,147, and 148. From these examples we get that*

$$\mathbf{A} = \begin{pmatrix} 1 & 3 & 4 & 2 \\ -2 & 3 & 4 & 1 \\ 4 & -6 & -8 & -2 \end{pmatrix},$$

and

$$\mathbf{A} = \mathbf{P}_e^{-1}\begin{pmatrix} \mathbf{I}_r & \mathbf{0} \\ \mathbf{0} & \mathbf{0} \end{pmatrix}\mathbf{Q}_e^{-1},$$

with

$$\mathbf{P}_e = \begin{pmatrix} \frac{1}{3} & -\frac{1}{3} & 0 \\ \frac{2}{9} & \frac{1}{9} & 0 \\ 0 & 2 & 1 \end{pmatrix} \quad \text{and} \quad \mathbf{Q}_e = \begin{pmatrix} 1 & 0 & 0 & -\frac{1}{3} \\ 0 & 1 & -\frac{4}{3} & -\frac{5}{9} \\ 0 & 0 & 1 & 0 \\ 0 & 0 & 0 & 1 \end{pmatrix}.$$

and

$$\mathbf{A} = \mathbf{A}_{\mathbf{P}_e}\mathbf{A}_{\mathbf{Q}_e},$$

with

$$\mathbf{A}_{\mathbf{P}_e} = \begin{pmatrix} 1 & 3 \\ -2 & 3 \\ 4 & -6 \end{pmatrix} \quad \text{and} \quad \mathbf{A}_{\mathbf{Q}_e} = \begin{pmatrix} 1 & 0 & 0 & \frac{1}{3} \\ 0 & 1 & \frac{4}{3} & \frac{5}{9} \end{pmatrix}$$

Now

$$\mathbf{A}^+ = \mathbf{A}^\top_{\mathbf{Q}_e}\left(\mathbf{A}_{\mathbf{Q}_e}\mathbf{A}^\top_{\mathbf{Q}_e}\right)^{-1}\left(\mathbf{A}^\top_{\mathbf{P}_e}\mathbf{A}_{\mathbf{P}_e}\right)^{-1}\mathbf{A}_{\mathbf{P}_e},$$

with

$$\left(\mathbf{A}^\top_{\mathbf{P}_e}\mathbf{A}_{\mathbf{P}_e}\right)^{-1}\begin{pmatrix} \frac{2}{15} & \frac{1}{15} \\ \frac{1}{15} & \frac{7}{135} \end{pmatrix} \quad \text{and} \quad \left(\mathbf{A}_{\mathbf{Q}_e}\mathbf{A}^\top_{\mathbf{Q}_e}\right)^{-1} = \begin{pmatrix} \frac{10}{11} & -\frac{3}{55} \\ -\frac{3}{55} & \frac{18}{55} \end{pmatrix},$$

and finally

$$\mathbf{A}^+ = \begin{pmatrix} \frac{16}{55} & -\frac{17}{275} & \frac{34}{275} \\ \frac{3}{55} & \frac{3}{275} & -\frac{6}{275} \\ \frac{4}{55} & \frac{4}{275} & -\frac{8}{275} \\ \frac{7}{55} & -\frac{4}{275} & \frac{8}{275} \end{pmatrix}.$$

2.11 The Drazin inverse and the group inverse

The generalized inverses presented in this chapter and especially the Moore-Penrose generalized inverse are primarily of interest because of their ability to produce solutions of consistent systems of linear equations. The right generalized inverse to use in connection with finite Markov chains (Meyer (1975)) is that of the Drazin generalized inverse (Drazin (1958)). This is more so with the special case of a Drazin generalized inverse, the group inverse (Erdélyi (1967)). If $\mathbf{P}$ is the one step transition probability matrix of a finite homogeneous Markov chain and if $\mathbf{A} = (\mathbf{I} - \mathbf{P})$ then it is singular. However, as it is shown in Meyer (1975) the group inverse $(\mathbf{I} - \mathbf{P})^\#$ plays a very important role in answering many important questions in the theory of finite Markov chains. We will firstly define the Drazin inverse:

Definition 155 *Let* $\mathbf{A} \in M_n(\mathbb{R})$ *then the Drazin inverse denoted by* $\mathbf{A}^D$ *has the following properties:*

$$(i)\ \mathbf{A}^{k+1}\mathbf{A}^D = \mathbf{A}^k\ .$$
$$(ii)\ \mathbf{A}^D\mathbf{A}\mathbf{A}^D = \mathbf{A}^D.$$
$$(iii)\ \mathbf{A}\mathbf{A}^D = \mathbf{A}^D\mathbf{A},$$

where k *is the index of* $\mathbf{A}$, *being the smallest non-negative integer such that* $rank\left(\mathbf{A}^k\right) = rank\left(\mathbf{A}^{k+1}\right)$.

Remark 156 *When* $\mathbf{A}$ *is non-singular then* $k = 0$, *using the convection that* $\mathbf{A}^0 = \mathbf{I}$, *and* $\mathbf{A}^D = \mathbf{A}^{-1}$. *From the definition it is apparent that* $\mathbf{A}^D$ *commutes with* $\mathbf{A}$, *whereas the Moore-Penrose and other inverses do not in general have this property. The Drazin inverse of a matrix is unique.*

Let $\mathbf{A} \in M_n$ then it could proved that it is similar to a blog matrix of the form

$$\mathbf{A} = \mathbf{S}\begin{pmatrix}\mathbf{B} & \mathbf{0}\\ \mathbf{0} & \mathbf{C}_k\end{pmatrix}\mathbf{S}^{-1},$$

where $\mathbf{B}$ is non-singular and $\mathbf{C}_k$ is nilpotent of index k, that is, $\mathbf{C}_k$ is such that $(\mathbf{C}_k)^k = 0$, $(\mathbf{C}_k)^{k-1} \neq 0$ for some positive k. Then

$$\mathbf{A}^D = \mathbf{S}\begin{pmatrix}\mathbf{B}^{-1} & \mathbf{0}\\ \mathbf{0} & \mathbf{0}\end{pmatrix}\mathbf{S}^{-1}.$$

It is easy to verify that the above matrix $\mathbf{A}^D$ satisfies the defining properties for the Drazin generalized inverse.

We will now define the group inverse of a square matrix:

Definition 157 *Let* $\mathbf{A} \in M_n$ *then the group inverse of* $\mathbf{A}$, *when it exists, is defined to be the matrix* $\mathbf{A}^\sharp$ *which satisfies the conditions*

$$\mathbf{A}\mathbf{A}^\sharp\mathbf{A} = \mathbf{A},\ \ \mathbf{A}^\sharp\mathbf{A}\mathbf{A}^\sharp = \mathbf{A}^\sharp\ \text{and}\ \ \mathbf{A}\mathbf{A}^\sharp = \mathbf{A}^\sharp\mathbf{A}.$$

Remark 158 *The matrix* $\mathbf{A}^\sharp$ *when it exists is unique if and only if*

$$rank\,(\mathbf{A}) = rank\left(\mathbf{A}^2\right).$$

For a proof see Erdélyi (1967). The term "group inverse" is used because due to the fact $rank\,(\mathbf{A}) = rank\left(\mathbf{A}^2\right)$ *then* $\mathbf{A}$ *belongs to a set* $\mathcal{G}(\mathbf{A}) \subseteq M_n(\mathbb{R})$ *such that* $\mathcal{G}(\mathbf{A})$ *forms a multiplicative group under ordinary matrix multiplication (see Rao and Mitra (1971)).*

The group inverse plays a very important role in the theory of Markov chains. The first fundamental fact is that for any transition matrix $\mathbf{P}$, which uniquely defines a Markov chain, then the group inverse $(\mathbf{I} - \mathbf{P})^\sharp$ always exists. This is established in the following lemma and theorem the proof of which could be found in Mayer (1975). The interested reader will find no difficulty in understanding the proofs with the knowledge he acquired so far on Matrix Analysis.

Lemma 159 *Let* $\mathbf{A} \in M_n$ *and assume that it has the form*

$$\mathbf{A} = \begin{pmatrix}\mathbf{B} & \mathbf{C}\\ \mathbf{0} & \mathbf{D}\end{pmatrix},$$

where $\mathbf{B}$ *and* $\mathbf{D}$ *are square matrices. If* $\mathbf{B}$ *is nonsingular and if* $\mathbf{D}^\sharp$ *exists, then* $\mathbf{A}^\sharp$ *exists and*

$$\mathbf{A}^\sharp = \begin{pmatrix}\mathbf{B}^{-1} & \mathbf{B}^{-2}\mathbf{C}\left(\mathbf{I} - \mathbf{D}\mathbf{D}^\sharp\right) - \mathbf{B}^{-1}\mathbf{C}\mathbf{D}^\sharp\\ \mathbf{0} & \mathbf{D}^\sharp\end{pmatrix}.$$

Theorem 160 *For any transition matrix* $\mathbf{P}$ *which uniquely defines a Markov chain then the group inverse* $(\mathbf{I}-\mathbf{P})^{\sharp}$ *always exists.*

In the present book, we choose to use Chapter 3 in Vassiliou (2010) as a guide in the theory of finite homogeneous Markov chains. The terminology and the results included in this chapter will be respectively adopted and assumed known. It will be very useful for the understanding of the present if the reader reads this chapter first. More on finite homogeneous Markov chains could be found in books such as Cox[1] and Miller (1968), Doob (1990), Dunkin (1965), Feller (1968, 1971), Howard (1971a, 1971b), Iosifescou (1980), Isaacson and Madsen (1976), Karlin and Taylor (1975), Kemeny and Snell (1976), Kijima (1997), Meyn and Tweedie (2009), Ross (1996), Sefrozo (2009), Seneta (1981) and Taylor and Karlin (1998). In the above books, different terminologies have been used for the same concepts creating sometimes some confusion. In order to avoid this we will clarify in the present some concepts.

For any homogeneous Markov chain the transition probability matrix $\mathbf{P}$ uniquely determines the chain. In addition, properties of the matrix $\mathbf{P}$ have a one to one correspondence with properties of the Markov chain since most of the important concepts are based on the transition probabilities among the states. We will start with the definition of a reducible stochastic matrix.

Definition 161 *Let* $\mathbf{P}$ *be a stochastic matrix. We say that* $\mathbf{P}$ *is reducible if it is possible by permutation of the rows and columns to be written in the form*

$$\mathbf{P}=\begin{pmatrix}\mathbf{P}_1 & \mathbf{0}\\ \mathbf{R} & \mathbf{Q}\end{pmatrix},$$

where $\mathbf{P}_1, \mathbf{Q}$ *are square matrices. In the opposite case the matrix is called irreducible.*

Note that the above definition could be generalized for any non-negative matrix $\mathbf{A}\in M_n$. A Markov chain is called irreducible if its transition probability matrix $\mathbf{P}$ is irreducible. It is rather apparent that in all generality a stochastic matrix $\mathbf{P}$ will be reducible. The study of Markov chains with an arbitrary stochastic matrix $\mathbf{P}$ as the transition probabilities matrix could be found in Sierksma (1999) for finite homogeneous Markov chains. The case of non-homogeneous Markov chains is studied in Vassiliou (2015). Let $\mathbf{P}$ be a stochastic matrix with state space $\{1,2,...,k\}$. Then two states i and j *communicate,* if there exists a positive probability to go from i to j in any number of steps and at the same time a positive probability to go from j to i in any number of steps. We say in this case that i leads to j and we denote it by $i\rightarrow j$ and $j\rightarrow i$ if also j leads to i (Seneta (1981)). When i and j communicate then we denote it by $i\longleftrightarrow j$. A state i is called *essential* if $i\rightarrow j$ implies $i\longleftrightarrow j$; and there is at least one j such that $i\rightarrow j$. If $i\rightarrow j$ but $j\nrightarrow i$ for some j, then the

[1]Sir David Cox (1924–2022) was well known for the ground breaking paths he created in statistics and applied probability. He is considered by many to be the statistician of the century. Cox proportional hazard model is one of the most cited papers of all time. His work on stochastic process started in 1955 with a paper published in JRSS, Part B. He was internationally renowned for Cox processes, Box-Cox transform, Matrix exponential distribution, method of supplementary variables, Design of experiments, Analysis of binary data and various parts of stochastic processes. He held Professorships at Birkbeck College London, Imperial College London and the University of Oxford. He also spent a remarkable 25 years editing the statistics journal *Biometrika,* where he read all the papers before sending them to the referees. He won many prizes and wrote almost 400 papers and books which received more than 100000 citations on the Web of science.

For those who knew him apart from his genius which was stunning he will be remembered for his energy, devotion and concentration to his creative work, his remarkable kindness and humility. He continued to be creative and in good mental form up to the end.

state i is called inessential. Therefore, all essential classes can be subdivided into essential classes in such a way that all the states belonging to one class communicate, but cannot lead to a state outside the class. Moreover, all inessential states which communicate with some state, may be divided into inessential classes such that all states communicate. Let there be a state i then if the probability of returning to i is 0, except if the number of steps required is multiple of a number d, then we call period of state i the maximum common divisor of such d's. In an essential communicating class all states have the same period, hence we may consider the concept of the period of an essential class. It is essential though to note, that all the above classifications and the concept of the period depends on the positions of the zero's and not on the actual size of the positive elements. Let $\mathbf{P}$ an arbitrary stochastic matrix, then according to our classification of states above its canonical form will be

$$\mathbf{P} = \begin{pmatrix} \mathbf{P}_1 & \mathbf{0} & \dots & \mathbf{0} & \mathbf{0} \\ \mathbf{0} & \mathbf{P}_2 & .. & \mathbf{0} & \mathbf{0} \\ \dots & \dots & \dots & \dots & \dots \\ \dots & \dots & \dots & \mathbf{P}_r & \mathbf{0} \\ \mathbf{T}_1 & \mathbf{T}_2 & \dots & \mathbf{T}_r & \mathbf{Q} \end{pmatrix}.$$

The matrix $\mathbf{P}_i$ for $i = 1, 2, ..., r$ is the matrix of transition probabilities among the class i of essential states. All states in the class i communicate for $i = 1, 2, ..., r$ and each class in general is assumed to be of different size. It is apparent that all the matrices $\mathbf{P}_i$ for $i = 1, 2, ..., r$ are square matrices. Therefore, all the matrices $\mathbf{P}_i$ for $i = 1, 2, ..., r$ are irreducible and consist of an essential classes, hence in all generality we may assume that their period is d_i. The matrix $\mathbf{T}_i$ for $i = 1, 2, ..., r$ is the transition probability matrix from the inessential states to the essential classes of states. The matrices $\mathbf{T}_i$ for $i = 1, 2, ..., r$ are not necessarily square matrices nor they of the same size. In general the matrix $\mathbf{Q}$ is substochastic.

We will now provide a useful Corollary of Theorem 160:

Corollary 162 *Let a finite homogeneous Markov chain with transition probability matrix* $\mathbf{P}$ *and state space* $S = \{1, 2, ..., k\}$, *and let* $\mathbf{A} = (\mathbf{I} - \mathbf{P})$. *Then there exists a nonsingular matrix* $\mathbf{L}$ *such that*

$$\mathbf{P} = \mathbf{L}^{-1} \begin{pmatrix} \mathbf{I} & \mathbf{0} \\ \mathbf{0} & \mathbf{Q} \end{pmatrix} \mathbf{L}, \quad \mathbf{A} = (\mathbf{I} - \mathbf{P}) = \mathbf{L}^{-1} \begin{pmatrix} \mathbf{I} & \mathbf{0} \\ \mathbf{0} & \mathbf{I} - \mathbf{Q} \end{pmatrix} \mathbf{L},$$

where 1 *is not an eigenvalue of* $\mathbf{Q}$ *and*

$$\mathbf{A}^{\#} = (\mathbf{I} - \mathbf{P})^{\#} = \mathbf{L}^{-1} \begin{pmatrix} \mathbf{0} & \mathbf{0} \\ \mathbf{0} & (\mathbf{I} - \mathbf{Q})^{-1} \end{pmatrix} \mathbf{L}.$$

We will now provide the following equivalent definition of an irreducible Markov chain (Vassiliou (2010)):

Definition 163 *A Markov chain is called irreducible if it has only one class of communicating states and not transient states.*

One of the most important results in the theory of finite homogeneous Markov chains is the fact that when the transition probability matrix $\mathbf{P}$ has one single communicating class, which is aperiodic then $\lim_{n \to \infty} \mathbf{P}^n$ exists and it is a stable matrix. That is, it has identical rows, expressing the physical meaning that in the long run the probability of being in a state is independent of the initial state. The following theorem exhibits one application of the concept of group inverse to the theory of Markov chains and its contribution to the study of asymptotic behavior.

Theorem 164 *Let a finite irreducible aperiodic homogeneous Markov chain with transition probability matrix* $\mathbf{P}$ *and state space* $S = \{1, 2, ..., k\}$. *Let* $\mathbf{A} = (\mathbf{I} - \mathbf{P})$. *Then*

$$\lim_{n\to\infty} \mathbf{P}^n = \mathbf{I} - \mathbf{A}\mathbf{A}^{\sharp}.$$

Proof. Since $\mathbf{P}$ is the transition probability matrix of a a Markov chain in which all states communicate, then we know that 1 is a simple eigenvalue and there exists a nonsingular matrix $\mathbf{L}$ such that

$$\mathbf{P} = \mathbf{L}^{-1}\begin{pmatrix} 1 & \mathbf{0} \\ \mathbf{0} & \mathbf{Q} \end{pmatrix}\mathbf{L}.$$

From the previous Corollary, we know that 1 is not an eigenvalue of $\mathbf{Q}$, and therefore $\mathbf{I} - \mathbf{Q}$ is nonsingular and in addition $\lim_{n\to\infty} \mathbf{Q}^n = \mathbf{0}$. Hence,

$$\lim_{n\to\infty} \mathbf{P}^n = \mathbf{L}^{-1}\begin{pmatrix} 1 & \mathbf{0} \\ \mathbf{0} & \mathbf{0} \end{pmatrix}\mathbf{L}.$$

Now from the previous Corollary, we have that

$$\mathbf{A}^{\sharp} = (\mathbf{I} - \mathbf{P})^{\sharp} = \mathbf{L}^{-1}\begin{pmatrix} \mathbf{0} & \mathbf{0} \\ \mathbf{0} & (\mathbf{I} - \mathbf{Q})^{-1} \end{pmatrix}\mathbf{L}.$$

It is easily verified that $\lim_{n\to\infty} \mathbf{P}^n = \mathbf{I} - \mathbf{A}\mathbf{A}^{\sharp}$. ■

We will now present how to calculate $(\mathbf{I} - \mathbf{P})^{\sharp}$, which is the basis for answering many important questions in the theory of finite Markov chains. The next theorem which concerns finite irreducible homogeneous Markov chains is the key result which allows one to compute $(\mathbf{I} - \mathbf{P})^{\sharp}$ without an undue amount of difficulty. The proof of the theorem is quite simple and could be found in Meyer (1975).

Theorem 165 *Let a finite homogeneous Markov chain with transition probability matrix* $\mathbf{P}$ *and state space* $S = \{1, 2, ..., k\}$. *Assume that all states communicate and let* $\mathbf{A} = (\mathbf{I} - \mathbf{P})$. *Every principal submatrix of* $\mathbf{A}$ *of size* $n = 1, 2, ..., k - 1$ *is nonsingular. Furthermore, the inverse of each principal submatrix of* $\mathbf{A}$ *is a non-negative matrix.*

It is now possible to prove the following theorem which provides a formula for finding $(\mathbf{I} - \mathbf{P})^{\sharp}$. The proof of the theorem could be found in Meyer (1975).

Theorem 166 *Let a finite irreducible homogeneous Markov chain with transition probability matrix* $\mathbf{P}$ *and state space* $S = \{1, 2, ..., k\}$. *Let* $\mathbf{A} = (\mathbf{I} - \mathbf{P})$. *Write* $\mathbf{A}$ *as*

$$\mathbf{A} = \begin{pmatrix} \mathbf{B} & \mathbf{c} \\ \mathbf{d}^{\top} & \mathbf{b} \end{pmatrix},$$

where $\mathbf{B}$ *is of order* $k - 1 \times k - 1$. *Define by*

$$\boldsymbol{\varphi}^{\top} = \mathbf{d}^{\top}\mathbf{B}^{-1}, \quad \gamma = -\boldsymbol{\varphi}^{\top}\mathbf{B}^{-1}\mathbf{1}^{\top}, \text{ where } \mathbf{1}^{\top} \text{is a column of 1's.}$$

$$\beta = 1 - \boldsymbol{\varphi}^{\top}\mathbf{1}^{\top}, \quad \mathbf{M} = \mathbf{B}^{-1} - \frac{\gamma}{\beta}\mathbf{I}.$$

The scalars $\beta > 1$ *and* $\gamma > 0$ *and*

$$\mathbf{A}^{\sharp} = \begin{pmatrix} \mathbf{B}^{-1} + \frac{\mathbf{B}^{-1}\mathbf{1}^{\top}\boldsymbol{\varphi}^{\top}\mathbf{B}^{-1}}{\gamma} - \frac{\mathbf{M}\mathbf{1}^{\top}\boldsymbol{\varphi}^{\top}\mathbf{M}}{\gamma} & -\frac{\mathbf{M}\mathbf{1}^{\top}}{\beta} \\ \frac{\boldsymbol{\varphi}^{\top}\mathbf{M}}{\beta} & \frac{\gamma}{\beta^2} \end{pmatrix}.$$

Example 167 *Consider the finite homogeneous Markov chain* $\{X_n\}_{n=0}^{\infty}$, *which is being uniquely determined by the transition probability matrix*

$$\mathbf{P}=\begin{pmatrix}0.8 & 0.1 & 0.1\\ 0.3 & 0.6 & 0.1\\ 0.1 & 0.4 & 0.5\end{pmatrix}.$$

Then

$$\mathbf{A}=\mathbf{I}-\mathbf{P}=\begin{pmatrix}0.2 & -0.1 & -0.1\\ -0.3 & 0.4 & -0.1\\ -0.1 & -0.4 & 0.5\end{pmatrix}.$$

We will find $(\mathbf{I}-\mathbf{P})^{\#}=\mathbf{A}^{\#}$. *We choose* $\mathbf{B}$ *to be the* 2×2 *matrix*

$$\mathbf{B}=\begin{pmatrix}0.2 & -0.1\\ -0.3 & 0.4\end{pmatrix} \text{ and } \mathbf{B}^{-1}=\begin{pmatrix}8.0 & 2.0\\ 6.0 & 4.0\end{pmatrix}.$$

Then we get that

$$\mathbf{d}^{\top}=\begin{pmatrix}-0.1 & -0.4\end{pmatrix} \quad \text{and} \quad \boldsymbol{\varphi}^{\top}=\mathbf{d}^{\top}\mathbf{B}^{-1}=\begin{pmatrix}-3.2 & -1.8\end{pmatrix},$$

$$\gamma=-\boldsymbol{\varphi}^{\top}\mathbf{B}^{-1}\mathbf{1}^{\top}=50 \quad \text{and} \quad \beta=1-\boldsymbol{\varphi}^{\top}\mathbf{1}^{\top}=6,$$

and

$$\mathbf{M}=\mathbf{B}^{-1}-\frac{\gamma}{\beta}\mathbf{I}=\begin{pmatrix}-0.33333 & 2.0\\ 6.0 & -4.3333\end{pmatrix}.$$

We are now ready to apply Theorem 153 in order to calculate $\mathbf{A}^{\#}$. *We will proceed by calculating each block separately. We have*

$$\mathbf{B}^{-1}+\frac{\mathbf{B}^{-1}\mathbf{1}^{\top}\boldsymbol{\varphi}^{\top}\mathbf{B}^{-1}}{\gamma}-\frac{\mathbf{M}\mathbf{1}^{\top}\boldsymbol{\varphi}^{\top}\mathbf{M}}{\gamma}=\begin{pmatrix}1.0444 & -0.76666\\ -0.95555 & 1.2333\end{pmatrix},$$

$$-\frac{\mathbf{M}\mathbf{1}^{\top}}{\beta}=\begin{pmatrix}-0.27778\\ -0.27778\end{pmatrix} \quad \text{and} \quad \frac{\gamma}{\beta^2}=\frac{25}{18},$$

and

$$\frac{\boldsymbol{\varphi}^{\top}\mathbf{M}}{\beta}=\begin{pmatrix}-1.6222 & 0.23332\end{pmatrix}.$$

Hence

$$\mathbf{A}^{\#}=\begin{pmatrix}1.0444 & -0.76666 & -0.27778\\ -0.95555 & 1.2333 & -0.27778\\ -1.6222 & 0.23332 & 1.34444\end{pmatrix}.$$

Now we have

$$\mathbf{I}-\mathbf{A}\mathbf{A}^{\#}=\begin{pmatrix}0.53335 & 0.29999 & 0.16222\\ 0.53332 & 0.30001 & 0.16222\\ 0.53332 & 0.29999 & 0.18889\end{pmatrix},$$

which is easily verified to be equal with $\lim_{n\to\infty}\mathbf{P}^n$ *a stable matrix as it should be.*

Remark 168 *It is well known that the Moore-Penrose inverse and the group inverse for a matrix* $\mathbf{A}$ *are equal, if and only if, the null space of* $\mathbf{A}$ *is equal with the null space of* $\mathbf{A}^{\top}$. *In our context it is of interest to know, for which type of Markov chains this is true. It is proved in Meyer (1975) that the Moore-Penrose inverse and the group inverse for a transition probability matrix* $\mathbf{P}$ *are equal, if and only if, it is doubly stochastic.*

2.12 The group inverse and Markov chains

In the previous section, we have founded an important link between an irreducible aperiodic finite homogeneous Markov chain and the group inverse of $\mathbf{A} = \mathbf{I} - \mathbf{P}$, where $\mathbf{P}$ the transition probability matrix of the Markov chain. In fact, as we will discuss in the present section their relation is a lot deeper. We start with concluding the relation between the group inverse of $\mathbf{A} = \mathbf{I} - \mathbf{P}$ and the asymptotic behavior of $\mathbf{P}$ the transition probability matrix of the Markov chain. It is well known that for an irreducible finite homogeneous Markov chain which is aperiodic there exists a vector of probabilities $\boldsymbol{\pi}$, which is the solution of the linear equations $\boldsymbol{\pi} = \boldsymbol{\pi}\ \mathbf{P}$. The following theorem relates the asymptotic probability distribution $\boldsymbol{\pi}$ with the group inverse $\mathbf{A}^{\sharp}$. The proof of the theorem is direct and simple (see Meyer (1975)).

Theorem 169 *Let an irreducible finite homogeneous Markov chain $\{X_n\}_{n=0}^{\infty}$ with state space $S = \{1, 2, ..., k\}$ and $\mathbf{A} = \mathbf{I} - \mathbf{P}$. Then the asymptotic probability distribution of $\{X_n\}_{n=0}^{\infty}$ is given by*

$$\boldsymbol{\pi} = \mathbf{e}_i - \mathbf{a}_i^{(\sharp)}\mathbf{A}, \ \ for\ i = 1, 2, ..., k,$$

where $\mathbf{e}_i$ is the i-th row of $\mathbf{I}$ and $\mathbf{a}_i^{(\sharp)}$ is the i-th row of $\mathbf{A}^{\sharp}$.

Note that in most cases it is preferable computationally to find the group inverse than solve a set of linear equations. It is known that the fundamental matrix

$$\mathbf{Z} = (\mathbf{I} - (\mathbf{P} - \boldsymbol{\Pi}))^{-1},$$

where $\boldsymbol{\Pi} = \lim_{t\to\infty} \mathbf{P}^t$ plays a central role in answering important problems for an irreducible finite homogeneous Markov chain. In this section, it is shown that although $\mathbf{Z} \neq \mathbf{A}^{\sharp}$, the group inverse in most cases may directly replace $\mathbf{Z}$ and in others may be used analogously. The advantage again in this case is that it is preferable computational to find the group inverse than finding $\mathbf{Z}$. The following theorem shows how the two matrices are related. The proof of the theorem is direct and simple (see Meyer (1975)).

Theorem 170 *Let an irreducible finite homogeneous Markov chain $\{X_n\}_{n=0}^{\infty}$ with state space $S = \{1, 2, ..., k\}$ and $\mathbf{A} = \mathbf{I} - \mathbf{P}$. Then*

$$\mathbf{Z} = \mathbf{I} + \mathbf{P}\mathbf{A}^{\sharp}.$$

In many cases, it is the diagonal elements of $\mathbf{A}^{\sharp}$ that play a very important role. Therefore, the next theorem which concerns the sign of the diagonal elements of $\mathbf{A}^{\sharp}$is important (see Meyer (1975)).

Theorem 171 *Let an irreducible finite homogeneous Markov chain $\{X_n\}_{n=0}^{\infty}$ with state space $S = \{1, 2, ..., k\}$ and $\mathbf{A} = \mathbf{I} - \mathbf{P}$. Then all the diagonal elements of $\mathbf{A}^{\sharp}$ are positive.*

It is of interest to examine the physical meaning of the elements of $\mathbf{A}^{\sharp}$ for an irreducible aperiodic finite homogeneous Markov chain. In this respect define by:

$R_{ij}^{(t)}$: *the number of times the chain is in state j given that it started in i at time zero.*

Define by $\mathbf{R}^{(t)}$ the matrix with elements $R_{ij}^{(t)}$. Then the following theorem by Meyer (1975) provides a physical meaning for the elements of $\mathbf{A}^{\sharp}$.

Theorem 172 *Let an irreducible, aperiodic finite homogeneous Markov chain $\{X_n\}_{n=0}^{\infty}$ with state space $S = \{1, 2, ..., k\}$ and $\mathbf{A} = \mathbf{I} - \mathbf{P}$. Then*

$$\mathbf{A}^{\sharp} = \lim_{n\to\infty} \left(\mathbb{E}\left(\mathbf{R}^{(t)}\right) - k\boldsymbol{\Pi}\right).$$

For an irreducible finite homogeneous Markov chain denote by:

m_{ij} : *the number of steps before entering state j for the first time given that the chain started in state i.*

Denote by $\mathbf{M}$ the matrix with elements m_{ij}. We call by $\mathbb{E}(\mathbf{M})$ the matrix of the *expected first passage times.* It is proved in Kemeny and Snell (1976) that $\mathbb{E}(\mathbf{M})$ has finite elements given by

$$\mathbb{E}(\mathbf{M}) = (\mathbf{I} - \mathbf{Z} + \mathbf{E}\mathbf{Z}_{dg})\,\mathbf{D},$$

where $\mathbf{E}$ is a matrix of ones, $\mathbf{Z}_{dg} = diag\{\mathbf{Z}\}$ and $\mathbf{D}$ is a diagonal matrix with diagonal elements $1/\pi_i$.

The following theorem in fact shows that actually in the computation of the matrix of the expected first passage times we may replace the matrix $\mathbf{Z}$ with $\mathbf{A}^{\sharp}$. The proof could be found in Meyer (1975) and is easily understood for the reader of the present who arrived up to here graciously.

Theorem 173 *Let an irreducible finite homogeneous Markov chain $\{X_n\}_{n=0}^{\infty}$ with state space $S = \{1, 2, ..., k\}$ and $\mathbf{A} = \mathbf{I} - \mathbf{P}$. Then*

$$\mathbb{E}(\mathbf{M}) = \left(\mathbf{I} - \mathbf{A}^{\sharp} + \mathbf{E}\mathbf{A}^{\sharp}_{dg}\right)\mathbf{D}.$$

2.13 Sensitivity of Markov chains

One of the important applications of the group inverse on the theory of Markov chains is in the study of the sensitivity of the asymptotic probability distribution of an irreducible finite homogeneous Markov chain. The problem of sensitivity is what happens to the asymptotic probability distribution of a Markov chain with transition probability matrix $\mathbf{P}$ if it is perturbed to a matrix $\tilde{\mathbf{P}}$. We start first with a formal definition of perturbation of a stochastic matrix.

Definition 174 *Let an irreducible aperiodic finite homogeneous Markov chain $\{X_n\}_{n=0}^{\infty}$ with state space $S = \{1, 2, ..., k\}$ and transition probability matrix $\mathbf{P}$. Then the matrix $\mathcal{E} \in M_k$ is called a perturbation matrix, if the matrix $\tilde{\mathbf{P}} = \mathbf{P} - \mathcal{E}$ is the transition probability matrix of an irreducible aperiodic finite homogeneous Markov chain.*

Apparently $\|\mathcal{E}\|$ is small relative to 1 for some appropriate norm when the physical meaning of perturbation concerns small errors in the estimation of the transition probabilities. It is known that for an irreducible aperiodic finite homogeneous Markov chain with transition probability matrix $\mathbf{P}$ there exists the asymptotic probability distribution $\boldsymbol{\pi}$ such that

$$\boldsymbol{\pi}\mathbf{P} = \boldsymbol{\pi}, \quad \boldsymbol{\pi}\mathbf{1}^{\top} = 1.$$

Let the perturbed matrix $\tilde{\mathbf{P}} = \mathbf{P} - \mathcal{E}$ which then is the transition probability matrix of another irreducible aperiodic finite homogeneous Markov chain and let $\tilde{\boldsymbol{\pi}}$ its asymptotic probability distribution, that is,

$$\tilde{\boldsymbol{\pi}}\mathbf{P} = \tilde{\boldsymbol{\pi}}, \quad \tilde{\boldsymbol{\pi}}\mathbf{1}^{\top} = 1.$$

The central problem in this area of research is to find the change $\boldsymbol{\pi} - \tilde{\boldsymbol{\pi}}$ in the asymptotic probability distributions for the two Markov chains in terms of the change $\tilde{\mathbf{P}} - \mathbf{P} = \mathcal{E}$ in the transition probability matrices. A large research effort was devoted to this important problem and the group inverse proved to play a central role. In the present we review some of the results which could be placed in a common frame. It was proved by many authors, that the bounds of change on the asymptotic probability distribution of the perturbed Markov chain, of the perturbed Markov chain is given by the general form

$$\|\boldsymbol{\pi} - \tilde{\boldsymbol{\pi}}\| \leq k_l \|\mathcal{E}\|,$$

where the norms are the appropriate ones depending on k_l. The parameter k_l is called the condition number of the Markov chain. Various condition numbers have been proposed by many authors but more or less all were functions of the group inverse of the Markov chain, or the fundamental matrix or the expected first passage matrix. A review and comparison of the various proposed condition numbers could be found in Cho and Meyer (2001). In the present, we will refer to only a few of what is useful for the purposes of the present. Historically the first condition number was proposed by Schweitzer (1968) based on the fundamental matrix of the Markov chain. It was proved that

$$\|\boldsymbol{\pi} - \tilde{\boldsymbol{\pi}}\| \leq \|\mathbf{Z}\| \, \|\mathcal{E}\| ,$$

where convenient norms for the vectors could be the L_1 or the L_∞ norms and for the matrices L_r or the L_∞ norms. Therefore, the condition number in this case was $k_1 = \|\mathbf{Z}\|_\infty$.

The second matrix related with perturbation was the group inverse of the matrix $\mathbf{A} = \mathbf{I} - \mathbf{P}$ by Meyer (1975, 1980) and his school of research. Meyer showed that the group inverse could be used in a similar way as $\mathbf{Z}$ and in fact he showed that

$$\tilde{\boldsymbol{\pi}} = \boldsymbol{\pi} \left(\mathbf{I} + \mathcal{E}\mathbf{A}^\sharp\right)^{-1},$$

and the resulting perturbation is

$$\|\boldsymbol{\pi} - \tilde{\boldsymbol{\pi}}\|_1 \leq \left\|\mathbf{A}^\sharp\right\|_\infty \|\mathcal{E}\|_\infty .$$

Therefore, a second condition number is

$$k_2 = \left\|\mathbf{A}^\sharp\right\|_\infty .$$

Funderlic and Meyer (1986) explores again the potential use of the group inverse of $\mathbf{A} = \mathbf{I} - \mathbf{P}$ for which it is known that

$$\boldsymbol{\pi}\mathbf{A} = \mathbf{0}.$$

They demonstrate that in many common situations it is only necessary to inspect the magnitudes of the transition probabilities in order to be certain that $\boldsymbol{\pi}$ is not sensitive to small perturbations in $\mathbf{P}$. In fact, they go on to prove that if there exists at least one state j which is directly accessible from every other state, and if p_{ij} is not close to zero for $i \neq j$, then $\boldsymbol{\pi}$ is not sensitive to small perturbations in $\mathbf{P}$.

The smallest condition number (Cho and Meyer (2001)) was proposed by Haviv and Van Heyde (1984) and Kirkland et al. (1998). They proved that

$$\|\boldsymbol{\pi} - \tilde{\boldsymbol{\pi}}\|_\infty \leq \frac{\max_j \left(a^\sharp_{jj} - \min_i a^\sharp_{ij}\right)}{2} \|\mathcal{E}\|_\infty .$$

Let $\mathbf{P}$ be a transition probability matrix then we define as the ergodicity coefficient and we denote it by $\tau_1(\mathbf{P})$

$$\tau_1(\mathbf{P}) = \frac{1}{2} \max_{i,j} \sum_{k=1}^{n} |p_{ik} - p_{jk}| .$$

That is, the ergodicity coefficient is a measure of how close the rows of a stochastic matrix are of being identical. In a similar way the ergodicity coefficient of $\mathbf{A}^\sharp = (\mathbf{I} - \mathbf{P})^\sharp$ could be defined. From Seneta (1993) we get that

$$\tau_1\left(\mathbf{A}^\sharp\right) \leq (1 - \tau_1(\mathbf{P}))^{-1} .$$

Seneta (1984, 1988a) used the coefficients of ergodicity $\tau_1(\mathbf{P})$ and $\tau_1(\mathbf{A}^{\#})$ to provide condition numbers

$$\|\boldsymbol{\pi} - \tilde{\boldsymbol{\pi}}\|_1 \leq (1 - \tau_1(\mathbf{P}))^{-1} \|\mathcal{E}\|_\infty \leq \tau_1(\mathbf{A}^{\#}) \|\mathcal{E}\|_\infty .$$

The approach so far on sensitivity of the asymptotic probability distribution under perturbation for a Markov chain was concerned about absolute perturbations in transition probabilities. That is, the size of the perturbation was always measured relative to 1. However, it was established in O'Cinneide (1993) that small relative entry-wise perturbations, that is, a perturbation to a transition probability p_{ij} is measured relative to the size of p_{ij}, do not unduly affect the asymptotic probability distribution of the Markov chain. On the other hand, even small absolute perturbations can drastically affect the asymptotic probability distribution. To deal with the problem Cho and Meyer (2000) introduced into the study of perturbations the matrix of the expected first passage times among the states of a Markov chain. They proved that

$$|\pi_j - \tilde{\pi}_j| \leq \frac{\|\mathcal{E}\|_\infty}{2} \frac{\max_{i \neq j} \mathbb{E}[m_{ij}]}{\mathbb{E}[m_{jj}]},$$

or equivalently

$$\frac{|\pi_j - \tilde{\pi}_j|}{\pi_j} \leq \frac{\|\mathcal{E}\|_\infty}{2} \max_{i \neq j} \mathbb{E}[m_{ij}] .$$

They also proved that if the j column of $\mathbf{P}$ is positive, then there exists a perturbation $\mathcal{E}$, that is, $\tilde{\mathbf{P}} = \mathbf{P} - \mathcal{E}$ such that equality can be attained in the above inequalities for the elements of column j.

We call a state *centrally located in an irreducible* aperiodic finite homogeneous Markov chain $\{X_n\}_{n=0}^{\infty}$ with state space $S = \{1, 2, ..., k\}$ and transition probability matrix $\mathbf{P}$, if every other state is "close" to it in the sense of expected first passage times. The above inequalities guarantees that if a state j is centrally located, then π_j has to be non-sensitive. On the other hand if at least one state is "far removed" from state j, then π_j will exhibit sensitivities.

Chapter 3

NON-HOMOGENEOUS MARKOV CHAINS; WEAK ERGODICITY

3.1 Introductory notes

In this chapter, we will study one of the most important stochastic process the non-homogeneous Markov chain. There are various degrees of generality and rigorousness by which we may define what a stochastic process is. Firstly, the word "stochastic" comes from the Greek and in this context means "random". The same word is used in Greek, maybe in similar context, to express the process of thought when someone is searching all the possible outcomes of a certain phenomenon that has important influence in one's life. Probably this fact is what makes the use of this word so successful as a terminology instead of the word "random process". Rather loosely a stochastic process is a family of random variables. The class of stochastic processes that are distinguished by the Markov property in their structure are called Markov processes. Uncertainty, complexity and dynamism have been continuing challenges to man's understanding and control of physical environment.

In Markov processes the probabilistic dependence on the past is only through the present state which contains all the necessary information for the evolution of the processes. This limited information or memory suffices to produce a great diversity of behaviors. Most probably this is the reason Markov processes have found applications in many diverse domains, including biology, medicine, sociology, stochastic finance, operations research, engineering and other, where they provide qualitative and quantitative answers as well as precious insights for systems design. One would rather say with confidence that in the development of logical structures to describe and control various physical phenomena, the model originated by A.A. Markov stands as a major accomplishment.

The reader would benefit by reading a review of conditional probabilities which could be found in many texts. The author of the present suggests Chapter 3 in Vassiliou (2010) as a feasible solution at hand.

3.2 Stochastic processes

Stochastic processes are probably the most important tools in many areas of science such as biology, operational research, social sciences, stochastic finance and other. Important characteristics in these areas evolve with time in more or less a random way and since stochastic processes are mainly sequences or families of random variables, in which their index represents time, they are the natural tool to use. The theory and applications of stochastic processes has its roots in the genesis of one of the richest one, that is, the Brownian motion. That was rather unexpected since Brownian motion is a beautiful object which is at the same time a martingale, a Gaussian processes, a diffusion, a Levy process, a Markov process etc.; concepts that were discovered quite later in the evolution of time.

Let a complete probability space $(\Omega, \mathcal{F}, \mathbb{P})$ on which a family of random variables $\{X_t\}_{t\in\mathcal{T}}$ over an appropriate index set $t \in \mathcal{T}$ is defined. We call such a family a *stochastic*

DOI: 10.1201/b23204-3

processes. The parameter t usually represents time, but in various applications if required it could be otherwise. If the members of the family are countable, then we denote the stochastic process by $\{X_t\}_{t=0}^{\infty}$, and the stochastic process is said to be in *discrete time.* If the members of the family are not countable, that is, $\mathcal{T} = \mathbb{R}_+ = [0, \infty)$ then we denote the process by $\{X_t\}_{t>0}$ and the stochastic process is said to be in *continuous time.*

We denote by $\mathbb{S}$ the set of all possible values of the stochastic process and we refer to it as the *state space* of the stochastic process. We also refer to each value of the stochastic process as its *state.* If $\mathbb{S} = \{0, 1, 2, ...\}$, then we say that we have a stochastic process with integer values or we have a stochastic process with *discrete state space.* If $\mathbb{S} = (-\infty, \infty)$, then we say that we have a stochastic process with real values or a stochastic process with *continuous state space.* If $\mathbb{S}$ is a subset of $\mathbb{R}^n$, we say that we have an n-dimensional stochastic process. There are cases where $\mathbb{S}$ is a space with some topology or a measurable space in which case we say that we have a stochastic process in *general state space.*

Example 175 *Consider X_t to represent the number of cars in a specific traffic light at time t. Then $\{X_t\}_{t>0}$ is a stochastic process in continuous time, since cars may arrive at any instant of time. It is also a discrete state space stochastic process since the number of cars in a traffic light is always an integer.*

Let a complete probability space $(\Omega, \mathcal{F}, \mathbb{P})$ on which a stochastic process $\{X_t\}_{t\in\mathcal{T}}$ is defined. For any $\omega \in \Omega$, a function $X_t(\omega)$ with respect to t, is called a *sample path or realization* of the process $\{X_t\}_{t\in\mathcal{T}}$. The realization of the process $\{X_t\}_{t\in\mathcal{T}}$ up to time t is $\{X_t(\omega), s \leq t\}$. The time before t, if t is the present, is called a past and the time after t a future.

The roots of the theory of stochastic process are considered to be back in the first ten years after 1900. In March 29, 1900, Bachelier[1] successfully defended at Sorbone his celebrated thesis, *Théorie de la Spéculation*, on the application of probability to stock markets. This was historically the first attempt to introduce stochastic thinking with the use of mathematics and gave the birth of what is now known as Stochastic Finance. Also, it can be viewed as the origin of stochastic calculus, the theory of Brownian motion, Markov processes, diffusion processes and other. What is astonishing, is that it was only in 1930 that the foundation of probability was developed. More on Bachelier his times, life and work could be found in Courtault (2000) and Taqqu (2001). Later was the work of Lundberg (1903), which introduced the theory of stochastic processes in insurance problems, the work of Einstein (1905) on Brownian motion, the work of Erlang (1909) who introduced the Poisson process for queuing problems, and the work of Rutherford and Geiger (1908) in radiation. The Russian mathematician A.A. Markov (1907) introduced the study of Markov processes, with the intension of introducing a first degree of dependency in a series of independent Bernoulli trials. For more on the life and work of A.A. Markov, see Basharin, Longville Naimov (2004). Later, in the present there will be a synopsis of his life and work as a small tribute to the man who founded the theory that was the main cause of the genesis of the present book. Note however that these early studies on stochastic processes were more intuitive than stochastically rigorous. It was after the work of Kolmogorov[2], in the

---

[1]Luis Jean Baptiste Alphonse Bachelier, March 11, 1870–April 28, 1946 was a French mathematician considered to be the forefather of Mathematical Finance and a pioneer in the study of stochastic processes. He laid the foundation for the study of Brownian motion process a stochastic process with amazing properties that shaped nowdays financial markets. He received his doctoral degree from the University of Paris and the main evaluator of his thesis was the famous worlwide mathematician Henri Poincare. Unfortunately, Bachelier and his work were much more appreciated after his death than during his life.

[2]Andrei Kolmogorov (1903–1987) was one of the most eminent mathematicians of the 20th century and the greatest probability theorist of recent times. In 1933, Kolmogorov published his book *Foundations of the Theory of Probability,* presenting the axiomatic foundations of probability theory. To this day, it has stood the test of time and constitutes the cornerstone around which the entire theory of probability and statistics flourished. His entire life was one of intense mathematical creativity spanning probability and stochastic

decade 1930–1940 that the study of stochastic process became the most mathematically and stochastically rigorous subject together with some parts of pure mathematics. (see Cramer (1976)).

3.3 Markov chain

In this chapter, we will study discrete time stochastic processes with discrete state space that their structure has the Markov property. Such stochastic processes are called *Markov chains.* Let a complete probability space $(\Omega, \mathcal{F}, \mathbb{P})$ on which a stochastic process $\{X_t\}_{t=0}^{\infty}$ is defined and let $\mathbb{S} = \{1, 2, ...\}$ be its state space which is countable. Then the Markov property is the following:

Markov Property

Definition 176 *We say that a stochastic process $\{X_t\}_{t=0}^{\infty}$ on a countable state space $\mathbb{S} = \{1, 2, ...\}$ has the Markov property if for any function $f : \mathbb{S} \to \mathbb{R}$ and for any $i, j \in \mathbb{S}$ and $t \geq 0$,*

$$\mathbb{E}_{\mathbb{P}}[f(X_{t+1}) \mid X_0, X_1, ..., X_t] = \mathbb{E}_{\mathbb{P}}[f(X_{t+1}) \mid X_t]. \tag{3.1}$$

or equivalently

$$\mathbb{P}(X_{t+1} = j \mid X_0 = i_0, X_1 = i_1, ..., X_t = i) = \mathbb{P}(X_{t+1} = j \mid X_t = i). \tag{3.2}$$

Conditions (3.1) and (3.2) says that, at any time t, the next state X_{t+1} is conditionally independent of the past $X_0, X_1, ..., X_{t-1}$ given the present state X_t. In other words, the future state is dependent on the present state and not on the more distant past states. Note that, in a later chapter we will study Markov chains for which the future state is dependent on a more distant past than the present. We call such Markov chains *high order Markov chains.*

We denote by

$$p_{ij}(t+1) = \mathbb{P}(X_{t+1} = j \mid X_t = i), \tag{3.3}$$

and call it the *transition probability* of going from state i at time t to the state j at time $t+1$. We collect the transition probabilities in the form of a matrix $\mathbf{P}(t) = \{p_{ij}(t)\}_{i,j \in \mathbb{S}}$, which we call the *transition probability matrix* or sometimes *the transition matrix* of the Markov chain $\{X_t\}_{t=0}^{\infty}$ in the time interval $[t-1, t)$. We call *Markov chains* for which the transition probabilities depend on the time of transition *non-homogeneous Markov chains* or sometimes *inhomogeneous Markov chains.* In addition, if the state space is finite we call them *finite non-homogeneous Markov chains.*

Note that every sequence $\{\mathbf{P}(t)\}_{t=0}^{\infty}$ of transition probability matrices defines uniquely a non-homogeneous Markov chain. The characteristic property of each member of the sequence $\{\mathbf{P}(t)\}_{t=0}^{\infty}$ is that is a non-negative matrix and that the sum of all its rows is equal to one. That is, the sequence $\{\mathbf{P}(t)\}_{t=0}^{\infty}$ is a sequence of stochastic matrices. In applications each element of $\mathbf{P}(t)$ for each t is estimated separately, which reduces significantly the amount of the least number of data needed to estimate the transition probabilities of a non-homogeneous Markov chain.

---

process, topology, intuitionistic logic, turbulence, classical mechanics, algorithmic information theory and computational complexity. An entire collection of limit theorems for sums of independent random variables were beautifully presented in his book with B.V. Gnedenko: *Limit Distributions for Sums of Independent Random variables* published in Russian in 1949 and translated in 1954. More on Kolmogorov's contributions in probability and stochastic process written in a compact way without missing important parts of the entire picture could be found in Parathasarathy (1988). He was an outdoor mathematician, to whom key ideas occurred while walking in the woods, swimming in the sea or in a lake, or skiing down the slopes of a mountain. The interaction between probability theory, measure theory, topology and classical mechanics constituted one of the most passionate pursuits in the life of Kolmogorov. His greatest achievement in this direction is his initiation of the so-called Kolmogorov-Arnold-Moser (KAM) theory which solved the long-standing problem of the stability of the solar system proposed by Henri Poincare.

3.3.1 A more advanced definition of a non-homogeneous Markov chain

In the present subsection, we will define a non-homogeneous Markov chain more generally with the use of σ-algebras and filtrations. The readers that they do not feel yet at home with these concepts they could probably read this subsection, the first time, superficially.

Let $\{X_t\}_{t=0}^{\infty}$ be a sequence of random variables on $(\Omega, \mathcal{G}, \mathbb{P})$ with values in $\mathbb{S} = \{1, 2, ...\}$, and let $\mathbb{F}^X$ be the natural filtration or history generated by the process $\{X_t\}_{t=0}^{\infty}$, that is, $\mathcal{F}_t^X = \sigma(X_s : s = 0, ..., t)$. We denote $\mathbb{G}$ to be some filtration in $(\Omega, \mathcal{G}, \mathbb{P})$ and we assume that $\mathbb{F}^X$is a subfiltration of $\mathbb{G} : \mathbb{F}^X \subseteq \mathbb{G}$. We now define

Definition 177 *The stochastic process $\{X_t\}_{t=0}^{\infty}$ is a discrete-time inhomogeneous Markov chain under $\mathbb{P}$ with respect to $\mathbb{G}$ or briefly a $\mathbb{G}$-Markov chain, if for any bounded and measurable function $f : \mathbb{S} \to \mathbb{R}$ we have*

$$\mathbb{E}_{\mathbb{P}}[f(X_{t+1}) \mid \mathcal{G}_t] = \mathbb{E}_{\mathbb{P}}[f(X_{t+1}) \mid \sigma(X_t)] \text{ for every } t. \tag{3.4}$$

Note that if $\{X_t\}_{t=0}^{\infty}$ is a Markov chain and $\mathcal{F}_t^X = \sigma(X_s : s = 0, ...t)$ is its history then it always is an $\mathbb{F}^X$-Markov chain. Equivalently, for any bounded and measurable function $f : \mathbb{S} \to \mathbb{R}$ and for all $0 \leq s \leq t$ we have

$$\mathbb{E}_{\mathbb{P}}\left[f(X_t) \mid \mathcal{F}_s^X\right] = \mathbb{E}_{\mathbb{P}}[f(X_t) \mid \sigma(X_s)] \text{ for every } t.$$

Another equivalent way for the definition of an $\mathbb{F}^X$-Markov chain under the probability measure $\mathbb{P}$ is the following:

Definition 178 *Let $\{X_t\}_{t=0}^{\infty}$ be a stochastic process defined on the complete probability space $(\Omega, \mathcal{F}, \mathbb{P})$, with state space $\mathbb{S} = \{1, 2, ...\}$ and adapted to its history $\mathcal{F}_t^X = \sigma(X_s : s = 0, ...t)$. It is called an $\mathbb{F}^X$-Markov chain under the probability measure $\mathbb{P}$ iff for all $t \geq 0$*

$$\sigma(X_s, s \geq t) \text{ and } \mathcal{F}_t^X \text{ are } \mathbb{P}\text{-independent given } X_t.$$

From the above definitions it is evident the difference in generality between a $\mathbb{G}$-Markov chain and an $\mathbb{F}^X$-Markov chain with $\mathbb{F}^X \subseteq \mathbb{G}$. That is, if $\{X_t\}_{t=0}^{\infty}$ is a $\mathbb{G}$-Markov chain under $\mathbb{P}$ then, of course it is also an $\mathbb{F}^X$-Markov chain under $\mathbb{P}$. However, the converse implication is not true, in general. When the state space $\mathbb{S}$ is finite, then condition (3.4) is equivalent to the following one: for every $t = 0, 1, ...$ and any $j \in \mathbb{S}$ we have

$$\mathbb{P}\{X_{t+1} = j \mid \mathcal{G}_t\} = \mathbb{P}\{X_{t+1} = j \mid \sigma(X_t)\}.$$

Especially, in the case of non-homogeneous Markov chains the concept of a $\mathbb{G}$-Markov chain is quite useful since the probabilities $p_{ij}(t)$ are functions of various parameters of the environment within which the phenomenon modeled as a non-homogeneous Markov chain takes place.

Some important stochastic precesses such as Brownian motion, Levy processes and others are characterized by the properties of their increments. Hence, at this point it is useful to state that roughly speaking a stochastic process $\{X_t\}_{t=0}^{\infty}$ is a Markov chain if its increments $X_{t+s} - X_t$ depend upon the past $\mathcal{F}_t^X$ only through X_t.

We will now make a small parenthesis by presenting a small section on the life and work of A.A. Markov. This section certainly is not a complete account of the work of A.A. Markov and it is not even a sufficient synopsis. However, it serves to provide some of the flavor of the era, the man's values in respect to his work and as a member of the society, to pinpoint sources where more extensive studies for Markov's life and work could be found and most of all it is an effort successful or not to pay a small tribute to this great probabilist.

3.4 The life and work of A.A. Markov

The main sources on the life and work of A.A. Markov are written in Russian. There are however some important sources in English, which where the main sources of information for what follows. These are: Basharin et al. (2004), Seneta (1982), Seneta (1996), Sheynin (1988) and Ondar (1981).

Andrey Andreevich Markov was born in the town of Ryazan, on June 14, 1856 and he died after an operation on his knee which was fatal; on July 20,1922 he died of sepsis.

Markov entered in 1874 the Faculty of Mechanics and Mathematics of St. Petesburg University, where Chebychev was a prominent figure. In 1877, he was awarded a gold metal for his research on the topic, proposed by the faculty, "On Solution of Differential equations with the help of continuous fractions". In 1884, he defended his Doctorate thesis, "On Certain Applications of the Algebraic Continuous Fractions" which was written under the supervision of Chebychev. At the age of 30, he became a professor at St. Petesberg University and a member of the St. Petesberg Academy of Sciences. He wrote a book which was classical by the title "Calculus of Probabilities", which was published four times in Russian and was translated into German. He published more than 120 scientific papers on number theory, continuous fraction theory, differential equations, probability theory, and statistics. Many of his papers created a new field of research, Markov chains. As a teacher Markov provided extensive and detailed numerical examples for all the concepts and theories that he taught. He believed that Mathematics to a considerable extent consists in solving problems and together with the proper discussion, he provided his students with many exercises.

Markov was very active in political issues and expressed openly his views with rebel tendencies for his era which were published to newspapers. Markov opposed honorary membership for the Academy of St. Petersburg for Royals. Also, refused to accept tsarist awards in protest against the exclusion of the writer A.M. Gorgy from the Academy. In 1912 in a letter, which is consider by some rather humorous, asked the Holy Synod to be excommunicated from the Russian Orthodox Church. This was his response to the excommunication of Tolstoy. The joke of the situation is that they excommunicated him. Many of the nowdays community of scientists will look at these attributes of Markov's character as very positive and admirable. Also, many will express scepticism to the fact that later in life by a letter Markov made accusations for his mentor Chebyshev.

In 1898 P.A. Nekrasov originally a theologian by training who later turned into mathematics and became professor at Moscow University published a paper. The title of the paper is "General properties of numerous independent events in connection with approximate calculation of functions of very large numbers". Later on in 1902 in a related paper Nekrasov erroneously claimed that "independence is a necessary condition for the law of large numbers".

Markov a very rigorous mathematician spotted the errors and began studying types of dependent random variables. Markov studied the Law of Large Numbers and Central Limit theorems for these type of dependent families of random variables and that gave the genesis of the Theory of Markov chains. Naturally, a war was on between the two adversaries and it is rather interesting to read a letter of Markov to Chuprov Basharin et al (2004 p. 15) in this respect. In there he states: "In this way a construction of highly general character was actually arrived at, which P.A. Nekrasov can not even dream about". That was actually the birth of Markov chains.

After the development of the Markov chain, with the advantage of hindsight, we can identify earlier ideas that are connected to the later idea of a Markov chain. The gamblers ruin problem (1654), various urn models 1769 and 1812, simple branching processes 1873, hydrology applications in 1846 et al. For more details see Seneta (1996) and Basharin et al (2004). However, most of these earlier studies were either unknown or unrecognized by Markov as his correspondence with long-time friend and fellow mathematician Chuprov shows.

In his many papers that followed Markov proved that the independence of random variables was not a necessary condition for the validity of the weak law of large numbers and the central limit theorem. He developed a large part of what is now known the Theory of Markov chains. His theory was developed further by a very large number of researchers all around the globe up to the present days, creating an amazing instrument for analyzing, predict and control various phenomena of the world.

3.5 Probability distribution in the states of an NHMC

Let a non-homogeneous Markov chain $\{X_t\}_{t=0}^{\infty}$ on a complete probability space $(\Omega, \mathcal{F}, \mathbb{P})$ with a finite state space $\mathbb{S} = \{1, 2, ..., k\}$. We define as the initial probability distribution and we denote it by $\mathbf{p}_0(0)$ the row vector of the probabilities

$$\mathbf{p}(0) = [p_1(0), p_2(0), ..., p_k(0)],$$

where for any state $j \in \mathbb{S}$, $p_j(0) = \mathbb{P}(X_0 = j)$. It is apparent that

$$\sum_{i=1}^{k} p_i(0) = 1$$

We also define as the probability distribution at time t of $\{X_t\}_{t=0}^{\infty}$ given the initial distribution $\mathbf{p}(0)$ and we denote it by $\mathbf{p}(0, t)$ the row vector of probabilities

$$\mathbf{p}(0, t) = [p_1(t \mid \mathbf{p}(0)), p_2(t \mid \mathbf{p}(0)), ..., p_k(t \mid \mathbf{p}(0))],$$

where for any state $j \in \mathbb{S}$, $p_j(t \mid \mathbf{p}(0)) = \mathbb{P}(X_t = j \mid \mathbf{p}(0))$. It is apparent that

$$\sum_{j=1}^{k} p_j(t \mid \mathbf{p}(0)) = 1$$

In the case that the probability distribution of $\{X_t\}_{t=0}^{\infty}$ is known at some time $s < t$, that is, the $\mathbf{p}(s)$ is known where

$$\mathbf{p}(s) = [p_1(s), p_2(s), ..., p_k(s)],$$

where for any state $j \in \mathbb{S}$, $p_j(s) = \mathbb{P}(X_s = j)$. Then we define as the probability distribution at time t of $\{X_t\}_{t=0}^{\infty}$, given the initial distribution $\mathbf{p}(s)$ and we denote it by $\mathbf{p}(s, t)$ the row vector of probabilities

$$\mathbf{p}(s, t) = [p_1(t \mid \mathbf{p}(s)), p_2(t \mid \mathbf{p}(s)), ..., p_k(t \mid \mathbf{p}(s))].$$

Note that when the state space $\mathbb{S} = \{1, 2, ...\}$ is countably infinite the above definitions and concepts are completely analogous. The following theorem shows that for the calculation of the probability distribution in the states of a non-homogeneous Markov chain at time t, it is sufficient to know the initial probability distribution and the sequence of transition probabilities matrices $\{\mathbf{P}(t)\}_{t=0}^{\infty}$. In this respect define by

$$\mathbf{P}(s, t) = \mathbf{P}(s+1)\mathbf{P}(s+2)...\mathbf{P}(t) \quad \text{for} \quad s < t.$$

Note that $\mathbf{P}(s,s) = \mathbf{I}$ and the entries $p_{ij}(s,t)$ of $\mathbf{P}(s,t)$ have the probabilistic interpretation that if $\mathbb{P}(X(s) = i) > 0$ then

$$p_{ij}(s,t) = \mathbb{P}(X(t) = j \mid X(s) = i).$$

Theorem 179 *Consider a non-homogeneous Markov chain $\{X_t\}_{t=0}^{\infty}$ on a complete probability space $(\Omega, \mathcal{F}, \mathbb{P})$ with a finite state space $\mathbb{S} = \{1, 2, ..., k\}$ with initial probability distribution $\mathbf{p}(0)$ and the sequence of transition probability matrices $\{\mathbf{P}(t)\}_{t=0}^{\infty}$. Then the following relations are true*

$$(i)\ \ \mathbf{p}(0,t) = \mathbf{p}(0)\mathbf{P}(0,t).$$

$$(ii)\ \ \mathbf{p}(s,t) = \mathbf{p}(s)\mathbf{P}(s,t).$$

$$(iii)\ \ p_{ij}(0,s+t) = \sum_{r=1}^{k} p_{rj}(s,s+t)\,p_{ir}(0,s).$$

or

$$(iv)\ \ \mathbf{P}(0,s+t) = \mathbf{P}(0,s)\mathbf{P}(s,s+t).$$

Proof. Let us consider the probability $p_j(0,t)$ and try to find a functional relationship with the probabilities $p_i(0,t-1)$ for all $i \in \mathbb{S}$. In Table 3.1 an analysis is provided of all possible paths, with which the Markov chain is possible to arrive at state j at time t given that it starts at any state at time 0.

In Table 3.1 in the first column are given all the possible states that the Markov chain is possible to occupy at time $t-1$. In the time interval $(t-1,t]$ the Markov chain makes the necessary transition to be at state j at time t. In the fourth column, we have the probability of each possible path. For example, the first path is for the Markov chain to be in state 1 at time $t-1$, which has probability $p_1(0,t-1)$ and then at the time interval $(t-1,t]$ to make a transition to state j which has probability $p_{1j}(t-1)$.

Table 3.1

Time $t-1$	Step $(t-1,t]$	Time t	Probability
1	$\rightarrow$	j	$p_1(0,t-1)\,p_{1j}(t)$
2	$\rightarrow$	j	$p_2(0,t-1)\,p_{2j}(t)$
....	.....	.....	......
k	$\rightarrow$	j	$p_k(0,t-1)\,p_{1j}(t)$

Due to the Markov property, we immediately get that the two parts of the path to be followed are independent events, and hence, the probability of following the entire path is $p_1(0,t-1)\,p_{1j}(t)$. Since all the paths in Table 3.1 constitute mutually exclusive events and at the same time are all the possible ones, we obtain

$$p_j(0,t) = p_1(0,t-1)\,p_{1j}(t) + p_2(0,t-1)\,p_{2j}(t) + ... + p_k(0,t-1)\,p_{kj}(t). \quad (3.5)$$

for all $j = 1, 2, ..., k$. Now, in matrix form, equation (3.5) could be written as

$$\mathbf{p}(0,t) = \mathbf{p}(0,t-1)\mathbf{P}(t). \quad (3.6)$$

Applying (3.6) recursively we obtain

$$\mathbf{p}(0,t) = \mathbf{p}(0,t-2)\mathbf{P}(t-1)\mathbf{P}(t) = \mathbf{p}(0,t-2)\mathbf{P}(t-2,t)$$
$$= ... = \mathbf{p}(0)\mathbf{P}(0,t). \tag{3.7}$$

In a completely similar way we could prove that

$$\mathbf{p}(s,t) = \mathbf{p}(s)\mathbf{P}(s,t). \tag{3.8}$$

What remains is to prove (iii). In this respect, we have

$$p_{ij}(0,s+t) = \mathbb{P}(X_{s+t} = j \mid X_0 = i)$$
$$= \sum_{r=1}^{k} \mathbb{P}(X_{s+t} = j, X_s = r \mid X_0 = i)$$
$$= \sum_{r=1}^{k} \mathbb{P}(X_{s+t} = j \mid X_s = r, X_0 = i)\mathbb{P}(X_s = r \mid X_0 = i)$$
$$= \text{(due to the Markov property)}$$
$$= \sum_{r=1}^{k} \mathbb{P}(X_{s+t} = j \mid X_s = r)\mathbb{P}(X_s = r \mid X_0 = i)$$
$$= \sum_{r=1}^{k} p_{rj}(s+t,s)\,p_{ir}(0,s).$$

Apparently,

$$\mathbf{P}(0,s+t) = \mathbf{P}(0,s)\mathbf{P}(s,s+t).$$

■

3.6 Examples

In what follows we will present some examples which are useful in clarifying the above concepts.

Example 180 *Non-homogeneous random walk.*

Let $\{X_t\}_{t=0}^{\infty}$ be a family of random variables with values in the set $\mathbb{S} = \{0,1,2,...,b\}$. Let also $\{Z_n\}_{n\geq 0}$ be a sequence of i.i.d. random variables, independent of $\{X_t\}_{t=0}^{\infty}$ taking the values of $+1, 0, -1$ and with probability distribution

$$\mathbb{P}(Z_n = +1 \mid X_{n-1} = i) = p(n), \quad \mathbb{P}(Z_n = -1 \mid X_{n-1} = i) = q(n),$$
$$\text{and} \quad \mathbb{P}(Z_n = 0 \mid X_{n-1} = i) = 1 - p(n) - q(n),$$

for $i = 1, 2, ..., b-1$ and $n = 1, 2, ...$ Also

$$\mathbb{P}(Z_n = +1 \mid X_{n-1} = 0) = 0, \quad \mathbb{P}(Z_n = -1 \mid X_{n-1} = 0) = 0,$$
$$\text{and} \quad \mathbb{P}(Z_n = 0 \mid X_{n-1} = 0) = 1,$$

and

$$\mathbb{P}(Z_n = +1 \mid X_{n-1} = b) = 0, \quad \mathbb{P}(Z_n = -1 \mid X_{n-1} = b) = 0\ ,$$
$$\text{and} \quad \mathbb{P}(Z_n = 0 \mid X_{n-1} = b) = 1,$$

for $n = 1, 2, ...$ The family of random variables defined by

$$X_{n+1} = X_n + Z_{n+1},$$

is called a *non-homogeneous random walk* on $\mathbb{S}$ with two absorbing barriers in states 0 and b. It is apparent that the process $\{X_t\}_{t=0}^{\infty}$ has the Markov property its state space is $\mathbb{S} = \{0, 1, ..., b\}$ and that the matrix of transition probabilities are

$$\mathbf{P}(n) = \begin{pmatrix} 1 & 0 & ... & 0 & 0 \\ q(n) & 1 - p(n) - q(n) & ... & 0 & 0 \\ ... & ..... & ... & ... & ... \\ 0 & 0 & ... & 1 - p(n) - q(n) & p(n) \\ 0 & 0 & ... & 0 & 1 \end{pmatrix}.$$

Let that the non-homogeneous random walk starts at state a where $0 < a < b$ then

$$\mathbf{p}(0) = (0, ..., 1, ...b)\ ,$$

where the 1 is in the ath position. Of interest are the probabilities of absorption at a certain time, the probabilities of being in a certain nonabsorbing state at time t given that absorption did not appear up to time t, the probability of ultimate absorption etc. The study of non-homogeneous random walks is done with methods especially devised for random walks or with general methods devised for absorbing non-homogeneous Markov chains.

Example 181 *Let* $\{X_t\}_{t=0}^{\infty}$ *be a non-homogeneous Markov chain with state space* $\mathbb{S} = \{1, 2, 3\}$ *and sequence of transition matrices*

$$\mathbf{P}(t) = \begin{pmatrix} 0 & 0 & 0.3 + \frac{1}{10t} & 0.7 - \frac{1}{10t} \\ 0 & 0 & 0.4 - \frac{1}{t^2+8} & 0.6 + \frac{1}{t^2+8} \\ 0.2 + \frac{1}{t^3+3} & 0.8 - \frac{1}{t^3+3} & 0 & 0 \\ 0.1 + \frac{3}{t^4+11} & 0.9 - \frac{3}{t^4+11} & 0 & 0 \end{pmatrix}.$$

Also let

$$\mathbf{p}(0) = (0.3, 0.2, 0.6, 0.1)$$

be the initial probability distribution. Then we have

$$\mathbf{P}(1) = \begin{pmatrix} 0 & 0 & 0.4 & 0.6 \\ 0 & 0 & 0.28888 & 0.71112 \\ 0.53333 & 0.46667 & 0 & 0 \\ 0.325 & 0.675 & 0 & 0 \end{pmatrix},$$

$$\mathbf{P}(2) = \begin{pmatrix} 0 & 0 & 0.35 & 0.65 \\ 0 & 0 & 0.31667 & 0.68333 \\ 0.2909 & 0.7091 & 0 & 0 \\ 0.21111 & 0.78889 & 0 & 0 \end{pmatrix},$$

$$\mathbf{P}(3) = \begin{pmatrix} 0 & 0 & 0.33333 & 0.66667 \\ 0 & 0 & 0.37142 & 0.62857 \\ 0.23333 & 0.76667 & 0 & 0 \\ 0.1326 & 0.8674 & 0 & 0 \end{pmatrix}.$$

From Theorem 179, we get that

$$\mathbf{p}(0,3) = \mathbf{p}(0)\,\mathbf{P}(0,3) = \mathbf{p}(0)\,\mathbf{P}(1)\,\mathbf{P}(2)\,\mathbf{P}(3)$$

$$= \begin{pmatrix} 0.116\,33 & 0.583\,67 & 0.181\,15 & 0.318\,85 \end{pmatrix}.$$

This example will be used further in illustrating new concepts.

Example 182 *Let $\{X_t\}_{t=0}^{\infty}$ be a non-homogeneous Markov chain with state space $\mathbb{S} = \{1,2,3\}$ and sequence of transition matrices*

$$\mathbf{P}(t) = \begin{pmatrix} 0.2 + \frac{2}{8t^3+4} & 0.4 - \frac{3}{8t^3+4} & 0.4 + \frac{1}{8t^3+4} \\ 0.1 + \frac{2}{10t^2} & 0.7 - \frac{3}{10t^2} & 0.2 + \frac{1}{10t^2} \\ 0.8 - \frac{4}{12t^2+8} & 0.1 + \frac{3}{12t^2+8} & 0.1 + \frac{1}{12t^2+8} \end{pmatrix}$$

Also let

$$\mathbf{p}(0) = (0.2, 0.3, 0.4, 0.1)$$

be the initial probability distribution. Then we have

$$\mathbf{P}(1) = \begin{pmatrix} 0.36666 & 0.15 & 0.48334 \\ 0.3 & 0.4 & 0.3 \\ 0.6 & 0.15 & 0.25 \end{pmatrix},$$

$$\mathbf{P}(2) = \begin{pmatrix} 0.22941 & 0.35588 & 0.41470 \\ 0.15 & 0.625 & 0.225 \\ 0.72857 & 0.15357 & 0.11786 \end{pmatrix},$$

$$\mathbf{P}(3) = \begin{pmatrix} 0.21818 & 0.38636 & 0.39546 \\ 0.12222 & 0.66666 & 0.21111 \\ 0.76551 & 0.12586 & 0.10863 \end{pmatrix}.$$

From Theorem 179, we get that

$$\mathbf{p}(0,3) = \mathbf{p}(0)\,\mathbf{P}(0,3) = \mathbf{p}(0)\,\mathbf{P}(1)\,\mathbf{P}(2)\,\mathbf{P}(3)$$

$$= \begin{pmatrix} 0.362\,20 & 0.461\,76 & 0.276\,04 \end{pmatrix}.$$

This example will be used further in illustrating new concepts.

3.6.1 A more advanced result on the relation between a $\mathbb{G}$-non-homogeneous Markov chain and martingales

In the present subsection we will establish a relation between a $\mathbb{G}$-non-homogeneous Markov chain and Martingales more generally with the use of σ-algebras and filtrations. The readers that they do not feel yet at home with these concepts they could probably read this subsection, the first time, superficially.

The roots of the theory of martingales goes back to the study of gambling. Actually, the term martingale comes from an old strategy in gambling from around 1815, where when playing a game of luck, each time the player lost his bet, he doubled, the money he just lost until he won. The reader is referred to chapter 5 in Vassiliou (2010) for an introduction on discrete time Martingales with many examples. Although, in the present subsection we will only need a formal definition of a discrete time martingale which we will provide. The word martingale outside the stochastic context is a part of a horse's harness, which goes into his mouth and stops the horse from throwing its head up. Martingales have been extensively studied since 1934 by Paul Levy[3]. Also Doob[4] made very lasting contributions from 1940 to the theory of martingales (see Bingham (2005)) and created links between martingale theory and mathematical analysis known as potential theory Doob (1981). More on his life and work could be found in Snell (1997, 2005). A classic book on martingales is by Doob (1990). Other very good texts are those by Karlin and Taylor (1975), Bremaud (1981), Kopp (1984), Williams (2000), Ross (1996), Rolski, Schmidili, Schmidt and Teugels (1999), Williams (2004), Grimmett and Stirzaker (2001) and Prabbu (2007). If one wishes to do research on the theory of martingales on a very beautiful advanced mathematical level, the two volumes by Rogers and Williams Vol 1,2 (2000) are indispensable.

We will now introduce some notation. For any $i, j \in \mathbb{S}$ we denote by

$$H_t^i = \mathbf{1}_{\{X_t=i\}} \text{ and } H_t^{ij} = \sum_{r=1}^{t} H_{r-1}^i H_t^j.$$

Note that, H_t^{ij} represents the number of one step transitions of the process $\{X_t\}$ from the state i to the state j that occur between time 0 and time t. Let the Kronecker constant δ then we have

$$\hat{H}_t^{ij} = H_t^{ij} - \delta_{ij} \sum_{r=1}^{t} H_{r-1}^i,$$

and

$$M_t^{ij} = \hat{H}_t^{ij} - \sum_{r=1}^{t} (p_{ij} - \delta_{ij}) H_{r-1}^i = H_t^{ij} - \sum_{r=1}^{t} p_{ij}(t) H_{r-1}^i$$

We will now prove the following theorem.

Theorem 183 *Let $(\Omega, \mathcal{G}, \mathbb{P})$ be a complete probability space and $\{X_t\}$ be a $\mathbb{G}$-non-homogeneous Markov chain on it. Then for $i, j \in \mathbb{S}$, the process $\left\{M_t^{ij}\right\}$ follows a $\mathbb{G}$-martingale under $\mathbb{P}$.*

Proof. Let that $i \neq j$ then we have that

$$\mathbb{E}_{\mathbb{P}}\left(M_{t+1}^{ij} - M_t^{ij} \mid \mathcal{G}_t\right)$$

$$= \mathbb{E}_{\mathbb{P}}\left(H_{t+1}^{ij} - \sum_{r=1}^{t+1} p_{ij}(t) H_{r-1}^i - H_t^{ij} - \sum_{r=1}^{t} p_{ij}(t) H_{r-1}^i \mid \mathcal{G}_t\right)$$

$$= \mathbb{E}_{\mathbb{P}}\left(\sum_{r=1}^{t+1} H_{r-1}^i H_t^j - \sum_{r=1}^{t} H_{r-1}^i H_t^j. - p_{ij}(t+1) H_t^i \mid \mathcal{G}_t\right)$$

$$= \mathbb{E}_{\mathbb{P}}\left(H_t^i H_{t+1}^j - p_{ij}(t+1) H_t^i \mid \mathcal{G}_t\right) = \mathbb{E}_{\mathbb{P}}\left[H_t^i \left(H_{t+1}^j - p_{ij}(t+1)\right) \mid X_t\right]$$

[3]Paule Levy was born in Paris on 15 September 1886 where he died on 15 December 1971. Levy attended the Ecole Polythechnique and in 1905 he published his first paper. He became Professor at the Ecole des Mines in 1913. In 1920, he was appointed Professor of Analysis at the Ecole Polytechnique where he remained until his retirment in 1959 with a small gap during the World War II because he was fired under the Vichy Statute for Jews. He is very well known among other research findings for the Levy Process, Levy flight, Levy measure, Levy constant, Levy distribution. He had very notable research students.

[4]Joseph Leonard Doob was born in Cincinnati on February 27th 1910 and died on June 4 2004 in Urbana Illinois. Doob was a mathematical analyst who turned to Probability theory to survive the Great Depression. In 1933, Kolmogorov provided the first axiomatic foundation for the theory of probability and Doob worked mainly to turn Probability Theory into almost a pure mathematics subject. He is famous for many things among which the foundation and many seminal theorems in martingales. His book on stochastic process is a classic. Aso his books on the measure theory and his last work on the link between martingales and potential theory are among the everlasting valuable books that he left to the Probabilist community.

$$= \left[\mathbb{P}\left(H_{t+1}^{j} \mid X_t = i \right) - p_{ij}\left(t+1\right) \right] = 0.$$

■

3.7 Weak and strong ergodicity

The problem of asymptotic behavior of non-homogeneous Markov chains is one of the most important problems as is in the case of homogeneous Markov chains (Vassiliou 2010, chapter 3). One important difference is that we have two types of asymptotic behaviors the *weak ergodicity and the strong ergodicity.* Let $\{\mathbf{P}\left(t\right)\}_{t=0}^{\infty}$ be the sequence of transition probabilities of the non-homogeneous Markov chain $\{X_t\}_{t=0}^{\infty}$ and assume a finite state space $S = \{1, 2, ..., k\}$ for the remaining of the chapter. One is concerned mainly with the asymptotic behavior of the *forward products*

$$\lim_{n\to\infty} \mathbf{P}\left(t, t+n\right) = \lim_{n\to\infty} \mathbf{P}\left(t+1\right) \mathbf{P}\left(t+2\right) ... \mathbf{P}\left(t+n\right),$$

and its dependence on t.

The concept of *weak ergodicity of* $\{X_t\}_{t=0}^{\infty}$ introduced by Kolmogorov in 1931, requires that for all i, j, r, t

$$p_{ir}\left(t, t+n\right) - p_{jr}\left(t, t+n\right) \to 0 \text{ as } n \to \infty. \tag{3.9}$$

Note that in (3.9) $\lim_{n\to\infty} p_{ir}\left(t, t+n\right)$ is not actually necessary to exist.

If weak ergodicity holds, and the $p_{ij}\left(t, t+n\right)$ themselves tend to a limit for all i, j, t as $n \to \infty$, then *strong ergodicity* holds.

The proposed definition is not only essential from the pure mathematics side for the foundation of the necessary and sufficient conditions for ergodicity weak and strong. It also has a strong physical meaning which clarifies important aspects of ergodicity from the point of view of applications. An ergodic chain, as here defined, has the property that knowledge of its state at any point of time has no influence on the state it will be in a sufficiently distant future. In other words, an ergodic chain is one which "forgets" the distant past and not only the initial state.

An early idea inherent in the work of Markov that of a finite stochastic matrix with an entirely positive column, called Markov matrices in the Russian language, triggered the work on the theory of both weak and strong ergodicity, as could be seen in the textbook of Bernstein (1946). More on the historical development of the theory of finite inhomogeneous Markov chains could be found in Seneta (1973). The first necessary and sufficient condition for weak ergodicity is ascribed in Hajnal (1956, 1958) and in parallel in the Russian school of probability by Sarymsakov (1956, 1958) and his students. In this respect an interesting paper is that of Seneta (2014).

Closely related with the study of weak ergodicity is the concept of a coefficient of ergodicity. This notion was partly crystallized in a recently discovered paper of Doeblin in 1937, but was put into its powerful form and was exploited in the theory of finite Markov chains by Dobrushin (1956) and Hajnal (1976).

There is a large amount of work done on the structure and properties of various coefficients of ergodicity and it is clear that the utility of a choice of a particular coefficient of ergodicity depends heavily on knowing its form as a function of the elements of $\mathbf{P}$.

However, as it will be evident in the next section in relation to weak ergodicity a very important part is played by the Dobrushin (1956) coefficient of ergodicity which in terms of the elements of $\mathbf{P}$ has the following form

$$\tau_1\left(\mathbf{P}\right) = \frac{1}{2} \max_{i,j} \sum_{r=1}^{n} \left|p_{ir} - p_{jr}\right| = 1 - \min_{i,j} \sum_{r=1}^{n} \min\left(p_{ir}, p_{jr}\right). \tag{3.10}$$

3.8 Structures for coefficients of ergodicity

Coefficients of ergodicity have proved to be valuable instruments for the study of both homogeneous and inhomogeneous Markov chains. There is a large number of coefficients of ergodicity which is rather due to their applications to different problems, the large diversity of research schools that studied them, and probably the fact that they have some properties which are not common to all of them and which are useful to certain problems. It has also been proved that they can be taken as upper bounds for the modulus of the largest non-unit eigenvalue of stochastic matrices. Hence, providing bounds for the rate of convergence of homogeneous and inhomogeneous Markov chains.

The reasons in the above paragraph caused, unfortunately, a certain ambiguity as to the notion coefficient of ergodicity itself (Lešanovský 1990). The following are the main definitions that have appeared:

Definition 184 *Seneta (1981). Any scalar function $\tau(.)$ continuous on the set of $(n \times n)$ stochastic matrices (treated as points in $\mathbb{R}^{n^2}$) and $0 \leq \tau(\mathbf{P}) \leq 1$, is called a coefficient of ergodicity. It is said to be proper if*

$$\tau(\mathbf{P}) = 0 \quad \text{if and only if} \quad \mathbf{P} = \mathbf{1}^{\mathsf{T}}\mathbf{p}, \tag{3.11}$$

where $\mathbf{p} \geq 0$ is any probability vector, that is, $\mathbf{P}$ has identical rows equal with $\mathbf{p}$, in which case we say that $\mathbf{P}$ is stable.

Remark 185 *The ergodicity coefficient is in fact a measure used to indicate how close a stochastic (or non-negative) matrix is to a stable matrix.*

The first ergodicity coefficient was in fact used by A.A. Markov himself and was

$$\mu(\mathbf{P}) = \max_j(\min_i p_{ij}). \tag{3.12}$$

It is sometimes more convenient to use $1 - \mu(\mathbf{P})$, that is

$$c(\mathbf{P}) = 1 - \max_j(\min_i p_{ij}). \tag{3.13}$$

It is easy to see that $\mu(\mathbf{P})$ and $c(\mathbf{P})$ are not proper.

Definition 186 *A stochastic matrix $\mathbf{P}$ is said to be a Markov matrix if and only if $\mu(\mathbf{P}) > 0$. This is equivalent to the fact that $\mathbf{P}$ has a column whose entries are positive.*

Another ergodicity coefficient was suggested by Doeblin (1937) see also Seneta (1973) and was

$$\delta(\mathbf{P}) = \sum_j \min_i p_{ij}, \tag{3.14}$$

where it is sometimes more convenient to use $1 - \delta(\mathbf{P})$, that is

$$b(\mathbf{P}) = 1 - \sum_j \min_i p_{ij}. \tag{3.15}$$

It is easy to see that $\delta(\mathbf{P})$ and $b(\mathbf{P})$ are proper. It is not difficult to prove that

$$\mu(\mathbf{P}) \leq \delta(\mathbf{P}). \tag{3.16}$$

Hence, if $\mu(\mathbf{P}) > 0$ then also $\delta(\mathbf{P}) > 0$, and the converse is also easily seen that it is true. Therefore Markov matrices could be characterized by either $\mu(\mathbf{P}) > 0$ or $\delta(\mathbf{P}) > 0$. The difference is that $\delta(\mathbf{P})$ is proper while $\mu(\mathbf{P})$ is not. Dobrushin (1956) a student of Markov introduced the ergodicity coefficient

$$a(\mathbf{P}) = \min_{i,j} \sum_k \min(p_{ik}, p_{jk}), \tag{3.17}$$

which is more often known as

$$\tau_1(\mathbf{P}) = 1 - \min_{i,j} \sum_k \min(p_{ik}, p_{jk}). \tag{3.18}$$

The coefficient of ergodicity $\tau_1(\mathbf{P})$ plays a vital role in non-homogeneous Markov chains and will be also derived in what follows also as one which is generated from the norm $\|.\|_1$. In the following Proposition, we provide some equivalent expressions for the coefficient of ergodicity $\tau_1(\mathbf{P})$.

Proposition 187 *Equivalent expressions for $\tau_1(\mathbf{P})$ are the following*

$$\tau_1(\mathbf{P}) = 1 - \min_{i,j} \sum_k \min(p_{ik}, p_{jk}) = 1 - \max_{i,k} \sum_k [p_{ik} - p_{jk}]^+ \tag{3.19}$$
$$= 1 - \frac{1}{2} \max_{i,k} \sum_k |p_{ik} - p_{jk}|.$$

Proof. By definition, we have that for any $a \in \mathbb{R}$

$$a^+ = \{ {}^{a, \text{ if } a \geq 0}_{0, \text{ if } a<0} \quad \text{and} \quad a^- = \{ {}^{0, \text{ if } a \geq 0}_{a, \text{ if } a<0}.$$

Let $i, j, k \in S$ then it is easy to see that

$$(p_{ik} - p_{jk})^+ = [p_{ik} - \min(p_{ik}, p_{jk})],$$

and since $\sum_k p_{ik} = 1$, we have

$$1 - \sum_k (p_{ik} - p_{jk})^+ = 1 - \sum_k [p_{ik} - \min(p_{ik}, p_{jk})] = \sum_k \min(p_{ik}, p_{jk}),$$

from which we get

$$\tau_1(\mathbf{P}) = 1 - \min_{i,j} \sum_k \min(p_{ik}, p_{jk}) = 1 - \max_{i,k} \sum_k [p_{ik} - p_{jk}]^+.$$

Now, we have that

$$\sum_k [p_{ik} - p_{jk}]^+ = -\sum_k [p_{ik} - p_{jk}]^- = \frac{1}{2} \sum_k |p_{ik} - p_{jk}|,$$

from which we get that

$$\tau_1(\mathbf{P}) = 1 - \frac{1}{2} \max_{i,k} \sum_k |p_{ik} - p_{jk}|.$$

■

It is easy to see that also with the use of (3.16) we get that

$$\mu(\mathbf{P}) \leq \delta(\mathbf{P}) \leq a(\mathbf{P}).$$

We now provide the following definition of an important class of matrices:

Definition 188 *A stochastic matrix* $\mathbf{P}$ *is said to be scrambling if and only if* $a(\mathbf{P}) > 0$ *or* $\tau_1(\mathbf{P}) < 1$*. This is equivalent with if and only if any submatrix of* $\mathbf{P}$ *consisting of two rows of it possesses a column whose entries are positive. Also this is equivalent with if and only if no two rows of* $\mathbf{P}$ *are orthogonal or again alternatively, any two rows intersect in the sense of having at least one positive element in a coincident position.*

Clearly, a Markov matrix is scrambling, too. The converse is not true, as is shown by the following example:

Example 189 *The stochastic matrix*

$$\mathbf{P} = \begin{pmatrix} 0 & 1/3 & 2/3 \\ 3/5 & 2/5 & 0 \\ 1/6 & 0 & 5/6 \end{pmatrix},$$

is scrambling although it is not Markov.

A fundamental assumption in homogeneous Markov chains where the concept of weak and strong ergodicity coincide is that the transition probability matrix is regular (Vassiliou (2010). There are various equivalent definitions of a regular stochastic matrix. One in terms of the eigenvalues of the stochastic matrix is that one is the only eigenvalue of modulus one. There are other definitions also in terms of the categorization of the states and their relation of communication (Seneta 1981, Vassiliou 2010). The essence of these definitions is that a stochastic matrix is regular depends on the pattern of its incidence matrix.

In non-homogeneous Markov chains, a fundamental assumption for forward or backward products for many theorems is that $\mathbb{P}(t, t+n)$ is regular for $t \geq 0$, $n \geq 0$. The conditions needed to satisfy this are not facilitated by the following facts: (i) The product of two non regular matrices could be regular and (ii) the product of two regular matrices could be a nonregular matrix. Fortunately, verification of this assumption is facilitated by a remarkable property of scrambling matrices which is proven in the following Lemma.

Lemma 190 *For two stochastic matrices* $\mathbf{P}$ *and* $\mathbf{Q}$*, if* $\mathbf{P}$ *is scrambling then so are* $\mathbf{PQ}$ *and* $\mathbf{QP}$.

Proof. Denote by $\mathbf{R} = \mathbf{PQ}$, we will prove that the matrix $\mathbf{R}$ is scrambling. It is easy to see that for any stochastic matrix $\mathbf{Q}$ then $a(\mathbf{Q}) = 0$ if and only if $\mathbf{Q}$ is not a scrambling matrix; $0 < a(\mathbf{Q}) < 1$ acts as measure of how close the rows of $\mathbf{Q}$ are identical; and $a(\mathbf{Q}) = 1$ if and only if all rows of $\mathbf{Q}$ are identical. Hence, it is sufficient to prove that $a(\mathbf{R}) > 0$. We have that by definition

$$a(\mathbf{R}) = \min_{i,j} \sum_s \min(r_{is}, r_{js}).$$

Without loss of generality assume that the minimum is achieved for $i = m$ and $j = l$. By the definition of the product of two matrices, we have that the element of the matrix $\mathbf{R}$ could be written as

$$r_{ij} = \sum_d p_{id} q_{dj},$$

from which we get that

$$\min(r_{ms}, r_{ls}) = \min\left(\sum_d p_{md} q_{ds}, \sum_d p_{ld} q_{ds}\right) \geq \sum_d q_{ds} \min(p_{md}, p_{ld}),$$

hence

$$\begin{aligned} \sum_s \min(r_{ms}, r_{ls}) &\geq \sum_s \sum_d q_{ds} \min(p_{md}, p_{ld}) \\ &\geq \sum_d \min(p_{md}, p_{ld}), \end{aligned}$$

Thus

$$a(\mathbf{R}) \geq a(\mathbf{P}) > 0,$$

since $\mathbf{P}$ is scrambling and therefore $\mathbf{R}$ is scrambling.

Let us now denote by $\mathbf{R}^* = \mathbf{QP}$ with $\mathbf{P}$ scrambling and $\mathbf{Q}$ stochastic. We will show that $\mathbf{R}^*$ is also scrambling. It is sufficient to show that for any rows i, j of $\mathbf{R}^*$ there exist a column s such that $r^*_{is} > 0$ and $r^*_{js} > 0$.

Since $\mathbf{Q}$ stochastic for any two rows i, j there is a positive element in its row and let us assume that they are $q_{id_1} > 0$ and $q_{jd_2} > 0$. Since $\mathbf{P}$ is scrambling, we have that for the rows d_1 and d_2 there exist a column s such that $p_{d_1 s} > 0$ and $p_{d_2 s} > 0$. Hence,

$$r^*_{is} = \sum_d q_{id} p_{ds} > q_{id_1} p_{d_1 s} > 0,$$

and

$$r^*_{js} = \sum_d q_{jd} p_{ds} > q_{jd_2} p_{d_2 s} > 0.$$

Therefore, $\mathbf{R}^*$ is scrambling. ∎

Remark 191 *The above Lemma actually underlines the fact that in a non-homogeneous Markov chain if the sequence $\{\mathbf{P}(t)\}_{t=0}^{\infty}$ of its transition probability matrices contains an infinite number of scrambling matrices then $\mathbf{P}(t, t+n)$ for any fixed t, and n sufficiently large is a regular matrix. Also from the above Lemma it is apparent that if we denote by G_3 the set of all scrambling stochastic matrices and by G_1 the set of all regular stochastic matrices then $G_3 \subset G_1$.*

We will now define another class of ergodicity coefficients which will be generated by different norms.

Definition 192 *Seneta (1979). Let the set*

$$D^n = \{\mathbf{x};\ \mathbf{x} \in \mathbb{R}^n,\ \mathbf{x} \geq 0, \mathbf{x}\mathbf{1}^\intercal = 1\}$$

and $d(.,.)$ be any metric on this set. Then the quantity

$$\tau(\mathbf{P}) = \sup_{\substack{\mathbf{x},\mathbf{y} \in D^n \\ \mathbf{x} \neq \mathbf{y}}} \frac{d(\mathbf{xP}, \mathbf{yP})}{d(\mathbf{x}, \mathbf{y})}, \tag{3.20}$$

for any stochastic matrix $\mathbf{P}$, is called a coefficient of ergodicity.

Any vector norm $\|.\|$ on $\mathbb{R}^n$ or $\mathbb{C}^n$ will yield an appropriate metric on D^n via $d(\mathbf{x}, \mathbf{y}) = \|\mathbf{x} - \mathbf{y}\|$, $\mathbf{x}, \mathbf{y} \in D$. Then for any stochastic matrix $\mathbf{P}$ we get

$$\tau_{\|.\|}(\mathbf{P}) = \sup_{\substack{\mathbf{x},\mathbf{y} \in D^n \\ \mathbf{x} \neq \mathbf{y}}} \frac{\|(\mathbf{x} - \mathbf{y})\mathbf{P}\|}{\|\mathbf{x} - \mathbf{y}\|}. \tag{3.21}$$

It is an elementary fact that any $\mathbf{z} \in \mathbb{R}^n$ with $\mathbf{z}\mathbf{1}^\intercal = 0$ may be written as $a(\mathbf{x} - \mathbf{y})$ with $\mathbf{x}, \mathbf{y} \in D^n$ with $a = (1/2)\sum_i |z_i|$. Hence,

$$\tau_{\|.\|}(\mathbf{P}) = \sup_{\substack{\|\mathbf{z}\|=1 \\ \mathbf{z}\mathbf{1}^\intercal=0}} \|\mathbf{zP}\|, \text{ over all } \mathbf{z} \in \mathbb{R}^n. \tag{3.22}$$

We will call $\tau_{\|.\|}(\mathbf{P})$ the coefficient of ergodicity induced by the norm $\|.\|$. We will denote by $\tau_{\|.\|_p}(\mathbf{P})$ any coefficient of ergodicity defined by (3.22) when the L_p norm is being used. When the L_1 norm is being used, then we denote by

$$\tau_1(\mathbf{P}) = \sup_{\substack{\|\mathbf{z}\|_1=1 \\ \mathbf{z}\mathbf{1}^\intercal=0}} \|\mathbf{zP}\|_1, \text{ over all } \mathbf{z} \in \mathbb{R}^n,$$

which is the Dobrushin coefficient which as a function of elements of $\mathbf{P}$ is given by (3.10). The proof of that could be found in Seneta (1981) Lemma 4.3 p.139.

Remark 193 *It is very easy to see that $\tau_1(\mathbf{P})$ is a proper coefficient as defined in definition 184. Also, it is obvious Lešanovský (1990), that a function $\tau(\mathbf{P}) = p_{ij}$ for any $i, j \in S$ is a coefficient of ergodicity by Definition 184, although not proper. However, it cannot be generated on the set of $n \times n$ stochastic matrices by any metric on D^n. Now consider the norm $\|.\|_\infty$ which is defined as $\|\mathbf{x}\|_\infty = \max\{|x_i|;\ i = 1, 2, ..., n\}$. The norm $\|.\|_\infty$ generates the coefficient of ergodicity*

$$\tau_\infty(\mathbf{P}) = \sup_{\substack{\|\mathbf{z}\|_\infty = 1 \\ \mathbf{z}\mathbf{1}^\intercal = 0}} \|\mathbf{z}\mathbf{P}\|_\infty, \quad \text{over all } \mathbf{z} \in \mathbb{R}^n,$$

for which Tan(1982) and Rhodious(1988) proved that

$$\sup_{\mathbf{P}} \tau_\infty(\mathbf{P}) = \begin{cases} n/2 & \text{for } n \text{ even}, \\ (n-1)/2 & \text{for } n \text{ odd}, \ n \geq 3. \end{cases}$$

Hence, for $n \geq 4$ there exists a $\mathbf{P}$ for which $\tau_\infty(\mathbf{P}) > 1$. Thus, according to definition 184 $\tau_\infty(\mathbf{P})$ is not a coefficient of ergodicity. Consequently, we showed by example that the two definitions 184 and 192 are not compatible.

We will now prove the following properties of coefficients of ergodicity.

Lemma 194 *The coefficient of ergodicity generated by any metric as in* (3.20) *or induced by any vector norm on $\mathbb{R}^n$ has the following properties:*

$$(1)\ \ \tau(\mathbf{P}_1\mathbf{P}_2) \leq \tau(\mathbf{P}_1)\,\tau(\mathbf{P}_2) \quad \text{for any } \mathbf{P}_1, \mathbf{P}_2 \text{ stochastic matrices.}$$

(2) *For any stochastic matrix $\mathbf{P}$, $\tau(\mathbf{P}) = 0$ if and only if $rank(\mathbf{P}) = 1$.*

Proof. We have that

$$\begin{aligned}
\tau(\mathbf{P}_1\mathbf{P}_2) &= \sup_{\substack{\|\mathbf{z}\|=1 \\ \mathbf{z}\mathbf{1}^\intercal=0}} \|\mathbf{z}\mathbf{P}_1\mathbf{P}_2\|, \text{ over all } \mathbf{z} \in \mathbb{R}^n \\
&\leq \sup_{\substack{\|\mathbf{z}\|=1 \\ \mathbf{z}\mathbf{1}^\intercal=0}} \|\mathbf{z}\mathbf{P}_1\|\,\|\mathbf{P}_2\|, \text{ over all } \mathbf{z} \in \mathbb{R}^n \\
&= \sup_{\substack{\|\mathbf{z}\|=1 \\ \mathbf{z}\mathbf{1}^\intercal=0}} \|\mathbf{z}\mathbf{P}_1\|\,\|\mathbf{z}\|\,\|\mathbf{P}_2\|, \text{over all } \mathbf{z} \in \mathbb{R}^n \\
&\leq \sup_{\substack{\|\mathbf{z}\|=1 \\ \mathbf{z}\mathbf{1}^\intercal=0}} \|\mathbf{z}\mathbf{P}_1\| \sup_{\substack{\|\mathbf{z}\|=1 \\ \mathbf{z}\mathbf{1}^\intercal=0}} \|\mathbf{z}\mathbf{P}_2\|, \text{over all } \mathbf{z} \in \mathbb{R}^n \\
&\leq \tau(\mathbf{P}_1)\,\tau(\mathbf{P}_2).
\end{aligned}$$

Now, what remains and it is not obvious is that when $\tau(\mathbf{P}) = 0$ then $rank(\mathbf{P}) = 1$. This is equivalent with the fact that the matrix $\mathbf{P}$ is stable. That is, it is sufficient to show that $\mathbf{P} = \mathbf{1}^\intercal\mathbf{p}$ for some stochastic vector $\mathbf{p}$ the common row of $\mathbf{P}$. Since $\tau(\mathbf{P}) = 0$ then from (3.22) we get that

$$\tau(\mathbf{P}) = \sup_{\substack{\|\mathbf{z}\|=1 \\ \mathbf{z}\mathbf{1}^\intercal=0}} \|\mathbf{z}\mathbf{P}\| = 0, \text{ over all } \mathbf{z} \in \mathbb{R}^n,$$

from which it is evident that $\mathbf{z}\mathbf{P} = \mathbf{0}$ for all $\mathbf{z}\mathbf{1}^\intercal = 0$ with $\mathbf{z} \in \mathbb{R}^n$. Thus, for any $i, j = 1, 2, ..., n$ we take $\mathbf{z}$ with 1 in the i-th position and -1 in the j-position and zeros elsewhere, which yields that the i-th and j-th rows of $\mathbf{P}$ are equal. ■

Let us denote by SM_n the set of $n \times n$ stochastic matrices. It is known that SM_n is a convex compact subset of $\mathbb{R}^{n^2}$ and if we denote by $Extr\,(SM_n)$ the set of all extremal points of SM_n then we have that

$$SM_n = conv\,(Extr\,(SM_n))\,.$$

We define by

$$H = \left\{\mathbf{x} = (x_1, x_2, ..., x_n) \in \mathbb{R}^n : \mathbf{x}\mathbf{1}^\top = 0\right\}.$$

Then (3.22), where the concept of ergodicity coefficient τ for $\mathbf{P} \in SM_n$ with respect to the norm $\|.\|$ is defined, could be written as

$$\tau\,(\mathbf{P}) = \sup\left\{\|\mathbf{xP}\| : \mathbf{x} \in H,\ \|\mathbf{x}\| = 1\right\}\ \ (\mathbf{P} \in SM_n)\,.$$

In the next theorem, we will study the relation of the maximal value for $\tau\,(\mathbf{P})$, $(\mathbf{P} \in SM_n)$, and the extremal points of SM_n.

Theorem 195 *(Rhodius 1988). Let $\mathbf{P} \in SM_n$ and let $\tau_{\|.\|}\,(\mathbf{P})$ be the ergodicity coefficient τ for $\mathbf{P}$ with respect to the norm $\|.\|$. Then*

$$(i)\ \max_{\mathbf{P}\in SM_n} \tau_{\|.\|}\,(\mathbf{P}) = \max_{\mathbf{x}\in H,\|\mathbf{x}\|=1}\left(\max_{\mathbf{Q}\in Extr(SM_n)} \|\mathbf{xQ}\|\right),$$

$$(ii)\quad \max_{\mathbf{P}\in SM_n} \tau_{\|.\|}\,(\mathbf{P}) = \max_{\mathbf{x}\in H,\|\mathbf{x}\|=1}\left(\max_{(\Lambda_1,\Lambda_2,...,\Lambda_n)\in\Lambda}\right)\left\|\left(\sum_{i\in\Lambda_1} x_i, ..., \sum_{i\in\Lambda_n} x_i\right)\right\|,$$

where Λ denotes the set of all decompositions $(\Lambda_1, \Lambda_2, ..., \Lambda_n)$ of the set of states $\mathbb{S} = \{1, 2, ..., n\}$, which satisfy the conditions

$$a)\ \mathbb{S} = \Lambda_1 \cup \Lambda_2 \cup ... \cup \Lambda_n.$$

$$b)\ \Lambda_i \cap \Lambda_j = \varnothing \text{ for } i \neq j.$$

In particular

$$(iii)\quad \max_{\mathbf{P}\in SM_n} \tau_{\|\cdot\|_1}\,(\mathbf{P}) = 1,$$

$$(iv)\ \max_{\mathbf{P}\in SM_n} \tau_{\|\cdot\|_\infty}\,(\mathbf{P}) = \begin{cases} n/2 & \text{for } n \text{ even,} \\ (n-1)/2 & \text{for } n \text{ odd, } n\geq 3. \end{cases}$$

Proof. Let $\mathbf{P} \in SM_n$ and let $\tau_{\|.\|}\,(\mathbf{P})$ be the ergodicity coefficient τ for $\mathbf{P}$ with respect to the norm $\|.\|$. Then we have that

$$\tau_{\|.\|}\,(\mathbf{P}) = \sup\left\{\|\mathbf{xP}\| : \mathbf{x} \in H,\ \|\mathbf{x}\| = 1\right\}\ \ (\mathbf{P} \in SM_n)\,.$$

Hence, we get that

$$\max_{\mathbf{P}\in SM_n} \tau_{\|.\|}\,(\mathbf{P}) = \max_{\mathbf{P}\in SM_n}\left(\max_{\mathbf{x}\in H,\|\mathbf{x}\|=1} \|\mathbf{xP}\|\right) = \max_{\mathbf{x}\in H,\|\mathbf{x}\|=1}\left(\max_{\mathbf{P}\in SM_n} \|\mathbf{xP}\|\right).$$

Since $\tau_{\|.\|}$ is a convex functional on the convex subset SM_n of $\mathbb{R}^{n^2}$ it assumes its maximum at an extermal point of SM_n. Hence, we obtain

$$\max_{\mathbf{P}\in SM_n} \|\mathbf{xP}\| = \max_{\mathbf{Q}\in Extr(SM_n)} \|\mathbf{xQ}\| \text{ for } \mathbf{x} \in H,\ \|\mathbf{x}\| = 1,$$

and

$$\max_{\mathbf{P}\in SM_n} \tau_{\|.\|}\,(\mathbf{P}) = \max_{\mathbf{x}\in H,\|\mathbf{x}\|=1}\left(\max_{\mathbf{Q}\in Extr(SM_n)} \|\mathbf{xQ}\|\right).$$

(ii) A stochastic matrix $\mathbf{Q}$ is an extremal point of SM_n if and only if $\mathbf{Q}$ has the number one in each of its rows. Note that such a matrix could not be scrambling, that is, $\tau_{\|.\|_1}(\mathbf{Q}) < 1$. This is so because for a stochastic matrix with the number 1 in each row only $\tau_{\|.\|_1}(\mathbf{Q}) = 1$ or $\tau_{\|.\|_1}(\mathbf{Q}) = 0$ is possible.

Now, the stochastic matrix $\mathbf{Q} \in Extr\,(SM_n)$ if and only if

$$(i)\, \mathbb{S} = \Lambda_1 \cup \Lambda_2 \cup ... \cup \Lambda_n,$$

$$(ii)\ \ \Lambda_i \cap \Lambda_j = \varnothing \text{ for } i \neq j,$$

$$(iii)\ \ \text{For each } j = 1, 2, ..., n, \text{ we have } q_{ij} = 1 \text{ if and only if } i \in \Lambda_j.$$

Thus it follows that

$$\mathbf{xQ} = \left(\sum_{i \in \Lambda_1} x_i, ..., \sum_{i \in \Lambda_n} x_i \right) \text{ for } \mathbf{x} \in \mathbb{R}^n.$$

Therefore we get that

$$\max_{\mathbf{P} \in SM_n} \tau_{\|.\|}(\mathbf{P}) = \max_{\mathbf{x} \in H, \|\mathbf{x}\|=1} \left(\max_{(\Lambda_1, \Lambda_2, ..., \Lambda_n) \in \Lambda} \right) \left\| \left(\sum_{i \in \Lambda_1} x_i, ..., \sum_{i \in \Lambda_n} x_i \right) \right\|,$$

where Λ denotes the set of all decompositions $(\Lambda_1, \Lambda_2, ..., \Lambda_n)$ of the set of states $\mathbb{S} = \{1, 2, ..., n\}$.

(iii) From the above we immediately get that

$$\begin{aligned} \|\mathbf{xQ}\|_1 &= \left\| \left(\sum_{i \in \Lambda_1} x_i, ..., \sum_{i \in \Lambda_n} x_i \right) \right\|_1 = \sum_{j=1}^{n} \left| \sum_{i \in \Lambda_j} x_i \right| \\ &\leq \sum_{j=1}^{n} \sum_{i \in \Lambda_j} |x_i| = \|\mathbf{x}\|_1 . \end{aligned}$$

Since we seek the maximum for $\|\mathbf{x}\| = 1$ we get that $\max_{\mathbf{P} \in SM_n} \tau_{\|.\|_1}(\mathbf{P}) \leq 1$. On the other hand $\tau_1(\mathbf{I}) = 1$ hence $\max_{\mathbf{P} \in SM_n} \tau_1(\mathbf{P}) = 1$.

(iv) For the l_∞−norm $\|.\|_\infty$ we get that

$$\|\mathbf{xQ}\|_\infty = \max_j \left| \sum_{i \in \Lambda_j} x_i \right| .$$

Therefore, we get that

$$\max_{\mathbf{P} \in SM_n} \tau_{\|.\|_\infty}(\mathbf{P}) = \max_{\mathbf{x} \in H, \|\mathbf{x}\|=1} \left(\max_j \left| \sum_{i \in \Lambda_j} x_i \right| \right) = \begin{cases} n/2 & \text{for } n \text{ even}, \\ (n-1)/2 & \text{for } n \text{ odd}, \ n \geq 3. \end{cases}$$

■

A class of similar coefficients of ergodicity, as the one given in Definition 192, for non-negative irreducible matrices were studied by Rothblum and Tan (1985). Let $\mathbf{Q}$ be an irreducible non-negative matrix with spectral radius $\rho(\mathbf{Q}) > 0$ which is a simple eigenvalue of $\mathbf{Q}$. Then it is well known that there exist positive right eigenvector $\boldsymbol{\omega}$ of $\mathbf{Q}$ corresponding to $\rho(\mathbf{Q})$. Then the similar class of ergodicity coefficients is defined as

$$\tau_{\|.\|}^{(\boldsymbol{\omega})}(\mathbf{Q}) = \sup\{\|\mathbf{xQ}\| : \|\mathbf{x}\| = 1, \mathbf{x}\boldsymbol{\omega} = 0, \mathbf{x} \in \mathbb{R}^n\},$$

$$\mu_{\|.\|}^{(\boldsymbol{\omega})}(\mathbf{Q}) = \sup\{\|\mathbf{xQ}\| : \|\mathbf{x}\| = 1, \mathbf{x}\boldsymbol{\omega} = 0, \mathbf{x} \in \mathbb{C}^n\}$$

Rothblum and Tan (1985) have proved that for every norm $\|.\|$ there exists a continuation norm $|||.|||$ on $\mathbb{C}^n$ such that

$$\mu^{(\omega)}_{|||\cdot|||}(\mathbf{Q}) = \tau^{(\omega)}_{\|\cdot\|}(\mathbf{Q}).$$

Rhodius (1993) proved the analogous result when $\mathbf{Q}$ is not necessarily irreducible.

3.9 Conditions for weak ergodicity for general products

As it is well documented in the books by Seneta (1981), Paz (1971), Iosifescu (1980) and Isaacson and Madsen (1976) and in the papers by Seneta (1973 and 2014) the problem of ergodicity for non-homogeneous Markov chains had started with Markov himself. The concept of weak ergodicity was provided by Kolmogorov (1931) and was put in its most powerful form and studied by his research student Dobrusin (1956). In the mean time the idea was also partly crystallized in a recently discovered paper by the brilliant Jewish mathematician, Wolfgang Doeblin (1937, 1938) who claimed that he had proved some additional results that were never found. In 1991, there was a conference titled *Doeblin and Modern Probability,* whose proceedings under the same title were published as volume 149 of *Contemporary Mathematics*, to honor his memory. In there Hajnal's (1993) paper is an overt tribute to Doeblin's work on non-homogeneous Markov chains.

3.9.1 A synopsis of the life and mathematical legacy of Wolfgang Doeblin

Wolfgang Doeblin is one among the greatest major contributors to probability theory in the first half of the 20th century up to World War II. He was a quite original mathematician, whose novel ways of looking at things took a long time to be accepted. For example, the Ito-Doeblin theorem as it is now known, which was hidden in the attic of the French Academie of Sciences for over 50 years and was proposed independently by Ito in 1944 took 25 years to be accepted. The date of acceptance being estimated to be the publication of McKean's (1969) marvellous little book: *Stochastic integrals.*

However, Doeblin's life and fate which are gripping and very remarkable is what makes one being stunned by what he accomplished to contribute in his very short career and life. Hence, in order to appreciate more what comments we will provide for his contributions we will start with some details on his life.

Wolfgang Doeblin was born on the 17th of March 1915, in Berlin and he died on the 21st June 1940 in Housseras France. It is there where he took his life in order not to become a prisoner of the Nazi Germany being a Jewish-German serving in the French army. He was the son of Alfred Döblin and Erna Reiss. His father a physician and a novelist he became famous in 1929 once his novel *Berlin Alexanderplatz* was published. The Döblin family was forced in exile in March 1933 after the burning of the Reichstag. Studying probability theory at the Institute Henri Poincaré under Maurice Frechet and Paul Levy he quickly made a name of himself with the research he did on Markov chains. More details on his life and accomplishments could be found in Bru and Yor (2002) and O'Connor and Robertson in the Mac Tutor of Mathematics archive. He took French citizenship in 1936 and defended his doctorate thesis in 1938. At the age of 23 years and only two years of active research behind him, Doeblin's performance must be considered unique, probably since Laplace. In Spring 1938, he was enlisted for two years military service in the French army. Getting depressed by the unchallenged mentally life he stopped all his mathematical work for four months. In order to cure himself of the lethargy and other symptoms of an idle exceptionally intelligent mind he resumed his work and started submitting it for publication. The papers were sent to Fréchet, Levy and some notes to his family. Immediately after the outbreak of the war, was moved to a new position for the defence of Ardennes. The monotones days of training were illuminated by his devotion to his work which he did whenever he could still time from his military duties that did not needed too much focus. He was trying very hard, as he wrote to Fréchet, "to fight against depression. As I am not interested in alcohol, I cannot resort to getting drunk". Mathematics as a therapeutic against the hardships of life, a wonderful recipe for those blessed with creative talent, especially for mathematicians that I hope I can speak of. The young telephonist of the French army taking night shifts to get the opportunity to do research work was getting more and more addicted to his mathematical work. His optimism about his future was at its highest during that period.

In the middle of January 1940, the dream was brutally replaced by reality, with the "alert on Belgium". A German plane was crashed nearby his battalion and from the papers found it was evident that the war had not taken the course the French propaganda was claiming. On April 1940 his regiment moved to the front line. Under enemy fire, the soldier Doeblin restored interrupted contacts, an act for which he was decorated with the Croix de Guerre and which occasion he proved his great physical courage and disdain of death.

The procedure of a "Pli cacheté" goes back to the very origin of the Académie des Sciences. One of the first known examples was that of the deposit by Johann Benoulli, on February 1701. A "Pli cacheté" allows an author to establish a priority, over the result. Doeblin used Pli cacheté (Nos 11445 and 11446) which he claimed back when his findings were accepted for publication. In February 1940 he sent his last research findings as Pli cacheté no 11668. More than ever, he was anguished his results would forever be unknown. As a precaution he sent a letter to Fréchet alerting him about sending the Pli. It is this very letter that was found on Fréchet's office since he had the habit not to throw away even a single memo. In May 2000 the sealed envelope was found and opened containing a treasure of important findings. The Pli is mostly concerned with Markov processes in general state spaces and the start of stochastic integration.

During the night of June 20th to 21st the remaining of his regiment are encircled by German troops and surrender is imminent. Doeblin after walking all night long, he finds himself into the German net in the village of Housseras. He destroys all his notes and his life stops here. His life was the subject of a 2007 movie by Agnes Handwerk and Harrie Willems entitled *A Mathematician Rediscovered.*

On the scientific work that Deblin's genius has produced the reader is referred to Lindvall (1991), Levy (1955), Cohn (1993), Chung (1964, 1992), Doob (1990), Feller (1971). As Paul Levy wrote

After all there can be no greater testimony to a man's work than its influence on others.

Fortunately, for Doeblin, this influence has been visible and is still continuing.

3.9.2 General products of stochastic matrices

Let a complete probability space $(\Omega, \mathcal{F}, \mathbb{P})$ and an inhomogeneous Markov chain $\{X_t\}_{t=0}^{\infty}$ on it with $\{\mathbf{P}(t)\}_{t=0}^{\infty}$ be the sequence of transition probability matrices. In all initial works on weak ergodicity Hajnal (1956, 1958, 1976), Mott (1957), Mott and Sneider (1957), Sarymsakov (1953, 1956) and the works documented in the books referred to in the previous paragraph the problem of weak ergodicity is set as follows:

For the inhomogeneous Markov chain $\{X_t\}_{t=0}^{\infty}$ the sequence of transition probability matrices is given in a prescribed order and the weak ergodicity problem is the study of *forward products*

$$\mathbf{P}(t, t+n) = \mathbf{P}(t+1)\,\mathbf{P}(t+2)\ldots\mathbf{P}(t+n),$$

as $n \to \infty$ and for every possible value of t if they converge in the way described in the previous section. Also, the same problem is of interest for the *backward products*

$$\mathbf{P}(t+n)\,\mathbf{P}(t+n-1)\ldots\mathbf{P}(t+1),$$

as $n \to \infty$ and for every possible value of of t.

In the case of weak ergodicity the probability of the occurrence of any of the k states at the $n-th$ time tends to independence of the initial probability distribution, but is in general dependent on n. In the case of strong ergodicity the probabilities as each state at time n tend to be independent of the initial probability distribution and of n. This characteristic is expressed by the fact that in the case of strong ergodicity $\lim_{n\to\infty} \mathbf{P}(t, t+n)$ tends to a matrix with identical rows. We call matrices with identical rows *stable*. That is, strong ergodicity is the tendency as $n \to \infty$ of $\mathbf{P}(t, t+n)$ to stability.

In the present we will start with a more general problem studied mainly by Leizarowitz (1992), Rhodius (1997), and Neumann and Schneider (1999). Let a sequence of matrices $\{\mathbf{A}_n\}_{n=1}^{\infty}$ with elements in general from the field $\mathbb{C}$ of complex numbers. Neumann and Schneider (1999) considered products of matrices obtained from this sequence in the following manner:

First consider some permutation p or a shuffling as you may visualize it of the matrices of the infinite sequence $\{\mathbf{A}_n\}_{n=1}^{\infty}$ and obtain a sequence of matrices $\mathbf{B}_1, \mathbf{B}_2, \ldots$. Let as call this shuffling of the sequence of matrices $\{\mathbf{A}_n\}_{n=1}^{\infty}$ the p-shuffling. Choose from the p-shuffling one matrix let say $\mathbf{A}_m$ and construct $\mathbf{C}_{p,1} = \mathbf{A}_m$. Choose from the p-shuffling a second matrix let say the matrix $\mathbf{A}_r$. Shuffle the matrices $\mathbf{A}_m, \mathbf{A}_r$ and construct the matrix let say $\mathbf{C}_{p,2} = \mathbf{A}_r\mathbf{A}_m$. Choose from the p-shuffling a third matrix let say the matrix $\mathbf{A}_s$. Shuffle the matrices $\mathbf{A}_m, \mathbf{A}_r, \mathbf{A}_s$ and construct the matrix let say $\mathbf{C}_{p,3} = \mathbf{A}_r\mathbf{A}_s\mathbf{A}_m$. In this way eventually you construct the sequence $\{\mathbf{C}_{p,1}, \mathbf{C}_{p,2}, \mathbf{C}_{p,3}, \ldots\}$, which we call a sequence of a general product of matrices. We will be interested in the study of existence of $\lim_{n\to\infty} \mathbf{C}_{p,n}$ for all $p \geq 0$. This definition of a sequence of a general product of matrices was introduced by Neumann and Schneider (1999).

For example for a sequence $\{\mathbf{A}_n\}_{n=1}^{\infty}$ let the p-shuffle has produced the order $\mathbf{A}_{44}, \mathbf{A}_7, \mathbf{A}_{84}, \mathbf{A}_{33}, \mathbf{A}_{48}, \mathbf{A}_4, \ldots$ We choose the first matrix(card) and let it be $c_{p,1} = \mathbf{A}_4$. In the next step let say that the matrix $\mathbf{A}_{84}$ is chosen. Scuffle the two cards $\mathbf{A}_4, \mathbf{A}_{84}$ and let us say that the order $\mathbf{A}_{84}, \mathbf{A}_4$ results. Then $\mathbf{C}_{p,2} = \mathbf{A}_{84}\mathbf{A}_4$. In the third step let say that the matrix $\mathbf{A}_{33}$ is chosen. Scuffle the three cards $\mathbf{A}_4, \mathbf{A}_{84}, \mathbf{A}_{33}$ and let us say that the order $\mathbf{A}_{84}, \mathbf{A}_{33}, \mathbf{A}_4$ results. Then $\mathbf{C}_{p,3} = \mathbf{A}_{84}\mathbf{A}_{33}\mathbf{A}_4$. Then the sequence of general product is $\{\mathbf{C}_{p,1}, \mathbf{C}_{p,2}, \mathbf{C}_{p,3}, \ldots\}$. Note that $\mathbf{C}_{p,n+1}$ has all the factors of $\mathbf{C}_{p,r}$ but not necessarily in the same order.

In Leizarowitz (1992), another kind of possible products of matrices from a sequence of stochastic matrices $\{\mathbf{Q}_n\}_{n=0}^{\infty}$ was defined. We start with the stochastic matrix $\mathbf{Q}_0$ which we multiply by $\mathbf{Q}_1$ from the right and from the left which produce two stochastic matrices $\mathbf{Q}_0\mathbf{Q}_1$ and $\mathbf{Q}_1\ \mathbf{Q}_0$. Then these two stochastic matrices we multiply by $\mathbf{Q}_2$ from the right and from the left, which produce four stochastic matrices $\mathbf{Q}_0\mathbf{Q}_1\mathbf{Q}_2$, $\mathbf{Q}_2\mathbf{Q}_0\mathbf{Q}_1$, $\mathbf{Q}_1\mathbf{Q}_0\mathbf{Q}_2$ and $\mathbf{Q}_2\ \mathbf{Q}_1\ \mathbf{Q}_0$. Hence, from a set of $n+1$ stochastic matrices we get 2^n different products of stochastic matrices. Among these are the forward and the backward product. It easy to see that up to now the set of different products produced so far is a subset of the general product of stochastic matrices from the infinite sequence $\{\mathbf{Q}_n\}_{n=0}^{\infty}$. Now, we consider infinite permutations of the integers, that is, 1-1 mappings $i \to \phi(i)$ from the set of nonnegative integers onto itself. We then consider infinite products of $\left\{\mathbf{Q}_{\phi(i)}\right\}_{n=0}^{\infty}$ taken in the same way as described for the sequence of stochastic matrices $\{\mathbf{Q}_n\}_{n=0}^{\infty}$. For all possible $\phi(i)'$s then we say that we have the set of *products taken in arbitrary orders of multiplication from the sequence of stochastic matrices* $\{\mathbf{Q}_n\}_{n=0}^{\infty}$. The sets of general products of stochastic matrices and the set of products taken in arbitrary orders of multiplication from the sequence of stochastic matrices $\{\mathbf{Q}_n\}_{n=0}^{\infty}$ coincide. The same definition of general products was used by Rhodius (1997).

We would say that a property hods for a sequence of matrices if it holds for every sequence of products constructed as in the previous two paragraphs. In this respect now we naturally arrive at the following definition of weak and strong ergodicity:

Definition 196 *For a given sequence of stochastic matrices $\{\mathbf{Q}_n\}_{n=0}^{\infty}$, defining uniquely a non-homogeneous Markov chain on a complete probability space $(\Omega, \mathcal{F}, \mathbb{P})$, we say that it is weakly ergodic if for any sequence of general products of stochastic matrices or for any product of stochastic matrices for any order of multiplication as the number factors goes to infinity then the product tends to a sequence of stable matrices. In addition, we say it is strongly ergodic if weak ergodicity obtains and the stable matrix is unique.*

Let $\|.\|_{\mu}$ be a matrix norm in $M_n(\mathbb{R})$ or $M_n(\mathbb{C})$ and $\mathbf{P} \in M_n(\mathbb{R})$ or $M_n(\mathbb{C})$. Then define by

$$\|\mathbf{P}\|_{\mu}^{+} = \max\left(\|\mathbf{P}\|_{\mu}, 1\right) \quad \text{and} \quad \|\mathbf{P}\|_{\mu}^{-} = \min\left(\|\mathbf{P}\|_{\mu}, 1\right). \tag{3.23}$$

Using the above generally defined norm and (3.20) we will prove the basic theorems that follows using classical results from mathematical analysis on the convergence and divergence of series of nonegative real numbers. In order to study series the non-zero terms of which are of the same sign it is sufficient to study series of which all terms are non-negative. Let a series

$$\sum_{n=1}^{\infty} a_n \quad \text{with } a_n \geq 0, \tag{3.24}$$

then it is well known that this series is either convergent or properly divergent, that is, its sum is equal with $+\infty$. In what follows we will use the following theorem from Hyslop (1965 p. 36):

Theorem 197 *Let the series (3.24) and let $\sum_{n=1}^{\infty} b_n$ be any series whose terms are those of the series $\sum_{n=1}^{\infty} a_n$ in any different order. If the series $\sum_{n=1}^{\infty} a_n$ converges to a then so does $\sum_{n=1}^{\infty} b_n$ and if $\sum_{n=1}^{\infty} a_n$ is properly divergent then so is $\sum_{n=1}^{\infty} b_n$.*

Let a sequence $\{a_n\}_{n=0}^{\infty}$ and consider the infinite product $\prod_{n=1}^{\infty}(1+a_n)$. If it tends to a finite non-zero limit, then we say that it *converges.* If it does not converge, then we say that the above infinite product is *divergent.* If it tends to zero, we say that it *diverges to zero.* The expression diverges to zero may seem curious, but it is not, if we observe that the behavior of the product $\prod_{n=1}^{\infty}(1+a_n)$ is closely related by

$$\log \prod_{n=1}^{k}(1+a_n) = \sum_{n=1}^{k} \log(1+a_n) \text{ for } a_n > 1,$$

and from which it is evident that the product diverges to zero when the series diverges to $-\infty$. In what follows we will use the following theorem from Hyslop (1965 p. 94):

Theorem 198 *If $a_n \geq 0$ the series $\sum a_n$ and the product $\prod(1+a_n)$ converge or diverge together.*

We now define the following two conditions (Neyman and Schneider (1999)).

Condition (C). We say that a sequence of matrices $\{\mathbf{A}_n\}_{n=1}^{\infty} \in M_n(\mathbb{R})$ or $M_n(\mathbb{C})$ satisfies condition (C) for the norm $\|.\|_{\mu}$ if

$$\sum_{i=1}^{\infty}\left(\|\mathbf{A}_i\|_{\mu}^{+} - 1\right), \text{ converges.}$$

Condition (D). We say that a sequence of matrices $\{\mathbf{A}_n\}_{n=1}^{\infty} \in M_n(\mathbb{R})$ or $M_n(\mathbb{C})$ satisfies condition (D) for the norm $\|.\|_\mu$ if

$$\sum_{i=1}^{\infty}\left(1-\|\mathbf{A}_i\|_\mu^-\right), \text{ diverges.}$$

We will now prove the following theorem:

Theorem 199 *(Neyman and Schneider, 1999). Let a sequence of matrices $\{\mathbf{A}_n\}_{n=1}^{\infty}$ and also a norm $\|.\|_\mu$ in $M_n(\mathbb{R})$ or $M_n(\mathbb{C})$. Assume that the sequence $\{\mathbf{A}_n\}_{n=1}^{\infty}$ satisfies condition **C** for the norm $\|.\|_\mu$. Then all general products or products taken in arbitrary orders of multiplication from the sequence of matrices $\{\mathbf{A}_n\}_{n=1}^{\infty}$ are bounded.*

Proof. Since condition **C** holds for the sequence $\{\mathbf{A}_n\}_{n=1}^{\infty}$ then we have that

$$\sum_{i=1}^{\infty}\left(\|\mathbf{A}_i\|_\mu^+-1\right), \text{ converges.}$$

Now, for any p-shuffle of the sequence $\{\mathbf{A}_n\}_{n=1}^{\infty}$ the sequence $\{\mathbf{C}_{p,i}\}_{i=1}^{\infty}$ has elements $\mathbf{C}_{p,r} = \mathbf{A}_{p,1}\mathbf{A}_{p,2}...\mathbf{A}_{p,r}$ were $\mathbf{A}_{p,s}$ is an element from the p-shuffle of the sequence $\{\mathbf{A}_n\}_{n=1}^{\infty}$ and whose position s in the product $\mathbf{C}_{p,r}$ is the result of the process described above and explained by an example. It follows that according to Theorem 197 we get that

$$\sum_{i=1}^{\infty}\left(\|\mathbf{A}_{p,i}\|_\mu^+-1\right), \text{ converges.}$$

Hence, by Theorem 198 the product

$$\prod_{i=1}^{\infty}\|\mathbf{A}_{p,i}\|_\mu^+, \quad \text{converges,}$$

and so there exist a positive constant b such that

$$\prod_{i=1}^{r}\|\mathbf{A}_{p,i}\|_\mu^+ \leq b \text{ for every } r \in \mathbb{N}.$$

Note that

$$\begin{aligned}
\text{If } \|\mathbf{A}_i\|_\mu^- &= \|\mathbf{A}_i\|_\mu \text{ then } \|\mathbf{A}_i\|_\mu^+ = 1 \text{ and} \\
\text{If } \|\mathbf{A}_i\|_\mu^+ &= \|\mathbf{A}_i\|_\mu \text{ then } \|\mathbf{A}_i\|_\mu^- = 1.
\end{aligned}$$

Hence, for each $r \in \{1, 2, ...\}$ it follows that

$$\begin{aligned}
\|\mathbf{C}_{p,r}\|_\mu &= \|\mathbf{A}_{p,1}\mathbf{A}_{p,2}...\mathbf{A}_{p,r}\|_\mu \leq \|\mathbf{A}_{p,1}\|_\mu\|\mathbf{A}_{p,2}\|_\mu...\|\mathbf{A}_{p,r}\|_\mu \\
&= \prod_{i=1}^{r}\|\mathbf{A}_{p,i}\|_\mu^+\prod_{i=1}^{r}\|\mathbf{A}_{p,i}\|_\mu^- \leq b\prod_{i=1}^{r}\|\mathbf{A}_{p,i}\|_\mu^- \leq b. \qquad (3.25)
\end{aligned}$$

■

We are now in a position to prove the following important theorem.

Theorem 200 *(Neyman and Schneider, 1999). Let a sequence of matrices $\{\mathbf{A}_n\}_{n=1}^{\infty}$ and also a norm $\|.\|_\mu$ in $M_n(\mathbb{R})$ or $M_n(\mathbb{C})$. Assume that the sequence $\{\mathbf{A}_n\}_{n=1}^{\infty}$ satisfies conditions **C** and **D** for the norm $\|.\|_\mu$. Then all general products or products taken in arbitrary orders of multiplication from the sequence of matrices $\{\mathbf{A}_n\}_{n=1}^{\infty}$ diverges to zero.*

Proof. Since the conditions of the previous theorem hold relation (3.25) is valid in here. Hence,

$$\left\|\mathbf{C}_{p,r}\right\|_{\mu} = \left\|\mathbf{A}_{p,1}\mathbf{A}_{p,2}...\mathbf{A}_{p,r}\right\|_{\mu} \leq b\prod_{i=1}^{r}\left\|\mathbf{A}_{p,i}\right\|_{\mu}^{-}. \tag{3.26}$$

By condition **D** the series

$$\sum_{i=1}^{\infty}\left(1-\left\|\mathbf{A}_{i}\right\|_{\mu}^{-}\right), \text{ diverges.}$$

Thus, by Theorem 197 the series

$$\sum_{i=1}^{\infty}\left(1-\left\|\mathbf{A}_{p,i}\right\|_{\mu}^{-}\right), \text{ also diverges.}$$

Hence, by Theorem 198 the product

$$\prod_{i=1}^{r}\left\|\mathbf{A}_{p,i}\right\|_{\mu}^{-}, \text{ also diverges.} \tag{3.27}$$

By the fact that $\left\|\mathbf{A}_{p,i}\right\|_{\mu}^{-} = \min\left(\left\|\mathbf{A}_{p,i}\right\|_{\mu}, 1\right) \leq 1$ and (3.26) we easily arrive at our needed conclusion. ■

3.9.3 Consequences of the above general theorems to weak ergodicity of inhomogeneous Markov chains

Let $\|.\|$ be an arbitrary vector norm in $M_{1n}(\mathbb{R})$, and $SM_n(\mathbb{R})$ the set of all $n\times n$ stochastic matrices. As we have seen in Definition 192 and relation (3.20) Seneta (1979) has introduced the coefficient of weak ergodicity $\tau_{\|,\|}(\mathbf{P})$ with $\mathbf{P}\in SM_n(\mathbb{R})$ induced by the norm $\|.\|$ as

$$\tau_{\|\cdot\|}(\mathbf{P}) = \sup\left\{\|\mathbf{xP}\| : \mathbf{x}\in\mathbb{H}, \ \|\mathbf{x}\|=1\right\}, \tag{3.28}$$

where

$$\mathbb{H} = \left\{\mathbf{x}\in M_{1n}(\mathbb{R}) : \mathbf{x1}^{\intercal}=0\right\}.$$

We now provide an analogous definition, as in relation (3.9) for all general products or products taken in arbitrary orders of multiplication from the sequence $\{\mathbf{P}_i\}_{i=1}^{\infty}$ with $\mathbf{P}_i\in SM_n(\mathbb{R})$ for every $i=1,2,...$ to be weakly ergodic.

Definition 201 *Let $\{\mathbf{P}_i\}_{i=1}^{\infty}$ with $\mathbf{P}_i\in SM_n(\mathbb{R})$ for every $i=1,2,....$ We shall say that all general products or products taken in arbitrary orders of multiplication from the sequence $\{\mathbf{P}_i\}_{i=1}^{\infty}$ are weakly ergodic if for all general products $\mathbf{A}_{p,i}$, $i=1,2,...$, we have that*

$$\lim_{i\to\infty}\mathbf{xA}_{p,i} = 0 \text{ for all } \mathbf{x}\in\mathbb{H}. \tag{3.29}$$

Remark 202 *Since $\mathbf{x}\in\mathbb{H}$ it is known that it can be written as $\mathbf{x}=c(\mathbf{u}-\mathbf{v})$ with $c\in\mathbb{R}$ and $\mathbf{u},\mathbf{v}\in M_{1,n}(\mathbb{R})$ and $\mathbf{u1}^{\top}=\mathbf{v1}^{\top}=1$. Then it is easily seen that, for each product $\mathbf{A}_{p,i}$ considered, our definition is equivalent to that in relation (3.9).*

From (3.28) we get that

$$\tau_{\|\cdot\|}(\mathbf{P}) = \sup_{\substack{\mathbf{x}\in\mathbb{H}\\ \|\mathbf{x}\|=1}}\|\mathbf{xP}\| \leq \sup_{\substack{\mathbf{x}\in\mathbb{H}\\ \|\mathbf{x}\|=1}}\|\mathbf{x}\|\,\|\mathbf{P}\| \leq \|\mathbf{P}\|, \tag{3.30}$$

from which by Theorem 200, we immediately obtain

Theorem 203 *(Neumann and Schneider 1999). Let* $\|\ .\|$ *be a norm on* $\mathbb{R}^n$ *and let* $\tau_{\|.\|}(\mathbf{P})$ *be the coefficient of ergodicity induced by this norm. Let a sequence* $\{\mathbf{P}_i\}_{i=1}^{\infty}$ *with* $\mathbf{P}_i \in SM_n(\mathbb{R})$ *for every* $i = 1, 2, ...$ *Then all general products or products taken in arbitrary orders of multiplication from the sequence* $\{\mathbf{P}_i\}_{i=1}^{\infty}$ *are weakly ergodic if*

$$\sum_{i=1}^{\infty} \left(\tau^{+}_{\|.\|}(\mathbf{P}_i) - 1\right) \quad \textit{converges,} \tag{3.31}$$

and

$$\sum_{i=1}^{\infty} \left(1 - \tau^{-}_{\|.\|}(\mathbf{P}_i)\right) \quad \textit{diverges.} \tag{3.32}$$

We will now provide the following important Corollary

Corollary 204 *(Neumann and Schneider 1999). Let* $\|\ .\|$ *be a norm on* $\mathbb{R}^n$ *and let* $\tau_{\|.\|}(\mathbf{P})$ *be the coefficient of ergodicity induced by this norm. Let a sequence* $\{\mathbf{P}_i\}_{i=1}^{\infty}$ *with* $\mathbf{P}_i \in SM_n(\mathbb{R})$ *for every* $i = 1, 2, ...$ *If* $\tau_{\|.\|}(\mathbf{P}_i) \leq 1$ *for all* $i = 1, 2, ...,$ *and there exists a point of accumulation* $\mathrm{w} \in \mathcal{W}$ *with* $\mathrm{w} < 1$ *for the sequence* $\left\{\tau_{\|.\|}(\mathbf{P}_i)\right\}_{i=1}^{\infty}$, *then all general products or products taken in arbitrary orders of multiplication from the sequence* $\{\mathbf{P}_i\}_{i=1}^{\infty}$ *are weakly ergodic.*

Proof. Since $\tau_{\|.\|}(\mathbf{P}_i) \leq 1$ for all $i = 1, 2, ...,$ we have that

$$\tau^{+}_{\|.\|}(\mathbf{P}_i) - 1 = \max\left(\tau_{\|.\|}(\mathbf{P}_i), 1\right) - 1 = 0,$$

hence condition (3.31) is satisfied. Since $w < 1$ is an accumulation point of the sequence $\left\{\tau_{\|,\|}(\mathbf{P}_i)\right\}_{i=1}^{\infty}$, then there exists an infinite subsequence $\left\{\tau_{\|.\|}\left(\mathbf{P}_{i_j}\right)\right\}_{j=1}^{\infty}$ such that

$$\tau_{\|.\|}\left(\mathbf{P}_{i_j}\right) < (1+w)/2 < 1 \text{ for } j = 1, 2, ...,$$

hence

$$\left(1 - \tau^{-}_{\|.\|}(\mathbf{P}_i)\right) > 0,$$

and thus condition (3.32) is satisfied. That completes our proof. ■

We have seen in section 3.8 that the coefficient of ergodicity induced by the l_1 is the Dobrushin coefficient of ergodicity given as a function of the elements of the matrix $\mathbf{P}$ by (3.10). Rhodius (1997) has proved the following theorem using the l_1 norm, that is, the Dobrushin ergodicity coefficient which is a special case of the above Corollary 204, but one that its basic condition could be easily checked:

Theorem 205 *(Rhodius 1997, Theorem 3, Part 1). Let a sequence* $\{\mathbf{P}_i\}_{i=1}^{\infty}$ *with* $\mathbf{P}_i \in SM_n(\mathbb{R})$ *for every* $i = 1, 2, ...,$ *and let* $\mathcal{W}$ *be the set of all its accumulation points. Then if there exists at least one* $\mathbf{P} \in \mathcal{W}$ *with* $\tau_{\|.\|_1}(\mathbf{P}) < 1$, *then weak ergodicity obtains for products taken in arbitrary orders of multiplication from the sequence* $\{\mathbf{P}_i\}_{i=1}^{\infty}$.

Remark 206 *The property* $\tau_{\|.\|_1}(\mathbf{P}) < 1$ *in fact defines a special class of matrices which we call scrambling. The physical characteristic of such matrices is that given any two rows, say* i *and* j, *there exist at least one column* r, *such that both* $p_{ir} > 0$ *and* $p_{jr} > 0$. *The name scrambling matrix is derived from the following fact: Suppose that after the scrambling matrix has generated the next step of the Markov chain we find our selves in state* r. *Then we cannot be sure whether it was in state* i *or* j *in the immediate past. In a way a scrambling matrix scrambles its past. Moreover, the "scrambling power" of a matrix* $\mathbf{P}$ *is a measure of the degree to which it approaches a matrix with identical rows which scrambles all traces of the past (Hajnal (1958).*

Another important Corollary of Theorem 203 is the following:

Corollary 207 *(Neumann and Schneider 1999). Let* $\|\ .\|$ *be a norm on* $\mathbb{R}^n$ *and let* $\tau_{\|.\|}(\mathbf{P})$ *be the coefficient of ergodicity induced by this norm of a matrix* $\mathbf{P}$. *Let a sequence* $\{\mathbf{P}_i\}_{i=1}^{\infty}$ *with* $\mathbf{P}_i \in SM_n(\mathbb{R})$ *for every* $i = 1, 2, ...$ *If all points of accumulation* $\mathrm{w} \in \mathcal{W}$ *of the sequence* $\{\tau_{\|.\|}(\mathbf{P}_i)\}_{i=1}^{\infty}$, *satisfy* $\mathrm{w} < 1$ *then all general products or products taken in arbitrary orders of multiplication from the sequence* $\{\mathbf{P}_i\}_{i=1}^{\infty}$ *are weakly ergodic.*

Proof. Any coefficient of ergodicity $\tau_{\|.\|}(\mathbf{P})$ induced by any norm $\|\ .\|$ is bounded when $\mathbf{P}$ is stochastic since all its elements are ≤ 1. Hence, the sequence $\{\tau_{\|.\|}(\mathbf{P}_i)\}_{i=1}^{\infty}$, is bounded and therefore $\mathcal{W}$ the set of its accumulation points is compact. Therefore, there exists a number $b < 1$ such that only a finite number of terms of the sequence $\{\tau_{\|.\|}(\mathbf{P}_i)\}_{i=1}^{\infty}$ exceeds it. Hence the sequence $\{\tau_{\|.\|}(\mathbf{P}_i)\}_{i=1}^{\infty}$ satisfies equations (3.31) and (3.32) and thus all general products or products taken in arbitrary orders of multiplication from the sequence $\{\mathbf{P}_i\}_{i=1}^{\infty}$ are weakly ergodic. ■

The study of weak ergodicity for general products of stochastic matrices or products taken in arbitrary orders of multiplication from the sequence $\{\mathbf{P}_i\}_{i=1}^{\infty}$ started with a general Theorem A by Leizarowitz (1992) the largest portion of which has already been stated as Corrollaries 201 and 204 and followed by Rhodius (1997) Theorem 3 the Part 1 of which is Corollary 204. However, in what follows we will state and prove Theorem A due to Leizarowitz (1992) the remaining parts of which are interesting and because the proof has an interest of its own. We will start by clarifying some concepts. For more details and examples see Meyer (2000).

(1) **Invariant Subspaces.** Let $\mathbf{T}$ be a linear operator on a vector space $\mathcal{V}$, and for $\mathcal{X} \subseteq \mathcal{V}$, define

$$\mathbf{T}(\mathcal{X}) = \{\mathbf{T}(\mathbf{x}) \mid \mathbf{x} \in \mathcal{X}\},$$

to be the set of all possible images of vectors from $\mathcal{X}$ under the transformation $\mathbf{T}$. It is easy to see that when $\mathcal{X}$ is a subspace of $\mathcal{V}$ then $\mathbf{T}(\mathcal{X})$ is also a subspace of $\mathcal{V}$. In general though $\mathbf{T}(\mathcal{X})$ is not related to $\mathcal{X}$. It is a special case when $\mathbf{T}(\mathcal{X}) \subseteq \mathcal{X}$. In this case the subspace $\mathcal{X} \subseteq \mathcal{V}$ is said to be an invariant subspace under $\mathbf{T}$. In such a situation $\mathbf{T}$ can be considered as a linear operator on $\mathcal{X}$ and restrict $\mathbf{T}$ to act only on vectors from $\mathcal{X}$. This restricted operator will be denoted by $\mathbf{T}_{|\mathcal{X}}$. For more on invariant subspaces see Meyer(2000). Let $\mathbf{x} \in \mathbb{H}$ and $\mathbf{Q} \in SM_n$ then $\mathbb{H}$ is easy to see that it is an invariant subspace of $\mathbb{R}^n$ under $\mathbf{Q} \in SM_n$ in the sense that

$$\mathbf{xQ1}^{\top} = \sum_{i=1}^{n} x_i \sum_{j=1}^{n} q_{ij} = 0.$$

Hence, there is a meaning to consider the restricted operator $\mathbf{Q}_{|\mathbb{H}}$.

(2) **Geometric properties of vector norms.** A large part of linear algebra is, in fact, geometric in nature because much of the subject grew out of the need to generalize the basic geometry at $\mathbb{R}^2$ and $\mathbb{R}^3$ to nonvisual higher-dimentional spaces. The primary geometric feature of a norm is its unit ball, through which considerable insight into properties of the norm may be gained. In fact the unit ball is a geometric summary of a norm, because of the homogeneity property, which characterizes the norm. Let $\|.\|$ a norm on a real or complex vector space $\mathcal{V}$, $\mathbf{x} \in \mathcal{V}$ and $r > 0$ be given. The ball of radius r around $\mathbf{x}$ is the set

$$B_{\|.\|}(r, x) = \{\mathbf{y} \in \mathcal{V} : \ \|\mathbf{y} - \mathbf{x}\| \leq r\}, \tag{3.33}$$

and the unit ball is actually $B_{\|.\|}(1, 0)$. To get a better feel for the l_1, l_2, and l_∞ norms it helps to know the shapes and relative sizes of their unit balls, that is $B_{\|.\|_1}(1, 0), B_{\|.\|_2}(1, 0)$ and $B_{\|.\|_\infty}(1, 0)$. To visualize these unit balls let us sketch them in $\mathbb{R}^3$. It is easy to see that $B_{\|.\|_1}(1,0)$ is an octahedron, $B_{\|.\|_2}(1,0)$ a sphere and $B_{\|.\|_\infty}(1,0)$ a cube. Also, it is easy to see visually that $B_{\|.\|_1}(1,0)$ fits inside $B_{\|.\|_2}(1,0)$ which in turn fits inside $B_{\|.\|_\infty}(1,0)$. This summarizes the fact that

$$\|\mathbf{x}\|_1 \geq \|\mathbf{x}\|_2 \geq \|\mathbf{x}\|_\infty, \quad \text{for all } \mathbf{x} \in \mathbb{R}^n. \tag{3.34}$$

Also, we observe that all three unit balls are convex sets. This also is a general result since it is not difficult to prove that unit ball of a norm is a convex set. The unit ball of a norm on a finite-dimensional vector space is compact although this is not always the case in the infinite dimentional case. Note that $B_{\|.\|_1}(1,0)$ has eight extreme points, $B_{\|.\|_2}(1,0)$ has an infinite number of extreme points and $B_{\|.\|_\infty}(1,0)$ has sixteen extreme points. A norm as a continuous real-valued function on a compact set is bounded and achieves both its supremum and infimum on the set.

(3) Let $\mathcal{U}$ and $\mathcal{V}$ be vector spaces over a field $\mathcal{F}$ ($\mathbb{R}$ and $\mathbb{C}$ for us). A linear transformation from $\mathcal{U}$ to $\mathcal{V}$ is defined to be a linear function $\mathbf{T}$ mapping $\mathcal{U}$ into $\mathcal{V}$. That is

$$\mathbf{T}(a\mathbf{x}+\mathbf{y}) = a\mathbf{T}(\mathbf{x}) + \mathbf{T}(\mathbf{y}) \text{ for all } \mathbf{x},\mathbf{y} \in \mathcal{U} \text{ and } a \in \mathcal{F}. \tag{3.35}$$

A linear operator on $\mathcal{U}$ is defined to be a linear transformation from $\mathcal{U}$ into itself. The connection between linear transformations and matrices is at the heart of Linear Algebra. It is known that Cayley's observation that the composition of two linear transformations can be represented by the multiplication of two matrices gave the opportunity for the subject of Linear Algebra to grow.

(4) Let $\mathcal{U}$ be a vector space, $\mathcal{B} = \{\mathbf{u}_1, \mathbf{u}_2, ..., \mathbf{u}_n\}$ a basis of $\mathcal{U}$ and $\mathbf{v} = a_1\mathbf{u}_1 + a_2\mathbf{u}_2 + ... + a_n\mathbf{u}_n \in \mathcal{U}$. We will denote by $[\mathbf{v}]_{\mathcal{B}} = (a_1, a_2, ..., a_n)$ the coordinates of $\mathbf{v}$ with respect to $\mathcal{B}$. For each pair of vector spaces $\mathcal{U}$ and $\mathcal{V}$ the set $\mathcal{L}(\mathcal{U},\mathcal{V})$ of all linear transformations from $\mathcal{U}$ to $\mathcal{V}$ is a vector space over $\mathcal{F}$. Let $\mathcal{B}_u = \{\mathbf{u}_1, \mathbf{u}_2, ..., \mathbf{u}_n\}$ and $\mathcal{B}_v = \{\mathbf{v}_1, \mathbf{v}_2, ..., \mathbf{v}_m\}$ be bases for $\mathcal{U}$ and $\mathcal{V}$ respectively. Consider $[\mathbf{u}]_{\mathcal{B}} = \{c_1, c_2, ..., c_n\}$ for any $\mathbf{u} \in \mathcal{U}$ and define the linear transformation

$$\mathbf{T}_{ji}(\mathbf{u}) = c_j \mathbf{v}_i. \tag{3.36}$$

Then

$$\mathcal{B}_{\mathcal{L}} = \{\mathbf{T}_{ji}\}_{j=1,...,n}^{i=1,...,m} \text{ is a basis for } \mathcal{L}(\mathcal{U},\mathcal{V}), \tag{3.37}$$

$$\text{and } \dim \mathcal{L}(\mathcal{U},\mathcal{V}) = \dim(\mathcal{U}) \dim(\mathcal{V}). \tag{3.38}$$

(5) The coordinate matrix of $\mathbf{T} \in \mathcal{L}(\mathcal{U},\mathcal{V})$ with respect to the bases $\mathcal{B}_u = \{\mathbf{u}_1, \mathbf{u}_2, ..., \mathbf{u}_n\}$ and $\mathcal{B}_v = \{\mathbf{v}_1, \mathbf{v}_2, ..., \mathbf{v}_m\}$ is defined to be the matrix

$$[\mathbf{T}]_{\mathcal{B}_u\mathcal{B}_v} = \begin{pmatrix} a_{11} & a_{12} & \cdots & a_{1m} \\ a_{21} & a_{22} & \cdots & a_{2m} \\ \cdots & \cdots & \cdots & \cdots \\ a_{n1} & a_{n2} & \cdots & a_{nm} \end{pmatrix}, \text{ with} \tag{3.39}$$

$$[\mathbf{T}(\mathbf{u}_j)]_{\mathcal{B}_v} = \begin{pmatrix} a_{j1} & a_{j2} & \cdots & a_{jm} \end{pmatrix}.$$

Now for each $\mathbf{u} \in \mathcal{U}$ the action of $\mathbf{T}$ on $\mathbf{u}$ is being represented by matrix multiplication in the sense that

$$[\mathbf{T}(\mathbf{u})]_{\mathcal{B}_v} = [\mathbf{u}]_{\mathcal{B}_u} [\mathbf{T}]_{\mathcal{B}_u\mathcal{B}_v}. \tag{3.40}$$

(6) Let $\mathbf{T}_1 \in \mathcal{L}(\mathcal{U},\mathcal{V})$ and $\mathbf{T}_2 \in \mathcal{L}(\mathcal{V},\mathcal{W})$, the composition of $\mathbf{T}_1$ with $\mathbf{T}_2$ is defined to be the function $\mathbf{C} : \mathcal{U} \to \mathcal{V}$ such that for $\mathbf{x} \in \mathcal{U}$ we have $\mathbf{C}(\mathbf{x}) = \mathbf{T}_2(\mathbf{T}_1(\mathbf{x}))$. It is not difficult to see that the composition $\mathbf{C} = \mathbf{T}_2\mathbf{T}_1$ is also a linear transformation. Let $\mathcal{B}_u$, $\mathcal{B}_v$, and $\mathcal{B}_w$ respectively be the basis of $\mathcal{U}, \mathcal{V}$ and $\mathcal{W}$. Since $\mathbf{C}$ is also a linear transformation must have a coordinate matrix representation with respect to $(\mathcal{B}_u, \mathcal{B}_w)$. As expected this is given by $[\mathbf{C}]_{\mathcal{B}_u\mathcal{B}_w} = [\mathbf{T}_2]_{\mathcal{B}_v\mathcal{B}_w} [\mathbf{T}_1]_{\mathcal{B}_u\mathcal{B}_v}$, which as we have discussed was the motive of Cayley to define the multiplication of two matrices.

Let us now state and prove Theorem A from Leizarovitz (1992), which was the first that dealt with weak ergodicity for general products from a sequence of stochastic matrices and in addition proved the equivalence with strong ergodicity of backward products.

Theorem 208 *(Leizarovitz (1992)). Let $\|.\|$ be the induced (operator) matrix norm induced on matrices in $M_{nn}(\mathbb{R})$ by the norm $\|.\|$ on $\mathbb{R}^n$, and let the constant C be defined by*

$$C = \max\left\{\left\|\mathbf{Q}_{|\mathbb{H}}\right\| : \mathbf{Q} \in SM_n\right\}. \tag{3.41}$$

It is apparent that $C \geq 1$. Then, (i) if there exists at least one matrix $\mathbf{Q} \in \mathcal{W}$ the set of all the accumulation points of the sequence $\{\mathbf{Q}_i\}_{i=0}^{\infty}$ which satisfies

$$\left\|\mathbf{Q}_{|\mathbb{H}}\right\| < 1/C, \tag{3.42}$$

then weak ergodicity obtains for general products or infinite products of $\{\mathbf{Q}_i\}_{i=0}^{\infty}$ taken in arbitrary order of multiplications, and strong ergodicity obtains for the backward products. (ii) In the case that we have

$$\left\|\mathbf{Q}_{|\mathbb{H}}\right\| < 1, \quad \textit{for every } \mathbf{Q} \in \mathcal{W}, \tag{3.43}$$

then the conclusions in (i) hold and the convergences have a geometric rate.

Proof. (i) For the sequence $\{\mathbf{Q}_i\}_{i=0}^{\infty}$ consider sequences $\left\{\mathbf{Q}_{i(n)}\right\}_{n=0}^{\infty}$ by infinite shuffling of $\{\mathbf{Q}_i\}_{i=0}^{\infty}$. As described in 3.9 for sequences $\left\{\mathbf{Q}_{i(n)}\right\}_{n=0}^{\infty}$ we consider sequences of forward products $\{\mathbf{\Gamma}(N)\}_{N=1}^{\infty}$ taken in various orders of multiplication where N is the number of multipliers in the product. The property of weak ergodicity for general products or for products taken in arbitrary orders of multiplication from $\{\mathbf{Q}_i\}_{i=0}^{\infty}$ it holds if it is true for every sequence of products $\{\mathbf{\Gamma}(N)\}_{N=1}^{\infty}$. We will consider a specific order of multiplication for the sequence $\{\mathbf{\Gamma}(N)\}_{N=1}^{\infty}$and prove weak ergodicity independent of the specific order which then in fact proves weak ergodicity for any general product. We define the following set

$$K_N = \{\mathbf{x}\mathbf{\Gamma}(N) : \mathbf{x} \in SM_{1n}\}. \tag{3.44}$$

In order to prove weak ergodicity for $\{\mathbf{\Gamma}(N)\}_{N=1}^{\infty}$, it is sufficient to show that

$$d(K_N) \to 0 \text{ as } N \to \infty, \tag{3.45}$$

where $d(.)$ is the diameter of a set which is known that for a set A is defined as

$$d(A) = \sup\{\|\mathbf{x} - \mathbf{y}\| : \mathbf{x}, \mathbf{y} \in A\}. \tag{3.46}$$

This is so since if $\mathbf{x} \neq \mathbf{y} \in \mathbb{H}$ then from (3.44) we get that

$$\lim_{N\to\infty} d(K_N) = 0 = \sup\left\{\lim_{N\to\infty} \|\mathbf{x}\mathbf{\Gamma}(N) - \mathbf{y}\mathbf{\Gamma}(N)\| = 0 : \mathbf{x} \neq \mathbf{y} \in S_{1n}\right\}, \tag{3.47}$$

and for which to be true it is sufficient for $\mathbf{\Gamma}(N)$ for N large to tend to matrices with identical rows.

If there exists at least one matrix $\mathbf{Q} \in \mathcal{W}$ the set of all the accumulation points of the sequence $\{\mathbf{Q}_i\}_{i=0}^{\infty}$ which satisfies

$$\left\|\mathbf{Q}_{|\mathbb{H}}\right\| < 1/C,$$

then there exists an increasing subsequence $\left\{\mathbf{Q}_{n_j}\right\}_{j=1}^{\infty}$ which converges to $\mathbf{Q} \in \mathcal{W}$ and a $0 < a < 1/C$ such that

$$\left\|\mathbf{Q}_{n_j}\right\| \leq a \text{ for every } j \geq 1. \tag{3.48}$$

An abstract representative of $\mathbf{\Gamma}(N)$ could be using in the product a large number of factors from the increasing subsequence $\left\{\mathbf{Q}_{n_j}\right\}_{j=1}^{\infty}$ each member of which will be multiplied from the left and from the right by stochastic matrices $\mathbf{R}_i$ to fill the gap between two members of the subsequence. If two members p_i and p_{i+1} are selected to be consecutive integers, then the matrix $\mathbf{R}_i$ is the identity matrix.

$$\mathbf{\Gamma}(N) = \mathbf{R}_1 \mathbf{Q}_{p_1} \mathbf{R}_2 \mathbf{Q}_{p_2} ... \mathbf{R}_l \mathbf{Q}_{p_l} \mathbf{R}_{l+1} \tag{3.49}$$

for some stochastic matrices $\mathbf{R}_1, ..., \mathbf{R}_{l+1}$, and where every integer p_i, $1 \leq i \leq l$, belongs to the sequence $\left\{\mathbf{Q}_{n_j}\right\}_{j=1}^{\infty}$. Since $\mathbf{Q} \in \mathcal{W}$ when $N \to \infty$ then also $l \to \infty$. From (3.48) and (3.49) we get that

$$\|\mathbf{\Gamma}(N)\| \leq C (aC)^l \quad \text{with } aC < 1. \tag{3.50}$$

From (3.50) we get $\|\mathbf{\Gamma}(N)\| \to_{N\to\infty} 0$. Now, since $\|\mathbf{x\Gamma}(N) - \mathbf{y\Gamma}(N)\| \leq \|\mathbf{x} - \mathbf{y}\| \, \|\mathbf{\Gamma}(N)\|$ we get that $d(K_N) \to 0$ as $N \to \infty$ which establishes weak ergodicity.

(*ii*) We will now establish strong ergodicity for the backward products. We consider sequences of backward products $\{\mathbf{B}(N)\}_{N=1}^{\infty}$ taken in various orders of multiplication where N is the number of multipliers in the product. The property of strong ergodicity for general products or for products taken in arbitrary orders of multiplication from $\{\mathbf{Q}_i\}_{i=0}^{\infty}$ it holds if it is true for every sequence of products $\{\mathbf{B}(N)\}_{N=1}^{\infty}$. We will consider a specific order of multiplication for the sequence $\{\mathbf{B}(N)\}_{N=1}^{\infty}$ and prove strong ergodicity independent of the specific order which then in fact proves strong ergodicity for any general product. Let in this respect

$$\mathbf{B}(N) = \mathbf{Q}_N \mathbf{Q}_{N-1} ... \mathbf{Q}_1 \mathbf{Q}_0,$$

be the representative general product for which we will prove strong ergodicity without leaning on its specific order of multipliers. Denote by

$$K_N^B = \{\mathbf{xB}(N) : \mathbf{x} \in SM_{1n}\}.$$

Let a vector $\mathbf{y} \in K_{N+1}^B$ then

$$\mathbf{y} \in \{\mathbf{xB}(N+1) : \mathbf{x} \in SM_{1n}\} \implies \mathbf{y} \in \{\mathbf{xQ_{N+1}B}(N) : \mathbf{x} \in SM_{1n}\},$$

and

$$\{\mathbf{xQ_{N+1}B}(N) : \mathbf{x} \in SM_{1n}\} \subseteq \{\mathbf{zB}(N) : \mathbf{z} \in SM_{1n}\} = K_N^B.$$

hence $\mathbf{y} \in K_N^B$ from which we get that

$$K_{N+1}^B \subseteq K_N^B.$$

On the other hand, since $\mathbf{B}(N)$ is also one of the general products, we have from (3.46) that

$$\lim_{N\to\infty} d\left(K_N^B\right) = 0 = \sup\left\{\lim_{N\to\infty} \|\mathbf{xB}(N) - \mathbf{yB}(N)\| = 0 : \mathbf{x} \neq \mathbf{y} \in S_{1n}\right\},$$

from which we get that

$$\cap_{N=1}^{\infty} K_N^B = \mathbf{q} \quad \text{with } \mathbf{q} \in S_{1n}.$$

Now from (3.46) since $\left\|\mathbf{Q}_{|\mathbb{H}}\right\| < 1$, for every $\mathbf{Q} \in \mathcal{W}$ we get that for some $0 < \beta < 1$ there exists a n_0 such that

$$\|\mathbf{Q}_n\| < 1 \quad \text{for } n \geq n_0, \tag{3.51}$$

from which we easily get that $\|\mathbf{B}(N)\| \to 0$ geometrically fast. ■

We will now derive Theorem 208 when the norm used is the l_1 norm in which case the induced coefficient of ergodicity is the Dobrushin coefficient of ergodicity $\tau_{\|\cdot\|_1}(\mathbf{P})$ for any stochastic matrix $\mathbf{P}$.

Theorem 209 *(Leizarowitz (1992)). Let* $\{X_t\}_{t=0}^{\infty}$ *be a non-homogeneous Markov chain on a complete probability space* $(\Omega, \mathcal{F}, \mathbb{P})$ *and let* $\{\mathbf{P}(t)\}_{t=0}^{\infty}$ *be its sequence of transition probability matrices. Suppose that* $\{\mathbf{P}(t)\}_{t=0}^{\infty}$ *is such that* $\mathcal{W}$ *the set of its accumulation points contains a scrambling matrix. Then weak ergodicity obtains for general products or infinite products of* $\{\mathbf{P}(t)\}_{t=0}^{\infty}$ *taken in arbitrary order of multiplications, and strong ergodicity obtains for the backward products. If every member of* $\mathcal{W}$ *is scrambling, then all the convergences have geometric rate.*

Proof. For the l_1 norm and a stochastic matrix $\mathbf{P}$ we have $\|\mathbf{P}\|_1 = 1$, thus in (3.42) we easily see that $C = 1$. Hence, in order that the conditions of Theorem 208 are valid it remains to show that

$$\left\|\mathbf{P}_{|\mathbb{H}}\right\|_1 < 1 \text{ for every scrambling matrix } \mathbb{P}. \tag{3.52}$$

Equivalently

$$\|\mathbf{xP}\|_1 < 1 \text{ if } \|\mathbf{x}\|_1 = 1 \text{ and } \sum_{i=1}^{n} x_i = 0. \tag{3.53}$$

Let us denote by $\mathbf{p}_i$ the i-th row of $\mathbf{P}$, then we have that

$$\|\mathbf{xP}\|_1 = \left\|\sum_{i=1}^{n} x_i \mathbf{p}_i\right\|_1 \leq \sum_{i=1}^{n} |x_i| = 1. \tag{3.54}$$

Since, $\sum_{i=1}^{n} |x_i| = 1$ and $\sum_{i=1}^{n} x_i = 0$ there exist an i and a j for which x_i and x_j have opposite signs and for these since $\mathbf{P}$ is scrambling there exist an $1 \leq n \leq k$ such that $p_{ik} > 0$ and $p_{jk} > 0$. Hence (3.54) is true as a strict inequality and consequently (3.51) holds which proves (3.52). ■

3.9.4 An application of backward products in non-homogeneous Markov chains

All too often one finds when teaching backward products that listeners wonder about the usefulness of studying such products since "backwards in time" sounds like science fiction. For this and very many other reasons we will present a well known in the members of the non-homogeneous Markov chain circles application of weak and strong ergodicity of backward product of a non-homogeneous Markov chain.

Let a group of k experts who are trying to evaluate a parameter θ. When the only prior information available to the expert is of an objective nature (i.e., consists of sample information), the choice of a prior distribution is immediate. If this is not the case, and the expert has some knowledge with regard to the parameters of interest, then, he should be able to quantify this knowledge and form a subjective probability. Savage (1954) shows that subjective probability distributions assessed in accordance with certain plausible behavioral postulates of consistency must conform mathematically to a probability measure. Perhaps the best way to fully appreciate the problems involved is to assess one's own subjective probability distributions. What appears in the surface to be a simple and forward procedure requires a tremendous amount of intellectual self-discipline. For example, one could try to assess the probabilities of his favorite basketball or football team to win a game or a series of games.

The value of θ is assumed to lie in an abstract space Ω that is endowed with a σ-field of measurable subsets for which probabilities can be specified. In a formal Bayesian analysis, it is necessary for the expert to quantify his judgements and express them in the form of a subjective probability distribution. Thus, $F_1, F_2, ..., F_k$ are subjective probability distributions over Ω which represent the prior beliefs about θ of k individuals. In other words, for any measurable set A in the parameter space Ω, $F_i(A)$ for $i = 1, 2, ..., k$ is the prior beliefs of the k experts that the value of θ will lie in A. Let us denote by

$$\mathbf{F}(0) = [F_1(0), F_2(0), ..., F_k(0)], \tag{3.55}$$

their initial estimates. It is assumed that each individual expert revises his subjective probability distribution each time he receives the subjective opinions of the others. Let us denote by $p_{ij}(1)$ the weight that individual i assigns to the opinion of individual j when he carries out his revision for the first time. It is assumed that $p_{ij}(1) > 0$ and that $\sum_{j=1}^{k} p_{ij}(1) = 1$ for $i = 1, 2, ..., k$. Hence, it is implied that the new subjective probability for individual i will be

$$F_i(1) = \sum_{j=1}^{k} p_{ij}(1) F_j(0). \tag{3.56}$$

Hence,

$$\mathbf{F}(1) = \mathbf{P}(1)\mathbf{F}(0). \tag{3.57}$$

After t interchanges of opinions and judgements we will have that

$$\mathbf{F}^{\top}(t) = \mathbf{P}(t)\mathbf{P}(t-1)...\mathbf{P}(1)\mathbf{F}^{\top}(0). \tag{3.58}$$

In (3.58) the genesis of a backward product in a non-homogeneous Markov chain is apparent. Our main concern is then the behavior of (3.58) as $t \to \infty$.

In actual fact (3.58) represents a procedure described by Feller (1968) as "repeated averaging" and in De Groot (1974) is known as "reaching a consensus". In both cases, the Markov chain studied was homogeneous. The problem was generalized into a non-homogeneous one as presented in here by Chatterjee and Seneta (1977). Later on Cohen, Hajnal and Newman (1986) studied consensus for the case when the position of experts is harden, that is, they eventually increase the weight they put on their own opinion. A recent review of the problem and perspective on future research could be found in Xia, Wang and Xuan (2011).

Chatterjee and Seneta (1977) generalized the problem by studying backward products of the form

$$\lim_{t\to\infty} \mathbf{M}(r,t) = \lim_{t\to\infty} \mathbf{P}(t)\mathbf{P}(t-1)...\mathbf{P}(r). \tag{3.59}$$

It is evident from (3.59) that since weak ergodicity means that $\lim_{t\to\infty} \mathbf{M}(r,t) = \mathbf{M}_n^*$ a sequence of stable matrices for n large, then $\mathbf{F}^{\top}(n)$ will be a vector with identical elements. That means that the opinions of the k experts for large n will coincide although it will change at each step. We say then that we have consensus. We also have consensus when we have strong ergodicity the difference being that the common opinion of the experts is stable for large n.

It is important however at this point to stress that Chatterjee and Seneta (1977) have proved that for backward products weak and strong ergodicity are equivalent.

Theorem 210 *(Chatterjee and Seneta (1977)). For the matrix backward products* $\mathbf{M}(r,t)$ *weak and strong ergodicity are equivalent.*

Proof. Since strong ergodicity implies weak ergodicity for all matrix products it suffices to prove that weak ergodicity for the backward product $\mathbf{M}(r,t)$ implies strong ergodicity also for it.

Assuming weak ergodicity we have for $r \geq 0$ and $\epsilon > 0$ that

$$-\epsilon \leq p_{il}(r,t) - p_{jl}(r,t) \leq \epsilon \ \text{ for } t \geq t_0, \tag{3.60}$$

uniformly for all $i,j,l=1,2,...,k$. We also have that

$$\mathbf{M}(r,t+1)=\mathbf{P}(r+t+1)\,\mathbf{M}(r,t),$$

from which we get that the $lm-$element of $\mathbf{M}(r,t+1)$ is equal with

$$p_{lm}(r,t+1)=\sum_{j=1}^{k}p_{lj}(r+t+1)\,p_{jm}(r,t). \tag{3.61}$$

Now, it easily seen that using (3.60) we get that

$$\begin{aligned}\sum_{j=1}^{k}p_{lj}(r+t+1)\left[p_{im}(r,t)-\epsilon\right] &\le \sum_{j=1}^{k}p_{lj}(r+t+1)\,p_{jm}(r,t) \\ &\le \sum_{j=1}^{k}p_{lj}(r+t+1)\left[p_{im}(r,t)+\epsilon\right].\end{aligned} \tag{3.62}$$

from which we arrive at

$$\left[p_{im}(r,t)-\epsilon\right]\le p_{lm}(r,t+1)\le\left[p_{im}(r,t)+\epsilon\right]\quad\text{for } t\ge t_0. \tag{3.63}$$

By induction we get that for any $t\ge t_0, n\ge 1$

$$\left|p_{jm}(r,t+n)-p_{im}(r,t)\right|\le\epsilon\ \text{ for any } i,j,m. \tag{3.64}$$

Hence, also for $j=i$

$$\left|p_{im}(r,t+n)-p_{im}(r,t)\right|\le\epsilon,$$

which according to Cauchy convergence criterion implies strong ergodicity for $\mathbf{M}(r,t)$. ∎

3.10 The dominant role of the Dobrushin ergodicity coefficient

In this section, we will improve the sufficient conditions for the weak ergodicity of the infinite products of $\{\mathbf{Q}_n\}_{n=0}^{\infty}$, on the basis of the set $\mathcal{W}$ of all accumulation points of $\{\mathbf{Q}_n\}_{n=0}^{\infty}$. This improvement will be based on the Dobrushin ergodicity coefficient with which one has no problem, in contrast to an arbitrary norm, in computing $\tau_{\|.\|_1}$ and also for which we have $C_{\|.\|_1}=1$. This will actually emphasize the dominant role of the Dobrushin ergodicity coefficient in the class of all ergodicity coefficients induced by a norm. Refreshing our notation and definitions let $\mathbb{H}=\{(x_1,x_2,...,x_n\}\in\mathbb{R}^n:\sum_{i=1}^{n}x_i=0\}$, SM_n be the set of all $n\times n$ stochastic matrices and let $n\ge 2$. Then as we have seen Seneta (1979) for any $\mathbf{P}\in SM_n$

$$\tau_{\|.\|}(\mathbf{P})=\max\left\{\|x\mathbf{P}\|:x\in\mathbb{H},\ \|x\|=1\right\},$$

is the ergodicity coefficient corresponding to the norm $\|.\|$. The Dobrushin ergodicity coefficient $\tau_{\|.\|_1}(\mathbf{P})$ it corresponds to the l_1- norm $\|x\|_1=\sum_{i=1}^{n}|x_i|$ and it has the simple explicit form

$$\tau_{\|.\|_1}(\mathbf{P})=\frac{1}{2}\max_{i,j}\sum_{l=1}^{n}\left|p_{il}-p_{jl}\right|.$$

The Dobrushin coefficient of ergodicity is the only coefficient of ergodicity corresponding to a norm $\|.\|$ with

$$\max\left\{\tau_{\|.\|_1}(\mathbf{P}):\mathbf{P}\in SM_n\right\}=1,$$

in all other cases one has

$$\max\left\{\tau_{\|.\|}(\mathbf{P}):\mathbf{P}\in SM_n\right\}>1,$$

see Kuhne and Rhodius (1990) and Lesanovsky (1990). We will start with two Lemmas which will prove useful in the foundation of the basic theorems that will follow:

Lemma 211 *Let $\mathcal{L}(\mathbb{H})$ be the set of linear operators $\mathbf{T} : \mathbb{H} \to \mathbb{H}$. Then $\mathbf{T} \in \mathcal{L}(\mathbb{H})$ if and only if there are a stochastic matrix $\mathbf{P} \in SM_n$ and a $R_T > 0$ such that*

$$\mathbf{T}(\mathbf{x}) = (\mathbf{xP}) R_T \text{ for all } x \in \mathbb{H}.$$

Proof. *Let $\mathbf{T} \in \mathcal{L}(\mathbb{H})$ then $\mathbf{T}(\mathbf{x}) \in \mathbb{H}$ for every $x \in \mathbb{H}$. Then $\mathbf{T}$ according to 5 page 169 the coordinate matrix of $\mathbf{T} \in \mathfrak{L}(\mathbb{H}, \mathbb{H})$ with respect to the base $B_{\mathbb{H}}$ in $\mathbb{H}$ is defined to be a matrix $[\mathbf{T}]_{B_{\mathbb{H}} B_{\mathbb{H}}}$ such that*

$$[\mathbf{T}(\mathbf{x})]_{B_{\mathbb{H}}} = [\mathbf{x}]_{B_{\mathbb{H}}} [\mathbf{T}]_{B_{\mathbb{H}} B_{\mathbb{H}}} \in \mathbb{H}.$$

It is not difficult to see that the above equation is satisfied if and only if

$$[\mathbf{T}]_{B_{\mathbb{H}} B_{\mathbb{H}}} \mathbf{1}^{\top} \mathbf{e}_i = c_T \text{ for all } i = 1, 2, ..., n,$$

with $[\mathbf{T}]_{B_{\mathbb{H}} B_{\mathbb{H}}} \geq 0$. Then $\mathbf{P} = \frac{1}{c_T}$ is a stochastic matrix and $\frac{1}{c_T} = R_T$. ■

Lemma 212 *Define by*

$$S^* = \left\{ x \in \mathbb{R}^n : \sum_{i=1}^{n} x_i^+ = \sum_{i=1}^{n} x_i^- = \frac{1}{2} \right\},$$

and $\|.\|$ a norm on $\mathbb{R}^n$. Then

$$(i)\, C_{\|.\|} = \max \left\{ \tau_{\|.\|}(\mathbf{P}) : \mathbf{P} \in SM_n \right\}$$
$$= \max \left\{ \tau_{\|.\|}(\mathbf{P}) : \mathbf{P} \in SM_n, \tau_{\|.\|_1}(\mathbf{P}) = 1 \right\}, \tag{3.65}$$

and

$$(ii)\, C_{\|.\|} \geq \frac{\max_{x \in S^*} \|x\|}{\min_{x \in S^*} \|x\|}. \tag{3.66}$$

Proof. (i). The set SM_n of all $n \times n$ stochastic matrices it is known to be a convex subset of $\mathbb{R}^{n^2}$. The norm $\|.\|$ on $\mathbb{R}^n$ is a convex functional hence the coefficient of ergodicity $\tau_{\|.\|}(\mathbf{P}) = \|x\mathbf{P}\|$ corresponding to the norm $\|.\|$ is a convex functional on the convex set SM_n. Therefore the maximum of $\tau_{\|.\|}$is one of the external points of SM_n. Let us denote by $Extr\,[SM_n]$ the set of all extremal points of S^*.

Now, it is known that a stochastic matrix $\mathbf{P} \in Extr\,[SM_n]$ if and only if has the number 1 as an element in each of its rows. Therefore, for such a $\mathbf{P} \in Extr\,[SM_n]$ only $\tau_{\|.\|_1}(\mathbf{P}) = 1$ is possible, i.e. $\mathbf{P}$ is nonscrambling. Naturally, it is possible that $\tau_{\|.\|_1}(\mathbf{P}) = 0$ in which case though this would imply $\tau_{\|.\|}(\mathbf{P}) = 0$. Hence, (3.65) is true.

(ii) We have

$$\begin{aligned} C_{\|.\|} &= \max \left\{ \tau_{\|.\|}(\mathbf{P}) : \mathbf{P} \in SM_n \right\} \\ &= \max_{x \in \mathbb{H}, \|x\|=1} \left(\max_{\mathbf{P} \in SM_n} \|x\mathbf{P}\| \right) \\ &= \max_{x \in \mathbb{H}, \|x\|=1} \left(\max_{\mathbf{P} \in Extr[SM_n]} \|x\mathbf{P}\| \right) \end{aligned} \tag{3.67}$$

Now, the stochastic matrix $\mathbf{P} \in Extr\,(SM_n)$ if and only if

$$(i)\;\; S = \Lambda_1 \cup \Lambda_2 \cup ... \cup \Lambda_n,$$

$$(ii)\;\; \Lambda_i \cap \Lambda_j = \varnothing \text{ for } i \neq j,$$

$$(iii)\;\; \text{For each } j = 1, 2, ..., n \text{ we have } q_{ij} = 1 \text{ if and only if } i \in \Lambda_j.$$

Thus it follows that

$$\mathbf{xP} = \left(\sum_{i \in \Lambda_1} x_i, ..., \sum_{i \in \Lambda_n} x_i \right) \text{ for } \mathbf{x} \in \mathbb{R}^n.$$

Therefore we get that

$$C_{\|\cdot\|} = \max_{\mathbf{P} \in SM_n} \tau(\mathbf{P}) = \max_{\mathbf{x} \in H, \|\mathbf{x}\|=1} \left(\max_{(\Lambda_1, \Lambda_2, ..., \Lambda_n) \in \Lambda} \right) \left\| \left(\sum_{i \in \Lambda_1} x_i, ..., \sum_{i \in \Lambda_n} x_i \right) \right\|, \tag{3.68}$$

where Λ denotes the set of all decompositions $(\Lambda_1, \Lambda_2, ..., \Lambda_n)$ of the set of states $\mathbb{S} = \{1, 2, ..., n\}$.

From all the $x \in \mathbb{H}$ with $\|x\| = 1$ we choose x_0 for which $\|x_0\| = 1$, and that one for which

$$\|x_0\|_1 = \max_{\mathbf{x} \in H, \|\mathbf{x}\|=1} \|x\|_1. \tag{3.69}$$

Since $z = x_0 / \|x_0\|_1 \in S^*$ we have $\sum_{i=1}^n z_i^+ = \sum_{i=1}^n z_i^- = 1/2$. Hence

$$C_{\|\cdot\|} \geq \|x_0\|_1 \left\| \left(\sum_{i \in \Lambda_1} z_i, ..., \sum_{i \in \Lambda_n} z_i \right) \right\|,$$

for all decompositions $(\Lambda_1, \Lambda_2, ..., \Lambda_n)$ of the set of states $\mathbb{S} = \{1, 2, ..., n\}$.

It is not difficult to see that (see also Rothblum and Schneider (1980))

$$Extr\ [S^*] = \left\{ \frac{1}{2} (e_i - e_j) : i \neq j\ i, j = 1, 2, ..., n \right\}.$$

It is apparent that

$$Extr\ [S^*] \subset \left\{ \left(\sum_{i \in \Lambda_1} z_i, ..., \sum_{i \in \Lambda_n} z_i \right) \text{ for all } (\Lambda_1, \Lambda_2, ..., \Lambda_n) \right\}. \tag{3.70}$$

Hence, it immediately follows that

$$C_{\|\cdot\|} \geq \max_{x \in \mathbb{H}, \|x\|=1} \|x\|_1 \max_{x \in \mathbb{H}, x \in S^*} \|x\|. \tag{3.71}$$

It is not difficult to see that

$$\max_{x \in \mathbb{H}, \|x\|=1} \|x\|_1 = \max_{x \in \mathbb{H}, \|x\|_1=1} \|x\|_1 = \max_{x \in S^*} \|x\| = \frac{1}{\min_{x \in S^*} \|x\|}. \tag{3.72}$$

From (3.71) and (3.72) we get that

$$C_{\|\cdot\|} \geq \frac{\max_{x \in S^*} \|x\|}{\min_{x \in S^*} \|x\|}.$$

■

It is known that on a finite dimensional space, all norms are equivalent. Hence, for any two matrix norms $\|.\|_a$ and $\|.\|_b$ there is a least finite positive constant C_{ab} such that $\|\mathbf{A}\|_a \leq C_{ab} \|\mathbf{A}\|_b$ for all $\mathbf{A} \in M_n$. The constant C_{ab} can be computed as

$$C_{ab} = \max_{\mathbf{A} \neq \mathbf{0}} \frac{\|\mathbf{A}\|_a}{\|\mathbf{A}\|_b}.$$

There is an apparent analogy if we reverse the rules of a and b, that is,

$$C_{ba} = \max_{\mathbf{A}\neq\mathbf{0}} \frac{\|\mathbf{A}\|_b}{\|\mathbf{A}\|_a}.$$

Although there is no apparent relation between C_{ab} and C_{ba} it has been observed that for the matrix norms $\|.\|_1, \|.\|_2$ and $\|.\|_\infty$ and for each pair $C_{ab} = C_{ba}$ for $a, b = 1, 2, \infty$ and for all $\mathbf{A} \in M_n$. All these norms are induced norms and thus the following theorem was proved in Horn and Johnson (2013, Theorem 5.6.18).

Theorem 213 *Let $\|.\|_a$ and $\|.\|_b$ be given norms on $\mathbb{R}^n$. Let $\|.\|_{ia}$ and $\|.\|_{ib}$ denote their respective induced norms on M_n, that is,*

$$\|\mathbf{A}\|_{ia} = \max_{x\neq 0} \frac{\|\mathbf{Ax}\|_a}{\|\mathbf{x}\|_a} \quad \textit{and} \quad \|\mathbf{A}\|_{ib} = \max_{x\neq 0} \frac{\|\mathbf{Ax}\|_b}{\|\mathbf{x}\|_b},$$

then define by

$$R_{ab} = \max_{\mathbf{x}\neq\mathbf{0}} \frac{\|\mathbf{x}\|_a}{\|\mathbf{x}\|_b} \quad \textit{and} \quad R_{ba} = \max_{\mathbf{x}\neq\mathbf{0}} \frac{\|\mathbf{x}\|_b}{\|\mathbf{x}\|_a}.$$

Then

$$\max_{\mathbf{A}\neq\mathbf{0}} \frac{\|\mathbf{A}\|_{ia}}{\|\mathbf{A}\|_{ib}} = R_{ab}R_{ba},$$

and

$$\max_{\mathbf{A}\neq\mathbf{0}} \frac{\|\mathbf{A}\|_{ia}}{\|\mathbf{A}\|_{ib}} = \max_{\mathbf{A}\neq\mathbf{0}} \frac{\|\mathbf{A}\|_{ib}}{\|\mathbf{A}\|_{ia}} = R_{ab}R_{ba}.$$

We will now state and prove the following basic theorem.

Theorem 214 *Let*

$$S^* = \left\{ \mathbf{x} \in \mathbb{R}^n : \sum_{i=1}^n x_i^+ = \sum_{i=1}^n x_i^- = \frac{1}{2} \right\},$$

then for any norm $\|.\|$ on $\mathbb{R}^n$ the following are true:

(1) *If $\mathbf{P} \in SM_n$ is scrambling, then $\tau_{\|.\|}(\mathbf{P}) < C_{\|.\|}$.*

(2) *A matrix $\mathbf{P} \in SM_n$ with $\tau_{\|.\|}(\mathbf{P}) < 1/C_{\|.\|}$ is scrambling.*

$$\begin{aligned}(3)\ C_{\|.\|} &= \max\left\{\tau_{\|.\|}(\mathbf{P}) : \mathbf{P} \in SM_n\right\} \\ &= \max\left\{\tau_{\|.\|}(\mathbf{P}) : \mathbf{P} \in SM_n, \tau_{\|.\|_1}(\mathbf{P}) = 1\right\} \\ &= \frac{\max_{\mathbf{x}\in S^*}\|\mathbf{x}\|}{\min_{\mathbf{x}\in S^*}\|\mathbf{x}\|}.\end{aligned}$$

$$(4)\ \frac{1}{C_{\|.\|}}\tau_{\|.\|_1}(\mathbf{P}) \leq \tau_{\|.\|}(\mathbf{P}) \leq C_{\|.\|}\tau_{\|.\|_1}(\mathbf{P})$$

and $C_{\|.\|}$ is the smallest such number.

Proof. We have that

$$R = \max_{x\in\mathbb{H}\backslash\{0\},\|x\|=1} \frac{\|\mathbf{x}\|}{\|\mathbf{x}\|_1} = \max_{\mathbf{x}\in S^*} \frac{\|\mathbf{x}\|}{\|\mathbf{x}\|_1} = \max_{\mathbf{x}\in S^*}\|\mathbf{x}\|, \tag{3.73}$$

and

$$R_1 = \max_{x\in\mathbb{H}\backslash\{0\},\|x\|=1} \frac{\|\mathbf{x}\|_1}{\|\mathbf{x}\|} = \max_{\mathbf{x}\in S^*} \frac{\|\mathbf{x}\|_1}{\|\mathbf{x}\|} = \max_{\mathbf{x}\in S^*} \frac{\mathbf{1}}{\|\mathbf{x}\|} = \frac{1}{\min_{\mathbf{x}\in S^*}\|\mathbf{x}\|}. \tag{3.74}$$

Now, for all $\mathbf{T} \in \mathcal{L}(\mathbb{H})$ the induced operator norms $\|\mathbf{T}\|_i$ and $\|\mathbf{T}\|_{i1}$ of the corresponding vector norms $\|.\|$ and $\|.\|_1$ are related analogously like the induced matrix norms and the corresponding matrix norms considered in the previous Theorem 213. Hence, by applying it we arrive at the conclusion that

$$\max_{\mathbf{T}\in\mathcal{L}(\mathbb{H}),\mathbf{T}\neq\mathbf{0}} \frac{\|\mathbf{T}\|_{i1}}{\|\mathbf{T}\|_i} = \max_{\mathbf{T}\in\mathcal{L}(\mathbb{H}),\mathbf{T}\neq\mathbf{0}} \frac{\|\mathbf{T}\|_i}{\|\mathbf{T}\|_{i1}} = RR_1.$$

Now, by Lemma 211 we have that

$$\|\mathbf{T}\|_{i1} = \max_{\mathbf{x}\in S^*} \{\|\mathbf{T}(\mathbf{x})\|_1\} = \max_{\mathbf{x}\in S^*,\mathbf{P}\in SM_n} \{\|\mathbf{xP}\|_1 R_T\} = \tau_{\|.\|_1}(\mathbf{P}) R_T,$$

and

$$\|\mathbf{T}\|_i = \max_{\mathbf{x}\in\mathbb{H},\|\mathbf{x}\|=1} \{\|\mathbf{T}(\mathbf{x})\|\} = \max_{\mathbf{x}\in\mathbb{H},\|\mathbf{x}\|=1,\mathbf{P}\in SM_n} \{\|\mathbf{xP}\| R_T\} = \tau_{\|.\|}(\mathbf{P}) R_T.$$

Therefore

$$\max_{\mathbf{P}\in SM_n} \frac{\tau_{\|.\|_1}(\mathbf{P})}{\tau_{\|.\|}(\mathbf{P})} = \max_{\mathbf{P}\in SM_n} \frac{\tau_{\|.\|}(\mathbf{P})}{\tau_{\|.\|_1}(\mathbf{P})} = RR_1,$$

from which we get that

$$\frac{1}{RR_1}\tau_{\|.\|_1}(\mathbf{P}) \leq \tau_{\|.\|}(\mathbf{P}) \leq RR_1\tau_{\|.\|_1}(\mathbf{P}). \tag{3.75}$$

Now, from Lemma 212 and (3.73) and (3.74) we get that

$$C_{\|.\|} \geq RR_1.$$

From (3.75) and since $\tau_{\|.\|_1}(\mathbf{P}) \leq 1$ we get that

$$C_{\|.\|} \leq \tau_{\|.\|}(\mathbf{P}) \leq RR_1\tau_{\|.\|_1}(\mathbf{P}) \leq RR_1,$$

and therefore $C_{\|.\|} = RR_1$. Hence, we get the following important inequality

$$\frac{1}{C_{\|.\|}}\tau_{\|.\|_1}(\mathbf{P}) \leq \tau_{\|.\|}(\mathbf{P}) \leq C_{\|.\|}\tau_{\|.\|_1}(\mathbf{P}). \tag{3.76}$$

From the right hand side of inequality (3.76) we get

$$\tau_{\|.\|}(\mathbf{P}) \leq C_{\|.\|}\tau_{\|.\|_1}(\mathbf{P}),$$

and if $\mathbf{P}$ is scrambling then $\tau_{\|.\|_1}(\mathbf{P}) < 1$ and therefore $\tau_{\|.\|}(\mathbf{P}) \leq C_{\|.\|}$.

From the left hand side of (3.76), we get that $\tau_{\|.\|_1}(\mathbf{P}) \leq C_{\|.\|}\tau_{\|.\|}(\mathbf{P})$. Therefore, if $\tau_{\|.\|}(\mathbf{P}) < \frac{1}{C_{\|.\|}}$ then $\tau_{\|.\|_1}(\mathbf{P}) < 1$ and $\mathbf{P}$ is scrambling. ■

Our next important result is a generalization of the basic Theorem 208 by Leizarowitz mainly in the sense that makes use only of the Dobrushin ergodicity coefficient. This fact is another indication of the dominant role of the Dobrushin ergodicity coefficient in weak ergodicity of non-homogeneous Markov chains.

Theorem 215 *(Rhodius (1997)). Let a non-homogeneous Markov chain* $\{X_t\}_{t=0}^{\infty}$ *and* $\{\mathbf{P}(t)\}_{t=0}^{\infty}$ *the sequence of transition probabilities. If the infinite products of* $\{\mathbf{P}(t)\}_{t=0}^{\infty}$ *are taken in arbitrary order of multiplication and there exists at least one* $\mathbf{P} \in \mathcal{W}$ *scrambling then weak ergodicity obtains for the non-homogeneous Markov chain* $\{X_t\}_{t=0}^{\infty}$.

Proof. Let $\mathbf{P} \in \mathcal{W}$ scrambling then $\tau_{\|.\|_1}(\mathbf{P}) < a < 1$. Then there exist a subsequence $\mathbf{P}_{n_j}$ with $\mathbf{P} = \lim_{j\to\infty} \mathbf{P}_{n_j}$. Therefore, there exists a j_0 such that

$$\tau_{\|.\|_1}(\mathbf{P}_{n_j}) < a \text{ for all } j > j_0.$$

Consider all possible products taken in arbitrary order of multiplication with factors taken from the sequence $\{\mathbf{P}(t)\}_{t=0}^{\infty}$. All such products containing a large number of factors are such that they contain at least r factors $\mathbf{P}_{n_j}$ for $j > j_0$. Let $\mathbf{H}_m$ such a product of m factors then there exists an m_0 such that

$$\tau_{\|.\|_1}(\mathbf{H}_m) < a^r \quad \text{for } m > m_0.$$

Therefore, $\lim_{m\to\infty} \tau_{\|.\|_1}(\mathbf{H}_m) = 0$ which implies weak ergodicity. ■

Now, we will refresh the concept of *equicontinuity* for a family of functions (Rudin (1987)). We will provide the following definition which usually appears in the context of topological vector spaces.

Definition 216 *When X is a topological space, a family of functions F from X to Y is said to be equicontinuous at x if for every $\epsilon > 0$, x has a neighborhood U_x such that*

$$d_Y(f(y), f(x)) < \epsilon,$$

for all $y \in U_x$ and $f \in F$, where $d_Y(.,.)$ is a distance in Y.

We will now refer to some known results which will be useful in what follows

(1) When X is compact, a set is uniformly equicontinuous if and only if it is equicontinuous at every point. On a compact space, the notions of pointwise equicontinuity and uniform equicontinuity coincide.

(2) A Banach space is a vector space X over any scalar field $\mathfrak{F}$, which is a complete normed vector space. That is, there exist a norm in X, and for every Caushy sequence in X there exist an element in X to which the sequence converges in norm.

(3) Let X and Y be Banach spaces, and $\mathbf{O} : X \to Y$ be a family of continuous linear operators. Then $\mathbf{O}$ is equicontinuous if and only if

$$\sup\{\|\mathbf{T}\| : \mathbf{T} \in \mathbf{O}\} < \infty.$$

We will now state and prove the following basic theorem:

Theorem 217 *(Rhodius (1997)) Let a non-homogeneous Markov chain $\{X_t\}_{t=0}^{\infty}$ and $\{\mathbf{P}(t)\}_{t=0}^{\infty}$ the sequence of transition probabilities. If the infinite products of $\{\mathbf{P}(t)\}_{t=0}^{\infty}$ are taken in arbitrary order of multiplication and there exist a natural number $m \geq 1$ and an $K < 1$ such that for all m-tuples $(\mathbf{P}_1, \mathbf{P}_2, ..., \mathbf{P}_m) \in \mathcal{W}^{(m)}$*

$$\tau_{\|.\|_1}(\mathbf{P}_1\mathbf{P}_2...\mathbf{P}_m) \leq K,$$

then weak ergodicity obtains for the non-homogeneous Markov chain $\{X_t\}_{t=0}^{\infty}$.

Proof. Let $SM_n^{(m)}$ be the set of all m-tuples of stochastic matrices. That is,

$$SM_n^{(m)} = \{(\mathbf{Q}_1, \mathbf{Q}_2, ..., \mathbf{Q}_m) : \mathbf{Q}_i \in SM_n \text{ for } i = 1, 2, ..., m\}.$$

Then $SM_n^{(m)}$ is a compact set. Then the functional

$$\tau_{\|.\|_1}(\mathbf{Q}_1\mathbf{Q}_2, ..., \mathbf{Q}_m) \quad \text{for} \quad (\mathbf{Q}_1, \mathbf{Q}_2, ..., \mathbf{Q}_m) \in SM_n^{(m)},$$

is equicontinuous (uniformly) on the compact set $SM_n^{(m)}$.

Our basic assumption in the theorem states that for every m-tuple $(\mathbf{P}_1, \mathbf{P}_2, ..., \mathbf{P}_m) \in \mathcal{W}^{(m)}$ we have

$$\tau_{\|\cdot\|_1}(\mathbf{P}_1\mathbf{P}_2...\mathbf{P}_m) \leq K.$$

Define by $U_a(\mathbf{P}) = \{\mathbf{Q} \in SM_n : d(\mathbf{Q}, \mathbf{P}) < a\}$, where $d(.)$ is the Euclidean metric on R^{n^2}, that is $U_a(\mathbf{P})$ is a neighborhood of $\mathbf{P} \in SM_n$ of radius a. Let $\epsilon > 0$ and $K + \epsilon < 1$. Then for any m-tuple $(\mathbf{P}_1, \mathbf{P}_2, ..., \mathbf{P}_m) \in \mathcal{W}^{(m)}$ there exists a $\mu > 0$ suh that the following is true:

For every $(\mathbf{Q}_1, \mathbf{Q}_2, ..., \mathbf{Q}_m) \in SM_n^{(m)}$ such that

$$\mathbf{Q}_i \in U_\mu(\mathbf{P}_i) \text{ for } i = 1, 2, ..., m,$$

we have that

$$\tau_{\|\cdot\|_1}(\mathbf{Q}_1\mathbf{Q}_2...\mathbf{Q}_m) \leq K + \epsilon < 1.$$

Now, since $\mathcal{W}$ is compact and $\mathcal{W} \subseteq SM_n$, there exists a finite set $\{\mathbf{T}_1, \mathbf{T}_2, ..., \mathbf{T}_r\} \subseteq \mathcal{W}$ such that

$$\mathcal{W} \subseteq \cup_{i=1}^r U_\mu(\mathbf{T}_i).$$

Now, let us consider the transition probability sequence $\{\mathbf{P}(t)\}_{t=0}^\infty$ of the non-homogeneous Markov chain $\{X_t\}_{t=0}^\infty$. Since $\mathbf{P}(t)$ as $t \to \infty$ converges to an element of $\mathcal{W}$ there exists a t_0 with

$$\mathbf{P}(t) \in \cup_{i=1}^r U_\mu(\mathbf{T}_i) \quad \text{for } t > t_0.$$

This is because otherwise it will exist a convergent subsequence $\{\mathbf{P}(t_i)\}_{i=1}^\infty$ of the sequence $\{\mathbf{P}(t)\}_{t=0}^\infty$ such that

$$\mathbf{P}(t_i) \notin \cup_{i=1}^r U_\mu(\mathbf{T}_i) \quad \text{for all } i = 1, 2, ...$$

and $\lim_{i\to\infty} \mathbf{P}(t_i) = \mathbf{P}^*$ and hence $\mathbf{P}^* \notin \mathcal{W}$, which is an apparent contradiction.

Consider all possible products taken in arbitrary order of multiplication with factors taken from the sequence $\{\mathbf{P}(t)\}_{t=0}^\infty$. All such products containing a large number of factors, are such that they contain as factors the matrices $\mathbf{P}(0), \mathbf{P}(1), ..., \mathbf{P}(t_0)$. Let $\mathbf{H}_s$ such a product of s factors then there exists an s_0 such that the matrices $\mathbf{P}(0), \mathbf{P}(1), ..., \mathbf{P}(t_0)$ are factors of $\mathbf{H}_s$ for $s \geq s_0$. Hence, we have

$$\tau_{\|\cdot\|_1}(\mathbf{H}_{s_0+km+v}) \leq \tau_{\|\cdot\|_1}(\mathbf{H}_{s_0}) . (K + \epsilon)^k,$$

for all $k \geq 2, v = 0, 1, 2, ..., m-1$ and therefore

$$\lim_{s\to\infty} \tau_{\|\cdot\|_1}(\mathbf{H}_s) = 0,$$

which implies weak ergodicity for the non-homogeneous Markov chain $\{X_t\}_{t=0}^\infty$. ■

It is not difficult see that Part 1 of Theorem 208 follows in a straight forward way from Theorem 215. Also, Part II of Theorem 208 follows from Theorem 217.

3.11 Transition probability matrices are in chronological order

The study of weak ergodicity of Markov chains when the transition probability matrices are in some usually chronological order is the oldest in the literature. It started with Markov itself and Kolmogorov (1931) and partly crystallized in a paper by Doeblin in 1937. However, it was revived and exploited thoroughly by Dobrushin (1956), Hajnal (1956, 1958, 1976, 1993), Mott and Shneider (1957). Also from the Russian school of probability by Sarysmakov (1953, 1956, 1958, 1961). More could be found on the thorough book by Seneta (2006) and in Seneta (2014). In the present, we will study two basic theorems which treat the problem.

Theorem 218 *(Mott and Schneider (1957)). Let a norm* $\|.\|$ *on* SM_n *and* $\mathbf{S}(1), \mathbf{S}(2), \ldots$, *be a sequence of stable matrices in* SM_n. *Also let the non-homogeneous Markov chain* $\{X_t\}_{t=0}^{\infty}$ *with* $\{\mathbf{P}(t)\}_{t=0}^{\infty}$ *the sequence of transition probabilities. Then the forward product* $\mathbf{P}(t, t+n)$ *is weakly ergodic if*

$$\lim_{n\to\infty} \prod_{t=1}^{n} \|\mathbf{P}(t) - \mathbf{S}(t)\mathbf{P}(t)\| = 0.$$

Proof. Consider the product

$$[\mathbf{P}(1) - \mathbf{S}(1)\mathbf{P}(1)][\mathbf{P}(2) - \mathbf{S}(2)\mathbf{P}(2)] =$$

$$\mathbf{P}(1)\mathbf{P}(2) - \mathbf{P}(1)\mathbf{S}(2)\mathbf{P}(2) - \mathbf{S}(1)\mathbf{P}(1)\mathbf{P}(2) + \mathbf{S}(1)\mathbf{P}(1)\mathbf{S}(2)\mathbf{P}(2). \tag{3.77}$$

Now, it easy to check that if $\mathbf{P}$ is stochastic and $\mathbf{S}$ stable then

$$\mathbf{PS} = \mathbf{S} \text{ and } \mathbf{SP} \text{ a stable stochastic matrix.} \tag{3.78}$$

From (3.78) we get that (3.77) could be written as

$$[\mathbf{P}(1) - \mathbf{S}(1)\mathbf{P}(1)][\mathbf{P}(2) - \mathbf{S}(2)\mathbf{P}(2)] = \mathbf{P}(1)\mathbf{P}(2) - \mathbf{S}(1)\mathbf{P}(1)\mathbf{P}(2),$$

and by induction we easily get that

$$\prod_{t=1}^{n} [\mathbf{P}(t) - \mathbf{S}(t)\mathbf{P}(t)] = \mathbf{P}(1,n) - \mathbf{S}(1)\mathbf{P}(1,n). \tag{3.79}$$

We now have

$$\begin{aligned} \prod_{t=1}^{n} \|\mathbf{P}(t) - \mathbf{S}(t)\mathbf{P}(t)\| &\geq \left\| \prod_{t=1}^{n} [\mathbf{P}(t) - \mathbf{S}(t)\mathbf{P}(t)] \right\| \\ &= \|\mathbf{P}(1,n) - \mathbf{S}(1)\mathbf{P}(1,n)\|. \end{aligned} \tag{3.80}$$

From the assertion of the theorem and (3.80) we get that

$$\lim_{n\to\infty} \|\mathbf{P}(1,n) - \mathbf{S}(1)\mathbf{P}(1,n)\| = 0,$$

which implies weak ergodicity. ■

We have mentioned many times before that a matrix $\mathbf{P} \in SM_n$ is scrambling, by definition, if and only if $\tau_{\|.\|_1}(\mathbf{P}) < 1$, or equivalently, that no two rows of $\mathbf{P}$ are orthogonal, or alternatively given two rows i and j there is at least one column k such that $p_{ik} > 0$ and $p_{jk} > 0$. A fundamental assumption for many theorems of ergodicity is that the products of matrices $\mathbf{P}(1,n)$ from the sequence of transition probability matrices is a regular stochastic matrix. This assumption is not easily checked in a practical way and from the point of view of utility we need conditions on individual matrices $\mathbf{P}(t)$. The difficulty increases by the following facts which can be easily checked: a) the product of two non-regular matrices may be regular and b) the products of regular matrices may not be regular. In order to overcome partially at least these difficulties we distinguish the following classes of matrices.

Definition 219 *a) A matrix* $\mathbf{P} \in \mathcal{M}$ *is called a Markov if at least one column of* $\mathbf{P}$ *is entirely positive. b)* $\mathbf{P} \in \mathrm{G}_2$ *if (i)* $\mathbf{P} \in \mathrm{G}_1$ *a regular matrix; (ii)* $\mathbf{QP} \in \mathrm{G}_1$ *for any* $\mathbf{Q} \in \mathrm{G}_1$. *c)* $\mathbf{P} \in \mathrm{G}_3$ *if* $\tau_{\|\cdot\|_1}(\mathbf{P}) < 1$, *i.e.,* $\mathbf{P}$ *is scrambling.*

We will now prove the following theorem on the relation between these classes of stochastic matrices.

Theorem 220 *For all* $\mathbf{P} \in SM_n$ *we have* $\mathcal{M} \subset \mathrm{G}_3 \subset \mathrm{G}_2 \subset \mathrm{G}_1$.

Proof. From the above definition we have that if $\mathbf{P} \in G_2$ then $\mathbf{P} \in G_1$ and thus $G_2 \subset G_1$. Also if $\mathbf{P} \in \mathcal{M}$ then it has at least one column entirely positive hence $\tau_{\|\cdot\|_1}(\mathbf{P}) < 1$ and $\mathbf{P} \in G_3$, hence $\mathcal{M} \subset G_3$.

Now, what remains is to prove that $G_3 \subset G_2$. Let a scrambling matrix $\mathbf{P} \in G_3$ we need to show that $\mathbf{P} \in G_2$, that is by definition we should prove that (i) $\mathbf{P} \in G_1$ a regular matrix; (ii) $\mathbf{QP} \in G_1$ for any $\mathbf{Q} \in G_1$.

Let us start with if $\mathbf{P} \in G_3$ then $\mathbf{P} \in G_1$ a regular matrix. We need to prove that $\mathbf{P} \in G_3$ cannot have more than one essential class, and if it has one cannot be periodic. Let that $\mathbf{P} \in G_3$ has two essential classes then its canonical form will be

$$\mathbf{P} = \begin{pmatrix} \mathbf{P}_1 & \mathbf{0} & \mathbf{0} \\ \mathbf{0} & \mathbf{P}_2 & \mathbf{0} \\ \mathbf{R}_1 & \mathbf{R}_2 & \mathbf{Q} \end{pmatrix},$$

where $\mathbf{P}_1$ and $\mathbf{P}_2$ are stochastic matrices. If we choose a row i in $\mathbf{P}_1$ and a row j in $\mathbf{P}_2$ then

$$\frac{1}{2}\sum_{k=1}^{n} |p_{ik} - p_{jk}| = 1,$$

hence the matrix $\mathbf{P} \notin G_3$. Therefore, $\mathbf{P} \in G_3$ consists of only one essential class. Let us now think if it is possible to be periodic. Assume periodicity and let that its cyclic subclasses are $\mathcal{C}_0, \mathcal{C}_1, \mathcal{C}_2$, then it could be written in the form

$$\mathbf{P} = \begin{pmatrix} \mathbf{0} & \mathbf{P}_1 & \mathbf{0} \\ \mathbf{0} & \mathbf{0} & \mathbf{P}_2 \\ \mathbf{P}_3 & \mathbf{0} & \mathbf{0} \end{pmatrix},$$

and with the same reasoning we conclude that it can not be periodic, thus $\mathbf{P} \in G_1$ a regular matrix.

It remains to prove that if $\mathbf{P} \in G_3$ scrambling then $\mathbf{QP} \in G_1$ for any $\mathbf{Q} \in G_1$. Take any two rows i and j of $\mathbf{QP}$ then the entries of column α are

$$\sum_{k=1}^{n} q_{ik}p_{ka} \quad \text{and} \quad \sum_{r=1}^{n} q_{jr}p_{ra}.$$

Since $\mathbf{Q}$ is stochastic then there exists at least one k, r such that $q_{ik} > 0$ and $q_{jr} > 0$. Since $\mathbf{P} \in G_3$ scrambling there exists α such that $p_{ka} > 0$ and $p_{ra} > 0$. Therefore, $\mathbf{QP} \in G_1$. ■

A natural extension of the above is the following Lemma.

Lemma 221 *If* $\mathbf{P}$ *is scrambling, then so are* $\mathbf{PQ}$ *and* $\mathbf{QP}$ *for any stochastic* $\mathbf{Q}$.

Proof. The proof could be done as an exercise easily in view of the previous theorem (see Seneta 2006). ■

We will now prove two useful Lemmas for the proof of the basic theorem that follows (see Seneta (2006)).

Lemma 222 *If* $\mathbf{P},\mathbf{Q} \in SM_n$, $\mathbf{Q}$ *regular and* $\mathbf{PQ}$ *or* $\mathbf{QP}$ *has the same incidence matrix as* $\mathbf{P}$,*that is,* $\mathbf{PQ} \sim \mathbf{P}$ *or* $\mathbf{QP} \sim \mathbf{P}$, *then* $\mathbf{P} \in \mathcal{M}$.

Proof. Since $\mathbf{Q} \in G_1$ regular, there exists a $r > 0$ such that $\mathbf{Q}^r \in M$ (see Vassiliou (2010)). Let us assume first that $\mathbf{PQ} \sim \mathbf{P}$ then $\mathbf{PQ}^2 \sim (\mathbf{PQ})\,\mathbf{Q} \sim \mathbf{PQ} \sim \mathbf{P}$ and so on and finally $\mathbf{PQ}^r \sim \mathbf{P}$. Therefore $\mathbf{P} \in \mathcal{M}$. ■

Lemma 223 *If* $\mathbf{P}(t, t+n) \in \mathrm{G}_1$ *with* $t \geq 0$, $n \geq 1$, *then* $\mathbf{P}(t, t+n) \in \mathrm{M}$ *for* $t + n \geq \mu$ *the number of distinct incidence matrices corresponding to* G_1.

Proof. Since μ is finite for a fixed t there are numbers m and k such that $1 \leq m < k < \mu+1$ and

$$\begin{aligned} &\mathbf{P}(t)\,\mathbf{P}(t+1)\ldots\mathbf{P}(t+m)\,\mathbf{P}(t+m+1)\ldots\mathbf{P}(t+k) \\ \sim\; &\mathbf{P}(t)\ldots\mathbf{P}(t+m). \end{aligned}$$

Therefore,

$$\mathbf{P}(t,t+m)\,\mathbf{P}(t+m,t+k) \sim \mathbf{P}(t,t+m).$$

Since, $\mathbf{P}(t+m,t+k) \in G_1$ from the previous Lemma we get $\mathbf{P}(t,t+m) \in M$. Therefore, $\mathbf{P}(t,t+n) \in M$ for $t+n \geq \mu$. ■

We will now ready to prove the following theorem which is slightly altered from Theorem 4.10 in Seneta (2006).

Theorem 224 *Let the non-homogeneous Markov chain* $\{X_t\}_{t=0}^{\infty}$ *and the sequence of transition probabilities* $\{\mathbf{P}(t)\}_{t=0}^{\infty}$. *If* $\mathbf{P}(t) \in \mathrm{G}_2$ *for every* $t = 1, 2, \ldots$ *and*

$$\min_{i,j}(p_{ij}(t), 0)^{+} \geq \gamma > 0, \tag{3.81}$$

uniformly for all $t \geq 1$, *then weak ergodicity obtains at a uniform geometric rate.*

Proof. Firstly, it is not difficult for the reader to show that if $\mathbf{P}, \mathbf{Q} \in G_2$, then $\mathbf{PQ}$, $\mathbf{Q}$ $\mathbf{P} \in G_2$. Therefore, $\mathbf{P}(t,t+n) \in G_2$ for every $t, n \geq 0$.

Now, let that $\mu > 0$ is the number of distinct incidence matrices corresponding to G_1. Let also that for arbitrary but fixed t and n "large enough" we have that $k\mu \leq t+n$ with k the largest such integer, then

$$\begin{aligned} \mathbf{P}(t,t+n) &= \mathbf{P}(t,\mu)\,\mathbf{P}(\mu+1,2\mu)\ldots\mathbf{P}(k-1)\mu+1,k\mu)\,\mathbf{P}(k\mu+1,t+n) \\ &= \mathbf{P}(t,k\mu)\,\mathbf{P}(k\mu+1,t+n). \end{aligned}$$

Since by Lemma 223 we have that $\mathbf{P}(i-1)\mu+1, i\mu)$ is Markov then by (3.81) we get that

$$c(\mathbf{P}(i-1)\mu+1, i\mu)) \leq 1-\gamma^{\mu},$$

with $c(P)$ the improper ergodicity coefficient

$$c(\mathbf{P}) = 1 - \max_{k}\left(\min_{i} p_{ik}\right).$$

Another, proper though ergodicity coefficient is

$$a(\mathbf{P}) = \max_{k}\max_{i,j}|p_{ik}-p_{jk}|,$$

for which it is known that $a(\mathbf{P}) \leq c(\mathbf{P})$. Therefore, we have that

$$a(\mathbf{P}(t,t+n)) \leq \prod_{i=0}^{k-1} c(\mathbf{P}(i-1)\mu+1, i\mu)) \leq (1-\gamma^{\mu})^{k},$$

which since $\gamma < 1$ as $n \to \infty$ then $a(\mathbf{P}(t,t+n)) \to 0$ which implies weak ergodicity at a uniform geometric rate. ■

3.12 Examples on the use of weak ergodicity theorems

Example 225 *(Theorem 224). Let the matrix*

$$\mathbf{P}(1) = \begin{pmatrix} 0.2 & 0.8 & 0 \\ 0.3 & 0.5 & 0.2 \\ 0 & 0.2 & 0.8 \end{pmatrix}.$$

Using the method presented in Vassiliou (2010) p. 101, we find that it consists of one essential class. Also using the method presented in Vassiliou (2010) p. 109, we find that the matrix is aperiodic. We continue creating the matrix $\mathbf{P}(2)$ *having the same incidence matrix as* $\mathbf{P}(1)$ *and let*

$$\mathbf{P}(2) = \begin{pmatrix} 0.7 & 0.3 & 0 \\ 0.4 & 0.2 & 0.4 \\ 0 & 0.3 & 0.7 \end{pmatrix}$$

Each of these matrices it is not difficult to prove that they are regular and also they belong to the set G_2*. Also, all of them satisfy the assumption of Theorem 224 that*

$$\min_{i,j} (p_{ij}(t), 0)^+ \geq 0.2.$$

With the same incidence matrix, we could create as many members of $\mathbf{P}(t)$ *needed with the same properties. However, since the convergence is at a geometric rate that will not be necessary. After seven and eight steps we get respectively*

$$\mathbf{P}(1,7) = \begin{pmatrix} 0.2 & 0.5 & 0.3 \\ 0.2 & 0.5 & 0.3 \\ 0.2 & 0.5 & 0.3 \end{pmatrix},$$

and

$$\mathbf{P}(1,8) = \begin{pmatrix} 0.16 & 0.58 & 0.26 \\ 0.16 & 0.58 & 0.26 \\ 0.16 & 0.58 & 0.26 \end{pmatrix},$$

hence, weak ergodicity is evident.

Example 226 *(Theorem 215). Let the non-homogeneous Markov chain* $\{X_t\}_{t=0}^{\infty}$ *and the sequence of transition probabilities* $\{\mathbf{P}(t)\}_{t=0}^{\infty}$ *which are constructed in the following way:*

$$\mathbf{P}(t) = \begin{pmatrix} 0.8 - \frac{1}{2t} & 0.2 + \frac{1}{2t} & 0 \\ 0.1 + \frac{1}{2t^2} & 0.7 - \frac{1}{2t^2} & 0.2 \\ 0.3 & 0.6 & 0.1 \end{pmatrix}, \text{ for } t = 1 + 3\pi, \ \pi = 0, 1, 2...,$$

$$\mathbf{P}(t) = \begin{pmatrix} 1 - \frac{1}{2t^3} & \frac{1}{2t^3} & 0 \\ 0.1 + \frac{1}{2t+3} & 0.9 - \frac{1}{2t+3} & 0 \\ 0 & 1 & 0 \end{pmatrix}, \text{ for } t = 2 + 3\pi, \ \pi = 0, 1, 2...,$$

$$\mathbf{P}(t) = \begin{pmatrix} 1 - \frac{1}{2+t^2} & \frac{1}{2+t^2} & 0 \\ 0 & 0.9 - \frac{1}{2t+3} & 0.1 + \frac{1}{2t+3} \\ 1 & 0 & 0 \end{pmatrix}, \text{ for } t = 3\pi, \ \pi = 1, 2....$$

The accumulation points of the sequence $\{\mathbf{P}(t)\}_{t=1}^{\infty}$ *are*

$$\mathcal{W} = \left\{ \begin{pmatrix} 0.8 & 0.2 & 0 \\ 0.1 & 0.7 & 0.2 \\ 0.3 & 0.6 & 0.1 \end{pmatrix}, \begin{pmatrix} 1 & 0 & 0 \\ 0.1 & 0.9 & 0 \\ 0 & 1 & 0 \end{pmatrix}, \begin{pmatrix} 1 & 0 & 0 \\ 0 & 0.9 & 0.1 \\ 1 & 0 & 0 \end{pmatrix} \right\}.$$

In $\mathcal{W}$ only the first matrix is scrambling with coefficient of ergodicity equal 0.7. *The other two have coefficients of ergodicity equal with* 1. *From the sequence* $\{\mathbf{P}(t)\}_{t=1}^{\infty}$ *we construct a general product as described in section 3.9.2. Let that*

$$\begin{aligned} C_{p,8} &= \mathbf{P}(3)\,\mathbf{P}(7)\,\mathbf{P}(1)\,\mathbf{P}(9)\,\mathbf{P}(5)\,\mathbf{P}(4)\,\mathbf{P}(16)\,\mathbf{P}(22) \\ &= \begin{pmatrix} 0.4 & 0.5 & 0.1 \\ 0.4 & 0.5 & 0.1 \\ 0.4 & 0.5 & 0.1 \end{pmatrix}, \end{aligned}$$

and

$$\begin{aligned} C_{p,9} &= \mathbf{P}(3)\,\mathbf{P}(7)\,\mathbf{P}(1)\,\mathbf{P}(9)\,\mathbf{P}(5)\,\mathbf{P}(4)\,\mathbf{P}(16)\,\mathbf{P}(22)\,\mathbf{P}(33) \\ &= \begin{pmatrix} 0.5 & 0.45 & 0.05 \\ 0.5 & 0.45 & 0.05 \\ 0.5 & 0.45 & 0.05 \end{pmatrix}. \end{aligned}$$

As it is apparent $C_{p,8}$ and $C_{p,9}$ are stable matrices as it is expected in weak ergodicity.

Chapter 4

NON-HOMOGENEOUS MARKOV CHAINS; STRONG ERGODICITY

In the previous chapter, we introduced the concept of an non-homogeneous Markov chain and provided necessary and sufficient conditions for weak ergodicity mainly for finite state space. The style of presentation we followed was to present first what we thought to be the most important general theorems and then the classical theorems that provided the first semantic answers to the problem of weak ergodicity. In the present chapter, we will study the problem of strong ergodicity for non-homogeneous Markov chains and we will in general assume that the state space is infinitely countable. We will adopt in this chapter the inverse than in the previous chapter style of presentation. That is, we will start with some classical results on the problem of strong ergodicity.

4.1 Strong ergodicity

Let an non-homogeneous Markov chain $\{X_t\}_{t=0}^{\infty}$ with state space $S = \{1, 2, ...\}$ and sequence of transition probabilities $\{\mathbf{P}(t)\}_{t=0}^{\infty}$. In the present chapter we will use the norm $\|.\|_{\infty}$ which for a matrix $\mathbf{A}$ with countably infinite size we have seen that is given by:

$$\|\mathbf{A}\|_{\infty} = \sup_i \sum_j |a_{ij}| \,.$$

For simplicity though we will use the symbol of $\|.\|$ for this norm in this chapter. Now, if $\mathbf{Q}$ is a stochastic matrix of infinitely countable size, that is, $S = \{1, 2, ...\}$ then the Dobruchin ergodicity coefficient it is known that is given by

$$\tau_1(\mathbf{Q}) = \frac{1}{2} \sup_{i,k} \sum_{j\in S} |q_{ij} - q_{kj}| \,.$$

We now provide the following definition of strong ergodicity:

Definition 227 *We say that a non-homogeneous Markov chain $\{X_t\}_{t=0}^{\infty}$ with state space $S = \{1, 2, ...\}$ and sequence of transition probabilities $\{\mathbf{P}(t)\}_{t=0}^{\infty}$ is strongly ergodic if there exists a stochastic matrix $\mathbf{Q}$ stable such that for every t*

$$\lim_{n\to\infty} \|\mathbf{P}(t, t+n) - \mathbf{Q}\| = 0. \tag{4.1}$$

If the limit is zero uniformly in t we say that $\{X_t\}_{t=0}^{\infty}$ or equivalently $\{\mathbf{P}(t)\}_{t=0}^{\infty}$ is uniformly strongly ergodic.

In here it is useful to remark that the mode of convergence above used to define strong ergodicity is stronger than coordinate-wise convergence. That is, it is not sufficient to have

DOI: 10.1201/b23204-4

$\lim_{n\to\infty} p_{ij}(t,t+n) = q_i$ in order to have strong ergodicity in the case of infinitely countable non-homogeneous Markov chains. In fact, it is necessary to have

$$\lim_{n\to\infty} \sup_{i\in S} \sum_{j=1}^{\infty} |p_{ij}(t,t+n) - q_j| = 0. \tag{4.2}$$

On the contrary in the case of finite non-homogeneous Markov chains coordinate convergence and convergence in norm are equivalent. It is also useful to remark that it is not difficult for the reader to prove that strong ergodicity implies weak ergodicity.

Example 228 *Consider the non-homogeneous Markov chain* $\{X_t\}_{t=0}^{\infty}$ *with transition matrices*

$$\mathbf{P}(2t) = \begin{pmatrix} 1-\frac{2}{t^2} & \frac{1}{t^2} & \frac{1}{t^2} \\ 1-\frac{2}{t^2} & \frac{1}{t^2} & \frac{1}{t^2} \\ 1-\frac{2}{t^2} & \frac{1}{t^2} & \frac{1}{t^2} \end{pmatrix} \text{ and } \mathbf{P}(2t+1) = \begin{pmatrix} 1-\frac{2}{t^2-1} & \frac{1}{t^2-1} & \frac{1}{t^2-1} \\ 1-\frac{2}{t^2-1} & \frac{1}{t^2-1} & \frac{1}{t^2-1} \\ 1-\frac{2}{t^2-1} & \frac{1}{t^2-1} & \frac{1}{t^2-1} \end{pmatrix},$$

for $t = 1, 2, \ldots$. *Since* $\mathbf{P}(2t)$ *and respectively* $\mathbf{P}(2t+1)$ *are stable stochastic matrices, that is, stochastic matrices with equal rows we easily see that*

$$\mathbf{P}(t,t+n) = \begin{pmatrix} 1-\frac{2}{t^2} & \frac{1}{t^2} & \frac{1}{t^2} \\ 1-\frac{2}{t^2} & \frac{1}{t^2} & \frac{1}{t^2} \\ 1-\frac{2}{t^2} & \frac{1}{t^2} & \frac{1}{t^2} \end{pmatrix} \text{ for } t,n \text{ both even or both odd,}$$

and

$$\mathbf{P}(t,t+n) = \begin{pmatrix} \frac{1}{t^2+1} & 1-\frac{2}{t^2+1} & \frac{1}{t^2+1} \\ \frac{1}{t^2+1} & 1-\frac{2}{t^2+1} & \frac{1}{t^2+1} \\ \frac{1}{t^2+1} & 1-\frac{2}{t^2+1} & \frac{1}{t^2+1} \end{pmatrix} \text{ for } t,n \text{ for either one of them odd .}$$

Hence the chain is weakly ergodic but not strongly ergodic.

For non-homogeneous Markov chains with infinitely countable state space $S = \{1, 2, \ldots\}$ we distinguish two more cases of ergodicity which are the analogues of the homogenous Markov chain case with infinitely countable state space $S = \{1, 2, \ldots\}$, see Isaacson (1979).

Definition 229 *Let a non-homogeneous Markov chain* $\{X_t\}_{t=0}^{\infty}$ *with state space* $S = \{1, 2, \ldots\}$ *and sequence of transition probabilities* $\{\mathbf{P}(t)\}_{t=0}^{\infty}$. *Also, let* $\mathbf{\Pi}$ *be a stable stochastic matrix of the same size as each* $\mathbf{P}(t)$ *and with row* $\boldsymbol{\pi} = [\pi_1, \pi_2, \pi_3, \ldots]$. *Then* $\{X_t\}_{t=0}^{\infty}$ *is said to be ergodic if and only if for each* $i \in S$ *and* $t \geq 0$

$$\lim_{n\to\infty} \sum_{j} |p_{ij}(t,t+n) - \pi_j| = 0, \quad \text{independent of } t. \tag{4.3}$$

Moreover, $\{X_t\}_{t=0}^{\infty}$ *is said to be geometrically ergodic if and only if for each* t *there exists constants* $c(t) \geq 0$ *and* $0 < b(t) < 1$ *such that*

$$\sup_{i} \sum_{j} |p_{ij}(t,t+n) - \pi_j| \leq c(t)\,[b(t)]^n. \tag{4.4}$$

Remark 230 *Strong ergodicity is a stronger requirement than geometric ergodicity.*

4.2 Ergodicity and geometric strong ergodicity

We will start by providing a necessary and sufficient condition for the ergodicity of an infinitely countable inhomogeneous Markov chain. Firstly though we will state what they are called the Lebesgue's Theorems for series which are well known and their proofs could be found in many texts, and for example we refer to Bremaud (2020 p. 530). These theorems will prove useful in what follows.

Theorem 231 ***(Dominated convergence theorem).*** *Let $\{a_{nk}\}_{n\geq 1,k\geq 1}$ be a double sequence of real numbers such that for some sequence $\{b_k\}_{k\geq 1}$ of non-negative real numbers which are such that*

$$\sum_{k=1}^{\infty} b_k < \infty,$$

and in addition let that

$$|a_{nk}| \leq b_k \text{ for } n,k \geq 1 \quad \text{and} \quad \lim_{n\to\infty} a_{nk} = a_k.$$

Then

$$\lim_{n\to\infty} \sum_{k=1}^{\infty} a_{nk} = \sum_{k=1}^{\infty} a_k.$$

Theorem 232 ***(Monotone convergence theorem).*** *Let $\{a_{nk}\}_{n\geq 1,k\geq 1}$ be a double sequence of non-negative real numbers which are such that*

$$\lim_{n\to\infty} a_{nk} = a_k \leq \infty.$$

Then

$$\lim_{n\to\infty} \sum_{k=1}^{\infty} a_{nk} = \sum_{k=1}^{\infty} a_k.$$

Let us now refresh the concept of lim inf for a sequence of real numbers, that is, limes inferior or limit infimum which we will use in the next theorem called also by the name Fatou's Lemma. Let $\{a_n\}_{n\geq 0}$ be a sequence of real numbers then the lim inf of this sequence is defined as

$$\liminf_{n\to\infty} a_n := \lim_{n\to\infty}\left(\inf_{m\geq n} a_m\right) = \sup\left\{\inf\left\{a_m : m\geq n\right\} : n\geq 0\right\}.$$

For any two sequences of real numbers $\{a_n\}_{n\geq 0}$ and $\{b_n\}_{n\geq 0}$, the limit inferior satisfies the following inequality if the right hand side is properly defined, that is, it is not infinity.

$$\liminf_{n\to\infty} (a_n + b_n) \geq \liminf_{n\to\infty} a_n + \liminf_{n\to\infty} b_n,$$

and if the right hand side is not of the form $0.\infty$ it satisfies the inequality

$$\liminf_{n\to\infty} (a_n b_n) \geq \left(\liminf_{n\to\infty} a_n\right)\left(\liminf_{n\to\infty} b_n\right).$$

Remark 233 *Define by Ψ the set of the limits of all possible subsequences of the sequence of real numbers $\{a_n\}_{n\geq 0}$ then*

$$\liminf_{n\to\infty} a_n = \inf \Psi.$$

The limes inferior always exist, as the extended real number line is complete.

Theorem 234 ***(Fatou's lemma).*** *Let $\{a_{nk}\}_{n\geq 1,k\geq 1}$ be a double sequence of non-negative real numbers. Then*

$$\sum_{k=1}^{\infty} \liminf_{n\to\infty} a_{nk} \leq \liminf_{n\to\infty} \sum_{k=1}^{\infty} a_{nk}.$$

Theorem 235 *(Isaacson and Seneta, 1982). Let a non-homogeneous Markov chain $\{X_t\}_{t=0}^{\infty}$ with state space $S=\{1,2,...\}$ and sequence of transition probabilities $\{\mathbf{P}(t)\}_{t=0}^{\infty}$. Then $\{X_t\}_{t=0}^{\infty}$ is said to be ergodic if and only if for all $i,r\in S$*

$$\sup_k \sum_j |p_{ij}(t,t+n) - p_{rj}(t,t+n+k)| \to 0 \quad \text{as } n\to\infty. \tag{4.5}$$

Proof. Let that $\{X_t\}_{t=0}^{\infty}$ is ergodic, that is

$$\sum_j |p_{ij}(t,t+n) - \pi_j| \to 0 \text{ as } n \to \infty, \tag{4.6}$$

we will firstly prove that then (4.5) is true. We have that

$$\begin{aligned}
&\sup_k \sum_j |p_{ij}(t,t+n) - p_{rj}(t,t+n+k)| \\
&= \text{(add and subtract } \pi_j \text{ within the absolute value)} \\
&\leq \sup_k \sum_j |p_{ij}(t,t+n) - \pi_j| + \sup_k \sum_j |p_{ij}(t,t+n+k) - \pi_j| \\
&= \sum_j |p_{ij}(t,t+n) - \pi_j| + \sup_k \sum_j |p_{ij}(t,t+n+k) - \pi_j| \\
&= \text{(by (4.6) for every } \epsilon > 0 \text{ there exists a } N(\epsilon) \text{ such that)} \\
&< \epsilon \quad \text{for } n \geq N(\epsilon).
\end{aligned}$$

Hence, we have proved (4.5). Assume now that (4.5) is true we will prove that then the Markov chain $\{X_t\}_{t=0}^{\infty}$ is ergodic. Since the relation (4.5) is true for all $i, r \in S$, it will hold for $r = i$ also. Therefore, we have

$$\sup_k \sum_j |p_{ij}(t,t+n) - p_{ij}(t,t+n+k)| \to 0 \text{ as } n \to \infty. \tag{4.7}$$

Hence, we get that for all $k \geq 1$ $|p_{ij}(t,t+n) - p_{ij}(t,t+n+k)| \to 0$ as $n \to \infty$, thus $p_{ij}(t,t+n)$ is a Caushy sequence in n and therefore,

$$p_{ij}(t,t+n) \to \pi_{ij}(t) \text{ as } n \to \infty. \tag{4.8}$$

Now, with the use of Fatou lemma we get

$$\begin{aligned}
\sum_j |p_{ij}(t,t+n) - \pi_{ij}(t)| &= \sum_j \liminf_{k\to\infty} |p_{ij}(t,t+n) - p_{ij}(t,t+n+k)| \\
&\leq \liminf_{k\to\infty} \sum_j |p_{ij}(t,t+n) - p_{ij}(t,t+n+k)| < \epsilon \quad \text{for} \quad k > K(\epsilon).
\end{aligned} \tag{4.9}$$

On the left hand side of (4.9) within the absolute value, we add and subtract $p_{rj}(t,t+n+1)$ and we get

$$\begin{aligned}
\sum_j |p_{ij}(t,t+n) - \pi_{rj}(t)| &\leq \sum_j |p_{ij}(t,t+n) - p_{rj}(t,t+n+1)| \\
&+ \sum_j |\pi_{rj}(t) - p_{rj}(t,t+n+1)| \to 0 \text{ as } n \to \infty,
\end{aligned} \tag{4.10}$$

due to the fact that the first term on the right hand side of (4.9) by (4.5) goes to zero as $n \to \infty$, and the second term on the right hand side of (4.9) by (4.8) goes to zero as $n \to \infty$.

From (4.8) and (4.10) we get that

$$\pi_{ij}(t) = \pi_{rj}(t) = \pi_j(t) \quad \text{for all } i, r \in S. \tag{4.11}$$

From the well known Chapman-Kolmogorov equation, we have

$$p_{ij}(t,t+n) = \sum_l p_{il}(t+1)\, p_{lj}(t+1.t+n). \tag{4.12}$$

Having in mind the Lebesgue's dominated-convergence theorem set as the double sequence

$$a_{nl}(t) = p_{il}(t+1)\, p_{lj}(t+1.t+n),\ b_l(t) = p_{il}(t+1),$$

then it is easy to see that

$$\begin{aligned} \sum_l b_l(t) &= 1 < \infty,\ a_{nl}(t) < b_l(t),\ \lim_{n\to\infty} a_{nl}(t) = a_l(t), \\ \text{where } a_l(t) &= \lim_{n\to\infty} p_{il}(t+1)\, p_{lj}(t+1.t+n) = p_{il}(t+1)\,\pi_j(t+1). \end{aligned}$$

and therefore

$$\pi_j(t) = \sum_j p_{il}(t+1)\,\pi_j(t+1),$$

hence $\pi_j(t) = \pi_j(t+1) = \pi_j$ for every t and $j \in S$. It remains to show that $\boldsymbol{\pi} = [\pi_1, \pi_2, ...]$ is a stochastic vector. In this respect, we have

$$\begin{aligned} \left|1 - \sum_j \pi_j\right| &= \left|\sum_j p_{ij}(t,t+n) - \sum_j \pi_j\right| \\ &\leq \sum_j |p_{ij}(t,t+n) - \pi_j| \to 0 \ \text{ as } n \to \infty. \end{aligned}$$

■

In complete analogy to Theorem 235 we could show the following theorem, which provides the respective conditions for geometric ergodicity and strong ergodicity.

Theorem 236 *(Isaacson and Seneta, 1982). Let a non-homogeneous Markov chain $\{X_t\}_{t=0}^{\infty}$ with state space $S = \{1, 2, ...\}$ and sequence of transition probabilities $\{\mathbf{P}(t)\}_{t=0}^{\infty}$. Then $\{X_t\}_{t=0}^{\infty}$ is said to be geometrically strongly ergodic if and only if for each $t \geq 0$ there exists $c(t) > 0$ and $0 < b(t) < 1$ such that*

$$\sup_{i,r} \sum_j |p_{ij}(t,t+n) - p_{rj}(t,t+n+1)| \leq c(t)\,[b(t)]^n \tag{4.13}$$

Also $\{X_t\}_{t=0}^{\infty}$ is said to be strongly ergodic if and only if for each $t \geq 0$, we have that

$$\sup_{k\geq 1} \sup_{i,r} \sum_j |p_{ij}(t,t+n) - p_{rj}(t,t+n+k)| \to 0 \ \ \text{as } n \to \infty. \tag{4.14}$$

4.3 Criteria for strong ergodicity for NHMC

In this section we will study various criteria for strong ergodicity of non-homogeneous Markov chains. One of our goals of selecting the present results is among others to provide easily checked conditions under which strong ergodicity holds. It is important to start having in mind the result in the following well known result which will be given as a Lemma:

Lemma 237 *(Bowerman (1974)). Let a non-homogeneous Markov chain $\{X_t\}_{t=0}^{\infty}$ with state space $S = \{1, 2, ...\}$ and sequence of transition probabilities $\{\mathbf{P}(t)\}_{t=0}^{\infty}$, which is strongly ergodic, that is, there exists a stable matrix $\mathbf{Q}$ such that $\lim_{n\to\infty} \|\mathbf{P}(t,t+n) - \mathbf{Q}\| = 0$. Then the row $\mathbf{q} = [q_1, q_2, ...]$ of $\mathbf{Q}$ has $q_i \geq 0$ and $\sum_i q_i = 1$.*

Proof. For any $t, n \geq 0$, we have that $\mathbf{P}(t, t+n)$ is a stochastic matrix hence we have that for every $i, j \in S$

$$p_{ij}(t, t+n) \geq 0 \text{ and } \sum_j p_{ij}(t, t+n) = 1 \text{ for every } i.$$

Since $\{\mathbf{P}(t)\}_{t=0}^{\infty}$ is strongly ergodic we have that

$$\lim_{n\to\infty} \|\mathbf{P}(t, t+n) - \mathbf{Q}\| = 0,$$

where $\mathbf{Q}$ is a stable stochastic matrix and let $\mathbf{q} = [q_1, q_2, ...]$ it's row. From equation (4.1), we have that for any $\epsilon > 0$ there exists an $N(\epsilon)$ such that for $n \geq N(\epsilon)$ we have

$$\sup_i \sum_j |p_{ij}(t, t+n) - q_i| < \epsilon. \tag{4.15}$$

Since for any t, n and every i, j we have that, $p_{ij}(t, t+n) \geq 0$ could not converge to a negative number and therefore $q_i \geq 0$ for every i. We will now prove that $\sum_i q_i = 1$ by contradiction. Assume that $\sum_i q_i \neq 1$ and therefore there exists $a \neq 0$ such that $\sum_i q_i = 1 - a$. From (4.15), we get that for $n \geq N(|a|/2)$

$$\sup_i \sum_j |p_{ij}(t, t+n) - q_i| < |a|/2. \tag{4.16}$$

We also have that for $n \geq N(|a|/2)$

$$\sup_i \sum_j |p_{ij}(t, t+n) - q_i| \geq \sup_i \left| \sum_j p_{ij}(t, t+n) - \sum_i q_i \right| = |a|, \tag{4.17}$$

which is a contradiction to (4.16). ■

The following Lemma the proof of which could be found in Paz (1971) will be useful in what follows:

Lemma 238 *If* $\mathbf{P}$ *is a stochastic matrix and if* $\mathbf{R}$ *is any real matrix such that* $\mathbf{R}\mathbf{1}^{\top} = 0$ *and in addition* $\|\mathbf{R}\| < \infty$, *then*

$$\|\mathbf{RP}\| \leq \|\mathbf{R}\| \tau_1(\mathbf{P}). \tag{4.18}$$

The following Lemma is easily to be proved as an exercise and the proof could be found in Isaacson and Madsen (1976).

Lemma 239 *If* $\mathbf{P}$ *and* $\mathbf{Q}$ *are stochastic matrices, then*

$$|\tau_1(\mathbf{P}) - \tau_1(\mathbf{Q})| \leq \|\mathbf{P} - \mathbf{Q}\|. \tag{4.19}$$

In what follows in a theorem sufficient conditions will be provided for a non-homogeneous Markov chain which is weakly ergodic to be strongly ergodic.

Theorem 240 *(Bremaud, 2020). Let a non-homogeneous Markov chain* $\{X_t\}_{t=0}^{\infty}$ *with state space* $S = \{1, 2, ...\}$ *and sequence of transition probabilities* $\{\mathbf{P}(t)\}_{t=0}^{\infty}$. *If the Markov chain* $\{X_t\}_{t=0}^{\infty}$ *is weakly ergodic and if there exist a* $\boldsymbol{\pi}(t) = [\pi_1(t), \pi_2(t), ...]$ *probability distribution which is the left eigenvector of* $\mathbf{P}(t)$ *for* $t = 0, 1, 2, ...$, *i.e.*

$$\boldsymbol{\pi}(t) = \boldsymbol{\pi}(t)\mathbf{P}(t),$$

and in addition

$$\sum_{t=0}^{\infty} \|\boldsymbol{\pi}(t+1) - \boldsymbol{\pi}(t)\| < \infty, \tag{4.20}$$

then the chain is strongly ergodic.

Proof. From condition (4.20), we get that there exist a stochastic vector $\boldsymbol{\pi}$ such that

$$\lim_{t\to\infty} \|\boldsymbol{\pi}(t) - \boldsymbol{\pi}\| = 0. \tag{4.21}$$

Denote by $\mathbf{\Pi}(t)$ to be the stable matrix with row $\boldsymbol{\pi}(t)$ and respectively $\mathbf{\Pi}$ the stable matrix with row $\boldsymbol{\pi}$. It is easy to see that

$$\|\mathbf{\Pi}(t+1) - \mathbf{\Pi}(t)\| = \|\boldsymbol{\pi}(t+1) - \boldsymbol{\pi}(t)\|, \tag{4.22}$$

and

$$\|\mathbf{\Pi}(t) - \mathbf{\Pi}\| = \|\boldsymbol{\pi}(t) - \boldsymbol{\pi}\|. \tag{4.23}$$

In order to prove the strong ergodicity of the non-homogeneous Markov chain $\{X_t\}_{t=0}^{\infty}$ it is sufficient to show that for every $t \geq 0$

$$\lim_{n\to\infty} \|\mathbf{P}(t, t+n) - \mathbf{\Pi}\| = 0. \tag{4.24}$$

We have that

$$\begin{aligned}\mathbf{P}(t, t+n) - \mathbf{\Pi} &= \mathbf{P}(t, t+k)\,\mathbf{P}(t+k, t+n) - \mathbf{\Pi}(t+k)\,\mathbf{P}(t+k, t+n) \\ &+ \mathbf{\Pi}(t+k)\,\mathbf{P}(t+k, t+n) - \mathbf{\Pi}(t+n-1) + \mathbf{\Pi}(t+n-1) - \mathbf{\Pi}.\end{aligned} \tag{4.25}$$

Taking in (4.25) the norm on both sides of the equation and using properties of the norm we get that

$$\begin{aligned}\|\mathbf{P}(t, t+n) - \mathbf{\Pi}\| &\leq \|\mathbf{P}(t, t+k)\,\mathbf{P}(t+k, t+n) - \mathbf{\Pi}(t+k)\,\mathbf{P}(t+k, t+n)\| \\ &\|\mathbf{\Pi}(t+k)\,\mathbf{P}(t+k, t+n) - \mathbf{\Pi}(t+n-1)\| + \|\mathbf{\Pi}(t+n-1) - \mathbf{\Pi}\| \\ &= N_1 + N_2 + N_3.\end{aligned} \tag{4.26}$$

We now have

$$\begin{aligned}N_1 &= \|\mathbf{P}(t, t+k)\,\mathbf{P}(t+k, t+n) - \mathbf{\Pi}(t+k)\,\mathbf{P}(t+k, t+n)\| \\ &\leq \|\mathbf{P}(t, t+k) - \mathbf{\Pi}(t+k)\|\,\tau_1(\mathbf{P}(t+k, t+n)) \leq 2\tau_1(\mathbf{P}(t, t+k)).\end{aligned} \tag{4.27}$$

In order to find an upper bound for N_2, we have that

$$\begin{aligned}\mathbf{\Pi}(t+k)\,\mathbf{P}(t+k, t+n) &= (\mathbf{\Pi}(t+k) - \mathbf{\Pi}(t+k+1))\,\mathbf{P}(t+k, t+n) \\ &+ \mathbf{\Pi}(t+k+1)\,\mathbf{P}(t+k+1, t+n).\end{aligned} \tag{4.28}$$

and using this equation recursively we get

$$\begin{aligned}\mathbf{\Pi}(t+k)\,\mathbf{P}(t+k, t+n) &= \sum_{j=t+k+1}^{t+n-1} (\mathbf{\Pi}(j-1) - \mathbf{\Pi}(j))\,\mathbf{P}(j, t+n) + \\ &\mathbf{\Pi}(t+n-1)\,\mathbf{P}(t+n-1, t+n),\end{aligned} \tag{4.29}$$

and therefore from (4.29) Lemma 238, condition (4.20) and the fact that $\mathbf{\Pi}(t+n-1)$ $\mathbf{P}(t+n-1, t+n) = \mathbf{\Pi}(t+n-1)$ we have that

$$\begin{aligned}N_2 &= \|\mathbf{\Pi}(t+k)\,\mathbf{P}(t+k, t+n) - \mathbf{\Pi}\mathbf{\Pi}(t+n-1)\| \\ &\leq \sum_{j=t+k+1}^{t+n-1} \|\mathbf{\Pi}(j-1) - \mathbf{\Pi}(j)\|\,\tau_1(\mathbf{P}(j, t+n)) \leq\end{aligned} \tag{4.30}$$

$$\sum_{j=t+k+1}^{t+n-1} \|\boldsymbol{\pi}(j-1) - \boldsymbol{\pi}(j)\| < \infty.$$

Now for every $\epsilon > 0$ we can fix $t+k$ such that $N_2 \leq \epsilon/3$ for $n-1 \geq k$ and we can always take k large enough so that from (4.27) to have $N_1 \leq \epsilon/3$ and from (4.24) $N_3 \leq \epsilon/3$ which concludes our proof. ■

We will now present a theorem which is one of the basic theorems for strong ergodicity of non-homogeneous Markov chains and its roots go back to Hajnal (1956, 1958), Mott (1957) and has taken various forms, see for example, Isaacson and Madsen (1976), Bowerman (1974). In order to prove this important result, we will be needing the following well known theorem from the theory of homogeneous Markov chains with countably infinite state space.

Theorem 241 *Let a homogeneous Markov chain $\{X_t\}_{t=0}^{\infty}$ with state space $S = \{1, 2, ...\}$ and transition probability matrix $\mathbf{P}$, which is irreducible and weakly ergodic. Then it is strongly ergodic with limiting matrix the stable matrix $\mathbf{\Pi}$ with row $\boldsymbol{\pi}$, which is the stochastic eigenvector of the eigenvalue one of the matrix $\mathbf{P}$. That is, $\boldsymbol{\pi}$ is the unique solution of the system of equations $\boldsymbol{\pi}\mathbf{P} = \boldsymbol{\pi}$ and $\boldsymbol{\pi}\mathbf{1}^{\top} = 1$.*

Theorem 242 *Let a non-homogeneous Markov chain $\{X_t\}_{t=0}^{\infty}$ with state space $S = \{1, 2, ...\}$ and sequence of transition probabilities $\{\mathbf{P}(t)\}_{t=0}^{\infty}$. If $\lim_{t\to\infty} \|\mathbf{P}(t) - \mathbf{P}\| = 0$ where $\mathbf{P}$ is an irreducible and weakly ergodic stochastic matrix then the non-homogeneous Markov chain $\{X_t\}_{t=0}^{\infty}$ is strongly ergodic with limit matrix the stable stochastic matrix $\mathbf{\Pi}$ which is the limit of the homogeneous Markov chain defined uniquely by the matrix $\mathbf{P}$.*

Proof. Since $\mathbf{P}$ is irreducible and weakly ergodic stochastic matrix then from Theorem 241 we get that it is strongly ergodic and consequently

$$\lim_{t\to\infty} \left\|\mathbf{P}^t - \mathbf{\Pi}\right\| = 0, \tag{4.31}$$

where $\mathbf{\Pi}$ is a stable stochastic matrix with row $\boldsymbol{\pi} = [\pi_1, \pi_2, ...]$, which is an eigenvector of $\mathbf{P}$ of the eigenvalue one, that is, is the unique solution of the system

$$\boldsymbol{\pi}\mathbf{P} = \boldsymbol{\pi} \quad \text{and} \quad \sum_i \pi_i = 1.$$

From (4.31), we have that for every $\epsilon > 0$ there exist a $t_0 = t_0(\epsilon)$ for which

$$\left\|\mathbf{P}^{t_0} - \mathbf{\Pi}\right\| \leq \epsilon/2 \text{ for every } t \geq t_0. \tag{4.32}$$

Now, we have that

$$\begin{aligned} &\|\mathbf{P}(t, t+n) - \mathbf{\Pi}\| = \\ &\|\mathbf{P}(t, t+n-t_0)\mathbf{P}(t+n-t_0, t+n) - \mathbf{P}(t, t+n-t_0)\mathbf{\Pi}\| \\ &\leq \|\mathbf{P}(t, t+n-t_0)\| \, \|\mathbf{P}(t+n-t_0, t+n) - \mathbf{\Pi}\| \\ &\leq \left\|\mathbf{P}(t+n-t_0, t+n) - \mathbf{P}^{t_0} + \mathbf{P}^{t_0} - \mathbf{\Pi}\right\| \\ &\leq \left\|\mathbf{P}(t+n-t_0, t+n) - \mathbf{P}^{t_0}\right\| + \left\|\mathbf{P}^{t_0} - \mathbf{\Pi}\right\|. \end{aligned} \tag{4.33}$$

Now, we will show by induction that the first term of the second part of equation (4.33) is less than $\epsilon/2$ for some $n > N(\epsilon)$. It is apparent that $\mathbf{P}(t, t+1) = \mathbf{P}(t+1)$ and since by the condition of the theorem

$$\lim_{t\to\infty} \|\mathbf{P}(t) - \mathbf{P}\| = 0$$

we get that it is true for $n = 1$. Assume that it is true for $n = r$, that is

$$\lim_{t\to\infty} \|\mathbf{P}(t, t+r) - \mathbf{P}^r\| = 0. \tag{4.34}$$

Then we have that

$$\lim_{t\to\infty}\left\|\mathbf{P}(t,t+r+1)-\mathbf{P}^{r+1}\right\| = \lim_{t\to\infty}\left\|\mathbf{P}(t,t+r)\mathbf{P}(t+r+1)-\mathbf{P}^{r}\mathbf{P}\right\|$$

$$\leq \lim_{t\to\infty}\|\mathbf{P}(t,t+r)\|\,\|\mathbf{P}(t+r+1)-\mathbf{P}\| + \lim_{t\to\infty}\|\mathbf{P}(t,t+r)-\mathbf{P}^{r}\|\,\|\mathbf{P}\| = 0. \quad (4.35)$$

From (4.32), (4.33) and (4.34), we arrive at the desired conclusion. ■

Remark 243 *When the non-homogeneous Markov chain in Theorem 242 is finite, that is,* $\mathbf{P}$ *is a finite stochastic matrix then the condition on* $\mathbf{P}$ *being irreducible and weakly ergodic could be replaced by* $\mathbf{P}$ *regular and then we get the classical theorem proved by Mott (1957).*

It is important for practical reasons in the applications that the convergence of a non-homogeneous Markov chain in the strong ergodicity case be very fast, that is, at a geometric rate. In what follows we will study the conditions under which we have geometric rate of convergence. We will start with a Lemma.

Lemma 244 *(Huang, Isaacson and Vinograde (1976)). Let a homogeneous Markov chain* $\{X_t\}_{t=0}^{\infty}$ *with state space* $S=\{1,2,...\}$ *and transition probability* $\mathbf{P}$ *which is strongly ergodic. Then there exists constants* c *and* b $(0<b<1)$ *such that*

$$\left\|\mathbf{P}^{t}-\mathbf{\Pi}\right\| \leq cb^{t}, \quad t=1,2,3,.... \quad (4.36)$$

Proof. From the strong ergodicity of $\mathbf{P}$ we get that there exists a positive integer M such that

$$0 \leq \tau_1\left(\mathbf{P}^{M}\right) < 1. \quad (4.37)$$

Let that for $t \geq M$, we have in general $t = Md+v$ where $v=1,2,...,d-1$. Then we have that

$$\left\|\mathbf{P}^{t}-\mathbf{\Pi}\right\| = \left\|\mathbf{P}^{Md+v}-\mathbf{\Pi}\right\|$$

$$= (\text{since } \mathbf{\Pi} \text{ is a stable matrix and } \mathbf{\Pi}=\mathbf{\Pi}\mathbf{P}=\mathbf{P}\mathbf{\Pi})$$

$$= \left\|\mathbf{P}^{Md+v}-\mathbf{\Pi}\mathbf{P}^{Md}\right\| = (\text{by Lemma 238})$$

$$\leq \left\|\mathbf{P}^{v}-\mathbf{\Pi}\right\|\tau_1\left(\mathbf{P}^{Md}\right) \leq 2\tau_1\left(\mathbf{P}^{M}\right)^{d} \leq \frac{2}{\tau_1\left(\mathbf{P}^{M}\right)}\left(\tau_1\left(\mathbf{P}^{M}\right)^{\frac{1}{M}}\right)^{t}, \quad (4.38)$$

From (4.37) and $t \geq 1$ we get that

$$0 < \tau_1\left(\mathbf{P}^{M}\right)^{\frac{1}{M}} < 1.$$

Therefore for $t \geq M$

$$\left\|\mathbf{P}^{t}-\mathbf{\Pi}\right\| \leq cb^{t}, \text{ where } b=\tau_1\left(\mathbf{P}^{M}\right)^{\frac{1}{M}} \text{ and } c=\frac{2}{\delta\left(\mathbf{P}^{M}\right)}. \quad (4.39)$$

It remains to prove the same for $t=1,2,...,M-1$. Quite easily we get that for $t<M$

$$\left\|\mathbf{P}^{t}-\mathbf{\Pi}\right\| \leq 2 = \frac{2}{\left(\tau_1\left(\mathbf{P}^{M}\right)^{\frac{1}{M}}\right)^{t}}\left(\tau_1\left(\mathbf{P}^{M}\right)^{\frac{1}{M}}\right)^{t} < cb^{t},$$

From (4.38) and (4.39), we get (4.36). ■

In the following Lemma, we will prove a few properties of a non-homogeneous Markov chain which converge in the strong sense to a homogeneous Markov chain, which is assumed to be strongly ergodic. These properties will prove to be very useful in proving the basic theorems that follow.

Lemma 245 *(Huang, Isaacson and Vinograde (1976)). Let a non- homogeneous Markov chain* $\{X_t\}_{t=0}^{\infty}$ *with state space* $S = \{1, 2, ...\}$ *and transition probability sequence of matrices* $\{\mathbf{P}(t)\}_{t=0}^{\infty}$. *Assume that*

$$lim_{t\to\infty} \|\mathbf{P}(t) - \mathbf{P}\| = 0, \quad then$$

(i) If $\tau_1(\mathbf{P}) < 1$, *then there exist a positive integer* M *such that*

$$0 \leq \tau_1(\mathbf{P}(t)) < 1, \; for \;\; t \geq M. \tag{4.40}$$

that is the tail of $\{\mathbf{P}(t)\}_{t=0}^{\infty}$ *behaves similarly to* $\mathbf{P}$.

(ii) For any positive integers t,n we have that

$$\|\mathbf{P}(t, t+n) - \mathbf{P}^n\| \leq \sum_{i=1}^{n} \|\mathbf{P}(i) - \mathbf{P}\|. \tag{4.41}$$

(iii) If $\tau_1(\mathbf{P}^n) < 1$ *for some* positive *integer n, then there exists a positive integer* t_0 *such that*

$$\tau_1(\mathbf{P}(t, t+n)) < 1 \quad for\, t \geq t_0.$$

Proof. (i). From Lemma 239 we get that

$$|\tau_1(\mathbf{P}(t)) - \tau_1(\mathbf{P})| \leq \|\mathbf{P}(t) - \mathbf{P}\|,$$

and since $\tau_1(\mathbf{P}) < 1$, then there exist a positive integer M such that

$$0 \leq \tau_1(\mathbf{P}(t)) < 1, \text{ for } \; t \geq M.$$

(ii) We have that

$$\begin{aligned}
\|\mathbf{P}(t, t+n) - \mathbf{P}^n\| &\leq \|\mathbf{P}(t, t+n-1)\| \, \|\mathbf{P}(t+n) - \mathbf{P}\| \\
&\quad + \left\|\mathbf{P}(t, t+n-1) - \mathbf{P}^{n-1}\right\| \|\mathbf{P}\| \\
&\leq \|\mathbf{P}(t+n) - \mathbf{P}\| + \left\|\mathbf{P}(t, t+n-1) - \mathbf{P}^{n-1}\right\| \\
\text{(working recursively)} &\leq \sum_{i=t+1}^{t+n} \|\mathbf{P}(i) - \mathbf{P}\|.
\end{aligned}$$

(iii). We have that

$$|\tau_1(\mathbf{P}(t, t+n)) - \tau_1(\mathbf{P}^n)| \leq \|\mathbf{P}(t, t+n) - \mathbf{P}^n\| \leq \sum_{i=t+1}^{t+n} \|\mathbf{P}(i) - \mathbf{P}\|,$$

and consequently there exists a t_0 for which for $t \geq t_0$

$$|\tau_1(\mathbf{P}(t, t+n)) - \tau_1(\mathbf{P}^n)| \leq \epsilon$$

that is,

$$\lim_{t\to\infty} |\tau_1(\mathbf{P}(t, t+n)) - \tau_1(\mathbf{P}^n)| = 0,$$

and since $\tau_1(\mathbf{P}^n) < 1$ we arrive at the desired conclusion. ■

In the following we will actually prove that if a non-homogeneous Markov chain converges asymptotically geometrically fast to a homogeneous Markov chain which is strongly ergodic then the non-homogeneous Markov chain is uniformly strongly ergodic and its convergence is also geometric. Actually the following theorem includes inherently other rates of convergences also but the above case is the one with the most practical importance. Now, let $g(t)$ be any monotone increasing function from the positive integers into the positive real numbers.

Theorem 246 *(Huang, Isaacson and Vinograde, 1976)). Let a non- homogeneous Markov chain* $\{X_t\}_{t=0}^{\infty}$ *with state space* $S = \{1, 2, ...\}$ *and transition probability sequence of matrices* $\{\mathbf{P}(t)\}_{t=0}^{\infty}$. *Assume that*

$$lim_{t\to\infty} \|\mathbf{P}(t) - \mathbf{P}\| = 0, \tag{4.42}$$

where $\mathbf{P}$ *is strongly ergodic. Let* $g(n)$ *be a monotone increasing function from the positive integers into the positive real numbers. If*

$$\lim_{t\to\infty} g(2t) \|\mathbf{P}(t) - \mathbf{P}\| = 0, \tag{4.43}$$

then not only is $\{\mathbf{P}(t)\}_{t=0}^{\infty}$ *uniformly strongly ergodic, but*

$$\lim_{n\to\infty} \sup_{t\geq o} \{\min(a^n, g(n)) \|\mathbf{P}(t, t+n) - \mathbf{\Pi}\|\} = 0, \tag{4.44}$$

where $\mathbf{\Pi} = \mathbf{\Pi P}$, $1 < a < \left(\frac{1}{b}\right)^{1/2}$ *and where* b *is such that* $\|\mathbf{P}^n - \mathbf{\Pi}\| \leq cb^n$ *with* $c > 0$ *and* $(0 < b < 1)$.

Proof. By assumption $\mathbf{P}$ is strongly ergodic, therefore there exists a positive integer M such that $0 \leq \tau_1(\mathbf{P}) < 1$. Therefore, by Lemma 245 (iii) there exists a positive integer t_0 and a number $\vartheta < 1$ such that

$$\tau_1(\mathbf{P}(t, t+M)) < \vartheta < 1 \quad \text{for} \quad t > t_0. \tag{4.45}$$

From (4.43) we get that for every $\epsilon > 0$ there exists a $t_1(\epsilon)$ such that

$$g(2t) \|\mathbf{P}(t) - \mathbf{P}\| < \epsilon \quad \text{for} \quad t > t_1(\epsilon). \tag{4.46}$$

To avoid unnecessary complications in notation we assume from now on that the integer n is even and that $n/2 > \max(M, t_0, t_1(\epsilon))$. For n odd the proof is strongly similar.

Now, we have the following

$$\|\mathbf{P}(t, t+n) - \mathbf{\Pi}\| \leq \left\|\mathbf{P}(t, t+n) - \mathbf{P}(t, t+n/2)\mathbf{P}^{n/2}\right\| +$$
$$\left\|\mathbf{P}(t, t+n)\mathbf{P}^{n/2} - \mathbf{\Pi}\right\| \leq \left\|\mathbf{P}(t+n/2, t+n) - \mathbf{P}^{n/2}\right\| + \left\|\mathbf{P}^{n/2} - \mathbf{\Pi}\right\| \tag{4.47}$$
$$= A + B.$$

From (4.47) we get that

$$A \leq \left\|\mathbf{P}(t+n/2, t+n) - \mathbf{P}^{n/2-M}\mathbf{P}(t+n-M, t+n)\right\|$$
$$+ \left\|\mathbf{P}^{n/2-M}\mathbf{P}(t+n-M, t+n) - \mathbf{P}^{n/2}\right\|$$
$$\text{(By Lemma 238)}$$
$$\leq \left\|\mathbf{P}(t+n/2, t+n-M) - \mathbf{P}^{n/2-M}\right\| \tau_1(\mathbf{P}(t+n-M, t+n))$$
$$+ \left\|\mathbf{P}(t+n-M, t+n) - \mathbf{P}^M\right\|$$
$$\text{(given that } \tau_1(\mathbf{P}(t+n-M, t+n)) < \vartheta < 1 \text{ for } t \geq t_0)$$
$$\leq \left\|\mathbf{P}(t+n/2, t+n-M) - \mathbf{P}^{n/2-M}\right\| \vartheta + \left\|\mathbf{P}(t+n-M, t+n) - \mathbf{P}^M\right\|, \tag{4.48}$$

for $t \geq t_0$.We will now reduce the number of factors in $\mathbf{P}(t+n/2, t+n-M)$ by M following again the same reasoning. In this respect, we have

$$\left\|\mathbf{P}(t+n/2, t+n-M) - \mathbf{P}^{n/2-M}\right\| \leq$$

$$\left\|\mathbf{P}\left(t+n/2,t+n-2M\right)-\mathbf{P}^{n/2-2M}\right\|\vartheta+ \tag{4.49}$$

$$\left\|\mathbf{P}\left(t+n-2M,t+n-M\right)-\mathbf{P}^{M}\right\|,\quad fort\geq t_0.$$

Hence from (4.48) and (4.49) we get that

$$A\leq\left\|\mathbf{P}\left(t+n/2,t+n-2M\right)-\mathbf{P}^{n/2-2M}\right\|\vartheta^2+$$

$$\left\|\mathbf{P}\left(t+n-2M,t+n-M\right)-\mathbf{P}^{M}\right\|\vartheta+\left\|\mathbf{P}\left(t+n-M,t+n\right)-\mathbf{P}^{M}\right\|, \tag{4.50}$$

for $t\geq t_0$. Following this procedure recursively then for $t\geq t_0$, we get

$$\left\|\mathbf{P}\left(t+n/2,t+n\right)-\mathbf{P}^{n/2}\right\|\leq$$

$$\left\|\mathbf{P}\left(t+n/2,t+n-\left(d-1\right)M\right)-\mathbf{P}^{n/2-(d-1)M}\right\|\vartheta^{d-1}+$$

$$\sum_{i=1}^{d-1}\left\|\mathbf{P}\left(t+n-iM,t+n-\left(i-1\right)M\right)-\mathbf{P}^{M}\right\|\vartheta^{i-1}=\mathbf{C}\left(n\right)+\mathbf{D}\left(n,i,\vartheta\right), \tag{4.51}$$

where $d=[n/2M]$, $[x]$ meaning the integer part of x.

We now have by Lemma 244: *(ii)*

$$\mathbf{C}\left(n\right)\leq g\left(n\right)\left(\left\|\mathbf{P}\left(t+n/2+1\right)-\mathbf{P}\right\|+...+\left\|\mathbf{P}\left(t+n-\left(d-1\right)M\right)-\mathbf{P}\right\|\right)$$

$$\text{(since } g\left(n\right) \text{ is an increasing function and in } (4.33)\text{, we have } g\left(2t\right))$$

$$\leq g\left(n+2t+2\right)\left\|\mathbf{P}\left(t+\frac{n}{2}+1\right)-\mathbf{P}\right\|+g\left(n+2t+4\right)\left\|\mathbf{P}\left(t+\frac{n}{2}+2\right)-\mathbf{P}\right\|$$

$$+...+g(2\left(t+n-\left(d-1\right)M\right)\left\|\mathbf{P}t+n-\left(d-1\right)M\right)-\mathbf{P}\right\|. \tag{4.52}$$

From (4.46) since $n+2t+2>n/2>\max(t_0,t_1,M)$ we get that for each of the terms in the right hand side of (4.52) for $t\geq 0$

$$g(2\left(t+n-\left(d-1\right)M\right)\left\|\mathbf{P}(t+n-\left(d-1\right)M)-\mathbf{P}\right\|<\frac{2\epsilon}{d-1}M, \tag{4.53}$$

hence

$$\mathbf{C}\left(n\right)<2M\epsilon. \tag{4.54}$$

In a similar way it could be proved that for $t\geq 0$

$$g\left(n\right)\left\|\mathbf{P}\left(t+n-iM,t+n-\left(i-1\right)M\right)-\mathbf{P}^{M}\right\|<M\epsilon\ \text{ for } i=1,...d-1. \tag{4.55}$$

Therefore from $(4.51), (4.54)$ and (4.55), we get that

$$g\left(n\right)\left\|\mathbf{P}\left(t+n/2,t+n\right)-\mathbf{P}^{n/2}\right\|<\frac{2M}{1-\vartheta}\epsilon\ \text{ for } t\geq 0. \tag{4.56}$$

That is, it approaches zero uniformly in t.

By Lemma 244 since $\mathbf{P}$ is strongly ergodic the second part of the right hand side of (4.47), that is, B goes to zero geometrically fast. Therefore, there exists a $c>0$ and a $0<b<1$ such that

$$\left\|\mathbf{P}^{n/2}-\mathbf{\Pi}\right\|\leq cb^{n/2}. \tag{4.57}$$

Thus, if a is such that $1<a<\left(\frac{1}{b}\right)^{1/2}$ then we have

$$\min\left(g\left(n\right),a^n\right)\left\|\mathbf{P}^{n/2}-\mathbf{\Pi}\right\|\leq a^n cb^{n/2}=c\left(\frac{a^2}{1/b}\right)^{n/2}, \tag{4.58}$$

which goes to zero geometrically fast as $n \to \infty$ since $c > 0$ and $0 < a^2/1/b < 1$. From (4.43) and (4.45) we get

$$\lim_{n\to\infty} \sup_{t\geq o} \{\min(a^n, g(n)) \|\mathbf{P}(t, t+n) - \mathbf{\Pi}\|\} = 0,$$

uniformly in t. ■

4.4 Convergence in the Cesaro sense

In this section we will study ergodicity or equivalently limiting behavior for non-homogeneous Markov chains that converge to homogeneous periodic Markov chains. The interested reader to gain some depth on the theory of finite homogeneous periodic Markov chains could do so by visiting Vassiliou (2010 p. 104) or Iosifescou (1980).

Let us refresh the fact that the period of a state i of a homogeneous Markov chain is the greatest common divisor of the set of integers $t \geq 1$ such that $p_{ii}(t) > 0$, that is,

$$d_i = \gcd\{t \geq 1 : p_{ii}(t) > 0\}.$$

If $d_i = 0$ then the state is aperiodic. From Proposition 3.29 in Vassiliou (2010) we know that for a homogeneous finite Markov chain all communicating states have the same period. It is very easy following its proof to show that the same is true for countable homogeneous Markov chains also.

4.4.1 Cyclic subclasses

Let an irreducible countable homogeneous Markov chain $\{X_t\}_{t=0}^{\infty}$ all the states of which communicate and have period d and state space $\mathbb{S} = \{1, 2, ...\}$. Then the movement of the Markov chain among its states follows a cyclic behavior which we shall describe. Let $i \in \mathbb{S}$ then taking this state as a basis we define

$$C_b(i) = \{j \in \mathbb{S} : p_{ij}(n) > 0 \text{ for all } n \equiv b \,(\mathrm{mod}\, d)\}.$$

Apparently, we have

$$C_0(i) \cup C_1(i) \cup ... \cup C_{d-1}(i) = \mathbb{S},$$
$$C_b(i) \cap C_r(i) = \varnothing \quad \text{for} \quad b \neq r.$$

Following the proof of Lemma 3.30 in Vassiliou (2010) it is easy to show that the above separation of the state space $\mathbb{S}$ is in fact independent of the state chosen as basis in the sense that the classes C_b for $b = 1, 2, ..., d-1$ have always the same states and only their numbering changes with the state taken as basis. Hence, we denote the subsets $C_b(i)$, $i \in \mathbb{S}$ for $b = 1, 2, ..., d-1$ merely C_b for $b = 1, 2, ..., d-1$. We call the sets C_b for $b = 1, 2, ..., d-1$ the *cyclic subclasses of the Markov chain.* The reason for that is that by Theorem 3.31 in Vassiliou (2010) if $i \in C_b$ and $p_{ij} > 0$ then $j \in C_{b+1}$. It is apparent that the transition probability matrix of an irreducible countable homogeneous Markov chain $\{X_t\}_{t=0}^{\infty}$ which has period d could be written in the following form:

$$\mathbf{P} = \begin{array}{c} \\ C_0 \\ C_1 \\ ... \\ C_{d-2} \\ C_{d-1} \end{array} \begin{array}{c} \begin{array}{ccccc} C_0 & C_1 & C_2 & ... & C_{d-1} \end{array} \\ \begin{pmatrix} \mathbf{0} & \mathbf{P}_0 & \mathbf{0} & ... & \mathbf{0} \\ \mathbf{0} & \mathbf{0} & \mathbf{P}_1 & ... & \mathbf{0} \\ ... & ... & ... & ... & ... \\ \mathbf{0} & \mathbf{0} & \mathbf{0} & ... & \mathbf{P}_{d-2} \\ \mathbf{P}_{d-1} & \mathbf{0} & \mathbf{0} & ... & \mathbf{0} \end{pmatrix} \end{array}, \tag{4.59}$$

where $\mathbf{P}_0, \mathbf{P}_1, ..., \mathbf{P}_{d-1}$ are stochastic matrices that in general do not have the same dimensions and each one of them is not necessarily a square matrix. In Vassiliou (2010 p.109) there is an algorithm for finding the cyclic subclasses of a stochastic matrix. It is easy to see that the matrix $\mathbf{P}^d$ will be of the form

$$\mathbf{P}^d = \begin{array}{c} \\ C_0 \\ C_1 \\ \dots \\ C_{d-2} \\ C_{d-1} \end{array} \begin{array}{c} \begin{array}{ccccc} C_0 & C_1 & \dots & C_{d-2} & C_{d-1} \end{array} \\ \begin{pmatrix} \mathbf{X}_0 & 0 & \dots & 0 & 0 \\ 0 & \mathbf{X}_1 & \dots & 0 & 0 \\ \dots & \dots & \dots & \dots & \dots \\ 0 & 0 & \dots & \mathbf{X}_{d-2} & 0 \\ 0 & 0 & 0 & \dots & \mathbf{X}_{d-1} \end{pmatrix} \end{array}. \tag{4.60}$$

It is known that if the irreducible periodic matrix $\mathbf{P}$ is finite, then each $\mathbf{X}_i$, for $i = 1, 2, ..., d-1$ is automatically strongly ergodic. However, if $\mathbf{P}$ is infinite, then each $\mathbf{X}_i$, for $i = 1, 2, ..., d-1$ there is not guaranty that all $\mathbf{X}_i^{\prime}$s are strongly ergodic. In this chapter, we will assume that we will work only with infinite irreducible periodic matrices for which all $\mathbf{X}_i^{\prime}$s, for $i = 1, 2, ..., d-1$ are strongly ergodic. We will refer to such an infinite irreducible periodic matrix as a *periodic strongly ergodic matrix* or simply as a PSE matrix. At this point it is useful to prove the following proposition which simply states that it is sufficient for one of the $\mathbf{X}_i^{\prime}$s to be strongly ergodic and then all the others are too.

Proposition 247 *Let a homogeneous Markov chain $\{X_t\}_{t=0}^{\infty}$ with a countable state space $\mathbb{S} = \{1, 2, ...\}$ and an infinite irreducible transition probability matrix $\mathbf{P}$. Let $\mathbf{X}_i$, for $i = 1, 2, ..., d-1$ be the diagonal square matrices of the matrix $\mathbf{P}^d$. Then if any of the $\mathbf{X}_i^{\prime}$s, for $i = 1, 2, ..., d-1$ is strongly ergodic so are all.*

Proof. It is not difficult to see that working with the various powers of $\mathbf{P}^d$ that the $\mathbf{X}_i^{\prime}$s have the following structure

$$\mathbf{X}_0 = \mathbf{P}_0\mathbf{P}_1...\mathbf{P}_{d-1}, \quad \mathbf{X}_1 = \mathbf{P}_1\mathbf{P}_2...\mathbf{P}_{d-1}\mathbf{P}_0,$$

$$\mathbf{X}_2 = \mathbf{P}_2\mathbf{P}_3...\mathbf{P}_{d-1}\mathbf{P}_0\mathbf{P}_1, \ ... \ , \ \mathbf{X}_{d-1} = \mathbf{P}_{d-1}\mathbf{P}_0...\mathbf{P}_{d-2}.$$

Without loss of generality assume that $\mathbf{X}_0$ is strongly ergodic, then there is a stable matrix $\mathbf{\Pi}_0$ with row $\boldsymbol{\pi}_0$ for which $\boldsymbol{\pi}_0 = \boldsymbol{\pi}_0\mathbf{X}_0$ and $\boldsymbol{\pi}_0\mathbf{1}^\top = 1$ and for which

$$\lim_{t\to\infty} \left\|\mathbf{X}_0^t - \mathbf{\Pi}_0\right\| = 0.$$

Now, let $\mathbf{X}_i^{t+1}$ then we can write

$$\mathbf{X}_i^{t+1} = \mathbf{P}_i\mathbf{P}_{i+1}...\mathbf{P}_{d-1}X_0^t\mathbf{P}_0\mathbf{P}_1...\mathbf{P}_{i-1},$$

and denote by

$$\mathbf{\Pi}_i = \mathbf{\Pi}_0\mathbf{P}_0\mathbf{P}_1...\mathbf{P}_{i-1},$$

then since $\mathbf{\Pi}_0$ is a stable stochastic matrix, we have that

$$\lim_{t\to\infty} \left\|\mathbf{X}_i^{t+1} - \mathbf{\Pi}_i\right\| =$$

$$\lim_{t\to\infty} \left\|\mathbf{P}_i\mathbf{P}_{i+1}...\mathbf{P}_{d-1}X_0^t\mathbf{P}_0\mathbf{P}_1...\mathbf{P}_{i-1} - \mathbf{\Pi}_0\mathbf{P}_0\mathbf{P}_1...\mathbf{P}_{i-1}\right\|$$
$$\leq \lim_{t\to\infty} \left\|\mathbf{P}_i\mathbf{P}_{i+1}...\mathbf{P}_{d-1}\right\| \left\|\mathbf{X}_0^t - \mathbf{\Pi}_0\right\| \left\|\mathbf{P}_0\mathbf{P}_1...\mathbf{P}_{i-1}\right\|$$
$$\leq \lim_{t\to\infty} \left\|\mathbf{X}_0^t - \mathbf{\Pi}_0\right\| = 0.$$

■

From the theory of finite homogeneous Markov chains the following theorem from Vassiliou (2010) p.123, Theorem 3.48, will play a useful guide for what to expect for countable inhomogeneous Markov chains.

Theorem 248 *Consider an irreducible finite homogeneous Markov chain* $\{X_t\}_{t=0}^{\infty}$ *with period* d*. Denote by* $\mathbf{P}$ *the probability transition matrix of the Markov chain. Then, the sequence* $\{\mathbf{P}^t\}_{t=0}^{\infty}$ *splits into* d *convergent subsequences with limits*

$$\mathbf{P}^*\mathbf{P}^r \quad \text{for } r = 0, 1, ...d-1,$$

where

$$\mathbf{P}^* = \lim_{t\to\infty} \mathbf{P}^{td}. \tag{4.61}$$

Also, let $\mathrm{C}_0, \mathrm{C}_1, ..., \mathrm{C}_{d-1}$ *be the cyclic subclasses of the Markov chain and* $\mathbf{P}_0, \mathbf{P}_1, ..., \mathbf{P}_{d-1}$ *the respective transition probability matrices. Then, define by*

$$\mathbf{X}_i = \mathbf{P}_i\mathbf{P}_{i+1}...\mathbf{P}_{d-1}\mathbf{P}_0\mathbf{P}_1...\mathbf{P}_{i-1} \quad \text{for } i = 0, 1, ..., d-1. \tag{4.62}$$

The matrices $\mathbf{X}_i$ $(i = 0, 1, ..., d-1)$ *define uniquely* d *regular Markov chains and we have then*

$$\mathbf{\Pi}_i = \lim_{t\to\infty} \mathbf{X}_i^t, \tag{4.63}$$

where for $i = 0, 1, ..., d-1$ *the matrices* $\mathbf{\Pi}_i$ *are stable stochastic matrices with rows respectively* $\boldsymbol{\pi}_i$ *the solution of the system*

$$\boldsymbol{\pi}_i\mathbf{X}_i = \boldsymbol{\pi}_i \quad \text{and} \quad \boldsymbol{\pi}_i\mathbf{1}^\top = 1.$$

Denote by

$$\mathbf{P}^* = \begin{array}{c} \mathrm{C}_0 \\ \mathrm{C}_1 \\ ... \\ \mathrm{C}_{d-2} \\ \mathrm{C}_{d-1} \end{array} \begin{pmatrix} \mathbf{\Pi}_0 & \mathbf{0} & ... & \mathbf{0} & \mathbf{0} \\ \mathbf{0} & \mathbf{\Pi}_1 & ... & \mathbf{0} & \mathbf{0} \\ ... & ... & ... & ... & ... \\ \mathbf{0} & \mathbf{0} & ... & \mathbf{\Pi}_{d-2} & \mathbf{0} \\ \mathbf{0} & \mathbf{0} & \mathbf{0} & ... & \mathbf{\Pi}_{d-1} \end{pmatrix}. \tag{4.64}$$

Then the sequence $\{\mathbf{P}^t\}_{t=0}^{\infty}$ *has a Cesaro convergence as* $t \to \infty$*, that is*

$$\lim_{t\to\infty} \frac{1}{t}\sum_{s=0}^{t-1} \mathbf{P}^s = \frac{1}{d}\sum_{r=0}^{d-1} \mathbf{P}^*\mathbf{P}^r = \mathbf{\Pi}^*. \tag{4.65}$$

where $\mathbf{\Pi}^*$ *is a stable stochastic matrix the row of which is equal with*

$$\boldsymbol{\pi}^* = \frac{1}{d}\left(\boldsymbol{\pi}_0, \boldsymbol{\pi}_1, ..., \boldsymbol{\pi}_{d-1}\right). \tag{4.66}$$

Remark 249 *Theorem 248 holds as it is in the case of a countable homogeneous Markov chain if we assume that the transition probability matrix* $\mathbf{P}$ *is a periodic strongly ergodic matrix (PSE). The proof is easily done by following the steps of the proof of Theorem 248.*

Remark 250 *Since each one of the matrices* $\mathbf{X}_i$ *for* $i = 0, 1, ..., d-1$ *is regular then there exists constants* c *and* $0 < b < 1$ *such that*

$$\left\|\mathbf{P}^{\pi d+r} - \mathbf{P}^*\mathbf{P}^r\right\| \leq cb^{\pi}. \tag{4.67}$$

4.4.2 The non-homogeneous case

In this subsection having as a guide Theorem 248 we will study the countable non-homogeneous Markov chain case when the sequence of matrices $\{\mathbf{P}(t)\}_{t=0}^{\infty}$ converge that is, $\lim_{t\to\infty} \|\mathbf{P}(t) - \mathbf{P}\| = 0$, where $\mathbf{P}$ is a PSE matrix. We will study three important problems. Firstly, we will consider the problem of the uniform convergence of subsequences of the form $\mathbf{P}(m, m + td + r)$ as $t \to \infty$ and their rate of convergence. Secondly, we will study the

Cesaro convergence for the non-homogeneous case, that is, we will establish the conditions under which

$$\lim_{t\to\infty}\sup_{m}\left\|\sum_{n=1}^{t}\mathbf{P}\left(m,m+n\right)/t-\mathbf{\Pi}^{*}\right\|=0, \tag{4.68}$$

where $\mathbf{\Pi}^*$ is a stable matrix which is closely related as we will prove with the PSE matrix $\mathbf{P}$. Thirdly, we will study the rate of convergence of Cesaro averages in relation with the rate of convergence of the sequence $\{\mathbf{P}(t)\}_{t=0}^{\infty}$ to the matrix $\mathbf{P}$. Note that in spite of the fact that subsequences do not in general converge, Cesaro averages do in general converge. We will start with a useful definition for the *incidence matrix* of a non-negative matrix.

Definition 251 *Let* $\mathbf{A}$ *a non-negative matrix, we define as its incidence matrix* $\tilde{\mathbf{A}}$ *the matrix which has* 1*'s in the place of the positive numbers while it retains the zeros.*

We will start with the basic theorem for the convergence of subsequences.

Theorem 252 *Let a countable non-homogeneous Markov chain* $\{X_t\}_{t=0}^{\infty}$ *with a countable state space* $\mathbb{S}=\{1,2,...\}$ *and a sequence of transition probability matrices* $\{\mathbf{P}(t)\}_{t=0}^{\infty}$. *Let the non-homogeneous Markov chain* $\{X_t\}_{t=0}^{\infty}$ *converge to a countable irreducible homogeneous Markov chain with transition probability matrix* $\mathbf{P}$ *which is a PSE matrix of period* d. *That is, there exists a PSE probability matrix* $\mathbf{P}$ *such that*

$$\lim_{t\to\infty}\left\|\mathbf{P}\left(t\right)-\mathbf{P}\right\|=0. \tag{4.69}$$

Assume in addition that $\tilde{\mathbf{P}}(t)=\tilde{\mathbf{P}}$ *for every* $t\geq 0$, *that is, all transition probability matrices have the same incidence matrix. Then the subsequences* $\mathbf{P}(m,m+td+r)$ *as* $t\to\infty$ *converge independent of* m *for* $r=0,1,...,d-1$ *to a matrix* $\mathbf{P}^*\mathbf{P}^r$ *where* $\mathbf{P}^*$ *is given in Theorem 248. That is,*

$$\lim_{t\to\infty}\left\|\mathbf{P}\left(m,m+td+r\right)-\mathbf{P}^{*}\mathbf{P}^{r}\right\|=0 \text{ uniformly in } m, \tag{4.70}$$

for $r=0,1,...,d-1$.

Proof. Since $\mathbf{P}$ is a PSE matrix of period d then it can take the form (4.59) and the matrix $\mathbf{P}^d$ the form (4.60). Then it is easy to see that

$$\mathbf{P}^{nd}=\begin{matrix}C_0\\C_1\\...\\C_{d-2}\\C_{d-1}\end{matrix}\begin{pmatrix}\mathbf{X}_0^n & \mathbf{0} & ... & \mathbf{0} & \mathbf{0}\\ \mathbf{0} & \mathbf{X}_1^n & ... & \mathbf{0} & \mathbf{0}\\ ... & ... & ... & ... & ...\\ \mathbf{0} & \mathbf{0} & ... & \mathbf{X}_{d-2}^n & \mathbf{0}\\ \mathbf{0} & \mathbf{0} & \mathbf{0} & ... & \mathbf{X}_{d-1}^n\end{pmatrix}, \tag{4.71}$$

and according to Theorem 248, we have

$$\lim_{n\to\infty}\mathbf{P}^{nd}=\mathbf{P}^{*}==\begin{matrix}C_0\\C_1\\...\\C_{d-2}\\C_{d-1}\end{matrix}\begin{pmatrix}\mathbf{\Pi}_0 & \mathbf{0} & ... & \mathbf{0} & \mathbf{0}\\ \mathbf{0} & \mathbf{\Pi}_1 & ... & \mathbf{0} & \mathbf{0}\\ ... & ... & ... & ... & ...\\ \mathbf{0} & \mathbf{0} & ... & \mathbf{\Pi}_{d-2} & \mathbf{0}\\ \mathbf{0} & \mathbf{0} & \mathbf{0} & ... & \mathbf{\Pi}_{d-1}\end{pmatrix}. \tag{4.72}$$

From (4.72) we get that we could choose $\epsilon>0$ and then there exists an $n>0$ such that

$$\left\|\mathbf{P}^{nd}-\mathbf{P}^{*}\right\|<\epsilon/2. \tag{4.73}$$

Since $\tilde{\mathbf{P}}(t) = \tilde{\mathbf{P}}$ for every $t \geq 0$ we get that the form of $\mathbf{P}(m, m+td)$ is the same as that of $\mathbf{P}^{td}$, that is, they have the same incidence matrix. Therefore, $\mathbf{P}(m, m+td)\mathbf{P}^* = \mathbf{P}^*$. We now have for fixed n

$$\|\mathbf{P}(m, m+td) - \mathbf{P}^*\| = (\text{since } \mathbf{P}(m, m+td)\mathbf{P}^* = \mathbf{P}^*)$$

$$\begin{aligned}
&\|\mathbf{P}(m, m+(t-n)d)\mathbf{P}(m+(t-n)d, m+td) - \mathbf{P}(m, m+(t-n)d)\mathbf{P}^*\| \\
&\leq \|\mathbf{P}(m, m+(t-n)d)\| \|\mathbf{P}(m+(t-n)d, m+td) - \mathbf{P}^*\| \\
&\leq \|\mathbf{P}(m+(t-n)d, m+td) - \mathbf{P}^*\| \\
&\leq \left\|\mathbf{P}(m+(t-n)d, m+td) - \mathbf{P}^{nd}\right\| + \left\|\mathbf{P}^{nd} - \mathbf{P}^*\right\|.
\end{aligned} \tag{4.74}$$

Now, we have

$$\begin{aligned}
&\left\|\mathbf{P}(m+(t-n)d, m+td) - \mathbf{P}^{nd}\right\| = \\
&\left\|\mathbf{P}(m+(t-n)d, m+td-1)\mathbf{P}(m+td) - \mathbf{P}^{nd-1}\mathbf{P}\right\| \\
&\leq \|\mathbf{P}(m+td) - \mathbf{P}\| + \left\|\mathbf{P}(m+(t-n)d, m+td-1) - \mathbf{P}^{nd-1}\right\| \\
&\leq \ldots \leq \sum_{k=1}^{n} \|\mathbf{P}(m+td-kd) - \mathbf{P}\|.
\end{aligned} \tag{4.75}$$

From (4.73), (4.74) and (4.75) we get that

$$\lim_{t\to\infty} \|\mathbf{P}(m, m+td) - \mathbf{P}^*\| = 0. \tag{4.76}$$

In addition for $r = 1, 2, \ldots, d-1$, we have that

$$\begin{aligned}
&\lim_{t\to\infty} \|\mathbf{P}(m, m+td+r) - \mathbf{P}^*\mathbf{P}^r\| \leq \\
&\lim_{t\to\infty} \|\mathbf{P}(m, m+td)\mathbf{P}(m+td, m+td+r) - \mathbf{P}^*\mathbf{P}(m+td, m+td+r)\| \\
&+ \lim_{t\to\infty} \|\mathbf{P}^*\mathbf{P}(m+td, m+td+r) - \mathbf{P}^*\mathbf{P}^r\| \\
&\leq \lim_{t\to\infty} \|\mathbf{P}(m, m+td) - \mathbf{P}^*\| + \lim_{t\to\infty} \|\mathbf{P}(m+td, m+td+r) - \mathbf{P}^r\| = 0.
\end{aligned}$$

■

We will now provide the following theorem for the Cesaro convergence for non-homogeneous Markov chains and its applications.

Theorem 253 *(Yang, 2002). Let a non-homogeneous Markov chain $\{X_t\}_{t=0}^{\infty}$ with a countable state space $\mathbb{S} = \{1, 2, \ldots\}$ and a sequence of transition probability matrices $\{\mathbf{P}(t)\}_{t=0}^{\infty}$. Let also a homogeneous Markov chain $\{Y_t\}_{t=0}^{\infty}$ with a countable state space $\mathbb{S} = \{1, 2, \ldots\}$ and transition probability matrix $\mathbf{P}$ which is assumed to be periodic strongly ergodic stochastic matrix. Define $\mathbf{\Pi}^*$ as in (4.65) then (i) If*

$$\lim_{t\to\infty} \frac{1}{t} \sum_{k=1}^{t} \|\mathbf{P}(k) - \mathbf{P}\| = 0, \tag{4.77}$$

then for any positive integer numbers m and v

$$\lim_{t\to\infty} \frac{1}{t} \sum_{k=1}^{t} \|\mathbf{P}(m+k, m+k+v) - \mathbf{P}^v\| = 0, \tag{4.78}$$

and

$$\lim_{t\to\infty}\left\|\frac{1}{t}\sum_{k=1}^{t}\mathbf{P}(m,m+k)-\mathbf{\Pi}^*\right\|=0. \tag{4.79}$$

with

$$\mathbf{\Pi}^*=\frac{1}{d}\sum_{r=0}^{d-1}\mathbf{P}^*\mathbf{P}^r$$

(ii) If

$$\lim_{t\to\infty}\sup_{m\geq 0}\frac{1}{t}\sum_{k=1}^{t}\|\mathbf{P}(m+k)-\mathbf{P}\|=0, \tag{4.80}$$

then

$$\lim_{t\to\infty}\sup_{m\geq 0}\left\|\frac{1}{t}\sum_{k=1}^{t}\mathbf{P}(m,m+k)-\mathbf{\Pi}^*\right\|=0. \tag{4.81}$$

Proof. Similarly as we have proved (4.75) we could prove that

$$\|\mathbf{P}(m+k,m+k+v)-\mathbf{P}^v\|\leq\sum_{n=1}^{v}\|\mathbf{P}(m+k+n)-\mathbf{P}\|. \tag{4.82}$$

Therefore from $(4.77), (4.78)$ and (4.82) we get that

$$\lim_{t\to\infty}\frac{1}{t}\sum_{k=1}^{t}\|\mathbf{P}(m+k,m+k+v)-\mathbf{P}^v\|=$$

$$\lim_{t\to\infty}\frac{1}{t}\sum_{k=1}^{t}\sum_{n=1}^{v}\|\mathbf{P}(m+k+n)-\mathbf{P}\|=0. \tag{4.83}$$

We will now prove an important inequality for the present and the next theorem. With no loss of generality let that $t=\pi d+v$ and let us write what is inside the norm in (4.79) in the following way, for reasons which will be apparent in what follows:

$$\sum_{k=1}^{t}(\mathbf{P}(m,m+k)-\mathbf{\Pi}^*)=\sum_{k=1}^{\Lambda d+v}(\mathbf{P}(m,m+k)-\mathbf{\Pi}^*) \tag{4.84}$$

$$+\sum_{k=\Lambda d+v+1}^{t}((\mathbf{P}(m,m+k)-\mathbf{\Pi}^*)), \tag{4.85}$$

where we select Λ such that the summation on the right hand side of (4.84) to have substantially fewer terms than the summation (4.85); the summation (4.85) to contain a number of terms that is a multiple of d, and finally is such that all terms in (4.85) are at least $\Lambda d+v+1$ for reason that will be apparent in the next few steps. From (4.84) and (4.85) we get that

$$\left\|\sum_{k=1}^{t}(\mathbf{P}(m,m+k)-\mathbf{\Pi}^*)\right\|\leq\left\|\sum_{k=1}^{\Lambda d+v}(\mathbf{P}(m,m+k)-\mathbf{\Pi}^*)\right\| \tag{4.86}$$

$$+\sum_{\lambda=\Lambda}^{\pi-1}\left\|\sum_{r=1}^{d}(\mathbf{P}(m,m+\lambda d+r+v)-\mathbf{\Pi}^*)\right\|. \tag{4.87}$$

Now, $\mathbf{\Pi}^*$ is a stable matrix and so if we multiply it from the left by a stochastic matrix we get again $\mathbf{\Pi}^*$. Therefore, from (4.87) we get

$$\sum_{\lambda=\Lambda}^{\pi-1}\left\|\sum_{r=1}^{d}\left(\mathbf{P}\left(m,m+\lambda d+r+v\right)-\mathbf{\Pi}^*\right)\right\| \leq$$

$$\sum_{\lambda=\Lambda}^{\pi-1} \| \mathbf{P}\left(m,m+\lambda d+v+1-M\right)\times$$

$$\sum_{r=1}^{d}\left(\mathbf{P}\left(m,m+\lambda d+v+1-M,m+\lambda d+r+v\right)-\mathbf{\Pi}^*\right) \| \leq$$

$$\sum_{\lambda=\Lambda}^{\pi-1}\left\|\sum_{r=1}^{d}\left(\mathbf{P}\left(m+\lambda d+v+1-M,m+\lambda d+r+v\right)-\mathbf{\Pi}^*\right)\right\|. \tag{4.88}$$

It is apparent from (4.88) that the number of terms in the last sum is no more than $M+d-1$ and that was the goal of the exercise. We now in (4.88) add and subtract $\sum_{r=1}^{d}\mathbf{P}^{M+r-1}$ and apply properties that we have already used in this section.

$$\sum_{\lambda=\Lambda}^{\pi-1}\left\|\sum_{r=1}^{d}\left(\mathbf{P}\left(m+\lambda d+v+1-M,m+\lambda d+r+v\right)-\mathbf{\Pi}^*\right)\right\| \leq$$

$$\sum_{\lambda=\Lambda}^{\pi-1}\left\|\sum_{r=1}^{d}\left(\mathbf{P}\left(m+\lambda d+v+1-M,m+\lambda d+r+v\right)-\mathbf{P}^{M+r-1}\right)\right\| +$$

$$\sum_{\lambda=\Lambda}^{\pi-1}\left\|\sum_{r=1}^{d}\left(\mathbf{P}^{M+r-1}-\mathbf{\Pi}^*\right)\right\| \leq \tag{4.89}$$

$$d\sum_{\lambda=\Lambda}^{\pi-1}\sum_{k=1}^{M+d-1}\left\|\mathbf{P}\left(m+\lambda d+v+1-M+k\right)-\mathbf{P}\right\|$$

$$+\left(\pi-\Lambda\right)\left\|\sum_{r=1}^{d}\left(\mathbf{P}^{M+r-1}-\mathbf{\Pi}^*\right)\right\|. \tag{4.90}$$

Finally from (4.85)–(4.90) we arrive at

$$\frac{1}{t}\left\|\sum_{k=1}^{t}\left(\mathbf{P}\left(m,m+k\right)-\mathbf{\Pi}^*\right)\right\| \leq \frac{1}{t}\left\|\sum_{k=1}^{\Lambda d+v}\left(\mathbf{P}\left(m,m+k\right)-\mathbf{\Pi}^*\right)\right\| \tag{4.91}$$

$$+\frac{d}{t}\sum_{\lambda=\Lambda}^{\pi-1}\sum_{k=1}^{M+d-1}\left\|\mathbf{P}\left(m+\lambda d+v+1-M+k\right)-\mathbf{P}\right\| \tag{4.92}$$

$$+\frac{\left(\pi-\Lambda\right)}{t}\left\|\sum_{r=1}^{d}\left(\mathbf{P}^{M+r-1}-\mathbf{\Pi}^*\right)\right\|. \tag{4.93}$$

We now have for the right hand side of (4.91)

$$\frac{1}{t}\left\|\sum_{k=1}^{\Lambda d+v}\left(\mathbf{P}\left(m,m+k\right)-\mathbf{\Pi}^*\right)\right\| \leq \frac{2\left(\Lambda d+v\right)}{t}. \tag{4.94}$$

From (4.93) for every $\epsilon > 0$ for sufficiently large M we get that

$$\frac{(\pi - \Lambda)}{t} \left\| \sum_{r=1}^{d} \left(\mathbf{P}^{M+r-1} - \mathbf{\Pi}^* \right) \right\| < \epsilon. \tag{4.95}$$

From (4.92) and for $\Lambda d > \lambda d > M$ we have that

$$\frac{d}{t} \sum_{\lambda=\Lambda}^{\pi-1} \sum_{k=1}^{M+d-1} \left\| \mathbf{P}(m + \lambda d + v + 1 - M + k) - \mathbf{P} \right\| \leq$$

$$d \sum_{k=1}^{M+d-1} \frac{1}{t} \sum_{s=1}^{t} \left\| \mathbf{P}(m + s + k) - \mathbf{P} \right\| < \epsilon \tag{4.96}$$

for sufficiently large t according to assumption (4.77). In (4.91), (4.92), (4.93) applying (4.94), (4.95) and (4.96) we easily arrive at the desired conclusion. Note that the proof of (*b*) is quite similar to (*a*). ■

We will now study the next natural important problem, that is, the rate of convergence in the Cesaro convergence case. We will show that the rate at which Cesaro convergence takes place depends on the rate of convergence at which $\mathbf{P}(t)$ converge to the matrix $\mathbf{P}$ which is a PSE matrix and it is a certain geometric rate which we will determine in close analytic form. In fact these results are summarized in the following theorem.

Theorem 254 *(Bowerman et al., 1977). Let a non-homogeneous Markov chain* $\{X_t\}_{t=0}^{\infty}$ *with a countable state space* $\mathbb{S} = \{1, 2, ...\}$ *and a sequence of transition probability matrices* $\{\mathbf{P}(t)\}_{t=0}^{\infty}$. *Let also a homogeneous Markov chain* $\{Y_t\}_{t=0}^{\infty}$ *with a countable state space* $\mathbb{S} = \{1, 2, ...\}$ *and transition probability matrix* $\mathbf{P}$ *which is assumed to be periodic strongly ergodic stochastic matrix. Define* $\mathbf{\Pi}^*$ *as in* (4.65) *then*

(*a*) *If there exists a finite constant* C *such that*

$$\left\| \mathbf{P}(t) - \mathbf{P} \right\| \leq C/t^{a} \quad \text{for } t \geq 1, \tag{4.97}$$

then

(1) *If* $a > 1$ *there is, for any* $\epsilon > 0$, *a* $D(\epsilon)$ *such that*

$$\sup_{m \geq 0} \left\| \frac{1}{t} \sum_{n=1}^{t} \left(\mathbf{P}(m, m+n) - \mathbf{\Pi}^* \right) \right\| \leq \frac{D(\epsilon)}{t^{1-\epsilon}} \quad \text{for } t \geq 1, \tag{4.98}$$

(2) *If* $a \in (0, 1]$ *there is, for any* $\epsilon > 0$, *a* $D(\epsilon)$ *such that*

$$\sup_{m \geq 0} \left\| \frac{1}{t} \sum_{n=1}^{t} \left(\mathbf{P}(m, m+n) - \mathbf{\Pi}^* \right) \right\| \leq \frac{D(\epsilon)}{t^{a-\epsilon}} \quad \text{for } t \geq 1. \tag{4.99}$$

Proof. Following the proof of the previous theorem we arrive at

$$\frac{1}{t} \left\| \sum_{k=1}^{t} \left(\mathbf{P}(m, m+k) - \mathbf{\Pi}^* \right) \right\| \leq \frac{1}{t} \left\| \sum_{k=1}^{\Lambda d + v} \left(\mathbf{P}(m, m+k) - \mathbf{\Pi}^* \right) \right\| \tag{4.100}$$

$$+ \frac{d}{t} \sum_{\lambda=\Lambda}^{\pi-1} \sum_{k=1}^{M+d-1} \left\| \mathbf{P}(m + \lambda d + v + 1 - M + k) - \mathbf{P} \right\| \tag{4.101}$$

$$+ \frac{(\pi - \Lambda)}{t} \left\| \sum_{r=1}^{d} \left(\mathbf{P}^{M+r-1} - \mathbf{\Pi}^* \right) \right\|.$$

From our hypothesis, we have that $\mathbf{P}$ is a periodic strongly ergodic matrix hence there are constants $c > 0$ and $0 < b < 1$ such that

$$\left\| \sum_{r=1}^{d} \left(\mathbf{P}^{M+r-1} - \mathbf{\Pi}^*\right) \right\| \leq cb^M, \quad \text{with } c > 0 \text{ and } 0 < b < 1. \tag{4.102}$$

It is easy to see that if we choose $M(t) = [-a \ln t / \ln b] + 1$ then (4.102) is equivalent with

$$\left\| \sum_{r=1}^{d} \left(\mathbf{P}^{M+r-1} - \mathbf{\Pi}^*\right) \right\| \leq ct^{-a}, \tag{4.103}$$

and from (4.102) we get that for M large enough

$$\left\| \sum_{r=1}^{d} \left(\mathbf{P}^{M+r-1} - \mathbf{\Pi}^*\right) \right\| \leq \epsilon / d. \tag{4.104}$$

Now from (4.97) we get that $\lim_{t\to\infty} \|\mathbf{P}(t) - \mathbf{P}\| = 0$ and therefore we can choose Λ large enough so that $\Lambda d > M$ and

$$\|\mathbf{P}(t) - \mathbf{P}\| \leq \epsilon / (M + d - 1) \quad \text{for} \quad t > \Lambda d - M. \tag{4.105}$$

Now, staying with the chosen M we select $\Lambda = \Lambda(t) = M(t) + 1$, we have for any $\epsilon > 0$, that, for large enough t,

$$\frac{d}{t} \sum_{\lambda=\Lambda}^{\pi-1} \sum_{k=1}^{M+d-1} \|\mathbf{P}(m + \lambda d + v + 1 - M + k) - \mathbf{P}\| \leq$$

$$\frac{d}{t} \sum_{\lambda=\Lambda}^{\pi-1} (M + d) C (\lambda d - M)^{-a} \leq \frac{Cd(M + d)}{t} \sum_{r=1}^{t} r^{-a} \leq \epsilon \tag{4.106}$$

Now from (4.104), (4.105), and (4.106) we get again (4.81). Also, $M + d$ is of order $\ln t$ we get that $M + d$ is of order t^{ϵ}. Next we observe that $\sum_{r=1}^{t} r^{-a}$ is of order t^{1-a} hence the right hand side of (4.106) is of order less than $t^{-a+\epsilon}$. Finally, since $\Lambda = M + 1$ is of order $\ln t$ the right hand side of (4.100) is of order no greater than $t^{-1+\epsilon}$. Finally, in the third part of (4.100) $(L - \Lambda)/t$ is of order $\ln t / t$ and from (4.103) we get that the third part is of order not less than t^{e-1-a}.Therefore, we arrive at the conclusion that the left hand side of (4.100) is of order not greater than

$$\min\left(\frac{1}{t^{a+1-\epsilon}}, \frac{1}{t^{a-\epsilon}}, \frac{1}{t^{1-\epsilon}}\right).$$

From which when $a > 1$ then we get (4.98) and when $\alpha \in (0, 1]$ then we get (4.99). ∎

4.5 Uniform strong ergodicity with the use of mean visit times

In this section, we aim at providing a criterion for uniform strong ergodicity with the use of mean visit times in the states of a non-homogeneous Markov chain. The roots on the use of mean visit times for uniform rate of convergence in Markov chains are in Pitman (1974). We will start by providing the definition of the mean visit time in a state.

Definition 255 *Let a non-homogeneous Markov chain $\{X_t\}_{t=0}^{\infty}$ with a countable state space $\mathbb{S}=\{1,2,...\}$ and a sequence of transition probability matrices $\{\mathbf{P}(t)\}_{t=0}^{\infty}$. The mean visit time from state i to state j is given by*

$$_t m_{ij} = \sum_{n=1}^{\infty} n f_{ij}(t, t+n), \tag{4.107}$$

where

$$f_{ij}(t,t+n) = \Pr\left\{\begin{matrix} \text{first visit to } j \text{ from } i \text{ at time } t+n \\ \text{given in } i \text{ at time } t+1 \end{matrix}\right\}.$$

We will start with two Lemmas which will help us to prove the main result of this section.

Lemma 256 *(Huang and Isaacson (1976). Let a non-homogeneous Markov chain $\{X_t\}_{t=0}^{\infty}$ with a countable state space $\mathbb{S}=\{1,2,...\}$ and a sequence of transition probability matrices $\{\mathbf{P}(t)\}_{t=0}^{\infty}$. If $\{\mathbf{P}(t)\}_{t=0}^{\infty}$ is weakly ergodic and $\lim_{n\to\infty} p_{jj}(t,t+n) = \pi_j$ then $\lim_{n\to\infty} p_{ij}(t,t+n) = \pi_j$ for all i.*

Proof. Since $\{\mathbf{P}(t)\}_{t=0}^{\infty}$ is weakly ergodic, we have that for all t and for any $\mathbf{p}(0)$ and $\hat{\mathbf{p}}(0)$ initial distributions

$$\lim_{n\to\infty} \sup_{\mathbf{p}(0),\hat{\mathbf{p}}(0)} \|\mathbf{p}(0)\mathbf{P}(t,t+n) - \hat{\mathbf{p}}(0)\mathbf{P}(t,t+n)\| = 0. \tag{4.108}$$

If we take as initial distributions the vectors $\mathbf{e}_i$ and $\mathbf{e}_k$, we get that

$$\lim_{n\to\infty} \|\mathbf{e}_i\mathbf{P}(t,t+n) - \mathbf{e}_k\mathbf{P}(t,t+n)\| = 0, \text{ uniformly in } i \text{ and } k. \tag{4.109}$$

From the hypothesis of the theorem we get that

$$|p_{jj}(t,t+n) - \pi_j| < \epsilon/2 \quad \text{for every} \quad n > N_1(\epsilon). \tag{4.110}$$

From (4.109) we get that

$$|p_{ij}(t,t+n) - p_{kj}(t,t+n)| < \epsilon/2 \text{ for all } i,k,j \text{ and for } n > N_2(\epsilon). \tag{4.111}$$

Let $N(\epsilon) = \max\{N_1(\epsilon), N_2(\epsilon)\}$ then for any i from (4.110) and (4.111) we get that

$$|p_{ij}(t,t+n) - \pi_j| \le |p_{jj}(t,t+n) - \pi_j| + |p_{ij}(t,t+n) - p_{kj}(t,t+n)| < \epsilon, \tag{4.112}$$

for all i and $n > N(\epsilon)$. This concludes the proof of the Lemma. ■

We will now proceed to state and prove the second Lemma that we need in order to present the main result of the present section.

Lemma 257 *(Huang and Isaacson (1976)). Let a non-homogeneous Markov chain $\{X_t\}_{t=0}^{\infty}$ with a countable state space $\mathbb{S}=\{1,2,...\}$ and a sequence of transition probability matrices $\{\mathbf{P}(t)\}_{t=0}^{\infty}$. Then $\{X_t\}_{t=0}^{\infty}$ is strongly ergodic if and only if $\{\mathbf{P}(t)\}_{t=0}^{\infty}$ is weakly ergodic and $\lim_{n\to\infty} p_{jj}(t,t+n) = \pi_j \ge 0$ for all j and t and in addition $\sum_{j=1}^{\infty} \pi_j = 1$.*

Proof. When $\{X_t\}_{t=0}^{\infty}$ is strongly ergodic then it is known that $\{\mathbf{P}(t)\}_{t=0}^{\infty}$ is weakly ergodic and $\lim_{n\to\infty} p_{jj}(t,t+n) = \pi_j \ge 0$ for all j and t and in addition $\sum_{j=1}^{\infty} \pi_j = 1$. Hence, it remains to prove that if $\{\mathbf{P}(t)\}_{t=0}^{\infty}$ is weakly ergodic and $\lim_{n\to\infty} p_{jj}(t,t+n) = \pi_j \ge 0$ for all j and t and in addition $\sum_{j=1}^{\infty} \pi_j = 1$ then $\{X_t\}_{t=0}^{\infty}$ is strongly ergodic.

From $\sum_{j=1}^{\infty} \pi_j = 1$ we get that for every $\epsilon > 0$ there exists a $K = K(\epsilon)$ such that

$$\sum_{j=k}^{\infty} \pi_j < \epsilon/4 \quad \text{for every } k \ge K. \tag{4.113}$$

We now have

$$\sum_{j=1}^{\infty} |p_{ij}(t,t+n) - \pi_j| \le \sum_{j=1}^{K-1} |p_{ij}(t,t+n) - \pi_j| + \sum_{j=K}^{\infty} p_{ij}(t,t+n) + \sum_{j=K}^{\infty} \pi_j. \tag{4.114}$$

Since $\{\mathbf{P}(t)\}_{t=0}^{\infty}$ is weakly ergodic and $\lim_{n\to\infty} p_{jj}(t,t+n) = \pi_j \ge 0$ for all j and t from Lemma 256 we get that for every $\epsilon > 0$ there exists an $N = N(\epsilon)$ such that

$$|p_{ij}(t,t+n) - \pi_j| < \epsilon/4(K-1) \text{ for all } i, \text{ for } j \le K-1, \text{ and for } n > N. \tag{4.115}$$

For the second sum on the right hand side of (4.114), we have that

$$\sum_{j=K}^{\infty} p_{ij}(t,t+n) = 1 - \sum_{j=1}^{K-1} p_{ij}(t,t+n) + \sum_{j=1}^{K-1} \pi_j - \sum_{j=1}^{K-1} \pi_j \le$$

$$\sum_{j=1}^{K-1} |p_{ij}(t,t+n) - \pi_j| + \sum_{j=1}^{K-1} \pi_j \le (\text{from } (4.113), (4115))$$

$$\le \epsilon/2 \text{ for all } i,t, \text{ for } j \le K-1, \text{ and for } n > N. \tag{4.116}$$

Therefore, from (4.113)–(4.116) we get that

$$\text{for all } t \ \sup_i \sum_{j=1}^{\infty} |p_{ij}(t,t+n) - \pi_j| \le \epsilon \text{ for } n > N, \tag{4.117}$$

Hence,

$$\text{for all } t \ \|\mathbf{P}(t,t+n) - \mathbf{\Pi}\| \le \epsilon \text{ for } n > N,$$

where $\mathbf{\Pi}$ is the stable matrix with row $\boldsymbol{\pi} = (\pi_1, \pi_2 ...)$. Therefore $\{X_t\}_{t=0}^{\infty}$ is strongly ergodic. ∎

Following the same path and taking similar steps we could prove the following Lemma which deals with the case of uniformly strongly ergodic.

Lemma 258 *Let a non-homogeneous Markov chain $\{X_t\}_{t=0}^{\infty}$ with a countable state space $\mathbb{S} = \{1, 2, ...\}$ and a sequence of transition probability matrices $\{\mathbf{P}(t)\}_{t=0}^{\infty}$. Then $\{X_t\}_{t=0}^{\infty}$ is uniformly strongly ergodic if and only if $\{\mathbf{P}(t)\}_{t=0}^{\infty}$ is uniformly weakly ergodic and $\lim_{n\to\infty} p_{jj}(t,t+n) = \pi_j \ge 0$ for all j and t and in addition $\boldsymbol{\pi}\mathbf{1}^{\top} = 1$, where $\boldsymbol{\pi} = (\pi_1, \pi_2 ...)$*

We will now establish the main result of the present section, which will be a characterization of uniform strong ergodicity using mean visit times.

Theorem 259 *(Huang and Isaacson, 1976). Let a non-homogeneous Markov chain $\{X_t\}_{t=0}^{\infty}$ with a countable state space $\mathbb{S} = \{1, 2, ...\}$ and a sequence of transition probability matrices $\{\mathbf{P}(t)\}_{t=0}^{\infty}$. Then $\{X_t\}_{t=0}^{\infty}$ is uniformly strongly ergodic if and only if for all diagonal elements*

$$\lim_{n\to\infty} p_{jj}(t,t+n) = \pi_j \ge 0 \ \text{ uniformly in } t, \ \ \boldsymbol{\pi}\mathbf{1}^{\top} = 1,$$

where $\boldsymbol{\pi} = (\pi_1, \pi_2 ...)$ and for some l for which $\pi_l > 0$ then $\sup_{i,s}({}_s m_{il}) < \infty$. In addition, uniform strong ergodicity implies that $\sup_{i,s}({}_s m_{il}) < \infty$ for all l with $\pi_l > 0$.

Proof. Let us start by assuming that $\{X_t\}_{t=0}^{\infty}$ is uniformly strongly ergodic. Then it only remains to prove that

$$\sup_{i,t}({}_t m_{il}) < \infty \text{ whenever } \pi_l > 0. \tag{4.118}$$

By the uniformly strong ergodicity of $\{\mathbf{P}(t)\}_{t=0}^{\infty}$ for a stable matrix $\mathbf{Q}$ with row $\boldsymbol{\pi} = (\pi_1, \pi_2 ...)$, we have that

$$\lim_{n\to\infty} \|\mathbf{P}(t, t+n) - \mathbf{Q}\| = 0 \text{ uniformly in } t,$$

therefore there exists a $N(\pi_l) = N$ such that

$$\|\mathbf{P}(t, t+n) - \mathbf{Q}\| < \pi_l/2 \text{ for } n \geq N(\pi_l).$$

consequently

$$|p_{il}(t, t+n) - \pi_l| < \pi_l/2 \text{ for } i, t, \text{ and } n \geq N(\pi_l).$$

from which we get $p_{il}(t, t+n) \geq \pi_l/2$ and finally

$$\inf_i p_{il}(t, t+n) \geq \pi_l/2 \text{ for all } t, \text{ and } n \geq N(\pi_l). \tag{4.119}$$

By writing $t + 2N(\pi_l) + r$ for $r = 0, 1, 2, ..., N(\pi_l) - 1$ we get all the numbers between $t + 2N(\pi_l)$ and $t + 3N(\pi_l) - 1$ and for these we get

$$\begin{aligned} f_{ij}(t, t+2N+r) &\leq \sum_{k\neq l} p_{ik}(t, t+N) f_{kl}(t+N, t+2N+r) \\ &\leq \sum_{k\neq l} p_{ik}(t, t+N) \leq 1 - p_{il}(t, t+N) \\ &\leq \text{ from (4.119) } \leq 1 - \pi_l/2. \end{aligned} \tag{4.120}$$

We repeat the same arguments for $f_{ij}(t, t+2N+r), ..., f_{ij}(t, t+mN+r)$ and we get

$$f_{ij}(t, t+mN+r) \leq (1 - \pi_l/2)^{m-1}. \tag{4.121}$$

Hence, from (4.107) and for all i and t

$$\begin{gathered} {}_t m_{il} = \sum_{n=1}^{\infty} n f_{il}(t, t+n) \leq \\ \sum_{n=1}^{2N-1} n f_{il}(t, t+n) + \sum_{n=2N}^{3N-1} n f_{il}(t, t+n) + \sum_{n=3N}^{4N-1} n f_{il}(t, t+n) + ... \leq \\ \leq N(1 - \pi_l/2) + \frac{N}{2}(5N-1)(1 - \pi_l/2) + \frac{N}{2}(7N-1)(1 - \pi_l/2)^2 + ... \\ N(1 - \pi_l/2) + \sum_{n=3}^{\infty} n N^2 (1 - \pi_l/2)^{n-2} < \infty, \end{gathered}$$

independent of i and t. Thus $\sup_{i,t}({}_t m_{il}) < \infty$.

We will now prove the converse. That is, we want to prove that if $\lim_{n\to\infty} p_{jj}(t, t+n) = \pi_j \geq 0$ uniformly in t, $\boldsymbol{\pi}\mathbf{1}^\top = 1$, and for some l for which $\pi_l > 0$ we have $\sup_{i,s}({}_s m_{il}) < \infty$ then $\{X_t\}_{t=0}^{\infty}$ is uniformly strongly ergodic. We will follow the opposite path, that is we will assume that when the previous conditions hold then $\{X_t\}_{t=0}^{\infty}$ is not uniformly strongly ergodic and look for a contradiction. By assuming $\{X_t\}_{t=0}^{\infty}$ is not uniformly strongly ergodic then by Lemma 258, we have also that it is not weakly ergodic, therefore, given n there exists $t = T$ such that

$$\tau_1(\mathbf{P}(T, T+n)) > 1 - \pi_l/4.$$

From the definition of the Dobrushin coefficient of ergodicity and by our hypothesis then for every n there exists a state $i(n)$ such that

$$p_{i(n)l}(T, T+n) < \pi_l/4. \tag{4.122}$$

From our hypothesis $\lim_{n\to\infty} p_{ll}(t, t+n) = \pi_l > 0$ uniformly in t hence there exists an N such that

$$p_{ll}(T, T+n) > \pi_l/2 \quad \text{for} \quad n \geq N. \tag{4.123}$$

Now, for $n \geq 2N$ we have that

$$p_{i(n)l}(T, T+n) = \sum_{k=1}^{n-1} f_{i(n)l}(T, T+k)\, p_{ll}(T+k, T+n)$$

(retaining the first $[n/2]$ (integer part) terms of the sum)

$$\geq \sum_{k=1}^{[n/2]} f_{i(n)l}(T, T+k)\, p_{ll}(T+k, T+n)$$

$$\geq \text{(from (4.123))} \geq \frac{\pi_l}{2} \sum_{k=1}^{[n/2]} f_{i(n)l}(T, T+k). \tag{4.124}$$

From (4.122) and (4.124) we get that

$$\sum_{k=1}^{[n/2]} f_{i(n)l}(T, T+k) < \frac{1}{2} \quad \text{for} \quad n \geq 2N. \tag{4.125}$$

Now, we have that for $n \geq 2N$

$${}_T m_{i(n)l} = \sum_{k=1}^{\infty} k f_{i(n)l}(T, T+k) >$$

(taking out the first $[n/2]$ terms and replacing k with $[n/2]+1$)

$$> ([n/2]+1) \sum_{k=[n/2]+1}^{\infty} f_{i(n)l}(T, T+k). \tag{4.126}$$

By hypothesis, we have ${}_T m_{i(n)l} < \infty$ therefore $\sum_{k=1}^{\infty} f_{i(n)l}(T, T+k) = 1$ and consequently from (4.126) we get that

$${}_T m_{i(n)l} > ([n/2]+1)(1-1/2) \geq \frac{n+1}{4} \quad \text{for } n \geq 2N.$$

Therefore,

$$\sup_{i,t}\left({}_t m_{il}\right) = \infty,$$

which contradicts our hypothesis, and hence $\{X_t\}_{t=0}^{\infty}$ is uniformly strongly ergodic. ■

4.6 Strong ergodicity for general products of matrices

In the present section we will consider general infinite products of stochastic matrices or infinite products of stochastic matrices taken in arbitrary order of multiplication. We will concentrate merely on forward products. That is, if $\{\mathbf{Q}(t)\}_{t=0}^{\infty}$ is the sequence of matrices from which we draw matrices we will study the asymptotic behavior of products of the type

$$\lim_{t\to\infty} \mathbf{Q}(0,t).$$

We will prove the following basic theorem:

Theorem 260 *(Leizarowitz, 1992). Let $\{X_t\}_{t=0}^{\infty}$ be a non-homogeneous Markov chain on a complete probability space $(\Omega, \mathcal{F}, \mathbb{P})$ and let $\{\mathbf{P}(t)\}_{t=0}^{\infty}$ be its sequence of transition probability matrices. Suppose that $\{\mathbf{P}(t)\}_{t=0}^{\infty}$ is such that $\mathcal{W}$ the set of its accumulation points contains a scrambling matrix. Assume also that all the matrices $\mathbf{P}(t)$ for every $t \geq 0$ share a common eigenvector of the eigenvalue one. That is, if $\mathbf{p}$ is the common eigenvector then*

$$\mathbf{pP}(t) = \mathbf{p} \quad \text{for every } t \geq 0.$$

Then we have strong ergodicity for forward products $\mathbf{P}(0,t)$ $t \geq 1$ as $t \to \infty$ taken in arbitrary orders of multiplication or general infinite products and in addition

$$\lim_{t\to\infty} \mathbf{P}(0,t) = \begin{pmatrix} \mathbf{p} \\ \cdots \\ \mathbf{p} \end{pmatrix}.$$

Proof. Let the l_1-norm in $\mathbb{R}^n$ and let the induced norm $\|.\|_1$ for matrices in SM_n. We will consider a specific order of multiplication for the sequences $\{\mathbf{P}(0,t)\}_{t=1}^{\infty}$ and prove strong ergodicity independent of the specific order, which in fact proves strong ergodicity for any infinity general product or infinite products taken in any order of multiplication. We define the following set

$$K_t = \{\mathbf{xP}(0,t) : \mathbf{x} \in SM_{1n}\}.$$

We will firstly show that as $t \to \infty$ the diameter of the set K_t goes to zero, that is for any two vectors $\mathbf{x}, \mathbf{y} \in SM_{1n}$ with $\mathbf{x} \neq \mathbf{y}$

$$\lim_{t\to\infty} d(K_t) = 0 = \sup\left\{\lim_{t\to\infty} \|\mathbf{xP}(0,t) - \mathbf{yP}(0,t)\|_1 = 0 : \mathbf{x} \neq \mathbf{y} \in SM_{1n}\right\},$$

or equivalently

$$\lim_{t\to\infty} \|\mathbf{xP}(0,t) - \mathbf{yP}(0,t)\|_1 = 0 \text{ for every } \mathbf{x}, \mathbf{y} \in SM_{1n}.$$

From the assumption of the theorem, we know that in the set of all accumulation points $\mathcal{W}$ there is a matrix which is scrambling. Let $\mathbf{Q} \in \mathcal{W}$ the scrambling matrix for which apparently

$$\|\mathbf{Q}_{|\mathbb{H}}\|_1 < 1.$$

Now following similar steps as in Theorem 208, we arrive at the conclusion that

$$\lim_{t\to\infty} \|\mathbf{P}(0,t)\|_1 = 0,$$

from which we get that

$$\lim_{t\to\infty} \|(\mathbf{x} - \mathbf{y})\mathbf{P}(0,t)\|_1 = 0 \text{ for every } \mathbf{x}, \mathbf{y} \in SM_{1n}.$$

Since the common eigenvector $\mathbf{p} \in SM_{1n}$ we obtain

$$\lim_{t\to\infty} \|\mathbf{xP}(0,t) - \mathbf{pP}(0,t)\|_1 = 0 \text{ for every } \mathbf{x} \in SM_{1n},$$

or equivalently

$$\lim_{t\to\infty} \|\mathbf{xP}(0,t) - \mathbf{p}\|_1 = 0 \text{ for every } \mathbf{x} \in SM_{1n}.$$

From which we get that

$$\mathbf{P}(0,t)\,\mathbf{e}_i \to \mathbf{p} \text{ as } t \to \infty \text{ for all } i = 1, 2, ..., n.$$

Hence, any general forward infinite product of matrices from the sequence $\{\mathbf{P}(t)\}_{t=0}^{\infty}$ is strongly ergodic. ■

The above could take a rather more general form which may be proved similarly if instead of the l_1-norm from which we constructed the induced norm we use any induced norm for matrices from a norm on $\mathbb{R}^n$. Then the theorem takes the following form.

Theorem 261 *(Leizarowitz (1992)). Let $\{X_t\}_{t=0}^{\infty}$ be a non-homogeneous Markov chain on a complete probability space $(\Omega, \mathcal{F}, \mathbb{P})$ and let $\{\mathbf{P}(t)\}_{t=0}^{\infty}$ be its sequence of transition probability matrices. Assume that there exists a norm on $\mathbb{R}^n$ from which we construct the induced matrix norm for matrices. Suppose that $\{\mathbf{P}(t)\}_{t=0}^{\infty}$ is such that $\mathcal{W}$ the set of its accumulation points is such that either*

$$\left\|\mathbf{Q}_{|\mathbb{H}}\right\| < 1/C \text{ for some } \mathbf{Q} \in \mathcal{W},$$

where C is as defined in Theorem 208, or

$$\left\|\mathbf{Q}_{|\mathbb{H}}\right\| < 1 \text{ for every } \mathbf{Q} \in \mathcal{W}.$$

In addition assume that all the matrices $\mathbf{P}(t)$ for every $t \geq 0$ share a common eigenvector of the eigenvalue one. That is, if $\mathbf{p}$ is the common eigenvector then

$$\mathbf{pP}(t) = \mathbf{p} \quad \text{for every } t \geq 0.$$

Then we have strong ergodicity for forward products $\mathbf{P}(0,t)$ $t \geq 1$ as $t \to \infty$ taken in arbitrary orders of multiplication or general infinite products and in addition

$$\lim_{t\to\infty} \mathbf{P}(0,t) = \begin{pmatrix} \mathbf{p} \\ \cdots \\ \mathbf{p} \end{pmatrix}.$$

We will now prove another basic theorem in which the condition that all matrices in the sequence $\{\mathbf{P}(t)\}_{t=0}^{\infty}$ share a common eigenvector of the eigenvalue 1 will be imposed instead on the set $\mathcal{W}$, which usually is a much smaller set.

Theorem 262 *(Leizarowitz (1992)). Let $\{X_t\}_{t=0}^{\infty}$ be a non-homogeneous Markov chain on a complete probability space $(\Omega, \mathcal{F}, \mathbb{P})$ and let $\{\mathbf{P}(t)\}_{t=0}^{\infty}$ be its sequence of transition probability matrices. Suppose that $\{\mathbf{P}(t)\}_{t=0}^{\infty}$ is such that $\mathcal{W}$ the set of its accumulation points contains only scrambling matrices. Assume also that all the matrices in $\mathcal{W}$ share a common eigenvector of the eigenvalue one. That is, if $\mathbf{p}$ is the common eigenvector then*

$$\mathbf{pP} = \mathbf{p} \quad \text{for every } \mathbf{P} \in \mathcal{W}.$$

Then we have strong ergodicity for forward products $\mathbf{P}(0,t)$ $t \geq 1$ as $t \to \infty$ taken in arbitrary orders of multiplication or general infinite forward products and in addition

$$\lim_{t\to\infty} \mathbf{P}(0,t) = \mathbf{P}^* = \begin{pmatrix} \mathbf{p} \\ \cdots \\ \mathbf{p} \end{pmatrix}.$$

Proof. We will assume that $\lim_{t\to\infty} \mathbf{P}(0,t)$ does not converge and we will show that this contradicts the conditions of the theorem. Since the sequence of transition probability matrices is not strongly ergodic, then we can find an increasing sequence of integers $\{n_k\}_{k=0}^{\infty}$ such that for every $k \geq 1$ we can find a finite number of selection times $t_1(k), t_2(k), ..., t_k(k)$ all grater than n_k, that is, $t_i(k) \geq n_k$ for every $i = 1, 2, ...k$ and consequently we get a finite sequence of matrices $\mathbf{P}(t_1(k))\mathbf{P}(t_2(k))...\mathbf{P}(t_k(k))$. This sequence of matrices is such that for every $\epsilon > 0$ and for any number of matrices from them using the induced l_1-norm $\mathbb{R}^n$

$$\|\mathbf{P}(t_1(k))\mathbf{P}(t_2(k))...\mathbf{P}(t_r(k)) - \mathbf{P}^*\| \geq \epsilon \quad \text{for } r = 1, 2, ..., k. \tag{4.127}$$

By taking values of k every time larger, we can apparently create an increasing sequence $\{k_m\}_{m=0}^{\infty}$ such that

$$\lim_{m\to\infty} \mathbf{P}(t_i(k_m)) = \mathbf{P}^*(i) \quad \text{for every fixed } i.$$

Apparently $\mathbf{P}^*(i) \in \mathcal{W}$ for every i and satisfy the conditions of Theorem 261 hence as a sequence is strongly ergodic which contradicts the fact that by construction the set of matrices $\mathbf{P}^*(i)$ for every $i \geq 1$ satisfy by construction a relation of the type (4.127). ■

The above could take a rather more general form which may be proved similarly if instead of the l_1-norm from which we constructed the induced norm we use any induced norm for matrices from a norm on $\mathbb{R}^n$. Then the theorem takes the following form.

Theorem 263 *(Leizarowitz (1992)). Let $\{X_t\}_{t=0}^{\infty}$ be a non-homogeneous Markov chain on a complete probability space $(\Omega, \mathcal{F}, \mathbb{P})$ and let $\{\mathbf{P}(t)\}_{t=0}^{\infty}$ be its sequence of transition probability matrices. Suppose that $\{\mathbf{P}(t)\}_{t=0}^{\infty}$ is such that $\mathcal{W}$ the set of its accumulation points is such that*

$$\left\|\mathbf{P}_{|\mathbb{H}}\right\| < 1 \quad \text{for every } \mathbf{P} \in \mathcal{W}.$$

Assume also that all the matrices in $\mathcal{W}$ share a common eigenvector of the eigenvalue one. That is, if $\mathbf{p}$ is the common eigenvector then

$$\mathbf{pP} = \mathbf{p} \quad \text{for every } \mathbf{P} \in \mathcal{W}.$$

Then we have strong ergodicity for forward products $\mathbf{P}(0,t)$ $t \geq 1$ as $t \to \infty$ taken in arbitrary orders of multiplication or general infinite forward products and in addition

$$\lim_{t\to\infty} \mathbf{P}(0,t) = \mathbf{P}^* = \begin{pmatrix} \mathbf{p} \\ ... \\ \mathbf{p} \end{pmatrix}.$$

A very important Corollary of Theorem 263 which is a generalization of Theorem 242 is the following.

Theorem 264 *Let $\{X_t\}_{t=0}^{\infty}$ be a non-homogeneous Markov chain on a complete probability space $(\Omega, \mathcal{F}, \mathbb{P})$ and let $\{\mathbf{P}(t)\}_{t=0}^{\infty}$ be its sequence of transition probability matrices. Suppose that $\{\mathbf{P}(t)\}_{t=0}^{\infty}$ is such that*

$$\lim_{t\to\infty} \mathbf{P}(t) = \mathbf{P} \quad \text{a regular matrix.}$$

Then we have strong ergodicity for forward products $\mathbf{P}(0,t)$ $t \geq 1$ as $t \to \infty$ taken in arbitrary orders of multiplication or general infinite forward products and in addition

$$\lim_{t\to\infty} \mathbf{P}(0,t) = \mathbf{P}^* = \begin{pmatrix} \mathbf{p} \\ ... \\ \mathbf{p} \end{pmatrix},$$

where

$$\mathbf{pP} = \mathbf{p}.$$

Proof. Since $\mathbf{P}$ is regular there is a norm in $\mathbb{R}^n$ induced for matrices such that $\left\|\mathbf{P}_{|\mathbb{H}}\right\| < 1$ and from the theorem, we have apparently that $\mathbf{P} \in \mathcal{W}$ the set of accumulation points for the sequence $\{\mathbf{P}(t)\}_{t=0}^{\infty}$ which in this case is a singlenton. Therefore the conditions of the previous theorem are satisfied and we arrive at the desired conclusion. ∎

4.7 Sets of matrices all products of which converge

There is a large amount of research published on infinite matrix products of matrices taken from any subset Σ of $M_n(\mathbb{R})$ or $M_n(\mathbb{C})$. Such a matrix set Σ usually contains a finite number of matrices $\mathbf{A}_i$ for $i = 0, 1, ..., m$ and it is said to be *bounded* if there is a positive constant a such that for some matrix norm $\|.\|$, $\|\mathbf{A}_i\| \leq a$ for all $\mathbf{A}_i \in \Sigma$. The set Σ is *product bounded* if there is a positive constant b such that $\|\mathbf{A}_1\mathbf{A}_2...\mathbf{A}_k\| \leq b$ for all k and all $\mathbf{A}_i \in \Sigma$. Work has also been done for the case when Σ is compact.

The set Σ is said to be an *RCP* set, i.e., Right Converging Product, or to have the *RCP* if all infinite products of members of Σ of the type

$$\lim_{n\to\infty} \mathbf{A}_1\mathbf{A}_2\mathbf{A}_3... \quad \text{converge.}$$

On the other hand the set Σ is said to be an *LCP* set, i.e., Left Converging Product, or to have the *LCP* if all infinite products of members of Σ of the type

$$\lim_{n\to\infty} \mathbf{A}_n\mathbf{A}_{n-1}\mathbf{A}_{n-2}... \quad \text{converge.}$$

It is easy to see that Σ is an *RCP* set if and only if $\Sigma^{\top}$ is an *LCP* set. For this reason it suffices to study *RCP* sets. Hartfiel (1974) observes that there are *RCP* sets that are not *LCP* sets. We denote with Σ^k the set of all products of k matrices from the set Σ and with Σ^{∞} the set of the limits of all infinite product of matrices from the set Σ.

A basic reference for the roots of the study of sets of matrices all infinite products of which converge is the work of Daubechies and Lagarias (1992). *RCP* sets of matrices arise in a large variety of contexts. For example, they appear in parametrizing various fractal-like objects; also can be similarly parametrized at tractors of hyperbolic iterated function systems on $\mathbb{R}^n$; also they arise in computer-aided geometric design in constructing parametrized curves and surfaces; also in the study of lattice two-scale difference equations. Another large area of study of such sets is the case when Σ consists in general from non-negative matrices. A very good and thorough book for the work done on this problem is by Hartfiel (2002). Naturally, sets of stochastic matrices with the *RCP* or *LCP* appear in the study of non-homogeneous Markov chains and in probabilistic automata. Apart from the very good book by Hartfiel (2002) basic results on *RCP* or *LCP* sets stochastic matrices exist in Daubechies and Lagarias (1992) as an application of his general results on infinite products of finite complex matrices and in Hartfiel and Rothblum (1998). The latter founds necessary and sufficient conditions for *LCP* sets of finite non-negative matrices which naturally are directly applicable in the case of finite stochastic matrices. The basic theorem in Daubechies and Lagarias (1992) is Theorem 6.1 which follows:

Theorem 265 *(Daubechies and Lagarias (1992)). For a finite set* $\Sigma = \{\mathbf{P}(0), \mathbf{P}(1), ..., \mathbf{P}(m)\}$ *of* $n \times n$ *column stochastic matrices, the following conditions are equivalent:* (1) Σ *is an RCP set whose left* 1*-eigenspace* $E_1(\Sigma)$ *is one dimensional spanned by* $(1, 1, ..., 1)$*.* (2) *All finite products* $\mathbf{P}(k_1), \mathbf{P}(k_2), ..., \mathbf{P}(k_i)$ *are irreducible and aperiodic.* (3) *There exists a finite* μ *such that all* $k \geq \mu$ *RCP products of matrices from* Σ *are scrambling.* (4) *There exists a finite* ν *such that all* $k \geq \nu$ *RCP products of matrices from* Σ *have a row with all entries nonzero.* (5) *All left infinite products from* Σ *are weakly ergodic.*

We now move on to the basic result on Hartfiel and Rothblum (1998). We assume that:

Assumption $\mathbf{E}_1$. *All the matrices in Σ have 1 as an eigenvalue and have a common set of independent non-negative eigenvectors corresponding to 1, say $\mathbf{v}_1, \mathbf{v}_2, ..., \mathbf{v}_s$.*

Let $\mathbf{V} = (\mathbf{v}_1, \mathbf{v}_2, ..., \mathbf{v}_s)$ be the matrix with columns $\mathbf{v}_i$ then it is easy to see that for every $\mathbf{A} \in \Sigma$ we have $\mathbf{AV} = \mathbf{V}$. We observe that if the matrices Σ are nonnegative and have a common strictly positive eigenvector corresponding to 1, this is the case when the assumption is satisfied with $\sum_{i=1}^{s} \mathbf{v}_i > 0$.

Let now Σ be a set of $n \times n$ matrices satisfying assumption $\mathbf{E}_1$. Let also $\|.\|$ be a norm on $\mathbb{R}^n$. For a matrix $\mathbf{A} \in \Sigma$ the $\|.\|$-*coefficient of ergodicity* of $\mathbf{A}$, denoted $\tau_{\|.\|}(\mathbf{A})$, is defined by

$$\tau_{\|.\|}(\mathbf{A}) := \max_{\substack{\mathbf{x}\in\mathbb{R}^n \\ \|\mathbf{x}\|=1 \\ \mathbf{xV}=\mathbf{0}}} \|\mathbf{xA}\|.$$

Also the $\|.\|$-*coefficient of ergodicity* of Σ denoted by $\tau_{\|.\|}(\mathbf{\Sigma})$, is defined by

$$\tau_{\|.\|}(\mathbf{\Sigma}) = \sup_{\mathbf{A}\in\Sigma} \tau_{\|.\|}(\mathbf{A}).$$

Now the main result in Hartfiel and Rothblum (1998), which provides a sufficient condition for the LCP property in terms of coefficient of ergodicity is the following

Theorem 266 *(Hartfiel and Rothblum (1998)). If Σ is bounded and for some norm $\|.\|$ on $\mathbb{R}^n$ $\tau_{\|.\|}(\mathbf{\Sigma}) < 1$, then Σ has the LCP property. In addition, the convergence of infinite sequences (LCP) from Σ to its limit is geometric with $\tau_{\|.\|}(\mathbf{\Sigma})$ as a bound on the geometric coefficient of convergence.*

4.8 A geometric approach to ergodic NHMC

In the present section we will present an interesting approach which is called geometric introduced by Shen (2000a) for the ergodicity of Non-homogeneous Markov chains with a finite state space. This approach was inspired from the work of Daubechies and Lagarias (1992). In this approach a stochastic matrix is understood as a transform of simplices.

For any ordered n vectors which usually geometrically are referred as vertices $\mathbf{v}_1, \mathbf{v}_2, ..., \mathbf{v}_n$ in $\mathbb{R}^n$, the associated n-simplex is their convex hull, denoted by $[\mathbf{v}_1\mathbf{v}_2...\mathbf{v}_n]$. That is,

$$[\mathbf{v}_1\mathbf{v}_2...\mathbf{v}_n] = \left\{ \sum_{i=1}^{n} a_i\mathbf{v}_i \mid a_i \geq 0, \ \sum_{i=1}^{n} a_i = 1 \right\}.$$

The simplex is called *non-degenerate* if the n vertices are linearly independent. Note that apparently a non-degenerate n-simplex is $n-1$ dimensional. Let

$$\mathbf{e}_i = [0, 0, ...0, 1, 0, ..., 0],$$

where 1 is in the *i-th* position for $i = 1, ..., n$ be the canonical basis of $\mathbb{R}^n$. Then the simplex $[\mathbf{e}_1\mathbf{e}_2...\mathbf{e}_n]$ is called the *canonical simplex.* For example, $[\mathbf{e}_1\ \mathbf{e}_2\ \mathbf{e}_3]$ is the triangle $(1, 0, 0), (0, 1, 0)$ and $(0, 0, 1)$ in $\mathbb{R}^3$ which is in fact 2-dimensional.

In this section given a set $\Sigma \subset SM_n$, we are interested on the asymptotic behavior of all Σ-Markov chains, that is, all non-homogeneous Markov chains whose transition probability matrices are drown from Σ.

For a given stochastic matrix $\mathbf{P}$ we define a transform of simplices

$$[\mathbf{v}_1\mathbf{v}_2...\mathbf{v}_n] \to \left[\mathbf{v}_1^{\mathbf{P}}\mathbf{v}_2^{\mathbf{P}}...\mathbf{v}_n^{\mathbf{P}}\right],$$

where

$$\mathbf{v}_i^{\mathbf{P}} = p_{i1}\mathbf{v}_1 + p_{i2}\mathbf{v}_2 + ... + p_{in}\mathbf{v}_n.$$

It is easy to see that

$$\left[\mathbf{v}_1^{\mathbf{P}}\mathbf{v}_2^{\mathbf{P}}...\mathbf{v}_n^{\mathbf{P}}\right] \subset \left[\mathbf{v}_1\mathbf{v}_2...\mathbf{v}_n\right].$$

Let $\mathbf{P}$ be a stochastic matrix and let $\mathbf{p}_i$ be its i-th row vector. Then it is apparent that

$$\left[\mathbf{e}_1^{\mathbf{P}}\mathbf{e}_2^{\mathbf{P}}...\mathbf{e}_n^{\mathbf{P}}\right] = \left[\mathbf{p}_1\mathbf{p}_2...\mathbf{p}_n\right] := \left[\mathbf{P}\right]. \tag{4.128}$$

Remark 267 *It is easily seen that there exist a one to one mapping between the set of* $\mathbf{P} \in SM_n$ *and the set of all ordered* n*-simplices contained in the canonical simplex.*

For any two $1 \times n$ vectors $\mathbf{p}_i$ and $\mathbf{p}_j$ define by

$$\mathbf{p}_i \wedge \mathbf{p}_j = \left(\min\left(p_{i1}, p_{j1}\right), ..., \min\left(p_{in}, p_{jn}\right)\right).$$

Now, it will be helpful for now on to follow the following example

Example 268 *Let the non-degenerate set of matrices in* $\mathbb{R}^3$

$$\mathbf{v}_1 = [0.1 \;\; 0.2 \;\; 0.7];\; \mathbf{v}_2 = [0.7 \;\; 0.2 \;\; 0.1];\; \mathbf{v}_3 = [0.1 \;\; 0.8 \;\; 0.1].$$

See Figure 4.1

Let $\mathbf{P}$ *be the following scrambling matrix*

$$\mathbf{P} = \begin{pmatrix} 0.4 & 0 & 0.6 \\ 0.3 & 0.7 & 0 \\ 0 & 0.2 & 0.8 \end{pmatrix},$$

then

$$\mathbf{v}_1^{\mathbf{P}} = \begin{pmatrix}0.1 & 0.56 & 0.34\end{pmatrix},\; \mathbf{v}_2^{\mathbf{P}} = \begin{pmatrix}0.52 & 0.20 & 0.28\end{pmatrix},$$
$$\text{and} \;\; \mathbf{v}_3^{\mathbf{P}} = \begin{pmatrix}0.22 & 0.68 & 0.1\end{pmatrix}.$$

In Figure 4.1 the triangle $(1,0,0), (0,1,0)$ *and* $(0,0,1)$ *in* $\mathbb{R}^3$*, that is,* $[\mathbf{e}_1\mathbf{e}_2\mathbf{e}_3]$ *is drawn. Also the triangle with vertices* $\mathbf{v}_1, \mathbf{v}_2,$ *and* $\mathbf{v}_3$ *is drawn and it is immediately evident that* $[\mathbf{v}_1\mathbf{v}_2\mathbf{v}_3] \subset [\mathbf{e}_1\mathbf{e}_2\mathbf{e}_3]$*. As it is seen from the Figure 4.1 the vertex* $\mathbf{v}_1^{\mathbf{P}}$ *is on the line* $\mathbf{v}_1\mathbf{v}_3$*,*

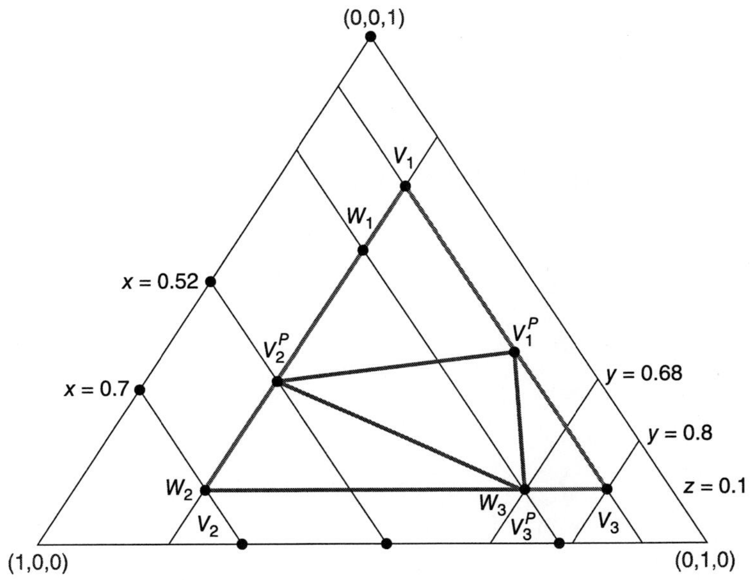

Figure 4.1

and the vertex $\mathbf{v}_2^{\mathbf{P}}$ *is on the line* $\mathbf{v}_1\mathbf{v}_2$ *expressing the fact which we will generalize later that for a scrambling matrix there exist a* $k = 1$ *such that*

$$\left[\mathbf{v}_1^{\mathbf{P}}\mathbf{v}_2^{\mathbf{P}}\right] \cap [\mathbf{v}_2\mathbf{v}_3] = \varnothing.$$

Also, the vertex $\mathbf{v}_3^{\mathbf{P}}$ *is on the line* $\mathbf{v}_2\mathbf{v}_3$ *and so*

$$\left[\mathbf{v}_1^{\mathbf{P}}\mathbf{v}_3^{\mathbf{P}}\right] \cap [\mathbf{v}_1\mathbf{v}_2] = \varnothing \quad \textit{and} \quad \left[\mathbf{v}_2^{\mathbf{P}}\mathbf{v}_3^{\mathbf{P}}\right] \cap [\mathbf{v}_1\mathbf{v}_3] = \varnothing.$$

In Figure 4.1 we also observe that the vertices $\mathbf{v}_1^{\mathbf{P}}$, $\mathbf{v}_2^{\mathbf{P}}$, $\mathbf{v}_3^{\mathbf{P}}$ *are all on the lines of the triangle with vertices* $\mathbf{v}_1, \mathbf{v}_2, \mathbf{v}_3$. *This is a characteristic of a scrambling matrix and it is a way of visually distinguish a scrambling matrix. Note that if* $\mathbf{P}$ *is for example a Markov matrix then the above property does not hold. Note also that*

$$\mathbf{p}_1 \wedge \mathbf{p}_2 = \begin{pmatrix}0.3 & 0 & 0\end{pmatrix}, \quad \mathbf{p}_2 \wedge \mathbf{p}_3 = \begin{pmatrix}0 & 0.2 & 0\end{pmatrix},$$

$$\textit{and} \quad \mathbf{p}_1 \wedge \mathbf{p}_3 = \begin{pmatrix}0 & 0 & 0.6\end{pmatrix}.$$

It is now easy to state the following general proposition of a geometric property for scrambling matrices.

Proposition 269 *(Scrambling matrices). A stochastic matrix* $\mathbf{P}$ *is scrambling if and only if for any non-degenerate* n*-simplex* $[\mathbf{v}_1\mathbf{v}_2...\mathbf{v}_n]$ *and any* i *and* j*, there exists* k*, such that*

$$\left[\mathbf{v}_i^{\mathbf{P}}\mathbf{v}_j^{\mathbf{P}}\right] \cap [\mathbf{v}_1...\mathbf{v}_{k-1}\hat{\mathbf{v}}_k\mathbf{v}_{k+1}...\mathbf{v}_n] = \varnothing,$$

where by $\hat{\mathbf{v}}_k$ *we denote the fact that the element* $\mathbf{v}_k$ *is dropped.*

To provide a kind of measure of how close is a matrix in being scrambling Hajnal (1958) defined

$$\lambda(\mathbf{P}) = \min_{i,j} \|\mathbf{p}_i \wedge \mathbf{p}_j\|_1.$$

It is apparent that $\mathbf{P}$ is scrambling if and only if $\lambda(\mathbf{P}) > 0$. Also, $\lambda(\mathbf{P}) = 1$ if and only if $\mathbf{P}$ is stable. Also, Hajnal (1958) defined the *Hajnal diameter* of a stochastic matrix $\mathbf{P}$ as

$$diam_\infty(\mathbf{P}) = \max_{i,j} \|\mathbf{p}_i - \mathbf{p}_j\|_\infty.$$

We will now provide a useful theorem proved by Shen (2000a,b), which is called the generalized Hajnal inequality.

Theorem 270 *For any norm in* $\mathbb{R}^n$

$$diam\left(\left[\mathbf{v}_1^{\mathbf{P}}...\mathbf{v}_n^{\mathbf{P}}\right]\right) \leq (1 - \lambda(\mathbf{P}))\, diam\left([\mathbf{v}_1...\mathbf{v}_n]\right),$$

where $[\mathbf{v}_1...\mathbf{v}_n]$ *is any non-degenerate simplex.*

Proof. For any states i and j, let that

$$\mathbf{p}_i \wedge \mathbf{p}_j = \begin{pmatrix}p_1^* & p_2^* & \cdots & p_n^*\end{pmatrix},$$

where as we defined it $p_k^* = \min(p_{ik}, p_{jk})$. Now construct

$$\mathbf{w}_k = \left(1 - \sum_{\nu \neq k} p_\nu^*\right)\mathbf{v}_k + \sum_{\nu \neq k} p_\nu^* \mathbf{v}_\nu \quad \text{for} \quad k = 1, 2, ..., n. \tag{4.129}$$

See also Example 271 in parallel for deeper understanding. It is easy to verify that

$$\mathbf{v}_i^{\mathbf{P}}, \mathbf{v}_j^{\mathbf{P}} \in [\mathbf{w}_1...\mathbf{w}_n].$$

Therefore it is immediate that

$$\left\|\mathbf{v}_i^{\mathbf{P}} - \mathbf{v}_j^{\mathbf{P}}\right\| \leq diam\left[\mathbf{w}_1...\mathbf{w}_n\right]. \tag{4.130}$$

It is known that

$$diam\left[\mathbf{w}_1...\mathbf{w}_n\right] = \max_{i,j}\left\|\mathbf{w}_i - \mathbf{w}_j\right\| \geq \max_{i,j}\left\|\mathbf{v}_i^{\mathbf{P}} - \mathbf{v}_j^{\mathbf{P}}\right\|.$$

We have that for any states i, j from (4.129) we get that

$$\mathbf{w}_i - \mathbf{w}_j = \left(1 - \sum_{\nu=1}^{n} p_\nu^*\right)(\mathbf{v}_i - \mathbf{v}_j),$$

therefore

$$diam\left[\mathbf{w}_1...\mathbf{w}_n\right] = \left(1 - \sum_{\nu=1}^{n} p_\nu^*\right) diam\left(\left[\mathbf{v}_1...\mathbf{v}_n\right]\right). \tag{4.131}$$

Now, since

$$\lambda(\mathbf{P}) = \left\|\mathbf{p}_i \wedge \mathbf{p}_j\right\|_1 \leq \sum_{\nu=1}^{n} p_\nu^*,$$

from (4.130) and (4.131) we get

$$diam\left(\left[\mathbf{v}_1^{\mathbf{P}}...\mathbf{v}_n^{\mathbf{P}}\right]\right) \leq (1 - \lambda(\mathbf{P}))\, diam\left(\left[\mathbf{v}_1...\mathbf{v}_n\right]\right).$$

■

Example 271 *Applying Theorem 270 for Example 268 which we continue in this present example and also see our findings in Figure 4.1. From* (4.129) *and taking as* $i = 2$ *and* $j = 3$ *we get that*

$$\mathbf{w}_1 = \left(1 - p_2^* - p_3^*\right)\mathbf{v}_1 + p_2^*\mathbf{v}_2 + p_3^*\mathbf{v}_3 = \begin{pmatrix}0.22 & 0.20 & 0.58\end{pmatrix},$$

$$\mathbf{w}_2 = \left(1 - p_1^* - p_3^*\right)\mathbf{v}_1 + p_1^*\mathbf{v}_1 + p_3^*\mathbf{v}_3 = \begin{pmatrix}0.7 & 0.20 & 0.1\end{pmatrix},$$

$$\mathbf{w}_3 = \left(1 - p_1^* - p_2^*\right)\mathbf{v}_1 + p_1^*\mathbf{v}_2 + p_3^*\mathbf{v}_3 = \begin{pmatrix}0.22 & 0.68 & 0.1\end{pmatrix}.$$

From Figure 4.1 *we see that*

$$\mathbf{v}_2^{\mathbf{P}}, \mathbf{v}_3^{\mathbf{P}} \in \left[\mathbf{w}_1\mathbf{w}_2\mathbf{w}_3\right],$$

we also observe that

$$\left[\mathbf{v}_1^{\mathbf{P}}\mathbf{v}_2^{\mathbf{P}}\mathbf{v}_3^{\mathbf{P}}\right] \subset \left(\left[\mathbf{v}_1\mathbf{v}_2\mathbf{v}_3\right]\right).$$

From example we get that $\lambda(\mathbf{P}) = 0.2$. *We could easily check that*

$$diam\left(\left[\mathbf{v}_1^{\mathbf{P}}\mathbf{v}_2^{\mathbf{P}}\mathbf{v}_3^{\mathbf{P}}\right]\right) = \max_{ij}\left\|\mathbf{v}_i^{\mathbf{P}} - \mathbf{v}_j^{\mathbf{P}}\right\| \leq 0.8\max_{ij}\left\|\mathbf{v}_i - \mathbf{v}_j\right\|$$

$$\leq 0.8 diam\left(\left[\mathbf{v}_1\mathbf{v}_2\mathbf{v}_3\right]\right).$$

It is apparent from Figure 4.1 and it could also be easily proved by applying Theorem 270 to the canonical simplex the following Corollary.

Corollary 272 *Under any norm in* $\mathbb{R}^n$ *and for any matrix* $\mathbf{P} \in SM_n$,

$$diam\left[\mathbf{P}\right] \leq (1 - \lambda(\mathbf{P}))\, diam\left(\left[\mathbf{e}_1...\mathbf{e}_n\right]\right).$$

Also if we use the norm $\|.\|_1$, we could easily obtain the following inequality of the generalized Hajnal inequality by Dobrushin (1956) and Paz and Reichaw (1967) as a corollary of Theorem 270.

Corollary 273 *Let* $\mathbf{P}(1), \mathbf{P}(2), ..., \mathbf{P}(k) \in SM_n$. *Then*

$$1-\lambda(\mathbf{P}(1)\mathbf{P}(2)...\mathbf{P}(k)) \leq (1-\lambda(\mathbf{P}(1)))(1-\lambda(\mathbf{P}(2)))...(1-\lambda(\mathbf{P}(k))).$$

We are now in a position to prove the following classical result as an application of Theorem 270.

Theorem 274 *Let a compact set* Σ *of scrambling stochastic matrices. Then any non-homogeneous Markov chain with transition probability matrices drawn from* Σ *is ergodic.*

Proof. Let $\mathbf{P}(t) \in \Sigma$ be any matrix drawn from the compact set then since it is scrambling, we have $\lambda(\mathbf{P}(t)) > 0$. Now, $\lambda(\mathbf{P}(t)) : \Sigma \to [0,1]$ and Σ being compact $\lambda(.)$ is a continuous function and also

$$\lambda(\Sigma) := \inf_{\mathbf{P}\in\Sigma} \lambda(\mathbf{P}) > 0.$$

Let $\mathbf{P}(1), \mathbf{P}(2), ..., \mathbf{P}(k) \in \Sigma$ any set of k matrices drawn from Σ, then applying Theorem 270, we get

$$diam([\mathbf{P}(1)\mathbf{P}(2)...\mathbf{P}(k)]) \leq (1-\lambda(\Sigma))^k \, diam([\mathbf{e}_1\mathbf{e}_2...\mathbf{e}_n])$$

from which we get that

$$\lim_{k\to\infty} diam([\mathbf{P}(1)\mathbf{P}(2)...\mathbf{P}(k)]) = 0,$$

which apparently implies that the Σ-chain is uniformly ergodic.

This could also be seen in another way. Let $[\mathbf{P}(1)\mathbf{P}(2)...\mathbf{P}(k)]$ be a set C_k which apparently is a compact subset of $\mathbb{R}^n$ a complete metric space. Also by direct application of Theorem 270, we get that

$$C_1 \supset C_2 \supset ... \supset C_k \supset ...$$

is a chain of nonempty compact sets such that the diameters tend to zero. Then there must exist a unique point $\mathbf{p} \in \mathbb{R}^n$ such that

$$\cap_{k=1}^{\infty} C_k = \mathbf{p}.$$

■

We have discussed that there are difficulties brought in finding conditions for ergodicity (weak or strong) by the fact that the product of two regular matrices may be regular and the product of two non regular matrices may be non-regular. Hence the following theorem may well considered central (see Shen (2000a,b)).

Theorem 275 *Let* Σ *be a compact set of regular matrices. Then any* Σ*-Markov chain is ergodic (weakly or strongly).*

Proof. According to Lemma 223 there exists a finite number μ such that

$$\Sigma_\mu = \{\mathbf{P}(1)...\mathbf{P}(\mu) \mid \mathbf{P}(i) \in \Sigma\},$$

is a Markov set. Due to Theorem 220, we know that a Markov set is in fact a subset of a scrambling set, hence Σ_μ is a scrambling set. Since Σ is compact therefore it is easy to see that Σ_μ is compact. We proved so far that Σ_μ is a scrambling compact set hence it is ergodic according to Theorem 274. Hence, for any initially chosen matrix from Σ let say $\mathbf{P}(t_0)$ we have that the sequence

$$\begin{aligned}\mathbf{Q}(1) &= \mathbf{P}(t_0, t_0+\mu), \mathbf{Q}(2) = \mathbf{P}(t_0+\mu, t_0+2\mu), ..., \\ \mathbf{Q}(\nu) &= \mathbf{P}(t_0+(\nu-1)\mu, t_0+\nu\mu), ...\end{aligned}$$

is a sequence of scrambling matrices and as such is ergodic.

Let $\mathbf{P}(s)$ any initial matrix drawn from Σ then without loss of generality, we may assume that

$$t_0 + (k-1)\mu \leq s \leq t_0 + k\mu$$

and consider the product $\mathbf{P}(s, s+n)$ as $n \to \infty$ with s naturally fixed. If now we prove that $\lim_{n\to\infty} \mathbf{P}(s, s+n)$ for every s

not necessarily independent of n then Σ is ergodic.

Since $\mathbf{Q}(\nu)$ is ergodic for every $\epsilon > 0$ there exists a $N(\epsilon) > 0$ and a stochastic vector $\mathbf{q}(K)$ such that for any norm

$$\left\|\mathbf{Q}(t_0 + k\mu, t_0 + K\mu) - \mathbf{q}(K)\mathbf{1}^\top\right\| \leq \epsilon \text{ for every } K \geq N(\epsilon).$$

Again without loss of generality assume that $s + n > t_0 + K\mu$

$$\mathbf{p}(n) = \mathbf{q}(K)\mathbf{P}(t_0 + K\mu, s+n)$$

We now have for any norm that

$$\left\|\mathbf{P}(s, s+n) - \mathbf{p}(n)\mathbf{1}^\top\right\| =$$

$$\left\|\mathbf{P}(s, t_0 + k\mu)\mathbf{Q}(t_0 + k\mu, t_0 + K\mu)\mathbf{P}(t_0 + K\mu, s+n) - \mathbf{p}(n)\mathbf{1}^\top\right\|$$

$$\leq (\text{since } \mathbf{p}(n)\mathbf{1}^\top \text{ is a stable stochastic matrix})$$

$$\leq a\left\|\mathbf{Q}(t_0 + k\mu, t_0 + K\mu) - \mathbf{q}(K)\mathbf{1}^\top\right\| b \leq \epsilon.$$

■

Consider a more general framework, that is, a set Σ of square matrices (real or complex) which is a right convergent product (RCP) set, which means any infinite product $M(1)M(2)...M(n)...$ drawn from Σ converges. In other words for any sequence $(M(1), M(2), ...) \in \Sigma_\infty$,

$$\lim_{n\to\infty} M(1)M(2)...M(n),$$

exists. Shen (2000a,b) proved that for such a set Σ the closure $\bar{\Sigma}$ in the matrix space shares almost all the properties.

4.9 Asymptotic behavior with arbitrary stochastic matrices

In the present section, we will study ergodicity for Markov chains for which the sequence of transition probability matrices are given in chronological order but they are arbitrary stochastic matrices. In this way we will achieve full generality for the sequence of transition probability matrices. Let $\{\mathbf{P}(t)\}_{t=0}^\infty$ the sequence of finite transition probability matrices of a Markov chain given in chronological order and assume that they are arbitrary stochastic matrices all of which have the same incidence matrix. That is, $\mathbf{P}(t) = \bar{\mathbf{P}}(t)$ for every t. Since they all have the same incidence matrix they will have the same canonical form and let it be of the form

$$\mathbf{P}(t) = \begin{pmatrix} \mathbf{P}_1(t) & \mathbf{0} & ... & \mathbf{0} & \mathbf{0} \\ \mathbf{0} & \mathbf{P}_2(t) & ... & \mathbf{0} & 0 \\ ... & ... & ... & ... & ... \\ ... & ... & ... & \mathbf{P}_r(t) & \mathbf{0} \\ \mathbf{T}_1(t) & \mathbf{T}_2(t) & ... & \mathbf{T}_r(t) & \mathbf{Q}(t) \end{pmatrix} \text{ for } t = 0, 1, ... \quad (4.132)$$

The matrix $\mathbf{P}_i(t)$ for $i = 1, 2, ..., r$ is the matrix of transition probabilities among the class i of essential states. All states in the class $i = 1, 2, ..., r$ communicate and each class in general is assumed to be of different size. It is apparent that all matrices $\mathbf{P}_i(t)$ for $i = 1, 2, ..., r$ are square matrices. Therefore, all the matrices $\mathbf{P}_i(t)$ for $i = 1, 2, ..., r$ are irreducible and consists of an essential class, hence in all generality we may assume that their period is d_i for all $t = 0, 1, 2, ...$ The matrix $\mathbf{T}_i(t)$ for $i = 1, 2, ..., r$ is the transition probability matrix, from the inessential states to the essential classes of states. The matrices $\mathbf{T}_i(t)$ for $i = 1, 2, ..., r$ are not necessarily square matrices not they are of the same size. Apparently, $\mathbf{T}_i(t)$ for all t have the same incidence matrix. In general $\mathbf{Q}(t)$ for all t is substochastic.

It is easy to see that

$$\mathbf{P}(s,t) = \begin{pmatrix} \mathbf{P}_1(s,t) & \mathbf{0} & ... & \mathbf{0} & \mathbf{0} \\ \mathbf{0} & \mathbf{P}_2(s,t) & ... & \mathbf{0} & 0 \\ ... & ... & ... & ... & ... \\ ... & ... & ... & \mathbf{P}_r(s,t) & \mathbf{0} \\ \mathbf{T}_1(s,t) & \mathbf{T}_2(s,t) & ... & \mathbf{T}_r(s,t) & \mathbf{Q}(s,t) \end{pmatrix} \quad \text{for } t = 0, 1, ... \tag{4.133}$$

and

$$\mathbf{T}_i(s,t) = \sum_{m=0}^{t-s-1} \mathbf{Q}(s-1, s+m-1)\, \mathbf{T}_i(s+m)\, \mathbf{P}_i(s+m, t) \ , \tag{4.134}$$

for $i = 1, 2, ..., r.$

We will now prove the following basic theorem:

Theorem 276 *(Vassiliou (2015)). Let $\{\mathbf{P}(t)\}_{t=0}^{\infty}$ be a sequence of arbitrary stochastic matrices with the same incidence matrix and let with no loss of generality the canonical form for every t be*

$$\mathbf{P}(t) = \begin{pmatrix} \mathbf{P}_1(t) & \mathbf{0} & ... & \mathbf{0} & \mathbf{0} \\ \mathbf{0} & \mathbf{P}_2(t) & ... & \mathbf{0} & 0 \\ ... & ... & ... & ... & ... \\ ... & ... & ... & \mathbf{P}_r(t) & \mathbf{0} \\ \mathbf{T}_1(t) & \mathbf{T}_2(t) & ... & \mathbf{T}_r(t) & \mathbf{Q}(t) \end{pmatrix} \quad \textit{for } t = 0, 1, ... \tag{4.135}$$

that is, the matrices $\mathbf{P}(t)$ for $t = 0, 1, ...$ have r essential classes of states. Assume that each essential class is periodic with period d_i respectively. Assume asymptotic homogeneity, that is,

$$\lim_{t\to\infty} \|\mathbf{P}(t) - \mathbf{P}\| = 0, \tag{4.136}$$

with $\mathbf{P}$ having the same incidence matrix as $\mathbf{P}(t)$. Now let

$$s = \pi_{si} d_i + u_i, \qquad t = \pi_{ti} + v_i \quad \textit{and} \quad \nu_i = v_i - u_i + 1. \tag{4.137}$$

Then the infinite product $\lim_{t\to\infty} \mathbf{P}(s,t)$ for every s splits into

$$d = \operatorname{lcm}\{d_1, d_2, ..., d_r\}$$

convergent subsequences with limits

$$\mathbf{P}(\nu_1, \nu_2, ..., \nu_r) = \begin{pmatrix} \mathbf{P}_1^*\mathbf{P}_1^{d_1+\nu_1} & \mathbf{0} & ... & \mathbf{0} & \mathbf{0} \\ \mathbf{0} & \mathbf{P}_2^*\mathbf{P}_2^{d_2+\nu_2} & ... & \mathbf{0} & \mathbf{0} \\ ... & ... & ... & ... & ... \\ \mathbf{0} & 0 & ... & \mathbf{P}_r^*\mathbf{P}_r^{d_r+\nu_r} & \mathbf{0} \\ \mathbf{T}_1^* & \mathbf{T}_2^* & ... & \mathbf{T}_r^* & \mathbf{0} \end{pmatrix}. \tag{4.138}$$

where

$$\mathbf{P}_i^* = \lim_{t\to\infty} \mathbf{P}_i^t,$$

$$\mathbf{T}_i^* = \left(\mathbf{I} - \mathbf{Q}^{d_i}\right)^{-1} \left[\sum_{l=0}^{d-1} \mathbf{Q}^l \mathbf{T}_i \mathbf{P}_i^{d_i-l}\right] \mathbf{P}_i^* \mathbf{P}_i^{d_i+\nu_i}. \tag{4.139}$$

Proof. From Theorem 252, we get that

$$\lim_{t\to\infty} \mathbf{P}_i(s,t) = \mathbf{P}_i^*\mathbf{P}_i^{d_i+v_i-u_i+1} = \mathbf{P}_i^*\mathbf{P}_i^{d_i+\nu_i} \quad \text{for } i=1,2,...,r. \tag{4.140}$$

and since the involved matrices are finite we get that

$$\lim_{t\to\infty} \left\| \mathbf{P}_i(s,t) - \mathbf{P}_i^*\mathbf{P}_i^{d_i+\nu_i} \right\| = 0 \text{ for } i=1,2,...,r. \tag{4.141}$$

Also it is well known that if $\{\mathbf{Q}(t)\}_{t=0}^{\infty}$ is a sequence of substochastic matrices then

$$\lim_{t\to\infty} \|\mathbf{Q}(s,t)\| = 0 \text{ for every } s\in\mathbb{Z}^+ \text{ geometrically fast,} \tag{4.142}$$

hence there exists a $c_q > 0$ and a $0 < b_q < 1$ such that

$$\|\mathbf{Q}(s,t)\| \leq c_q b_q^{t-s}. \tag{4.143}$$

Now let that

$$\mathbf{M}(s,t) = \mathbf{P}_i(s,t) - \mathbf{P}_i^*\mathbf{P}_i^{d_i+\nu_i\cdot},$$

then equation (4.134) could be written in the form

$$\mathbf{T}_i(s,t) = \sum_{m=0}^{t-s-1} \mathbf{Q}(s-1,s+m-1)\,\mathbf{T}_i(s+m)\,\mathbf{M}_i(s+m,t)$$

$$= \sum_{m=0}^{t-s-1} \mathbf{Q}(s-1,s+m-1)\,\mathbf{T}_i(s+m)\,\mathbf{P}_i^*\mathbf{P}_i^{(t-s-m) \bmod d_i}. \tag{4.144}$$

$$= \langle\mathbf{A}\rangle + \langle\mathbf{B}\rangle.$$

We now have that

$$\|\langle\mathbf{A}\rangle\| = \left\| \sum_{m=0}^{t-s-1} \mathbf{Q}(s-1,s+m-1)\,\mathbf{T}_i(s+m)\,\mathbf{M}_i(s+m,t) \right\|$$

$$\leq \sum_{m=0}^{t-s-1} \|\mathbf{Q}(s-1,s+m-1)\| \, \|\mathbf{T}_i(s+m)\| \, \|\mathbf{M}_i(s+m,t)\|. \tag{4.145}$$

We have that

$$\left\| \mathbf{M}(s,t) = \mathbf{P}_i(s,t) - \mathbf{P}_i^*\mathbf{P}_i^{d_i+\nu_i\cdot} \right\| \leq 2, \tag{4.146}$$

and in general

$$\|\mathbf{T}_i(s+m)\| \leq 1. \tag{4.147}$$

From $(4.145), (4.146)$ and (4.147) we get that

$$\|\langle\mathbf{A}\rangle\| \leq 2c_q \sum_{m=0}^{t-s-1} b_q^{m-1}. \tag{4.148}$$

Hence

$$\lim_{t\to\infty} \|\langle\mathbf{A}\rangle\| = 0 \text{ for every } s\in\mathbb{Z}^+. \tag{4.149}$$

We now proceed to find the limit as $t\to\infty$ of $\langle\mathbf{B}\rangle$ in (4.144) assuming without loss of generality (4.137). In this respect we get

$$\langle\mathbf{B}\rangle = \sum_{m=0}^{t-s-1} \mathbf{Q}(s-1,s+m-1)\,\mathbf{T}_i(s+m)\,\mathbf{P}_i^*\mathbf{P}_i^{(t-s-m) \bmod d_i}$$

$$= \sum_{\tau=0}^{\pi_{ti}-\pi_{si}-2} \mathbf{Q}(s-1, s+\tau d_i - 1) \times$$

$$[\sum_{l=0}^{d_i+\nu_i-1} \mathbf{Q}(s+\tau d_i, s+\tau d_i + l - 1)\, \mathbf{T}_i(s+\tau d_i + l)\, \mathbf{P}_i^* \mathbf{P}_i^{(d_i+v_i-u_i-l)}$$

$$+ \sum_{l=d_i+\nu_i}^{d_i-1} \mathbf{Q}(s-1+\tau d_i, s+\tau d_i + l - 1)\, \mathbf{T}_i(s+\tau d_i + l)\, \mathbf{P}_i^* \mathbf{P}_i^{d_i+(d_i+v_i-u_i-l)}]$$

$$+ \sum_{r=0}^{d_i+v_i-u_i-1} \mathbf{Q}(s-1, s+(\pi_{ti}-\pi_{si}-1)\, d_i + r)$$

$$\times \mathbf{T}_i(s+((\pi_{ti}-\pi_{si}-1))\, d_i + r)\, \mathbf{P}_i^* \mathbf{P}_i^{(d_i+v_i-u_i+r)} = \langle \mathbf{C} \rangle + \langle \mathbf{D} \rangle. \tag{4.150}$$

It is straightforward to see that

$$\lim_{t\to\infty} \langle \mathbf{D} \rangle = \mathbf{0}. \tag{4.151}$$

Now, from (4.150), (4.151) we get that

$$\lim_{t\to\infty} \langle \mathbf{B} \rangle = \lim_{t\to\infty} \sum_{\tau=0}^{\pi_{ti}-\pi_{si}-2} \mathbf{Q}(s-1, s+\tau d_i - 1) \times$$

$$\left[\sum_{l=0}^{d_i-1} \mathbf{Q}(s-1+\tau d_i, s+\tau d_i + l - 1)\, \mathbf{T}_i(s+\tau d_i + l)\, \mathbf{P}_i^{d-l}\right] \mathbf{P}_i^* \mathbf{P}_i^{d_i+\nu_i}. \tag{4.152}$$

We now have

$$\mathbf{L} = \| \lim_{t\to\infty} \sum_{\tau=0}^{\pi_{ti}-\pi_{si}-2} \mathbf{Q}(s-1, s+\tau d_i - 1) \times$$

$$\left[\sum_{l=0}^{d_i-1} \mathbf{Q}(s-1+\tau d_i, s+\tau d_i + l - 1)\, \mathbf{T}_i(s+\tau d_i + l)\, \mathbf{P}_i^{d_i-l}\right] \mathbf{P}_i^* \mathbf{P}_i^{d_i+\nu_i}$$

$$- \left(\mathbf{I} - \mathbf{Q}^{d_i}\right)^{-1} \left[\sum_{l=0}^{d_i-1} \mathbf{Q}^l \mathbf{T}_i \mathbf{P}_i^{d_i-l}\right] \mathbf{P}_i^* \mathbf{P}_i^{d_i+\nu_i} \|$$

$$\leq \sum_{\tau=0}^{\infty} \|\mathbf{Q}(s-1, s+\tau d_i - 1)\, [\sum_{l=0}^{d_i-1} \mathbf{Q}(s+\tau d_i, s+\tau d_i + l - 1)$$

$$\times \mathbf{T}_i(s+\tau d_i + l)\, \mathbf{P}_i^{d_i-l}] - \left(\mathbf{Q}^{d_i}\right)^{\tau} \left[\sum_{l=0}^{d_i-1} \mathbf{Q}^l \mathbf{T}_i \mathbf{P}_i^{d_i-l}\right] \|. \tag{4.153}$$

Now, let us define by

$$\mathbf{Y}(t) = \sum_{l=0}^{d_i-1} \mathbf{Q}(s-1+\tau d_i, s+\tau d_i + l - 1)\, \mathbf{T}_i(s+\tau d_i + l)\, \mathbf{P}_i^{d_i-l}$$

$$- \sum_{l=0}^{d_i-1} \mathbf{Q}^l \mathbf{T}_i \mathbf{P}_i^{d_i-l}. \tag{4.154}$$

Then we get that

$$\|\mathbf{Y}(t)\| = \sum_{l=0}^{d_i-1} \left\|\mathbf{Q}(s-1+\tau d_i, s+\tau d_i+l-1)\mathbf{T}_i(s+\tau d_i+l) - \mathbf{Q}^l\mathbf{T}_i\right\| =$$

$$\sum_{l=0}^{d_i-1} \left\|\mathbf{Q}(s-1+\tau d_i, s+\tau d_i+l-1) - \mathbf{Q}^l\right\|$$

$$+\sum_{l=0}^{d_i-1} \|\mathbf{T}_i(s+\tau d_i+l) - \mathbf{T}_i\|. \tag{4.155}$$

Since the sequence $\{\mathbf{P}(t)\}_{t=0}^{\infty}$ is asymptotically homogeneous by (4.137), and in addition all matrices in $\{\mathbf{P}(t)\}_{t=0}^{\infty}$ have the same incidence matrix and $\mathbf{Q}(t), \mathbf{Q}$ are in general sub-stochastic matrices we get that

$$\left\|\mathbf{Q}(s-1+\tau d_i, s+\tau d_i+l-1) - \mathbf{Q}^l\right\| \leq \epsilon/2d_i \text{ and}$$

$$\|\mathbf{T}_i(s+\tau d_i+l) - \mathbf{T}_i\| \leq \epsilon/2d_i \text{ for } \tau \geq t_0. \tag{4.156}$$

Therefore,

$$\lim_{t\to\infty} \|\mathbf{Y}(t)\| = 0. \tag{4.157}$$

From (4.153) and (4.154) we get that

$$\mathbf{L} \leq \sum_{\tau=0}^{\infty} \|\mathbf{Q}(s-1, s+\tau d_i-1)\mathbf{Y}(t) + \mathbf{Q}(s-1, s+\tau d_i-1)\sum_{l=0}^{d_i-1} \mathbf{Q}^l\mathbf{T}_i\mathbf{P}_i^{d-l}$$

$$-\left(\mathbf{Q}^{d_i}\right)^{\tau}\sum_{l=0}^{d_i-1} \mathbf{Q}^l\mathbf{T}_i\mathbf{P}_i^{d-l}\|$$

$$\leq \sum_{\tau=0}^{\infty} \|\mathbf{Q}(s-1, s+\tau d_i-1)\| \, \|\mathbf{Y}(t)\| + \sum_{\tau=0}^{\infty} \left\|\mathbf{Q}(s-1, s+\tau d_i-1) - \left(\mathbf{Q}^{d_i}\right)^{\tau}\right\| d_i. \tag{4.158}$$

Now from (4.154) we get that $\|\mathbf{Y}(t)\| \leq 4d_i$, therefore, we have

$$\mathbf{L} \leq \sum_{\tau=0}^{\infty} \|\mathbf{Q}(s-1, s+\tau d_i-1)\| \, 4d_i + \sum_{\tau=0}^{\infty} \|\mathbf{Q}(s-1, s+\tau d_i-1)\| \, d_i - \sum_{\tau=0}^{\infty} \left\|\left(\mathbf{Q}^{d_i}\right)^{\tau}\right\| d_i$$

$$\leq 5d_i c_q b_q^{d_i} \sum_{\tau=0}^{\infty} b_q^{\tau} + c d_i b_i^{d_i} \sum_{\tau=0}^{\infty} b_i^{\tau} \leq d_i(5c_q + c_i)\frac{\min(b_q, b_i)^{d_i}}{1 - \min(b_q, b_i)}. \tag{4.159}$$

It is not difficult to see that the number of different subsequences $\mathbf{P}(\nu_1, \nu_2, ...\nu_r)$ is $d = \operatorname{lcm}\{d_1, d_2, ..., d_r\}$. We will now prove that the matrix $\mathbf{P}(\nu_1, \nu_2, ...\nu_r)$ is a stochastic matrix. In order to do that it is apparent that it is sufficient to show that

$$\left(\sum_{i=1}^{r} \mathbf{T}_i^*\right)\mathbf{1}^{\top} = \mathbf{1}^{\top},$$

that is,

$$\sum_{i=1}^{r}\left(\left(\mathbf{I} - \mathbf{Q}^{d_i}\right)^{-1}\left[\sum_{l=0}^{d-1} \mathbf{Q}^l\mathbf{T}_i\mathbf{P}_i^{d_i-l}\right]\mathbf{P}_i^*\mathbf{P}_i^{d_i+\nu_i\cdot}\right)\mathbf{1}^{\top} = \mathbf{1}^{\top}.$$

We have that

$$\sum_{i=1}^{r}\left(\left(\mathbf{I}-\mathbf{Q}^{d_i}\right)^{-1}\left[\sum_{l=0}^{d-1}\mathbf{Q}^{l}\mathbf{T}_i\mathbf{P}_i^{d_i-l}\right]\mathbf{P}_i^{*}\mathbf{P}_i^{d_i+\nu_i\cdot}\right)\mathbf{1}^{\top}=$$

$$\sum_{i=1}^{r}\left(\mathbf{I}-\mathbf{Q}^{d_i}\right)^{-1}\left(\sum_{l=0}^{d-1}\mathbf{Q}^{l}\right)\mathbf{T}_i\mathbf{1}^{\top}=$$

$$\sum_{i=1}^{r}\left(\mathbf{I}-\mathbf{Q}^{d_i}\right)^{-1}\left(\mathbf{I}-\mathbf{Q}^{d_i}\right)\left(\mathbf{I}-\mathbf{Q}\right)^{-1}\mathbf{T}_i\mathbf{1}^{\top}=\sum_{i=1}^{r}\left(\mathbf{I}-\mathbf{Q}\right)^{-1}\mathbf{T}_i\mathbf{1}^{\top}$$

$$=\left(\text{since }\sum_{i=1}^{r}\mathbf{T}_i+\mathbf{Q}=\mathbf{1}^{\top}=\mathbf{1}^{\top}\right)=\mathbf{1}^{\top}.$$

■

We will now prove the following theorem:

Theorem 277 *Let $\{X_t\}_{t=0}^{\infty}$ be a non-homogeneous Markov chain and let $\{\mathbf{P}(t)\}_{t=0}^{\infty}$ be a sequence of $k\times k$ arbitrary stochastic matrices that uniquely determine $\{X_t\}_{t=0}^{\infty}$. Assume that*

$$\lim_{t\to\infty}\|\mathbf{P}(t)-\mathbf{P}\|=0,\quad \bar{\mathbf{P}}(t)=\bar{\mathbf{P}}\ ,$$

that is,with the same incidence matrix and let with no loss of generality the canonical form for every t be

$$\mathbf{P}(t)=\begin{pmatrix}\mathbf{P}_1(t) & \mathbf{0} & \dots & \mathbf{0} & \mathbf{0}\\ \mathbf{0} & \mathbf{P}_2(t) & \dots & \mathbf{0} & 0\\ \dots & \dots & \dots & \dots & \dots\\ \dots & \dots & \dots & \mathbf{P}_r(t) & \mathbf{0}\\ \mathbf{T}_1(t) & \mathbf{T}_2(t) & \dots & \mathbf{T}_r(t) & \mathbf{Q}(t)\end{pmatrix}\quad \textit{for } t=0,1,\dots .$$

where $\mathbf{P}_i(t)\in SM_k(\mathbb{R})$, $\mathbf{T}_i(t)\in M_{k_q.k_i}(\mathbb{R})$, and as a consequence $\mathbf{Q}(t)\in SM_{k_q}(\mathbb{R})$. Let also the state probability vector

$$\mathbf{p}(t,1,k)=\{\Pr(X_t=j)\}_{j=1}^{k}=[p_1(t),p_2(t),\dots,p_k(t)]\,.$$

and the blocks of state probability vectors defined by

$$\mathbf{p}(t,m_{i-1},m_i)=\left[p_{m_{i-1}+1}(t),p_{m_{i-1}+2}(t),\dots,p_{m_i}(t)\right],$$

$$\textit{with}\ \ m_i=\sum_{j=1}^{i}k_j\ \ \textit{and}\ \ m_0=1.$$

Then the sequence of asymptotic state probability vectors $\mathbf{p}(t,m_{i-1},m_i)$ splits into $d=\operatorname{lcm}\{d_1,d_2,\dots,d_r\}$ convergent subsequences with limits given by

$$\mathbf{p}(\infty,1,k)=\left[\mathbf{p}_1(\infty,1,m_1),\mathbf{p}_2(\infty,m_1,m_2),\dots,\mathbf{p}_r(\infty,m_{r-1},m_r),\mathbf{0}\right],$$

where

$$\mathbf{p}(\infty,m_{i-1},m_i)=[\mathbf{p}_i(s,m_{i-1},m_i)+\mathbf{p}_{r+1}(s,m_r+1,k)\left(\mathbf{I}-\mathbf{Q}^{d_i}\right)^{-1}$$

$$\times\left(\sum_{l=0}^{d_i-1}\mathbf{Q}^{l}\mathbf{T}_i\mathbf{P}_i^{d_i-l}\right)]\mathbf{P}_i^{*}\mathbf{P}_i^{d_i+\nu_i}.$$

In addition

$$\mathbf{p}(\infty,1,k)\,\mathbf{1}_k^{\top}=1.$$

Proof. Let s the initial time as it is evident in the theorem. Then, it is known that

$$\mathbf{p}(t,1,k)=\mathbf{p}(0,1,k)\lim_{t\to\infty}\mathbf{P}(0,t),$$

from which we get

$$\lim_{t\to\infty}\mathbf{p}(t,1,k)=\mathbf{p}(0,1,k)\lim_{t\to\infty}\mathbf{P}(0,t),$$

and from the previous theorem, we get that

$$\lim_{t\to\infty}\mathbf{p}(t,1,k)=\mathbf{p}(\infty,1,k)$$

$$=\left[\mathbf{p}_1(0,1,m_1),\mathbf{p}_2(0,m_1,m_2),...,\mathbf{p}_r(0,m_{r-1},m_r),\mathbf{0}\right]$$
$$\times\begin{pmatrix}\mathbf{P}_1^*\mathbf{P}_1^{d_1+\nu_1} & \mathbf{0} & ... & \mathbf{0} & \mathbf{0}\\ \mathbf{0} & \mathbf{P}_2^*\mathbf{P}_2^{d_2+\nu_2} & ... & \mathbf{0} & \mathbf{0}\\ ... & ... & ... & ... & ...\\ \mathbf{0} & 0 & ... & \mathbf{P}_r^*\mathbf{P}_r^{d_r+\nu_r} & \mathbf{0}\\ \mathbf{T}_1^* & \mathbf{T}_2^* & ... & \mathbf{T}_r^* & \mathbf{0}\end{pmatrix}. \tag{4.160}$$

From the previous theorem and (4.160), we get that

$$\mathbf{p}(\infty,1,k)=\left[\mathbf{p}_1(\infty,1,m_1),\mathbf{p}_2(\infty,m_1,m_2),...,\mathbf{p}_r(\infty,m_{r-1},m_r),\mathbf{0}\right], \tag{4.161}$$

where

$$\mathbf{p}(\infty,m_{i-1},m_i)=[\mathbf{p}_i(0,m_{i-1},m_i)+\mathbf{p}_{r+1}(0,m_r+1,k)\left(\mathbf{I}-\mathbf{Q}^{d_i}\right)^{-1}$$
$$\times\left(\sum_{l=0}^{d_i-1}\mathbf{Q}^l\mathbf{T}_i\mathbf{P}_i^{d_i-l}\right)]\mathbf{P}_i^*\mathbf{P}_i^{d_i+\nu_i}.$$

Now, we need to show that $\mathbf{p}(\infty,1,k)$ is a stochastic vector, that is, $\mathbf{p}(\infty,1,k)\mathbf{1}^\top=\mathbf{1}^\top$. We have that

$$\mathbf{p}(\infty,1,k)\mathbf{1}^\top=\sum_{i=1}^{r}\mathbf{p}_i(\infty,m_{i-1},m_i)\mathbf{1}_{k_i}^\top$$
$$=\sum_{i=1}^{r}\mathbf{p}_i(0,m_{i-1},m_i)\mathbf{P}_i^*\mathbf{P}_i^{d_i+\nu_i}\mathbf{1}_{k_i}^\top+$$
$$\mathbf{p}_{r+1}(0,m_r+1,k)\sum_{i=1}^{r}\left[\left(\mathbf{I}-\mathbf{Q}^{d_i}\right)^{-1}\sum_{l=0}^{d_i-1}\mathbf{Q}^l\mathbf{T}_i\mathbf{P}^{d_i-l}\right]\mathbf{P}_i^*\mathbf{P}_i^{d_i+\nu_i}\mathbf{1}_{k_i}^\top$$
$$\sum_{i=1}^{r}\mathbf{p}_i(0,m_{i-1},m_i)\mathbf{1}_{k_i}^\top+\mathbf{p}_{r+1}(0,m_r+1,k)\mathbf{1}_{k_i}^\top=1.$$

■

4.9.1 An illustrative example

In the present subsection we will provide an illustrative example for the ergodicity of a non-homogeneous Markov chain with the sequence of transition probability matrices $\{\mathbf{P}(t)\}_{t=0}^{\infty}$ consisting of arbitrary stochastic matrices. Let $\{X_t\}_{t=0}^{\infty}$ be a non-homogeneous Markov chain with state space $\mathbb{S}=\{1,2,3,4,5,6,7,8\}$. Let that X_t has two essential classes, the first one consisting of the states $\{1,2,3,4\}$ and the second one consisting of the states

$\{5,6\}$. Finally the remaining states $\{7,8\}$ are transient. Then the sequence of transition probabilities $\{\mathbf{P}(t)\}_{t=0}^{\infty}$ having the same incidence matrix will have the following form:

$$\mathbf{P}(t)=\begin{pmatrix}\mathbf{P}_1(t) & \mathbf{0} & \mathbf{0}\\ \mathbf{0} & \mathbf{P}_2(t) & \mathbf{0}\\ \mathbf{T}_1(t) & \mathbf{T}_2(t) & \mathbf{Q}(t)\end{pmatrix},$$

where

$$\mathbf{P}_1(t)=\begin{pmatrix}0 & 0.9-\frac{3}{4t^2} & 0 & 0.1+\frac{3}{4t^2}\\ 0.8-\frac{2}{t^3+4} & 0 & 0.2+\frac{2}{t^3+4} & 0\\ 0 & 0.7-\frac{8}{t^4+11} & 0 & 0.3+\frac{8}{t^4+11}\\ 0.05+\frac{1}{3t} & 0 & 0.95-\frac{1}{3t} & 0\end{pmatrix},$$

$$\mathbf{P}_2(t)=\begin{pmatrix}0 & 1\\ 1 & 0\end{pmatrix},$$

$$\mathbf{T}_1(t)=\begin{pmatrix}0 & 0.2+\frac{1}{5t} & 0 & 0.1+\frac{2}{11t^4}\\ 0 & 0 & 0 & 0.2+\frac{3}{21t}\end{pmatrix},$$

$$\mathbf{T}_2(t)=\begin{pmatrix}0.1 & 0.3-\frac{2}{11t^4}\\ 0.3+\frac{1}{8t^3} & 0.3-\frac{3}{21t}\end{pmatrix},$$

$$\mathbf{Q}(t)=\begin{pmatrix}0.3-\frac{1}{5t} & 0\\ 0.1-\frac{1}{8t^3} & 0.1\end{pmatrix}.$$

Now, it is easy to check that $\mathbf{P}_1(t)$ and $\mathbf{P}_2(t)$ are periodic with periods respectively $d_1=d_2=2$ and consequently $d=\operatorname{lcm}\{d_1,d_2\}=2$. We have that $\lim_{t\to\infty}\|\mathbf{Q}(t)-\mathbf{Q}\|=0$ with

$$\mathbf{Q}=\begin{pmatrix}0 & 0.9 & 0 & 0.1 & 0 & 0 & 0 & 0\\ 0.8 & 0 & 0.2 & 0 & 0 & 0 & 0 & 0\\ 0 & 0.7 & 0 & 0.3 & 0 & 0 & 0 & 0\\ 0.05 & 0 & 0.95 & 0 & 0 & 0 & 0 & 0\\ 0 & 0 & 0 & 0 & 0 & 1 & 0 & 0\\ 0 & 0 & 0 & 0 & 1 & 0 & 0 & 0\\ 0 & 0.1 & 0 & 0.1 & 0.1 & 0.4 & 0.3 & 0\\ 0 & 0 & 0 & 0.2 & 0.2 & 0.2 & 0.3 & 0.1\end{pmatrix}.$$

Now applying Theorem 276 and simulating the process, we find that the results coincide and we get that for $t=2\pi$ for π large

$$\lim_{t\to\infty}\mathbf{Q}(0,t)=\begin{pmatrix}0 & 0.84 & 0 & 0.16 & 0 & 0 & 0 & 0\\ 0.68 & 0 & 0.32 & 0 & 0 & 0 & 0 & 0\\ 0 & 0.84 & 0 & 0.16 & 0 & 0 & 0 & 0\\ 0.68 & 0 & 0.32 & 0 & 0 & 0 & 0 & 0\\ 0 & 0 & 0 & 0 & 0 & 1 & 0 & 0\\ 0 & 0 & 0 & 0 & 1 & 0 & 0 & 0\\ 0.045 & 0.184 & 0.021 & 0.036 & 0.241 & 0.473 & 0 & 0\\ 0.06 & 0.191 & 0.029 & 0.038 & 0.373 & 0.309 & 0 & 0\end{pmatrix},$$

and for $t=2\pi+1$ as $\pi\to\infty$

$$\lim_{t\to\infty}\mathbf{Q}(0,t)=\begin{pmatrix}0.68 & 0 & 0.32 & 0 & 0 & 0 & 0 & 0\\ 0 & 0.84 & 0 & 0.16 & 0 & 0 & 0 & 0\\ 0.68 & 0 & 0.32 & 0 & 0 & 0 & 0 & 0\\ 0 & 0.84 & 0 & 0.16 & 0 & 0 & 0 & 0\\ 0 & 0 & 0 & 0 & 1 & 0 & 0 & 0\\ 0 & 0 & 0 & 0 & 0 & 1 & 0 & 0\\ 0.149 & 0.055 & 0.071 & 0.011 & 0.473 & 0.241 & 0 & 0\\ 0.155 & 0.074 & 0.074 & 0.015 & 0.31 & 0.373 & 0 & 0\end{pmatrix}.$$

Now, let that the initial probabilities be

$$\mathbf{p}(0,1,8) = \begin{pmatrix}0.1 & 0.1 & 0.3 & 0.2 & 0.1 & 0.2 & 0 & 0\end{pmatrix},$$

then applying Theorem 277, we get that for $t = 2\pi$

$$\mathbf{p}(\infty,1,8) = \begin{pmatrix}0.204 & 0.336 & 0.096 & 0.064 & 0.2 & 0.1 & 0.0 & 0.0\end{pmatrix}.$$

Also for $t = 2\pi + 1$, we have

$$\mathbf{p}(\infty,1,8) = \begin{pmatrix}0.272 & 0.252 & 0.128 & 0.048 & 0.1 & 0.2 & 0.0 & 0.0\end{pmatrix}.$$

Chapter 5

THE NON-HOMOGENEOUS MARKOV SYSTEM

5.1 Introductory notes

The concept of Non-homogeneous Markov system process (NHMS) was first introduced in Vassiliou (1982). Since then the evolution of the theory of NHMSs has realized an interesting growth and it still does. As we will present in what follows actually an NHMS process is a generalization of the non-homogeneous Markov chain. The roots of the motive for the development of the theory of NHMS is much older and could be summarized in the use of Markov models in manpower systems. This started with the work of Young and Almond (1961), Young (1965,1971) and Batholomew (1963). Young's[1] motive was the application of homogeneous Markov chain models in the British University system. Batholomew created important multiple renewal theory models for various social processes and his first related book Bartholomew[2] (1967) among other things, provided an important theoretical reference

[1]Andrew Young (1919-1992) was born on Edentown, Lancashire and raised with 11 other siblings as the third in the row by a working class family. As a young boy he distinguished very early himself by his ability in mathematics although he had at the same time to work from very early in the morning to contribute to the survival of the family. For more details see Shannon (1995). In 1938, he won the Harkness scholarship to read Mathematics at St Andrews University. At St Andrew he won the Carstairs Prize for mathematics in 1939 and the Miller prize in 1940. He married and immediately was called up for the war. As a combat officer in the Border regiment he was wounded at the battle of Kohima. The horrors of the war ultimately led him to pacifism and the Society of Friends. While working at Liverpool University he received his Ph.D entitled "The rotation of the earth axis". This was the first area of research domain for him and he quickly made an impact publishing a large number of papers mainly in *Monthly notices of the RAS.* In 1955, he gained a Leverhulme Research award. In 1958 he was invited to participate in the Xth Genearal Assembly of the IAU, Moscow. This first visit to Moscow had an influence on Young's political beliefs and although he rarely talked politics with his research students I recall many such discussions with him. I would say that this prolonged visit to Moscow was the end of some romantic political views for him. It is worth mentioning probably that he became an active member of the Labor Party and he played a significant part in its political campaigns.

He very rapidly became Professor at Liverpool University and Head of the Department of Computational and Statistical Sciences in 1966. Since 1961, he entered in the second face in his research mainly on Numerical Analysis and Mathematical Modeling. He wrote seminal papers on Linear Programming and was one of the main founders of Manpower Planning. Also Young played a major role in the Association of University Teachers. Over the years he supervised thirty-two successful Ph.D students. Many of his Ph.D students are now professors in countries throughout the world.

[2]Prof. David Bartholomew (1931–2017) was a British statistician who was president of the Royal Statistical Society between 1993 and 1995. He was one of the founders of the research area known as manpower planning. An Historical Review of the birth of this domain area of research exist in Smith and Bartholomew (1988). He was educated at University College London, where he earned his BSc and PhD. He was appointed professor of statistics at the University of Kent in 1967 before being made professor of statistics at the London School of Economics in 1973. Bartolomew was elected a Fellow of the British Academy in 1987. He was one of the founders of the Manpower society. He made a founder's impact on a vast area that could be labeled as "Stochastic Models for social processes" which has inspired a large number of younger promissing probabilists to devote their research efforts in this area. He has authored many seminal books

DOI: 10.1201/b23204-5

of applied probability style for everyone. The subject was named manpower planning, embracing a large variety of people with research backgrounds and studying relevant problems from, almost the whole spectrum. Also for human populations special cases of NHMS have appeared before in the literature under the name closed or open Markov models see Pollard (1966, 1967, 1973) and Bartholomew (1982). This terminology is carried also up to the present days and we find studies of open Markov models while in fact are mostly special cases of an NHMS.

The NHMS is a stochastic system which has a population of members which increase or fluctuate with some restrictions at every point in time. In addition the initial members need not to be the same entities at different time points since there is wastage from the system. The members of the population move among the different states, exit from the system (population) and new members are coming into the population (system) as replacements or to expand the system. In the case of a non-homogeneous Markov chain, we studied in previous chapters, we have only one particle in the population, which never leaves the system and no procedure to replace this particle exists. Mathematically now the NHMS has more elements in the one step in time equation of the expected relative population structures introduced in the stochastic difference equation. As the equation is applied recursively we end up with series of components which interact together. Hence, the problems to be solved are a lot harder, and new strategies and tools must be used, than the simple case of the Markov chain. The introduction of the concept of a membership is crucial in dealing with the different individual members as time progress. A very important new characteristic is now the fact that in every state we have an expected number of memberships and hence it arises the problem of the variances and covariances among the sizes of the states.

In the present chapter, we will start with an NHMS in discrete time and discrete state space and we will study in later chapters NHMS in continuous time and discrete state space, NHMS in discrete time and general state space, cyclic NHMS in discrete time and space, perturbed NHMS in discrete time and space and Non-homogeneous Markov set systems where the basic parameters of the NHMS are in intervals.

The results presented in the remaining of the book, which by some could be seen as a part two of the book, has been written in the desire that practitioners will use them. In this respect we will make the effort to illustrate the use of the theory in a systematic and accessible way. So the book will not be concentrated only on the theory of NHMS, but also in describing representative examples, which are paradigms or suggested ones of real applications. Real detailed applications will not be included due to space considerations and other restrictions concerning data which include humans.

5.2 The non-homogeneous Markov system in discrete time and space

Consider a population which has $T(t)$ memberships at time t, that is, at the end of the interval $(t-1,t]$, which are being held by its members. An example which help in understanding this concept is if you have in mind a hospital where the memberships are the available beds and the members are the various patients coming and going in the hospital. These memberships could be held by any kind of entities, i.e., human beings, animals, insects, fish in an organized area in the sea, biological micro-organisms, particles in a physical phenomenon, various types of machines, T-cells in a biological entity, cars in a highway etc. We will assume for the most part of the remaining book that $T(t)$ is known for every t, for example in a hospital the numbers of available beds for patients, i.e., the memberships

---

and papers in all three of the main research areas that he had devoted his time, that is, Statistical Inference under order restrictions, Stochastic models for social processes, Latent variable models and factor analysis.. Bartholomew also authored several books defending the existence of God arguing that randomness all around is part of God's wisdom to achive his purpose while retaining the freedom of individual elements.

are known in advance. Also, $\{T(t)\}_{t=0}^{\infty}$ could be a realization of a known stochastic process depending on the application.

Let that the population is stratified into classes which we call states and let that there are a finite number of states, i.e., the state space $\mathbb{S} = \{1, 2, ..., k\}$. The set of states are assumed to be exclusive and exhaustive so that each member of the population may be in one and only one state at any given time. We assume that the evolution of the population is in discrete time, i.e., $t = 1, 2, ...$ and we call the vector of random variables

$$\mathbf{N}(t) = [N_1(t), N_2(t), ..., N_k(t)]$$

where $N_i(t)$ is the number of memberships in state i at time t, the *population structure* of the NHMS at time t. Define by $\mathbf{q}(t) = \mathbf{N}(t)/T(t)$ to be the *relative population structure.* At every time instant $t = 1, 2, ...,$ we have internal transitions of members among the states in $\mathbb{S}$ and let $p_{ij}(t)$ be the probability of a member of the system moving from state i to state j at time t, that is, during the tth interval, i.e., during the period $(t-1, t]$. We collect these probabilities in the $k \times k$ matrix $\mathbf{P}(t)$. Also we have wastage from all the states with probabilities $p_{i,k+1}(t)$ from states $i = 1, 2, ..., k$ at the tth interval, where the state $k+1$ represents the external environment, and let the $1 \times k$ vector $\mathbf{p}_{k+1}(t) = [p_{1,k+1}(t), p_{2,k+1}(t), ..., p_{k,k+1}(t)]$ be the wastage vector of probabilities. Finally, we have recruitment or allocation probabilities of replacements or new entrants in the various states at time t, i.e., $p_{0j}(t)$ for $j = 1, 2, ..., k$ being the probability of a recruitment entering the population at time t (at the end of the interval $(t-1, t]$) to go to state j. We collect these probabilities in the $1 \times k$ stochastic vector $\mathbf{p}_0(t) = [p_{01}(t), ..., p_{0k}(t)]$. In an NHMS, we assume that the system is either expanding, i.e., $\Delta T(t) = T(t) - T(t-1) \geq 0$ or that the system is allowed to decrease in size only through voluntary wastage. The reader may well understand the reasons we do not develop the mathematics of decreasing the size of the population through forced wastage although it is not a difficult extension of the present results.

During the time interval $(t-1, t]$ a member of the system in state i either moves internally to another state j of the system with probability $p_{ij}(t)$ or leave the system to the external environment and his membership remains at the exit of the system. New entrants of the system are of two types, those to replace leavers and those needed to be added in the system to meet the target of $T(t)$ total memberships. The new entrant gets his membership at the entrance and he is being allocated or recruited at state j with probability $p_{0j}(t)$. There are times where it will be needed to define separately by $r_{0j}(t)$ the probabilities of allocating new memberships into the population and by $\mathbf{r}_0(t)$ the relative recruitment vector. However, most of the times will not be needed. Hence, the probability of a membership's transition from state i to state j at time t is $q_{ij}(t) = p_{ij}(t) + p_{0j}(t)p_{i,k+1}(t)$. We collect these probabilities in the $k \times k$ matrix $\mathbf{Q}(t) = \mathbf{P}(t) + \mathbf{p}_{k+1}^{\top}(t)\mathbf{p}_0(t)$, which is easily be seen to be a stochastic matrix. We call the non-homogeneous Markov chain defined by the sequence of matrices $\{\mathbf{Q}(t)\}_{t=0}^{\infty}$ the *imbedded or inherent non-homogeneous Markov chain* of the NHMS.

The sequences

$$\{\mathbf{P}(t)\}_{t=0}^{\infty}, \quad \{\mathbf{p}_0(t)\}_{t=0}^{\infty}, \quad \{\mathbf{P}(t)\}_{t=0}^{\infty}, \quad \{\mathbf{r}_0(t)\}_{t=0}^{\infty}, \quad \text{and} \quad \{\mathbf{Q}(t)\}_{t=0}^{\infty},$$

and the initial structure $\mathbf{N}(0)$ constitute sufficient information for answering the important questions for the evolution of the population. A population or any physical phenomenon that could be modeled in the above described way is defined to be a *non-homogeneous Markov system (NHMS).*

5.3 The expected and relative expected population structure

5.3.1 An expanding NHMS

Let us define by $N_{i.k+1}(t)$ the random variable representing the members of the system who leave the system at time t. Also define by $N_{k+1}(t)$ the total number of members leaving the system during the tth interval from all the states. Now, given $N_i(t)$ the random variable $N_{i.k+1}(t)$ is distributed Binomially, that is, $N_{i.k+1}(t) \sim B(N_i(t-1); p_{i,k+1}(t))$ therefore we have

$$\mathbb{E}\left[N_{i.k+1}(t) \mid N_i(t-1)\right] = N_i(t-1)\, p_{i,k+1}(t), \tag{5.1}$$

hence

$$\mathbb{E}\left[N_{i.k+1}(t)\right] = \mathbb{E}\left[N_i(t-1)\right] p_{i,k+1}(t). \tag{5.2}$$

Since $N_{k+1}(t) = \sum_{i=1}^{k} N_{i.k+1}(t)$ we get that

$$\mathbb{E}\left[N_{k+1}(t)\right] = \mathbb{E}\left[\mathbf{N}(t-1)\right] \mathbf{p}_{k+1}^{\top}(t). \tag{5.3}$$

Let us define by $R(t)$ the total number of memberships entering the system at time t then for an expanding system we have

$$R(t) = \Delta T(t) + N_{k+1}(t)$$

$$\mathbb{E}\left[R(t)\right] = \mathbb{E}\left[\mathbb{E}\left[R(t) \mid \mathbf{N}(t)\right]\right] = \Delta T(t) + \mathbb{E}\left[N_{k+1}(t)\right]. \tag{5.4}$$

Now, the expected number of memberships entering the system at time t, $\mathbb{E}[R(t)]$ will be distributed in the various states according to the probability distribution $\mathbf{p}_0(t)$. Therefore we have

$$\begin{aligned}
\mathbb{E}\left[\mathbf{N}(t)\right] = \mathbb{E}\left[\mathbb{E}\left[\mathbf{N}(t) \mid \mathbf{N}(t-1)\right]\right] &= \mathbb{E}\left[\mathbf{N}(t-1)\right] \mathbf{P}(t) + \mathbb{E}\left[R(t)\right] \mathbf{p}_0(t) \qquad (5.5)\\
&= \mathbb{E}\left[\mathbf{N}(t-1)\right] \mathbf{P}(t) + \left[\Delta T(t) + \mathbb{E}\left[N_{k+1}(t)\right]\right] \mathbf{p}_0(t)\\
&= (\text{ from 5.2 we get}) =\\
\mathbb{E}\left[\mathbf{N}(t-1)\right] &\left[\mathbf{P}(t) + \mathbf{p}_{k+1}^{\top}(t)\,\mathbf{p}_0(t)\right] + \Delta T(t)\,\mathbf{p}_0(t) =\\
&= \mathbb{E}\left[\mathbf{N}(t-1)\right] \mathbf{Q}(t) + \Delta T(t)\,\mathbf{p}_0(t)
\end{aligned}$$

Now define in general by

$$\mathbb{E}\left[\mathbf{N}(s,t)\right] = \mathbb{E}\left[\mathbb{E}\left[\mathbf{N}(t) \mid \mathbf{N}(s)\right]\right],$$

that is $\mathbb{E}[\mathbf{N}(s,t)]$ is the expected population structure at time t given that the system started at time s with known population structure $\mathbf{N}(s)$. From (5.3) and (5.5), we get that

$$\begin{aligned}
\mathbb{E}\left[\mathbf{N}(s,t)\right] &= \mathbb{E}\left[\mathbf{N}(s,t-1)\right] \mathbf{P}(t) + \left(\Delta T(t) + \mathbb{E}\left[\mathbf{N}(s,t-1)\right] \mathbf{p}_{k+1}^{\top}(t)\right) \mathbf{p}_0(t)\\
&= \mathbb{E}\left[\mathbf{N}(s,t-1)\right] \mathbf{Q}(t) + \Delta T(t)\,\mathbf{p}_0(t).
\end{aligned} \tag{5.6}$$

Applying (5.6) recursively we get that

$$\mathbb{E}\left[\mathbf{N}(s,t)\right] = \mathbf{N}(s)\,\mathbf{Q}(s,t) + \sum_{\tau=1}^{t} \Delta T(s+\tau)\,\mathbf{p}_0(s+\tau)\,\mathbf{Q}(s+\tau,t), \tag{5.7}$$

where $\mathbf{Q}(s,t) = \mathbf{Q}(s+1)\,\mathbf{Q}(s+2) \ldots \mathbf{Q}(t)$ for $s \leq t$ and $\mathbf{Q}(t,t) = \mathbf{I}$, $\mathbf{Q}(s,t) = 0$ for $s > t$. When $s = 0$ then we omit it if confusion is not created and we write

$$\mathbb{E}\left[\mathbf{N}(t)\right] = \mathbf{N}(0)\,\mathbf{Q}(0,t) + \sum_{\tau=1}^{t} \Delta T(\tau)\,\mathbf{p}_0(\tau)\,\mathbf{Q}(\tau,t). \tag{5.8}$$

Now the relative population structure of a system which started at time s and is at time t is defined to be $\mathbf{q}(s,t) = \mathbf{N}(s,t)/T(t)$ or if the system starts at 0 we usually write simply $\mathbf{q}(t) = \mathbf{N}(t)/T(t)$. By dividing both parts of equation (5.6) by $T(t)$ we get

$$\mathbb{E}[\mathbf{q}(s,t)] = \mathbb{E}[\mathbf{q}(s,t-1)]\, a(t-1)\,\mathbf{Q}(t) + b(t-1)\,\mathbf{p}_0(t), \tag{5.9}$$

where

$$a(t-1) = \frac{T(t-1)}{T(t)} \text{ and } b(t-1) = \frac{T(t) - T(t-1)}{T(t)}.$$

When $s = 0$ we may simply write

$$\mathbb{E}[\mathbf{q}(t)] = \mathbb{E}[\mathbf{q}(t-1)]\, a(t-1)\,\mathbf{Q}(t) + b(t-1)\,\mathbf{p}_0(t). \tag{5.10}$$

Now applying (5.9) recursively we get

$$\mathbb{E}[\mathbf{q}(s,t)] = \mathbf{q}(s)\frac{T(s)}{T(t)}\mathbf{Q}(s,t) + \frac{1}{T(t)}\sum_{\tau=1}^{t}\Delta T(s+\tau)\,\mathbf{p}_0(s+\tau)\,\mathbf{Q}(s+\tau,t), \tag{5.11}$$

and for the case that the initial time is zero we get

$$\mathbb{E}[\mathbf{q}(t)] = \mathbf{q}(0)\frac{T(0)}{T(t)}\mathbf{Q}(0,t) + \frac{1}{T(t)}\sum_{\tau=1}^{t}\Delta T(\tau)\,\mathbf{p}_0(\tau)\,\mathbf{Q}(\tau,t). \tag{5.12}$$

In each state $i = 1, 2, ..., k$ there are $N_i(t)$ memberships of the total population $T(t)$. We already have found the expected values of the population structure hence the natural question is what happens with the variances and covariances among the state sizes. As the literature shows the variability of the state sizes is an important problem which was firstly studied for manpower systems, that is human populations. The earliest work on the variability of the class sizes for the very-well-known models of population growth was introduced by Leslie (1945, 1948). An interesting treatment of a closed social system that is of an NHMS or HMS (Homogeneous Markov System) with $\Delta T(t) = 0$ was that of Prais (1955). The variability of the class sizes for these systems was studied by Bartholomew (1982) by applying the results given in Pollard (1966, 1967, 1973). These results were extended in Bartholomew (1982) for the model proposed for University teachers by Young et al. (1961), that is for Homogeneous Markov systems where the basic parameters are stationary. The study of the variability of the state sizes for an NHMS started with the work described in Vassiliou et al., (1985, 1989) and Tsaklidis et al., (1988). As we will show in what follows in fact the results are obtained in a straightforward way by just replacing in the corresponding equations for the homogeneous Markov system the basic stationary sequences with the respective non-homogeneous ones.

Let in what follows assume that the system starts at time $t = 0$, which for simplicity we omit, and let the vector $\boldsymbol{\mu}(t)$ consist of k means of the population of the states followed by the k^2 covariances in dictionary order, i.e.:

$$\begin{aligned}\boldsymbol{\mu}(t) = [&\mathbb{E}[N_1(t)], ..., \mathbb{E}[N_k(t)], cov(N_1(t), N_2(t)), ..., cov(N_1(t), N_k(t)),\\ &cov(N_2(t), N_1(t)), cov(N_2(t), N_2(t)), ..., cov(N_2(t), N_k(t)),\\ &cov(N_k(t), N_1(t)), ..., cov(N_k(t), N_k(t))].\end{aligned} \tag{5.13}$$

Let us assume as a first step that $\Delta T(t) = 0$ that is the system only replace leavers. Define by $N_{ij}(t-1)$ the random variable which represents the number of memberships moving from state i to state j in the time interval $(t-1, t]$. Then we have that

$$N_j(t) = \sum_{i=1}^{k} N_{ij}(t-1) \text{ for } j = 1, 2, ..., k. \tag{5.14}$$

From (5.14) we get that

$$cov\left(N_j(t), N_r(t)\right) = \mathbb{E}\left[N_j(t) N_r(t)\right] - \mathbb{E}\left[N_j(t)\right]\mathbb{E}\left[N_r(t)\right]$$

$$= \sum_{i=1}^{k}\sum_{s=1}^{k}\left\{\mathbb{E}\left[N_{ij}(t-1) N_{sr}(t-1)\right] - \mathbb{E}\left[N_{ij}(t-1)\right]\mathbb{E}\left[N_{sr}(t-1)\right]\right\}. \tag{5.15}$$

Since

$$N_{ij}(t) \sim B\left(N_i(t-1), p_{ij}(t)\right), \tag{5.16}$$

we get see also Bartholomew $(1982, p.21)$ that

$$\text{1) If } i \neq s \quad \mathbb{E}\left[N_{ij}(t-1) N_{sr}(t-1) \mid N_i(t-1), N_s(t-1)\right]$$

$$= N_i(t-1) N_s(t-1) p_{ij}(t) p_{sr}(t),$$

and consequently

$$\text{1) If } i \neq s \quad \mathbb{E}\left[N_{ij}(t-1) N_{sr}(t-1)\right] = \tag{5.17}$$

$$= \mathbb{E}\left[N_i(t-1)\right]\mathbb{E}\left[N_s(t-1)\right] p_{ij}(t) p_{sr}(t).$$

Now, if $i = s$ it could be proved straightforwardly that

$$\mathbb{E}\left[N_{ij}(t-1) N_{ir}(t-1) \mid N_i(t-1)\right] = \tag{5.18}$$

$$N_i(t-1)\left[N_i(t-1) - 1\right] p_{ij}(t) p_{ir}(t).$$

and if also $j = r$ then

$$\mathbb{E}\left[N_{ij}^2(t-1) \mid N_i(t-1)\right] = var\left[N_{ij}(t-1) \mid N_i(t-1)\right]$$

$$+ \left(\mathbb{E}\left[N_{ij}(t-1) \mid N_i(t-1)\right]\right)^2. \tag{5.19}$$

Given that

$$var\left[N_{ij}(t-1) \mid N_i(t-1)\right] = N_i(t-1) p_{ij}(t)\left(1 - p_{ij}(t)\right), \tag{5.20}$$

we get from $(5.18), (5.19)$ and (5.20) that

$$\text{2) If } i = s,\ \mathbb{E}\left[N_{ij}(t-1) N_{ir}(t-1)\right] =$$

$$\left[\mathbb{E}\left[N_i^2(t-1)\right] - \mathbb{E}\left[N_i(t-1)\right]\right] p_{ij}(t) p_{ir}(t) + \delta_{jr}\mathbb{E}\left[N_i(t-1)\right] p_{ij}(t). \tag{5.21}$$

Now, from (5.15) exploiting equations $(5.17), (5.21)$ we get that

$$cov\left\{N_j(t), N_r(t)\right\} = \sum_{i=1}^{k}\sum_{s=1}^{k} p_{ij}(t) p_{sr}(t)\, cov\left\{N_j(t-1), N_r(t-1)\right\}$$

$$+ \sum_{i=1}^{k}\left\{\delta_{jr} p_{ij}(t) - p_{ij}(t) p_{ir}(t)\right\}\mathbb{E}\left[N_i(t-1)\right]. \tag{5.22}$$

When the system is expanding, i.e. $\Delta T(t) > 0$ then we assume that the memberships of the expansion will be allocated to the states independent of the replacement memberships; then these flows will be independent of all other flows, so their contribution to the variances and covariances can simply be added. It is important to have in mind that the memberships entering the system in the time interval $(t-1, t]$ they are entering at the end of the interval at exactly the time t. Let $\boldsymbol{\mu}_0(t)$ be the vector of expectations, variances and covariances attributable to the expansion, then since they are distributed in the states multinomially we have that

$$\Delta T(t)\,\boldsymbol{\mu}_0(t) = \Delta T(t)\,[p_{01}(t), p_{02}(t), ..., p_{0k}(t), p_{01}(t)(1-p_{01}(t)), -p_{01}(t)\,p_{02}(t), ..., p_{0k}(t)(1-p_{0k}(t))]. \tag{5.23}$$

From $(5.13), (5.22)$ and the above reasoning after some careful algebra we arrive at the following matrix equation

$$\boldsymbol{\mu}(t) = \boldsymbol{\mu}(t-1)\,\mathbf{V}_q(t) + \Delta T(t)\,\boldsymbol{\mu}_0(t), \tag{5.24}$$

where the matrix $\mathbf{V}_q(t)$ is given by

$$\mathbf{V}_q(t) = \begin{bmatrix} \mathbf{Q}(t) & \mathbf{Q}(t)\,\mathbf{J} - \mathbf{J}(\mathbf{Q}(t) \otimes \mathbf{Q}(t)) \\ \mathbf{0} & \mathbf{Q}(t) \otimes \mathbf{Q}(t) \end{bmatrix}, \tag{5.25}$$

where $\mathbf{Q}(t) \otimes \mathbf{Q}(t)$ is the Kronecker product of the two matrices as defined in Chapter 2, $\mathbf{J}$ is a kk^2 matrix, the ith row of which has a "1" in column $(i-1)\,k+i$ and zeros elsewhere. The matrix $\mathbf{V}_q(t)$ is a $k(k+1) \times k(k+1)$ matrix which has some negative elements. In later chapters, we will study infinite products of such matrices.

If the initial time of the population is at time s then we define by $\boldsymbol{\mu}(s,t)$ the analogue of $\boldsymbol{\mu}(t)$ in a rather apparent way from equation (5.13) and following the above steps we get that

$$\boldsymbol{\mu}(s,t) = \boldsymbol{\mu}(s,t-1)\,\mathbf{V}_q(s,t) + \Delta T(t)\,\boldsymbol{\mu}_0(t), \tag{5.26}$$

where the matrix $\mathbf{V}_q(s,t)$ is of the form

$$\mathbf{V}_q(s,t) = \begin{bmatrix} \mathbf{Q}(s,t) & \mathbf{Q}(s,t)\,\mathbf{J} - \mathbf{J}(\mathbf{Q}(s,t) \otimes \mathbf{Q}(s,t)) \\ \mathbf{0} & \mathbf{Q}(s,t) \otimes \mathbf{Q}(s,t) \end{bmatrix}. \tag{5.27}$$

5.3.2 A fluctuating NHMS without forced wastage

In this subsection we will consider the case where we will be allowed mathematically to decrease the size of the NHMS with the condition that we will rely only to voluntary wastage. All the probabilities of internal movements at time t from a state i and the probabilities of leaving the system from that state also at time t apparently add up to one, i.e.:

$$\sum_{i=1}^{k+1} p_{ij}(t) = 1.$$

This equation could be written equivalently in matrix form as

$$\mathbf{p}_{k+1}^{\top}(t) = (\mathbf{I} - \mathbf{P}(t))\,\mathbf{1}^{\top}. \tag{5.28}$$

We know that the total number of memberships which will enter the population at the end of interval$(t-1,t]$ is

$$\mathbb{E}[R(t)] = T(t) - T(t-1) + \mathbb{E}[\mathbf{N}(t-1)]\,\mathbf{p}_{k+1}^{\top}(t),$$

which we could write it as follows

$$\mathbb{E}[R(t)] = \mathbb{E}[\mathbf{N}(t-1)]\left[\frac{T(t)}{T(t-1)}\mathbf{I} - \mathbf{P}(t)\right]\mathbf{1}^{\top}. \tag{5.29}$$

Now, from (5.5) we get that

$$\mathbb{E}\left[\mathbf{N}(t)\right]=\mathbb{E}\left[\mathbf{N}(t-1)\right]\left[\mathbf{P}(t)+\left[\frac{T(t)}{T(t-1)}\mathbf{I}-\mathbf{P}(t)\right]\mathbf{1}^{\top}\mathbf{p}_0(t)\right].$$

Since from (5.28) we get $\mathbf{p}_{k+1}^{\top}(t)\,\mathbf{p}_0(t))=(\mathbf{I}-\mathbf{P}(t))\,\mathbf{1}^{\top}\mathbf{p}_0(t))$,

we have that

$$\mathbb{E}\left[\mathbf{N}(t)\right]=\mathbb{E}\left[\mathbf{N}(t-1)\right]\left[\mathbf{Q}(t)+\frac{T(t)-T(t-1)}{T(t-1)}\mathbf{1}^{\top}\mathbf{p}_0(t)\right]$$

$$=\frac{T(t)}{T(t-1)}\mathbb{E}\left[\mathbf{N}(t-1)\right]\mathbf{M}(t), \tag{5.30}$$

where

$$\mathbf{M}(t)=\frac{T(t-1)}{T(t)}\left[\mathbf{Q}(t)+\frac{T(t)-T(t-1)}{T(t-1)}\mathbf{1}^{\top}\mathbf{p}_0(t)\right]. \tag{5.31}$$

We now make the following assumption:
Assumption 5.1.

$$\frac{T(t)}{T(t-1)}\geq\max\left[\sum_{j=1}^{k}p_{ij}(t)-p_{ij}(t)/p_{0j}(t)\right].$$

It could be proved that Assumption 5.1 is a necessary and sufficient condition for matrix $\mathbf{M}(t)$ to be a stochastic matrix. It is easy to show that $\mathbf{M}(t)\,\mathbf{1}^{\top}=\mathbf{1}^{\top}$, hence what remains is that

$$\mathbf{P}(t)+\left[\frac{T(t)}{T(t-1)}\mathbf{I}-\mathbf{P}(t)\right]\mathbf{1}^{\top}\mathbf{p}_0(t)\geq 0,$$

which from Assumption **5.1** is true. In terms of physical meaning it actually is a condition that the population could rely only on voluntary wastage in order to decrease its size. Hence, no forced wastage is allowed in the population.

Now, writing equation (5.30) recursively we get

$$\mathbb{E}\left[\mathbf{N}(0,t)\right]=\frac{T(t)}{T(0)}\mathbf{N}(0)\,\mathbf{M}(0,t). \tag{5.32}$$

When the initial time is s then (5.32) will take the form

$$\mathbb{E}\left[\mathbf{N}(s,t)\right]=\frac{T(t)}{T(s)}\mathbf{N}(s)\,\mathbf{M}(s,t).$$

For the relative population structure from (5.32) we get

$$\mathbb{E}\left[\mathbf{q}(t)\right]=\mathbf{q}(0)\,\mathbf{M}(0,t), \tag{5.33}$$

and

$$\mathbb{E}\left[\mathbf{q}(s,t)\right]=\mathbf{q}(s)\,\mathbf{M}(s,t). \tag{5.34}$$

5.4 A range of NHMS's environment

The following synopsis of applications illustrate the breath of application of NHMS, and a little of the reason why the theoretical results that follow are central for this processes.

1. Manpower planning in corporate organizations

Manpower planning can be defined as the problem of determining the number of personnel and their skills that best meet the future operational requirements of an enterprise. Steven Vajda[3] wrote one of the pioneering works on manpower planning Vajda (1947). Also of great interest is his paper Vajda (1975) and his book Vajda (1978). In Smith and Bartholomew(1988) there exist an historical review for the evolution of Manpower planning in the United Kingdom. A basic paper on wastage from manpower systems is McClean (1976b).

Humanity made the first attempts on manpower planning way back from ancient times, when civilized nations were preparing their naval power well in advance for reasons of future defence or aggressive policies in order to acquire assets for growth and survival. For example, it is well known the proposal of general Themistocles back in 478 BC in the assembly of Athenian democracy to invest all resources and the training of young people to create a powerful fleet to defend the city from the Persian aggressiveness. They did so and won the naval battle of Salamis in 480 BC although they were severely outnumbered. In the Admiralty in the UK, we meet the first serious effort for manpower planning back in 1779. Naturally the economic climate and the social conditions plays a vital role in the need for manpower planning. A strong example is the decision in the UK during World War II to plan to use single young women initially and later on the vast majority of women in the war effort. Immediately after the War the most senior staff of the Government Departments mainly concerned, having recognized the value of manpower planning decided that work should be done.

The early part was one of growth and optimism, where the emphasis was on manpower as a scarce resource. Wastage was damaging and costly and promotion was available in growing organizations. Lately, we have passed years of recession with increasing unemployment; organizations use voluntary wastage and promotion blockages to offer the prospect of easing the painful process of shedding labor. These changes nevertheless did not weakened the value and need for manpower planning.

Explosive growth of manpower planning appeared in 1965–1970. In April 1965, the Royal Statistical Society and the Operational Research Society sponsored a one day conference on manpower planning (see Young (1965))

The use of stochastic models in planning and management of human resources increased in the 1970's. Stochastic models have dominated due to the inherent stochastic nature of social and behavioral processes. More could be found on the Introductory Notes of the present chapter. In this example, we illustrate applications of NHMS in hierarchical structured management staffs in commercial and industrial organizations (see Young and Vassiliou (1974), Vassiliou (1976)). In fact the above work has been the basis for the theory of NHMS. The corporate organization is seen as a graded social system through which people move by recruitment, promotion and wastage. The career histories of managers who served with a large British firm between 1955 and 1966 have been analyzed and the model presented in the

[3]Steven Vajda (1901–1995) is known as the British father of linear programming. He was born in Budapest then moved to Vienna in 1902. Between the two world wars, Vienna was a focal point of intelllectual activity. During the Nazis occupation of Austria the Vajda familly felt compelled to leave. After many adventurous rather bad experiences which included some months in concentration camp the statistician Hilary Seal apointed him in the Admiralty Research Laboratory at Teddington. Steven work for the navy spanned manpower planning, military OR, and the provision of mathematical, statistical, and computational services. He is recognized one of the founders of the science of OR in Brittain and the rest of the world. Prof. Vajda continued his research up to the end and his name is closely linked with the duallity property in many mathematical problems.

above mentioned work has been tested against observed movements in the firm. We retain the notation introduced for an NHMS and a member of the system is now any member of the organization.

$N_i^{'}(t)$ the expected number of staff who remain in grade i at the end of the interval $(t-1,t]$ having been in the grade at the start of the interval.

$\alpha_{ij}(t)$ the "ecological parameter" (see explanation below).

$N_i^{*}(t)$ the expected number of staff in grade i in the inherent profile (see Young and Vassiliou 1974) at the start of the interval.

$p_{i,k+1}^{R}(t)$ the probability of an individual leaving the organization from grade i in the time interval $(t-1,t]$ because of resignation or redundancy.

$p_{i,k+1}^{s}(t)$ the probability of a person leaving the system from grade i in the time interval $(t-1,t]$ because of resignation under "normal" circumstances.

$p_{i,k+1}^{N}(t)$ the probability of a person leaving the system from grade i in the time interval $(t-1,t]$ because of death or medical retirement.

b_i the coefficient which expresses quantitatively the effect of chang ing rates on the number of people resigning from grade i in the time interval t.

Also define the cumulative acceleration $D^2(t)$ given by the following rules: If

$$\Delta^2 T(t) \leq 0 \ \text{ and } \ \Delta^2 T(t-k) \leq 0 \text{ for } k = 1,2,...,n,$$

or

$$\Delta^2 T(t) \leq 0 \ \text{ and } \Delta^2 T(t-k) > 0 \text{ for } k = 1,2,...,n \ \text{ and for at least } k = 1,2,$$

or

$$\Delta^2 T(t) > 0 \ \text{ and } \Delta^2 T(t-k) \leq 0 \ \text{ for } k = 1,2,...,n \ \text{ and for at least } k = 1,2,$$

then

$$D^2(t) = \sum_{k=0}^{n} \Delta^2 T(t-k). \tag{5.35}$$

In all other cases

$$D^2(t) = \Delta^2 T(t). \tag{5.36}$$

The ecological principle states that the rate at which promotions occur is proportional to the number available for promotion and to the number of vacancies. The ecological parameter $a_{ij}(t)$ expresses this relation quantitatively in the equations below. However, Bhat (1984) in his book on stochastic processes provided a nice interpretation by defining $a_{ij}(t)$ to be the probability that an individual in grade i will be acceptable as a candidate for one of the vacancies in grade j in the interval $(t-1,t]$. Now again from Young and Vassiliou (1974) and Vassiliou (1976) we get that

$$p_{ij}(t) N_i(t) = a_{ij}(t) \left[N_j^{*}(t+1) - N_i^{'}(t) \right] N_i(t) \quad \text{for } i,j = 1,2,...,k, \tag{5.37}$$

and

$$p_{i.k+1}(t) N_i(t) = p_{i,k+1}^{N} + p_{i,k+1}^{S} - \alpha_i D^2(t). \tag{5.38}$$

Hence, we have for a graded corporate organization the sequence of matrices $\{\mathbf{Q}(t)\}_{t=0}^{\infty}$ which are being calculated from (5.37) and (5.38). Examples of other applications of NHMS in manpower planning could be found many in the literature and it is not possible to refer to all of them. However, we will try to roughly categorize them as follows:

1.1. Hierarchical manpower systems

The basic work of manpower planning in hierarchical organizations was done for homogeneous Markov systems and up to 1992 they are well documented in the books by Vajda (1975), Bartholomew (1982), Bartholomew et. al., (1991). However there are some seminal papers that are worth reading such as Young and Almond (1961), Young (1965, 1971), McClean (1976). There is a large literature since then in a great diversity of journals. We will only provide some such applications in here from which the reader could start and uncover many more. With motive from hierarchical systems of the public service we refer the works of Georgiou and coworkers such as Georgiou and Tsantas (2002), Georgiou (1999), Dimitriou et al. (2013,2015). For hierarchical grade structures are also the works McClean (1977), Konta and Stewman (1980), Stewman and Konta(1983), Bruderl et al. (1993), Ugwuowo and McClean (2000), Yadavalli et al. (2006), Nilakantan and Raghavendra (2005), Yadavalli et al. (2002), Barbu et al. (2016), McClean (2021).

1.2. Sociology and Labor Markets

For various populations of specific characteristics problems of sociological nature have been studied through homogeneous Markov systems mainly. Examples of such studies which carry the flavor of the stochastic models used in such cases are Stewman (1975, 1986, 1988). There are also a number of known non-homogeneous Models for occupational mobility such as Stewman (1976a, 1976b), Hodge (1966), intragenerational mobility such as Mayer (1972), interregional mobility such Henry et al. (1971), job insecurity in Europe such as Symeonaki et al. such as (2014,2019), gender distances in education and employment such as Symeonaki et al. (2017). These studies and many others are special cases of NHMS the theory of which could be applied and answer important questions.

1.3 Military manpower planning

Homogeneous and Non-Homogeneous Markov systems theory had and is enjoying wide popularity through its use as a tool in military manpower planning. Among other reasons for that is the fact that military population systems are strictly hierarchical and that training of large subsections of their systems (populations) are going through strict specialized training from time to time as technology advances in various important defence systems.

The British Admiralty has been a pioneer in the use of manpower planning even before 1945. Manpower planning had reinforced political arguments in favor of abolition of compulsory military service, and National Service in Britain came to an end soon after 1960. The value of the UK experience was quickly evident throughout the NATO countries and as publications show in the entire globe.

Examples of such studies are the Markov models for the manpower in the police forces in UK, in Leeson (1981); the workforce planning and modeling of military training in Wang (2005) for the Department of Defence of the Australian Government; the military manpower planning models proposed by Gass (1991); the military reserve manpower planning model proposed by Reeves et al. (1999); the Markov chain modeling in the Slovenian Armed Forces (2008); the modeling of personnel sustainability by an NHMS for military force structure analysis in Filinkov et al. (2011); in Zais et al. (2016) the military personnel dynamics are studied with the use of a homogeneous Markov system model; that is, the size of the US Army is a function of competing budget, mission requirements, sentiments toward the retention (reenlistment) of enlisted soldiers; and most recently the military manpower planning via Markov system modeling in the Belgian Armed forces in Ousama et al. (2020).

2. Ecological modeling

Homogeneous Markov models have played traditionally an important role in biological problems and have proved an important tool. There are scientific journals devoted to Mathematical Biology and Markov models have been created for various ecological processes.

That is, such as energy flow in ecosystems, ecological succession, population dynamics, etc. However, various ecological processes are time dependent (see Colebrook (1979), Beers et al., (1982), Lindley (1987), Kasymov (1988), Ambland (1988), Moran et al. (1989), Patoucheas and Stamou (1993)). Even for benthic organisms living in very deep waters where the conditions are consider to be stable, seasonal changes do occur due to the seasonal cycles taking place near the surface. It has been reported that seasonality characterizes phenomena like succession (Ambland (1988)) and diversity ((Moran et al. (1989)). An application of NHMS under cyclic behavior has been published (Patoucheas and Stamou (1993)) for the study of zoobenthos dynamics in the Thessaloniki Gulf, Greece. Seasonal variations have been reported not only in the zoobenthos of the Thermaikos Gulf but also other biotic and antibiotic parameters like air and water temperature, algal biomass, hydrographic conditions, phytoplankton abundance, etc. Data concerning zoobenthos have been selected by 26 stations randomly dispersed in the study area. The number of species varied from 5 to 29 per sampling area and the density of population from 40 to 1300. Hence, the states for the number of species A_i and the population density B_i were grouped as follows:

A_1 : up to 10 species B_1 : up to 90 individuals

A_2 : from 11 to 15 species B_2 : from 91 to 140 individuals

A_3 : from 16 to 20 species B_3 : from 411 to 200 individuals

A_4 : more than 20 species B_4 : more than 200 individuals

Two NHMS have been fitted in (Patoucheas and Stamou (1993)) for the number of species and for the population density. The transition matrices among the A_i $(i = 1, 2, 3, 4)$ states $\mathbf{P}_A(t)$where reported to be cyclic, i.e., $\mathbf{P}_A(4\pi + \upsilon) = \mathbf{P}(\upsilon)$ for $\pi = 1, 2, ...$ and $\upsilon = 0, 1, 2, 3$ following the seasons. The same has been reported for the transition matrices among the B_i $(i = 1, 2, 3, 4)$ states $\mathbf{P}_B(t)$. This work is a classical application of the theory of NHMS under cyclic behavior (see Vassiliou (1984) and Vassiliou (1986)) which will be presented in detail in Chapter 8.

3. Healthcare systems

There is a huge literature on healthcare systems which is an important area of planning admissions, resources, lives of human beings, especially the most vulnerable ones the elderly, strategies of optimizing care services, in the face of all greater pressures for cuts in the budgets from the political system. Long waiting lists are classical symptoms all over the glob for inefficient care systems. The latest pandemic surfaced many more problems and proved beyond doubt the need for serious heath care planning with the use of stochastic modeling. In the long list of studies of such work there is the dominating figure of Sally McClean[4] and her school of research.

An integrated care system is a system in which patient care is provided in both hospital and the community. The states of the system are the different phases a particular illness goes through in a hospital or community care such as for example acute, treatment, rehabilitation or long stay, or conceptual, representing periods in hospital and/or community care depending on the patient phase of care or stage of the illness being treated.

The core of the models studied see for example Garg, McClean et al. (2010) in all studies using as instrument stochastic Markov models and which they have been most successful is

[4]Sally Ida McClean was born in Belfast and took her first degree an MA in mathematics from the University of Oxford in 1970. She earned an MSc in mathematical statistics and operations research from Cardiff University in 1972. She completed a Ph.D. in 1976 at the Ulster University at Coleraine. She is very well known for her huge contribution for Healthcare planning and especially her studies to improve the well being of the elderly. She is currently Professor of Mathematics at Ulster University. Her main research interests are in Stochastic Modeling and Optimization, particularly for Healthcare Planning, and Computer Science, specifically Databases, Internet of Things, Sensor Technology and Telecommunications.She has published over three hundred papers and was previously a recipient of Ulster University's Senior Distinguished Research Fellowship.

in compatibility of the terminology of the present book an NHMS in continuous time and discrete state with one absorbing state. Naturally, for an issue of such a magnitude and importance, from the humanitarian aspect, there are many forms that the transition matrices take in relation with the goals of the study, the illness under study, the covariates used for the transition inensities, and the type connections of the hospital with the community care phases. For the interested reader to collect the vast majority of these studies a search in any engine of publications which are under the name of Prof McClean will be sufficient for an almost perfect start.

In an integrated care system patient pathway for such a system is defined as the way a patient sequentially moves from one hospital phase to the next and similarly one community phase to the next. Assume that there are n hospital phases (HP) and m community phases (CP) and the death state, that is our NHMS has in all $n+m+1$ states. Patients can be discharged from any hospital or community phase due to death. Admission into the hospital is done only in the first hospital phase (state). Readmission into the first hospital phase is possible from any community phase. Transition from any hospital phase is possible to only the first community phase state.

Let $\lambda_i(t)$ represents the rate of transition or transition intensity from hospital phase i to hospital phase $i+1$ at time t; $\nu_i(t)$ represents the transition intensity from hospital i to first community phase at time t and $\mu_i(t)$ is the transition intensity from the hospital phase i to the absorbing state death. The remaining transition intensities are $a_i(t)$ the transition intensity from community phase i to community phase $i+1$ at time t; $\gamma_i(t)$ is the transition intensity from community phase i to the first hospital phase and $\beta_i(t)$ the transition intensity from community phase i to death. By time t it is meant that the transition intensities are assumed to depend log-linearly on the time dependent covariates $\mathbf{c}=(c_1,c_2,...,c_n)$. The log-linear function is assumed to take the form

$$\exp\left(\sigma+\mathbf{d}^{\top}\mathbf{c}\right),$$

where the coefficient parameters σ and vector $\mathbf{d}^{\top}$ are estimated for each of the transition rates using maximum likelihood estimates (Faddy and McClean (1999, 2005)). The most common covariates are two the patient age and the present calendar year. Hence the transition intensity matrix $\mathbf{Q}^*$ is given by

$$\begin{bmatrix}
q_{11}^* & q_{12}^* & \cdots & 0 & q_{1,n+1}^* & 0 & \cdots & 0 & q_{1,n+m+1}^* \\
0 & q_{22}^* & \cdots & 0 & q_{2,n+1}^* & 0 & \cdots & 0 & q_{2,n+m+1}^* \\
\cdots & \cdots & \cdots & . & \cdots & . & \cdots & . & \cdots \\
0 & 0 & \cdots & q_{n,n}^* & q_{n,n+1}^* & 0 & \cdots & 0 & q_{n,n+m+1}^* \\
q_{n+1,1}^* & 0 & \cdots & 0 & q_{n+1,n+1}^* & q_{n+1,n+2}^* & \cdots & 0 & q_{n+1,n+m+1}^* \\
q_{n+2,1}^* & 0 & \cdots & 0 & 0 & q_{n+2,n+2}^* & \cdots & 0 & q_{n+2,n+m+1}^* \\
. & . & \cdots & . & . & . & \cdots & . & . \\
q_{n+m,1}^* & 0 & \cdots & 0 & 0 & 0 & \cdots & q_{n+m,n+m}^* & q_{n+m,n+m+1}^* \\
0 & 0 & \cdots & 0 & 0 & 0 & \cdots & 0 & 0
\end{bmatrix},$$

where by $\mathbf{Q}^* = \mathbf{Q}(t)$ we denote the non-homogeneous transition intensity matrix. Now $q_{ii}^* = q_{ii}(t) = -\lambda_i(t) - \mu_i(t) - \nu_i(t)$ for $i = 1,2,...,n-1$ and $q_{nn}^* = q_{nn}(t) = -\mu_i(t) - \nu_i(t)$; $q_{i,i+1}^* = q_{i,i+1}(t) = \lambda_i(t)$ for $i = 1,2,...,n-1$; $q_{i,n+1}^* = q_{i,n+1}(t) = \nu_i(t)$ for $i = 1,2,...,n$; $q_{i,n+m+1}^* = q_{i,n+m+1}(t) = \mu_i(t)$ with $i = 1,2,...,n$; $q_{n+j,i}^*(t) = q_{n+j,1}(t) = \gamma_j(t)$ for $j = 1,2,...,m$; $q_{i,n+m+1}^* = q_{i,n+m+1}(t) = \mu_i(t)$ for $i = 1,2,...,n$; $q_{n+i,n+i}^* = q_{n+i,n+i}(t) = -a_i(t) - \beta_i(t) - \gamma_i(t)$ for $i = 1,2,...,m-1$; $q_{n+m,n+m}^* = q_{n+m,n+m}(t) = -\beta_m(t) - \gamma_m(t)$; $q_{n+i,n+i+1}^* = q_{n+i,n+i+1}(t) = a_i(t)$ for $i = 1,2,...,m-1$; $q_{n+j,n+m+1}^*(t) = q_{n+j,n+m+1}(t) = \beta_j$ for $j = 1,2,...,m$.

Models just as the one described where the states are put in some order and the process moves from one state only to the next and the absorbing state give rise to the well known Coxian distribution (Cox (1955), Marshall et al. (2004, 2004b)).

We observe that $\mathbf{Q}(t)\mathbf{1}^{\top} = 0$ for all t, which we call a conservative Markov chain, and also that $\sup_i \sup_t |q_{ii}(t)| < \infty$ which means that

$$\int_0^t q_{ii}(u)\,du < \infty \text{ for all } t \geq 0.$$

Under these conditions for the measurable transition matrix function $\mathbf{Q}(t)$ there exists a unique transition probability matrix $\mathbf{P}(s,t)$ which in addition satisfies the integral matrix equations (see Iosifescu (1980))

$$\mathbf{P}(s,t) = \mathbf{I} + \int_s^t \mathbf{Q}(u)\,\mathbf{P}(u,t)\,du,$$

for all $t \notin E$,where $E \subset [0,\infty)$ is a set of Lebesque measure zero. If $\mathbf{Q}(t)$ is continuous then E is the empty set, and

$$\mathbf{P}(s,t) = \mathbf{I} + \int_s^t \mathbf{P}(s,u)\,\mathbf{Q}(u)\,du,$$

for all $0 \leq s \leq t$. It follows also that $\mathbf{P}(.,.)$ satisfies the matrix differential equations

$$\frac{\partial \mathbf{P}(s,t)}{\partial s} = -\mathbf{Q}(s)\,\mathbf{P}(s,t)\,,\ 0 \leq s \leq t,\ s \notin E,$$

which is called the backward Kolmogorov equations, and

$$\frac{\partial \mathbf{P}(s,t)}{\partial t} = \mathbf{P}(s,t)\,\mathbf{Q}(t)\,,\ 0 \leq s \leq t,\ s \notin E,$$

which is called the forward Kolmogorov equation. We will study more on the subject of non-homogeneous Markov systems in continuous time in Chapter 11 of the present.

Let now according to our usual notation $\mathbf{N}(0)$ be the initial number of patients in each state/paths then $\mathbf{N}(0)$ is a $1 \times n+m+1$

$$\mathbf{N}(0) = (N_1(0),..., N_n(0), N_{n+1}(0),..., N_{n+m}(0), N_{n+m+1}(0)),$$

where $N_k(0)$ for $k = 1,2,...,n$ are the initial numbers of patients in hospital path k, $N_j(0)$ for $j = n+1,...,n+m$ are the initial number of patients in community phase j and finally $N_{n+m+1}(0)$ are the initial number of deaths. Let us also define by $\mathbf{N}(t)$ the vector with the number of patients in each hospital path, community phase, and deaths after t days then we have that

$$\mathbb{E}[\mathbf{N}(t)] = \mathbf{N}(0)\prod_{u=1}^{t}\mathbf{P}(u-1,u).$$

Now using the general theory of NHMS in continuous time and assuming various type of admissions like regular admissions, i.e., a constant number of admissions each day, or targeted admissions, i.e., admissions to complete the target number of patients in the hospital, or complete random admissions, i.e., the number of admissions follow a certain stochastic process for example Poisson or any type of time series we have the tools to answer important questions.

As we mentioned before there is a vast literature in the subject and there exist many important and interesting extensions and important reformed versions in order to study various aspects of the healthcare process. All of these use the above described non-homogeneous

Markov system as a shell for their development. At this point, we will refer to some of these work that I believe they constitute a probable minimum to start with in order to get the breath of the subject. In Jones, McClean et al. (2019) it is offered the maturity of working many years on the subject. In McClean et al. (2012), there exists solutions for the management and resource planning for complex stochastic healthcare systems. In Garg, McClean, et al. (2011) there is the extension of using survival trees and mixed distribution survival trees for clustering patient's hospital length of stay. In McClean et al. (2011) there is an interesting combination of Markov models and discrete-event simulation for stroke patient care. Finally, in Garg, McClean et al. (2009) an NHMS is proposed for sequential pattern mining of healthcare data.

4. Fish populations

There is a rather large number of studies of fish populations or populations restricted in fisheries that apply stochastic models especially Markov models to study problems of large diversity in these populations. The large majority of these applications of Markov modeling have the potential to be expanded using more advanced techniques and results on Markov theory like the ones present in the present book. We will refer to a few examples of such studies and the interested reader could start his research from there. Examples with a variety of problems tackled are Carreras et al. (2007) for self recruitment levels in a Mediterranean littoral fish population; Zhang et al. (2011) were they use Markov chains to analyze changes in wetland trends in Yinchuan plain, China; Dans (2012) studying the effects of tour boats on dolphin activity with the use of Markov chains; studying the dynamics of fish stock using Markov chain Monte Carlo simulation algorithm based on Markov chains on general state space which we will study in a later Chapter. Interesting studies at the management of fisheries are these of Clark (2010) and Hilker and Liz (2019).

5. Non-homogeneous Markov jump systems (NHJMS)

Since the mid-1960s, Markov jump systems were proposed and have attracted much attention for their practical applications. In recent years, the research on the behavior and control of MJS has been very active. Generally, the evolutions in time of NHMSs are appropriate to model the phenomenon that occur in practical systems for example in such areas such as aerospace industry, manufacturing systems, economic systems, wheeled mobile manipulators (see Kang et al., (2012)), electrical systems see (Costa et al. (2005)) and positive systems (see Yin et al. (2018a)).

Most theoretical results and applications related to MJSs are focused on homogeneous MJSs, that is with constant transition probability matrices. However, the assumption of homogeneous transition probabilities is not realistic in many real situations (see Yin and Lin (2017)).

As it will be apparent below NHJMS are closely related with NHMS and methodologies, directions of research and common problems would be possible to be shared by both areas. This fact has not been yet exploited in full by both areas. The world "Jump" probably is the heritage that came from the main applications of the theory which was network of modes. The multiple modes of NHMJS are switched on stochastically from on to another according to jumps of a Markov chain whose transition probabilities determine the stochastic evolution of the NHMJS.

Let $(\Omega, \mathcal{F}, \mathbb{P})$ be a probability space and consider the following discrete-time NHMJS with state variable

$$\mathbf{N}(t) = [N_1(t), N_2(t), ..., N_n(t)], \text{ that is } \mathbf{N}(t) \in \mathbb{R}^n.$$

Then for a NHMJS the evolution of the state variable is given by a number of possible stochastic difference equations depending on the assumptions made on the parameters, the assumptions made on the driven non-homogeneous Markov chain, and naturally the application involved see for examples Zhao et al. (2013), Liu et al. (2019), Luan et al. (2014), Yin et al. (2018).

$$\mathbf{N}(t+1) = \mathbf{A}(X_t)\mathbf{N}(t) + \mathbf{B}(X_t)\sigma(\mathbf{u}_t) + \mathbb{E}(X_t)\boldsymbol{\omega}_t,$$

where $\{X_t\}_{t=0}^{\infty}$ is a non-homogeneous Markov chain which drives the system which has state space $\mathbb{S} = \{1, 2, ..., k\}$ and could be in discrete or continuous time; $\mathbf{u}_t$ is the input which is in general a $1 \times m$ vector; $\boldsymbol{\omega}_t$ is the bounded disturbance of the system with $\boldsymbol{\omega}_t \in \{\boldsymbol{\omega}_t^{\top}\boldsymbol{\omega}_t \leq 1\}$; and

$$\sigma(\mathbf{u}_t) = [\sigma(u_{1t}), \sigma(u_{2t}), ..., \sigma(u_{mt})]^{\top}.$$

The non-homogeneous Markov chain with state space $\mathbb{S} = \{1, 2, ..., k\}$ governs the transitions among the k modes with transition probability matrices $\mathbf{P}(t)$ with elements

$$p_{ij}(t) = \mathbb{P}[X_{t+1} = j \mid X_t = i]$$

denoting the transition probability from mode i to mode j at the tth interval. The matrices $\mathbf{A}(X_t)$, $\mathbf{B}(X_t)$ are known matrices with the appropriate dimensions. They are numerous studies with various models for the type of transition probabilities. For example in Liu et al. (2019) it is assumed that they follow a truncated Gaussian processes; in Luan et al. (2014) they are unknown but assume that they follow a Gaussian process; in Zhao et al. (2013) they are assume to take values randomly in a finite set and a scheme of Bayesian estimation is proposed; in Chen (2018) time delay is introduced which is a similar idea as with NHMS of high order firstly introduced in Vassiliou and Moysiadis (2010) for which a chapter is devoted later on.

Important problems studied for NHMJS are, for example, those of control see for example Do Val and Costa (2000) and Luan (2011), probabilistic control by scenario optimization approach see for example Yin et al. (2018b), Liu et al. (2019), stochastic stability analysis see for example Yin et al. (2018a), filtering see for example Dong (2011) and H_{∞} filtering Luan (2014); there are also some studies where fuzzy sets theory is introduced in order to study certain uncertainties in the systems.

6. Biological Markov population models

There is an immense literature on biological Markov population models which are closely related, or potentially related or partially related with the present NHMS theory. In order to categorize these models in details would probably take another volume of the present and probably distract us for the main theme of the book. We will only try to describe in general lines these class of models, provide some examples of different kinds of populations that have been studied to uncover their large diversity, and give some references for the interest reader to have vague directions to start his search. These studies are mainly in mathematical biological journals but not exclusively.

Their main common characteristics are firstly that they study a structured population. The world structured is used in here in its most general meaning and it could mean for example different patches of some species, different states of the same species, biological populations in randomly varying environments, harvesting and seeding populations which creates new states and so on. The second characteristic is that stochastic immigration is present among the different states and to the external environment in various forms and these are equivalent in many cases with transitions of memberships among the states of an NHMS and the wastage from it from its states. Another characteristic is emigration in the various states of the biological Markov population models for example emigration from the external environment into the different patches of the population, seeding in the various states etc., which actually are equivalent with recruitment in the states of NHMS. Finally the problem of the growth of a biological population is an interesting problem and there are many studies dealing with this in biological Markov population models. In the case of an NHMS this corresponds to the number of memberships in the system and both areas of research could benefit from existing studies on this problem.

Let us start with the first example that of studies of metapopulations a review of which exist in the book Giplin and Hanski (1991). Natural populations of most species have a spatial structure (Gyllenberg and Sylvestrov (1994)). There are several habitat patches that can support local populations which are geographically distributed in a probably large area. Local populations in the various patches and of the mainland, which is the largest habitat constitute a metapopulation and are connected with stochastic migration. Population in the habitat patches may go extinct while the metapopulation persist. An empty habitat patch may be colonized by migrants from other patches and extinction and recolonization are important problems of a metapopulation. A basic question in a metapopulation as in the present book is its limiting behavior see for example Mc Vinish and Pollet (2012), Barbour and Pugliese (2005) and Buckley and Pollet (2010). As a second example we refer to the general class of Markov population models formulated as stochastic difference equations. Many such studies are related to the limiting behavior of biological populations such as analyses of gene frequency dynamics and competition for space in randomly varying environments. Most frequently the structure of the population refers to densities of the population collected in a vector with components which may represent different phenotypes, genotypes, age classes, species or sub-populations. Hence, their framework is closely related with the NHMS. The interest reader might find useful the study by Ellner (1984) which also provides an extensive bibliography. Another class of Markov population models are the ones which study to what extent density and resource shortage must be taken in explaining population growth. In order to succeed in this endeavor researches extended the theory of population dynamics to encompass stochastic environments. That is, the models involved such as carrying capacity undergo a stochastic perturbation see for example Gyllenberg et al. (1994).

A third example which is closely related to NHMS is the problem of harvesting and seeding of stochastic populations Hening and Tran (2020). In this class of models, we have an ecosystem of d species that interact, possibly non-linearly, due to competition for resources, predation, cooperation, mutualism etc., are influenced by random environmental fluctuations, and can be controlled through seeding and harvesting. There are examples where environmental fluctuations can drive a species extinct as well as examples where the environmental fluctuations create rescue effect that serves species from extinction. In this respect there is a powerful theory of stochastic persistence (Hening and Nguyen (2018)). In addition species are also harvested by humans. Excessive harvesting and hunting can lead species to become locally or globally extinct. In specific situations in restricted conservation habitats one can increase the population size of a certain species by bringing in individuals from a different location. Naturally there are economic implications, that is, there is a cost whenever one seeds and a gain when one harvest. The similarities with wastage and recruitment in an NHMS are apparent and strongly linked. The basic problem is a control one of how we harvest/seed in order to maximize revenue while ensuring species do not go extinct. The similarities with the problem of control in an NHMS, that is, maintainability and attainability of the state sizes studied in subsequent chapters is apparent. At this point let us describe a little more this class of models which will uncover a little the similarities with the theory of NHMS.

Assume we have a probability space $(\Omega, \mathcal{F}, \mathbb{P})$ and an appropriate filtration $\{\mathcal{F}_t\}_{t\geq 0}$ containing all the necessary and available information up to time t. Let that in our population coexist k species interacting nonlinearly in a stochastic environment. Let

$$\mathbf{N}(t) = [N_1(t), N_2(t), ..., N_k(t)]^\top ,$$

be the population structure of the various species in the population. Under the assumption that the environment mainly affects the growth/death rates of the population then the approach of Scheiber et al. (2011) and Evans et al. (2013) leads us to consider (Hening et al. (2020)) the following stochastic differential equation

$$d\mathbf{N}(t) = \mathbf{b}(\mathbf{N}(t))\, dt + \boldsymbol{\sigma}(\mathbf{N}(t))\, d\mathbf{W}(t),$$

where $\mathbf{W}(t)$ is the k-dimensional Brownian motion adapted to the filtration $\{\mathcal{F}_t\}_{t\geq 0}$ and $\mathbf{b}, \boldsymbol{\sigma}:[0,\infty)^k \to \mathbb{R}^k$ are locally Lipshitz continuous functions.

Let

$$\mathbf{Y}(t) = [Y_1(t), Y_2(t), ..., Y_k(t)]^\top \text{ and } \mathbf{Z}(t) = [Z_1(t), Z_2(t), ..., Z_k(t)]^\top,$$

be the vectors with $Y_i(t)$ for $i = 1, 2, ..., k$ denote the number of species i that have been harvested up to time t, that is wastage from the population; $Z_i(t)$ for $i = 1, 2, ..., k$ denote the number of species i seeded into the population up to time t, that is recruitment into the population. Then the population structure at time $t \geq 0$ will be given by

$$\mathbf{N}(t) = \mathbf{N}(0) + \int_0^t \mathbf{b}(\mathbf{N}(s))\, ds + \int_0^t \boldsymbol{\sigma}(\mathbf{N}(s))\, d\mathbf{W}(s) - \mathbf{Y}(t) + \mathbf{Z}(t),$$

where apparently $\mathbf{N}(0)$ is the initial population structure before any seeding or harvesting.

The basic important problem in this class of models is the *optimal harvesting-seeding problem*. That is, given any kind of value function to find the optimal harvesting-seeding strategy $(\mathbf{Y}^*(t), \mathbf{Z}^*(t))$ for every t that maximizes the value function.

For the interested reader the book by Ewens (1979) discusses biological issues and their history in a manner accessible to the mathematician.

7. Open Markov populations

In the first edition of his book Bartholomew (1967) used the terminology "open" and "closed" Markov manpower models. The reason was to distinguish the case where there were no recruitment of new members and wastage of members in the population which he called closed, with the case where there was recruitment and wastage which he called "open". Examples of closed systems are many in the literature such as the well known demographic models see Pollard (1966), Pollard (1967). There is a good selection of demographic models since their start in Smith and Keyfitz (1977). These homogeneous Markov systems are in fact special cases of an NHMS. In the literature there are a large number of Markov chain models used for populations consisting of various types of members where they use the terminology of "Open Markov population" for which apparently the theory of NHMS's is directly applicable. For instance Guerreiro et al. (2004, 2013) have studied a population which consists of contracts for bonus malus system. Their concern is to find the limiting distribution of the relative population structure among the different categories of contracts. A homogeneous Markov system with Poisson recruitment was studied in Pollard (1967) and extensions of these in continuous and discrete time were studied in McClean (1976, 1978) and Vassiliou (1982b). Open Markov population models fed by second order stationary processes have also been studied in Esquivel et al. (2014), where it is studied a special case of an NHMS for which the recruitment process of new memberships is modeled by a time series, coming from a second order stationary process, that is, a stationary process with a deterministic bias. Also Open Markov population models where the recruitment process which is not independent is considered in Stadje (1999). Open Markov population models have been also used for different populations to study consumption portfolios (Esquivel et al. (2017)); open automobile portfolios (Afonso et al. (2020), Des Lourdes Centeno et al. (2001); traffic flows by means of random networks (Floriani et al. (2016)). Gurerreiro and colleagues have also studied a class of open Markov chains subjected to periodical re-classifications see for example Guerreiro et al. (2010). Open Markov population processes in continuous time have also been studied in Esquivel (2021a, 2021b).

8. Social physics and statistical mechanics

Another interesting class of Open Markov population models in discrete space and continuous time are the ones which are closely related with statistical physics. Social physics, biological physics and physics of particles are related probably through the strong concepts of entropy, stability, energy which exists in all kind of populations. See for example for the concept of stability Feightinger (1976) for homogeneous Markov populations; Vassiliou (1981b) for the concept of stability in NHMS; for the concept of asymptotic variability for NHMS's Vassiliou (1986); for the concept of entropy see McClean et al. (1978), McClean (1986) for homogeneous Markov systems, Vassiliou (1984) for NHMS's. These initial studies gave the sparkle for a stronger connection of social physics of populations with statistical mechanics in the works of Tsaklidis and co-workers. For example, in Tsaklidis and Soldatos (2003) homogeneous Markov systems in continuous time and fixed size are seen as an elastic solid; energy is defined and studied for homogeneous Markov systems in continuous time and fixed size in Tsaklidis (1999); as a Newtonian fluid is seen a homogeneous Markov system in continuous time and fixed size in Tsaklidis (1997); the interpretation of the evolution of the homogeneous Markov system as a deformation of viscoelastic medium for the three dimensional case is given in Loumponias and Tsaklidis (2017). For Open Markov population models entropy and other physics concepts are discussed in some interesting depth in the works of Baez et al. (2016), Pollard (2016, 2017) and Salgado-Garcia (2021).

9. Non-Homogeneous Semi-Markov systems

The theory of NHMS was very successfully extended to the theory of non-homogeneous semi-Markov systems (NHSMS) by Vassiliou and Papadopoulou (1992) where basically the inherent Markov chain or process is now a semi-Markov process (Howard (1971a and b). Naturally, the difficulty in dealing with the mathematical problems which logically surface is greater see for example Papadopoulou and Vassiliou (1994, 2014), Papadopoulou (1997) in discrete time and Papadopoulou and Vassiliou (1999) in continuous time. The number of parameters to be estimated are increased in the NHSMS but not dramatically to reduce significantly its applications. Only a few such examples will be referred to, such as in credit risk a research area in stochastic analysis and financial mathematics see for example Vasileiou et al. (2006), D' Amico, et al (2007, 2010), Vassiliou, et al. (2013c), Vassiliou (2013c, 2014a, 2021). Also DNA sequences Papadopoulou (2013); in reward models McClean et al. (2004), Papadopoulou (2004); Healthcare systems Papadopoulou, et al. (2012); in reward paths in a stochastic environment Papadopoulou (2007).

5.5 The $\mathcal{G}$-non-homogeneous Markov system

At this point having already an idea of the range of NHMS's we will step up the definition of an NHMS with the help of some measure theory, not much more than the part already stated in the present book. That is, we will extend the definition of the NHMS, given in section 5.2, in order to accommodate the case, where the information needed to the observer for the evolution of the NHMS, is a larger set than the history of the multidimensional process NHMS.

As usual in probability theory, a complete filtered probability space $(\Omega, \mathcal{G}, \mathcal{G}_t, \mathbb{P})$ is given. Consider a family of sub-σ-algebras, which we call *filtration* or *history*–since it is assumed that $\mathcal{G}_s \subseteq \mathcal{G}_t$ for $s \leq t$ – which is assumed to contain all the information available up to time t for the evolution of the phenomenon (physical problem) at hand. As it is already evident, we will work in the present in discrete time and we will write $(\mathcal{G}_t,\ t = 0, 1, ...)$. By definition, $\mathcal{G} = \vee_{t=\infty}^{\infty} \mathcal{G}_t$ the smallest σ-algebra containing all the events of $\mathcal{G}_t,\ t = 0, 1, ...$ In association with any stochastic process X_t we will define for each t a sub-σ-algebra of $\mathcal{G}$, denoted $\mathcal{F}_t^X$, by $\mathcal{F}_t^X = \sigma(X_s, s \in [0, t])$. The family $\mathcal{F}_t^X$ is called the internal history of the process X_t. Assume that any stochastic process which comes into play from now on, its internal history is a subfiltration of $\mathcal{G}$.

Let $S = \{1, 2, ..., k\}$ be the set of states and since it is finite, it is clear that any function $h : S \to \mathbb{R}$ is bounded and measurable provided that we endow the state space with the σ-field of all its subsets. Let also $X_t^{(i)}$ be the stochastic process which describes the state the i-member is in at time t, with values on $S^* = \{1, 2, ..., k, k+1\}$, were $k+1$ is the state that contains in the external environment the memberships of leaving members. Let $\mathcal{F}_t^{X_t^{(i)}}$ be the internal history generated by the process $X_t^{(i)}$ assuming that $\mathcal{F}_t^{X_t^{(i)}} \subset \mathcal{G}_t$ for $i = 1, 2, ..., T(t)$. For a particular i assume that $X_t^{(i)}$ follows a $\mathcal{G}$-non-homogeneous Markov chain of first order in discrete time. That means, for any function $h : S \to \mathbb{R}$ we have

$$\mathbb{E}_{\mathbb{P}}\left(h\left(X_{t+1}^{(i)}\right) \mid \mathcal{G}_t\right) = \mathbb{E}_{\mathbb{P}}\left(h\left(X_{t+1}^{(i)}\right) \mid \sigma\left(X_t^{(i)}\right)\right). \tag{5.39}$$

Note that by the terminology of "first order" we mean that the Markov property is valid with only the dependence on the present state and not on the more distance past as in Markov chains of higher order. If a process $X_t^{(i)}$ is a $\mathcal{G}$-non-homogeneous Markov chain of first order, then of course, it is also an $\mathcal{F}_t^{X_t^{(i)}}$-non-homogeneous Markov chain under $\mathbb{P}$, but the converse implication is not true in general. Since the state space S^* is finite, it is apparent that condition (5.39) is equivalent to the following

$$\mathbb{P}\left(X_{t+1}^{(i)} = j \mid \mathcal{G}_t\right) = \mathbb{P}\left(X_{t+1}^{(i)} = j \mid \sigma\left(X_t^{(i)}\right)\right). \tag{5.40}$$

Assume that $X_t^{(i)}$ for $i = 1, 2, ..., T(t)$ are independent for every $t \in \mathbb{N}^*$. Then for any member i of the system (population) we define by

$$\mathbb{P}\left(X_t^{(i)} = j \mid X_{t-1}^{(i)} = r\right) = p_{rj}(t), \tag{5.41}$$

for $i \in T(t)$ and $r, j \in S$. Let $R_t^{(i)}$ be the random variable that expresses the state a new entrant (i) enter the system in the time interval $(t-1, t]$, (t-th interval) either as a replacement for a membership or as a new membership. Assume that $R_t^{(i)}$ for $i = 1, 2, ..., T(t)$ are independent for every $t \in \mathbb{N}^*$ and let

$$\mathbb{P}\left(R_t^{(i)} = j \mid X_{t-1}^{(i)} = 0\right) = p_{0j}(t), \tag{5.42}$$

where 0 expresses the pool of available memberships for replacement and new memberships for expansion.

Now let $Y_t^{(i)}$ be the stochastic process with values on S, which describes the state where the i-membership is at time t. From above we have that $X_t^{(i)}$ for $i = 1, 2, ..., T(t)$ are independent for every $t \in \mathbb{N}^*$. Then for any membership i of the system (population) we define by

$$\mathbb{P}\left(Y_t^{(i)} = j \mid Y_{t-1}^{(i)} = r\right) = q_{rj}(t). \tag{5.43}$$

It is easy to see that

$$q_{rj}(t) = p_{rj}(t) + p_{0j}(t)\, p_{rk+1}(t). \tag{5.44}$$

Obviously $q_{rj}(t) \geq 0$ and it could be proved that

$$\sum_{j=1}^{n} q_{rj}(t) = 1. \tag{5.45}$$

Thus Y_t is a $\mathcal{G}$-non-homogeneous Markov chain under $\mathbb{P}$ with values on $S = \{1, 2, ..., k\}$ of first order which we call the *inherent $\mathcal{G}$-non-homogeneous Markov chain under $\mathbb{P}$* of first order.

Now define by

$$\mathbf{Q}(t) = \{q_{ij}(t)\}_{i,j \in S},$$

which is the transition matrix for the inherent $\mathcal{G}$-non-homogeneous Markov chain under $\mathbb{P}$ of first order.

Define by $N_j(t)$ the stochastic process at time t which represents the number of memberships who are in state j at time t. We call the vector of stochastic processes

$$\mathbf{N}(t) = [N_1(t), N_2(t), ..., N_k(t)],$$

the random population structure of the $\mathcal{G}$-non-homogeneous Markov system.

Define by

$$\sigma(\mathbf{N}(t)) = \vee_j \sigma(N_j(t)), \tag{5.46}$$

that is, the minimum σ-algebra of the σ-algebras $\sigma(N_j(t))$ for $j = 1, 2, ..., k$. Assume that as we already mentioned $\sigma(\mathbf{N}(t)) \subset \mathcal{G}_t$. Now we have for an arbitrary chosen t

$$\mathbb{E}_{\mathbb{P}}[\mathbf{N}(t)] = \mathbb{E}_{\mathbb{P}}[\mathbb{E}_{\mathbb{P}}[\mathbf{N}(t) \mid \mathcal{G}_{t-1}]]. \tag{5.47}$$

Define by $\mathbf{p}_0(t) = [p_{01}(t), p_{02}(t), ..., p_{0k}(t)]$ and $\Delta T(t) = T(t) - T(t-1)$, then as we already have seen in this chapter

$$\mathbb{E}_{\mathbb{P}}[\mathbf{N}(t)] = \mathbb{E}_{\mathbb{P}}[\mathbf{N}(t-1)]\mathbf{Q}(t) + \Delta T(t)\mathbf{p}_0(t). \tag{5.48}$$

Define by $q_j(t) = N_j(t)/T(t)$ provided that $T(t) < \infty$. Then we define by

$$\mathbf{q}(t) = [q_1(t), q_2(t), ..., q_k(t)], \tag{5.49}$$

the random relative population structure of the $\mathcal{G}$-non-homogeneous Markov system.

5.6 Change of measure in $\mathcal{G}$-non-homogeneous Markov systems

In the present section, we will study what we chose to call **the first exotic property of an NHMS.** We will start by explaining the choice of the terminology *exotic*. Firstly, we must make it clear that the term does not imply that it is not a vital or important property, in fact quite the opposite is true. It has been observed that the large majority of researchers in the area, has a mental image, of the stochastic methods and pure mathematics background needed for tackling successfully research problems in the area. The terminology exotic stems from the fact that the mathematical background in what follows, is unusual for this research area. The interplay between stochastic theory of NHMS and other branches of probability, stochastic processes and mathematics I believe is a fascinating one, apart from being important. We borrowed the term exotic from stochastic finance, where in there it represents options with an unusual structure (see for example Vassiliou (2010)).

A central problem in Markov chains is what happens if we change the measure in a probability space to a Markov chain defined in this space. The solution to this problem for a homogeneous Markov chain is that under certain conditions on the relation of the two measures, it becomes a non-homogeneous Markov chain under the new measure see Vassiliou (2010). Change of measure problems is vital in Mathematical Finance and hence when non-homogeneous Markov chains are used as models then the problem arises and solutions have been found see Vassiliou (2010). Also the problem has been solved and successfully used in Financial problems for non-homogeneous semi-Markov chains and non-homogeneous Markov renewal processes, see for example Vasileiou et al. (2006), Vassiliou (2021). In many areas of applications of NHMS's as for example in most financial applications it is enough to study the behavior of an NHMS up to some horizon date $t^* < \infty$. In such an application for example $\mathbf{N}(t)$ or $\mathbf{q}(t)$ could represent a portfolio of some financial assets or derivatives. In such applications and others the main question to be answered is what happens if we change measure in a $\mathcal{G}$-NHMS. Does it remains a $\mathcal{G}$-NHMS, and if yes what is the influence of the change of measure on the sequence of transition probabilities.

Let a complete filtered probability space $(\Omega, \mathcal{G}, \mathcal{G}_t, \mathbb{P})$ and a $\mathcal{G}$-NHMS defined in it. Let also that we want to change the probability measure $\mathbb{P}$ to the probability measure $\tilde{\mathbb{P}}$. To change $\mathbb{P}$ to $\tilde{\mathbb{P}}$, we need to reassign probabilities in Ω using a random variable Z. The process is described by the following theorem the proof of which could be found for example in Shreve (2004).

Theorem 278 *Let a complete filtered probability space $(\Omega, \mathcal{G}, \mathcal{G}_t, \mathbb{P})$ and let Z be an almost surely non-negative random variable with $\mathbb{E}[Z] = 1$. For $A \in \mathcal{G}$, define*

$$\tilde{\mathbb{P}}(A) = \int_A Z(\omega)\, d\mathbb{P}(\omega). \tag{5.50}$$

Then $\tilde{\mathbb{P}}(A)$ with $A \in \mathcal{G}$ is a probability measure and in addition if X is a non-negative random variable, then

$$\mathbb{E}_{\tilde{\mathbb{P}}}[X] = \mathbb{E}_{\mathbb{P}}[XZ]. \tag{5.51}$$

If Z is almost surely strictly positive, then for any non-negative random variable X

$$\mathbb{E}_{\mathbb{P}}[X] = \mathbb{E}_{\tilde{\mathbb{P}}}\left[\frac{X}{Z}\right].$$

We are now in a position to extend the definition of two equivalent probability measures given in Definition 6.1., Vassiliou (2010).

Definition 279 *Let a complete filtered probability space $(\Omega, \mathcal{G}, \mathcal{G}_t, \mathbb{P})$. Two probability measures $\mathbb{P}$ and $\tilde{\mathbb{P}}$ on $(\Omega, \mathcal{G}, \mathcal{G}_t, \mathbb{P})$ are said to be equivalent if they agree which sets (events) have probability zero.*

Remark 280 *Under the assumptions of Theorem 278, including the assumption that $Z > 0$ almost surely, $\mathbb{P}$ and $\tilde{\mathbb{P}}$ are equivalent. Actually this is easy to see. Let $A \in \mathcal{G}$ with $\mathbb{P}(A) = 0$. Then from Theorem 278, we have*

$$\tilde{\mathbb{P}}(A) = \int_A \mathbf{1}_A(\omega)\, Z(\omega)\, d\mathbb{P}(\omega) = 0,$$

since $\mathbb{P}(A) = 0$. Inversely, assume that $A \in \mathcal{G}$ with $\tilde{\mathbb{P}}(A) = 0$. Then

$$\mathbb{P}(A) = \mathbb{E}_{\mathbb{P}}[\mathbf{1}_A] = \mathbb{E}_{\tilde{\mathbb{P}}}\left[\mathbf{1}_A \frac{1}{Z}\right] = 0,$$

since $\mathbf{1}_A \frac{1}{Z} = 0$ almost surely under $\tilde{\mathbb{P}}$.

In fact theorem 278, Definition 279 and Remark 280 are being related with the following very important theorem of Radon-Nikodým (Shreve 2004) which we provide below.

Theorem 281 ***(Radon-Nikodým)****. Let a complete filtered probability space $(\Omega, \mathcal{G}, \mathcal{G}_t, \mathbb{P})$ and two probability measures $\mathbb{P}$ and $\tilde{\mathbb{P}}$ on $(\Omega, \mathcal{G}, \mathcal{G}_t, \mathbb{P})$. There exists an almost surely positive random variable Z such that $\mathbb{E}_{\mathbb{P}}[Z] = 1$ and*

$$\tilde{\mathbb{P}}(A) = \int_A Z(\omega)\, d\mathbb{P}(\omega), \text{ for every } A \in \mathcal{G}. \tag{5.52}$$

We now provide the definition of an integrable random variable.

Definition 282 *Let a complete filtered probability space $(\Omega, \mathcal{G}, \mathcal{G}_t, \mathbb{P})$ and $X(\omega)$ a random variable on it. The random variable X is integrable if and only if*

$$\int_{\Omega} |X(\omega)| \, d\mathbb{P}(\omega) < \infty. \tag{5.53}$$

We write $\mathcal{L}^1(\Omega, \mathcal{G}, \mathbb{P})$ the space of all integrable random variables $(\Omega, \mathcal{G}, \mathbb{P})$.

Remark 283 *Let the vector space $\mathcal{L}^1(\Omega, \mathcal{G}, \mathbb{P})$. The almost sure equality of two random variables in $\mathcal{L}^1$ is an equivalence relation. In this respect a random variable X is a representative of all almost surely equals to it random variables and in fact they all have the same expectation. We may thus consider the space L^1 of all such representatives of the random variables in $\mathcal{L}^1$. Then since the addition of random variables preserve almost surely equality, it is known that the set L^1 is a vector space (see Jacod and Protter (2004)).*

We may easily generalize this concept by letting a parameter $1 < p < \infty$ and define the space $\mathcal{L}^p$ to be the space of all random variables such that $|X|^p \in \mathcal{L}^1$. Then the space L^p is defined analogously to L^1, as the space of representatives random variables that are almost surely equal and belong to $\mathcal{L}^p$.

Let now $\mathbb{P}$ and $\tilde{\mathbb{P}}$ be two equivalent probability measures defined on the common measurable space $(\Omega, \mathcal{G})$. Then from (5.52), we get that Radon-Nikodým derivative of $\tilde{\mathbb{P}}$ with respect to $\mathbb{P}$ could be written as

$$\frac{d\tilde{\mathbb{P}}}{d\mathbb{P}} = \zeta, \ \mathbb{P}\text{-almost surely.} \tag{5.54}$$

Note that the random variable ζ is strictly positive $\mathbb{P}$-almost surely, moreover ζ is $\mathbb{P}$ integrable, with $\mathbb{E}_{\mathbb{P}}[\zeta] = 1$. Then the proof of the following theorem could be found in Musiela and Rutkowski (1997).

Theorem 284 ***(Abstract version of Bayes's formula).*** *Let $\mathcal{F}$ be a sub-σ-field of the σ-field $\mathcal{G}$, and let ψ be a random variable integrable with respect to $\tilde{\mathbb{P}}$. Then the following holds*

$$\mathbb{E}_{\tilde{\mathbb{P}}}[\psi \mid \mathcal{F}] = \frac{\mathbb{E}_{\mathbb{P}}[\psi\zeta \mid \mathcal{F}]}{\mathbb{E}_{\mathbb{P}}[\zeta \mid \mathcal{F}]}. \tag{5.55}$$

Now, let $\tilde{\mathbb{P}}$ a probability measure equivalent to the probability measure $\mathbb{P}$ on the measurable space $(\Omega, \mathcal{G}_{t^*})$, set

$$\frac{d\tilde{\mathbb{P}}}{d\mathbb{P}} |_{\mathcal{G}_{t^*}} = \mathbb{E}_{\mathbb{P}}[\zeta \mid \mathcal{G}_{t^*}] = \zeta_{t^*}, \ \mathbb{P}\text{-almost surely,} \tag{5.56}$$

where the $\mathcal{G}_{t^*}$-measurable random variable is strictly positive $\mathbb{P}$-almost surely and $\mathbb{E}_{\mathbb{P}}[\zeta_{t^*}] = 1$.

It could be proved that the process

$$\zeta_t = \mathbb{E}_{\mathbb{P}}[\zeta \mid \mathcal{G}_t], \ \text{for } t = 0, 1, ..., t^*, \tag{5.57}$$

follows a strictly positive martingale (see Vassiliou (2010)) for an introduction on martingales) under $\mathbb{P}$.

We assume that the following which is labeled as **Condition M.1** is valid (Vassiliou (2013c)):

Condition M.1. The random variable

$$\zeta_t^{-1}\zeta_{t+1} \text{ is } \sigma(Y_t, Y_{t+1})\text{- measurable for any } t = 0, 1, ..., t^* - 1.$$

Equivalently, for every $t = 0, 1, ..., t^* - 1$, we have that

$$\zeta_t^{-1}\zeta_{t+1} = g_t\left(Y_t, Y_{t+1}\right),$$

for some function $g_t : S \cup \{0\} \times S \to \mathbb{R}$. Where it is assumed that $Y_t = 0$ has the meaning that a new membership is going to enter the system.

We will now provide the following theorem which answers the problem of the change of measure in a $\mathcal{G}$-NHMS.

Theorem 285 *(Vassiliou (2013c)). Let a $\mathcal{G}$-NHMS under the real probability measure $\mathbb{P}$. Let also $\tilde{\mathbb{P}}$ be an equivalent probability measure on the measurable space $(\Omega, \mathcal{G}_{t^*})$ given by (5.56), where ζ_{t^*} is the $\mathcal{G}_{t^*}$-measurable random variable, strictly positive $\mathbb{P}$-almost surely and $\mathbb{E}_{\mathbb{P}}\left[\zeta_{t^*}\right] = 1$.*

Assume that Condition M.1 is satisfied. Then $\{Y_t\}_{t=0}^{\infty}$ follows a $\mathcal{G}$-NHMS also under the probability measure $\tilde{\mathbb{P}}$. In addition we have that:

$$\tilde{q}_{ij}(t) = \tilde{\mathbb{P}}\left(Y_{t+1} = j \mid Y_t = i\right) = q_{ij}(t)\, g_t(i, j), \tag{5.58}$$

for arbitrary states $i, j \in S$ and $t = 0, 1, ..., t^* - 1$. Also

$$\tilde{p}_{0j}(t) = \tilde{\mathbb{P}}\left(Y_{t+1} = j \mid Y_t = 0\right) = p_{0j}(t)\, g_t(0, j). \tag{5.59}$$

Proof. Using the Abstract Bayes formula, we fix a $t \in \mathbb{N}^*$ and for any $j \in S$, we get

$$\begin{aligned}
\tilde{\mathbb{P}}\left(Y_{t+1} = j \mid \mathcal{G}_t\right) &= \mathbb{E}_{\tilde{\mathbb{P}}}\left[\mathbf{1}_{\{Y_{t+1}=j\}} \mid \mathcal{G}_t\right] = \frac{\mathbb{E}_{\mathbb{P}}\left[\zeta_{t^*}\mathbf{1}_{\{Y_{t+1}=j\}} \mid \mathcal{G}_t\right]}{\mathbb{E}_{\mathbb{P}}\left[\zeta_{t^*} \mid \mathcal{G}_t\right]} \\
&= (\zeta_t \text{ is a martingale under } \mathbb{P} \text{ and } \mathcal{G}_t\text{-measurable}) \\
&= \mathbb{E}_{\mathbb{P}}\left[\zeta_t^{-1}\zeta_{t^*}\mathbf{1}_{\{Y_{t+1}=j\}} \mid \mathcal{G}_t\right] = (\text{tower property}) \\
&= \mathbb{E}_{\mathbb{P}}\left[\mathbb{E}_{\mathbb{P}}\left[\zeta_t^{-1}\zeta_{t^*}\mathbf{1}_{\{Y_{t+1}=j\}} \mid \mathcal{G}_{t+1}\right] \mid \mathcal{G}_t\right] = \mathbb{E}_{\mathbb{P}}\left[\zeta_t^{-1}\zeta_{t+1}\mathbf{1}_{\{Y_{t+1}=j\}} \mid \mathcal{G}_t\right] \\
&= \mathbb{E}_{\mathbb{P}}\left[\mathbb{E}_{\mathbb{P}}\left[\zeta_t^{-1}\zeta_{t^*}\mathbf{1}_{\{Y_{t+1}=j\}} \mid \mathcal{G}_{t+1}\right] \mid \mathcal{G}_t\right] == \mathbb{E}_{\mathbb{P}}\left[\zeta_t^{-1}\zeta_{t+1}\mathbf{1}_{\{Y_{t+1}=j\}} \mid \mathcal{G}_t\right] \\
&= \mathbb{E}_{\mathbb{P}}\left[g_t\left(Y_t, Y_{t+1}\right)\mathbf{1}_{\{Y_{t+1}=j\}} \mid \mathcal{G}_t\right] = \mathbb{E}_{\mathbb{P}}\left[g_t\left(Y_t, Y_{t+1}\right)\mathbf{1}_{\{Y_{t+1}=j\}} \mid \sigma\left(Y_t\right)\right] \\
&= \tilde{\mathbb{P}}\left(Y_{t+1} = j \mid \sigma\left(Y_t\right)\right).
\end{aligned} \tag{5.60}$$

The equality before the last is due to (5.39) with

$$h\left(Y_{t+1}\right) = g_t\left(Y_t, Y_{t+1}\right)\mathbf{1}_{\{Y_{t+1}=j\}},$$

and it shows that the conditional probability is a $\sigma(Y_t)$-measurable random variable and since $\sigma(Y_t) \subseteq \mathcal{G}_t$ we get the last equality. This in fact shows that $\{Y_t\}_{t=0}^{\infty}$ follows a $\mathcal{G}$-NHMS under the probability measure $\tilde{\mathbb{P}}$.

Now we have that

$$\begin{aligned}
\tilde{q}_{ij}(t) &:= \tilde{\mathbb{P}}\left(Y_{t+1} = j \mid Y_t = i\right) = \mathbb{E}_{\mathbb{P}}\left[\zeta_t^{-1}\zeta_{t+1}\mathbf{1}_{\{Y_{t+1}=j\}} \mid Y_t = i\right] \\
&= \mathbb{E}_{\mathbb{P}}\left[g_t\left(Y_t, Y_{t+1}\right)\mathbf{1}_{\{Y_{t+1}=j\}} \mid Y_t = i\right] = g_t(i, j)\, q_{ij}(t).
\end{aligned} \tag{5.61}$$

Relation (5.59) is proved in a similar way as relation (5.61). ∎

Now let us denote by

$$\mathbf{G}(t) = \{g_t(i, j)\}_{i,j \in S} \text{ and } \mathbf{G}_0(t) = \{g_t(0, j)\}_{j \in S}. \tag{5.62}$$

From section 5.6 we know that $\{Y_t\}_{t=0}^{\infty}$ follows a $\mathcal{G}$-NHMS under the probability measure $\tilde{\mathbb{P}}$ thus

$$\mathbb{E}_{\tilde{\mathbb{P}}}\left[\mathbf{N}(t)\right] = \mathbb{E}_{\tilde{\mathbb{P}}}\left[\mathbf{N}(t-1)\right]\left[\mathbf{G}(t) \vartriangle \mathbf{Q}(t)\right] + \Delta T(t)\left[\mathbf{G}_0(t) \vartriangle \mathbf{p}_0(t)\right], \tag{5.63}$$

for $t = 0, 1, ..., t^* - 1$.

Remark 286 *The assumption $\zeta_t^{-1}\zeta_{t+1} = g_t(Y_t, Y_{t+1})$ is very general and includes a large number of cases. Well known cases are those in credit risk (Bielecki and Rutckowski (2004), Vasileiou et al (2006), Vassiliou (2021)).*

5.7 The space of random population structures as a Hilbert space

We suppose as previously a given complete probability space $(\Omega, \mathcal{G}, \mathcal{G}_t, \mathbb{P})$. The question which has been in the mind of researchers for some time is "What type of space constitute the random population structures of an NHMS?". A successful answer to this question apparently could provide a wealth of mathematical tools to our disposal.

We start by denoting by PS^2 all equivalent classes for almost surely equality of random population structures $\mathbf{N} = [N_1, N_2, ..., N_k]$ of a $\mathcal{G}$-NHMS such that $\mathbb{E}_{\mathbb{P}}\left[\mathbf{N}^2\right] < \infty$, that is, $\mathbb{E}_{\mathbb{P}}\left[N_i^2\right] < \infty$ for $i = 1, 2, ..., k$. We henceforth identify all random population structures $\mathbf{N}$ and $\hat{\mathbf{N}}$ in PS^2 that are equal almost surely and consider them to be representatives of the same variable. We will now prove the following theorem.

Theorem 287 *(Vassiliou (2013c)). Let PS^2 the space of all equivalent classes for almost surely equality of random population structures $\mathbf{N} = [N_1, N_2, ..., N_k]$ of a $\mathcal{G}$-NHMS such that $\mathbb{E}_{\mathbb{P}}\left[\mathbf{N}^2\right] < \infty$, that is, $\mathbb{E}_{\mathbb{P}}\left[N_i^2\right] < \infty$ for $i = 1, 2, ..., k$. Define in PS^2 the inner product of two random population structures to be*

$$\left\langle \mathbf{N}, \hat{\mathbf{N}} \right\rangle = \sum_{i=1}^{k} \mathbb{E}_{\mathbb{P}}\left[N_i \hat{N}_i\right]. \tag{5.64}$$

Also define as possible norm for an element of the space PS^2 the following:

$$\|\mathbf{N}\| = \langle \mathbf{N}, \mathbf{N} \rangle^{\frac{1}{2}} = \left\{ \sum_{i=1}^{k} \mathbb{E}_{\mathbb{P}}\left[N_i^2\right] \right\}^{\frac{1}{2}}. \tag{5.65}$$

Then PS^2 is a normed linear space with an inner product given by (5.64) *and moreover it has the norm* (5.65).

Proof. Let $\mathbf{N}$ and $\hat{\mathbf{N}}$ in PS^2 and $a, b \in \mathbb{R}$. We have that

$$\left(a\mathbf{N}+b\hat{\mathbf{N}}\right)^2 \leq 2a^2\mathbf{N}^2 + 2b^2\hat{\mathbf{N}}^2,$$

and by taking expectations on both sides of the inequality we get that

$$\mathbb{E}_{\mathbb{P}}\left[\left(a\mathbf{N}+b\hat{\mathbf{N}}\right)^2\right] \leq 2a^2\mathbb{E}_{\mathbb{P}}\left[\mathbf{N}^2\right] + 2b^2\mathbb{E}_{\mathbb{P}}\left[\hat{\mathbf{N}}^2\right] < \infty,$$

which proves that $a\mathbf{N}+b\hat{\mathbf{N}} \in PS^2$. It is not difficult to prove that the inner product defined in (5.64) is linear in each component: For example,

$$\left\langle a\mathbf{N}+b\hat{\mathbf{N}}, \tilde{\mathbf{N}} \right\rangle = \sum_{i=1}^{k} \mathbb{E}_{\mathbb{P}}\left[\left(aN_i + b\hat{N}_i\right)\tilde{N}_i\right] =$$

$$a\sum_{i=1}^{k} \mathbb{E}_{\mathbb{P}}\left[N_i\tilde{N}_i\right] + b\sum_{i=1}^{k} \mathbb{E}_{\mathbb{P}}\left[\hat{N}_i\tilde{N}_i\right] = a\left\langle \mathbf{N}, \tilde{\mathbf{N}} \right\rangle + b\left\langle \hat{\mathbf{N}}, \tilde{\mathbf{N}} \right\rangle. \tag{5.66}$$

Observe that

$$\begin{aligned} \langle \mathbf{N}, \mathbf{N} \rangle &= \sum_{i=1}^{k} \mathbb{E}_{\mathbb{P}}\left[N_i^2\right] \geq 0; \qquad (5.67) \\ \text{and } \langle \mathbf{N}, \mathbf{N} \rangle &= 0 \text{ if and only if } \mathbf{N} = \mathbf{0} \text{ almost surely.} \end{aligned}$$

Note that, $\mathbf{N} = \mathbf{0}$ almost surely implies $\mathbf{N} = \mathbf{0}$ by our convention of identifying almost surely equal random population structures.

It remains to show that the norm proposed in (5.65) is indeed a norm. It is not difficult to show that

$$1)\ \|\mathbf{N}\| = \left\{ \sum_{i=1}^{k} \mathbb{E}_{\mathbb{P}}\left[N_i^2\right] \right\}^{\frac{1}{2}} \geq 0.$$

$$\begin{aligned} 2)\ \|a\mathbf{N}\| &= \left\{ \sum_{i=1}^{k} \mathbb{E}_{\mathbb{P}}\left[a^2 N_i^2\right] \right\}^{\frac{1}{2}} = |a| \left\{ \sum_{i=1}^{k} \mathbb{E}_{\mathbb{P}}\left[N_i^2\right] \right\}^{\frac{1}{2}} \\ &= |a|\,\|\mathbf{N}\| \text{ for all } a \in \mathbb{R}, \end{aligned}$$

and by bilinearity and the Cauchy-Schwarz inequality we could prove the inequality of Minkowski.

$$3)\ \left\|\mathbf{N} + \hat{\mathbf{N}}\right\| \leq \|\mathbf{N}\| + \left\|\hat{\mathbf{N}}\right\|. \tag{5.68}$$

Therefore, our norm is a true norm. ■

We now provide two well known theorems (see Jacod and Protter (2004)), which will be used in what follows.

Theorem 288 *(Monotone Convergence Theorem). If a sequence of random variables* $\{X_n\}$ *are positive and increasing almost surely to* X*, then* $lim_{n\to\infty}\mathbb{E}_{\mathbb{P}}[X_n] = \mathbb{E}_{\mathbb{P}}[X]$.

Theorem 289 *(Lebesgue dominated convergence theorem). If a sequence of random variables* $\{X_n\}$ *converges almost surely to* X *and if* $|X_n| < Y \in L^1$ *almost surely, all* n*, then* $X_n \in L^1, X \in L^1$*, and* $\lim_{n\to\infty}\mathbb{E}_{\mathbb{P}}[X_n] = \mathbb{E}_{\mathbb{P}}[X]$.

We will now prove the following:

Theorem 290 *(Vassiliou (2013c)). The normed linear space* PS^2 *is complete and hence it is a Hilbert space.*

Proof. In order to prove completeness of PS^2 it is sufficient to show that if

$$\mathbf{N}_n = [N_{1n}, N_{2n}, ..., N_{kn}]$$

is a sequence of random population structures which is a Caushy sequence under the norm (5.65), then $\{\mathbf{N}_n\}_{n=0}^{\infty}$ converges in PS^2.

Let $\{\mathbf{N}_n\}_{n=0}^{\infty}$ be a Cauchy sequence in PS^2, that is, for every $\epsilon > 0$, there exists a n^* such that $\|\mathbf{N}_n - \mathbf{N}_m\| \leq \epsilon$ for every $n, m \geq n^*$. Then there exists a subsequence $\{\mathbf{N}_{n_r}\}_{r=1}^{\infty}$ such that

$$\left\|\mathbf{N}_{n_r} - \mathbf{N}_{n_{r+1}}\right\| \leq \frac{1}{2^r}. \tag{5.69}$$

Define the row vector

$$\begin{aligned} \mathbf{S}_n &= \sum_{\nu=1}^{n} \left|\mathbf{N}_{n_\nu} - \mathbf{N}_{n_{\nu+1}}\right| \qquad (5.70) \\ &= \left[\sum_{\nu=1}^{n} \left|N_{n_\nu,1} - N_{n_{\nu+1},1}\right|, \sum_{\nu=1}^{n} \left|N_{n_\nu,2} - N_{n_{\nu+1},2}\right|, ..., \sum_{\nu=1}^{n} \left|N_{n_\nu,k} - N_{n_{\nu+1},k}\right| \right]. \end{aligned}$$

Then we have that

$$\|\mathbf{S}_n\| \leq \sum_{\nu=1}^{n} \left\|\mathbf{N}_{n_\nu} - \mathbf{N}_{n_{\nu+1}}\right\|. \tag{5.71}$$

From (5.69) and (5.71) we get that

$$\|\mathbf{S}_n\| \leq \sum_{\nu=1}^{n} \|\mathbf{N}_{n_\nu} - \mathbf{N}_{n_{\nu+1}}\| = \sum_{\nu=1}^{n} \frac{1}{2^\nu} \leq 1. \tag{5.72}$$

Since $\mathbf{S}_n(\omega)$ is a nondecreasing sequence almost surely for each ω, then there exists a random variable $\mathbf{S} = \lim_{n\to\infty} \mathbf{S}_n$. By (5.72) since $\|\mathbf{S}_n\| \leq 1$ we get also that $\|\mathbf{S}_n\|^2 \leq 1$ and therefore $\mathbb{E}_{\mathbb{P}}\left[\mathbf{S}_n^2\right] < 1$. Due to the Monotone Convergent Theorem, we get $\mathbb{E}_{\mathbb{P}}\left[\mathbf{S}^2\right] < 1$. Hence, $\mathbf{S} < \infty$ almost surely. Consequently the sequence

$$\mathbf{N}_{n_1} + \sum_{\nu=1}^{n} \left(\mathbf{N}_{n_\nu} - \mathbf{N}_{n_{\nu+1}}\right),$$

converges absolutely almost surely, and since it is a telescoping series we conclude $\mathbf{N}_{n_\nu}(\omega)$ as $\nu \to \infty$ converges to a limit let say $\mathbf{N}(\omega)$ and in addition

$$|\mathbf{N}(\omega)| \leq |\mathbf{N}_{n_1}(\omega)| + \mathbf{S}(\omega).$$

Thus, since we know that $\mathbf{N}_{n_1}, \mathbf{S} \in PS^2$ we get also that $\mathbf{N} \in PS^2$.

Consider the following sequence in m and r

$$\sum_{\nu=r}^{m} \left(\mathbf{N}_{n_\nu} - \mathbf{N}_{n_{\nu+1}}\right),$$

then it is easy to see that

$$\mathbf{N} - \mathbf{N}_{n_r} = \lim_{m\to\infty} \sum_{\nu=r}^{m} \left(\mathbf{N}_{n_\nu} - \mathbf{N}_{n_{\nu+1}}\right). \tag{5.73}$$

Now we have that

$$\begin{aligned} \left|\sum_{\nu=r}^{m} \left(\mathbf{N}_{n_\nu} - \mathbf{N}_{n_{\nu+1}}\right)\right| &\leq \sum_{\nu=r}^{m} \left|\mathbf{N}_{n_\nu} - \mathbf{N}_{n_{\nu+1}}\right| \\ &\leq \sum_{\nu=r}^{\infty} \left|\mathbf{N}_{n_\nu} - \mathbf{N}_{n_{\nu+1}}\right| = \mathbf{S}. \end{aligned} \tag{5.74}$$

Since (5.73) and (5.74) is true for every m, r by Lebesgue's dominated convergence theorem, we have that

$$\|\mathbf{N} - \mathbf{N}_{n_r}\| \leq \lim_{m\to\infty} \sum_{\nu=r}^{m} \|\mathbf{N}_{n_\nu} - \mathbf{N}_{n_{\nu+1}}\| \leq \frac{1}{2^{r-1}},$$

and hence we conclude that $\lim_{r\to\infty} \|\mathbf{N} - \mathbf{N}_{n_r}\| = 0$. Therefore, $\mathbf{N}_{n_r}$ converges as $r \to \infty$ to $\mathbf{N} \in PS^2$

Now, as the last step we have

$$\|\mathbf{N}_n - \mathbf{N}\| \leq \|\mathbf{N}_n - \mathbf{N}_{n_r}\| + \|\mathbf{N}_{n_r} - \mathbf{N}\|,$$

from which we conclude that as $n, r \to \infty$ then $\mathbf{N}_n \to \mathbf{N} \in PS^2$. ■

5.8 Estimation of the transition probabilities of an NHMS

One of the questions which naturally arise for the practitioner in any application of the NHMS is that of how to estimate, from the available data, the real world probabilities of the inherent non-homogeneous Markov chain. Naturally, the problems encounter depend on the environment of the application. However, there is a common problem to almost all applications and that is the right censoring. In the environment of manpower planning, one of the important variables is the completion of length of service before an event such as leaving the job or getting a promotion. For the time span of our data there will surely be some cases where although they are present in the data the time event does not occur. If we ignore this fact then we lose information from the available data (see McCLean and Gribbin (1987, 1991), McClean (1991) and Bartholomew et al. (1991, p. 184)). It is also essentially the same as the competing risk model in the medical literature, as discussed by Kalbfleisch and Prentice (2002, Chapter 8), Lee (1992, p. 26) and Cox and Oakes (1984, Chapter 4), where the event of interest is frequently time elapsed prior to either the appearance of various symptoms or death. Similar problems appear also in reliability theory, where the event of interest is time elapsed until a component breaks down, and actuarial studies, as discussed in Elandt-Johnson and Johnson (1980, Chapter 7). Finally, we will also refer to the environment of credit risk where the states are the rating categories, which are distinguished as: Aaa, Aa, A, Baa, Ba, B, Caa, and Default. However, there is an entry with rating WR that reports the issuers had withdrawn at the date referred to. A senior rating might be withdrawn for any number of reasons, from retirement of all rated debt to completion of an exchange offer for all dated debt. In this case the "withdrawn" flow is treated as right censoring. We will assume that our sencoring scheme is independent, which in the case of credit risk means that the reason for the rating WR is not that they understand that their possibilities of default have increased.

For all the above kinds of data much work has been done on both parametric and nonparametric estimation. For our present purposes, we will provide a nonparametric estimation method for the transition probabilities following mainly the work by Anderson and Goodman (1957), Vassiliou (1976), and Vasileiou et.al (2006). Define by $W_i(t)$ the number of members of the NHMS right censored in state i in the interval $(t-1,t]$. Then the maximum likelihood estimate for a particular year of the transition probabilities is

$$\hat{p}_{ij}(t) = N_{ij}(t) / (N_i(t) - W_i(t)).$$

Under the assumption that the system is homogeneous in the time interval $[0,T]$ then the maximum likelihood estimate is

$$\hat{p}_{ij} = \sum_{t=1}^{T} N_{ij}(t) / \sum_{t=1}^{T} (N_i(t) - W_i(t)).$$

If we consider the null hypothesis that the transition probabilities from grade i remain constant over time:

$$H_0 : p_{ij}(t) = p_{ij} \text{ for all } j \in F(i) \text{ and } t, \text{ given } i,$$

where $F(i)$ is the set of all values of j for which $p_{ij}(t) > 0$. We denote by $n(i)$ the number of elements of the set $F(i)$. Under this hypothesis the statistic

$$\chi^2(i) = \sum_{t=1}^{T} \sum_{F(i)} (N_i(t) - W_i(t)) \frac{[\hat{p}_{ij}(t) - \hat{p}_{ij}]^2}{\hat{p}_{ij}},$$

is approximately chi-square distributed with $(T-1) \times [n(i) - 1]$, degrees of freedom.

For testing the hypothesis if a particular flow (given j) is independent of time we can use the statistic

$$\chi^2(i,j) = \sum_{t=1}^{T} (N_i(t) - W_i(t)) \frac{[\hat{p}_{ij}(t) - \hat{p}_{ij}]^2}{\hat{p}_{ij}},$$

which is approximately chi-square distributed with $(T-1)$ degrees of freedom. Statistics based on the generalized likelihood ratio statistic may also be derived to test each of the foregoing hypothesis. Note that the above statistics and the generalized likelihood ratio statistics are asymptotically equivalent.

5.9 Research notes

The problem of embeddability in continuous time Markov chains is quite old (Elving (1937)). Guerry (2014) studied the problem for discrete time HMS. Let us assume that the available data allows us to estimate the transition probability matrix in one time unit and denote this by $\mathbf{P}(1)$. The problem studied is whether there exists a transition matrix $\mathbf{P}(1/m)$, that is, a transition probability matrix with time step $1/m$ which is compatible with $\mathbf{P}(1)$. After existence is established the next step is how to estimate $\mathbf{P}(1/m)$ from the available data especially when the solution for $\mathbf{P}(1/m)$ is not unique.

In the study by Symeonaki et al. (2002) the Fuzzy Non-Homogeneous Markov System (F-NHMS) was introduced. The idea was to combine the concepts of fuzzy sets and their theory introduced by Zadeh (1968) with the theory of NHMS. The main instrument of the *membership function* of a fuzzy set A being the dominating concept. The elements of the theory of *Fuzzy Logic* are sets whose boundaries are not precise. The membership in a fuzzy set is not a matter of affirmation or denial, but rather a matter of degree. More specifically, when A is a fuzzy set and x is a relevant element, the proposition "x is a member of A" is not necessarily either true or false, but it may be true only to some degree, being usually a number in the interval $[0,1]$. This degree is represented by the membership function of the fuzzy set A. Note that membership functions are not probabilities. Moreover, they are not subjective probabilities although intuitively in many cases act like ones. With these instruments Symeonaki et al. (2002) estimate the transition probabilities of the inherent Markov chain using the experience of experts on the population parameters using a Fuzzy Inference System. In Symeonaki and Stamou (2004) they progress to study NHMS's with fuzzy states. That is, the asymptotic behavior of the expected population structure of an NHMS when the states are fuzzy. Latter on in Symeonaki and Stamou (2006) they study the rate of convergence and the sensitivity of the limiting behavior. In Symeonaki (2015) the theory of F-NHMS is developed when the states are also fuzzy.

Papadopoulou and Tsaklidis (2007) introduced the theory of inhomogeneous semi-Markov systems (NHSMS) with fuzzy states. Also, Vassiliou (2013) studied in Credit Risk, the fuzzy semi-Markov migration process. In this way the phenomenon of different commercial rating companies attributing ratings is rather vague and overlapping rating classes have found with the help of fuzzy sets theory a successful way to be modeled. An interesting collection of papers related with the theory and applications of NHMS cound be found in Vassiliou and Georgiou (2021).

Chapter 6

ASYMPTOTIC BEHAVIOR OF A NON-HOMOGENEOUS MARKOV SYSTEM

6.1 Introductory notes

In the present chapter we will study the asymptotic behavior of the expected population structure and the relative expected population structure. That is, we want to find $\lim_{t\to\infty} \mathbb{E}[\mathbf{N}(t)]$ if it exists, and if it is, to find the conditions that makes it possible. The asymptotic behavior of Markov chains as we have seen is a problem of central importance and so does the asymptotic behavior of NHMS. The first basic theorem answering this question appeared in Vassiliou (1981a) and we will start with this. The expected population structure of an NHMS is given in equation (5.8). As we have seen in Chapter 5, where we defined an NHMS with a finite number of states and also we noticed this was the consequence of the fact that from the large number of various areas of applications, that we referred to, all the various NHMS had a finite number of states. Hence, we will study in the remaining of the book finite NHMS.

6.2 Asymptotic behavior of the expected population structure

We will use the following norm of a matrix:

Let a matrix $\mathbf{A} \in M_n(\mathbb{R})$ and a matrix $\mathbf{P} \in SM_n(\mathbb{R})$ then we defined the norm $\|.\|_\infty$ which we called the L_∞ norm for a matrix $\mathbf{A} \in M_n(\mathbb{R})$ as the maximum of the absolute values of its elements. Now this norm has in addition the following properties

$$\|\mathbf{PA}\|_\infty \leq \|\mathbf{A}\|_\infty \quad \text{and} \quad \|\mathbf{AP}\|_\infty \leq n\|\mathbf{A}\|_\infty. \tag{6.1}$$

We will prove the following basic theorem which to a large extent follows the proof of Theorem 3.2 in Vassiliou (1981a) and Theorem 1 in Vassiliou and Georgiou (1990).

Theorem 291 *Let a complete probability space $(\Omega, \mathcal{F}, \mathbb{P})$ and consider a finite NHMS defined on it as in Chapter 5 in discrete time. Assume that*

$$\lim_{t\to\infty} \|\mathbf{Q}(t) - \mathbf{Q}\|_\infty = 0 \quad \text{and} \quad \lim_{t\to\infty} \|\mathbf{p}_0(t) - \mathbf{p}_0\|_\infty = 0, \tag{6.2}$$

where $\mathbf{Q}$ is regular, that is, irreducible and aperiodic. In addition assume that $\lim_{t\to\infty} T(t) = T$ *and* $\Delta T(t) \geq 0$ *then*

$$\lim_{t\to\infty} \mathbb{E}[\mathbf{N}(t)] = T\mathbf{q}^*, \tag{6.3}$$

where $\mathbf{q}^*$ is the row of the stable stochastic matrix $\mathbf{Q}^* = \lim_{t\to\infty} \mathbf{Q}^t$. Also the asymptotic expected relative limiting structure is given by $\lim_{t\to\infty} \mathbb{E}[\mathbf{q}(t)] = \mathbf{q}^*$.

Proof. Since $\mathbf{Q}$ is a regular stochastic matrix then it is well known that $\lim_{t\to\infty} \|\mathbf{Q}^t - \mathbf{Q}^*\|_\infty = 0$ where $\mathbf{Q}^*$ is a stable stochastic matrix with row $\mathbf{q}^*$such that

$$\mathbf{q}^*\mathbf{Q}^* = \mathbf{q}^* \quad \text{and} \quad \mathbf{q}^*\mathbf{1}^\top = \mathbf{1}^\top. \tag{6.4}$$

DOI: 10.1201/b23204-6

Now, it is apparent that we could always choose m such that

$$\left\| \mathbf{Q}^{m-1} - \mathbf{Q}^{*} \right\|_{\infty} < \frac{1}{2}\epsilon \text{ where } \epsilon > 0. \tag{6.5}$$

Since $\mathbf{Q}^{*}$ is a stable stochastic matrix it is easy to see that

$$\mathbf{Q}(\tau, t) - \mathbf{Q}^{*} = (\mathbf{Q}(\tau, t-m)) \left(\prod_{s=1}^{m} \mathbf{Q}(t-m+s) - \mathbf{Q}^{*} \right), \tag{6.7}$$

from which we get that

$$\left\| \mathbf{Q}(\tau, t) - \mathbf{Q}^{*} \right\|_{\infty} \leq \left\| \prod_{s=1}^{m} \mathbf{Q}(t-m+s) - \mathbf{Q}^{*} \right\|_{\infty} \leq$$

$$\left\| \prod_{s=1}^{m} \mathbf{Q}(t-m+s) - \mathbf{Q}^{m} \right\|_{\infty} + \left\| \mathbf{Q}^{m} - \mathbf{Q}^{*} \right\|_{\infty}. \tag{6.8}$$

Let $\mathbf{A}_u = \mathbf{Q}(t-m+u)$. Then

$$\left\| \prod_{u=1}^{m} \mathbf{A}_u - \mathbf{Q}^{m} \right\|_{\infty} \leq \left\| \mathbf{A}_1 \left(\prod_{u=2}^{m} \mathbf{A}_u - \mathbf{Q}^{m-1} \right) \right\|_{\infty} + \left\| (\mathbf{A}_1 - \mathbf{Q}) \mathbf{Q}^{m-1} \right\|_{\infty},$$

then using the properties given in (6.1) we get

$$\left\| \prod_{u=1}^{m} \mathbf{A}_u - \mathbf{Q}^{m} \right\|_{\infty} \leq \left\| \left(\prod_{u=2}^{m} \mathbf{A}_u - \mathbf{Q}^{m-1} \right) \right\|_{\infty} + n \left\| (\mathbf{A}_1 - \mathbf{Q}) \right\|_{\infty}$$

$$\leq n \sum_{u=1}^{m} \left\| \mathbf{A}_u - \mathbf{Q} \right\|_{\infty}. \tag{6.9}$$

From (6.2) it is apparent that we could always choose a t_0 such that

$$\left\| \mathbf{Q}(t-m) - \mathbf{Q} \right\|_{\infty} \leq \epsilon / (2nm) \quad \text{for } t \geq t_0. \tag{6.10}$$

Then we easily see that

$$\left\| \mathbf{Q}(\tau, t) - \mathbf{Q}^{*} \right\|_{\infty} < \epsilon \text{ for } t \geq t_0, \ \tau \leq t-m. \tag{6.11}$$

Now, let us define from (5.8) the second part on the right hand side as

$$\mathbf{U}(t) = \sum_{\tau=1}^{t} \Delta T(\tau) \mathbf{p}_0(\tau) \mathbf{Q}(\tau, t). \tag{6.12}$$

Then

$$\left\| \mathbf{U}(t) - [T(t) - T(0)] \mathbf{p}_0 \mathbf{Q}^{*} \right\|_{\infty} = \tag{6.13}$$

$$\left\| \sum_{\tau=1}^{t} \Delta T(\tau) \mathbf{p}_0(\tau) \mathbf{Q}(\tau, t) - \sum_{\tau=1}^{t} \Delta T(\tau) \mathbf{p}_0 \mathbf{Q}^{*} \right\|_{\infty} \leq$$

$$\sum_{\tau=1}^{t} \Delta T(\tau) \left\| \mathbf{p}_0(\tau) [\mathbf{Q}(\tau, t) - \mathbf{Q}^{*}] + \mathbf{p}_0(\tau) \mathbf{Q}^{*} - \mathbf{p}_0 \mathbf{Q}^{*} \right\|_{\infty}$$

Since the stochastic matrix $\mathbf{Q}^*$is the limit of the powers of a regular stochastic matrix we know that it is a stable stochastic matrix, that is, a matrix with identical rows let say $\mathbf{q}^*$. It is easy then to see that

$$\mathbf{p}_0(\tau)\,\mathbf{Q}^* = \mathbf{p}_0\mathbf{Q}^* = \mathbf{q}^*. \tag{6.14}$$

Thus from (6.13) and (6.14) we get

$$\|\mathbf{U}(t) - [T(t) - T(0)]\,\mathbf{p}_0\mathbf{Q}^*\|_\infty \le \sum_{\tau=1}^{t} \Delta T(\tau)\,\|\mathbf{p}_0(\tau)\,[\mathbf{Q}(\tau,t) - \mathbf{Q}^*]\|_\infty \le$$

$$\sum_{\tau=1}^{t} \Delta T(\tau)\,\|\mathbf{Q}(\tau,t) - \mathbf{Q}^*\|_\infty . \tag{6.15}$$

From (6.11), (6.13) and (6.15) we get that for $t \ge t_0$

$$\|\mathbf{U}(t) - [T(t) - T(0)]\,\mathbf{p}_0\mathbf{Q}^*\|_\infty \le$$

$$\sum_{\tau=1}^{t-m} \Delta T(\tau)\,\|\mathbf{Q}(\tau,t) - \mathbf{Q}^*\|_\infty + \sum_{\tau=t-m+1}^{t} \Delta T(\tau)\,\|\mathbf{Q}(\tau,t) - \mathbf{Q}^*\|_\infty$$

$$\le [T(t-m) - T(0)] \in +2\,[T(t) - T(t-m)]\,. \tag{6.16}$$

Hence, since $\lim_{t\to\infty} T(t) = T$, we have that

$$\lim_{t\to\infty} \|\mathbf{U}(t) - [T(t) - T(0)]\,\mathbf{p}_0\mathbf{Q}^*\|_\infty = 0. \tag{6.17}$$

Now, we also have using (6.11)

$$\lim_{t\to\infty} \|\mathbf{N}(0)\,\mathbf{Q}(0,t) - T(0)\,\mathbf{q}^*\|_\infty = \lim_{t\to\infty} \left\|\mathbf{N}(0)\,\mathbf{Q}(0,t) - \mathbf{N}(0)\,\mathbf{1}^\top\mathbf{q}^*\right\|_\infty$$

$$\le \lim_{t\to\infty} \|\mathbf{N}(0)\|\,\|\mathbf{Q}(0,t) - \mathbf{Q}^*\|_\infty = 0. \tag{6.18}$$

Hence, from $(5.8), (6.17)$ and (6.18) we have that

$$\lim_{t\to\infty} \|\mathbb{E}[\mathbf{N}(t)] - T\mathbf{q}^*\|_\infty = 0. \tag{6.19}$$

From (6.19) we easily get that $\lim_{t\to\infty} \mathbb{E}[\mathbf{q}(t)] = \mathbf{q}^*$. ■

6.3 The relative expected population structure

It is of interest what happens if we replace the condition of $\lim_{t\to\infty} T(t) = T$ by the less restrictive

$$\lim_{t\to\infty} \frac{\Delta T(t)}{T(t)} = 0, \tag{6.20}$$

which is a general assumption with the physical interpretation that the relative growth rate of the population must vanish in the limit (Bartholomew (1982)). Such a condition is met in many real systems. It is important to note that this assumption actually allows $\lim_{t\to\infty} T(t) = T$.

We have seen in Chapter 4 that the rate of convergence for a strongly ergodic non-homogeneous Markov chain is an important problem and many research efforts have been devoted to it. In an NHMS, the study of asymptotic behavior is also important from the practical point of view, since although the limits may be approached too slowly, in some cases, to be attained in practice, they give valuable information about the inherent tendencies in the system (Bartholomew (1973). One of the most useful things to know is the "direction" in which the structure is changing. For example, the statement that an NHMS tends to become top heavy summarizes an important type of change. Thus, it will be very important to study under what conditions the relative expected population structure of an NHMS converges geometrically fast, since this is actually achieved in 5–6 time steps.

We will start the study of geometrical rate of convergence for an NHMS with the following Lemmas.

Lemma 292 *(Vassiliou and Tsaklidis (1989). For an NHMS with growth rate $T(t)$ the following statements are equivalent:*

(i) The sequence $\{\Delta T(t)\}_{t=0}^{\infty}$ converges to 0 with geometrical rate.

(ii) The sequence $\{T(t)\}_{t=0}^{\infty}$ converges ($\lim_{t\to\infty} T(t) = T$) with geometrical rate.

Proof. $(i) \to (ii)$ From (i), we get that there exist $c > 0$ and $0 < b < 1$ such that

$$|\Delta T(t)| \leq cb^t \text{ for } t = 0, 1, 2, ... \tag{6.21}$$

Then

$$\sum_{t=0}^{\infty} |\Delta T(t)| \leq c \sum_{t=0}^{\infty} b^t = c(1-b)^t, \tag{6.22}$$

therefore the series $\sum_{t=0}^{\infty} \Delta T(t)$ converges and hence the limit

$$\lim_{t\to\infty} T(t) = \sum_{t=0}^{\infty} \Delta T(t) + T(0) = T, \quad \text{exists.} \tag{6.23}$$

In addition

$$|T - T(t)| = \left| \sum_{t=t+1}^{\infty} \Delta T(t) \right| \leq \sum_{t=t+1}^{\infty} |\Delta T(t)| \leq c(1-b)^{-1} b^t, \tag{6.24}$$

hence, the rate of convergence of the sequence $\{T(t)\}_{t=0}^{\infty}$ is geometrical.

$(ii) \to (i)$ From (ii) we get that there are $c_1 > 0$ and $0 < b_1 < 1$ such that

$$\left| \sum_{t=t+1}^{\infty} \Delta T(t) \right| \leq c_1 b_1^t \text{ for } \ t = 0, 1, 2, ... \tag{6.25}$$

Then

$$\left| \Delta T(t+1) + \sum_{t=t+2}^{\infty} \Delta T(t) \right| \leq c_1 b_1^t, \text{ for } t = 0, 1, 2, ..., \tag{6.26}$$

from which we get

$$-c_1 b_1^t - \sum_{t=t+2}^{\infty} \Delta T(t) \leq \Delta T(t+1) \leq c_1 b_1^t - \sum_{t=t+2}^{\infty} \Delta T(t) \text{ for every } t. \tag{6.27}$$

Now, if $\sum_{t=t+2}^{\infty} \Delta T(t) \geq 0$ then from (6.27), we have

$$-\left[c_1 b_1^t + \sum_{t=t+2}^{\infty} \Delta T(t) \right] \leq \Delta T(t+1) \leq c_1 b_1^t + \sum_{t=t+2}^{\infty} \Delta T(t),$$

or

$$|\Delta T(t+1)| \leq c_1 (1 + b_1) b_1^t. \tag{6.28}$$

Similarly, we arrive at the same result if $\sum_{t=t+2}^{\infty} \Delta T(t) \leq 0$. ■

Lemma 293 *(Vassiliou and Georgiou (1990)). Consider an NHMS and suppose that the sequence*

$$\left\{\frac{\Delta T(t)}{T(t)}\right\}_{t=0}^{\infty},$$

converges to zero as $t \to \infty$ geometrically fast with $\Delta T(t) \geq 0$. Then $\{T(t)\}_{t=0}^{\infty}$ converges geometrically fast.

Proof. There exists $c > 0$ and $0 < b < 1$ such that

$$\left|\frac{\Delta T(t)}{T(t)}\right| \leq cb^t, \text{ for } t = 0, 1, 2, ..., \tag{6.29}$$

From (6.29) it is easy to see that

$$T(t+1) \leq T(0) \prod_{n=0}^{t} (1 + cb^n). \tag{6.30}$$

Now the sequence on the write hand side of (6.30) converges. In order to prove that just take its logarithm and then the series which will be the outcome is convergent. The rate of convergence in addition is the same with the rate of convergent of the sequence $1 + cb^t$ which converges with geometrical rate to 1.Hence the sequence $\{T(t)\}_{t=0}^{\infty}$ converges with geometrical rate. ■

Remark 294 *From the above Lemma it is evident that when $\left\{\frac{\Delta T(t)}{T(t)}\right\}_{t=0}^{\infty}$ converges geometrically fast then it is not possible that the growth of the NHMS will be infinite.*

We will now prove a basic theorem for the convergence of the expected relative population structure and its rate of convergence. In what follows we will use the $\|.\|_r$ norm defined in section 2.5 but for reasons of simplicity, we will omit the subscript r. Note that we could have used this norm in the proof of the previous theorem too, but we used instead the norm $\|.\|_\infty$ in order to exhibit some of its properties and its use.

Theorem 295 *(Vassiliou and Georgiou (1990)). Let an NHMS and let it's basic parameters be $\{\mathbf{P}(t)\}_{t=0}^{\infty}$, $\{T(t)\}_{t=0}^{\infty}$, $\{\mathbf{p}_{k+1}(t)\}_{t=0}^{\infty}$, $\{\mathbf{p}_0(t)\}_{t=0}^{\infty}$ and $\{\mathbf{Q}(t)\}_{t=0}^{\infty}$ where $\mathbf{Q}(t) = \mathbf{P}(t) + \mathbf{p}_{k+1}^\top(t)\,\mathbf{p}_0(t)$. Assume that*

$$\lim_{t\to\infty} \|\mathbf{P}(t) - \mathbf{P}\| = 0, \ \lim_{t\to\infty} \|\mathbf{p}_{k+1}(t) - \mathbf{p}_{k+1}\| = 0, \ \lim_{t\to\infty} \|\mathbf{p}_0(t) - \mathbf{p}_0\| = 0, \tag{6.31}$$

geometrically fast in all three cases and

$$\lim_{t\to\infty} \frac{\Delta T(t)}{T(t)} = 0 \text{ geometrically fast with } \Delta T(t) \geq 0 \tag{6.32}$$

$$\text{then } \lim_{t\to\infty} \|\mathbf{Q}(t) - \mathbf{Q}\| = 0, \tag{6.33}$$

and assume that $\mathbf{Q} = \mathbf{P} + \mathbf{p}_{k+1}^\top \mathbf{p}_0$ is a regular stochastic matrix. Then

$$\lim_{t\to\infty} \mathbb{E}[\mathbf{q}(t)] = \mathbf{q}^*, \text{ geometrically fast.}$$

where $\mathbf{q}^*$ is the row of the stable stochastic matrix $\mathbf{Q}^* = \lim_{t\to\infty} \mathbf{Q}^t$.

Proof. It is sufficient to show that there are constants $c > 0$ and $0 < b < 1$ such that

$$\|\mathbb{E}[\mathbf{q}(t)] - \mathbf{q}^*\| \leq cb^t \text{ for } t = 0, 1, 2, ... \tag{6.34}$$

From (5.12) we get that

$$\|\mathbb{E}\left[\mathbf{q}\left(t\right)\right]-\mathbf{q}^{*}\|=\|\mathbf{q}\left(0\right)\frac{T\left(0\right)}{T\left(t\right)}\mathbf{Q}\left(0,t\right)+$$

$$\frac{1}{T\left(t\right)}\sum_{\tau=1}^{t}\Delta T\left(\tau\right)\mathbf{p}_{0}\left(t\right)\mathbf{Q}\left(\tau,t\right)-\mathbf{q}^{*}\|\leq$$

$$\left\|\mathbf{q}\left(0\right)\frac{T\left(0\right)}{T\left(t\right)}\mathbf{Q}\left(0,t\right)-\mathbf{q}\left(0\right)\frac{T\left(0\right)}{T}\mathbf{Q}^{*}\right\|+$$

$$\left\|\frac{1}{T\left(t\right)}\sum_{\tau=1}^{t}\Delta T\left(\tau\right)\mathbf{p}_{0}\left(t\right)\mathbf{Q}\left(\tau,t\right)-\left[1-\frac{T\left(0\right)}{T}\right]\mathbf{p}_{0}\mathbf{Q}^{*}\right\| \tag{6.35}$$

From Theorem 246, (6.32) and since $\mathbf{Q}$ is a regular stochastic matrix, we have that the sequence of matrices $\{\mathbf{Q}\left(t\right)\}_{t=0}^{\infty}$ are such that there are constants $c_1>0$ and $0<b_1<1$ such that

$$\|\mathbf{Q}\left(t,t+n\right)-\mathbf{Q}^{*}\|\leq c_1 b_1^{n}\ \text{ for every } t,n=0,1,2,.... \tag{6.37}$$

From (6.32) and Lemma 292, we get that there are constants $c_2>0$ and $0<b_2<1$ such that

$$|T\left(t\right)-T|\leq c_2 b_2^{t}\ \text{ and }\ \left|\frac{1}{T\left(t\right)}-\frac{1}{T}\right|\leq c_2 b_2^{t}\text{ for } t=0,1,.... \tag{6.38}$$

We will take separately the first part of the last relation in (6.35),

$$A=\left\|\mathbf{q}\left(0\right)\frac{T\left(0\right)}{T\left(t\right)}\mathbf{Q}\left(0,t\right)-\mathbf{q}\left(0\right)\frac{T\left(0\right)}{T}\mathbf{Q}^{*}\right\|\leq$$

$$T\left(0\right)\left\|\frac{1}{T\left(t\right)}\mathbf{Q}\left(0,t\right)-\frac{1}{T}\mathbf{Q}^{*}\right\|\leq\frac{T\left(0\right)}{TT\left(t\right)}\|T\mathbf{Q}\left(0,t\right)-T\left(t\right)\mathbf{Q}^{*}\|$$

$$\leq\frac{T\left(0\right)}{T}\left(\frac{1}{T}+c_2 b_2^{t}\right)\|T\mathbf{Q}\left(0,t\right)-T\left(t\right)\mathbf{Q}^{*}\|\leq$$

$$T\left(0\right)\left(\frac{1}{T}+c_2 b_2^{t}\right)\|\mathbf{Q}\left(0,t\right)-\mathbf{Q}^{*}\|+$$

$$\frac{T\left(0\right)}{T}\left(\frac{1}{T}+c_2 b_2^{t}\right)|T\left(t\right)-T|\leq$$

$$T\left(0\right)\left(\frac{1}{T}+c_2 b_2^{t}\right)c_1 b_1^{t}+\frac{T\left(0\right)}{T}\left(\frac{1}{T}+c_2 b_2^{t}\right)c_2 b_2^{t}\leq$$

$$(\text{ we can always choose } b_1>b_2)\leq c_3 b_1^{t}. \tag{6.39}$$

Now, we will take separately the second part of the last relation in (6.35),

$$B=\left\|\frac{1}{T\left(t\right)}\sum_{\tau=1}^{t}\Delta T\left(\tau\right)\mathbf{p}_{0}\left(t\right)\mathbf{Q}\left(\tau,t\right)-\left[1-\frac{T\left(0\right)}{T}\right]\mathbf{p}_{0}\mathbf{Q}^{*}\right\|$$

$$\leq\frac{1}{T}\left(\frac{1}{T}+c_2 b_2^{t}\right)[T\left\|\sum_{\tau=1}^{t}\Delta T\left(\tau\right)\mathbf{p}_{0}\left(t\right)\mathbf{Q}\left(\tau,t\right)-\{T-T\left(0\right)\}\mathbf{p}_{0}\mathbf{Q}^{*}\right\|$$

$$+\{T-T\left(0\right)\}|T\left(t\right)-T|\,]. \tag{6.40}$$

We now have

$$\Gamma = \left\| \sum_{\tau=1}^{t} \Delta T(\tau)\, \mathbf{p}_0(t)\, \mathbf{Q}(\tau,t) - \{T - T(0)\}\, \mathbf{p}_0 \mathbf{Q}^* \right\| =$$

$$\left\| \sum_{\tau=1}^{t} \Delta T(\tau)\, \mathbf{p}_0(t)\, \mathbf{Q}(\tau,t) - \sum_{\tau=1}^{t} \Delta T(\tau)\, \mathbf{p}_0 \mathbf{Q}^* - \sum_{\tau=1}^{\infty} \Delta T(\tau)\, \mathbf{p}_0 \mathbf{Q}^* \right\|$$

$$\leq \sum_{\tau=1}^{t} \Delta T(\tau) \left\| \mathbf{p}_0(t)\, \mathbf{Q}(\tau,t) - \mathbf{q}^* \right\| + |T - T(t)|. \tag{6.41}$$

Now, from (6.40) and (6.41) we get that

$$B \leq \frac{1}{T}\left(\frac{1}{T} + c_2 b_2^t\right)\left[T \sum_{\tau=1}^{t} c_2 b_2^{\tau} c_1 b_1^{t-\tau} + c_2 b_2^t\right] + \{T - T(0)\}\, c_2 b_2^t$$

$$\left(\text{set } T c_2 c_1 = c_3, \text{ choose } b_1 > b_2,\ \{T - T(0)\}\, c_2 = c_4,\ \frac{1}{T} = c_5\right) \leq$$

$$c_5\left(c_5 + c_2 b_2^t\right) c_3 b_1^t \left(1 - (b_2/b_1)^t\right)\left(1 - (b_2/b_1)\right)^{-1} + c_4 c_2 b_2^t \leq c_6 b_1^t. \tag{6.42}$$

From (6.39), (6.40), (6.41), (6.42) we get that there are constants $c > 0$ and $0 < b < 1$ such that

$$\left\| \mathbb{E}\left[\mathbf{q}(t)\right] - \mathbf{q}^* \right\| \leq c b^t \quad \text{for} \quad t = 0, 1, 2, \ldots. \tag{6.43}$$

■

6.4 A contracting NHMS without forced wastage

In this section we will study the asymptotic behavior of an NHMS, whose size fluctuates, but decrease in size is allowed only through voluntary wastage of memberships. As we have discussed previously, Assumption 5.1 is a necessary and sufficient condition for that. Under this condition, we have established in subsection 5.3.2 the equations needed to find the expected population structure and the relative expected population structure. Having these at hand, we will study now, under what conditions there exists convergence as $t \to \infty$ that is, we will study the strong ergodicity problem for an NHMS which relies only to voluntary wastage to contract. We will provide the answer to this problem in the theorem that follows which is a combination in theorems in Vassiliou (1981a, 1982a) and a part is new.

Theorem 296 *Let a contracting NHMS and let that*

$$(a)\ \frac{T(t)}{T(t-1)} \geq \max\left[\sum_{j=1}^{k} p_{ij}(t) - p_{ij}(t)/p_{0j}(t)\right],$$

$$(b)\ \lim_{t\to\infty} \frac{T(t-1)}{T(t)} = \tau, \quad (c)\ \lim_{t\to\infty} \left\|\mathbf{P}(t) - \mathbf{P}\right\| = 0,$$

$$(d)\ \lim_{t\to\infty} \left\|\mathbf{p}_0(t) - \mathbf{p}_0\right\| = 0, \quad (e)\ \lim_{t\to\infty} \left\|\mathbf{p}_{k+1}(t) - \mathbf{p}_{k+1}\right\| = 0.$$

(i) then if the matrix $\mathbf{L} = \tau\mathbf{P} + (\mathbf{I} - \tau\mathbf{P})\,\mathbf{1}^{\top}\mathbf{p}_0$ *is a regular stochastic matrix, the expected limiting relative population structure exists and is given by*

$$\mathbb{E}\left[\mathbf{q}(\infty)\right] = \lim_{t\to\infty} \mathbb{E}\left[\mathbf{q}(0,t)\right] = \mathbf{q}(0)\,\mathbf{L}^* = \mathbf{l}^*,$$

where $\mathbf{L}^* = \lim_{t\to\infty} \mathbf{L}^t$ *and* $\mathbf{l}^*$ *is the row of the stable stochastic matrix* $\mathbf{L}^*$. *Also we get that the NHMS is strongly ergodic since*

$$\lim_{t\to\infty} \|\mathbb{E}[\mathbf{q}(s,t)] - \mathbf{l}^*\| = 0, \text{ for every } s.$$

(ii) Now, if (a), (b), (c), and (d) are true and

(2) *The matrix* $\mathbf{L} = \tau\mathbf{P} + (\mathbf{I} - \tau\mathbf{P})\mathbf{1}^\top\mathbf{p}_0$ *is an irreducible stochastic matrix,*

with one essential class of period d,

(3) *The matrices* $\{\mathbf{M}(t)\}_{t=0}^{\infty}$ *all have the same incidence matrix,*

and so does $\mathbf{L}$,

then the sequence of the expected relative population structure $\{\mathbb{E}[\mathbf{q}(0,t)]\}_{t=0}^{\infty}$ *splits into d subsequences with limits*

$$\mathbf{q}(0)\mathbf{A}\mathbf{L}^v \text{ for } v = 0, 1, ..., d-1,$$

where $\mathbf{A} = \lim_{n\to\infty} \mathbf{L}^{nd}$. *(iii) Assume now that*

$$\lim_{t\to\infty} \frac{1}{t}\sum_{k=1}^{t} \|\mathbf{M}(t) - \mathbf{L}\| = 0,$$

Then in addition, we have convergence in the Cesaro sense, that is,

$$\lim_{t\to\infty} \frac{1}{t}\sum_{s=1}^{t} \mathbb{E}[\mathbf{q}(0,s)] = \mathbf{q}(0)\frac{1}{d}\sum_{v=0}^{d-1} \mathbf{A}\mathbf{L}^v = \boldsymbol{\pi}^*,$$

were $\boldsymbol{\pi}^*$ *is the row of the stable stochastic matrix* $\boldsymbol{\Pi}^* = \frac{1}{d}\sum_{v=0}^{d-1} \mathbf{A}\mathbf{L}^v$.

Proof. *(i)*. In (5.33) we have established that

$$\mathbb{E}[\mathbf{q}(0,t)] = \mathbf{q}(0)\mathbf{M}(0,t), \tag{6.44}$$

where due to *(a)* the matrices $\mathbf{M}(t)$ for $t = 1, 2, ...$ are stochastic matrices. Now, due to *(b)*,*(c)* and *(d)* and relation (5.31), we get that

$$\lim_{t\to\infty} \mathbf{M}(t) = \mathbf{L} = \tau\mathbf{P} + (\mathbf{I} - \tau\mathbf{P})\mathbf{1}^\top\mathbf{p}_0, \tag{6.45}$$

an irreducible regular stochastic matrix. Note that since $\mathbf{M}(t)$ is finite the convergence in norm is equivalent with the elementwise convergence. Since the matrix $\mathbf{L}$ is regular and a finite stochastic matrix by assumption then due to Theorem 242, we get that

$$\lim_{t\to\infty} \|\mathbf{M}(0,t) - \mathbf{L}^*\| = 0, \tag{6.46}$$

where $\mathbf{L}^* = \lim_{t\to\infty} \mathbf{L}^t$ a stable stochastic matrix. From (6.44), (6.45) and $\mathbf{L}^*$ being a stable stochastic matrix we arrive at

$$\mathbb{E}[\mathbf{q}(\infty)] = \lim_{t\to\infty} \mathbb{E}[\mathbf{q}(0,t)] = \mathbf{l}^*. \tag{6.47}$$

(ii) Now the matrix $\mathbf{L}$ is an irreducible periodic stochastic matrix with period d. The matrices $\mathbf{M}(t)$ are finite and they all have the same incidence matrix with $\mathbf{L}$, that is,

$\bar{\mathbf{M}}(t) = \bar{\mathbf{L}}$ for every $t \geq 0$. Hence the conditions of Theorem 252 are satisfied and so we get that

$$\lim_{t\to\infty} \|\mathbf{M}(s, s+td+v) - \mathbf{A}\mathbf{L}^{v}\| = 0 \text{ uniformly in } s, \text{ for } v = 0, 1, ..., d-1. \tag{6.48}$$

were $\mathbf{A} = \lim_{n\to\infty} \mathbf{L}^{nd}$. Hence from (6.45) and (6.48), we immediately get that as $t \to \infty$ the sequence $\{\mathbb{E}[\mathbf{q}(0,t)]\}_{t=0}^{\infty}$ splits into d subsequences with limits

$$\mathbf{q}(0)\,\mathbf{A}\mathbf{L}^{v} \text{ for } v = 0, 1, ..., d-1.$$

Also, similarly we that $\{\mathbb{E}[\mathbf{q}(s,t)]\}_{t=0}^{\infty}$ splits into d subsequences with limits

$$\mathbf{q}(s)\,\mathbf{A}\mathbf{L}^{v} \text{ for } v = 0, 1, ..., d-1, \text{ uniformly in } s. \tag{6.49}$$

(iii) From the conditions of the theorem, we have that the conditions of Theorem 253 are valid hence we get that

$$\lim_{t\to\infty} \left\| \frac{1}{t} \sum_{k=1}^{t-1} \mathbf{M}(0,k) - \mathbf{\Pi}^{*} \right\| = 0. \tag{6.50}$$

We now have that

$$\left\| \frac{1}{t} \sum_{s=1}^{t} \mathbb{E}[\mathbf{q}(0,s)] - \boldsymbol{\pi}^{*} \right\| = \left\| \frac{1}{t} \sum_{s=1}^{t} \mathbf{q}(0)\,\mathbf{M}(0,s) - \mathbf{q}(0)\,\mathbf{\Pi}^{*} \right\|$$

$$\leq \left\| \frac{1}{t} \sum_{s=1}^{t} \mathbf{q}(0)\,\mathbf{M}(0,s) - \frac{1}{t} \sum_{s=1}^{t} \mathbf{q}(0)\,\mathbf{\Pi}^{*} \right\|. \tag{6.51}$$

or finally

$$\left\| \frac{1}{t} \sum_{s=1}^{t} \mathbb{E}[\mathbf{q}(0,s)] - \boldsymbol{\pi}^{*} \right\| \leq \frac{1}{t} \sum_{s=1}^{t} \|\mathbf{M}(0,s) - \mathbf{\Pi}^{*}\|. \tag{6.52}$$

then from (6.50) and (6.51) we arrive at the desired conclusion. In a similar way we can prove that ■

6.5 NHMS with non-homogeneous Poisson recruitment

An interesting problem in NHMS's through the years has been, what happens if the flow of new members in the system follows a non-homogeneous Poisson process. The roots of the problem has been the work of Pollard (1967) for discrete time homogeneous Markov populations (systems) and homogeneous Poisson distribution for the number of new members, followed by McClean (1976) for continuous time homogeneous Markov systems with homogeneous Poisson process as input flow, Mehlmann (1977) for homogeneous Markov systems with non-homogeneous Poisson process as input flow, McClean (1980) for semi-Markov homogeneous Markov systems, Vassiliou (1982b) for non-homogeneous Markov systems in discrete time but with a stationary probability distribution of the new entrants in the various states and with non-homogeneous Poisson process as input flow. On the Research Notes section of this chapter we will describe some later developments and some other works with other stochastic processes as input flow. In this section, we will provide a basic theorem for the asymptotic behavior of the expected population structure of an NHMS and also for the asymptotic distribution when the input flow is a non-homogeneous Poisson process. We will start with the following Lemma

Lemma 297 *(Feller (1968), p. 216). Let a complete probability space $(\Omega, \mathcal{F}, \mathbb{P})$. Let a population of elements the number of which T is a random variable which is Poisson distributed, that is, $T \sim \mathcal{P}(\lambda)$. Assume that the population has k categories to which the elements of the population are distributed multinomially. That is, they are distributed to the categories $1, 2, ..., k$ with probabilities $p_1, p_2, ..., p_k$. Let X_i be the random variable which represents the number of elements from the population T in category i. Then the random variables X_i for $i = 1, 2, ..., k$ are independent Poisson distributed random variables, that is, $X_i \sim \mathcal{P}(\lambda p_i)$ for $i = 1, 2, ..., k$.*

Proof. We have that

$$\mathbb{P}[X_1 = \tau_1, X_2 = \tau_2, ..., X_k = \tau_k, T = \tau] =$$

$$= (\text{where } \tau = \tau_1 + \tau_2 + ... + \tau_k) =$$

$$\mathbb{P}[X_1 = \tau_1, X_2 = \tau_2, ..., X_k = \tau_k \mid T = \tau] \mathbb{P}[T = \tau] =$$

$$\frac{\tau!}{\tau_1! \tau_2! ... \tau_k!} p_1^{\tau_1} p_2^{\tau_2} ... p_k^{\tau_k} e^{-\lambda} \frac{\lambda^\tau}{\tau!} =$$

$$e^{-\lambda \tau_1} \frac{(\lambda p_1)^{\tau_1}}{\tau_1!} \times e^{-\lambda \tau_2} \frac{(\lambda p_1)^{\tau_2}}{\tau_2!} \times ... \times e^{-\lambda \tau_k} \frac{(\lambda p_1)^{\tau_k}}{\tau_k!}, \tag{6.53}$$

which proves the Lemma. ■

Theorem 298 *Let a complete probability space $(\Omega, \mathcal{F}, \mathbb{P})$ and an NHMS and let it's basic parameters be $\{\mathbf{P}(t)\}_{t=0}^{\infty}$, $\{T(t)\}_{t=0}^{\infty}$, $\{\mathbf{p}_{k+1}(t)\}_{t=0}^{\infty}$, $\{\mathbf{p}_0(t)\}_{t=0}^{\infty}$ and $\{\mathbf{Q}(t)\}_{t=0}^{\infty}$. Let that $\{R(t)\}_{t=0}^{\infty}$ be the random process representing the new entrants into the system at time t. Assume that $\{R(t)\}_{t=0}^{\infty}$ is a non-homogeneous Poisson process with parameter $\lambda(t)$ and in addition*

$$(i) \quad \lim_{t\to\infty} \|\mathbf{P}(t) - \mathbf{P}\| = 0, \quad \lim_{t\to\infty} \|\mathbf{p}_0(t) - \mathbf{p}_0\| = 0, \ \lim_{t\to\infty} \lambda(t) = \lambda, \tag{6.54}$$

$$(ii) \quad \sup_t \|\mathbf{P}(t)\| = a \text{ and } \sup_t [\lambda(t) a] = b < 1. \tag{6.55}$$

Then asymptotically the expected population structure of members converges to

$$\mathbb{E}[\mathbf{N}(\infty)] = \lim_{t\to\infty} \mathbb{E}[\mathbf{N}(t)] = \lambda \mathbf{p}_0 (\mathbf{I} - \mathbf{P})^{-1}. \tag{6.56}$$

Also the number of members in the various states are asymptotically mutually independent Poisson variates.

Proof. From relation (5.5), we get that the expected population structure of members is given by

$$\mathbb{E}[\mathbf{N}(t)] = \mathbb{E}[\mathbf{N}(t-1)] \mathbf{P}(t) + \mathbb{E}[R(t)] \mathbf{p}_0(t). \tag{6.57}$$

From (6.57) since the stochastic process $\{R(t)\}_{t=0}^{\infty}$ is non-homogeneous Poisson with parameter $\lambda(t)$ we get that

$$\mathbb{E}[\mathbf{N}(t)] = \mathbb{E}[\mathbf{N}(t-1)] \mathbf{P}(t) + \lambda(t) \mathbf{p}_0(t) =$$

$$\mathbf{N}(0) \mathbf{P}(0,t) + \sum_{n=0}^{t-1} \lambda(t-n) \mathbf{p}_0(t-n) \mathbf{P}(t-n,t). \tag{6.58}$$

Now the matrices $\mathbf{P}(t)$ for every t are substochastic and so $a < 1$ and

$$\lim_{t\to\infty} \mathbf{P}(0,t) = \mathbf{0}. \tag{6.59}$$

Also

$$\|\Sigma(\infty)\| = \lim_{t\to\infty} \left\| \sum_{n=0}^{t-1} \lambda(t-n)\,\mathbf{p}_0(t-n)\,\mathbf{P}(t-n,t) \right\| \leq$$

$$\lim_{t\to\infty} \sum_{n=0}^{t-1} \lambda(t-n)\,\|\mathbf{p}_0(t-n)\|\,\|\mathbf{P}(t-n,t)\| \leq (\text{from } (ii)) \leq$$

$$\sum_{n=0}^{\infty} \lambda(t-n)\,a^n \leq \sum_{n=0}^{\infty} b^n = (1-b)^{-1}, \tag{6.60}$$

hence. $\|\Sigma(\infty)\|$ is bounded and since it is also a monotonically increasing function it converges. From (6.59) and the last conclusion from (6.58) we get that $\lim_{t\to\infty}\mathbb{E}[\mathbf{N}(t)] = \mathbb{E}[\mathbf{N}(\infty)]$ exists therefore from (6.60) we get

$$\mathbb{E}[\mathbf{N}(\infty)] = \mathbf{p}_0(\mathbf{I}-\lambda\mathbf{P})^{-1}. \tag{6.61}$$

We will now show that asymptotically we need only find expectations since the number of members in each state are asymptotically independent Poisson variates. Firstly observe that from (6.59), we get that $\mathbf{N}(0)\,\mathbf{P}(0,t) \to 0$ as $t \to \infty$. Therefore, the initial members of the population asymptotically vanish from the system. From Lemma 297, we get that, for large t the new members in the system at each point in time are distributed in the k states, such that the number of members in each state are independent Poisson variates. From each state now at a large time t given the members being Poisson variates they are at time $t+1$ again multinomially distributed to the k grades. The $\{R(t)\}$ are mutually independent and the total number of members recruited in each state in a large time interval will be the sum of independent Poisson variates and hence a Poisson variate.

Attentively there are different proofs for the last part in Vassiliou (1982a) and Esquirel (2014). ■

6.6 Asymptotic stability in NHMS's

The concept of stability is of central importance in theoretical and applied study of populations especially the human ones. An NHMS is said to be stable (see Feichtinger (1976), Conlisk (1976) and Vassiliou (1981b)) if for its sequence of expected population structure $\{\mathbb{E}[\mathbf{N}(t)]\}_{t=0}^{\infty}$, we have $\mathbb{E}[\mathbf{N}(t)] = x\mathbb{E}[\mathbf{N}(t-1)]$. The positive scalar x is called the growth factor per unit time of the NHMS. We will now provide the definition of an asymptotically stable NHMS.

Definition 299 *A non-homogeneous Markov system for which*

$$\lim_{t\to\infty}\|\mathbf{P}(t)-\mathbf{P}\| = 0 \quad and \quad \lim_{t\to\infty}\|\mathbf{p}_0(t)-\mathbf{p}_0\| = 0$$

is said to be asymptotically stable if for any initial population structure $\mathbf{N}(0)$

$$\lim_{t\to\infty}\left\{\mathbb{E}[\mathbf{N}(t)]/x^t\right\} = C\mathbf{p}_0(x\mathbf{I}-\mathbf{P})^{-1},$$

for some constant $C>0$ *and some* $x > a = \sup_t \|\mathbf{P}(t)\|$ *which is called the asymptotically stable growth factor.*

We will now prove a theorem from which it will be apparent the motivation to define asymptotic stability. We will start with a lemma from Nikaido (1968):

Lemma 300 *Let* $\mathbf{P}$ *be a non-negative square matrix, and let* a *be the largest eigen value of* $\mathbf{P}$. *The matrix* $x\mathbf{I}-\mathbf{P}$ *is non-negatively invertible if and only if* $x > a$.

Theorem 301 *(Vassiliou (1981b)). Let a complete probability space* $(\Omega, \mathcal{F}, \mathbb{P})$ *and an NHMS and let it's basic parameters be* $\{\mathbf{P}(t)\}_{t=0}^{\infty}$, $\{T(t)\}_{t=0}^{\infty}$, $\{\mathbf{p}_{k+1}(t)\}_{t=0}^{\infty}$, $\{\mathbf{p}_0(t)\}_{t=0}^{\infty}$ *and* $\{\mathbf{Q}(t)\}_{t=0}^{\infty}$. *Assume that*

$$\lim_{t\to\infty} \|\mathbf{P}(t) - \mathbf{P}\| = 0, \quad \lim_{t\to\infty} \|\mathbf{p}_0(t) - \mathbf{p}_0\| = 0, \quad \sup_t \|\mathbf{p}_0(t)\| = b$$

$$\text{and } \sup_t \|\mathbf{P}(t)\| = a\ .$$

If the expected input process is exponential, that is, $\mathbb{E}[R(t)] = Rx^t$ *with* $x > a$ *and* $R > 0$ *then*

$$\lim_{t\to\infty} \left\{\mathbb{E}[\mathbf{N}(t)]/x^t\right\} = R\mathbf{p}_0 (x\mathbf{I} - \mathbf{P})^{-1}.$$

Proof. From 6.57 working recursively we get

$$\mathbb{E}[\mathbf{N}(t)] = \mathbf{N}(0)\mathbf{P}(0,t) + \sum_{n=0}^{t-1} R(t-n)\mathbf{p}_0(t-n)\mathbf{P}(t-n,t). \tag{6.62}$$

Since the matrices $\mathbf{P}(t)$ for every t are in general substochastic we get that

$$\lim_{t\to\infty} \mathbf{N}(0)\mathbf{P}(0,t) = 0. \tag{6.63}$$

We now have from (6.62) and (6.63)

$$\lim_{t\to\infty} \left\{\mathbb{E}[\mathbf{N}(t)]/x^t\right\} = \left\|\sum_{n=0}^{t-1} \frac{R(t-n)}{x^t}\mathbf{p}_0(t-n)\mathbf{P}(t-n,t)\right\| \le$$

$$\sum_{n=0}^{t-1} \frac{Rx^{t-n}}{x^t} \|\mathbf{p}_0(t-n)\| \|\mathbf{P}(t-n,t)\| \le Rb\sum_{n=0}^{t-1}\left(\frac{a}{x}\right)^n \le$$

$$(\text{since } x > a) \le R(1-a/x)^{-1}, \tag{6.64}$$

so it is bounded, and also monotonically increasing function and thus it converges. Consequently from (6.62), we get

$$\lim_{t\to\infty} \left\{\mathbb{E}[\mathbf{N}(t)]/x^t\right\} = 1/x \lim_{t\to\infty} \left\{\mathbb{E}[\mathbf{N}(t-1)]/x^{t-1}\right\}\mathbf{P} + R/x \lim_{t\to\infty} \mathbf{p}_0(t),$$

from which since $x > a$ we know that by Lemma 300 $(x\mathbf{I} - \mathbf{P})$ is non-negatively invertible, hence we get the desired result. ∎

We now provide the following result which characterizes asymptotic stability in non-homogeneous Markov systems.

Theorem 302 *(Vassiliou (1981b)). Let a complete probability space* $(\Omega, \mathcal{F}, \mathbb{P})$ *and an NHMS and let it's basic parameters be* $\{\mathbf{P}(t)\}_{t=0}^{\infty}$, $\{T(t)\}_{t=0}^{\infty}$, $\{\mathbf{p}_{k+1}(t)\}_{t=0}^{\infty}$, $\{\mathbf{p}_0(t)\}_{t=0}^{\infty}$ *and* $\{\mathbf{Q}(t)\}_{t=0}^{\infty}$. *Assume that*

$$\lim_{t\to\infty} \|\mathbf{P}(t) - \mathbf{P}\| = 0, \quad \lim_{t\to\infty} \|\mathbf{p}_0(t) - \mathbf{p}_0\| = 0, \quad \sup_t \|\mathbf{p}_0(t)\| = b$$

$$\text{and } \sup_t \|\mathbf{P}(t)\| = a\ .$$

Then the system is asymptotically stable with growth factor x *if and only if the input sequence is asymptotically exponential, i.e.,* $\lim_{t\to\infty}\{R(t)/x^t\} = R$ *where* $x > a$.

Proof. (*i*) We will first prove that if the input flow is asymptotically exponential then the NHMS is asymptotically stable. That is if

$$\lim_{t\to\infty}\left\{R(t)/x^t\right\} = R \text{ where } x > a, \tag{6.65}$$

then

$$\lim_{t\to\infty}\left\{\mathbf{N}(t)/x^t\right\} = C\mathbf{p}_0(x\mathbf{I}-\mathbf{P})^{-1}. \tag{6.66}$$

Due to (6.63) and from the proof of the previews theorem, we get

$$\lim_{t\to\infty}\left\{\mathbb{E}\left[\mathbf{N}(t)\right]/x^t\right\} = \lim_{t\to\infty}\sum_{n=0}^{t-1}\frac{R(t-n)}{x^t}\mathbf{p}_0(t-n)\mathbf{P}(t-n,t) =$$

$$R\mathbf{p}_0(x\mathbf{I}-\mathbf{P})^{-1}.$$

(*ii*) We will now prove that if the NHMS is asymptotically stable then the input flow is asymptotically exponential. That is,

$$\lim_{t\to\infty}\left\{\mathbf{N}(t)/x^t\right\} = C\mathbf{p}_0(x\mathbf{I}-\mathbf{P})^{-1} \text{ then } \lim_{t\to\infty}\left\{R(t)/x^t\right\} = R. \tag{6.67}$$

From (6.53) and asymptotic stability we get

$$\lim_{t\to\infty}\left\{R(t)\,\mathbf{p}_0(t)/x^t-1\right\} = \lim_{t\to\infty}\left\{x\frac{\mathbf{N}(t)}{x^t}-\frac{\mathbf{N}(t-1)}{x^{t-1}}\mathbf{P}(t)\right\}$$

$$= C\mathbf{p}_0(x\mathbf{I}-\mathbf{P})^{-1}((x\mathbf{I}-\mathbf{P})) = C\mathbf{p}_0. \tag{6.68}$$

Postmultiplication with the vector $\mathbf{1}^{\top}$ gives

$$\lim_{t\to\infty}\left\{R(t)/x^t\right\} = C.$$

■

6.7 NHMS as a martingale with Poisson input

In this section we will again take one more step up in the scale, by continuing our theory for NHMS with the help of measure theory. In fact, this section acts somehow as a link between the theory of NHMS's and the theory of martingales. A short but sufficient introduction on martingales could be found in Vassiliou (2010, Chapter 5). However, we will start at this point with some useful definitions and insights about martingales.

The roots of the theory of martingales go back to the study of gambling. Actually the term martingale comes from an old strategy in gambling from around 1815, where when playing a game of luck, each time the player lost his bet, he doubled, the money he just lost until he won. Games of luck have been very often the motive for the first steps in probability theory and this is a very well known fact from the correspondence in this respect between Pascal and Fermat around 1654. The term martingale for this class of stochastic process was due to Ville (1939). The word martingale outside the mathematics environment is a part of a horse's harness, which goes into its mouth and stops the horse from throwing its head up. Martingales have been extensively studied since 1934 by Paul Levy (1886–1971).

Also Doob[1](1910–2002) made very lasting contributions from 1940 to the theory of martingales and a classic book on the subject is Doob (1990). Other very good texts are Williams (2000), Ross (1996) and Williams (2004). If one wishes to do research on the theory of martingales on a very high mathematical level, the two volumes by Rogers and Williams (2000) are indispensable.

We will now provide a definition of a martingale.

Definition 303 *Let a complete filtered probability space* $(\Omega, \mathcal{F}, \mathcal{F}_t, \mathbb{P})$ *and a stochastic process* $\{X_t\}_{t=0}^{\infty}$ *in discrete time. Then* $\{X_t\}_{t=0}^{\infty}$ *is a martingale with respect to the filtration* $\mathcal{F}_t$ *if the following conditions are satisfied*

$$(i)\ \mathbb{E}\left[|X_t|\right] < \infty, \quad \text{for all } t \in \mathbb{N},$$

$$(ii)\ X_t \text{ is } \mathcal{F}_t\text{-measurable},$$

$$(iii)\ \mathbb{E}\left[X_{t+1} \mid \mathcal{F}_t\right] = X_t.$$

It is important to remark that in practice and theory proving that a stochastic process is a martingale is not particularly difficult. The real challenge is to intuitively find which process is a martingale. Let us at this point provide a well known example of a martingale:

Example 304 *The Doob martingale: Consider a complete filtered probability space* $(\Omega, \mathcal{F}, \mathcal{F}_t, \mathbb{P})$ *and a random variable* X *for which* $\mathbb{E}\left[|X|\right] < \infty$*. Then the stochastic process defined as*

$$X_t = \mathbb{E}\left[X \mid \mathcal{F}_t\right],$$

is a martingale in relation with the filtration $\{\mathcal{F}_t\}_{t=0}^{\infty}$*. The martingale* $\{X_t\}_{t=0}^{\infty}$ *is called a Doob martingale. In order to show that it is a martingale we must prove that it satisfies the three conditions of the definition of a martingale.. We already have condition* $\mathbb{E}\left[|X|\right] < \infty$ *as an assumption. It is known that for any random variable* Y

$$|\mathbb{E}(Z)| \leq \mathbb{E}(|Z|).$$

---

[1]Joseph Doob registered at Harvard in 1926 to study physics but by the end of the first year he had already taken by mathematics. After a few weeks as a sophomore he complaint to his advisor that the course was going too slow. In 1930 Doob was awarded his degree and started working for his Ph.D. degree with J.L. Walsh. Doob wrote:

Walsh accepted me and we had a fine relationship; he never bothered me, and conversely.

He obtained a Master's degree in 1931, the same year he married Elsie Haviland Field and had three children. He was awarded his doctorate in 1932 for a thesis entitled *Boundary Values of Analytic Functions.* Doob was a looner in research and he had discussions mainly with his friend Wladimir Seidel who was a postgraduade assistant at Harvard at the time. Later in life wrote that he had missed firtile contact with Birkhoff, Kellogg and Morse all three at Harvard and leaders in their fields.

After completing his doctorate, Doob was forced by the Great Depression to turn to probability theory to survive while continuing being a researcher in mathematics. He was appointed as Associate Professor at the University of Illinois which he never left despite the many offers. Doob had famous Ph.D. students such as Paul Halmos, David Blackwell, Donald Blackwell, Yuan-Shih Chow, M. Thompson and others. Doob was influenced by the axiomatic foundation of probability by Andrey Kolmogorov. Doob's approach in probability, as a branch of measure theory for him, was evident in his first probability paper, in which he proved theorems related to the law of large numbers, using a probabilistic interpretation on Birkhoff's ergodic theorem. Doob wrote a series of papers including martingales, Markov processes, and stationary processes. Then he wrote a book *Stochastic Processes* in 1953 which is now in the series of classic books by John Wiley. Doob's also studied the relations between probability and potential theory. After his early retirement in 1978 he continued to be very active writting books and research papers. A break appeared due to the death of his wife when he stated that mathematics was over for him. He bounced back and among others wrote another classic text, *Classical Potential Theory and its Probabilistic Counterpart* first published in 1984 which was 800 pages. He recieved several prestigious awards for his outstanding achievments in particular the Steele Prize of the American Mathematical Society in 1984. The Steel Prize was awarded for his fundamental work in establishing probability as a branch of mathematics and his continuing profound influence on its development.

Therefore, we have

$$\mathbb{E}\left[|X_t|\right] = \mathbb{E}\left\{|\mathbb{E}\left[X \mid \mathcal{F}_t\right]|\right\} \leq \mathbb{E}\left\{\mathbb{E}\left[|X| \mid \mathcal{F}_t\right]\right\} = \mathbb{E}\left[|X|\right] < \infty.$$

We now move to prove the following key condition

$$\mathbb{E}\left[X_{t+1} \mid \mathcal{F}_t\right] = \mathbb{E}\left\{\mathbb{E}\left[X \mid \mathcal{F}_{t+1}\right] \mid \mathcal{F}_t\right\} =$$
$$= (\textit{using the tower property}) = \mathbb{E}\left[X \mid \mathcal{F}_t\right] = X_t.$$

The third condition is apparently satisfied, hence $\{X_t\}_{t=0}^{\infty}$ *is a martingale.*

Now, define the *asymptotically compensated population structure for an NHMS to be for* $s \leq t$,

$$\mathbf{M}\left(t\right) = \mathbf{N}\left(t\right) - \lambda t \mathbf{p}_0 \left(\mathbf{I} - \mathbf{P}\right)^{-1},$$

where we assume that the input process for the NHMS is a Poisson non-homogeneous process with parameter $\lambda\left(t\right)$ which asymptotically tends to λ in a finite number of steps. We will now prove the following theorem which links NHMS's with the theory of martingales.

Theorem 305 *(Vassiliou (2013c)). Let a complete filtered probability space* $(\Omega, \mathcal{G}, \mathcal{G}_t, \mathbb{P})$ *and a* $\mathcal{G}$*-NHMS in it. Assume instead that* $\mathcal{R}\left(t\right)$ *the number of new memberships entering the system at the end of the interval* $(t-1, t]$*, is* $\mathcal{G}_t$*-measurable and that for* $t = 1, 2, ...$ *is a non-homogeneous Poisson process with intensity* $\lambda\left(t\right)$*. Assume that the following sequences converge*

$$\lim_{t\to\infty} \left\|\mathbf{P}\left(t\right) - \mathbf{P}\right\| = 0;\ \lim_{t\to\infty} \left\|\mathbf{p}_0\left(t\right) - \mathbf{p}_0\right\| = 0 \text{ and } \lim_{t\to\infty} \left|\lambda\left(t\right) - \lambda\right| = 0, \tag{6.69}$$

and in addition assume that convergence take place in a finite time and let that

$$\sup_t \lambda\left(t\right) = \lambda^{*};\ \sup_t \left\|\mathbf{p}_0\left(t\right)\right\| = b;\ \sup_t \left\|\mathbf{P}\left(t\right)\right\| = a. \tag{6.70}$$

Then the asymptotically compensated population structure for $t \geq t^{*}$*, where* t^{*} *is the minimum time at which all sequences* (6.69) *have converged,* $\mathbf{M}\left(t\right)$ *is a multidimensional martingale in respect to the filtration* $\mathcal{G}_t$*.*

Proof. In the previous chapter, we have shown that, the number of members in the various grades of an NHMS under the assumptions of the present theorem are mutually independent Poisson variates. Therefore, asymptotically the number of members in each state of a $\mathcal{G}$-NHMS are Poisson processes with intensity $\lambda \mathbf{p}_0 \left(\mathbf{I} - \mathbf{P}\right)^{-1}$. Then so far we have proved that for any $t \geq t^{*}$, we have that

$$N_i\left(t\right) \sim \text{Poisson with intensity } \lambda \mathbf{p}_0 \left(\mathbf{I} - \mathbf{P}\right)^{-1} \mathbf{e}_i^{\top} \text{ independent}, \tag{6.71}$$

for $i = 1, 2, ..., k$, where $\mathbf{e}_i$ is a $1 \times k$ vector with 1 in the i-th position and zeros elsewhere. Therefore, the population structure $\mathbf{N}\left(t\right)$, for $t \geq t^{*}$ consists of k independent Poisson processes with intensity $\lambda \mathbf{p}_0 \left(\mathbf{I} - \mathbf{P}\right)^{-1}$. Due to the above conclusion we have consequently that, for $t^{*} \leq s \leq t$

$$\mathbb{E}_{\mathbb{P}}\left[\mathbf{N}\left(t\right) - \mathbf{N}\left(s\right)\right] = \lambda\left(t - s\right) \mathbf{p}_0 \left(\mathbf{I} - \mathbf{P}\right)^{-1}. \tag{6.72}$$

Let $t^{*} \leq s \leq t$ then

$$\begin{aligned}
\mathbb{E}_{\mathbb{P}}\left[\mathbf{M}\left(t\right) \mid \mathcal{G}_s\right] &= \mathbb{E}_{\mathbb{P}}\left[\mathbf{M}\left(t\right) - \mathbf{M}\left(s\right) \mid \mathcal{G}_s\right] + \mathbb{E}_{\mathbb{P}}\left[\mathbf{M}\left(s\right) \mid \mathcal{G}_s\right] \\
&= \mathbb{E}_{\mathbb{P}}\left[\mathbf{N}\left(t\right) - \mathbf{N}\left(s\right) - \lambda\left(t-s\right)\mathbf{p}_0\left(\mathbf{I} - \mathbf{P}\right)^{-1} \mid \mathcal{G}_s\right] + \mathbf{M}\left(s\right) \\
&= (\text{since } \mathbf{N}\left(t\right) \sim \text{Poisson, } \mathbf{N}\left(t\right) - \mathbf{N}\left(s\right) \text{ is independent of } \mathcal{G}_s) \\
&= \mathbb{E}_{\mathbb{P}}\left[\mathbf{N}\left(t\right) - \mathbf{N}\left(s\right)\right] - \lambda\left(t-s\right)\mathbf{p}_0\left(\mathbf{I} - \mathbf{P}\right)^{-1} + \mathbf{M}\left(s\right) = \mathbf{M}\left(s\right).
\end{aligned} \tag{6.73}$$

■

6.8 Research notes

A classical problem in many types of Human populations is the age distribution of their members. The asymptotic behavior of the age distribution in Manpower Systems were studied in Vassiliou (1981c). In this study the states of the NHMS are taken to be the age classes within the system and the time step to be compatible with the age states. The transition probability matrices now have special forms but the general theory presented in this chapter does apply. The results obtained provide the asymptotic behavior of the age distribution within the manpower system and the results are quite general to include as corollaries other attempts to solve the problem as for example Keenay et al. (1975). Using the methods of the present chapter, it is straight forward to find the conditions under which the convergence of the age distribution is geometric and also to find the relative variability vector and its asymptotic behavior.

The concept of entropy has found a large number of applications in a variety of fields. Entropy has been used as a measure in statistical mechanics, thermodynamics, communication theory, mathematical statistics and population studies. In manpower systems, entropy was introduced by McClean and Abodunde (1978b). It is a well known result that, for a discrete finite sample space $\Omega = \{\omega_1, \omega_2, ..., \omega_k\}$ and probabilities $P_i = \mathbb{P}(\omega_i)$ for $i = 1, 2, ..., k$ the entropy is given by

$$H = -c\sum_{i=1}^{k} P_i \log P_i,$$

where c is a constant. This is a measure of the degree of randomness in the system. In a human population, the length of service profile is defined as the proportion of staff in each length of service class, that is, the state which consists of all members with the same amount of time in the population. Apparently, this is an NHMS with states the length of service classes and we may take the time step to be compatible with the time step of the length of service. Using then the terminology of the NHMS theory we get that the relative length of service profile is $\mathbf{q}(t) = \mathbf{N}(t)/T(t)$. An NHMS adapted to predict the length of service distribution has special forms of transition probability matrices that were provided in Vassiliou (1981c). Also the relative limiting length of service distribution is found from which it is possible to find the limiting inertia of the length of service system which is given by

$$H(\infty) = -(1/\log k)\sum_{i=1}^{k} q_i(\infty)\log q_i(\infty).$$

It is not difficult to see that $0 \leq H(\infty) \leq 1$. When $q_1(\infty) = q_2(\infty) = ... = q_k(\infty) = 1/k$, then $H(\infty) = 1$ and represents the case were there is equal distribution of experience in the system. When $q_1(\infty) = 1$, $q_2(\infty) = ... = q_k(\infty) = 0$, then $H(\infty) = 0$ and represents the situation that all members of the population have minimum experience on the processes within the system and leave after one step of time in the population. For intermediate values, the entropy measures the degree of experience which will be present in the steady state or the direction into which the population of members is heading. Using the methods of the present chapter, it is straight forward to find the conditions under which the convergence is geometric and also to find the relative variability vector and its asymptotic behavior.

An interesting problem in some populations is to retain a desired order structure in the expected population structure. For example, in hierarchical organizations it is desired for apparent reasons to have a pyramid structure and moreover to retain it for a long time, that is, to retain it asymptotically. For closed homogeneous Markov systems Kipouridis and Tsaklidis (2001) provided a theorem where under the conditions of $\mathbf{P}$ regular and given that the limiting population structure has an acceptable size order then they prove that there exist a time t_0 such that for $t \geq t_0$ the columns of the matrix $\mathbf{P}^t$ exhibit the same size order. They also provide a method for evaluating the crucial time t_0 as a function of the eigenvalues of the matrix **P**. Finally, they extend their results for the case were **P** is periodic.

The evolution of an HMS when we restrict the expected population structure only in integer values for its elements is studied by Guerry (1997, 2008).

An interesting application of an NHMS to manpower planning was given in Dimitriou and Tsantas (2010). The inherent non-homogeneous Markov chain has a special form and richness in order to accomodate the possible real application. The authors go on to study the limiting behavior of the expected population structure under various conditions. Dimitriou et al. (2015) study the asymptotic behavior of the expected population structure for a Multivariate homogeneous Markov system. That is, a homogeneous Markov system with a finite number of various department with internal transitions and transitions among the departments. The various departments have known total number of members and there are out flows of members from the population to the external environment, and inflows of new members.

Chapter 7

ASYMPTOTIC VARIABILITY OF NHMS

7.1 Introductory notes

In Chapter 6, we have seen under what conditions the expected population structure asymptotically converges to a stationary expected population structure. We also have studied the problem of convergence of the relative expected population structure. The next important logical question is since we have convergence after a large time of the expectations of the number of memberships in the various states, is it intuitively expected that the variances and covariances would also be possible under certain conditions to present analogous behavior? The answer to this important problem will be the theme of the present chapter.

In Chapter 5 and equation (5.24), we established the difference equation from which it is possible to find the $\lim_{t\to\infty}\mu(t)$ of the vector of expected values, variances and covariances of the population of memberships in the various states of the system. We call this the problem of asymptotic variability for an NHMS. In section 2 and Theorem 313, we will establish the existence of this limit under certain conditions when the sequences $\{\mathbf{Q}(t)\}_{t=0}^{\infty}$ of the inherent Markov chain of an NHMS consists of general stochastic matrices.

7.2 The V matrices and their properties

A vital part in the study of $\lim_{t\to\infty}\mu(t)$ as it is apparent from equation (5.24) plays the matrix $\mathbf{V}_q(t)$, see relation (5.25), which is of the form

$$\mathbf{V}_q(t)=\begin{bmatrix}\mathbf{Q}(t) & \mathbf{Q}(t)\mathbf{J}-\mathbf{J}(\mathbf{Q}(t)\otimes\mathbf{Q}(t))\\ \mathbf{0} & \mathbf{Q}(t)\otimes\mathbf{Q}(t)\end{bmatrix}. \tag{7.1}$$

This is a special type of a class of matrices with also negative elements the general form of which is called a $\mathbf{V}$-matrix and they were firstly defined and studied in Tsaklidis and Vassiliou (1988) and given in the following definition:

Definition 306 *A matrix $\mathbf{V}\in M_n(\mathbb{R})$ is called a $\mathbf{V}$-matrix if it can take the form*

$$\mathbf{V}=\begin{bmatrix}\mathbf{Q} & \mathbf{X}\\ \mathbf{0} & \mathbf{Y}\end{bmatrix}, \tag{7.2}$$

where $\mathbf{Q}\in SM_{k_1}$ is a $k_1\times k_1$ stochastic matrix, $\mathbf{X}$ is a $k_1\times k_2$ matrix with real elements with the property that $\mathbf{X}\mathbf{1}^\top=0$, that is, its row sums are zero implying that there are negative elements in it, and $\mathbf{Y}\in SM_{k_2}$ is a $k_2\times k_2$ stochastic matrix. Hence, apparently a $\mathbf{V}$-matrix has its row sums equal to one but it is not a stochastic matrix since it is not a non-negative matrix.

Infinite powers of a $\mathbf{V}$-matrix have been studied in Tsaklidis and Vassiliou (1992), while infinite non-homogeneous products of $\mathbf{V}$-matrices have been studied in Tsaklidis and Vassiliou (1989, 1992b). While there is a huge literature for infinite powers and infinite products of non-negative matrices there are only a few about real matrices with negative elements.

DOI: 10.1201/b23204-7

More specifically Tsaklidis and Vassiliou (1989, 1992b) and Daubechies and Lagarias (1992) considered products of matrices with row sums one but not necessarily having non-negative elements. In the environment of numerical analysis later on Tarazaga et al. (2001) studied symmetric matrices with some small negative elements and followed by Noutsos (2006) for more general matrices which possess a Perron-Frobenious eigenpair. The reason might be its difficulty or the somehow different background needed for that. For more details see the section Reasearch Notes of the present chapter. We will now prove some Propositions which will be very helpful in the proofs of theorems that follow:

Firstly it would be helpful to provide two examples of the $\mathbf{J}$ matrix. For $k = 2$ then

$$\mathbf{J} = \begin{pmatrix} 1 & 0 & 0 & 0 \\ 0 & 0 & 0 & 1 \end{pmatrix},$$

and for $k = 3$

$$\mathbf{J} = \begin{pmatrix} 1 & 0 & 0 & 0 & 0 & 0 & 0 & 0 & 0 \\ 0 & 0 & 0 & 0 & 1 & 0 & 0 & 0 & 0 \\ 0 & 0 & 0 & 0 & 0 & 0 & 0 & 0 & 1 \end{pmatrix}.$$

Proposition 307 *(Vassiliou (2015)). For the matrix*

$$\mathbf{X}(t) = \mathbf{Q}(t)\mathbf{J} - \mathbf{J}(\mathbf{Q}(t) \otimes \mathbf{Q}(t))$$

we have that $\mathbf{X}(t)\mathbf{1}_{k^2}^\top = 0$, hence the matrix $\mathbf{V}_q(t)$ is a $\mathbf{V}$-matrix. Also if $\mathbf{Q}^*$ is a stable stochastic matrix then $\mathbf{XQ}^* = 0$.

Proof. We have that

$$\begin{aligned} \mathbf{X}(t)\mathbf{1}_{k^2}^\top &= \mathbf{Q}(t)\mathbf{J}\mathbf{1}_{k^2}^\top - \mathbf{J}(\mathbf{Q}(t) \otimes \mathbf{Q}(t))\mathbf{1}_{k^2}^\top \\ &= \mathbf{Q}(t)\mathbf{1}_{k^2}^\top - \mathbf{J}(\mathbf{Q}(t) \otimes \mathbf{Q}(t))(\mathbf{1}_k^\top \otimes \mathbf{1}_k^\top) \\ &= \mathbf{1}_{k^2}^\top - \mathbf{J}\left(\mathbf{Q}(t)\mathbf{1}_k^\top \otimes \mathbf{Q}(t)\mathbf{1}_k^\top\right) = \mathbf{1}_{k^2}^\top - \mathbf{J}\left(\mathbf{1}_k^\top \otimes \mathbf{1}_k^\top\right) \\ &= \mathbf{1}_{k^2}^\top - \mathbf{J}\mathbf{1}_{k^2}^\top = \mathbf{0}. \end{aligned} \tag{7.3}$$

Now, let $\mathbf{Q}^*$be a $k^2 \times k^2$ stable stochastic matrix, that is, a stochastic matrix with identical rows. Then, we get that

$$\begin{aligned} \mathbf{XQ}^* &= \mathbf{Q}(t)\mathbf{JQ}^* - \mathbf{J}(\mathbf{Q}(t) \otimes \mathbf{Q}(t))\mathbf{Q}^* \\ &= (\text{since } \mathbf{Q}(t) \otimes \mathbf{Q}(t) \text{ is a stochastic matrix}) \\ &= \mathbf{Q}(t)\mathbf{JQ}^* - \mathbf{JQ}^* = \mathbf{0}. \end{aligned} \tag{7.4}$$

■

Proposition 308 *(Vassiliou (2015)). If $\mathbf{V}$ and $\mathbf{V}^*$ are matrices of the form*

$$\mathbf{V} = \begin{pmatrix} \mathbf{Q} & \mathbf{X} \\ \mathbf{0} & \mathbf{Q} \otimes \mathbf{Q} \end{pmatrix} \text{ and } \mathbf{V}^* = \begin{pmatrix} \mathbf{Q}^* & \mathbf{X}^* \\ \mathbf{0} & \mathbf{Q}^* \otimes \mathbf{Q}^* \end{pmatrix},$$

where $\mathbf{Q}$ is a $k \times k$ stochastic matrix, $\mathbf{X}$ is a $k \times k^2$ matrix with $\mathbf{X1}_{k^2}^\top = \mathbf{0}$, $\mathbf{Q}^*$ is a stable stochastic matrix, $\mathbf{X}^*$ a matrix with identical rows and $\mathbf{X}^*\mathbf{1}_{k^2}^\top = \mathbf{0}$, then $\mathbf{VV}^* = \mathbf{V}^*$.

Proof. We have

$$\mathbf{VV}^* = \begin{pmatrix} \mathbf{Q} & \mathbf{X} \\ \mathbf{0} & \mathbf{Q} \otimes \mathbf{Q} \end{pmatrix} \begin{pmatrix} \mathbf{Q}^* & \mathbf{X}^* \\ \mathbf{0} & \mathbf{Q}^* \otimes \mathbf{Q}^* \end{pmatrix} =$$

$$\begin{pmatrix} \mathbf{QQ}^* & \mathbf{QX}^* + \mathbf{X}(\mathbf{Q}^* \otimes \mathbf{Q}^*) \\ \mathbf{0} & (\mathbf{Q} \otimes \mathbf{Q})(\mathbf{Q}^* \otimes \mathbf{Q}^*) \end{pmatrix} = \begin{pmatrix} \mathbf{Q}^* & \mathbf{X}^* \\ \mathbf{0} & \mathbf{Q}^* \otimes \mathbf{Q}^* \end{pmatrix} = \mathbf{V}^*,$$

since if $\mathbf{X}$ is a matrix with row sums equal to zero and $\mathbf{Q}^*$ a stable stochastic matrix then $\mathbf{XQ}^* = \mathbf{0}$. ■

Proposition 309 *(Vassiliou (2015)). If $\{\mathbf{V}(t)\}_{t=0}^{\infty}$ is a sequence of* $\mathbf{V}$*-matrices then*

$$\mathbf{V}(t,t+2)=\begin{pmatrix}\mathbf{Q}(t,t+2) & \mathbf{Q}(t,t+2)\mathbf{J}-\mathbf{J}(\mathbf{Q}(t,t+2)\otimes\mathbf{Q}(t,t+2))\\ \mathbf{0} & \mathbf{Q}(t,t+2)\otimes\mathbf{Q}(t,t+2)\end{pmatrix}.$$

Proof.

$$\mathbf{V}(t,t+2)=\begin{pmatrix}\mathbf{Q}(t+1) & \mathbf{Q}(t+1)\mathbf{J}-\mathbf{J}(\mathbf{Q}(t+1)\otimes\mathbf{Q}(t+1))\\ \mathbf{0} & \mathbf{Q}(t+1)\otimes\mathbf{Q}(t+1)\end{pmatrix}\times$$

$$\begin{pmatrix}\mathbf{Q}(t+2) & \mathbf{Q}(t+2)\mathbf{J}-\mathbf{J}(\mathbf{Q}(t+2)\otimes\mathbf{Q}(t+2))\\ \mathbf{0} & \mathbf{Q}(t+2)\otimes\mathbf{Q}(t+2)\end{pmatrix}=$$

$$=\text{(using the properties of the Kronecker product)}=$$

$$\begin{pmatrix}\mathbf{Q}(t+1)\mathbf{Q}(t+2) & \begin{matrix}\mathbf{Q}(t+1)\mathbf{Q}(t+2)\mathbf{J}-\mathbf{J}(\mathbf{Q}(t+1)\mathbf{Q}(t+2))\otimes\\ \mathbf{Q}(t+1)\mathbf{Q}(t+2))\end{matrix}\\ \mathbf{0} & (\mathbf{Q}(t+1)\mathbf{Q}(t+2))\otimes\mathbf{Q}(t+1)\mathbf{Q}(t+2)\end{pmatrix}.$$

■

Similarly, we can prove that

$$\mathbf{V}(t,t+n)=\begin{pmatrix}\mathbf{Q}(t,t+n) & \mathbf{Q}(t,t+n)\mathbf{J}-\mathbf{J}(\mathbf{Q}(t,t+n)\otimes\mathbf{Q}(t,t+n))\\ \mathbf{0} & \mathbf{Q}(t,t+n)\otimes\mathbf{Q}(t,t+n)\end{pmatrix}, \tag{7.5}$$

and

$$\mathbf{V}^t=\begin{pmatrix}\mathbf{Q}^t & \mathbf{Q}^t\mathbf{J}-\mathbf{J}\left(\mathbf{Q}^t\otimes\mathbf{Q}^t\right)\\ \mathbf{0} & \mathbf{Q}^t\otimes\mathbf{Q}^t\end{pmatrix}. \tag{7.6}$$

The following Lemma is well known for sequences of numbers (see Isaacson et al. (1976)) and following the same steps it can be proved for matrices.

Lemma 310 *Let $\{a_n\}$ be a sequence of non-negative numbers such that $a_n/\sum_{k=0}^{n}a_k\to 0$ as $n\to\infty$. Let also $\{\mathbf{B}_n\}$ be a sequence of finite matrices converging to $\mathbf{B}$; then*

$$\lim_{n\to\infty}\frac{\mathbf{B}_0a_n+\mathbf{B}_1a_{n-1}+\ldots+\mathbf{B}_na_0}{\sum_{k=0}^{n}a_k}=\mathbf{B}.$$

The following Lemma interconnects the Hadamard and Kronecker products introduced to the reader in Chapter 2.

Lemma 311 *(Vassiliou (2015)). Let $\mathbf{A},\mathbf{C}\in M_m(\mathbb{R})$ and $\mathbf{B},\mathbf{F}\in M_n(\mathbb{R})$ then*

$$(\mathbf{A}\otimes\mathbf{B})\vartriangle(\mathbf{C}\otimes\mathbf{F})=(\mathbf{A}\vartriangle\mathbf{C})\otimes(\mathbf{B}\vartriangle\mathbf{F}). \tag{7.7}$$

Proof. Let $\mathbf{A}=\{a_{ij}\}_{i.j=1}^{m}$, $\mathbf{C}=\{c_{ij}\}_{i,j=1}^{m}$ and $\mathbf{B}=\{b_{ij}\}_{i,j=1}^{n}$, $\mathbf{F}=\{f_{ij}\}_{i,j=1}^{n}$. The (r,k)-element of the matrix $(\mathbf{A}\vartriangle\mathbf{C})$ is $a_{rk}c_{rk}$ and the (d,ν)-element of the matrix $(\mathbf{B}\vartriangle\mathbf{F})$ is $b_{d\nu}f_{d\nu}$. Hence, the element $a_{rk}c_{rk}b_{d\nu}f_{d\nu}$ is the (i,j)-element of the matrix

$$(\mathbf{A}\vartriangle\mathbf{C})\otimes(\mathbf{B}\vartriangle\mathbf{F}),$$

where $i=(r-1)n+d$ and $j=(k-1)n+\nu$. Now, from the first part of equation (7.7), we have that $a_{rk}b_{d\nu}$ is the (i,j)-element of the matrix $(\mathbf{A}\otimes\mathbf{B})$, where $i=(r-1)n+d$ and $j=(k-1)n+\nu$. Also, $c_{rk}f_{d\nu}$ is the (i,j)-element of the matrix $(\mathbf{C}\otimes\mathbf{F})$, where $i=(r-1)n+d$ and $j=(k-1)n+\nu$. Therefore, $a_{rk}c_{rk}b_{d\nu}f_{d\nu}$ is the (i,j)-element of the matrix

$$(\mathbf{A}\otimes\mathbf{B})\vartriangle(\mathbf{C}\otimes\mathbf{F}),$$

where $i=(r-1)n+d$ and $j=(k-1)n+\nu$ and thus it coincides with the (i,j)-element of the matrix $(\mathbf{A}\vartriangle\mathbf{C})\otimes(\mathbf{B}\vartriangle\mathbf{F})$. ■

We will now prove another useful theorem concerning convergence of the infinite products of sequence of $\mathbf{V}$-matrices when the inherent sequence of the NHMS $\{\mathbf{Q}(t)\}_{t=0}^{\infty}$ are general stochastic matrices.

Theorem 312 *(Vassiliou (2015)). Let an NHMS and let that the sequence of arbitrary stochastic matrices* $\{\mathbf{Q}(t)\}_{t=0}^{\infty}$*, which determine uniquely the inherent Markov chain of the NHMS. Let also, that all* $\mathbf{Q}(t)$*'s have the same incidence matrix for all* $t \in \mathbb{Z}^+$ *and that each* $\mathbf{Q}(t)$ *have r essential classes with periods* $d_1, d_2, ..., d_r$ *respectively. Assume in addition, that* $\lim_{t\to\infty} \|\mathbf{Q}(t) - \mathbf{Q}\| = 0$*, with* $\bar{\mathbf{Q}}(t) = \bar{\mathbf{Q}}$ *and let with no loss of generality that the canonical form be*

$$\mathbf{Q}(t) = \begin{pmatrix} \mathbf{Q}_1(t) & \mathbf{0} & ... & \mathbf{0} & \mathbf{0} \\ \mathbf{0} & \mathbf{Q}_2(t) & ... & \mathbf{0} & \mathbf{0} \\ ... & ... & ... & ... & ... \\ \mathbf{0} & \mathbf{0} & ... & \mathbf{Q}_r(t) & \mathbf{0} \\ \mathbf{T}_1(t) & \mathbf{T}_2(t) & ... & \mathbf{T}_r(t) & \mathbf{T}(t) \end{pmatrix}$$

Then

$$\begin{aligned} \lim_{t\to\infty} \|\mathbf{Q}_i(t) - \mathbf{Q}_i\| &= 0, \lim_{t\to\infty} \|\mathbf{T}_i(t) - \mathbf{T}_i\| = 0 \textit{ for } i = 1, 2, ..., r, \\ \textit{and} \lim_{t\to\infty} \|\mathbf{T}(t) - \mathbf{T}\| &= 0. \end{aligned}$$

Define by $\mathbf{Q}_i^* = \lim_{t\to\infty} \mathbf{Q}_i^t$*. Then* $\mathbf{V}(s,t)$ *as* $t \to \infty$ *splits into* $d = \operatorname{lcm}\{d_1, d_2, ..., d_r\}$ *converging subsequences with limits*

$$\mathbf{V}(\boldsymbol{\nu}) = \mathbf{V}(\nu_1, \nu_2, ..., \nu_r) = \mathbf{V}_{\mathbf{Q}^*}\mathbf{V}_{\mathbf{Q}(\boldsymbol{\nu})}\mathbf{W}_{\mathbf{T}\triangle\mathbf{Q}_{\mathbf{T}}(\boldsymbol{\nu})},$$

where

$$\mathbf{V}_{\mathbf{Q}^*} = \begin{pmatrix} \hat{\mathbf{Q}}^* & \hat{\mathbf{Q}}^*\mathbf{J} - \mathbf{J}\left(\hat{\mathbf{Q}}^* \times \hat{\mathbf{Q}}^*\right) \\ 0 & \hat{\mathbf{Q}}^* \times \hat{\mathbf{Q}}^* \end{pmatrix},$$

$$\mathbf{V}_{\mathbf{Q}(\boldsymbol{\nu})} = \begin{pmatrix} \hat{\mathbf{Q}}(\boldsymbol{\nu}) & \hat{\mathbf{Q}}(\boldsymbol{\nu})\mathbf{J} - \mathbf{J}\left(\hat{\mathbf{Q}}(\boldsymbol{\nu}) \times \hat{\mathbf{Q}}(\boldsymbol{\nu})\right) \\ 0 & \hat{\mathbf{Q}}(\boldsymbol{\nu}) \times \hat{\mathbf{Q}}(\boldsymbol{\nu}) \end{pmatrix},$$

$$\mathbf{W}_{\mathbf{T}\triangle\mathbf{Q}_{\mathbf{T}}(\boldsymbol{\nu})} =$$

$$\begin{pmatrix} \hat{\mathbf{T}} \triangle \hat{\mathbf{Q}}_{\mathbf{T}}(\boldsymbol{\nu}) & \hat{\mathbf{T}} \triangle \hat{\mathbf{Q}}_{\mathbf{T}}(\boldsymbol{\nu})\mathbf{J} - \mathbf{J}\left(\hat{\mathbf{T}} \triangle \hat{\mathbf{Q}}_{\mathbf{T}}(\boldsymbol{\nu}) \times \hat{\mathbf{T}} \triangle \hat{\mathbf{Q}}_{\mathbf{T}}(\boldsymbol{\nu})\right) \\ 0 & \left(\hat{\mathbf{T}} \triangle \hat{\mathbf{Q}}_{\mathbf{T}}(\boldsymbol{\nu}) \times \hat{\mathbf{T}} \triangle \hat{\mathbf{Q}}_{\mathbf{T}}(\boldsymbol{\nu})\right) \end{pmatrix},$$

and $\hat{\mathbf{Q}}^*, \hat{\mathbf{Q}}(\boldsymbol{\nu}), \hat{\mathbf{T}} \triangle \hat{\mathbf{Q}}_{\mathbf{T}}$ *are defined below in* (7.8)*. The matrix* $\mathbf{W}_{\mathbf{T}\triangle\mathbf{Q}_{\mathbf{T}}(\boldsymbol{\nu})}$ *is not a V-matrix.*

Proof. Following Theorem 276 and replacing $\mathbf{P}(t)$, $\mathbf{P}$, $\mathbf{P}_i(t)$, $\mathbf{P}_i$, $\mathbf{Q}(t)$, $\mathbf{Q}$ respectively with $\mathbf{Q}(t)$, $\mathbf{Q}$, $\mathbf{Q}_i(t)$, $\mathbf{Q}_i$, $\mathbf{T}(t)$, $\mathbf{T}$, we immediately arrive at the result that the sequence $\mathbf{Q}(s,t)$ as $t \to \infty$ splits into $d = \operatorname{lcm}\{d_1, d_2, ..., d_r\}$ subsequences with limits

$$\mathbf{Q}(\nu_1, \nu_2, ..., \nu_r) = \begin{pmatrix} \mathbf{Q}_1^*\mathbf{Q}_1^{d_1+\nu_1} & \mathbf{0} & ... & \mathbf{0} & \mathbf{0} \\ \mathbf{0} & \mathbf{Q}_2^*\mathbf{Q}_2^{d_2+\nu_2} & ... & \mathbf{0} & \mathbf{0} \\ ... & ... & ... & ... & ... \\ \mathbf{0} & \mathbf{0} & ... & \mathbf{Q}_r^*\mathbf{Q}_r^{d_r+\nu_r} & \mathbf{0} \\ \mathbf{T}_1^* & \mathbf{T}_2^* & ... & \mathbf{T}_r^* & \mathbf{0} \end{pmatrix}.$$

It is not difficult to see that

$$\begin{pmatrix} \mathbf{Q}_1^*\mathbf{Q}_1^{d_1+\nu_1} & \mathbf{0} & ... & \mathbf{0} & \mathbf{0} \\ \mathbf{0} & \mathbf{Q}_2^*\mathbf{Q}_2^{d_2+\nu_2} & ... & \mathbf{0} & \mathbf{0} \\ ... & ... & ... & ... & ... \\ \mathbf{0} & \mathbf{0} & ... & \mathbf{Q}_r^*\mathbf{Q}_r^{d_r+\nu_r} & \mathbf{0} \\ \mathbf{T}_1^* & \mathbf{T}_2^* & ... & \mathbf{T}_r^* & \mathbf{0} \end{pmatrix} =$$

$$\begin{pmatrix} \mathbf{Q}_1^* & \mathbf{0} & \dots & \mathbf{0} & \mathbf{0} \\ \mathbf{0} & \mathbf{Q}_2^* & \dots & \mathbf{0} & \mathbf{0} \\ \dots & \dots & \dots & \dots & \dots \\ \mathbf{0} & \mathbf{0} & \dots & \mathbf{Q}_r^* & \mathbf{0} \\ \mathbf{0} & \mathbf{0} & \dots & \mathbf{0} & \mathbf{I} \end{pmatrix} \begin{pmatrix} \mathbf{Q}_1^{d_1+\nu_1} & \mathbf{0} & \dots & \mathbf{0} & \mathbf{0} \\ \mathbf{0} & \mathbf{Q}_2^{d_2+\nu_2} & \dots & \mathbf{0} & \mathbf{0} \\ \dots & \dots & \dots & \dots & \dots \\ \mathbf{0} & \mathbf{0} & \dots & \mathbf{Q}_r^{d_r+\nu_r} & \mathbf{0} \\ \mathbf{0} & \mathbf{0} & \dots & \mathbf{0} & \mathbf{I} \end{pmatrix}.$$

$$\times \begin{pmatrix} \mathbf{I} & \mathbf{0} & \dots & \mathbf{0} & \mathbf{0} \\ \mathbf{0} & \mathbf{I} & \dots & \mathbf{0} & \mathbf{0} \\ \dots & \dots & \dots & \dots & \dots \\ \mathbf{0} & \mathbf{0} & \dots & \mathbf{I} & \mathbf{0} \\ \mathbf{T}_1 & \mathbf{T}_2 & \dots & \mathbf{T}_r & \mathbf{0} \end{pmatrix} \triangle \tag{7.8}$$

$$\begin{pmatrix} \mathbf{I} & \mathbf{0} & \dots & \mathbf{0} & \mathbf{0} \\ \mathbf{0} & \mathbf{I} & \dots & \mathbf{0} & \mathbf{0} \\ \dots & \dots & \dots & \dots & \dots \\ \mathbf{0} & \mathbf{0} & \dots & \mathbf{I} & \mathbf{0} \\ \mathbf{Q}_1^*\mathbf{Q}_1^{d_1+\nu_1} & \mathbf{Q}_2^*\mathbf{Q}_2^{d_2+\nu_2} & \dots & \mathbf{Q}_r^*\mathbf{Q}_r^{d_r+\nu_r} & \mathbf{0} \end{pmatrix} = \hat{\mathbf{Q}}^*\hat{\mathbf{Q}}(\nu)\hat{\mathbf{T}} \triangle \hat{\mathbf{Q}}_{\mathbf{T}}(\nu),$$

where $\mathbf{T}_i = \left(\mathbf{I} - \mathbf{T}^{d_i}\right)^{-1} \left[\sum_{l=0}^{d-1} \mathbf{T}^l \mathbf{T}_i \mathbf{P}_i^{d_i-l-1}\right]$.

From Proposition 309 and relation (7.8) we get that $\lim_{t\to\infty} \mathbf{V}(s,t)$ splits into $d = \text{lcm}\{d_1, d_2, ..., d_r\}$ converging subsequences with limits

$$\begin{pmatrix} \hat{\mathbf{Q}}^*\hat{\mathbf{Q}}(\nu)\hat{\mathbf{T}}\hat{\mathbf{Q}}_{\mathbf{T}}(\nu) & \begin{matrix} \hat{\mathbf{Q}}^*\hat{\mathbf{Q}}(\nu)\hat{\mathbf{T}} \triangle \hat{\mathbf{Q}}_{\mathbf{T}}(\nu)\mathbf{J} - \mathbf{J}(\hat{\mathbf{Q}}^*\hat{\mathbf{Q}}(\nu)\hat{\mathbf{T}} \triangle \\ \hat{\mathbf{Q}}_{\mathbf{T}}(\nu) \times \hat{\mathbf{Q}}^*\hat{\mathbf{Q}}(\nu)\hat{\mathbf{T}} \triangle \hat{\mathbf{Q}}_{\mathbf{T}}(\nu)) \end{matrix} \\ \mathbf{0} & \hat{\mathbf{Q}}^*\hat{\mathbf{Q}}(\nu)\hat{\mathbf{T}} \triangle \hat{\mathbf{Q}}_{\mathbf{T}}(\nu) \times \hat{\mathbf{Q}}^*\hat{\mathbf{Q}}(\nu)\hat{\mathbf{T}} \triangle \hat{\mathbf{Q}}_{\mathbf{T}}(\nu) \end{pmatrix}$$

$$= \begin{pmatrix} \hat{\mathbf{Q}}^* & \hat{\mathbf{Q}}^*\mathbf{J} - \mathbf{J}\left(\hat{\mathbf{Q}}^* \times \hat{\mathbf{Q}}^*\right) \\ \mathbf{0} & \hat{\mathbf{Q}}^* \times \hat{\mathbf{Q}}^* \end{pmatrix} \tag{7.9}$$

$$\times \begin{pmatrix} \hat{\mathbf{Q}}(\nu) & \hat{\mathbf{Q}}(\nu)\mathbf{J} - \mathbf{J}\left(\hat{\mathbf{Q}}(\nu) \times \hat{\mathbf{Q}}(\nu)\right) \\ \mathbf{0} & \hat{\mathbf{Q}}(\nu) \times \hat{\mathbf{Q}}(\nu) \end{pmatrix}$$

$$\times \begin{pmatrix} \hat{\mathbf{T}} \triangle \hat{\mathbf{Q}}_{\mathbf{T}}(\nu) & \hat{\mathbf{T}} \triangle \hat{\mathbf{Q}}_{\mathbf{T}}(\nu)\mathbf{J} - \mathbf{J}\left(\hat{\mathbf{T}} \triangle \hat{\mathbf{Q}}_{\mathbf{T}}(\nu) \times \hat{\mathbf{T}} \triangle \hat{\mathbf{Q}}_{\mathbf{T}}(\nu)\right) \\ \mathbf{0} & \hat{\mathbf{T}} \triangle \hat{\mathbf{Q}}_{\mathbf{T}}(\nu) \times \hat{\mathbf{T}} \triangle \hat{\mathbf{Q}}_{\mathbf{T}}(\nu) \end{pmatrix}$$

$$= \mathbf{V}_{\mathbf{Q}^*}\mathbf{V}_{\mathbf{Q}(\nu)}\mathbf{W}_{\mathbf{T}\triangle\mathbf{Q}_{\mathbf{T}}(\nu)}.$$

The matrix $\mathbf{W}_{\mathbf{T}\triangle\mathbf{Q}_{\mathbf{T}}(\nu)}$ is not a V-matrix since the matrix

$$\left[\hat{\mathbf{T}} \triangle \hat{\mathbf{Q}}_{\mathbf{T}}(\nu)\mathbf{J} - \mathbf{J}\left(\hat{\mathbf{T}} \triangle \hat{\mathbf{Q}}_{\mathbf{T}}(\nu) \times \hat{\mathbf{T}} \triangle \hat{\mathbf{Q}}_{\mathbf{T}}(\nu)\right)\right]\mathbf{1}^\top \neq \mathbf{0}.$$

This is also the reason why it cannot be written as a Hadamard product of two V-matrices. ■

We will now prove the following theorem concerning the asymptotic behavior of the variability of population structure.

Theorem 313 *(Vassiliou (2015)). Let an NHMS and let that the sequence of arbitrary stochastic matrices* $\{\mathbf{Q}(t)\}_{t=0}^{\infty}$*, which determine uniquely the inherent Markov chain of the NHMS. Let also, that all* $\mathbf{Q}(t)$*'s have the same incidence matrix for all* $t \in \mathbb{Z}^+$ *and that each* $\mathbf{Q}(t)$ *have* r *essential classes with periods* $d_1, d_2, ..., d_r$ *respectively. Assume in addition,* *that* $\lim_{t\to\infty} \|\mathbf{Q}(t) - \mathbf{Q}\| = 0$, *with* $\bar{\mathbf{Q}}(t) = \bar{\mathbf{Q}}$ *and let with no loss of generality that the canonical form be*

$$\mathbf{Q}(t) = \begin{pmatrix} \mathbf{Q}_1(t) & \mathbf{0} & \dots & \mathbf{0} & \mathbf{0} \\ \mathbf{0} & \mathbf{Q}_2(t) & \dots & \mathbf{0} & \mathbf{0} \\ \dots & \dots & \dots & \dots & \dots \\ \mathbf{0} & \mathbf{0} & \dots & \mathbf{Q}_r(t) & \mathbf{0} \\ \mathbf{T}_1(t) & \mathbf{T}_2(t) & \dots & \mathbf{T}_r(t) & \mathbf{T}(t) \end{pmatrix}$$

Then

$$\begin{aligned} \lim_{t\to\infty} \|\mathbf{Q}_i(t) - \mathbf{Q}_i\| &= 0, \lim_{t\to\infty} \|\mathbf{T}_i(t) - \mathbf{T}_i\| = 0 \text{ for } i = 1, 2, ..., r, \\ \text{and } \lim_{t\to\infty} \|\mathbf{T}(t) - \mathbf{T}\| &= 0. \end{aligned}$$

Define by $\mathbf{Q}_i^* = \lim_{t\to\infty} \mathbf{Q}_i^t$. *Assume also that* $\lim_{t\to\infty} \|\boldsymbol{\mu}_0(t) - \boldsymbol{\mu}_0\| = 0$; $\Delta T(t) \geq 0$ *and that the total population of memberships* $T(t)$ *as* $t \to \infty$ *converges to* T. *Let in general* $t = \pi_i d_i + \upsilon_i$ *for* $i = 1, 2, ..., r$ *and let* $\boldsymbol{\upsilon} = (\upsilon_1, \upsilon_2, ..., \upsilon_r)$, *then the sequence* $\{\boldsymbol{\mu}(t)\}_{t=0}^{\infty}$ *of means variances and covariances of the population structure splits in* $d = \operatorname{lcm}\{d_1, d_2, ..., d_n\}$ *converging subsequence given by*

$$\begin{aligned} \boldsymbol{\mu}(\boldsymbol{\upsilon}) &= [\boldsymbol{\mu}(0) + (T - T(0))\boldsymbol{\mu}_0(\infty)] \mathbf{V}_{\mathbf{Q}^*} \\ &\quad \times \left[\mathbf{V}_{\mathbf{Q}(\boldsymbol{\upsilon})} \mathbf{W}_{\mathbf{T}\triangle\mathbf{Q}_{\mathbf{T}}(\boldsymbol{\upsilon})} + \mathbf{V}_{\mathbf{Q}(\mathbf{S}(\boldsymbol{\upsilon}))} \mathbf{W}_{\mathbf{T}\triangle\mathbf{Q}_{\mathbf{T}}((\mathbf{S}(\boldsymbol{\upsilon})))}\right], \end{aligned}$$

where

$$\mathbf{V}_{\mathbf{Q}^*} = \begin{pmatrix} \hat{\mathbf{Q}}^* & \hat{\mathbf{Q}}^*\mathbf{J} - \mathbf{J}\left(\hat{\mathbf{Q}}^* \otimes \hat{\mathbf{Q}}^*\right) \\ \mathbf{0} & \hat{\mathbf{Q}}^* \times \hat{\mathbf{Q}}^* \end{pmatrix},$$

with $\hat{\mathbf{Q}}^*$ *given in* (7.8),

$$\mathbf{V}_{\mathbf{Q}(\boldsymbol{\upsilon})} = \begin{pmatrix} \hat{\mathbf{Q}}(\boldsymbol{\upsilon}) & \hat{\mathbf{Q}}(\boldsymbol{\upsilon})\mathbf{J} - \mathbf{J}\left(\hat{\mathbf{Q}}(\boldsymbol{\upsilon}) \otimes \hat{\mathbf{Q}}(\boldsymbol{\upsilon})\right) \\ \mathbf{0} & \hat{\mathbf{Q}}(\boldsymbol{\upsilon}) \otimes \hat{\mathbf{Q}}(\boldsymbol{\upsilon}) \end{pmatrix},$$

with $\hat{\mathbf{Q}}(\boldsymbol{\upsilon})$ *given in* (7.8).

$$\mathbf{W}_{\mathbf{T}\triangle\mathbf{Q}_{\mathbf{T}}(\boldsymbol{\upsilon})} = \begin{pmatrix} \hat{\mathbf{T}} \triangle \hat{\mathbf{Q}}_{\mathbf{T}}(\boldsymbol{\upsilon}) & \begin{array}{c} \hat{\mathbf{T}} \triangle \hat{\mathbf{Q}}_{\mathbf{T}}(\boldsymbol{\upsilon})\mathbf{J} - \mathbf{J}(\hat{\mathbf{T}} \triangle \hat{\mathbf{Q}}_{\mathbf{T}}(\boldsymbol{\upsilon}) \\ \otimes \hat{\mathbf{T}} \triangle \hat{\mathbf{Q}}_{\mathbf{T}}(\boldsymbol{\upsilon})) \end{array} \\ \mathbf{0} & \hat{\mathbf{T}} \triangle \hat{\mathbf{Q}}_{\mathbf{T}}(\boldsymbol{\upsilon}) \otimes \hat{\mathbf{T}} \triangle \hat{\mathbf{Q}}_{\mathbf{T}}(\boldsymbol{\upsilon}) \end{pmatrix}$$

with $\hat{\mathbf{T}}\triangle\, \hat{\mathbf{Q}}_{\mathbf{T}}(\nu)$ *given in* (7.8),

$$\mathbf{V}_{\mathbf{Q}(\mathbf{S}(\boldsymbol{\upsilon}))} = \begin{pmatrix} \hat{\mathbf{Q}}(\mathbf{S}(\boldsymbol{\upsilon})) & \hat{\mathbf{Q}}(\mathbf{S}(\boldsymbol{\upsilon}))\mathbf{J} - \mathbf{J}\left(\hat{\mathbf{Q}}(\mathbf{S}(\boldsymbol{\upsilon})) \otimes \hat{\mathbf{Q}}(\mathbf{S}(\boldsymbol{\upsilon}))\right) \\ \mathbf{0} & \hat{\mathbf{Q}}(\mathbf{S}(\boldsymbol{\upsilon})) \otimes \hat{\mathbf{Q}}(\mathbf{S}(\boldsymbol{\upsilon})) \end{pmatrix}$$

with $\hat{\mathbf{Q}}(\mathbf{S}(\boldsymbol{\upsilon}))$ *given in* (7.24) *and* (7.25). *In addition*

$$\mathbf{W}_{\mathbf{T}\triangle\mathbf{Q}_{\mathbf{T}}((\mathbf{S}(\boldsymbol{\upsilon})))} = \begin{pmatrix} \hat{\mathbf{T}}\triangle\hat{\mathbf{Q}}_{\mathbf{T}}(\mathbf{S}(\boldsymbol{\upsilon})) & \begin{array}{c} \hat{\mathbf{T}}\triangle\hat{\mathbf{Q}}_{\mathbf{T}}(\mathbf{S}(\boldsymbol{\upsilon}))\mathbf{J} - \mathbf{J}(\hat{\mathbf{T}}\triangle \\ \hat{\mathbf{Q}}_{\mathbf{T}}(\mathbf{S}(\boldsymbol{\upsilon})) \otimes \hat{\mathbf{T}}\triangle\hat{\mathbf{Q}}_{\mathbf{T}}(\mathbf{S}(\boldsymbol{\upsilon}))) \end{array} \\ 0 & \hat{\mathbf{T}}\triangle\hat{\mathbf{Q}}_{\mathbf{T}}(\mathbf{S}(\boldsymbol{\upsilon})) \otimes \hat{\mathbf{T}}\triangle\hat{\mathbf{Q}}_{\mathbf{T}}(\mathbf{S}(\boldsymbol{\upsilon})) \end{pmatrix}$$

with $\hat{\mathbf{T}}\triangle\, \hat{\mathbf{Q}}_{\mathbf{T}}(\mathbf{S}(\boldsymbol{\upsilon}))$ *given in* (7.24) *and* (7.25). *The matrix* $\mathbf{W}_{\mathbf{T}\triangle\mathbf{Q}_{\mathbf{T}}((\mathbf{S}(\boldsymbol{\upsilon})))}$ *is not a V-matrix.*

Proof. From (5.24) we get that

$$\boldsymbol{\mu}(t) = \boldsymbol{\mu}(0)\,\mathbf{V}(0,t) + \sum_{\tau=1}^{t} \Delta T(\tau)\,\boldsymbol{\mu}_0(\tau)\,\mathbf{V}(\tau,t). \tag{7.10}$$

Let $t = \pi_t d_i + v_i$ then from (7.10) and Theorem 312 we get that

$$\lim_{t\to\infty} \boldsymbol{\mu}(0)\,\mathbf{V}(0,t),$$

splits into $d = \operatorname{lcm}\{d_1, d_2, ..., d_r\}$ converging subsequences with limits

$$\boldsymbol{\mu}(0)\,\mathbf{V}_{\mathbf{Q}^*}\mathbf{V}_{\mathbf{Q}(v)}\mathbf{W}_{\mathbf{T}\Delta\mathbf{Q}_{\mathbf{T}}(v)}. \tag{7.11}$$

Hence, what remains is to find the limit

$$\lim_{t\to\infty} \mathbf{U}(t) = \lim_{t\to\infty} \sum_{\tau=1}^{t} \Delta T(\tau)\,\boldsymbol{\mu}_0(\tau)\,\mathbf{V}(\tau,t) =$$

$$\lim_{t\to\infty} \sum_{\tau=0}^{t} \Delta T(\tau)\,\boldsymbol{\mu}_0(\tau) \begin{pmatrix} \mathbf{Q}(\tau,t) & \mathbf{Q}(\tau,t)\,\mathbf{J} - \mathbf{J}\,(\mathbf{Q}(\tau,t)\times\mathbf{Q}(\tau,t)) \\ \mathbf{0} & \mathbf{Q}(\tau,t)\times\mathbf{Q}(\tau,t) \end{pmatrix}. \tag{7.12}$$

We will firstly start with the limit that follows, which is part of the limit in (7.12). Consider

$$\begin{aligned} \lim_{t\to\infty} \mathbf{U}_1(t) &= \sum_{\tau=0}^{t} \Delta T(\tau)\,\boldsymbol{\mu}_0(\tau) \begin{pmatrix} \mathbf{Q}(\tau,t) \\ 0 \end{pmatrix} \qquad (7.13) \\ &= \sum_{\tau=0}^{t} \Delta T(\tau)\,\mathbf{p}_0(\tau)\,\mathbf{Q}(\tau,t) \\ &= \sum_{\tau=0}^{t} \Delta T(\tau)\,\mathbf{p}_0(\tau) \begin{pmatrix} \mathbf{Q}_1(\tau,t) & \mathbf{0} & ... & \mathbf{0} & \mathbf{0} \\ \mathbf{0} & \mathbf{Q}_2(\tau,t) & ... & \mathbf{0} & \mathbf{0} \\ ... & ... & ... & ... & ... \\ \mathbf{0} & \mathbf{0} & ... & \mathbf{Q}_r(\tau,t) & \mathbf{0} \\ \mathbf{T}_1(\tau,t) & \mathbf{T}_2(\tau,t) & ... & \mathbf{T}_r(\tau,t) & \mathbf{T}(\tau,t) \end{pmatrix} \end{aligned}$$

Define

$$\mathbf{p}_{0i}(t, m_{i-1}, m_i) = \left[p_{0,m_{i-1}+1}(t), p_{0,m_{i-1}+2}(t), ..., p_{0,m_i}(t)\right],$$

where

$$m_i = \sum_{j=1}^{i} k_j \text{ with } m_0 = 1.$$

Since we have already assumed that $t = \pi_{ti} d_i + v_i$ then

$$\sum_{\tau=0}^{t} \Delta T(\tau)\,\mathbf{p}_0(\tau, m_{i-1}, m_i)\,\mathbf{Q}_i(\tau,t) =$$

$$\sum_{m=0}^{d_i-1} \sum_{\tau=0}^{\pi_{ti}-1} \Delta T(\tau d_i + m)\,\mathbf{p}_{0i}(\tau d_i + m, m_{i-1}, m_i)\,\mathbf{Q}_i(\tau d_i + m, \pi_{ti} d_i + v_i) \tag{7.14}$$

$$+ \sum_{m=0}^{v_i} \Delta T(\pi_{ti} d_i + m)\,\mathbf{p}_{0i}(\pi_{ti} d_i + m, m_{i-1}, m_i)\,\mathbf{Q}_i(\pi_{ti} d_i + m, \pi_{ti} d_i + v_i)$$

$$= \mathbf{U}_{11}(t) + \mathbf{U}_{12}(t).$$

Since, $\lim_{\pi_{ti}\to\infty} \Delta T(\pi_{ti} d_i + m) = 0$ from (7.14) we get that

$$\lim_{\pi_{ti}\to\infty} \|\mathbf{U}_{12}(t)\| \leq$$

$$\lim_{\pi_{ti}\to\infty} \| \sum_{m=0}^{v_i} \Delta T(\pi_{ti} d_i + m) \mathbf{p}_{0i}(\pi_{ti} d_i + m, m_{i-1}, m_i)$$
$$\times \mathbf{Q}_i(\pi_{ti} d_i + m, \pi_{ti} d_i + v_i) \| \leq$$
$$\lim_{\pi_{ti}\to\infty} \sum_{m=0}^{v_i} |\Delta T(\pi_{ti} d_i + m)| \, \|\mathbf{p}_{0i}(\pi_{ti} d_i + m, , m_{i-1}, m_i)\| \tag{7.15}$$
$$\times \|\mathbf{Q}_i(\pi_{ti} d_i + m, \pi_{ti} d_i + v_i)\| = 0$$

We will now proceed to find $\lim_{t\to\infty} \mathbf{U}_{11}(t)$. Since, $\Delta T(t) \geq 0$ we have that

$$\sum_{\tau=0}^{\pi_{ti}-1} \Delta T(\tau d_i + m) \leq \sum_{\tau=0}^{\pi_{ti} d_i + m} \Delta T(\tau) = T(\pi_{ti} d_i + m) - T(0).$$

Hence, the series $\sum_{\tau=0}^{\pi_{ti}-1} \Delta T(\tau d_i + m)$ is a series of non-negative elements and is bounded by $T - T(0)$ and thus it converges. We define by

$$w_m(d_i) = \frac{\lim_{\pi_{ti}\to\infty} \sum_{\tau=0}^{\pi_{ti}-1} \Delta T(\tau d_i + m)}{\lim_{t\to\infty} \sum_{\tau=0}^{t} \Delta T(\tau)}. \tag{7.16}$$

Now, from (7.14) we have that

$$\lim_{t\to\infty} \mathbf{U}_{11}(t) = \sum_{m=0}^{d_i-1} \lim_{\pi_{ti}\to\infty} \sum_{\tau=0}^{\pi_{ti}-1} \Delta T(\tau d_i + m) \mathbf{p}_{0i}(\tau d_i + m, m_{i-1}, m_i)$$
$$\times \mathbf{Q}_i(\tau d_i + m, (\pi_{ti} - 1) d_i + d_i + v_i). \tag{7.17}$$

From (7.16) and (7.17) we get that

$$\lim_{t\to\infty} \mathbf{U}_{11}(t) = (T - T(0)) \sum_{m=0}^{d_i-1} w_m(d_i) \sum_{\tau=0}^{\pi_{ti}-1} \frac{\Delta T(\tau d_i + m)}{\sum_{\tau=0}^{\pi_{ti}-1} \Delta T(\tau d_i + m)} \tag{7.18}$$
$$\times \mathbf{p}_{0i}(\tau d_i + m, m_{i-1}, m_i) \mathbf{Q}_i(\tau d_i + m, (\pi_{ti} - 1) d_i + d_i + v_i).$$

By Lemma 310 and (7.18), we get that $\lim_{t\to\infty} \mathbf{U}_{11}(t)$ splits into $d_i - 1$ converging subsequences with limits

$$(T - T(0)) \mathbf{p}_{0i}(\infty, m_{i-1}, m_i) \sum_{m=0}^{d_i-1} w_m(d_i) \mathbf{Q}_i^* \mathbf{Q}_i^{d_i+v_i-m+1}. \tag{7.19}$$

Following similar steps we could prove that

$$\lim_{t\to\infty} \sum_{\tau=0}^{t} \Delta T(\tau) \mathbf{p}_{0r}(\tau, m_{r-1}, m_r) \mathbf{T}_i(\tau, t), \tag{7.20}$$

splits into $d_i - 1$ converging subsequences with limits

$$(T - T(0))\,\mathbf{p}_{0r}(\infty, m_{i-1}, m_i)\left(\mathbf{I} - \mathbf{T}^{d_i}\right)^{-1}\left[\sum_{l=0}^{d_i-1}\mathbf{T}^l\mathbf{T}_i\mathbf{Q}_i^{d_i-l-1}\right]$$

$$\times\left[\sum_{l=0}^{d_i-1}\mathbf{T}^l\mathbf{T}_i\mathbf{Q}_i^{d_i-l-1}\right]\sum_{m=0}^{d_i-1} w_m(d_i)\,\mathbf{Q}_i^*\mathbf{Q}_i^{d_i+v_i-m+1}. \tag{7.21}$$

Also, it not difficult to show that

$$\lim_{t\to\infty}\sum_{\tau=0}^{t}\Delta T(\tau)\,\mathbf{p}_{0r+1}(\tau, m_r+1, k)\,\mathbf{T}(\tau, t) = \mathbf{0}. \tag{7.22}$$

Hence, from (7.13), (7.15), (7.19), (7.20), (7.21) and (7.22), we get that $\lim_{t\to\infty}\mathbf{U}_1(t)$ splits into $d = \operatorname{lcm}\{d_1, d_2, ..., d_r\}$ converging subsequences with limits

$$(T - T(0))\,\mathbf{p}_0(\infty)\begin{pmatrix} \mathbf{Q}_1^*\mathbf{S}_1(v_1) & \mathbf{0} & \cdots & \mathbf{0} & \mathbf{0} \\ \mathbf{0} & \mathbf{Q}_2^*\mathbf{S}_2(v_2) & \cdots & \mathbf{0} & \mathbf{0} \\ \cdots & \cdots & \cdots & \cdots & \cdots \\ \mathbf{0} & \mathbf{0} & \cdots & \mathbf{Q}_r^*\mathbf{S}_r(v_r) & \mathbf{0} \\ \mathbf{T}_1\mathbf{S}_1(v_1) & \mathbf{T}_2\mathbf{S}_2(v_2) & \cdots & \mathbf{T}_r\mathbf{S}_r(v_r) & \mathbf{0} \end{pmatrix} \tag{7.23}$$

$$= (T - T(0))\,\mathbf{p}_0(\infty)\,\mathbf{L}(\boldsymbol{v})$$

where

$$\mathbf{S}_i(v_i) = \sum_{m=0}^{d_i-1} w_m(d_i)\,\mathbf{Q}_i^{d_i+v_i-m+1} \quad\text{and} \tag{7.24}$$

$$\mathbf{T}_i = \left(\mathbf{I} - \mathbf{T}^{d_i}\right)^{-1}\left[\sum_{l=0}^{d_i-1}\mathbf{T}^l\mathbf{T}_i\mathbf{Q}_i^{d_i-l-1}\right].$$

Now, (7.23) could be written as

$$\mathbf{L}(\boldsymbol{v}) = \begin{pmatrix} \mathbf{Q}_1^* & \mathbf{0} & \cdots & \mathbf{0} & \mathbf{0} \\ \mathbf{0} & \mathbf{Q}_2^* & \cdots & \mathbf{0} & \mathbf{0} \\ \cdots & \cdots & \cdots & \cdots & \cdots \\ \mathbf{0} & \mathbf{0} & \cdots & \mathbf{Q}_r^* & \mathbf{0} \\ \mathbf{0} & \mathbf{0} & \cdots & \mathbf{0} & \mathbf{I} \end{pmatrix}\begin{pmatrix} \mathbf{S}_1(v_1) & \mathbf{0} & \cdots & \mathbf{0} & \mathbf{0} \\ \mathbf{0} & \mathbf{S}_2(v_2) & \cdots & \mathbf{0} & \mathbf{0} \\ \cdots & \cdots & \cdots & \cdots & \cdots \\ \mathbf{0} & \mathbf{0} & \cdots & \mathbf{S}_r(v_r) & \mathbf{0} \\ \mathbf{0} & \mathbf{0} & \cdots & \mathbf{0} & \mathbf{I} \end{pmatrix} \times \tag{7.25}$$

$$\begin{pmatrix} \mathbf{I} & \mathbf{0} & \cdots & \mathbf{0} & \mathbf{0} \\ \mathbf{0} & \mathbf{I} & \cdots & \mathbf{0} & \mathbf{0} \\ \cdots & \cdots & \cdots & \cdots & \cdots \\ \mathbf{0} & \mathbf{0} & \cdots & \mathbf{I} & \mathbf{0} \\ \mathbf{T}_1 & \mathbf{T}_2 & \cdots & \mathbf{T}_r & \mathbf{0} \end{pmatrix} \vartriangle \begin{pmatrix} \mathbf{I} & \mathbf{0} & \cdots & \mathbf{0} & \mathbf{0} \\ \mathbf{0} & \mathbf{I} & \cdots & \mathbf{0} & \mathbf{0} \\ \cdots & \cdots & \cdots & \cdots & \cdots \\ \mathbf{0} & \mathbf{0} & \cdots & \mathbf{I} & \mathbf{0} \\ \mathbf{Q}_1^*\mathbf{S}_1(v_1) & \mathbf{Q}_2^*\mathbf{S}_2(v_2) & \cdots & \mathbf{Q}_r^*\mathbf{S}_r(v_r) & \mathbf{0} \end{pmatrix}$$

$$= \hat{\mathbf{Q}}^*\hat{\mathbf{Q}}(\mathbf{S}(\boldsymbol{v}))\,\hat{\mathbf{T}} \vartriangle \hat{\mathbf{Q}}_{\mathbf{T}}(\mathbf{S}(\boldsymbol{v})).$$

Following similar steps as above, we could show that the

$$\lim_{t\to\infty}\mathbf{U}(t) = \lim_{t\to\infty}\sum_{\tau=0}^{t}\Delta T(\tau)\,\boldsymbol{\mu}_0(\tau)\,\mathbf{V}(\tau+1, t)$$

splits into $d = \operatorname{lcm}\{d_1, d_2, ..., d_r\}$ converging subsequences given by

$$(T - T(0))\,\boldsymbol{\mu}_0(\infty)\begin{pmatrix} \mathbf{L}(\boldsymbol{v}) & \mathbf{L}(\boldsymbol{v})\mathbf{J} - \mathbf{J}(\mathbf{L}(\boldsymbol{v}) \times \mathbf{L}(\boldsymbol{v})) \\ \mathbf{0} & \mathbf{L}(\boldsymbol{v}) \times \mathbf{L}(\boldsymbol{v}) \end{pmatrix} = \tag{7.26}$$

$$(T - T(0))\,\boldsymbol{\mu}_0(\infty)\begin{pmatrix}\hat{\mathbf{Q}}^* & \hat{\mathbf{Q}}^*\mathbf{J} - \mathbf{J}\left(\hat{\mathbf{Q}}^* \times \hat{\mathbf{Q}}^*\right) \\ \mathbf{0} & \hat{\mathbf{Q}}^* \times \hat{\mathbf{Q}}^*\end{pmatrix} \times$$

$$\begin{pmatrix}\hat{\mathbf{Q}}(\mathbf{S}(v)) & \hat{\mathbf{Q}}(\mathbf{S}(v))\mathbf{J} - \mathbf{J}\left(\hat{\mathbf{Q}}(\mathbf{S}(v)) \times \hat{\mathbf{Q}}(\mathbf{S}(v))\right) \\ \mathbf{0} & \hat{\mathbf{Q}}(\mathbf{S}(v)) \times \hat{\mathbf{Q}}(\mathbf{S}(v))\end{pmatrix} \times$$

$$\begin{pmatrix}\hat{\mathbf{T}} \vartriangle \hat{\mathbf{Q}}_{\mathbf{T}}(\mathbf{S}(v)) & \begin{matrix}\hat{\mathbf{T}} \vartriangle \hat{\mathbf{Q}}_{\mathbf{T}}(\mathbf{S}(v))\mathbf{J} - \\ \mathbf{J}\left(\hat{\mathbf{T}} \vartriangle \hat{\mathbf{Q}}_{\mathbf{T}}(\mathbf{S}(v)) \times \hat{\mathbf{T}} \vartriangle \hat{\mathbf{Q}}_{\mathbf{T}}(\mathbf{S}(v))\right)\end{matrix} \\ \mathbf{0} & \left(\hat{\mathbf{T}} \vartriangle \hat{\mathbf{Q}}_{\mathbf{T}}(\mathbf{S}(v)) \times \hat{\mathbf{T}} \vartriangle \hat{\mathbf{Q}}_{\mathbf{T}}(\mathbf{S}(v))\right)\end{pmatrix}$$

$$= (T - T(0))\,\boldsymbol{\mu}_0(\infty)\,\mathbf{V}_{\mathbf{Q}^*}\mathbf{V}_{\mathbf{Q}(\mathbf{S}(v))}\mathbf{W}_{\mathbf{T}\vartriangle\mathbf{Q}_{\mathbf{T}}((\mathbf{S}(v)))}. \tag{7.27}$$

The matrix $\mathbf{W}_{\mathbf{T}\vartriangle\mathbf{Q}_{\mathbf{T}}(\mathbf{S}(v))}$ is not a V-matrix since the matrix

$$\left[\hat{\mathbf{T}} \vartriangle \hat{\mathbf{Q}}_{\mathbf{T}}(\mathbf{S}(v))\mathbf{J} - \mathbf{J}\left(\hat{\mathbf{T}} \vartriangle \hat{\mathbf{Q}}_{\mathbf{T}}(\mathbf{S}(v)) \times \hat{\mathbf{T}} \vartriangle \hat{\mathbf{Q}}_{\mathbf{T}}(\mathbf{S}(v))\right)\right]\mathbf{1}^\top \neq \mathbf{0}.$$

This is also the reason, why it cannot be written as a Hadamard product of two V-matrices.

In conclusion from (7.11) and (7.25) we get that the variability population structure $\boldsymbol{\mu}(t)$ as $t \to \infty$ splits into $d = \mathrm{lcm}\{d_1, d_2, ..., d_r\}$ converging subsequences given by

$$\boldsymbol{\mu}(0)\,\mathbf{V}_{\mathbf{Q}^*}\mathbf{V}_{\mathbf{Q}(v)}\mathbf{W}_{\mathbf{T}\vartriangle\mathbf{Q}_{\mathbf{T}}(v)} + (T - T(0))\,\boldsymbol{\mu}_0(\infty)\,\mathbf{V}_{\mathbf{Q}^*}\mathbf{V}_{\mathbf{Q}(\mathbf{S}(v))}$$
$$\times\mathbf{W}_{\mathbf{T}\vartriangle\mathbf{Q}_{\mathbf{T}}((\mathbf{S}(v)))},$$

and

$$[\boldsymbol{\mu}(0) + (T - T(0))\,\boldsymbol{\mu}_0(\infty)]\,\mathbf{V}_{\mathbf{Q}^*}\times$$
$$\left[\mathbf{V}_{\mathbf{Q}(v)}\mathbf{W}_{\mathbf{T}\vartriangle\mathbf{Q}_{\mathbf{T}}(v)} + \mathbf{V}_{\mathbf{Q}(\mathbf{S}(v))}\mathbf{W}_{\mathbf{T}\vartriangle\mathbf{Q}_{\mathbf{T}}((\mathbf{S}(v)))}\right].$$

■

7.3 An illustrative probable application

In the present section an illustrative probable application of the findings of the previous section will be given from the area of manpower planning. The example is taken from Vassiliou (2015) and serves only to illustrate the previous results. This area is well known from personal experience of the researches, which are under employment and the special technicalities of this area of applications are minimal. Consider a firm with 8 grades which are divided into two large sections. We chose the number 8 in order that our transition probability matrices will be presentable within the limit of space of the present book. The first section is assumed to have grades $\{1, 2, 3, 4\}$, which for example could be the section of traders in an investment bank. The second section has grades $\{5, 6\}$, which for example could be the section of "back office" staff in an investment bank (Quants, IT, etc.). The two sections do not interchange staff, due to different skills created after years of experience. The remaining grades $\{7, 8\}$ are mainly the training grades. In those are set new graduates who are being recruited in grade 8, and then after some time make the appropriate transition according to proven abilities into the two basic sections. In such a system with this structure, the sequence of transition probability matrices $\{\mathbf{Q}(t)\}_{t=0}^{\infty}$ having the same incidence matrix are of the form

$$\mathbf{Q}(t) = \begin{pmatrix}\mathbf{Q}_1(t) & 0 & 0 \\ 0 & \mathbf{Q}_2(t) & 0 \\ \mathbf{T}_1(t) & \mathbf{T}_2(t) & \mathbf{T}(t)\end{pmatrix}, \tag{7.28}$$

where

$$\mathbf{Q}_1(t) = \begin{pmatrix} 0 & 0.9-\frac{1}{3t^4+1} & 0 & 0.1+\frac{1}{3t^4+1} \\ 0.8-\frac{1}{5t} & 0 & 0.2+\frac{1}{5t} & 0 \\ 0 & 0.7-\frac{3}{7t} & 0 & 0.3+\frac{3}{7t} \\ 0.05+\frac{1}{8t^2} & 0 & 0.95-\frac{1}{8t^2} & 0 \end{pmatrix}, \tag{7.29}$$

$$\mathbf{Q}_2(t) = \begin{pmatrix} 0 & 1 \\ 1 & 0 \end{pmatrix}, \tag{7.30}$$

$$\mathbf{T}_1(t) = \begin{pmatrix} 0 & 0.1+\frac{1}{3t^2} & 0 & 0.1+\frac{1}{6t+1} \\ 0 & 0 & 0 & 0.2+\frac{2}{11t^2} \end{pmatrix}, \tag{7.31}$$

$$\mathbf{T}_2(t) = \begin{pmatrix} 0.1 & 0.4-\frac{1}{3t^2} \\ 0.2-\frac{3}{14t^3} & 0.2+\frac{1}{7t^3} \end{pmatrix}, \tag{7.32}$$

and

$$\mathbf{T}(t) = \begin{pmatrix} 0.3-\frac{1}{6t+1} & 0 \\ 0.3-\frac{2}{11t^2} & 0.1+\frac{1}{14t^3} \end{pmatrix} \tag{7.33}$$

It is easy to see that $\mathbf{Q}_1(t)$ is periodic with period $d_1 = 2$ and $\mathbf{Q}_2(t)$ is periodic with period $d_2 = 2$ and consequently $d = \operatorname{lcm}\{d_1, d_2\} = 2$.

Let that the initial number of memberships in the system is $T(0) = 1100$ and that

$$\begin{aligned} T(1) &= 1150, T(2) = 1190, T(3) = 1220, T(4) = 1250, \\ T(5) &= 1310,\ T(6) = 1400,\ \text{and } T = 1500. \end{aligned} \tag{7.34}$$

The entrance probabilities are the following

$$\mathbf{p}_0 = \left[0, 0.2-\frac{1}{35t}, 0, 0.2+\frac{1}{35t}, 0, 0.2-\frac{1}{7t}, 0, 0.4+\frac{1}{7t}\right]. \tag{7.35}$$

From (7.28) to (7.35), we have that $\lim_{t\to\infty} \|\mathbf{Q}(t) - \mathbf{Q}\| = 0$ with

$$\mathbf{Q} = \begin{pmatrix} 0 & 0.9 & 0 & 0.1 & 0 & 0 & 0 & 0 \\ 0.8 & 0 & 0.2 & 0 & 0 & 0 & 0 & 0 \\ 0 & 0.7 & 0 & 0.3 & 0 & 0 & 0 & 0 \\ 0.05 & 0 & 0.95 & 0 & 0 & 0 & 0 & 0 \\ 0 & 0 & 0 & 0 & 0 & 1 & 0 & 0 \\ 0 & 0 & 0 & 0 & 1 & 0 & 0 & 0 \\ 0 & 0.1 & 0 & 0.1 & 0.1 & 0.4 & 0.3 & 0 \\ 0 & 0 & 0 & 0.2 & 0.2 & 0.2 & 0.3 & 0.1 \end{pmatrix}. \tag{7.36}$$

From (3.8), we have that $\lim_{t\to\infty} \|\mathbf{p}_0(t) - \mathbf{p}_0(\infty)\| = 0$ with

$$\mathbf{p}_0(\infty) = [0, 0.2, 0, 0.2, 0, 0.2, 0, 0.4]. \tag{7.37}$$

Now, by simulating the system we find that, the results coincided with the results of Theorem 312 and relation (313) and we get that, for $t = 2\pi$ for π large

$$\lim_{t\to\infty} \mathbf{Q}(0,t) = \begin{pmatrix} 0 & 0.84 & 0 & 0.16 & 0 & 0 & 0 & 0 \\ 0.68 & 0 & 0.32 & 0 & 0 & 0 & 0 & 0 \\ 0 & 0.84 & 0 & 0.16 & 0 & 0 & 0 & 0 \\ 0.68 & 0 & 0.32 & 0 & 0 & 0 & 0 & 0 \\ 0 & 0 & 0 & 0 & 0 & 1 & 0 & 0 \\ 0 & 0 & 0 & 0 & 1 & 0 & 0 & 0 \\ 0.045 & 0.184 & 0.021 & 0.036 & 0.241 & 0.473 & 0 & 0 \\ 0.06 & 0.191 & 0.029 & 0.038 & 0.373 & 0.309 & 0 & 0 \end{pmatrix}, \tag{7.38}$$

and for $t = 2\pi + 1$ for π large

$$\lim_{t\to\infty} \mathbf{Q}(0,t) = \begin{pmatrix} 0.68 & 0 & 0.32 & 0 & 0 & 0 & 0 & 0 \\ 0 & 0.84 & 0 & 0.16 & 0 & 0 & 0 & 0 \\ 0.68 & 0 & 0.32 & 0 & 0 & 0 & 0 & 0 \\ 0 & 0.84 & 0 & 0.16 & 0 & 0 & 0 & 0 \\ 0 & 0 & 0 & 0 & 1 & 0 & 0 & 0 \\ 0 & 0 & 0 & 0 & 0 & 1 & 0 & 0 \\ 0.149 & 0.055 & 0.071 & 0.011 & 0.473 & 0.241 & 0 & 0 \\ 0.155 & 0.074 & 0.074 & 0.015 & 0.31 & 0.373 & 0 & 0 \end{pmatrix}. \tag{7.39}$$

Now, let that the initial population structure of the system is

$$\boldsymbol{\mu}(0) = [50, 80, 120, 150, 200, 300, 50, 150, 0, 0, ..., 0]. \tag{7.40}$$

Now, from (7.37) and (7.40) we get that

$$\begin{aligned} & [0, 0.2, 0, 0.2, 0, 0.2, 0, 0.4, 0, 0, 0, 0, 0, 0, 0, 0, \\ & 0, 0.16, 0, -0.04, 0, -0.04, 0, -0.08, 0, 0, 0, 0, 0, 0, 0, 0, \\ \boldsymbol{\mu}_0 = \ & 0, -0.04, 0, 0.16, 0, -0.04, 0, -0.08, 0, 0, 0, 0, 0, 0, 0, 0, \\ & 0, -0.04, 0, -0.04, 0, 0.16, 0, -0.08, 0, 0, 0, 0, 0, 0, 0, 0, \\ & 0, -0.08, 0, -0.08, 0, -0.08, 0, 0.24]. \end{aligned} \tag{7.41}$$

By simulating the system and from Theorem 313 we get that, for $t = 2\pi$ for π large

$$\mathbf{L}(\mathbf{0}) = \mathbf{L}(0,0) = \begin{pmatrix} 0.37 & 0.39 & 0.17 & 0.07 & 0 & 0 & 0 & 0 \\ 0.31 & 0.45 & 0.15 & 0.09 & 0 & 0 & 0 & 0 \\ 0.37 & 0.39 & 0.17 & 0.07 & 0 & 0 & 0 & 0 \\ 0.31 & 0.45 & 0.15 & 0.09 & 0 & 0 & 0 & 0 \\ 0 & 0 & 0 & 0 & 0.53 & 0.47 & 0 & 0 \\ 0 & 0 & 0 & 0 & 0.47 & 0.53 & 0 & 0 \\ 0.10 & 0.12 & 0.05 & 0.02 & 0.37 & 0.34 & 0 & 0 \\ 0.11 & 0.13 & 0.05 & 0.03 & 0.34 & 0.34 & 0 & 0 \end{pmatrix}, \tag{7.42}$$

and for $t = 2\pi + 1$ for π large we get that

$$\mathbf{L}(\mathbf{1}) = \mathbf{L}(1,1) = \begin{pmatrix} 0.31 & 0.45 & 0.15 & 0.09 & 0 & 0 & 0 & 0 \\ 0.37 & 0.39 & 0.17 & 0.07 & 0 & 0 & 0 & 0 \\ 0.31 & 0.45 & 0.15 & 0.09 & 0 & 0 & 0 & 0 \\ 0.37 & 0.39 & 0.17 & 0.07 & 0 & 0 & 0 & 0 \\ 0 & 0 & 0 & 0 & 0.47 & 0.53 & 0 & 0 \\ 0 & 0 & 0 & 0 & 0.53 & 0.47 & 0 & 0 \\ 0.09 & 0.13 & 0.04 & 0.03 & 0.35 & 0.36 & 0 & 0 \\ 0.1 & 0.14 & 0.05 & 0.03 & 0.34 & 0.34 & 0 & 0 \end{pmatrix}. \tag{7.43}$$

Now, from (7.34) we have that $(T - T(0)) = 400$ and by simulating the system and by Theorem 313 we get that for $t = 2\pi$ for π large

$$\begin{aligned} \boldsymbol{\mu}(\mathbf{0}) \ = \ & [233.25, 275.05, 111.0, 53.9, 460.0, 366.8, 0, 0, \\ & 115.5, -17.61, -55.61, -3.57, -19.0, -19.75, 0, 0, \\ & -17.61, 125.65, -8.57, -28.8, -34.61, -36.06, 0, 0, \\ & -55.61, -8.57, 84.74, -1.74, -9.24, -9.58, 0, 0, \\ & -3.57, -28.8, -1.74, 48.35, -6.98, -7.26, 0, 0, \\ & -19.0, -34.61, -9.24, -6.98, 115.07, -45.25, 0, 0, \\ & -19.75, -36.05, -9.58, -7.26, -45.25, 117.86, 0, 0, \\ & 0, 0, 0, 0, 0, 0, 0, 0, 0, 0, 0, 0, 0, 0, 0, 0]. \end{aligned} \tag{7.44}$$

Note that in $(7,44)$, we have $cov\left(N_i\left(\infty\right), N_j\left(\infty\right)\right) = cov\left(N_j\left(\infty\right), N_i\left(\infty\right)\right)$ for every $i, j = 1, 2, ..., 8$ as it should be. Now, for for $t = 2\pi + 1$ for π large we get that

$$\begin{aligned}\boldsymbol{\mu}\left(\mathbf{1}\right) \;=\; & [275.05, 233.25, 53.9, 111.0, 366.8, 460.0, 0, 0, \\ & 125.65, \; -17.61, \; -28.8, \; -8.57, \; -36.06, \; -34.61, 0, 0, \\ & -17.61, 115.5, \; -3.57, \; -55.61, -19.75, \; -19.0, 0, 0, \\ & -28.8, \; -3.57, 48.35, \; -1.74, \; -7.26, \; -6.98, 0, 0, \\ & -8.57, \; -55.61, \; -1.74, 84.74, \; -9.58, \; -9.24, 0, 0, \\ & -36.05, \; -19.75, \; -7.26, \; -9.58, 117,86, \; -45.25, 0, 0, \\ & -34.61, \; -19.0, \; -6.98, \; -9.24, \; -45.25, 115.07, 0, 0, \\ & 0, 0, 0, 0, 0, 0, 0, 0, 0, 0, 0, 0, 0, 0, 0, 0].\end{aligned} \tag{7.45}$$

Hence, the sequence $\boldsymbol{\mu}(t)$ splits into two converging subsequences as $t \to \infty$ given in (7.44) and (7.45) as was proved in Theorems 312 and 313. Note that the first eight values in (7.44) and (7.45) are the expected number of memberships in the various states of the system, the positive elements that follow are their variances and the negative are the covariances.

7.4 Rate of convergence of the variability vector

In the previous section, we studied the convergence of the vector of means, variances and covariances of the state sizes in an NHMS in the most general case. That is, when the sequence of matrices $\{\mathbf{Q}(t)\}_{t=0}^{\infty}$ of the inherent Markov chain consists of arbitrary stochastic matrices all of which have the same incidence matrix. In this section, we will study another practical problem. Our goal will be to find the conditions under which the rate of convergence of the vector of means, variances and covariances of the state sizes in an NHMS is geometrical. The importance of answering this problem for practical purposes is apparent in Bartholomew (1973) and (1982) for the homogeneous Markov system which is a very special case. It is also includes as corollaries the studies of special cases in as in Feichtinger (1976), Mehlmann (1977), Feichtinger and Mehlmann (1976), Bartholomew and Forbes (1979) in manpower systems.

We will now start with a theorem concerning the convergence of the sequence $\{\mathbf{V}(t)\}_{t=0}^{\infty}$ in a geometrical rate in an NHMS when the inherent Markov chain uniquely defined by the sequence of matrices $\{\mathbf{Q}(t)\}_{t=0}^{\infty}$ converges to a regular homogeneous Markov chain.

Theorem 314 *(Vassiliou and Tsaklidis (1989)). Let an NHMS and assume that the sequence* $\{\mathbf{Q}(t)\}_{t=0}^{\infty}$ *of the inherent non-homogeneous Markov chain converges to a regular homogeneous Markov chain with transition probability matrix* $\mathbf{Q}$ *then the respective products from the non-homogeneous sequence of* $\mathbf{V}$*-matrices* $\{\mathbf{V}_q(t)\}_{t=0}^{\infty}$ *converges to a homogeneous* $\mathbf{V}_q$*-matrix*

$$\mathbf{V}_q(\infty) = \begin{pmatrix} \mathbf{Q}(\infty) & \mathbf{Q}(\infty)\mathbf{J} - \mathbf{J}(\mathbf{Q}(\infty) \otimes \mathbf{Q}(\infty)) \\ \mathbf{0} & \mathbf{Q}(\infty) \otimes \mathbf{Q}(\infty) \end{pmatrix}, \tag{7.46}$$

where $\mathbf{Q}(\infty) = \lim_{t\to\infty} \mathbf{Q}^t$, *geometrically fast and uniformly in* t *the initial time point of the product.*

Proof. From the properties of the $\mathbf{V}$-matrices we easily see that

$$\mathbf{V}_q(t, t+n) = \begin{pmatrix} \mathbf{Q}(t, t+n) & \mathbf{Q}(t, t+n)\mathbf{J} - \mathbf{J}(\mathbf{Q}(t, t+n) \otimes \mathbf{Q}(t, t+n)) \\ \mathbf{0} & \mathbf{Q}(t, t+n) \otimes \mathbf{Q}(t, t+n) \end{pmatrix}, \tag{7.47}$$

Denote by

$$\mathbf{D}_1(t, n) = \mathbf{Q}(t, t+n) - \mathbf{Q}(\infty), \tag{7.48}$$

$$\mathbf{D}_2(t,n) = \mathbf{Q}(t,t+n) \otimes \mathbf{Q}(t,t+n) - \mathbf{Q}(\infty) \otimes \mathbf{Q}(\infty), \tag{7.49}$$

then we get that

$$\begin{aligned} &\|\mathbf{V}_q(t,t+n) - \mathbf{V}_q(\infty)\| = \\ &\max(\|\mathbf{D}_1(t,n) + \mathbf{D}_1(t,n)\mathbf{J} - \mathbf{J}\mathbf{D}_2(t,n)\|, \|\mathbf{D}_2(t,n)\|) \\ &\leq \max\{2\|\mathbf{D}_1(t,n)\| + \|\mathbf{D}_2(t,n)\|, \|\mathbf{D}_2(t,n)\|\} \\ &= 2\|\mathbf{D}_1(t,n)\| + \|\mathbf{D}_2(t,n)\|. \end{aligned} \tag{7.50}$$

According to Theorem 246 since

$$\lim_{t\to\infty} \|\mathbf{Q}(t) - \mathbf{Q}\| = 0 \text{ with } \mathbf{Q} \text{ regular uniformly in } t, \tag{7.51}$$

there are constants $c > 0$ and $0 < b < 1$ such that

$$\|\mathbf{D}_1(t,n)\| \leq cb^{n+1} \text{ for every } t, n \in \mathbb{N}. \tag{7.52}$$

Also we have that

$$\begin{aligned} &\|\mathbf{D}_2(t,n)\| \leq \|(\mathbf{Q}(t,t+n) - \mathbf{Q}(\infty)) \otimes \mathbf{Q}(\infty)\| + \\ &\|\mathbf{Q}(t,t+n) \otimes (\mathbf{Q}(t,t+n) - \mathbf{Q}(\infty))\| = 2\|\mathbf{Q}(t,t+n) - \mathbf{Q}(\infty)\|. \end{aligned} \tag{7.53}$$

From (7.53), (7.52), and (7.50) we get that

$$\|\mathbf{V}_q(t,t+n) - \mathbf{V}_q(\infty)\| \leq 4cb^{n+1} \text{ for every } t, n.$$

■

Lemma 315 *Let an NHMS and assume that the sequence $\{\mathbf{Q}(t)\}_{t=0}^{\infty}$ of the inherent non-homogeneous Markov chain converges to a regular homogeneous Markov chain with transition probability matrix $\mathbf{Q}$ then*

$$\begin{aligned} &\boldsymbol{\mu}(0)\mathbf{V}_q(\infty) + (T - T(0))\boldsymbol{\mu}_0\mathbf{V}_q(\infty) = \\ &T\left[\mathbf{q}(\infty) \quad \mathbf{q}(\infty)\mathbf{J} - \mathbf{J}(\mathbf{q}(\infty) \otimes \mathbf{q}(\infty))\right] = T\mathbf{v}_q(\infty), \end{aligned}$$

where $\mathbf{q}(\infty)$ is the row of the stable matrix $\mathbf{Q}(\infty)$.

Proof. We know that

$$\boldsymbol{\mu}(0) = \left[\mathbf{N}(0) \quad \mathbf{0}\right] \text{ where } \mathbf{N}(0) \text{ is } 1 \times k \text{ and } \mathbf{0} \text{ is } k(1+k).$$

Now, we have that

$$\begin{aligned} \boldsymbol{\mu}(0)\mathbf{V}_q(\infty) &= \boldsymbol{\mu}(0)\begin{pmatrix} \mathbf{Q}(\infty) & \mathbf{Q}(\infty)\mathbf{J} - \mathbf{J}(\mathbf{Q}(\infty) \otimes \mathbf{Q}(\infty)) \\ \mathbf{0} & \mathbf{Q}(\infty) \otimes \mathbf{Q}(\infty) \end{pmatrix} \\ &= T(0)\left[\mathbf{q}(\infty) \quad \mathbf{q}(\infty)\mathbf{J} - \mathbf{J}(\mathbf{q}(\infty) \otimes \mathbf{q}(\infty))\right] = T(0)\mathbf{v}_q(\infty). \end{aligned} \tag{7.54}$$

The input vector $\boldsymbol{\mu}_0$ could be written as

$$\boldsymbol{\mu}_0 = \begin{pmatrix}\mathbf{p}_0 & \mathbf{p}_0 \otimes \mathbf{p}_0\end{pmatrix},$$

hence, we have that

$$\begin{aligned} &(T - T(0))\boldsymbol{\mu}_0\mathbf{V}_q(\infty) = \\ &(T - T(0))\begin{pmatrix}\mathbf{p}_0 & \mathbf{p}_0 \otimes \mathbf{p}_0\end{pmatrix}\begin{pmatrix} \mathbf{Q}(\infty) & \mathbf{Q}(\infty)\mathbf{J} - \mathbf{J}(\mathbf{Q}(\infty) \otimes \mathbf{Q}(\infty)) \\ \mathbf{0} & \mathbf{Q}(\infty) \otimes \mathbf{Q}(\infty) \end{pmatrix}, \end{aligned}$$

due to Theorem 131, we get that

$$(T - T(0))\boldsymbol{\mu}_0\mathbf{V}_q(\infty) = (T - T(0))\mathbf{v}_q(\infty). \tag{7.55}$$

Now, from (7.54) and (7.55) we get what we want to prove. ■

We are now in a position to provide and prove under what conditions the rate of convergence of the vector $\boldsymbol{\mu}(t)$ for an NHMS is geometrical in the form of the following theorem:

Theorem 316 *(Vassiliou and Tsaklidis (1989)). Let a complete probability space* $(\Omega, \mathcal{F}, \mathbb{P})$ *and an NHMS. Let that the sequence* $\{\Delta T(t)\}_{t=0}^{\infty}$ *be nonnegative and that it converges to zero geometrically fast. Also let that the transition probability sequence* $\{\mathbf{Q}(t)\}_{t=0}^{\infty}$ *is such that*

$$\lim_{t\to\infty} \|\mathbf{Q}(t) - \mathbf{Q}\| = 0 \text{ where } \mathbf{Q} \text{ is a regular matrix,}$$

$$\text{and } \lim_{t\to\infty} \|\mathbf{p}_0(t) - \mathbf{p}_0\| = 0,$$

and let that the convergence be geometrically fast. Then the sequence $\boldsymbol{\mu}(t)$ *as* $t \to \infty$ *converges geometrically fast to a vector* $\boldsymbol{\mu}(\infty)$ *given by*

$$\boldsymbol{\mu}(\infty) = (T - T(0)) \mathbf{v}_q(\infty).$$

Proof. According to the previous theorem, it is sufficient to show that

$$\lim_{t\to\infty} \boldsymbol{\mu}(t) = \boldsymbol{\mu}(\infty) = \boldsymbol{\mu}(0) \mathbf{V}_q(\infty) + (T - T(0)) \boldsymbol{\mu}_0 \mathbf{V}_q(\infty)$$

geometrically fast. Firstly, since the sequence $\{\Delta T(t)\}_{t=0}^{\infty}$ is nonnegative and it converges to zero geometrically fast we get that there are $c > 0$ and $0 < b < 1$ such that

$$|\Delta(t)| \leq cb^t. \tag{7.56}$$

We have that

$$T - T(0) = \sum_{\tau=0}^{\infty} \Delta T(t) = \sum_{\tau=0}^{t} \Delta T(t) + \sum_{\tau=t+1}^{\infty} \Delta T(t),$$

hence

$$\begin{aligned} &\|\boldsymbol{\mu}(t) - \boldsymbol{\mu}(\infty)\| \leq \|\boldsymbol{\mu}(0)\| \|\mathbf{V}_q(0,t) - \mathbf{V}_q(\infty)\| + \\ &\sum_{\tau=0}^{t} \|\Delta T(\tau) \boldsymbol{\mu}_0(\tau) \mathbf{V}_q(\tau,t) - \Delta T(\tau) \boldsymbol{\mu}_0 \mathbf{V}_q(\infty)\| + \\ &\sum_{\tau=t+1}^{\infty} |\Delta T(t)| \|\boldsymbol{\mu}_0 \mathbf{V}_q(\infty)\| = \mathbf{A}(t) + \mathbf{B}(t) + \mathbf{C}(t). \end{aligned} \tag{7.57}$$

We now have from Theorem 314 that there are $c_1 > 0$ and $0 < b_1 < t$ such that

$$\|\mathbf{V}_q(0,t) - \mathbf{V}_q(\infty)\| \leq c_1 b_1^t,$$

therefore

$$\mathbf{A}(t) \leq \|\boldsymbol{\mu}(0)\| c_1 b_1^t \leq c_2 b_2^t. \tag{7.58}$$

From (7.57), theorem 314, and the conditions of the present theorem, we get

$$\begin{aligned} &\mathbf{B}(t) \leq \sum_{\tau=0}^{t} |\Delta T(\tau)| \|\boldsymbol{\mu}_0(\tau) \mathbf{V}_q(\tau,t) - \boldsymbol{\mu}_0 \mathbf{V}_q(\infty)\| \leq \\ &\sum_{\tau=0}^{t} |\Delta T(\tau)| \|\boldsymbol{\mu}_0\| \|\mathbf{V}_q(\tau,t) - \mathbf{V}_q(\infty)\| + \\ &\sum_{\tau=0}^{t} |\Delta T(\tau)| \|\boldsymbol{\mu}_0(\tau) - \boldsymbol{\mu}_0\| \|\mathbf{V}_q(\tau,t)\| \leq \\ &\sum_{\tau=0}^{t} cb^\tau \|\boldsymbol{\mu}_0\| c_1 b_1^{t-\tau} + \sum_{\tau=0}^{t} cb^\tau c_3 b_3^{t-\tau} \|\mathbf{V}_q(\tau,t)\| \leq \end{aligned}$$

$$\text{(with no loss of generality assume } b < b_1 \text{ and } b < b_3)$$
$$\leq \|\boldsymbol{\mu}_0\| \, cc_1 b_1^t \left(1 - b/b_1\right)^{-1} + \|\mathbf{V}_q(\tau, t)\| \, cc_3 b_3^t \left(1 - b/b_3\right)^{-1}$$
$$\leq c_1^* b_1^t + c_2^* b_3^t \leq (\text{assuming } b_1 < b_3) \leq \left(c_1^* + c_2^*\right) b_3^t \tag{7.59}$$

for $t = 1, 2, ...$ where apparently $\left(c_1^* + c_2^*\right) > 0$ and $0 < b_3 < 1$. Now for the remaining sum, we have

$$\mathbf{C}(t) \leq \sum_{\tau=0}^{\infty} |\Delta T(\tau)| \, \|\boldsymbol{\mu}_0 \mathbf{V}_q(\infty)\| \leq c \, \|\boldsymbol{\mu}_0\| \, \|\mathbf{V}_q(\infty)\| \sum_{\tau=0}^{\infty} b^{\tau}$$
$$\leq c_3^* \left(1 - b\right)^{-1} \quad \text{for } t = 1, 2, ... \tag{7.60}$$

From (7.57), (7.58), (7.59), (7.60) we arrive easily at the desired conclusion. ■

7.5 Research notes

Related work of some interest is also the study of the factorial moments of the individual state sizes in a homogeneous Markov system (HMS) and also of the factorial moments jointly (mixed) of all the state sizes which appeared in Vasiliadis and Tsaklidis (2008). This is done for both cases when the transition probability matrix **B** is regular and periodic. The insertion of a new matrix product which is closely related with the Kronecker product provides the necessary instrument to find the factorial and joint factorial moments in a matrix form. As a consequence the asymptotic behavior of the moments is possible to be found in closed elegant form. Also the p.d.f.'s of the state sizes follow directly by means of the moments.

Chapter 8

CYCLIC BEHAVIOR OF NON-HOMOGENEOUS MARKOV SYSTEMS

8.1 Introductory notes

In this chapter, we will introduce and study cyclic behavior of an NHMS. The importance of cyclic behavior was firstly stressed in Bartholomew (1982), p. 71 in the context of manpower planning. The motive of this concept arose from the study in Gani (1963) of student enrolment at Michigan State University. Note that all too often in practice random variables in various models exhibit a cyclic behavior. Examples of that apart from the study of student enrolment could be found in modeling ecological phenomena where the sequence of the four epochs create naturally a cycle in the random variables involved. Such an example was mentioned in Chapter 5 in the study of Patoucheas and Stamou (1993) were seasonal variations has been reported not only the zoobenthos which actually evaluates the ecological conditions in the Thermaikos Golf but also to other biotic and antibiotic parameters involved in the ecology of the entire region.

The central problem of the present chapter will be the asymptotic behavior of the expected population structure and the relative population structure of cyclic non-homogeneous Markov systems (Cy-NHMS) and the behavior of their asymptotic variability. We will provide a series of theorems were under various conditions there exist a stationary distribution for the expected relative population structure and also for the variability vector of variance and covariances, which will be found in closed analytic form. We will also study the important question under what conditions the rate of convergence of the expected population structure and the variability vector of variances and covariances is geometrical.

8.2 Cyclic non-homogeneous Markov systems

We will start with the definition of a Cy-NHMS as was given for the first time in Vassiliou (1986):

Definition 317 *A non-homogeneous Markov system undergoes a cyclic behavior with period d if and only if*

$$\mathbf{P}(nd+r) = \mathbf{P}(r), \ \mathbf{p}_{k+1}^{\top}(nd+r) = \mathbf{p}_{k+1}^{\top}(r), \ \mathbf{p}_0(nd+r) = \mathbf{p}_0(r) \qquad (8.1)$$
$$\text{and} \quad T(nd+r) = T(r).$$

That is, all the parameters that uniquely determine an NHMS exhibit a cyclic behavior of period d. It is apparent also that the inherent non-homogeneous Markov chain will exhibit also a cyclic behavior with transition probability sequence $\{\mathbf{Q}(t)\}_{t=0}^{\infty}$ given by

$$\mathbf{Q}(nd+r) = \mathbf{P}(nd+r) + \mathbf{p}_{k+1}^{\top}(nd+r)\,\mathbf{p}_0(nd+r) = \mathbf{Q}(r). \qquad (8.2)$$

DOI: 10.1201/b23204-8

Also for the sequence of $\{\mathbf{V}_q(t)\}_{t=0}^{\infty}$ matrices, we have

$$\mathbf{V}_q(nd+r) = \begin{pmatrix} \mathbf{Q}(nd+r) & \mathbf{Q}(nd+r)\mathbf{J} - \mathbf{J}(\mathbf{Q}(nd+r)\otimes\mathbf{Q}(nd+r)) \\ \mathbf{0} & \mathbf{Q}(nd+r)\otimes\mathbf{Q}(nd+r) \end{pmatrix}$$

$$= \begin{pmatrix} \mathbf{Q}(r) & \mathbf{Q}(r)\mathbf{J} - \mathbf{J}(\mathbf{Q}(r)\otimes\mathbf{Q}(r)) \\ \mathbf{0} & \mathbf{Q}(r)\otimes\mathbf{Q}(r) \end{pmatrix} = \mathbf{V}_q(r). \tag{8.3}$$

Now let us define by

$$\begin{aligned} \mathbf{Q}_0 &= \mathbf{Q}(0)\mathbf{Q}(1)...\mathbf{Q}(d-1), \\ \mathbf{Q}_1 &= \mathbf{Q}(1)\mathbf{Q}(2)...\mathbf{Q}(d-1)\mathbf{Q}(0), \\ &\ldots\ldots\ldots\ldots\ldots\ldots \\ \mathbf{Q}_{d-1} &= \mathbf{Q}(d-1)\mathbf{Q}(0)...\mathbf{Q}(d-2), \end{aligned} \tag{8.4}$$

and

$$\begin{aligned} \mathbf{V}_{q0} &= \mathbf{V}_q(0)\mathbf{V}_q(1)...\mathbf{V}_q(d-1), \\ \mathbf{V}_{q1} &= \mathbf{V}_q(1)\mathbf{V}_q(2)...\mathbf{V}_q(d-1)\mathbf{V}_q(0), \\ &\ldots\ldots\ldots\ldots\ldots\ldots \\ \mathbf{V}_{qd-1} &= \mathbf{V}_q(d-1)\mathbf{V}_q(0)...\mathbf{V}_q(d-2), \end{aligned} \tag{8.5}$$

Now, from Proposition 309 we immediately get that

$$\mathbf{V}_q^t = \begin{pmatrix} \mathbf{Q}^t & \mathbf{Q}^t\mathbf{J} - \mathbf{J}(\mathbf{Q}^t\otimes\mathbf{Q}^t) \\ \mathbf{0} & \mathbf{Q}^t\otimes\mathbf{Q}^t \end{pmatrix} \text{ for } t \geq 1. \tag{8.6}$$

We are now in a position to prove the following Proposition:

Proposition 318 *For the sequence of* $\mathbf{V}$*-matrices* $\{\mathbf{V}_q(t)\}_{t=0}^{\infty}$*, we have*

$$\mathbf{V}_{qm} = \begin{pmatrix} \mathbf{Q}_m & \mathbf{Q}_m\mathbf{J} - \mathbf{J}(\mathbf{Q}_m\otimes\mathbf{Q}_m) \\ \mathbf{0} & \mathbf{Q}_m\otimes\mathbf{Q}_m \end{pmatrix}. \tag{8.7}$$

Proof. From (8.4) and (8.5), we get that

$$\mathbf{V}_{qm} = \mathbf{V}_q(m)\mathbf{V}_q(m+1)...\mathbf{V}_q(0)...\mathbf{V}_q(m-1)$$

$$= \begin{pmatrix} \mathbf{Q}(m) & \mathbf{Q}(m)\mathbf{J} - \mathbf{J}(\mathbf{Q}(m)\otimes\mathbf{Q}(m)) \\ \mathbf{0} & \mathbf{Q}(m)\otimes\mathbf{Q}(m) \end{pmatrix} \times$$

$$\begin{pmatrix} \mathbf{Q}(m+1) & \mathbf{Q}(m+1)\mathbf{J} - \mathbf{J}(\mathbf{Q}(m+1)\otimes\mathbf{Q}(m+1)) \\ \mathbf{0} & \mathbf{Q}(m+1)\otimes\mathbf{Q}(m+1) \end{pmatrix} \times$$

$$... \times \begin{pmatrix} \mathbf{Q}(0) & \mathbf{Q}(0)\mathbf{J} - \mathbf{J}(\mathbf{Q}(0)\otimes\mathbf{Q}(0)) \\ \mathbf{0} & \mathbf{Q}(0)\otimes\mathbf{Q}(0) \end{pmatrix} \times ... \times$$

$$\begin{pmatrix} \mathbf{Q}(m-1) & \mathbf{Q}(m-1)\mathbf{J} - \mathbf{J}(\mathbf{Q}(m-1)\otimes\mathbf{Q}(m-1)) \\ \mathbf{0} & \mathbf{Q}(m-1)\otimes\mathbf{Q}(m-1) \end{pmatrix} =$$

$$\begin{pmatrix} \mathbf{Q}_m & \mathbf{Q}_m\mathbf{J} - \mathbf{J}(\mathbf{Q}_m\otimes\mathbf{Q}_m) \\ \mathbf{0} & \mathbf{Q}_m\otimes\mathbf{Q}_m \end{pmatrix}.$$

■

We will now prove the following basic theorem:

Theorem 319 *(Vassiliou (1986)). Let a complete probability $(\Omega, \mathcal{F}, \mathbb{P})$ and a non-homogeneous Markov system defined on it, with its basic parameters to undergo a cyclic behavior, that is, we consider a Cy-NHMS. Let also that $\boldsymbol{\mu}(t)$ be the vector of variances and covariances defined previously. In addition assume that the system is expanding, i.e., $T(t) \geq 0$ for every t and that $\lim_{t\to\infty} T(t) = T$. Then if $\mathbf{Q}_0$ as defined in the set of equations (8.4) is a regular stochastic matrix, then the sequence $\{\boldsymbol{\mu}(t)\}_{t=0}^{\infty}$ splits into d subsequences with limits*

$$\boldsymbol{\mu}_r(\infty) = \boldsymbol{\mu}(0)\mathbf{V}_{q,r+1} + [T - T(0)]\{\sum_{m=0}^{r} w_m \boldsymbol{\mu}_0(m)\mathbf{V}^*_{q,m+1}\mathbf{V}_q(m+1,r)$$

$$+ \sum_{m=r+1}^{d-1} w_m \boldsymbol{\mu}_0(m)\mathbf{V}^*_{q,m+1}\mathbf{V}_q(m+1,d-1)\mathbf{V}_q(0,r)\}. \tag{8.8}$$

Where w_m is defined as

$$w_m = \frac{\lim_{t\to\infty}\sum_{\tau=0}^{[(\pi d+r)/d]-1}\Delta(\tau d+m)}{\lim_{t\to\infty}\sum_{\tau=0}^{t}\Delta(\tau)}, \tag{8.9}$$

and $\mathbf{Q}_0^* = \lim_{t\to\infty}\mathbf{Q}_0^t$;

$$\mathbf{V}^*_{q,0} = \lim_{t\to\infty}\mathbf{V}^t_{q0} = \lim_{t\to\infty}\begin{pmatrix}\mathbf{Q}_0^t & \mathbf{Q}_0^t\mathbf{J} - \mathbf{J}(\mathbf{Q}_0^t \times \mathbf{Q}_0^t) \\ \mathbf{0} & \mathbf{Q}_0^t \times \mathbf{Q}_0^t\end{pmatrix} \tag{8.10}$$

$$= \begin{pmatrix}\mathbf{Q}_0^* & \mathbf{Q}_0^*\mathbf{J} - \mathbf{J}(\mathbf{Q}_0^* \times \mathbf{Q}_0^*) \\ \mathbf{0} & \mathbf{Q}_0^* \times \mathbf{Q}_0^*\end{pmatrix}$$

and

$$\mathbf{V}^*_{q,i} = \mathbf{V}^*_{q,0}\left(\prod_{j=0}^{i-1}\mathbf{V}_q(j)\right) \quad \textit{for } i = 1,2,...,d-1. \tag{8.11}$$

Proof. Since $\mathbf{Q}_0$ is a regular stochastic matrix then $\mathbf{Q}_0^* = \lim_{t\to\infty}\mathbf{Q}_0^t$ exists and is a stable stochastic matrix. From this fact it is evident that (8.10) is valid. Also after reading carefully section 7.2 it will be easily evident that we may multiply the second part of the equation from the left by a $\mathbf{V}_q$ matrix without altering the result. Hence we get

$$\mathbf{V}^*_{q,i} = \left(\prod_{j=i}^{d-1}\mathbf{V}_q(j)\right)\mathbf{V}^*_{q,0}\left(\prod_{j=0}^{i-1}\mathbf{V}_q(j)\right) \quad \text{for } i = 1,2,...,d-1. \tag{8.12}$$

We have already seen that when the NHMS undergoes a cyclic behavior so does also the sequence of $\{\mathbf{V}_q(t)\}_{t=0}^{\infty}$ matrices, hence from (8.5) we have that

$$\mathbf{V}^t_{qi} = \mathbf{V}_{qi}\mathbf{V}_{qi}...\mathbf{V}_{qi} =$$

$$\{\mathbf{V}_q(i)\mathbf{V}_q(i+1)...\mathbf{V}_q(d-1)\mathbf{V}_q(0)\mathbf{V}_q(1)...\mathbf{V}_q(i-1)\} \times$$

$$\{\mathbf{V}_q(i)\mathbf{V}_q(i+1)...\mathbf{V}_q(d-1)\mathbf{V}_q(0)\mathbf{V}_q(1)...\mathbf{V}_q(i-1)\} \times ...$$

$$\times\{\mathbf{V}_q(i)\mathbf{V}_q(i+1)...\mathbf{V}_q(d-1)\mathbf{V}_q(0)\mathbf{V}_q(1)...\mathbf{V}_q(i-1)\}$$

$$= \left(\prod_{j=i}^{d-1}\mathbf{V}_q(j)\right)\mathbf{V}^{t-1}_{q,0}\left(\prod_{j=0}^{i-1}\mathbf{V}_q(j)\right). \tag{8.13}$$

Now from (8.12) and (8.13) we get that

$$\left\|\mathbf{V}_{qi}^{t}-\mathbf{V}_{q,i}^{*}\right\| = \| \left(\prod_{j=i}^{d-1}\mathbf{V}_q(j)\right)\mathbf{V}_{q,0}^{*}\left(\prod_{j=0}^{i-1}\mathbf{V}_q(j)\right) -$$

$$\left(\prod_{j=i}^{d-1}\mathbf{V}_q(j)\right)\mathbf{V}_{q,0}^{t-1}\left(\prod_{j=0}^{i-1}\mathbf{V}_q(j)\right) \| \leq$$

$$\left\|\left(\prod_{j=i}^{d-1}\mathbf{V}_q(j)\right)\right\|\left\|\mathbf{V}_{q,0}^{t-1}-\mathbf{V}_{q,0}^{*}\right\|\left\|\left(\prod_{j=0}^{i-1}\mathbf{V}_q(j)\right)\right\|. \tag{8.14}$$

Since we have $\lim_{t\to\infty}\mathbf{V}_{q,0}^{t-1}=\mathbf{V}_{q,0}^{*}$ we immediately get that

$$\lim_{t\to\infty}\left\|\mathbf{V}_{qi}^{t}-\mathbf{V}_{q,i}^{*}\right\|=0. \tag{8.15}$$

Now, without loss of generality assume that $t=\pi d+r$ then we have

$$\mathbf{V}_q(0,t)=\mathbf{V}_q(0,\pi d+r)=\mathbf{V}_q(0)\mathbf{V}_q(1)...\mathbf{V}_q(d-1)\times$$

$$\mathbf{V}_q(0)\mathbf{V}_q(1)...\mathbf{V}_q(d-1)\times...\times$$

$$\mathbf{V}_q(0)\mathbf{V}_q(1)...\mathbf{V}_q(d-1)\times \qquad (\pi d \ \text{ times})$$

$$\mathbf{V}_q(0)\mathbf{V}_q(1)...\mathbf{V}_q(r)=\mathbf{V}_{q,0}^{\pi d}\mathbf{V}_q(0,r). \tag{8.16}$$

Taking limits in both sides, it follows

$$\lim_{t\to\infty}\mathbf{V}_q(0,t)=\lim_{\pi\to\infty}\mathbf{V}_q(0,\pi d+r)=\lim_{\pi\to\infty}\mathbf{V}_{q,0}^{\pi d}\mathbf{V}_q(0,r)=$$

$$\mathbf{V}_{q,0}^{*}\mathbf{V}_q(0,r)=\mathbf{V}_{q,r+1}^{*}. \tag{8.17}$$

We go now in the second part of the right hand side of equation (7.10) and we write

$$U(t)=\sum_{\tau=1}^{t}\Delta T(\tau)\boldsymbol{\mu}_0(\tau)\mathbf{V}_q(\tau,t)=$$

$$U(\pi d+r)=\sum_{\tau=1}^{\pi d+r}\Delta T(\tau)\boldsymbol{\mu}_0(\tau)\mathbf{V}_q(\tau,\pi d+r)=$$

$$\sum_{m=0}^{d-1}\sum_{\tau=0}^{[(\pi d+r)/d]-1}\Delta T(\tau d+m)\boldsymbol{\mu}_0(\tau d+m)\mathbf{V}_q(\tau d+m,\pi d+r)$$

$$+\sum_{m=0}^{r}\Delta T(\pi d+m)\boldsymbol{\mu}_0(\pi d+m)\mathbf{V}_q(\pi d+m,\pi d+r)= \tag{8.18}$$

$$=U_1(\pi d+r)+U_2(\pi d+r)$$

where in the summation [.] means the greatest integer function. Now, since $\lim_{t\to\infty}T(t)=T$ then $\lim_{t\to\infty}\Delta T(t)=\lim_{t\to\infty}[T(t)-T(t-1)]=0$. Hence

$$\lim_{\pi\to\infty}\left\|\sum_{m=0}^{r}\Delta T(\pi d+m)\boldsymbol{\mu}_0(\pi d+m)\mathbf{V}_q(\pi d+m,\pi d+r)\right\|\leq$$

$$\lim_{\pi\to\infty}\sum_{m=0}^{r}\Delta T(\pi d+m)\left\|\boldsymbol{\mu}_0(\pi d+m)\right\|\left\|\mathbf{V}_q(\pi d+m,\pi d+r)\right\|=0. \tag{8.19}$$

Hence, what remains is to find the limit of the first summation in the second part of equation (8.18). Let that $m\leq r$ then we have

$$\begin{aligned}\mathbf{V}_q(\tau d+m,\pi d+r)&=\mathbf{V}_q(\tau d+m)\mathbf{V}_q(\tau d+m+1)\ldots\mathbf{V}_q(\tau d+d-1)\times\\ &\mathbf{V}_q((\tau+1)d+m)\mathbf{V}_q((\tau+1)d+m+1)\ldots\mathbf{V}_q((\tau+1)d+d-1)\times\ldots\\ &\times\mathbf{V}_q(\pi d+1)\mathbf{V}_q(\pi d+2)\ldots\mathbf{V}_q(\pi d+r).\end{aligned} \tag{8.20}$$

Since the system is under cyclic behavior

$$\begin{gathered}\mathbf{V}_q(\tau d+m,\pi d+r)=\mathbf{V}_q(m)\mathbf{V}_q(m+1)\ldots\\ \times\mathbf{V}_q(d-1)\mathbf{V}_q(0)\mathbf{V}_q(1)\ldots\mathbf{V}_q(m-1)\times\\ \mathbf{V}_q(m)\mathbf{V}_q(m+1)\ldots\mathbf{V}_q(d-1)\mathbf{V}_q(0)\mathbf{V}_q(1)\ldots\mathbf{V}_q(m-1)\times\\ \ldots\ldots\ldots\ldots\ldots\ldots\ldots\ldots\\ \mathbf{V}_q(m)\mathbf{V}_q(m+1)\ldots\mathbf{V}_q(d-1)\mathbf{V}_q(0)\mathbf{V}_q(1)\ldots\mathbf{V}_q(m-1)\times\\ \text{(i.e., }[(\pi-\tau)d+r-m]d\quad\text{times)}\\ \times\mathbf{V}_q(m)\mathbf{V}_q(m+1)\ldots\mathbf{V}_q(r).\end{gathered} \tag{8.21}$$

Hence, by the the notation introduced in (8.17) we have

$$\mathbf{V}_q(\tau d+m,\pi d+r)=\mathbf{V}_{qm}^{[(\pi-\tau)d+r-m]d}\,\mathbf{V}_q(m,r). \tag{8.22}$$

Similarly when $m<r$ we get

$$\mathbf{V}_q(\tau d+m,\pi d+r)=\mathbf{V}_{qm}^{[(\pi-\tau)d+r-m]d}\,\mathbf{V}_q(m,d-1)\mathbf{V}_q(0,r), \tag{8.23}$$

where

$$\begin{aligned}\mathbf{V}_q(k+1,k)&=\mathbf{I}\text{ for every integer }k,\ \mathbf{V}_{qd}=\mathbf{V}_{q0}\\ \text{and}\sum_{m=i}^{j}(.)&=0\ \text{ for every }i>j.\end{aligned} \tag{8.24}$$

Since $\Delta T(t)\geq 0$ we have

$$\sum_{\tau=0}^{[(\pi d+r)/d]-1}\Delta T(\tau d+m)\leq\sum_{\tau=0}^{\pi d+r}\Delta T(\tau)=T(\pi d+r)-T(0),$$

and thus the series

$$\lim_{\pi\to\infty}\sum_{\tau=0}^{[(\pi d+r)/d]-1}\Delta T(\tau d+m)$$

is a series of non-negative elements and it is bounded by $T-T(0)$ and thus it converges. Define by

$$w_m=\frac{\lim_{\pi\to\infty}\sum_{\tau=0}^{[(\pi d+r)/d]-1}\Delta T(\tau d+m)}{\sum_{\tau=0}^{\infty}\Delta T(\tau)}, \tag{8.25}$$

Now from (8.18) and assuming $m \leq r$ we get that

$$\lim_{t\to\infty} U_1(\pi d + r) = \sum_{m=0}^{r} \boldsymbol{\mu}_0(m) \lim_{\pi\to\infty} \sum_{\tau=0}^{[(\pi d+r)/d]-1} \Delta T(\tau d + m) \times \tag{8.26}$$

$$\lim_{\pi\to\infty} \sum_{\tau=0}^{[(\pi d+r)/d]-1} \frac{\Delta T(\tau d + m)}{\sum_{\tau=0}^{[(\pi d+r)/d]-1} \Delta T(\tau d + m)} \mathbf{V}_{qm}^{[(\pi-\tau)d+r-m]d} \mathbf{V}_q(m, r).$$

From (8.25) we get

$$\lim_{\pi\to\infty} \sum_{\tau=0}^{[(\pi d+r)/d]-1} \Delta T(\tau d + m) = w_m [T - T(0)]. \tag{8.27}$$

Hence, (8.26) could be written as

$$\lim_{t\to\infty} U_1(\pi d + r) = [T - T(0)] \sum_{m=0}^{r} w_m \boldsymbol{\mu}_0(m) \times$$

$$\lim_{\pi\to\infty} \sum_{\tau=0}^{[(\pi d+r)/d]-1} \frac{\Delta T(\tau d + m)}{\sum_{\tau=0}^{[(\pi d+r)/d]-1} \Delta T(\tau d + m)} \mathbf{V}_{qm}^{[(\pi-\tau)d+r-m]d} \mathbf{V}_q(m, r). \tag{8.28}$$

Lemma 310 is directly applicable in (8.28) since its conditions are satisfied, thus we get

$$\lim_{t\to\infty} U_1(\pi d + r) = [T - T(0)] \sum_{m=0}^{r} w_m \boldsymbol{\mu}_0(m) \mathbf{V}_{qm}^{*} \mathbf{V}_q(m, r). \tag{8.29}$$

Following the same steps for $m > r$ we could find

$$\lim_{t\to\infty} U_1(\pi d + r) = [T - T(0)] \sum_{m=0}^{r} w_m \boldsymbol{\mu}_0(m) \mathbf{V}_{qm}^{*} \mathbf{V}_q(m, d-1) \mathbf{V}_q(0, r). \tag{8.30}$$

Combining (8.18), (8.19), (8.29), (8.30) we arrive at the result needed to be proved. ■

8.3 An illustrative example from a manpower system

In the present section the previous results are illustrated with a typical example from a manpower system and a guide of how to apply the results is also supplied. It will help to have in mind the mental image of a University system with 3 grades, that is probably, Professor as grade 1, Associate Professor as grade 2, and Assistant Professor as grade 3. The last year in the records is the present for our study and let for example that the initial population structure being also the present is for example:

$$\boldsymbol{\mu}(0) = \begin{pmatrix} 100 & 300 & 500 & 0 & 0 & 0 & 0 & 0 & 0 \end{pmatrix},$$

that is, the University system starts with $T(0) = 900$ and assume that

$$\Delta T(1) = 100; \Delta T(2) = 50; \Delta T(3) = 40; \Delta T(4) = 30;$$

$$\Delta T(5) = 40; \Delta T(6) = 40; \Delta T(7) = 0,$$

that is $T = 1200$. Assume that from the data available we estimate the transition probabilities using the maximum likelihood estimators given in Chapter 5. Assume also that we discover the following cyclic behavior for $\pi = 0, 1, 2, ...$

$$\mathbf{P}(3\pi) = \begin{pmatrix} 0.9 & 0 & 0 \\ 0.15 & 0.65 & 0 \\ 0.05 & 0.15 & 0.60 \end{pmatrix}, \quad \mathbf{P}(3\pi+1) = \begin{pmatrix} 0.8 & 0 & 0 \\ 0.07 & 0.68 & 0 \\ 0.02 & 0.08 & 0.75 \end{pmatrix},$$

$$\mathbf{P}(3\pi+2) = \begin{pmatrix} 0.82 & 0 & 0 \\ 0.09 & 0.70 & 0 \\ 0.03 & 0.11 & 0.70 \end{pmatrix},$$

$$\mathbf{p}_4(3\pi) = \begin{pmatrix} 0.1 & 0.2 & 0.2 \end{pmatrix}; \quad \mathbf{p}_0(3\pi) = \begin{pmatrix} 0.18 & 0.37 & 0.45 \end{pmatrix},$$

$$\mathbf{p}_4(3\pi+1) = \begin{pmatrix} 0.01 & 0.02 & 0.03 \end{pmatrix}; \mathbf{p}_0(3\pi+1) = \begin{pmatrix} 0.2 & 0.3 & 0.5 \end{pmatrix},$$

$$\mathbf{p}_4(3\pi+2) = \begin{pmatrix} 0.18 & 0.21 & 0.16 \end{pmatrix}; \mathbf{p}_0(3\pi+2) = \begin{pmatrix} 0.15 & 0.35 & 0.5 \end{pmatrix}.$$

Then it is easy to get the transition probabilities of the inherent Markov chain from $\mathbf{Q}(3\pi+r) = \mathbf{P}(3\pi+r) + \mathbf{p}_4^\top(3\pi+r)\,\mathbf{p}_0(3\pi+r)$ and find that

$$\mathbf{Q}(3\pi) = \begin{pmatrix} 0.918 & 0.037 & 0.045 \\ 0.186 & 0.724 & 0.09 \\ 0.086 & 0.224 & 0.69 \end{pmatrix}, \quad \mathbf{Q}(3\pi+1) = \begin{pmatrix} 0.84 & 0.06 & 0.1 \\ 0.12 & 0.755 & 0.125 \\ 0.05 & 0.125 & 0.825 \end{pmatrix},$$

$$\mathbf{Q}(3\pi+2) = \begin{pmatrix} 0.847 & 0.063 & 0.09 \\ 0.1215 & 0.7735 & 0.105 \\ 0.054 & 0.166 & 0.78 \end{pmatrix}.$$

The above matrices and data constitute sufficient information from our manpower system in order to predict its long term behavior and have at least an idea about the direction which is going. We will describe the steps which should be taken for that purpose.

STEP 1: Find $\boldsymbol{\mu}_0(0)$, $\boldsymbol{\mu}_0(1)$ and $\boldsymbol{\mu}_0(2)$ from the respective entrance probabilities;
Find w_0, w_1, w_2 from the total number of memberships in each year.

STEP 2: Find $\mathbf{Q}_0 = \mathbf{Q}(0)\,\mathbf{Q}(1)\,\mathbf{Q}(2)$ and proceed to find $\mathbf{Q}_0^* = \lim_{t\to\infty} \mathbf{Q}_0^t$ which apparently exist since $\mathbf{Q}_0$ is easy to see that it is regular.

STEP 3: Find

$$\mathbf{V}_{q0}^* = \begin{pmatrix} \mathbf{Q}_0^* & \mathbf{Q}_0^*\mathbf{J} - \mathbf{J}(\mathbf{Q}_0^* \otimes \mathbf{Q}_0^*) \\ 0 & \mathbf{Q}_0^* \otimes \mathbf{Q}_0^* \end{pmatrix}.$$

STEP 4: Find $\mathbf{Q}_1 = \mathbf{Q}(1)\,\mathbf{Q}(2)\,\mathbf{Q}(0)$ and $\mathbf{Q}_2 = \mathbf{Q}(2)\,\mathbf{Q}(0)\,\mathbf{Q}(1)$ and

$$\mathbf{V}_q(i) = \begin{pmatrix} \mathbf{Q}(i) & \mathbf{Q}(i)\,\mathbf{J} - \mathbf{J}(\mathbf{Q}(i) \otimes \mathbf{Q}(i)) \\ 0 & \mathbf{Q}(i) \otimes \mathbf{Q}(i) \end{pmatrix},$$

for $i = 0, 1, 2$. Evaluate

$$\mathbf{V}_{qi}^* = \left(\prod_{j=i}^{d-1} \mathbf{V}_q(j)\right) \mathbf{V}_{q0}^* \left(\prod_{j=0}^{i-1} \mathbf{V}_q(j)\right),$$

for $i = 0, 1, 2$.

STEP 5: Use (8.8) to find $\boldsymbol{\mu}_r(\infty)$ for $r = 0, 1, 2$.

8.4 Rate of convergence of $\{\boldsymbol{\mu}(t)\}_{t=0}^{\infty}$ in an NHMS under cyclic behavior

One important question for cyclic NHMS's is how fast does the variability vector $\{\boldsymbol{\mu}(t)\}_{t=0}^{\infty}$ converges to its limit found in the previous section as $t \to \infty$. However, probably most important is to find the conditions under which the rate of convergence is geometric. In Chapter 7, we studied the problem of the rate of convergence for a non-cyclic NHMS. The properties of the **V**-matrices which we studied in section 7.2 are rather easily transferrable in the case of Cyclic NHMS, and hence following the steps of the proof of Theorem 313, it is a good exercise to prove the following theorem which could be found in Vassiliou and Tsaklidis (1988, section 3).

Theorem 320 *(Vassiliou and Tsaklidis (1988)). Let a cyclic NHMS with period d. If (i)* $\Delta T(t) \geq 0$ *for every* $t \geq 0$*, (ii) the sequence* $\{T(t)\}_{t=0}^{\infty}$ *converges to* T *geometrically fast, (iii) the matrix* $\mathbf{Q}_0$ *(as defined in the previous section) is regular, then* $\{\boldsymbol{\mu}(t)\}_{t=0}^{\infty}$ *splits into d subsequences which converge to the limits* (8.8) *geometrically fast.*

8.5 Research notes

The concept of the nonstationary cyclic behavior was introduced by Georgiou and Tsantas (1996). This concept is a generalization of the cyclic behavior described in the present chapter. This idea allows the transition matrix sequences of an NHMS for example the transition matrix $\mathbf{P}(t_n)$ is provided by $r = 0, 1, ..., d-1$ different time dependent functions, where d the time period of the cycle. Naturally, an application of such a model there is a need for a large number of data from which to estimate the parameters. The limiting expected population structure is established under the asumptions easily met in practice. Also the sets of asymptotically attainable structures are found under the assumption of the input policies taking all possible values. The question of the rate of convergence to these sets is also addressed. Finally, the close connection between nonstationary cyclic behavior and the concept of stochastic environment is studied in some detail.

Chapter 9

STOCHASTIC CONTROL IN NHMS'S

9.1 Introductory notes

An important class of problems in various populations is concerned with the control of the expected population structure or the relative expected population structure. Both structures are crucially important for the well being and constructive productivity, in whatever way these are defined, of these systems. For example when the system consists of human beings then, as it is well known, almost surely in manpower planning it is very important the control of expected population structure of the grades or the age or the service distributions. Certainly the most difficult populations to stochastically control are the human populations due to the freedom of decisions available to the individuals, the many restrictions put by the management in terms of economic budgets, inspirations of individuals, social implications etc. A useful mental image is for the reader to have in mind the management of a team in any sport. This mental image is proposed for the reader due to the fact that almost all decisions for teams in any sport are open to debate in the social media hence too many people have an intuition of management control of their favorite team at least.

The origins of these problems for manpower planning or equivalently for homogeneous Markov systems are in Bartholomew (1973, Chapter 4), Bartholomew (1975, 1977, 1979), Davies (1973, 1975,1981), Vajda (1975, 1978b), Grinold and Stanford (1974). In these studies the number of members, in the various categories defined, change through wastage, promotion and recruitment. Many factors often dictate a desirable population structure and the task of the planner is then to determine whether this can be achieved with the available controlled variables and, if so, how. This is known as the attainability problem in populations. Naturally, a desired population structure needs to be maintained and the question is if this is possible. This is the maintainability problem for planners in human populations. In theory one could argue that control could be exercised through any variable of the NHMS. However, for the most delicate case, that of manpower systems the literature shows that the input to the system offers the most attractive and practical means for control. Hence, in the present chapter we will study stochastic control of NHMS's by means of control of the input flow.

Stochastic control for the general NHMS's by means of the input flow were studied by Vassiliou and Tsantas (1984a, 1984b), while Vassiliou and Georgiou (1990), Georgiou and Vassiliou (1992), Vassiliou et al. (1990) introduced and studied the concept of asymptotic attainability. This concept contains the problem of control of the asymptotic behavior of an NHMS by controlling the input flow.

9.2 Maintainability in NHMS by input control

For various reasons too often in NHMS there are desired structures. For example in hierarchical manpower systems a pyramid structure of the population structure is desired for obvious reasons. In this respect assume that the desired population structure is $\mathbf{N}^*$ or the desired relative expected population structure $\mathbf{q}^*$. The problem is to control our system in

DOI: 10.1201/b23204-9

such way as to maintain this structure. Naturally, this must be exercised through the flows of members of the system. The flows are:

i) Wastage from the system, i.e. the sequence of vectors $\{\mathbf{p}_{k+1}(t)\}_{t=0}^{\infty}$.

ii) Internal transitions among the states of the system, that is the sequence of the transition probabilities matrices $\{\mathbf{P}(t)\}_{t=0}^{\infty}$.

iii) The input flow or as is used in the context of manpower planning the recruitment flow, which has two parts, first the total input number of memberships new and replacements, and second their allocation probabilities to the various states $\{\mathbf{p}_0(t)\}_{t=0}^{\infty}$.

As the literature in manpower systems shows the input to the system offers the most attractive means to control. Apparently, the wastage flow implies discharge which apart that we are not willing to study it also has inherent serious social implications. Internal transitions have implications of fairness, which amounts in the years to come and creates the phenomenon of increased pressure for internal movements (Kalamatianou (1987), (1988)). In fact in some applications of the NHMS this is the only control variable. Hence, in this chapter we will exercise control in an NHMS by controlling the input sequence $\{\mathbf{p}_0(t)\}_{t=0}^{\infty}$. The natural question is if all the structures are possible to be maintained. As the reader may already intuitively felt the answer is no. Thus the next logical question is which expected population structures are maintainable under input control. The answer to this question will be the first step in this chapter. We provide the following definition:

Definition 321 *Let an NHMS with given $\{\mathbf{P}(t)\}_{t=0}^{\infty}$, $\{\mathbf{p}_{k+1}(t)\}_{t=0}^{\infty}$, and $\{T(t)\}_{t=0}^{\infty}$. We define by B_t the set of expected relative structures that could be maintained in the time interval $[t-1,t]$, i.e.*

$$B_t = \{\mathbf{q}^*: \mathbf{q}^* = \mathbb{E}[\mathbf{q}(t)] = \mathbb{E}[\mathbf{q}(t-1)]\}. \tag{9.1}$$

We also define by B_t^m the set of expected relative structures that could be maintained in the time interval $[t, t+m]$, i.e.

$$B_t^m = \{\mathbf{q}^*: \mathbf{q}^* = \mathbb{E}[\mathbf{q}(t+m)] = \mathbb{E}[\mathbf{q}(t)]\} \text{ for } i = 1, 2, ..., m. \tag{9.2}$$

We will now try to find a characterization of the set B_t. We will do so step by step and we will conclude with the theorem which represents the result.

From relation (5.30) we get that

$$\mathbb{E}[\mathbf{q}(t)] = \mathbb{E}[\mathbf{q}(t-1)]\,\mathbf{M}(t). \tag{9.3}$$

where from (5.31) we have

$$\begin{aligned}\mathbf{M}(t) &= \frac{T(t-1)}{T(t)}\left[\mathbf{Q}(t) + \frac{T(t)-T(t-1)}{T(t-1)}\mathbf{1}^{\top}\mathbf{p}_0(t)\right] \\ &= a(t)\,\mathbf{P}(t) + \left[a(t)\,\mathbf{p}_{k+1}^{\top}(t) + b(t)\,\mathbf{1}^{\top}\right]\mathbf{p}_0(t),\end{aligned} \tag{9.4}$$

where

$$a(t) = \frac{T(t-1)}{T(t)} \quad \text{and} \quad b(t) = \frac{T(t)-T(t-1)}{T(t)}. \tag{9.5}$$

The relative expected population structure $\mathbb{E}[\mathbf{q}(t-1)]$ is maintainable in the time interval $[t-1,t]$ if we can find acceptable values of the control variables $\{\mathbf{p}_0(t)\}$ such that $\mathbf{q}^* = \mathbb{E}[\mathbf{q}(t)] = \mathbb{E}[\mathbf{q}(t-1)]$ and therefore

$$\mathbf{q}^* = \mathbf{q}^*\left\{a(t)\,\mathbf{P}(t) + \left[a(t)\,\mathbf{p}_{k+1}^{\top}(t) + b(t)\,\mathbf{1}^{\top}\right]\mathbf{p}_0(t)\right\}. \tag{9.6}$$

Solving (9.6) we get that

$$\mathbf{p}_0(t) = \mathbf{q}^*\left[\mathbf{I} - a(t)\,\mathbf{P}(t)\right]/\mathbf{q}^*\left[a(t)\,\mathbf{p}_{k+1}^{\top}(t) + b(t)\,\mathbf{1}^{\top}\right]. \tag{9.7}$$

The solution in (9.7) is acceptable only if

$$(i)\ \mathbf{p}_0(t)\mathbf{1}^\top = 1 \ \text{ and } \ (ii)\ \mathbf{p}_0(t) \geq 0. \tag{9.8}$$

Condition (i) in (9.8) is always true since

$$\begin{aligned}\mathbf{q}^*\left[\mathbf{I} - a(t)\mathbf{P}(t)\right]\mathbf{1}^\top &= \mathbf{q}^*\left[\mathbf{1}^\top - a(t)\mathbf{1}^\top + a(t)\mathbf{p}_{k+1}^\top(t)\right] \\ &= \mathbf{q}^*\left[a(t)\mathbf{p}_{k+1}^\top(t) + b(t)\mathbf{1}^\top\right].\end{aligned} \tag{9.9}$$

Apparently condition (ii) in (9.8) holds if

$$\mathbf{q}^* \geq \mathbf{q}^* a(t)\mathbf{P}(t). \tag{9.10}$$

Relation (9.10) could surely serve as a simple characterization of the set B_t. However, in what follows we will try to improve on that. From (9.6) we get that

$$\mathbf{q}^*\left[\mathbf{I} - a(t)\mathbf{P}(t)\right] = \mathbf{q}^*\left[a(t)\mathbf{p}_{k+1}^\top(t) + b(t)\mathbf{1}^\top\right]\mathbf{p}_0(t). \tag{9.11}$$

Now, we make the following assumption:

Assumption 9.1. Let that

$$a(t)\max_i \sum_{j=1}^{k} p_{ij}(t) < 1 \quad \text{for } t = 0, 1, ... \tag{9.12}$$

It is important to note that when Assumption 9.1 holds then the matrix $\mathbf{M}(t)$ is stochastic but in addition the matrix $\left[\mathbf{I} - a(t)\mathbf{P}(t)\right]$ is invertible. Hence, from (9.12) we get that

$$\mathbf{q}^* = \mathbf{q}^*\left[a(t)\mathbf{p}_{k+1}^\top(t) + b(t)\mathbf{1}^\top\right]\mathbf{p}_0(t)\left[\mathbf{I} - a(t)\mathbf{P}(t)\right]^{-1}. \tag{9.13}$$

Now, the stochastic vector $\mathbf{p}_0(t)$ could be written as $\mathbf{p}_0(t) = \sum_{i=1}^{k} p_{0i}(t)\mathbf{e}_i$, where $\mathbf{e}_i$ is the $1 \times k$ vector with 1 in the i-th position and zeros elsewhere. Therefore (9.13) takes the form:

$$\mathbf{q}^* = \mathbf{q}^*\left[a(t)\mathbf{p}_{k+1}^\top(t) + b(t)\mathbf{1}^\top\right]\sum_{i=1}^{k} p_{0i}(t)\mathbf{e}_i\left[\mathbf{I} - a(t)\mathbf{P}(t)\right]^{-1}. \tag{9.14}$$

From (9.14) we get by multiplying both sides by the column vector of one's $\mathbf{1}^\top$

$$\mathbf{q}^*\left[a(t)\mathbf{p}_{k+1}^\top(t) + b(t)\mathbf{1}^\top\right]\sum_{i=1}^{k} p_{0i}(t)\mathbf{e}_i\lambda_i(t) = 1. \tag{9.15}$$

where $\lambda_i(t)$ is the sum of the elements of the i-row of the matrix $\left[\mathbf{I} - a(t)\mathbf{P}(t)\right]^{-1}$. From (9.14) and (9.15) we have that

$$\mathbf{q}^* = \sum_{i=1}^{k}\left\{p_{0i}(t)\mathbf{e}_i\lambda_i(t) \Big/ \sum_{j=1}^{k} p_{0j}(t)\lambda_j(t)\right\}\lambda_i^{-1}(t)\mathbf{e}_i\left[\mathbf{I} - a(t)\mathbf{P}(t)\right]^{-1}. \tag{9.16}$$

Therefore we have proved the following theorem:

Theorem 322 *(Vassiliou and Tsantas (1984a)). Let a complete probability space* $(\Omega, \mathcal{F}, \mathbb{P})$ *and an NHMS in it with given basic sequences of parameters* $\{\mathbf{P}(t)\}_{t=0}^{\infty}$, $\{\mathbf{p}_{k+1}(t)\}_{t=0}^{\infty}$ *and* $\{T(t)\}_{t=0}^{\infty}$. *If*

$$\frac{T(t-1)}{T(t)}a(t)\max_i \sum_{j=1}^{k} p_{ij}(t) < 1 \quad \textit{for } t = 0, 1, ...,$$

then the set B_t *of maintainable expected population structures is a convex set with vertices*

$$\lambda_i^{-1}(t)\,\mathbf{e}_i\,[\mathbf{I}-a(t)\,\mathbf{P}(t)]^{-1},\ i=1,2,...,k,$$

where $\lambda_i(t)$ *is the sum of the elements of the i-th row of the matrix*

$$[\mathbf{I}-a(t)\,\mathbf{P}(t)]^{-1}.$$

We will now find a characterization for the set B_t^m, which will be established in the following theorem:

Theorem 323 *(Vassiliou and Tsantas (1984a)). Let a complete probability space* $(\Omega,\mathcal{F},\mathbb{P})$ *and an NHMS in it with given basic sequences of parametrs* $\{\mathbf{P}(t)\}_{t=0}^{\infty}$, $\{\mathbf{p}_{k+1}(t)\}_{t=0}^{\infty}$ *and* $\{T(t)\}_{t=0}^{\infty}$. *Then the set* B_t^m *of maintainable expected relative structures in the interval* $[t,t+m]$ *is the convex set*

$$B_t^m=\cap_{s=1}^{m-1}B_{t+s}.$$

Proof. Let $\mathbf{q}^*\in B_t^m$ then that means that it is maintainable in the time interval $[t,t+m]$, i.e.,

$$\mathbf{q}^*=\mathbf{q}(t)=\mathbb{E}[\mathbf{q}(t+1)]=...=\mathbb{E}[\mathbf{q}(t+m)]. \tag{9.17}$$

Since $\mathbf{q}^*=\mathbf{q}(t)=\mathbb{E}[\mathbf{q}(t+1)]$ it means that $\mathbf{q}^*\in B_t$. Similarly, from (9.17) we get that

$$\mathbf{q}^*\in B_{t+1},\ \mathbf{q}^*\in B_{t+2},\ ...,\ \mathbf{q}^*\in B_{t+m-1}. \tag{9.18}$$

From (9.18) it is apparent that

$$\mathbf{q}^*\in\cap_{s=1}^{m-1}B_{t+s},$$

which is a convex set as the intersection of m convex sets.

The converse that any element of $\cap_{s=1}^{m-1}B_{t+s}$ is an m maintainable set is easily proved. ■

We will now make the following useful remark:

Remark 324 *From the above theorem it is obvious that if someone wants to find out if a specific expected relative structure* $\mathbf{q}^*$ *is maintainable for m consecutive times it suffices to see if the inequality* $\mathbf{q}^*\geq\mathbf{q}^*a(t+s)\,\mathbf{P}(t+s)$ *is satisfied for* $s=1,2,...,m$.

9.3 Attainability in NHMS by input control

We will proceed now to solve the next logical problem, that is, how to attain a desirable expected population structure which is also maintainable. Apparently, there is no meaning to attain a structure which is not possible to be maintained.

Let us denote by $\mathbf{N}^*$ the desirable population structure which we wish to attain by input control. That is, we need to find the appropriate input policy $\mathbf{p}_0(1)$, $\mathbf{p}_0(2)$, ... which will lead to the desirable population structure in the minimum of time. Note that the number of newcomers is not in our control since they are the number of expected leavers plus the new memberships needed in order to reach the appropriately put target of total memberships each time. Naturally, it is always desirable that, apart from the minimum time consideration, the attainability of the structure $\mathbf{N}^*$ should be accomplished in a way that will minimize a given cost function. Assume that we start with initial population structure $\mathbf{N}(0)$ and we want to arrive at time T^* the desired population structure $\mathbf{N}^*$. That is, we will have the following sequence of expected population structures

$$\mathbf{N}(0),\ \mathbb{E}[\mathbf{N}(1)],...,\mathbb{E}[\mathbf{N}(T^*-1)],\mathbb{E}[\mathbf{N}(T^*)]=\mathbf{N}^*. \tag{9.19}$$

As the research literature shows a desirable cost function is the following:

$$\sum_{t=1}^{T^*} \left[\mathbb{E}\left[\mathbf{N}(t)\right] \mathbf{c}^\top(t) + \mathbb{E}\left[R(t)\right] \mathbf{p}_0(t) \mathbf{s}^\top(t)\right], \tag{9.20}$$

where T^* is the minimum time at which the desirable structure $\mathbf{N}^*$ is attained; $\mathbf{c}(t)$ is the $1 \times k$ row vector with elements $c_i(t)$ being the mean cost of a member of the system in state i in the t-th interval for $i = 1, 2, ..., k$; $\mathbf{s}(t)$ is the $1 \times k$ row vector with elements $s_i(t)$, the mean cost of a new member of the system in state i. Note that $c_i(t)$ and $s_i(t)$ could also be the mean productivity indices, and in such a case we wish to maximize (9.20) which constitutes an equivalent problem. The constraints imposed by the evolution of the population through time are the following:

$$\mathbb{E}\left[\mathbf{N}(t)\right] = \mathbb{E}\left[\mathbf{N}(t-1)\right] \mathbf{P}(t) + \mathbb{E}\left[R(t)\right] \mathbf{p}_0(t) \quad \text{for } t = 1, 2, ..., T^* - 1, \tag{9.21}$$

$$\mathbf{N}^* = \mathbb{E}\left[\mathbf{N}(T^*-1)\right] \mathbf{P}(T^*) + \mathbb{E}\left[R(T^*)\right] \mathbf{p}_0(T^*), \tag{9.22}$$

$$\mathbb{E}\left[\mathbf{N}(t)\right] \mathbf{1}^\top = T(t) \quad \text{for } t = 1, 2, ..., T^* - 1, \tag{9.23}$$

$$\mathbb{E}\left[\mathbf{N}(t)\right] \geq 0, \text{ integers for } i = 1, 2, ..., k \text{ and } t = 1, 2, ..., T^* - 1. \tag{9.24}$$

Equations (9.20), (9.21) and (9.22) could be written as

$$\sum_{t=1}^{T^*} \left\{\mathbb{E}\left[\mathbf{N}(t)\right] \mathbf{c}^\top(t) + \left[\Delta T(t) + \mathbb{E}\left[\mathbf{N}(t)\right] \mathbf{p}_{k+1}^\top(t)\right] \mathbf{p}_0(t) \mathbf{s}^\top(t)\right\}, \tag{9.25}$$

$$\mathbb{E}\left[\mathbf{N}(t)\right] = \mathbb{E}\left[\mathbf{N}(t-1)\right] \mathbf{P}(t) + \left[\Delta T(t) + \mathbb{E}\left[\mathbf{N}(t-1)\right] \mathbf{p}_{k+1}^\top(t)\right] \mathbf{p}_0(t), \tag{9.26}$$

$$\mathbf{N}^* = \mathbb{E}\left[\mathbf{N}(T^*-1)\right] \mathbf{P}(T^*) + \left[\Delta T(t) + \mathbb{E}\left[\mathbf{N}(T^*-1)\right] \mathbf{p}_{k+1}^\top(T^*)\right] \mathbf{p}_0(T^*). \tag{9.27}$$

Apparently, (9.25), (9.26) and (9.27) are linear in the variables $\mathbf{N}(t)$ and $\mathbf{p}_0(t)$.

We will now provide an Algorithm for the solution of our problem with the help of dynamic programming and integer programming techniques. Let the optimal function be defined as:

$$\begin{aligned} F\left[\mathbb{E}\left[\mathbf{N}(t+1)\right], \mathbf{p}_0(t+1), t+1\right] = \min_{\mathbb{E}[\mathbf{N}(t+1)], \mathbf{p}_0(t+1)} &\left\{\mathbb{E}\left[\mathbf{N}(t)\right] \mathbf{c}^\top(t+1) + \right. \\ \left[\Delta T(t+1) + \mathbb{E}\left[\mathbf{N}(t)\right] \mathbf{p}_{k+1}^\top(t+1)\right] &\left. \mathbf{p}_0(t+1) \mathbf{s}^\top(t+1)\right\} \\ &+ F\left[\mathbb{E}\left[\mathbf{N}(t)\right], \mathbf{p}_0(t), t\right], \end{aligned} \tag{9.28}$$

with obvious boundary condition

$$F\left[\mathbf{N}(0), \mathbf{p}_0(0), 0\right] = 0. \tag{9.29}$$

The Algorithm for solving the attainability problem in a minimum time and cost is the following:

Algorithm 9.2. (Vassiliou and Tsantas (1984))
Step 1. **If** $\mathbf{N}^* - \mathbf{N}(0)\mathbf{P}(1) > 0$
then $\mathbf{p}_0(1) = \left[\mathbf{N}^* - \mathbf{N}(0)\mathbf{P}(1)\right] / \left[\Delta T(1) + \mathbf{N}(0)\mathbf{p}_{k+1}^\top(1)\right]$.
Set $T^* = 1$ and go to Step 3.
else Solve the mixed integer programming problem with $2k$ variables and $k+1$ equations

$$F\left[\mathbb{E}\left[\mathbf{N}(1)\right], \mathbf{p}_0(1), 1\right] = \min_{[\mathbf{N}(1)], \mathbf{p}_0(1)} \left\{\mathbb{E}\left[\mathbf{N}(1)\right] \mathbf{c}^\top(1)\right.$$

$$\left[\Delta T(1) + \mathbf{N}(0)\,\mathbf{p}_{k+1}^{\top}(1)\right]\mathbf{p}_0(1)\,\mathbf{s}^{\top}(1),$$

subject to

$$\mathbb{E}[\mathbf{N}(1)] = \mathbf{N}(0)\,\mathbf{P}(1) + \left[\Delta T(1) + \mathbf{N}(0)\,\mathbf{p}_{k+1}^{\top}(1)\right]\mathbf{p}_0(1),$$

$$\mathbb{E}[\mathbf{N}(1)]\,\mathbf{1}^{\top} = T(1),$$

$$\mathbb{E}[\mathbf{N}(1)] \geq 0, \;\; \mathbf{p}_0(1) \geq 0,$$

$$\mathbb{E}[N_i(1)], \quad i = 1, 2, ..., k.$$

Results

$$\mathbb{E}[\mathbf{N}(1)], \;\; \mathbf{p}_0(1) \quad \text{optimal.}$$

$$\text{Set} \;\; t = 1.$$

Step 2.

$$\text{Set} \;\; t = t + 1$$

If $\mathbf{N}^* - \mathbb{E}[\mathbf{N}(t-1)]\,\mathbf{P}(t) > 0$
then $\mathbf{p}_0(t) = \left[\mathbf{N}^* - \mathbb{E}[\mathbf{N}(t-1)]\,\mathbf{P}(t)\right] / \left[\Delta T(t) + \mathbb{E}[\mathbf{N}(t-1)]\,\mathbf{p}_{k+1}^{\top}(t)\right]$.
Set $T^* = t$ and go to Step 3.
else Solve the mixed integer programming problem with $2k$ variables and $k+1$ equations

$$F\left[\mathbb{E}[\mathbf{N}(t)], \mathbf{p}_0(t), t\right] = \min_{[\mathbf{N}(1)], \mathbf{p}_0(1)} \{\mathbb{E}[\mathbf{N}(t)]\,\mathbf{c}^{\top}(t) +$$

$$\left[\Delta T(t) + \mathbb{E}[\mathbf{N}(t-1)]\,\mathbf{p}_{k+1}^{\top}(t)\right]\mathbf{p}_0(t)\,\mathbf{s}^{\top}(t)\}$$

$$+F\left[\mathbb{E}[\mathbf{N}(t-1)], \mathbf{p}_0(t-1), t-1\right],$$

subject to

$$\mathbb{E}[\mathbf{N}(t)] = \mathbb{E}[\mathbf{N}(t-1)]\,\mathbf{P}(t) +$$

$$\left[\Delta T(t) + \mathbb{E}[\mathbf{N}(t-1)]\,\mathbf{p}_{k+1}^{\top}(t)\right]\mathbf{p}_0(t),$$

$$\mathbb{E}[\mathbf{N}(t)]\,\mathbf{1}^{\top} = T(t),$$

$$\mathbb{E}[\mathbf{N}(t)] \geq 0, \;\; \mathbf{p}_0(t) \geq 0,$$

$$\mathbb{E}[N_i(t)], \quad i = 1, 2, ..., k.$$

Results

$$\mathbb{E}[\mathbf{N}(t)], \;\; \mathbf{p}_0(t) \quad \text{optimal.}$$

Go to step 3 if $\mathbf{N}^* - \mathbb{E}[\mathbf{N}(t-1)]\,\mathbf{P}(t) = 0$ else **Go to step 2**
Step 3 Results $t = T^*$. STOP.

9.4 Asymptotically attainable structures

In this section we formulate and study a new aspect of the classical problem of providing conditions under which a limiting relative structure of an NHMS exists and finding this relative structure. Our interest is in finding which relative structures are possible as limiting ones provided that we control the sequence of input probabilities in the sense that we control the limiting vector of input probabilities. In other words, we are interested in the set of asymptotically attainable structures by control of the limiting input probability vector.

Again in here our instrument of control is the input limiting probability vector as the most convenient and in many ways the most logical for a large number of applications means for control. We start with the following basic Definition:

Definition 325 *Let a complete probability space* $(\Omega, \mathcal{F}, \mathbb{P})$ *and an NHMS in it. We say that it has an asymptotically attainable expected relative population structure* $\mathbb{E}[\mathbf{q}(\infty)]$ *under asymptotic recruitment control, if there exists a sequence of recruitment vectors* $\{\mathbf{p}_0(t)\}_{t=0}^{\infty}$ *with* $\lim_{t\to\infty}\|\mathbf{p}_0(t)-\mathbf{p}_0\|=0$ *such that* $\lim_{t\to\infty}\|\mathbb{E}[\mathbf{q}(t)]-\mathbb{E}[\mathbf{q}(\infty)]\|=0$.

Denote by analogy B^{∞} the set of expected asymptotically relative population structures $\mathbb{E}[\mathbf{q}(\infty)]$ under asymptotic recruitment control. In the following theorem we will provide a characterization of this set.

Theorem 326 *(Vassiliou and Georgiou (1990)). Let a complete probability space* $(\Omega, \mathcal{F}, \mathbb{P})$ *and an NHMS in it with given basic sequences of parameters* $\{\mathbf{P}(t)\}_{t=0}^{\infty}$, $\{\mathbf{p}_{k+1}(t)\}_{t=0}^{\infty}$ *and* $\{T(t)\}_{t=0}^{\infty}$. *Assume that the following conditions are satisfied*

$$(i)\ \lim_{t\to\infty}\|\mathbf{P}(t)-\mathbf{P}\|=0,\quad (ii)\ \lim_{t\to\infty}\left\|\mathbf{p}_{k+1}^{\top}(t)-\mathbf{p}_{k+1}^{\top}\right\|=0, \tag{9.30}$$

$$(iii)\ \lim_{t\to\infty}\|\mathbf{p}_0(t)-\mathbf{p}_0\|=0,$$

$$\mathbf{Q}=\mathbf{P}+\mathbf{p}_{k+1}^{\top}\mathbf{p}_0 \quad \text{is a regular stochastic matrix.} \tag{9.31}$$

Then if

$$\lim_{t\to\infty}\frac{\Delta T(t)}{T(t)}=0 \quad \text{with} \quad \Delta T(t)\geq 0 \quad \text{for every } t\in\mathbb{N}, \tag{9.32}$$

or

$$\lim_{t\to\infty}T(t)=T \quad \text{with} \quad \Delta T(t)\geq 0 \quad \text{for every } t\in\mathbb{N}, \tag{9.33}$$

then set B^{∞} *is the convex hull of the points*

$$\mathbf{z}=\lambda_i^{-1}\left[\mathbf{e}_i[\mathbf{I}-\mathbf{P}]^{-1}\right],\ i=1,2,...,k,$$

where λ_i *is the sum of the elements of the i-th row of the matrix* $[\mathbf{I}-\mathbf{P}]^{-1}$.

Proof. From Theorem 291 we get that under the conditions (9.30), (9.31) and (9.33) the asymptotic expected relative limiting structure of the population is given from the system of equations

$$\mathbb{E}[\mathbf{q}(\infty)]=\mathbb{E}[\mathbf{q}(\infty)]\mathbf{Q} \quad \text{and} \quad \mathbb{E}[\mathbf{q}(\infty)]\mathbf{1}^{\top}=1. \tag{9.34}$$

Now, from (9.31) and (9.34) we get that

$$\mathbb{E}[\mathbf{q}(\infty)]=\mathbb{E}[\mathbf{q}(\infty)]\left(\mathbf{P}+\mathbf{p}_{k+1}^{\top}\mathbf{p}_0\right)=\mathbb{E}[\mathbf{q}(\infty)]\mathbf{P}+\mathbb{E}[\mathbf{q}(\infty)]\mathbf{p}_{k+1}^{\top}\mathbf{p}_0$$

or

$$\mathbb{E}[\mathbf{q}(\infty)][\mathbf{I}-\mathbf{P}]=\mathbb{E}[\mathbf{q}(\infty)]\mathbf{p}_{k+1}^{\top}\mathbf{p}_0. \tag{9.35}$$

The matrix $[\mathbf{I}-\mathbf{P}]$ is nonnegatively invertible because $\mathbf{P}$ is, in general, substochastic (Nikaido (1968)). Thus

$$\mathbb{E}[\mathbf{q}(\infty)]=\mathbb{E}[\mathbf{q}(\infty)]\mathbf{p}_{k+1}^{\top}\mathbf{p}_0[\mathbf{I}-\mathbf{P}]^{-1}$$

$$=\mathbb{E}[\mathbf{q}(\infty)]\mathbf{p}_{k+1}^{\top}\sum_{i=1}^{k}p_{0i}\mathbf{e}_i[\mathbf{I}-\mathbf{P}]^{-1}. \tag{9.36}$$

Multiplying both sides of (9.36) by $\mathbf{1}^\top$ we get

$$\mathbb{E}\left[\mathbf{q}(\infty)\right]\mathbf{p}_{k+1}^\top\sum_{i=1}^{k}p_{0i}\lambda_i=1, \tag{9.37}$$

where λ_i is the sum of the $i-$th row of the matrix $\left[\mathbf{I}-\mathbf{P}\right]^{-1}$. Hence from (9.36) and (9.37) we get

$$\mathbb{E}\left[\mathbf{q}(\infty)\right]=\sum_{i=1}^{k}\frac{p_{0i}\lambda_i}{\sum_{j=1}^{k}p_{0j}\lambda_j}\lambda_i^{-1}\left[\mathbf{e}_i\left[\mathbf{I}-\mathbf{P}\right]^{-1}\right]. \tag{9.38}$$

So, $\mathbb{E}\left[\mathbf{q}(\infty)\right]$ is a convex combination of the vertices $\lambda_i^{-1}\left[\mathbf{e}_i\left[\mathbf{I}-\mathbf{P}\right]^{-1}\right]$ for $i=1,2,...,k$ since if we denote by

$$c_i=\frac{p_{0i}\lambda_i}{\sum_{j=1}^{k}p_{0j}\lambda_j},$$

then we have

$$\sum_{i=1}^{k}c_i=\sum_{i=1}^{k}\frac{p_{0i}\lambda_i}{\sum_{j=1}^{k}p_{0j}\lambda_j}=1.$$

If conditions (9.32), hold then from Theorem 294, we get again relation (9.34) from which we repeat the reasoning. ■

Remark 327 *The assumption* $\lim_{t\to\infty}\frac{\Delta T(t)}{T(t)}=0$ *allows for* $T(t)=\infty$.

We are now going to establish the characterization of the set B^∞ for an NHMS which has a fluctuating size without forced wastage.

Theorem 328 *Let a complete probability space* $(\Omega,\mathcal{F},\mathbb{P})$ *and an NHMS in it with given basic sequences of parameters* $\{\mathbf{P}(t)\}_{t=0}^{\infty}$, $\{\mathbf{p}_{k+1}(t)\}_{t=0}^{\infty}$ *and* $\{T(t)\}_{t=0}^{\infty}$. *Assume that the following conditions are satisfied*

$$(i)\ \lim_{t\to\infty}\left\|\mathbf{P}(t)-\mathbf{P}\right\|=0,\quad (ii)\ \lim_{t\to\infty}\left\|\mathbf{p}_{k+1}^\top(t)-\mathbf{p}_{k+1}^\top\right\|=0, \tag{9.39}$$

$$(iii)\ \lim_{t\to\infty}\left\|\mathbf{p}_0(t)-\mathbf{p}_0\right\|=0,$$

$$(iv)\ \frac{T(t)}{T(t-1)}\geq\max\left[\sum_{j=1}^{k}p_{ij}(t)-p_{ij}(t)/p_{oj}(t)\right],\ (v)\ \lim_{t\to\infty}\frac{T(t-1)}{T(t)}=\tau, \tag{9.40}$$

$$\tau\rho(\mathbf{P})<1. \tag{9.41}$$

Then if the matrix $\mathbf{L}=\tau\mathbf{P}+(\mathbf{I}-\tau\mathbf{P})\mathbf{1}^\top\mathbf{p}_0$ *is a regular stochastic matrix the set* B^∞ *is the convex hull of the points*

$$\mathbf{z}=\mu_i^{-1}\left[\mathbf{e}_i\left[\mathbf{I}-\tau\mathbf{P}\right]^{-1}\right],\ i=1,2,...,k, \tag{9.42}$$

where μ_i *is the sum of the elements of the* $i-$*th row of the matrix* $\left[\mathbf{I}-\tau\mathbf{P}\right]^{-1}$.

Proof. From Theorem 296, we get that the conditions (9.39) and (9.40) certify that the limiting expected relative structure of the population is given as the solution of the system of equations

$$\mathbb{E}\left[\mathbf{q}(\infty)\right]=\mathbb{E}\left[\mathbf{q}(\infty)\right]\mathbf{L}\ \text{ and }\mathbb{E}\left[\mathbf{q}(\infty)\right]\mathbf{1}^\top=1. \tag{9.43}$$

Now from (9.43) we get that

$$\mathbb{E}\left[\mathbf{q}(\infty)\right]=\mathbb{E}\left[\mathbf{q}(\infty)\right]\left(\tau\mathbf{P}+(\mathbf{I}-\tau\mathbf{P})\mathbf{1}^\top\mathbf{p}_0\right)$$

or

$$\mathbb{E}\left[\mathbf{q}(\infty)\right](\mathbf{I}-\tau\mathbf{P})=\mathbb{E}\left[\mathbf{q}(\infty)\right](\mathbf{I}-\tau\mathbf{P})\,\mathbf{1}^{\top}\mathbf{p}_{0}. \tag{9.44}$$

Now from (9.41) we get that the matrix $(\mathbf{I}-\tau\mathbf{P})$ is non-negatively invertible (Nikaido (1968)), hence

$$\begin{aligned}\mathbb{E}\left[\mathbf{q}(\infty)\right]&=\mathbb{E}\left[\mathbf{q}(\infty)\right](\mathbf{I}-\tau\mathbf{P})\,\mathbf{1}^{\top}\mathbf{p}_{0}(\mathbf{I}-\tau\mathbf{P})^{-1}\\&=\mathbb{E}\left[\mathbf{q}(\infty)\right](\mathbf{I}-\tau\mathbf{P})\,\mathbf{1}^{\top}\sum_{\mathbf{i}=\mathbf{1}}^{\mathbf{k}}p_{0i}\mathbf{e}_{i}(\mathbf{I}-\tau\mathbf{P})^{-1},\end{aligned} \tag{9.45}$$

we multiply (9.45) by $\mathbf{1}^{\top}$ in both parts and we get

$$\mathbb{E}\left[\mathbf{q}(\infty)\right](\mathbf{I}-\tau\mathbf{P})\,\mathbf{1}^{\top}\sum_{\mathbf{i}=\mathbf{1}}^{\mathbf{k}}p_{0i}\mathbf{e}_{i}\mu_{i}=1, \tag{9.46}$$

where μ_i is the sum of the i–th row of the matrix $(\mathbf{I}-\tau\mathbf{P})^{-1}$. Now from (9.45) and (9.46) we get that

$$\mathbb{E}\left[\mathbf{q}(\infty)\right]=\sum_{i=1}^{k}\frac{p_{0i}\mu_{i}}{\sum_{j=1}^{k}p_{0j}\mu_{i}}\mu_{i}^{-1}\left[\mathbf{e}_{i}(\mathbf{I}-\tau\mathbf{P})^{-1}\right]$$

So, $\mathbb{E}\left[\mathbf{q}(\infty)\right]$ is a convex combination of the vertices $\mu_i^{-1}\left[\mathbf{e}_i\left[\mathbf{I}-\tau\mathbf{P}\right]^{-1}\right]$ for $i=1,2,...,k$ since if we denote by

$$\gamma_{i}=\frac{p_{0i}\mu_{i}}{\sum_{j=1}^{k}p_{0j}\mu_{j}},$$

then we have

$$\sum_{i=1}^{k}\gamma_{i}=\sum_{i=1}^{k}\frac{p_{0i}\mu_{i}}{\sum_{j=1}^{k}p_{0j}\mu_{j}}=1.$$

■

9.5 Periodicity of asymptotically attainable structures

In this section, we will relax the basic assumption of the two previous theorems that the inherent(imbedded) non-homogeneous Markov chain of an NHMS asymptotically converges in a homogeneous regular Markov chain. In fact, we will study the problem when the inherent non-homogeneous Markov chain converges to a periodic homogeneous Markov chain. Under the assumption of asymptotic periodicity we will study the important issue of the control of the expected relative structure of the NHMS by control of the input flow or recruitment control. We will do this by finding all the d-periodic asymptotically expected population structures which are possible to be attained by asymptotic recruitment control. We will start with the following theorem for non-homogeneous Markov chains in general:

Theorem 329 *(Georgiou et al. (1992)). Let a finite non-homogeneous Markov chain $\{X_t\}_{t=0}^{\infty}$ with state space $\mathbb{S}=\{1,2,...,k\}$ and with sequence of transition probability matrices $\{\mathbf{Q}(t)\}_{t=0}^{\infty}$. Assume that*

$$\lim_{t\to\infty}\left\|\mathbf{Q}(t)-\mathbf{Q}\right\|=0,$$

with geometric rate of convergence, where $\mathbf{Q}$ is a periodic stochastic matrix with period d and let that the sequence of matrices $\{\mathbf{Q}(t)\}_{t=0}^{\infty}$ and $\mathbf{Q}$ have the same incidence matrix, that is, $\hat{\mathbf{Q}}(t)=\hat{\mathbf{Q}}$. Then (i) if $r\geq m$ where $r,m\in\mathbb{N}$, $0\leq r\leq d-1$ and $0\leq m\leq d-1$, then

$$\lim_{n\to\infty}\left\|\mathbf{Q}(sd+m,nd+r)-\mathbf{A}\mathbf{Q}^{r-m}\right\|=0\text{ for every }s\in\mathbb{N}, \tag{9.47}$$

with geometric rate of convergence. (ii) if $r < m$,

$$\lim_{n\to\infty}\left\|\mathbf{Q}\left(sd+m,nd+r\right)-\mathbf{A}\mathbf{Q}^{d+r-m}\right\|=0 \text{ for every } s\in\mathbb{N}, \tag{9.48}$$

with geometric rate of convergence. That is, the product $\mathbf{Q}\left(sd+m,nd+r\right)$ converges to its limit at a geometric rate, uniformly in s.

Proof. Let $r \geq m$. Then we have that

$$\left\|\mathbf{Q}\left(sd+m,nd+r\right)-\mathbf{A}\mathbf{Q}^{r-m}\right\|=$$

$$\left\|\mathbf{Q}\left(sd+m,nd+m\right)\mathbf{Q}\left(nd+m,nd+r\right)-\mathbf{A}\mathbf{Q}^{r-m}\right\|$$
$$\leq\left\|\mathbf{Q}\left(sd+m,nd+m\right)-\mathbf{A}\right\|+\left\|\mathbf{Q}\left(nd+m,nd+r\right)-\mathbf{Q}^{r-m}\right\|. \tag{9.49}$$

From the theory of homogeneous Markov chains (Vassiliou (2010)) we get that since $\mathbf{Q}$ is a periodic Markov chain we get that there exist $c_1 > 0$ and $0 < b < 1$ and $\nu > 0$ such that

$$\left\|\mathbf{Q}^{\nu d}-\mathbf{A}\right\|\leq c_1 b_1^t \text{ for every } n>\nu. \tag{9.50}$$

Now, using also Theorem 252 and from (9.49) we get that

$$\left\|\mathbf{Q}\left(sd+m,nd+m\right)-\mathbf{A}\right\|\leq$$

$$\left\|\mathbf{Q}\left(sd+m,nd-\nu d+m\right)\right\|\cdot\left\|\mathbf{Q}\left(nd-\nu d+m,nd+m\right)-\mathbf{A}\right\|\leq$$
$$\left\|\mathbf{Q}\left(nd-\nu d+m,nd+m\right)-\mathbf{Q}^{\nu d}\right\|+\left\|\mathbf{Q}^{\nu d}-\mathbf{A}\right\|. \tag{9.51}$$

Since by assumption the sequence $\{\mathbf{Q}(t)\}_{t=0}^{\infty}$ converge to the matrix $\mathbf{Q}$ with geometrical rate, there exist $c > 0$ and $0 < b < 1$ such that

$$\left\|\mathbf{Q}\left(t\right)-\mathbf{Q}\right\|\leq cb^t \text{ for every } t\in\mathbb{N}. \tag{9.52}$$

From Lemma 245 (*ii*) and (9.51) we get that

$$\left\|\mathbf{Q}\left(sd+m,nd+m\right)-\mathbf{A}\right\|\leq$$

$$\sum_{u=0}^{\nu d}\left\|\mathbf{Q}\left(nd-\nu d+m+u\right)-\mathbf{Q}\right\|+\left\|\mathbf{Q}^{\nu d}-\mathbf{A}\right\|. \tag{9.53}$$

Hence, using (9.52) and (9.53) we get that

$$\left\|\mathbf{Q}\left(sd+m,nd+m\right)-\mathbf{A}\right\|\leq$$

$$\sum_{u=0}^{\nu d}cb^{nd-\nu d+m}+c_1b_1^{\nu}\leq csdb^{nd-\nu d+m}+c_1b_1^{\nu}\leq csdb^{(n-\nu)d}+c_1b_1^{\nu}. \tag{9.54}$$

Without loss of generality, we can always choose $b > b_1$ and n large enough so that $\nu \leq nd-r$. Equation (9.54) becomes

$$\left\|\mathbf{Q}\left(sd+m,nd+m\right)-\mathbf{A}\right\|\leq c_2b^{(n-\nu)d}+cb^{nd-r},$$

where $c_2 = c\nu d$. Now, defining $c_3 = c_2 b^{-\nu d} + c_1 b^{-r}$ yields

$$\left\|\mathbf{Q}\left(sd+m,nd+m\right)-\mathbf{A}\right\|\leq c_3b^n \text{ for } c_3>0 \text{ and } 0<b<1. \tag{9.55}$$

Now, from Theorem 252 and (9.52) we get that

$$\left\|\mathbf{Q}(nd+m,nd+r)-\mathbf{Q}^{r-m}\right\| \le \sum_{u=1}^{r-m}\left\|\mathbf{Q}(nd+m+u)-\mathbf{A}\mathbf{Q}^{r-m+1}\right\|$$

$$\le \sum_{u=1}^{r-m} cb^{nd+m} \le c_4 b^n. \tag{9.56}$$

From the above we get that

$$\left\|\mathbf{Q}(sd+m,nd+r)-\mathbf{A}\mathbf{Q}^{r-m}\right\| \le c_3 b^n + c_4 b^n \le c_5 b^n,$$

for every n, with $c_5>0$ and $0<b<1$. The case for $r<m$ could be proved similarly. ■

We will now proceed with the following definition:

Definition 330 *Let a complete probability space $(\Omega,\mathcal{F},\mathbb{P})$ and an NHMS in it. We say that the NHMS has an expected d-periodic relative asymptotically attainable structures $\mathbb{E}[\mathbf{q}_r(\infty)]$, $r=0,1,...,d-1$, under asymptotic recruitment control if there exists a sequence of recruitment vectors $\{\mathbf{p}_0(t)\}_{t=0}^{\infty}$ with $\lim_{t\to\infty}\|\mathbf{p}_0(t)-\mathbf{p}_0\|=0$ such that the sequence of the expected relative structures $\{\mathbf{q}(t)\}_{t=0}^{\infty}$ splits into d-converging subsequences with limits $\mathbb{E}[\mathbf{q}_r(\infty)]$, for $r=0,1,...,d-1$.*

We denote by $\mathbb{A}_r^{\infty}$ $(r=0,1,...,d-1)$ the set of expected d-periodic relative asymptotically attainable structures $\mathbb{E}[\mathbf{q}_r(\infty)]$. We now provide the following theorem which describes the set $\mathbb{A}_r^{\infty}$.

Theorem 331 *(Georgiou and Vassiliou (1992)). Let a complete probability space $(\Omega,\mathcal{F},\mathbb{P})$ and an NHMS which is defined by the sequences $\{\mathbf{P}(t)\}_{t=0}^{\infty}$, $\{\mathbf{p}_{k+1}(t)\}_{t=0}^{\infty}$, $\{T(t)\}_{t=0}^{\infty}$, but with an arbitrary chosen sequence of input probabilities $\{\mathbf{p}_0(t)\}_{t=0}^{\infty}$. Assume that the following conditions are satisfied:*

$$(i)\ \lim_{t\to\infty}\frac{\Delta T(t)}{T(t)}=0,\quad (ii)\ \lim_{t\to\infty}\mathbf{P}(t)=\mathbf{P}\quad (iii)\ \lim_{t\to\infty}\mathbf{p}_{k+1}(t)=\mathbf{p}_{k+1} \tag{9.57}$$

$$(iv)\ \lim_{t\to\infty}\mathbf{p}_0(t)=\mathbf{p}_0,\quad (v)\ \mathbf{Q}=\mathbf{P}+\mathbf{p}_{k+1}^{\top}\mathbf{p}_0. \tag{9.58}$$

Let also $\hat{\mathbf{Q}}(t)=\hat{\mathbf{Q}}$ (incidence matrices) for every t, where $\mathbf{Q}$ is periodic with period d, and $T(t)\ge T(t-1)$. Then the following are true: (i) The sequence $\{\mathbf{q}(t)\}_{t=0}^{\infty}$ of the expected relative structure splits into d convergent subsequences with limits

$$\mathbb{E}[\mathbf{q}_r(\infty)]=\mathbf{p}_0\mathbf{A}\left[\sum_{m=1}^{r}s_{m-1}\mathbf{Q}^{r-m}+\sum_{m=r+1}^{d}s_{m-1}\mathbf{Q}^{d+r-m}\right] \tag{9.59}$$

where s_r for $r=0,1,...,d-1$ is defined in (9.62), or

$$\mathbb{E}[\mathbf{q}_r(\infty)]=\mathbf{p}_0\mathbf{A}\left[\sum_{m=1}^{d}s_{m-1}\mathbf{Q}^{d+r-m}\right]\quad \textit{for}\ \ r=0,1,...,d-1. \tag{9.60}$$

(ii) The limiting expected relative structures $\mathbb{E}[\mathbf{q}_r(\infty)]$ $(r=0,1,...,d-1)$ are left eigenvectors for the matrix $\mathbf{Q}^d$.
(iii) Let $\mathcal{C}_0,\mathcal{C}_1,...,\mathcal{C}_{d-1}$ be the cyclic subclasses of the imbedded (inherent) non-homogeneous Markov chain and assume that the elements of the cycle $\mathcal{C}_i$ are the states $1,2,...,k_i$. Then the matrix $\mathbf{P}^d$ can be written as

$$\mathbf{P}^d=\begin{matrix}\mathcal{C}_0\\ \mathcal{C}_1\\ \cdot\\ \mathcal{C}_{d-1}\end{matrix}\begin{pmatrix}\mathbf{P}_0\mathbf{P}_1...\mathbf{P}_{d-1} & 0 & ... & 0\\ 0 & \mathbf{P}_1...\mathbf{P}_{d-1}\mathbf{P}_0 & ... & 0\\ ... & ... & ... & ...\\ 0 & 0 & ... & \mathbf{P}_{d-1}\mathbf{P}_0...\mathbf{P}_{d-2}\end{pmatrix}$$

where $\mathbf{P}_i$ *is the submatrix of* $\mathbf{P}$ *which contains the transition probabilities from the states of the cyclic subclasses* C_i *to the states of the cyclic subclasses* C_{i+1}*. For* $i = d-1$*, the* $(i+1)th$ *cyclic subclass is considered to be the* C_0*. Let the elements of the vector* $\mathbb{E}\left[\mathbf{q}_r(\infty)\right]$ *which corresponds to the states of the cyclic subclass* C_i *be*

$$\left(\mathbb{E}\left[q_{i1}^r(\infty)\right], \mathbb{E}\left[q_{i2}^r(\infty)\right], ..., \mathbb{E}\left[q_{ik_i}^r(\infty)\right]\right)$$

and let $\mathbb{E}\left[\mu_i^r\right]$ *be their sum; then the sets of all asymptotically attainable elements*

$$\left(\mathbb{E}\left[q_{i1}^r(\infty)\right], \mathbb{E}\left[q_{i2}^r(\infty)\right], ..., \mathbb{E}\left[q_{ik_i}^r(\infty)\right]\right)/\mathbb{E}\left[\mu_i^r\right]$$

for $i, r = 0, 1, ..., d-1$ *are convex sets with vertices that are scalar multiples of the rows of the matrix*

$$\left[\mathbf{I} - \mathbf{P}_i\mathbf{P}_{i+1}...\mathbf{P}_0...\mathbf{P}_{i-1}\right]^{-1}. \tag{9.61}$$

(iv) The $\mathbb{E}\left[\mu_i^r\right]$*'s for* $r = 0, 1, ..., d-1$ *are convex combinations of the* s_r $(r = 0, 1, ..., d-1)$*, where*

$$s_r = \lim_{n\to\infty}\sum_{k=0}^{n-1}\frac{\Delta T(kd+r)}{T(nd+r)}. \tag{9.62}$$

For example, for $i, r = 0, 1, ..., d-1$ *we have*

$$\mathbb{E}\left[\mu_i^r\right] = \sum_{m=0}^{d-1}\sum_{j=1}^{k_m} p_{m_j}^0 s_v, \quad where \tag{9.63}$$

$$v = \begin{cases} r-(i+m) & if \;\; i+m \leq r \\ d+r-(i+m) & if \;\; i+m > r \end{cases},$$

with

$$\sum_{m=0}^{d-1}\sum_{j=1}^{k_m} p_{m_j}^0 = 1.$$

(v) The $\mathbb{E}\left[\mu_i^r\right]$ $(i, r = 0, 1, .., d-1)$ *have the following periodic behavior:*

$$\mathbb{E}\left[\mu_i^r\right] = \mathbb{E}\left[\mu_{i+1}^{r+1}\right] = \mathbb{E}\left[\mu_{i+2}^{r+2}\right] = ... = \mathbb{E}\left[\mu_{d-1}^{d-i-1}\right] \tag{9.64}$$

$$= \mathbb{E}\left[\mu_0^{d-i}\right] = ... = \mathbb{E}\left[\mu_{i-1}^{d-1}\right].$$

(iv) The sets of all asymptotically attainable $\mathbb{E}\left[\mathbf{q}_r(\infty)\right]$ *are the convex sets with vertices that are scalar multiples of the rows of the matrix* $\left[\mathbf{I} - \mathbf{P}^d\right]^{-1}$.

Proof. Without loss of generality assume that $t = nd + r$ (for $0 \leq r \leq d-1$). Then equation (5.12) could be written as

$$\mathbb{E}\left[\mathbf{q}(nd+r)\right] = \mathbf{q}(0)\frac{T(0)}{T(nd+r)}\mathbf{Q}(0, nd+r)$$

$$+\frac{1}{T(nd+r)}\left[\sum_{m=1}^{d}\sum_{s=0}^{n-1}\Delta T(sd+m)\,\mathbf{p}_0(sd+m)\,\mathbf{Q}(sd+m, nd+r)\right]$$

$$+\sum_{u=0}^{r}\frac{\Delta T(nd+u)}{T(nd+r)}\mathbf{p}_0(sd+u)\,\mathbf{Q}(sd+u, nd+r). \tag{9.65}$$

We will now proceed to study the convergence of the second term on the right-hand side of Equation (9.65) as $n \to \infty$. We have

$$U = \| \sum_{m=1}^{d} \sum_{s=0}^{n-1} \frac{\Delta T(sd+m)}{T(nd+r)} \mathbf{p}_0 (sd+m) \mathbf{Q}(sd+m, nd+r)$$

$$- \sum_{m=1}^{r} \sum_{s=0}^{n-1} \frac{\Delta T(sd+m)}{T(nd+r)} \mathbf{p}_0 \mathbf{A} \mathbf{Q}^{r-m}$$

$$- \sum_{m=r+1}^{d} \sum_{s=0}^{n-1} \frac{\Delta T(sd+m)}{T(nd+r)} \mathbf{p}_0 \mathbf{A} \mathbf{Q}^{d+r-m} \| \leq$$

$$\sum_{m=1}^{r} \sum_{s=0}^{n-1} \frac{\Delta T(sd+m)}{T(nd+r)} \| \mathbf{p}_0 (sd+m) \mathbf{Q}(sd+m, nd+r)$$

$$- \mathbf{p}_0 \mathbf{A} \mathbf{Q}^{r-m} \|$$

$$+ \sum_{m=r+1}^{d} \sum_{s=0}^{n-1} \frac{\Delta T(sd+m)}{T(nd+r)} \| \mathbf{p}_0 (sd+m) \mathbf{Q}(sd+m, nd+r)$$

$$- \mathbf{p}_0 \mathbf{A} \mathbf{Q}^{d+r-m} \|. \tag{9.66}$$

Now we start with the first part on the right hand size of (9.66)

$$\| \mathbf{p}_0 (sd+m) \mathbf{Q}(sd+m, nd+r) - \mathbf{p}_0 \mathbf{A} \mathbf{Q}^{r-m} \| \leq$$

$$\| \mathbf{p}_0 (sd+m) \| \left\| \mathbf{Q}(sd+m, nd+r) - \mathbf{p}_0 \mathbf{A} \mathbf{Q}^{r-m} \right\| +$$

$$\| \mathbf{p}_0 (sd+m) - \mathbf{p}_0 \| \left\| \mathbf{A} \mathbf{Q}^{r-m} \right\|. \tag{9.67}$$

From assumption (9.58), Theorem 252 and (9.67) we get that

$$\lim_{n \to \infty} \| \mathbf{p}_0 (sd+m) \mathbf{Q}(sd+m, nd+r) - \mathbf{p}_0 \mathbf{A} \mathbf{Q}^{r-m} \| = 0 \text{ if } r \geq m. \tag{9.68}$$

In a similar way we can prove that

$$\lim_{n \to \infty} \| \mathbf{p}_0 (sd+m) \mathbf{Q}(sd+m, nd+r) - \mathbf{p}_0 \mathbf{A} \mathbf{Q}^{d+r-m} \| = 0 \text{ if } r < m. \tag{9.69}$$

In fact relation (9.69) summarizes also relation (9.68) since although the parameter d is required as in (9.69) it has no effect in (9.68) and it is a surplus. The series

$$\sum_{s=0}^{n-1} \frac{\Delta T(sd+m)}{T(nd+r)} \text{ for } m = 0, 1, ..., d-1,$$

is bounded above by $\sum_{\tau=0}^{t} \Delta T(\tau) / T(t) \leq 1$, and as a consequence

$$\lim_{n \to \infty} \sum_{s=0}^{n-1} \frac{\Delta T(sd+m)}{T(nd+r)} \leq 1. \tag{9.70}$$

Now, $\{T(t)\}_{t=0}^{\infty}$ is a monotonically increasing function of t,

$$\Delta T(sd+m) / T(nd+r) \geq 0.$$

Consequently, the series in the first part of (9.70) is converging, and we define its limit by

$$s_m = \lim_{n\to\infty} \sum_{s=0}^{n-1} \frac{\Delta T(sd+m)}{T(nd+r)}. \tag{9.71}$$

From relations (9.66) and (9.68) we get that

$$\begin{aligned} &\| \sum_{s=0}^{n-1} \frac{\Delta T(sd+m)}{T(nd+r)} \mathbf{p}_0(sd+m)\mathbf{Q}(sd+m,nd+r) \\ &- \sum_{s=0}^{n-1} \frac{\Delta T(sd+m)}{T(nd+r)} \mathbf{p}_0\mathbf{AQ}^{r-m} \| \leq \\ &\sum_{s=0}^{n-1} \left| \frac{\Delta T(sd+m)}{T(nd+r)} \right| \times \\ &\left[\left\| \mathbf{Q}(sd+m,nd+r) - \mathbf{AQ}^{r-m} \right\| + \left\| \mathbf{p}_0(sd+m) - \mathbf{p}_0 \right\| \right], \end{aligned} \tag{9.72}$$

which combined with (9.67), (9.68) and (9.71) we arrive at

$$\begin{aligned} &\lim_{n\to\infty} \sum_{s=0}^{n-1} \frac{\Delta T(sd+m)}{T(nd+r)} \mathbf{p}_0(sd+m)\mathbf{Q}(sd+m,nd+r) = \\ &\begin{cases} s_m \mathbf{p}_0 \mathbf{AQ}^{r-m} & \text{if } r \geq m, \\ s_m \mathbf{p}_0 \mathbf{AQ}^{d+r-m} & \text{if } r < m \end{cases}, \end{aligned} \tag{9.73}$$

From (9.66) and (9.73) we get that

$$\begin{aligned} &\lim_{n\to\infty} \sum_{m=1}^{d} \sum_{s=0}^{n-1} \frac{\Delta T(sd+m)}{T(nd+r)} \mathbf{p}_0(sd+m)\mathbf{Q}(sd+m,nd+r) = \\ &\sum_{m=1}^{r} s_m \mathbf{p}_0 \mathbf{AQ}^{r-m} + \sum_{m=r+1}^{d} s_m \mathbf{p}_0 \mathbf{AQ}^{d+r-m} = \\ &\sum_{m=1}^{d} s_m \mathbf{p}_0 \mathbf{AQ}^{d+r-m}. \end{aligned} \tag{9.74}$$

We will now prove that the third term in the right-hand side of equation (9.65) converges to zero as $n \to \infty$.

$$\begin{aligned} &\left\| \sum_{u=0}^{r} \frac{\Delta T(nd+u)}{T(nd+r)} \mathbf{p}_0(sd+u)\mathbf{Q}(sd+u,nd+r) \right\| \leq \\ &\sum_{u=0}^{r} \frac{\Delta T(nd+u)}{T(nd+r)} \left\| \mathbf{p}_0(sd+u)\mathbf{Q}(sd+u,nd+r) \right\| \\ &\leq \sum_{u=0}^{r} \frac{\Delta T(nd+u)}{T(nd+r)}, \end{aligned}$$

which is evident from the assumptions on the total number of memberships that goes to zero as $t \to \infty$. Hence, we have that

$$\mathbb{E}[\mathbf{q}_r(\infty)] = \sum_{m=1}^{r} s_m \mathbf{p}_0 \mathbf{AQ}^{r-m} + \sum_{m=r+1}^{d} s_m \mathbf{p}_0 \mathbf{AQ}^{d+r-m} =$$

$$\sum_{m=1}^{d} s_m \mathbf{p}_0 \mathbf{A} \mathbf{Q}^{d+r-m}. \tag{9.75}$$

(*ii*) For $r = 0, 1, ..., d-1$, we have that

$$\mathbb{E}[\mathbf{q}_r(\infty)]\mathbf{Q} = \mathbf{p}_0 \mathbf{A}\left(\sum_{m=1}^{d} s_m \mathbf{Q}^{d+r-m}\right)\mathbf{Q}$$

$$= \sum_{m=1}^{d} s_m \mathbf{p}_0 \mathbf{A} \mathbf{Q}^{d+r+1-m} = \mathbb{E}[\mathbf{q}_{r+1}(\infty)]. \tag{9.76}$$

Hence, from (9.76) we get that

$$\mathbb{E}[\mathbf{q}_{r+1}(\infty)] = \mathbb{E}[\mathbf{q}_r(\infty)]\mathbf{Q},\ \ \mathbb{E}[\mathbf{q}_{r+2}(\infty)] = \mathbb{E}[\mathbf{q}_{r+1}(\infty)]\mathbf{Q} = \mathbb{E}[\mathbf{q}_r(\infty)]\mathbf{Q}^2,$$

$$...,\ \mathbb{E}[\mathbf{q}_{d-1}(\infty)] = \mathbb{E}[\mathbf{q}_r(\infty)]\mathbf{Q}^{d-r-1},$$

and finally

$$\mathbb{E}[\mathbf{q}_r(\infty)] = \mathbb{E}[\mathbf{q}_r(\infty)]\mathbf{Q}^d \ \text{ for } r = 0, 1, ..., d-1. \tag{9.77}$$

Hence, each limiting structure $\mathbb{E}[\mathbf{q}_r(\infty)]$ is a left eigenvector of the matrix $\mathbf{Q}^d$ which corresponds to the eigenvalue one.
Let the canonical form of the matrix $\mathbf{Q}$ which is an irreducible stochastic matrix of period d with cyclic subclasses $\mathcal{C}_0, \mathcal{C}_1, ..., \mathcal{C}_{d-1}$ be of the form

$$\mathbf{Q} = \begin{matrix} \mathcal{C}_0 \\ \mathcal{C}_1 \\ ... \\ \mathcal{C}_{d-1} \end{matrix} \begin{pmatrix} \mathbf{0} & \mathbf{Q}_0 & 0 & ... & 0 \\ 0 & \mathbf{0} & \mathbf{Q}_1 & ... & 0 \\ ... & ... & ... & ... & ... \\ \mathbf{Q}_{d-1} & 0 & ... & ... & \mathbf{0} \end{pmatrix}.$$

Since $\mathbf{Q} = \mathbf{P} + \mathbf{p}_{k+1}^{\top}\mathbf{p}_0$, we have that the matrix $\mathbf{P}$ will be of the following form

$$\mathbf{Q} = \mathbf{P} + \mathbf{p}_{k+1}^{\top}\mathbf{p}_0 = \begin{matrix} \mathcal{C}_0 \\ \mathcal{C}_1 \\ ... \\ \mathcal{C}_{d-1} \end{matrix} \begin{pmatrix} \mathbf{0} & \mathbf{P}_0 & 0 & ... & 0 \\ 0 & \mathbf{0} & \mathbf{P}_1 & ... & 0 \\ ... & ... & ... & ... & ... \\ \mathbf{P}_{d-1} & 0 & ... & ... & \mathbf{0} \end{pmatrix} + \mathbf{p}_{k+1}^{\top}\mathbf{p}_0. \tag{9.78}$$

It is easily seen that

$$\mathbf{Q}^d = \mathbf{P}^d + \mathbf{B}$$

where $\mathbf{B}$ is the matrix containing the remaining powers of $\mathbf{P}^r$ for $r = 0, 1, ..., d-1$ and which will play no part in our analysis, while $\mathbf{P}^d$ is of the form

$$\mathbf{P}^d = \begin{matrix} \mathcal{C}_0 \\ \mathcal{C}_1 \\ ... \\ \mathcal{C}_{d-1} \end{matrix} \begin{pmatrix} \mathbf{X}_{\mathbf{P}0} & \mathbf{0} & ... & 0 \\ 0 & \mathbf{X}_{\mathbf{P}1} & ... & 0 \\ ... & ... & ... & ... \\ \mathbf{0} & 0 & ... & \mathbf{X}_{\mathbf{P}d-1} \end{pmatrix}, \tag{9.79}$$

where

$$\mathbf{X}_{\mathbf{P}r} = \mathbf{P}_r \mathbf{P}_{r+1} ... \mathbf{P}_{d-1} \mathbf{P}_0 ... \mathbf{P}_{r-1}. \tag{9.80}$$

Assume that the subclass $\mathcal{C}_i$ has k_i elements, that is, states for $i = 1, 2, ..., d-1$ and let us denote by $\mathbb{E}\left[q_{ij}^r(\infty)\right]$ the j-th element of the subclass that corresponds to the vector $\mathbb{E}[\mathbf{q}_r(\infty)]$ with $r = 0, 1, ..., d-1$, $i = 0, 1, ..., d-1$, and $j = 1, 2, ..., k_i$. Also denote by

$$\mathbb{E}[\mu_i^r] = \sum_{j=1}^{k_i} \mathbb{E}\left[q_{ij}^r(\infty)\right], \tag{9.81}$$

the sum of the elements of $\mathbb{E}\left[q_{ij}^{r}(\infty)\right]$ that correspond to the subclass $\mathcal{C}_i$.

We will prove that the set of all asymptotic attainable vectors

$$\left(\mathbb{E}\left[q_{i1}^{r}(\infty)\right], \mathbb{E}\left[q_{i2}^{r}(\infty)\right], ..., \mathbb{E}\left[q_{ik_i}^{r}(\infty)\right]\right) / \mathbb{E}\left[\mu_i^r\right], \tag{9.82}$$

for $r = 0, 1, ..., d-1$ is convex, and we will determine its vertices. From equation (9.77) we get that

$$\mathbb{E}\left[\mathbf{q}_r(\infty)\right] = \mathbb{E}\left[\mathbf{q}_r(\infty)\right]\mathbf{Q}^d = \mathbb{E}\left[\mathbf{q}_r(\infty)\right]\mathbf{P}^d + \mathbf{B}_q \quad \text{for } r = 0, 1, ..., d-1. \tag{9.83}$$

Now define by

$$\mathbb{E}\left[\mathbf{q}_i^r(\infty)\right] = \left(\mathbb{E}\left[q_{i1}^{r}(\infty)\right], \mathbb{E}\left[q_{i2}^{r}(\infty)\right], ..., \mathbb{E}\left[q_{ik_i}^{r}(\infty)\right]\right), \tag{9.84}$$

then (9.83) implies that

$$\left(\mathbb{E}\left[\mathbf{q}_0^r(\infty)\right], \mathbb{E}\left[\mathbf{q}_1^r(\infty)\right], ..., \mathbb{E}\left[\mathbf{q}_i^r(\infty)\right]_{d-1}\right) = \tag{9.85}$$

$$\left(\mathbb{E}\left[\mathbf{q}_0^r(\infty)\right], \mathbb{E}\left[\mathbf{q}_1^r(\infty)\right], ..., \mathbb{E}\left[\mathbf{q}_{d-1}^r(\infty)\right]\right)\begin{pmatrix} \mathbf{X}_{\mathbf{P}0} & \mathbf{0} & ... & 0 \\ 0 & \mathbf{X}_{\mathbf{P}1} & ... & 0 \\ ... & ... & ... & ... \\ \mathbf{0} & 0 & ... & \mathbf{X}_{\mathbf{P}d-1} \end{pmatrix} + \mathbf{B}_q$$

from which we get

$$\begin{aligned} \mathbb{E}\left[\mathbf{q}_i^r(\infty)\right] &= \mathbb{E}\left[\mathbf{q}_i^r(\infty)\right]\mathbf{X}_{\mathbf{P}i} + \mathbb{E}\left[\mathbf{q}_i^r(\infty)\right]\mathbf{B}_q \\ &= \mathbb{E}\left[\mathbf{q}_i^r(\infty)\right]\mathbf{X}_{\mathbf{P}i} + \mathbf{b}_{qi} \quad \text{for} \quad i = 0, 1, ..., d-1. \end{aligned} \tag{9.86}$$

Hence, for $r, i = 0, 1, ..., d-1$ we get that

$$\mathbb{E}\left[\mathbf{q}_i^r(\infty)\right]\left[\mathbf{I}_{k_i} - \mathbf{X}_{\mathbf{P}i}\right] = \mathbf{b}_{qi}, \tag{9.87}$$

where $\mathbf{I}_{k_i}$ is the $k_i \times k_i$ unit matrix. Now, from (9.87) we obtain

$$\mathbb{E}\left[\mathbf{q}_i^r(\infty)\right] = \mathbf{b}_{qi}\left[\mathbf{I}_{k_i} - \mathbf{X}_{\mathbf{P}i}\right]^{-1}, \tag{9.88}$$

where the matrix $\left[\mathbf{I}_{k_i} - \mathbf{X}_{\mathbf{P}i}\right]^{-1}$ is nonnegatively invertible, since each $\mathbf{P}_i$ is substochastic. Equation (9.88) can be written as

$$\mathbb{E}\left[\mathbf{q}_i^r(\infty)\right] = \sum_{j=1}^{k_i}\mathbf{b}_{qi}(j)\,\mathbf{e}_j\left[\mathbf{I}_{k_i} - \mathbf{X}_{\mathbf{P}i}\right]^{-1}, \tag{9.89}$$

where apparently $\mathbf{b}_{qi}(j)$ is the j-coordinate of the vector $\mathbf{b}_{qi}$. Now, if we multiply both sides of equation (9.89) from the right by the vector $\mathbf{1}^\top$ and divide by $\mathbb{E}\left[\mu_i^r\right]$ we get

$$\mathbb{E}\left[\mathbf{q}_i^r(\infty)\right]\mathbf{1}^\top/\mathbb{E}\left[\mu_i^r\right] = 1 = \sum_{j=1}^{k_i}\mathbf{b}_{qi}(j)\,\lambda_j/\mathbb{E}\left[\mu_i^r\right], \tag{9.90}$$

where λ_j is the sum of the elements of the $j-$row of the matrix $\left[\mathbf{I}_{k_i} - \mathbf{X}_{\mathbf{P}i}\right]^{-1}$. Now, from (9.89) we get that

$$\mathbb{E}\left[\mathbf{q}_i^r(\infty)\right]/\mathbb{E}\left[\mu_i^r\right] = \sum_{j=1}^{k_i}\frac{\mathbf{b}_{qi}(j)\,\lambda_j}{\mathbb{E}\left[\mu_i^r\right]}\lambda_j^{-1}\mathbf{e}_j\left[\mathbf{I}_{k_i} - \mathbf{X}_{\mathbf{P}i}\right]^{-1}$$

$$= \sum_{j=1}^{k_i} c_j \lambda_j^{-1} \mathbf{e}_j \left[\mathbf{I}_{k_i} - \mathbf{X}_{\mathbf{P}i}\right]^{-1}, \tag{9.91}$$

where from (9.90) it is evident that

$$0 < c_i < 1 \text{ and } \sum_{j=1}^{k_i} c_j = 1. \tag{9.92}$$

From (9.91) and (9.92), we get that all possible vectors $\mathbb{E}\left[\mathbf{q}_i^r(\infty)\right]/\mathbb{E}\left[\mu_i^r\right]$ for $r.i = 0, 1, ...,$ $d-1$ is a convex set with vertices scalar multiples of the rows of the matrix

$$\left[\mathbf{I}_{k_i} - \mathbf{X}_{\mathbf{P}i}\right]^{-1}.$$

(*iv*) We will now show that the $\mathbb{E}\left[\mu_i^r\right]$ for $i, r = 0, 1, ..., d-1$ are convex combinations of the s_r for $r = 0, 1, ..., d-1$ which were defined in (9.62). From (9.60) for $r = 0, 1, ..., d-1$, we have that

$$\begin{aligned}\mathbb{E}\left[\mathbf{q}_r(\infty)\right] &= \mathbf{p}_0\mathbf{A}\left[\sum_{m=1}^{d} s_{m-1}\mathbf{Q}^{d+r-m}\right] \\ &= \mathbf{p}_0\mathbf{A}\left(s_0\mathbf{Q}^{r-1} + s_1\mathbf{Q}^{r-2} + \cdots + s_{d-1}\mathbf{Q}^r\right) \\ &= \mathbf{p}_0\mathbf{A}(s_{r-1} + s_{r-2}\mathbf{Q} + \cdots + s_0\mathbf{Q}^{r-1} \\ &\quad + s_{d-1}\mathbf{Q}^r + s_{d-2}\mathbf{Q}^{r+1} + \cdots + s_r\mathbf{Q}^{d-1}).\end{aligned} \tag{9.93}$$

For $r = 0$, the s_{r-1} becomes s_{d-1} due to the cyclic movement of the subclasses. It is well known that for a homogeneous Markov chain (Isaacson and Madsen (1976), Vassiliou (2010, Chapter 3) with transition probability matrix $\mathbf{Q}$ which is irreducible and periodic with period d then the sum $\sum_{r=0}^{d-1} \mathbf{AQ}^r$ is a stable stochastic matrix $\mathbf{\Pi}_q$. Assume that the cyclic subclasses $\mathcal{C}_0, \mathcal{C}_1, ..., \mathcal{C}_{d-1}$ are as previously then the row of the stable stochastic matrix will be of the form

$$\boldsymbol{\pi}_q = (\boldsymbol{\pi}_0, \boldsymbol{\pi}_1, ..., \boldsymbol{\pi}_{d-1}) \text{ with } \boldsymbol{\pi}_i = (\pi_{1i}, \pi_{2i}, ..., \pi_{k_i i}). \tag{9.94}$$

We need a more descriptive way of expressing the matrix $\mathbf{\Pi}_q$. Let us denote by $\mathbf{\Pi}_{qi}^{(k_j)}$ the

$$\mathbf{\Pi}_{qi}^{(k_j)} = \begin{matrix} - \\ \uparrow \\ k_j \\ \downarrow \\ - \end{matrix} \begin{pmatrix} \pi_1^{(i)} & \pi_2^{(i)} & \dots & \pi_{k_i}^{(i)} \\ \pi_1^{(i)} & \pi_2^{(i)} & \dots & \pi_{k_i}^{(i)} \\ \dots & \dots & \dots & \dots \\ \pi_1^{(i)} & \pi_2^{(i)} & \dots & \pi_{k_i}^{(i)} \end{pmatrix}, \tag{9.95}$$

which is a stable matrix with k_j rows equal with $\boldsymbol{\pi}_i$. Then the matrix $\mathbf{\Pi}_q$ could be written in the form

$$\mathbf{\Pi}_q = \begin{pmatrix} \mathbf{\Pi}_{q0}^{(k_0)} & \mathbf{\Pi}_{q1}^{(k_0)} & \dots & \mathbf{\Pi}_{qd-1}^{(k_0)} \\ \mathbf{\Pi}_{q0}^{(k_1)} & \mathbf{\Pi}_{q1}^{(k_1)} & \dots & \mathbf{\Pi}_{qd-1}^{(k_1)} \\ \dots & \dots & \dots & \dots \\ \mathbf{\Pi}_{q0}^{(k_{d-1})} & \mathbf{\Pi}_{q1}^{(k_{d-1})} & \dots & \mathbf{\Pi}_{qd-1}^{(k_{d-1})} \end{pmatrix}. \tag{9.96}$$

Given the notation in (9.95) and (9.96) it is not difficult to show in a similar way that

$$\mathbf{\Pi}_q(s) = \sum_{m=1}^{d} s_{m-1}\mathbf{AQ}^{d+r-m} = \tag{9.97}$$

$$\begin{pmatrix} s_{r-1}\mathbf{\Pi}_{q0}^{(k_0)} & s_{r-2}\mathbf{\Pi}_{q1}^{(k_0)} & .. & s_0\mathbf{\Pi}_{qr-1}^{(k_0)} & ... & s_r\mathbf{\Pi}_{qd-1}^{(k_0)} \\ s_r\mathbf{\Pi}_{q0}^{(k_1)} & s_{r-1}\mathbf{\Pi}_{q1}^{(k_1)} & ... & s_1\mathbf{\Pi}_{qr-1}^{(k_1)} & ... & s_{r+1}\mathbf{\Pi}_{qd-1}^{(k_1)} \\ ... & ... & ... & ... & ... & ... \\ s_{d-1}\mathbf{\Pi}_{q0}^{(k_{d-r})} & s_{d-2}\mathbf{\Pi}_{q1}^{(k_{d-r})} & ... & s_r\mathbf{\Pi}_{qr-1}^{(k_{d-r})} & ... & s_0\mathbf{\Pi}_{qd-1}^{(k_{d-r})} \\ s_0\mathbf{\Pi}_{q0}^{(k_{d-r+1})} & s_{d-1}\mathbf{\Pi}_{q1}^{(k_{d-r+1})} & ... & s_{r+1}\mathbf{\Pi}_{qr-1}^{(k_{d-r+1})} & ... & s_1\mathbf{\Pi}_{qd-1}^{(k_{d-r+1})} \\ ... & ... & ... & ... & ... & ... \\ s_{r-2}\mathbf{\Pi}_{q0}^{(k_{d-1})} & s_{r-3}\mathbf{\Pi}_{q1}^{(k_{d-1})} & ... & s_{d-1}\mathbf{\Pi}_{qr-1}^{(k_{d-1})} & ... & s_{r-1}1\mathbf{\Pi}_{qd-1}^{(k_{d-1})} \end{pmatrix}.$$

Let us denote by

$$\mathbf{p}_0^{(i)} = \left(p_{01}^{(i)}, p_{02}^{(i)}, ..., p_{0k_i}^{(i)}\right) \text{ then } \mathbf{p}_0 = \left(\mathbf{p}_0^{(0)}, \mathbf{p}_0^{(1)}, ..., \mathbf{p}_0^{(d-1)}\right),$$

that is, we divide the limiting input vector into the cyclic subclasses of the transition matrix $\mathbf{Q}$ of the imbedded limiting Markov chain. We will try now to express the elements of the vector $\mathbb{E}\left[\mathbf{q}_r(\infty)\right]$ in terms of the parameters s_r for $r = 0, 1, ..., d-1$ and the elements of the vector $\mathbf{p}_0$. The reader that will find difficulties to follow this part of the proof might as well read first the example 335. Now, from (9.97) we get that

$$\mathbb{E}\mathbf{q}_r(\infty) = \mathbf{p}_0 \sum_{m=1}^{d} s_{m-1}\mathbf{A}\mathbf{Q}^{d+r-m} = \left(\mathbf{p}_0^{(0)}, \mathbf{p}_0^{(1)}, ..., \mathbf{p}_0^{(d-1)}\right) \times \tag{9.98}$$

$$\begin{pmatrix} s_{r-1}\mathbf{\Pi}_{q0}^{(k_0)} & s_{r-2}\mathbf{\Pi}_{q1}^{(k_0)} & .. & s_0\mathbf{\Pi}_{qr-1}^{(k_0)} & ... & s_r\mathbf{\Pi}_{qd-1}^{(k_0)} \\ s_r\mathbf{\Pi}_{q0}^{(k_1)} & s_{r-1}\mathbf{\Pi}_{q1}^{(k_1)} & ... & s_1\mathbf{\Pi}_{qr-1}^{(k_1)} & ... & s_{r+1}\mathbf{\Pi}_{qd-1}^{(k_1)} \\ ... & ... & ... & ... & ... & ... \\ s_{d-1}\mathbf{\Pi}_{q0}^{(k_{d-r})} & s_{d-2}\mathbf{\Pi}_{q1}^{(k_{d-r})} & ... & s_r\mathbf{\Pi}_{qr-1}^{(k_{d-r})} & ... & s_0\mathbf{\Pi}_{qd-1}^{(k_{d-r})} \\ s_0\mathbf{\Pi}_{q0}^{(k_{d-r+1})} & s_{d-1}\mathbf{\Pi}_{q1}^{(k_{d-r+1})} & ... & s_{r+1}\mathbf{\Pi}_{qr-1}^{(k_{d-r+1})} & ... & s_1\mathbf{\Pi}_{qd-1}^{(k_{d-r+1})} \\ ... & ... & ... & ... & ... & ... \\ s_{r-2}\mathbf{\Pi}_{q0}^{(k_{d-1})} & s_{r-3}\mathbf{\Pi}_{q1}^{(k_{d-1})} & ... & s_{d-1}\mathbf{\Pi}_{qr-1}^{(k_{d-1})} & ... & s_{r-1}1\mathbf{\Pi}_{qd-1}^{(k_{d-1})} \end{pmatrix},$$

from we which we get that

$$\mathbb{E}\left[q_{1r}^{(0)}(\infty)\right] = [s_{r-1}\left(p_{01}^{(0)} + p_{02}^{(0)} + \cdots + p_{0k_0}^{(0)}\right) + s_r\left(p_{01}^{(1)} + p_{02}^{(1)} + \cdots + p_{0k_1}^{(1)}\right) +$$
$$\cdots + s_{r-2}\left(p_{01}^{(d-1)} + p_{02}^{(d-1)} + \cdots + p_{0k_{d-1}}^{(d-1)}\right)\pi_1^{(0)},$$
$$\mathbb{E}\left[q_{2r}^{(0)}(\infty)\right] = [s_{r-1}\left(p_{01}^{(0)} + p_{02}^{(0)} + \cdots + p_{0k_0}^{(0)}\right) + s_r\left(p_{01}^{(1)} + p_{02}^{(1)} + \cdots + p_{0k_1}^{(1)}\right) +$$
$$\cdots + s_{r-2}\left(p_{01}^{(d-1)} + p_{02}^{(d-1)} + \cdots + p_{0k_{d-1}}^{(d-1)}\right)\pi_2^{(0)},$$
$$\cdots\cdots\cdots\cdots\cdots\cdots\cdots\cdots\cdots$$
$$\mathbb{E}\left[q_{k_0r}^{(0)}(\infty)\right] = [s_{r-1}\left(p_{01}^{(0)} + p_{02}^{(0)} + \cdots + p_{0k_0}^{(0)}\right) + s_r\left(p_{01}^{(1)} + p_{02}^{(1)} + \cdots + p_{0k_1}^{(1)}\right) +$$
$$\cdots + s_{r-2}\left(p_{01}^{(d-1)} + p_{02}^{(d-1)} + \cdots + p_{0k_{d-1}}^{(d-1)}\right)\pi_{k_0}^{(0)}. \tag{9.99}$$

Analogous results in a similar way we get for the remaining elements of the subclasses of $\mathbb{E}\left[\mathbf{q}_r(\infty)\right]$ that correspond for the subclasses $\mathcal{C}_0, \mathcal{C}_1, ..., \mathcal{C}_{d-1}$. Now, if we sum up all the elements of $\mathbb{E}\left[q_{ir}^{(0)}(\infty)\right]$ we get:

$$\sum_{i=1}^{k_0}\mathbb{E}\left[q_{ir}^{(0)}(\infty)\right] = [s_{r-1}\left(\sum_{i=1}^{k_0} p_{0i}^{(0)}\right) + s_r\left(\sum_{i=1}^{k_1} p_{0i}^{(1)}\right) + \cdots +$$

$$s_{r-2}\left(\sum_{i=1}^{k_{d-1}} p_{0i}^{(d-1)}\right)]\left(\pi_1^{(0)}+\pi_2^{(0)}+\cdots+\pi_{k_0}^{(0)}\right)=\mathbb{E}\left[\mu_0^r\right], \tag{9.100}$$

from which we get that

$$\mathbb{E}\left[\mu_0^r\right]=\sum_{i=0}^{d-1} a_i s_i \text{ with } \sum_{i=0}^{d-1} a_i=\sum_{j=1}^{k} p_{0j}=1. \tag{9.101}$$

In general, from the above it is not difficult to show that

$$\mathbb{E}\left[\mu_u^r\right]=\sum_{m=0}^{d-1}\sum_{i=1}^{k_m} p_{0i}^{(m)} s_l \text{ where } l=\left\{\begin{smallmatrix} r-(u+m) & \text{if} & u+m\leq r \\ d+r-(u+m) & \text{if} & u+m>r\end{smallmatrix}\right., \tag{9.102}$$

with apparently

$$\sum_{m=0}^{d-1}\sum_{i=1}^{k_m} p_{0i}^{(m)}=1,$$

since it is the sum of the elements of the stochastic vector $\mathbf{p}_0$. Hence, we showed that the sum of the elements in each subclass is a convex combination of the sums s_r for $r=0,1,...,d-1$ with $\sum_{r=0}^{d-1} s_r=1$. Now, from (9.102) we get that

$$\mathbb{E}\left[\mu_{u+1}^{r+1}\right]=\sum_{m=0}^{d-1}\sum_{i=1}^{k_m} p_{0i}^{(m)} s_l \text{ where } l=\left\{\begin{smallmatrix} r-(u+m) & \text{if} & u+m\leq r \\ d+r-(u+m) & \text{if} & u+m>r\end{smallmatrix}\right., \tag{9.103}$$

hence from (9.102) and (9.103) we get that $\mathbb{E}\left[\mu_u^r\right]=\mathbb{E}\left[\mu_{u+1}^{r+1}\right]$. Therefore, we have the following cyclic behavior

$$\begin{aligned}\mathbb{E}\left[\mu_u^r\right]&=\mathbb{E}\left[\mu_{u+1}^{r+1}\right]=\mathbb{E}\left[\mu_{u+2}^{r+2}\right]= \\ &...=\mathbb{E}\left[\mu_0^{d-u}\right]= \\ &...=\mathbb{E}\left[\mu_{u-1}^{d-1}\right].\end{aligned} \tag{9.104}$$

From (9.104) it is evident that it is sufficient to find the above quantities for only some value of r and from them we get the values for every r. ■

The following Corollary is straight forward to be proved for the reader and can be checked in Georgiou (1989):

Corollary 332 *Let a complete probability space* $(\Omega,\mathcal{F},\mathbb{P})$ *and an NHMS which is defined by the sequences* $\{\mathbf{P}(t)\}_{t=0}^{\infty}$, $\{\mathbf{p}_{k+1}(t)\}_{t=0}^{\infty}$, $\{T(t)\}_{t=0}^{\infty}$, *but with an arbitrary chosen sequence of input probabilities* $\{\mathbf{p}_0(t)\}_{t=0}^{\infty}$. *Assume that the following conditions are satisfied:*

$$(i)\ \lim_{t\to\infty} T(t)=T<\infty,\ \ (ii)\ \lim_{t\to\infty}\mathbf{P}(t)=\mathbf{P}\ \ (iii)\ \lim_{t\to\infty}\mathbf{p}_{k+1}(t)=\mathbf{p}_{k+1}, \tag{9.105}$$

$$(iv)\ \lim_{t\to\infty}\mathbf{p}_0(t)=\mathbf{p}_0,\ \ (v)\ \mathbf{Q}=\mathbf{P}+\mathbf{p}_{k+1}^{\top}\mathbf{p}_0. \tag{9.106}$$

Let also $\hat{\mathbf{Q}}(t)=\hat{\mathbf{Q}}$ *for every* t, *where* $\mathbf{Q}$ *is periodic with period* d, *and* $T(t)\geq T(t-1)$. *Then the following are true: The sequence* $\{\mathbf{N}(t)\}_{t=0}^{\infty}$ *of the expected structure splits into* d *convergent subsequences with limits*

$$\mathbb{E}\left[\mathbf{N}_r(\infty)\right]=\mathbf{N}(0)\mathbf{A}\mathbf{Q}^r+\mathbf{p}_0\mathbf{A}\left[\sum_{m=1}^{r} s_{m-1}\mathbf{Q}^{r-m}+\sum_{m=r+1}^{d} s_{m-1}\mathbf{Q}^{d+r-m}\right] \tag{9.107}$$

or

$$\mathbb{E}\left[\mathbf{N}_r(\infty)\right]=\mathbf{N}(0)\mathbf{A}\mathbf{Q}^r+\mathbf{p}_0\mathbf{A}\left[\sum_{m=1}^{d-1} s_{m-1}\mathbf{Q}^{d+r-m}\right]\ \ \textit{for}\ \ r=0,1,...,d-1. \tag{9.108}$$

We will now study the conditions under which the convergences of the subsequences in the above subsequences are with geometrical rate:

Theorem 333 *Let a complete probability space* $(\Omega, \mathcal{F}, \mathbb{P})$ *and an NHMS which is defined by the sequences* $\{\mathbf{P}(t)\}_{t=0}^{\infty}$, $\{\mathbf{p}_{k+1}(t)\}_{t=0}^{\infty}$, $\{T(t)\}_{t=0}^{\infty}$, *and* $\{\mathbf{p}_0(t)\}_{t=0}^{\infty}$. *Assume that the following conditions are satisfied:*

$$(i)\ \lim_{t\to\infty} T(t) = T < \infty,\ \ (ii)\ \lim_{t\to\infty} \mathbf{P}(t) = \mathbf{P}\ \ (iii)\ \lim_{t\to\infty} \mathbf{p}_{k+1}(t) = \mathbf{p}_{k+1}, \tag{9.109}$$

$$(iv)\ \lim_{t\to\infty} \mathbf{p}_0(t) = \mathbf{p}_0,\ \ (v)\ \mathbf{Q} = \mathbf{P} + \mathbf{p}_{k+1}^{\top}\mathbf{p}_0. \tag{9.110}$$

Assume also that all the above convergences are at a geometric rate. Let also $\hat{\mathbf{Q}}(t) = \hat{\mathbf{Q}}$ *for every* t, *where* $\mathbf{Q}$ *is periodic with period* d, *and* $T(t) \geq T(t-1)$. *Then the sequence* $\{\mathbb{E}[\mathbf{N}(t)]\}_{t=0}^{\infty}$ *of the expected structure splits into* d *convergent subsequences with limits given in* (9.107) *and* (9.108) *with geometrical rate of convergence.*

Proof. The proof follows closely the techniques of Theorems 295, 314. The reader will benefit if he tries to do the proof as an exercise. If he feels that is needed to check it the proof is available in Georgiou (1989). ■

Also for the analogous problem for the expected relative structure, we have the following:

Theorem 334 *Let a complete probability space* $(\Omega, \mathcal{F}, \mathbb{P})$ *and an NHMS which is defined by the sequences* $\{\mathbf{P}(t)\}_{t=0}^{\infty}$, $\{\mathbf{p}_{k+1}(t)\}_{t=0}^{\infty}$, $\{T(t)\}_{t=0}^{\infty}$, *and* $\{\mathbf{p}_0(t)\}_{t=0}^{\infty}$. *Assume that the following conditions are satisfied:*

$$(i)\ \lim_{t\to\infty} \frac{\Delta T(t)}{T(t)} = 0,\ \ (ii)\ \lim_{t\to\infty} \mathbf{P}(t) = \mathbf{P}\ \ (iii)\ \lim_{t\to\infty} \mathbf{p}_{k+1}(t) = \mathbf{p}_{k+1}, \tag{9.111}$$

$$(iv)\ \lim_{t\to\infty} \mathbf{p}_0(t) = \mathbf{p}_0,\ \ (v)\ \mathbf{Q} = \mathbf{P} + \mathbf{p}_{k+1}^{\top}\mathbf{p}_0. \tag{9.112}$$

Assume also that all the above convergences are at a geometric rate. Let also $\hat{\mathbf{Q}}(t) = \hat{\mathbf{Q}}$ *for every* t, *where* $\mathbf{Q}$ *is periodic with period* d, *and* $T(t) \geq T(t-1)$. *Then the sequence* $\{\mathbb{E}[\mathbf{q}(t)]\}_{t=0}^{\infty}$ *of the expected relative structure splits into* d *convergent subsequences with limits given by*

$$\mathbb{E}[\mathbf{q}_r(\infty)] = \mathbf{q}(0)\,\mathbf{A}\mathbf{Q}^r + \mathbf{p}_0\mathbf{A}\left[\sum_{m=1}^{d-1} s_{m-1}\mathbf{Q}^{d+r-m}\right] \quad \text{for } r = 0, 1, ..., d-1, \tag{9.113}$$

with geometrical rate of convergence.

Proof. Taking into account Lemma 293 where we prove that $\lim_{t-\infty} \Delta T(t)/T(t) = 0$ with geometrical rate of convergence is equivalent with $\lim_{t\to\infty} T(t) = T$ with geometrical rate of convergence from Theorem 331 we immediately arrive at the proof. ■

Example 335 *Let a periodic stochastic matrix* $\mathbf{Q}$ *which is* 5×5 *and which has period* $d = 3$. *Also let that the cyclic subclasses be* $\mathcal{C}_0 = \{1, 2\}$, $\mathcal{C}_1 = \{3, 4\}$ *and* $\mathcal{C}_2 = \{5\}$. *Then the matrix* $\mathbf{Q}$ *could be written in the following bloc form*

$$\mathbf{Q} = \begin{matrix} \mathcal{C}_0 \\ \mathcal{C}_1 \\ \mathcal{C}_2 \end{matrix} \begin{pmatrix} 0 & \mathbf{Q}_0 & 0 \\ 0 & 0 & \mathbf{Q}_1 \\ \mathbf{Q}_2 & 0 & 0 \end{pmatrix},$$

where the matrix $\mathbf{Q}_i$ *contains the transition probabilities between the cyclic subclass* $\mathcal{C}_i$ *and* $\mathcal{C}_{i+1}$ *for* $i = 0, 1, 2$. *Apparently due to their cyclic nature* $\mathcal{C}_3 = \mathcal{C}_0$. *The matrices* $\mathbf{Q}_i$ *are regular stochastic matrices with positive elements at least in the positions which are necessary and sufficient to make all the states among the subclass i to communicate. Now it is known that*

$$\mathbf{Q}^{td} = \begin{matrix} \mathcal{C}_0 \\ \mathcal{C}_1 \\ \mathcal{C}_2 \end{matrix} \begin{pmatrix} \mathbf{X}_0 & \mathbf{0} & \mathbf{0} \\ \mathbf{0} & \mathbf{X}_1 & 0 \\ \mathbf{0} & \mathbf{0} & \mathbf{X}_2 \end{pmatrix},$$

where $\mathbf{X}_0 = \mathbf{Q}_0\mathbf{Q}_1\mathbf{Q}_2$, $\mathbf{X}_1 = \mathbf{Q}_1\mathbf{Q}_2\mathbf{Q}_0$ *and* $\mathbf{X}_2 = \mathbf{Q}_2\mathbf{Q}_1\mathbf{Q}_0$. *Then it is known that*

$$\lim_{t\to\infty} \mathbf{Q}^{td} = \mathbf{A} = \begin{matrix} \mathcal{C}_0 \\ \mathcal{C}_1 \\ \mathcal{C}_2 \end{matrix} \begin{pmatrix} \mathbf{\Pi}_0 & \mathbf{0} & \mathbf{0} \\ \mathbf{0} & \mathbf{\Pi}_1 & 0 \\ \mathbf{0} & \mathbf{0} & \mathbf{\Pi}_2 \end{pmatrix},$$

where $\mathbf{\Pi}_i = \lim_{t\to\infty} \mathbf{X}_i^t$ *a stable stochastic matrix with row* $\boldsymbol{\pi}^{(i)} = \left(\pi_1^{(i)}, \pi_2^{(i)}\right)$ *for* $i = 0, 1$ *and* $\boldsymbol{\pi}^{(2)} = \left(\pi_1^{(2)}\right)$. *Now, it is known that*

$$\sum_{r=0}^{2} \mathbf{AQ}^r = \begin{pmatrix} \pi_1^{(0)} & \pi_2^{(0)} & \pi_1^{(1)} & \pi_2^{(1)} & \pi_1^{(2)} \\ \pi_1^{(0)} & \pi_2^{(0)} & \pi_1^{(1)} & \pi_2^{(1)} & \pi_1^{(2)} \\ \pi_1^{(0)} & \pi_2^{(0)} & \pi_1^{(1)} & \pi_2^{(1)} & \pi_1^{(2)} \\ \pi_1^{(0)} & \pi_2^{(0)} & \pi_1^{(1)} & \pi_2^{(1)} & \pi_1^{(2)} \\ \pi_1^{(0)} & \pi_2^{(0)} & \pi_1^{(1)} & \pi_2^{(1)} & \pi_1^{(2)} \end{pmatrix}$$

$$= \begin{pmatrix} \mathbf{\Pi}_{q0}^{(2)} & \mathbf{\Pi}_{q1}^{(2)} & \mathbf{\Pi}_{q2}^{(2)} \\ \mathbf{\Pi}_{q0}^{(2)} & \mathbf{\Pi}_{q1}^{(2)} & \mathbf{\Pi}_{q2}^{(2)} \\ \mathbf{\Pi}_{q0}^{(1)} & \mathbf{\Pi}_{q1}^{(1)} & \mathbf{\Pi}_{q2}^{(1)} \end{pmatrix} = \mathbf{\Pi}_q.$$

It is also known that $1/3\mathbf{\Pi}_q$ *it is a stable stochastic matrix and that the sequence* $\{\mathbf{Q}^t\}_{t=0}^{\infty}$ *converges in Cezaro to it. Now, we have shown that*

$$\mathbb{E}\left[\mathbf{q}_r(\infty)\right] = \mathbf{p}_0 \sum_{m=1}^{d} s_{m-1}\mathbf{AQ}^{d+r-m} \text{ for } r = 0, 1, 2.$$

That is,

$$\mathbb{E}\left[\mathbf{q}_0(\infty)\right] = s_0\mathbf{p}_0\mathbf{AQ}^2 + s_1\mathbf{p}_0\mathbf{AQ} + s_2\mathbf{p}_0\mathbf{A},$$
$$\mathbb{E}\left[\mathbf{q}_1(\infty)\right] = s_0\mathbf{p}_0\mathbf{A} + s_1\mathbf{p}_0\mathbf{AQ}^2 + s_2\mathbf{p}_0\mathbf{AQ},$$
$$\mathbb{E}\left[\mathbf{q}_2(\infty)\right] = s_0\mathbf{p}_0\mathbf{AQ} + s_1\mathbf{p}_0\mathbf{A} + s_2\mathbf{p}_0\mathbf{AQ}^2.$$

Let us now work with $\mathbb{E}\left[\mathbf{q}_1(\infty)\right]$ *to illustrate our formulas. We have*

$$s_0\mathbf{p}_0\mathbf{A} = s_0\left(\mathbf{p}_0^{(0)} \quad \mathbf{p}_0^{(1)} \quad \mathbf{p}_0^{(2)}\right)\mathbf{A} = s_0\left(p_{01}^{(0)} \quad p_{02}^{(0)} \quad p_{01}^{(1)} \quad p_{02}^{(1)} \quad p_{01}^{(2)}\right)$$

$$\times \begin{pmatrix} \pi_1^{(0)} & \pi_2^{(0)} & 0 & 0 & 0 \\ \pi_1^{(0)} & \pi_2^{(0)} & 0 & 0 & 0 \\ 0 & 0 & \pi_1^{(1)} & \pi_2^{(1)} & 0 \\ 0 & 0 & \pi_1^{(1)} & \pi_2^{(1)} & 0 \\ 0 & 0 & 0 & 0 & \pi_1^{(2)} \end{pmatrix} =$$

$$(s_0\left[p_{01}^{(0)} + p_{02}^{(0)}\right]\pi_1^{(0)} \quad s_0\left[p_{01}^{(0)} + p_{02}^{(0)}\right]\pi_2^{(0)} \quad s_0\left[p_{01}^{(1)} + p_{02}^{(1)}\right]\pi_1^{(1)}$$
$$s_0\left[p_{01}^{(1)} + p_{02}^{(1)}\right]\pi_2^{(1)} \quad s_0 p_{01}^{(2)}\pi_1^{(2)}),$$

similarly

$$s_0\mathbf{p}_0\mathbf{AQ}^2 = s_0\left(\mathbf{p}_0^{(0)} \quad \mathbf{p}_0^{(1)} \quad \mathbf{p}_0^{(2)}\right)\mathbf{AQ}^2 = s_0\left(p_{01}^{(0)} \quad p_{02}^{(0)} \quad p_{01}^{(1)} \quad p_{02}^{(1)} \quad p_{01}^{(2)}\right)$$

$$\times\begin{pmatrix} 0 & 0 & 0 & 0 & \pi_1^{(2)} \\ 0 & 0 & 0 & 0 & \pi_1^{(2)} \\ \pi_1^{(0)} & \pi_2^{(0)} & 0 & 0 & 0 \\ \pi_1^{(0)} & \pi_2^{(0)} & 0 & 0 & 0 \\ 0 & 0 & \pi_1^{(1)} & \pi_2^{(1)} & 0 \end{pmatrix} =$$

$$(s_1\left[p_{01}^{(0)} + p_{02}^{(0)}\right]\pi_1^{(0)} \quad s_1\left[p_{01}^{(0)} + p_{02}^{(0)}\right]\pi_2^{(0)} \quad s_1 p_{01}^{(2)}\pi_1^{(1)}$$

$$s_1 p_{01}^{(2)}\pi_2^{(1)} \quad s_1\left[p_{01}^{(1)} + p_{02}^{(1)}\right]\pi_1^{(2)}),$$

and finally

$$s_0\mathbf{p}_0\mathbf{AQ} = s_0\left(\mathbf{p}_0^{(0)} \quad \mathbf{p}_0^{(1)} \quad \mathbf{p}_0^{(2)}\right)\mathbf{AQ} = s_0\left(p_{01}^{(0)} \quad p_{02}^{(0)} \quad p_{01}^{(1)} \quad p_{02}^{(1)} \quad p_{01}^{(2)}\right)$$

$$\times\begin{pmatrix} 0 & 0 & \pi_1^{(1)} & \pi_2^{(1)} & 0 \\ 0 & 0 & \pi_1^{(1)} & \pi_2^{(1)} & 0 \\ 0 & 0 & 0 & 0 & \pi_1^{(2)} \\ 0 & 0 & 0 & 0 & \pi_1^{(2)} \\ \pi_1^{(0)} & \pi_2^{(0)} & 0 & 0 & 0 \end{pmatrix} =$$

$$(s_2 p_{01}^{(2)}\pi_1^{(0)} \quad s_2 p_{01}^{(2)}\pi_2^{(0)} \quad s_2\left[p_{01}^{(0)} + p_{02}^{(0)}\right]\pi_1^{(1)}$$

$$s_2\left[p_{01}^{(0)} + p_{02}^{(0)}\right]\pi_2^{(1)} \quad s_2\left[p_{01}^{(1)} + p_{02}^{(1)}\right]\pi_1^{(2)}).$$

It is apparent now from the above that

$$\mathbb{E}\left[\mathbf{q}_r(\infty)\right] = \left(\mathbb{E}\left[\mathbf{q}_r^{(0)}(\infty)\right] \quad \mathbb{E}\left[\mathbf{q}_r^{(1)}(\infty)\right] \quad \mathbb{E}\left[\mathbf{q}_r^{(2)}(\infty)\right]\right) =$$

$$= \left(\mathbb{E}\left[q_{1r}^{(0)}(\infty)\right] \quad \mathbb{E}\left[q_{2r}^{(0)}(\infty)\right] \quad \mathbb{E}\left[q_{1r}^{(1)}(\infty)\right] \quad \mathbb{E}\left[q_{2r}^{(1)}(\infty)\right] \quad \mathbb{E}\left[q_{2r}^{(1)}(\infty)\right]\right),$$

and for $r = 1$

$$\mathbb{E}\left[q_{11}^{(0)}(\infty)\right] = \left[s_0\left(p_{01}^{(0)} + p_{02}^{(0)}\right) + s_1\left(p_{01}^{(0)} + p_{02}^{(0)}\right) + s_2 p_{01}^{(2)}\right]\pi_1^{(0)},$$

$$\mathbb{E}\left[q_{21}^{(0)}(\infty)\right] = \left[s_0\left(p_{01}^{(0)} + p_{02}^{(0)}\right) + s_1\left(p_{01}^{(0)} + p_{02}^{(0)}\right) + s_2 p_{02}^{(2)}\right]\pi_2^{(0)},$$

$$\mathbb{E}\left[q_{11}^{(1)}(\infty)\right] = \left[s_0\left(p_{01}^{(1)} + p_{02}^{(1)}\right) + s_1 p_{01}^{(2)} + s_2\left(p_{01}^{(0)} + p_{02}^{(0)}\right)\right]\pi_1^{(1)},$$

$$\mathbb{E}\left[q_{21}^{(1)}(\infty)\right] = \left[s_0\left(p_{01}^{(1)} + p_{02}^{(1)}\right) + s_1 p_{01}^{(2)} + s_2\left(p_{01}^{(0)} + p_{02}^{(0)}\right)\right]\pi_2^{(1)},$$

$$\mathbb{E}\left[q_{21}^{(2)}(\infty)\right] = \left[s_0 p_{01}^{(2)} + s_1\left(p_{01}^{(0)} + p_{02}^{(0)}\right) + s_2\left(p_{01}^{(1)} + p_{02}^{(1)}\right)\right]\pi_1^{(2)}.$$

Now adding the elements that belong to different subclasses we get for subclass (0):

$$\mathbb{E}\left[q_{11}^{(0)}(\infty)\right] + \mathbb{E}\left[q_{21}^{(0)}(\infty)\right] = \left[s_0\left(p_{01}^{(0)} + p_{02}^{(0)}\right) + s_1\left(p_{01}^{(1)} + p_{02}^{(1)}\right) + s_2 p_{01}^{(2)}\right]\times$$

$$\left(\pi_1^{(0)} + \pi_2^{(0)}\right) = \left[s_0\left(p_{01}^{(0)} + p_{02}^{(0)}\right) + s_1\left(p_{01}^{(1)} + p_{02}^{(1)}\right) + s_2 p_{01}^{(2)}\right] = c\left(1,0,2\right),$$

respectively for subclasses (1)*:*

$$\mathbb{E}\left[q_{11}^{(1)}\left(\infty\right)\right] + \mathbb{E}\left[q_{21}^{(1)}\left(\infty\right)\right] = \left[s_0\left(p_{01}^{(1)} + p_{02}^{(1)}\right) + s_1 p_{01}^{(2)} + s_2\left(p_{01}^{(0)} + p_{02}^{(0)}\right)\right] \times$$

$$\left(\pi_1^{(0)} + \pi_2^{(0)}\right) = \left[s_0\left(p_{01}^{(1)} + p_{02}^{(1)}\right) + s_1 p_{01}^{(2)} + s_2\left(p_{01}^{(0)} + p_{02}^{(0)}\right)\right] = c\left(1,1,2\right),$$

and finally for subclass (2)*:*

$$\mathbb{E}\left[q_{21}^{(2)}\left(\infty\right)\right] = \left[s_0 p_{01}^{(2)} + s_1\left(p_{01}^{(0)} + p_{02}^{(0)}\right) + s_2\left(p_{01}^{(1)} + p_{02}^{(1)}\right)\right]\pi_1^{(2)} =$$

$$\left[s_0 p_{01}^{(2)} + s_1\left(p_{01}^{(0)} + p_{02}^{(0)}\right) + s_2\left(p_{01}^{(1)} + p_{02}^{(1)}\right)\right] = c\left(1,2,1\right).$$

It is easy to see that

$$c\left(1,0,2\right) = \sum_{i=0}^{2} s_i a_i \text{ with } \sum_{i=0}^{2} a_i = 1,$$

$$c\left(1,1,2\right) = \sum_{i=0}^{2} s_i \beta_i \text{ with } \sum_{i=0}^{2} \beta_i = 1,$$

$$c\left(1,2,1\right) = \sum_{i=0}^{2} s_i \gamma_i \text{ with } \sum_{i=0}^{2} \gamma_i = 1.$$

9.6 An illustrative example

In the present section, we will provide an illustrative example and it will be benefit the reader to follow the instructions given complement the calculations in order to get a full insight of the theoretical results. With this effort the reader will reach a better deepening in the above results which started with the above given example. To check his results the interested reader could consult Georgiou and Vassiliou (1992).

Assume an NHMS modeling any real population which is convenient for the reader to visualize which has a state space $k = \{1,2,4,5,6,7\}$ and for which the non-homogeneous Markov chain has the sequence of matrices $\{\mathbf{P}(t)\}_{t=0}^{\infty}$ all with the same incidence matrix which is periodic with period $d = 3$ and cyclic subclasses $\mathcal{C}_0 = \{1,2\}$, $\mathcal{C}_1 = \{3,4,5\}$ and $\mathcal{C}_2 = \{6,7\}$. Let the transition probability matrices among the subclasses be respectively:

$$\mathcal{C}_0 \to \mathcal{C}_1 : \mathbf{P}_0\left(t\right) = \begin{pmatrix} \frac{1}{3} + \frac{1}{t^6+1} & \frac{1}{6} & \frac{1}{2} - \frac{1}{t^6+1} \\ \frac{1}{4} & \frac{1}{2} - \frac{1}{t^5+10} & \frac{1}{4} + \frac{1}{t^5+10} \end{pmatrix},$$

$$\mathcal{C}_1 \to \mathcal{C}_2 : \mathbf{P}_1\left(t\right) = \begin{pmatrix} \frac{1}{10} & \frac{9}{10} \\ \frac{8}{10} & \frac{2}{10} \\ \frac{3}{10} & \frac{7}{10} \end{pmatrix},$$

$$\mathcal{C}_2 \to \mathcal{C}_0 : \mathbf{P}_1\left(t\right) = \begin{pmatrix} \frac{4}{10} - \frac{1}{t^4+8} & \frac{2}{10} + \frac{1}{t^4+8} \\ \frac{2}{10} & \frac{7}{10} \end{pmatrix}.$$

Then $\lim_{t\to\infty} \mathbf{P}\left(t\right) = \mathbf{P}$ is

$$\mathbf{P} = \begin{pmatrix} 0 & 0 & \frac{1}{3} & \frac{1}{6} & \frac{1}{2} & 0 & 0 \\ 0 & 0 & \frac{1}{4} & \frac{1}{2} & \frac{1}{4} & 0 & 0 \\ 0 & 0 & 0 & 0 & 0 & \frac{1}{10} & \frac{9}{10} \\ 0 & 0 & 0 & 0 & 0 & \frac{8}{10} & \frac{2}{10} \\ 0 & 0 & 0 & 0 & 0 & \frac{3}{10} & \frac{7}{10} \\ \frac{4}{10} & \frac{2}{10} & 0 & 0 & 0 & 0 & 0 \\ \frac{2}{10} & \frac{7}{10} & 0 & 0 & 0 & 0 & 0 \end{pmatrix}$$

Assume that the recruitment policies or the input process into the population be

$$\mathbf{p}_0(t) = \left[\frac{1}{3} - \frac{1}{(t^6+10)}, \frac{2}{3} + \frac{1}{(t^6+10)}, 0,0,0,0,0\right],$$

then with the information, we have so far we are in a position to evaluate:

$$\mathbf{p}_0,\ \{\mathbf{p}_{k+1}(t)\}_{t=0}^{\infty},\ \mathbf{p}_{k+1},\ \{\mathbf{Q}(t)\}_{t=0}^{\infty},\ \mathbf{Q}.$$

The reader may choose any total population sequence $\{T(t)\}_{t=0}^{\infty}$ for his system and let that his choices are such that the parameters s_r $(r = 0, 1, ..., d-1)$ given by (9.62) are evaluated to be

$$s_0 = 0.222,\quad s_1 = 0.421,\quad s_2 = 0.357.$$

Now, execute the following steps using the above given data:

Step 1. Evaluate $\mathbb{E}[\mathbf{q}_r(\infty)]$ for $r = 0, 1, ..., d-1$ given by (9.60).

Step 2. Find the matrix

$$[\mathbf{I} - \mathbf{P}_0\mathbf{P}_1\mathbf{P}_2]^{-1},$$

then the set of all possible asymptotically attainable expected relative structures of the part of $\mathbb{E}[\mathbf{q}_r(\infty)]$ that corresponds with the subclass $\mathcal{C}_0$, that is,

$$\left(\mathbb{E}[q_{01}^r(\infty)], \mathbb{E}[q_{02}^r(\infty)], ..., \mathbb{E}[q_{0k_i}^r(\infty)]\right)/\mathbb{E}[\mu_0^r]$$

is a convex set with vertices that are scalar multiples of the rows of this matrix.

Find the matrices

$$[\mathbf{I} - \mathbf{P}_1\mathbf{P}_2\mathbf{P}_0]^{-1} \text{ and } [\mathbf{I} - \mathbf{P}_2\mathbf{P}_0\mathbf{P}_1]^{-1},$$

for which analogous results are valid for the set of all possible asymptotically attainable expected relative structures of the part of $\mathbb{E}[\mathbf{q}_r(\infty)]$ that corresponds with the subclasses $\mathcal{C}_1$ and $\mathcal{C}_2$ respectively.

Step 3. Evaluate $\mathbb{E}[\mu_i^r]$ for $i, r = 0, 1, 2, ..., d-1$ given in (9.81) and verify (9.64).

Step 4. Find $[\mathbf{I} - \mathbf{P}^3]^{-1}$ and sketch the set of all asymptotically attainable expected relative structures.

9.7 Control of asymptotic variability in NHMS

In Chapter 5, we have founded equation (5.24) which provides for the vector $\mu(t)$ of the expected number of memberships and their variances and covariances in a lexicographic order in the various states of the expected population structure a difference equation. From (5.24), we may calculate $\mu(t)$ for every t given the initial $\mu(0)$ and the basic sequences of an NHMS. In Chapter 7, we studied the asymptotic variability problem of an NHMS, that is, finding the conditions under which $\lim_{t\to\infty}\mu(t)$ exists and finding it in closed analytic form as a function of the basic parameters of the system.

As we have seen the problem of asymptotic variability of an NHMS has its roots as one of the important problem in manpower systems, starting with the work of Pollard (1966) who studied the variability of the class sizes for the very well-known models for the growth of populations introduced by Leslie (1945, 1948). The variability of the class sizes in the social mobility problem (Prais (1955) was studied by Bartholomew (1982, p. 20) by applying the results given by Pollard (1966, 1973). These results were extended by Bartholomew (1982, pp. 66, 78) for homogeneous Markov manpower systems.

In the present section we extend the idea of control, by controlling the asymptotic input process in an NHMS, when together with the expected population structure we also control the variances and covariances of the expected number of memberships in the various states.

Apparently, the control of variances and covariances is in general an important statistical problem. Among other uses, these methods could be used to provide the user with strategies to attain and consequently to maintain limiting expected population structures with reasonably small variances. This is of particular interest for manpower planning where as it has been noted in Bartholomew (1982) large variance is a common pathological situation in manpower systems.

We refresh in here for the reader the following notation:

$$\mathbf{V}_q(t) = \begin{pmatrix} \mathbf{Q}(t) & \mathbf{X}_q(t) \\ 0 & \mathbf{Q}(t) \otimes \mathbf{Q}(t) \end{pmatrix},$$

where $\{\mathbf{Q}(t)\}_{t=0}^{\infty}$ is a sequence of $k \times k$ stochastic matrices defining uniquely the inherent non-homogeneous Markov chain for the NHMS, $\mathbf{X}_q(t)$ is a $k \times k^2$ matrix given by $\mathbf{X}_q(t) = \mathbf{Q}(t)\mathbf{J} - \mathbf{J}(\mathbf{Q}(t) \otimes \mathbf{Q}(t))$ where $\mathbf{J}$ is a $k \times k^2$ matrix the i-th row of which has 1 in the column $(i-1)k+i$ and zeros elsewhere. In what remains we assume that $\lim_{t\to\infty} \mathbf{Q}(t) = \mathbf{Q}$ a regular stochastic matrix and thus $\lim_{t\to\infty} \mathbf{Q}^t = \mathbf{Q}^*$ is a stable stochastic matrix with row $\mathbf{q}_v^* = \begin{pmatrix} q_1^* & q_2^* & \dots & q_k^* \end{pmatrix}$. Now we will prove the following Proposition.

Proposition 336 *Let a complete probability space* $(\Omega, \mathcal{F}, \mathbb{P})$ *and an NHMS defined on it. Assume the above notation and assumption then the following is true:* *(i) If*

$$\begin{aligned} \mathbf{q}_v^* &= \begin{pmatrix} q_1^* & \dots & q_k^* & q_1^*(1-q_1^*) & -q_1^*q_2^* & \dots & q_k^*(1-q_k^*) \end{pmatrix} \\ &= \begin{pmatrix} \mathbf{q}^* & \mathbf{q}^*\mathbf{J} - \mathbf{J}(\mathbf{q}^* \otimes \mathbf{q}^*) \end{pmatrix} \end{aligned} \tag{9.114}$$

then

$$\boldsymbol{\mu}(0)\mathbf{V}_q^{\infty} = T(0)\,\mathbf{q}_v^* \text{ where } \mathbf{V}_q^{\infty} = \begin{pmatrix} \mathbf{Q}^* & \mathbf{Q}^*\mathbf{J} - \mathbf{J}(\mathbf{Q}^* \otimes \mathbf{Q}^*) \\ 0 & \mathbf{Q}^* \otimes \mathbf{Q}^* \end{pmatrix}. \tag{9.115}$$

(ii) If $\mathbf{q}$ *is any* $1 \times k$ *stochastic vector and* $\mathbf{q}_v = \begin{pmatrix} \mathbf{q} & \mathbf{q}\mathbf{J} - \mathbf{J}(\mathbf{q} \otimes \mathbf{q}) \end{pmatrix}$ *then it is true that* $\mathbf{q}_v\mathbf{V}_q^{\infty} = \mathbf{q}_v^*$. *(iii) The following holds for any* $t > 0$

$$\boldsymbol{\mu}_0(t)\mathbf{V}_q^{\infty} = \boldsymbol{\mu}_0(t+k)\mathbf{V}_q^{\infty} \quad \text{for } k = 0, 1, 2....$$

Proof. (i) We have

$$\begin{aligned} \boldsymbol{\mu}(0)\mathbf{V}_q^{\infty} &= T(0)\begin{pmatrix} \mathbf{q}(0)\mathbf{Q}^* & \mathbf{q}(0)\mathbf{Q}^*\mathbf{J} - \mathbf{q}(0)\mathbf{J}(\mathbf{Q}^* \otimes \mathbf{Q}^*) \end{pmatrix} \\ &= T(0)\begin{pmatrix} \mathbf{q}^* & \mathbf{q}^*\mathbf{J} - \mathbf{q}(0)\mathbf{J}(\mathbf{Q}^* \otimes \mathbf{Q}^*) \end{pmatrix}. \end{aligned} \tag{9.116}$$

It is apparent that $\mathbf{Q}^* \otimes \mathbf{Q}^*$ is a stable stochastic matrix with rows equal to $\mathbf{q}^* \otimes \mathbf{q}^*$ and consequently we get that

$$\boldsymbol{\mu}(0)\mathbf{V}_q^{\infty} = T(0)\begin{pmatrix} \mathbf{q}^* & \mathbf{q}^*\mathbf{J} - \mathbf{q}^* \otimes \mathbf{q}^* \end{pmatrix}$$

(ii) We have that

$$\begin{aligned} \mathbf{q}_v\mathbf{V}_q^{\infty} &= \begin{pmatrix} \mathbf{q}\mathbf{Q}^* & \mathbf{q}(\mathbf{Q}^*\mathbf{J} - \mathbf{J}(\mathbf{Q}^* \otimes \mathbf{Q}^*)) + (\mathbf{q}\mathbf{J} - \mathbf{q} \otimes \mathbf{q})(\mathbf{Q}^* \otimes \mathbf{Q}^*) \end{pmatrix} \\ &= \begin{pmatrix} \mathbf{q}^* & \mathbf{q}^*\mathbf{J} - \mathbf{q}\mathbf{J}(\mathbf{Q}^* \otimes \mathbf{Q}^*) + \mathbf{q}\mathbf{J}(\mathbf{Q}^* \otimes \mathbf{Q}^*) - (\mathbf{q} \otimes \mathbf{q})(\mathbf{Q}^* \otimes \mathbf{Q}^*) \end{pmatrix} \\ &= \begin{pmatrix} \mathbf{q}^* & \mathbf{q}^*\mathbf{J} - (\mathbf{q}\mathbf{Q}^*) \otimes (\mathbf{q}\mathbf{Q}^*) \end{pmatrix} = \begin{pmatrix} \mathbf{q}^* & \mathbf{q}^*\mathbf{J} - \mathbf{q}^* \otimes \mathbf{q}^* \end{pmatrix}. \end{aligned}$$

(iii) We have already seen that

$$\boldsymbol{\mu}_0(t) = \begin{pmatrix} \mathbf{p}_0(t) & \mathbf{p}_0(t)\mathbf{J} - \mathbf{p}_0(t) \otimes \mathbf{p}_0(t) \end{pmatrix},$$

hence using (ii) we arrive at the desired result. ■

Now, for an NHMS we introduce the following definition:

Definition 337 *We say that an NHMS has an asymptotically attainable structure* $\boldsymbol{\mu}(\infty)$ *under asymptotically recruitment control if there exists a limiting recruitment vector* $\lim_{t\to\infty}\mathbf{p}_0(t)=\mathbf{p}_0$ *such that* $\lim_{t\to\infty}\boldsymbol{\mu}(t)=\boldsymbol{\mu}(\infty)$*. We denote by* M^{∞} *the set of asymptotically attainable structures under input flow control or recruitment control.*

The following theorem characterize the set M^{∞}.

Theorem 338 *(Vassiliou et al. (1990)). Let a complete probability space* $(\Omega,\mathcal{F},\mathbb{P})$ *and an NHMS which is defined by the sequences* $\{\mathbf{P}(t)\}_{t=0}^{\infty}$, $\{\mathbf{p}_{k+1}(t)\}_{t=0}^{\infty}$, $\{T(t)\}_{t=0}^{\infty}$, *and* $\{\mathbf{p}_0(t)\}_{t=0}^{\infty}$. *Assume that the following conditions are satisfied:*

$$(i)\ \lim_{t\to\infty}T(t)=T\ \ (t),\ \ (T(t)\geq T(t-1))\ \ (ii)\ \lim_{t\to\infty}\mathbf{P}(t)=\mathbf{P} \tag{9.117}$$

$$(iii)\ \lim_{t\to\infty}\mathbf{p}_{k+1}(t)=\mathbf{p}_{k+1},\ (iv)\lim_{t\to\infty}\mathbf{p}_0(t)=\mathbf{p}_0,\ (v)\ \mathbf{Q}=\mathbf{P}+\mathbf{p}_{k+1}^{\top}\mathbf{p}_0, \tag{9.118}$$

and $\mathbf{Q}$ *is a regular stochastic matrix. If we denote by* $\mathbf{q}_{2v}^{*}=\mathbf{q}^{*}\mathbf{J}-\mathbf{q}^{*}\otimes\mathbf{q}^{*}$, *that is,the limiting covariances of the state sizes then:*
The set of all asymptotically attainable $\mathbf{q}_{2v}^{*}$ *is a convex set with vertices the rows of the matrix* $[\mathbf{I}-\mathbf{P}\otimes\mathbf{P}]^{-1}$. *(ii) The set of all asymptotically attainable* $\mathbf{q}_{v}^{*}$ *is a convex set of the form*

$$\{\textstyle\sum_{i=1}^{k}a_i s_i^{-1}\mathbf{e}_i[\mathbf{I}-\mathbf{P}]^{-1}\quad \sum_{i=1}^{k}a_i s_i^{-1}\mathbf{e}_i[\mathbf{I}-\mathbf{P}]^{-1}\mathbf{J}-$$

$$\sum_{i=1}^{k^2}b_i\lambda_i^{-1}\mathbf{e}_i[\mathbf{I}-\mathbf{P}\otimes\mathbf{P}]^{-1}\}, \tag{9.119}$$

where $0<a_i,\ b_i<1$ *for* $i=1,2,\ldots,$ $\sum_{i=1}^{k}a_i$, $\sum_{i=1}^{k^2}b_i$ *and* s_i,λ_i *are the sums of the i-th row of the matrices* $[\mathbf{I}-\mathbf{P}]^{-1}$ *and* $[\mathbf{I}-\mathbf{P}\otimes\mathbf{P}]^{-1}$ *respectively.*

Proof. From the results of Chapter 7 on the asymptotic variability of NHMS, we easily get that

$$\lim_{t\to\infty}\boldsymbol{\mu}(t)=\boldsymbol{\mu}(\infty)=T\mathbf{q}_v^{*}. \tag{9.120}$$

Now, in order to characterize the set of all asymptotically attainable $\boldsymbol{\mu}(\infty)$ under the control of the limiting input vector $\mathbf{p}_0$, it is sufficient to find the set of all possible values of $\mathbf{q}_v^{*}$ for all possible values of $\mathbf{p}_0$. Following an analogous reasoning as in Theorem 326 and Proposition 336, we get that

$$\mathbf{q}_v^{*}=\begin{pmatrix}\mathbf{q}_{1v}^{*} & \mathbf{q}_{2v}^{*}\end{pmatrix}=\begin{pmatrix}\mathbf{q}^{*} & \mathbf{q}^{*}\mathbf{J}-\mathbf{q}^{*}\otimes\mathbf{q}^{*}\end{pmatrix}=$$

$$\begin{pmatrix}\mathbf{q}^{*} & \mathbf{q}^{*}\mathbf{J}-\mathbf{q}^{*}\otimes\mathbf{q}^{*}\end{pmatrix}\begin{pmatrix}\mathbf{Q} & \mathbf{QJ}-\mathbf{J}(\mathbf{Q}\otimes\mathbf{Q})\\ 0 & \mathbf{Q}\otimes\mathbf{Q}\end{pmatrix}, \tag{9.121}$$

from which we get

$$\mathbf{q}_{1v}^{*}=\mathbf{q}^{*}=\mathbf{q}^{*}\mathbf{Q}, \tag{9.122}$$

$$\mathbf{q}_{2v}^{*}=\mathbf{q}^{*}\mathbf{J}-\mathbf{q}^{*}\otimes\mathbf{q}^{*}=\mathbf{q}^{*}\mathbf{J}-(\mathbf{q}^{*}\otimes\mathbf{q}^{*})(\mathbf{Q}\otimes\mathbf{Q}). \tag{9.123}$$

From (9.123) we get

$$\mathbf{q}^{*}\otimes\mathbf{q}^{*}=(\mathbf{q}^{*}\otimes\mathbf{q}^{*})(\mathbf{Q}\otimes\mathbf{Q}). \tag{9.124}$$

Now, we have

$$(\mathbf{Q}\otimes\mathbf{Q})=\left(\mathbf{P}+\mathbf{p}_{k+1}^{\top}\mathbf{p}_0\right)\otimes\left(\mathbf{P}+\mathbf{p}_{k+1}^{\top}\mathbf{p}_0\right)=$$

$$\mathbf{P}\otimes\mathbf{P}+\mathbf{P}\otimes(\mathbf{p}_{k+1}^{\top}\mathbf{p}_0)+(\mathbf{p}_{k+1}^{\top}\mathbf{p}_0)\otimes\mathbf{P}+(\mathbf{p}_{k+1}^{\top}\mathbf{p}_0)\otimes(\mathbf{p}_{k+1}^{\top}\mathbf{p}_0). \tag{9.125}$$

We know that $\mathbf{P}$ is a substochastic matrix, hence we have that also $\mathbf{P}\otimes\mathbf{P}$ is a substochastic matrix and thus the matrix $(\mathbf{I}-\mathbf{P}\otimes\mathbf{P})$ is a non-negatively invertible matrix, thus from (9.124) and (9.125) we get

$$\mathbf{q}^*\otimes\mathbf{q}^* =$$

$$\mathbf{q}^*\otimes\mathbf{q}^*\left[\mathbf{P}\otimes\left(\mathbf{p}_{k+1}^{\top}\mathbf{p}_0\right)+\left(\mathbf{p}_{k+1}^{\top}\mathbf{p}_0\right)\otimes\mathbf{P}+\left(\mathbf{p}_{k+1}^{\top}\mathbf{p}_0\right)\otimes\left(\mathbf{p}_{k+1}^{\top}\mathbf{p}_0\right)\right]\times$$

$$(\mathbf{I}-\mathbf{P}\otimes\mathbf{P})^{-1}=\sum_{i=1}^{k^2}\beta_i\mathbf{e}_i(\mathbf{I}-\mathbf{P}\otimes\mathbf{P})^{-1}. \tag{9.126}$$

From (9.126) we get that

$$(\mathbf{q}^*\otimes\mathbf{q}^*)\,\mathbf{1}^{\top}=\sum_{i=1}^{k^2}\beta_i\lambda_i=1, \tag{9.127}$$

where λ_i is the sum of the i-th row of the matrix $(\mathbf{I}-\mathbf{P}\otimes\mathbf{P})^{-1}$. Hence,

$$\mathbf{q}^*\otimes\mathbf{q}^*=\sum_{i=1}^{k^2}\left(\frac{\beta_i\lambda_i}{\sum_{j=1}^{k^2}\beta_j\lambda_j}\right)\lambda_i^{-1}\mathbf{e}_i(\mathbf{I}-\mathbf{P}\otimes\mathbf{P})^{-1}=$$

$$\sum_{i=1}^{k^2}b_i\lambda_i^{-1}\mathbf{e}_i(\mathbf{I}-\mathbf{P}\otimes\mathbf{P})^{-1}, \tag{9.128}$$

where $0<b_i<1$ and $\sum_{i=1}^{k^2}b_i=1$. That is, the set of all possible vectors $\mathbf{q}^*\otimes\mathbf{q}^*$ is a convex set with vertices scalar multiples of the rows of the matrix $(\mathbf{I}-\mathbf{P}\otimes\mathbf{P})^{-1}$. Taking also into account Theorem 326 where it is founded that the set of all possible vectors $\mathbf{q}^*$ is a convex set with vertices scalar multiples of the rows of the matrix $(\mathbf{I}-\mathbf{P})^{-1}$ and (9.121) we get that the set of all possible $\boldsymbol{\mu}(\infty)$ is a convex set described by (9.119). ■

9.8 The NHMS in a stochastic environment

In the present section, we modify one of the basic assumptions of an NHMS as it was defined in Chapter 5. Consider the set $\mathfrak{P}$ of all possible transitions matrices $\{\mathbf{P}(t)\}_{t=0}^{\infty}$ and let $\mathfrak{P}_{\nu}(t)=\{\mathbf{P}_1(t),\mathbf{P}_2(t),...,\mathbf{P}_{\nu}(t)\}$ be a finite subset of it such that

$$(\mathbf{I}-\mathbf{P}_m(t))\,\mathbf{1}^{\top}=\mathbf{p}_{k+1}^{\top}(t)\quad\text{for } m=1,2,...,\nu \text{ and } t=0,1,2,.... \tag{9.129}$$

Assume that at every time point $t=0.1,2,...$ the NHMS has the choice of selecting a transition matrix from the pool $\mathfrak{P}_{\nu}(t)$. In addition, assume that it makes its choice in a stochastic way and more specifically let

$$c_{ij}(t)=\mathbb{P}\{\mathbf{P}(t)=\mathbf{P}_j(t)\mid\mathbf{P}(t)=\mathbf{P}_i(t)\}\ \text{with } i,j=1,2,...,\nu, \tag{9.130}$$

be the probability that the transition matrix for the $\mathbf{P}(t)$ for the $(t-1)$th time interval to be $\mathbf{P}_j(t)$ given that at the $(t-2)$th time interval was $\mathbf{P}(t-1)=\mathbf{P}_i(t-1)$.

Apparently the matrix $\mathbf{C}(t)=\{c_{ij}(t)\}_{i,j=1}^{\nu}$ is a stochastic one and hence it defines uniquely a non-homogeneous Markov chain. We call the sequence $\{\mathbf{C}(t)\}_{t=1}^{\infty}$ the *compromise non-homogeneous Markov chain,* in the sense that it is the outcome of the choice of strategy under the various pressures in the environment. We call such an NHMS a *non-homogeneous Markov system in a stochastic environment (S-NHMS).*

The roots of the idea of having a pool $\mathfrak{P}_{\nu}(t)$ of transition probability matrices to choose from is also included in Choen (1976, 1977) for demography problems where the transition matrices are Leslie matrices. It is mentioned there that "... the empirical stimulus for such a step is a dissatisfaction with assuming fixed vital rates at all times before t". In conclusion, the idea of having to choose stochastically from a pool $\mathfrak{P}_\nu(t)$ $(t = 1, 2, ...)$ of transition matrices seems to be a step toward reality.

Let us discuss for a while some real systems where the present generalization is almost certain to provide a more realistic model. That is, where the S-NHMS would be most likely to be applicable. Most often for any population we start the modeling process by the estimation of the sequence of transition probabilities $\{\mathbf{P}(t)\}_{t=0}^{\infty}$ from historical data. Assume that during that process we observe a certain random repetition of matrices, that is, without a certain deterministic pattern as for example is the cyclic repetition where a Cy-NHMS is applicable. For example, assume that an estimated from historical data sequence of transition matrices may look as follows

$$\mathbf{P}_1 \to \mathbf{P}_2 \to \mathbf{P}_1 \to \mathbf{P}_2 \to \mathbf{P}_2 \to \mathbf{P}_3 \to \mathbf{P}_1 \to \mathbf{P}_2 \to \mathbf{P}_3 \to \mathbf{P}_3, ...$$

Then it is apparent that the theory of NHMS as it is developed so far in the present is rather difficult to deal with the situation satisfactorily, and a preferable way to proceed is to employ the model of the S-NHMS with the use of the compromising non-homogeneous Markov chain $\{\mathbf{C}(t)\}_{t=1}^{\infty}$.

It is moreover important to note that the concept of an S-NHMS is an interesting generalization of the cyclic NHMS the theory of which was presented in Chapter 8. For example, a cyclic NHMS with cycle of period d is an S-NHMS with compromise sequence of matrices which always has the following form:

$$\mathbf{C}(t) = \begin{pmatrix} 0 & 1 & 0 & ... & 0 \\ 0 & 0 & 1 & ... & 0 \\ ... & ... & ... & ... & ... \\ 0 & 0 & 0 & ... & 1 \\ 1 & 0 & 0 & ... & 0 \end{pmatrix} \quad \text{for } t = 1, 2, ... \tag{9.131}$$

9.8.1 *The expected population structure of the S-NHMS*

We will firstly start with the important problem of finding the expected population structure, that is the expected distribution of memberships of the population among the k states as a function of the basic parameters of the NHMS. It is apparent already from the previous paragraphs that we can no longer speak of a known deterministically transition matrix $\mathbf{P}(t)$ for the time interval $(t-1, t]$. The transition matrix for the time interval $(t-1, t]$ will be selected from the pool $\mathfrak{P}_\nu(t)$ by a stochastic mechanism which is described by the compromise non-homogeneous Markov chain $\{\mathbf{C}(t)\}_{t=1}^{\infty}$. Hence, it is natural and correct to introduce $\mathbb{E}[\mathbf{P}(t)]$, i.e., the idea of the expected transition matrix at the time interval $(t-1, t]$ all over the set $\mathfrak{P}_\nu(t)$.

9.8.2 *Evaluating the expected transition probability matrix* $\mathbb{E}[\mathbf{P}(t)]$

In the theorem that follows, we will evaluate the matrix $\mathbb{E}[\mathbf{P}(t)]$ as a function of the basic parameters of the NHMS:

Theorem 339 *(Tsantas and Vassiliou (1993)). Let a complete probability space* $(\Omega, \mathcal{F}, \mathbb{P})$ *and an NHMS which is defined by the sequences* $\{P(t)\}_{t=0}^{\infty}, \{\mathbf{p}_{k+1}(t)\}_{t=0}^{\infty}, \{T(t)\}_{t=0}^{\infty}$, *and* $\{\mathbf{p}_0(t)\}_{t=0}^{\infty}$. *Then the expected transition matrix at time interval* $(t-1, t]$ *is given by*

$$\mathbb{E}[\mathbf{P}(t)] = \sum_{j=1}^{\nu} \mathbf{e}_j \left[\mathbf{e}_{i_0} \prod_{s=1}^{t} \mathbf{C}(s)\right]^{\top} \mathbf{P}_j(t), \tag{9.132}$$

where i_0 *is the initial state of the compromise non-homogeneous Markov chain.*

Proof. We have that

$$\mathbb{P}\left[\mathbf{P}(t)=\mathbf{P}_j(t)\right]=\sum_{i=1}^{\nu}\mathbb{P}\left[\mathbf{P}(t)=\mathbf{P}_j(t)\mid\mathbf{P}(t-1)=\mathbf{P}_i(t-1)\right]\times$$
$$\mathbb{P}\left[\mathbf{P}(t-1)=\mathbf{P}_i(t-1)\right]= \tag{9.133}$$
$$\sum_{i=1}^{\nu}c_{ij}(t)\,\mathbb{P}\left[\mathbf{P}(t-1)=\mathbf{P}_i(t-1)\right].$$

Applying the above reasoning recursively and

$$\mathbb{P}\left[\mathbf{P}(t)=\mathbf{P}_{i_t}(t)\right]=\sum_{i_1=1}^{\nu}\sum_{i_2=1}^{\nu}\ldots\sum_{i_{t-1}=1}^{\nu}c_{i_0i_1}(1)\,c_{i_1i_2}(2)\ldots c_{i_{t-1}i_t}(t). \tag{9.134}$$

Now for the expected value $\mathbb{E}\left[\mathbf{P}(t)\right]$ from (9.134), we have that

$$\mathbb{E}\left[\mathbf{P}(t)\right]=\sum_{i_t=1}^{\nu}\mathbf{P}_{i_t}(t)\,\mathbb{P}\left[\mathbf{P}(t)=\mathbf{P}_{i_t}(t)\right]$$
$$=\sum_{i_1=1}^{\nu}\sum_{i_2=1}^{\nu}\ldots\sum_{i_{t-1}=1}^{\nu}\sum_{i_t=1}^{\nu}\mathbf{P}_{i_t}(t)\,c_{i_0i_1}(1)\,c_{i_1i_2}(2)\ldots c_{i_{t-1}i_t}(t).$$
$$\sum_{i=1}^{\nu}\mathbf{e}_i\left[\mathbf{e}_{i_0}\prod_{s=1}^{t}\mathbf{C}(s)\right]^{\top}\mathbf{P}_i(t),\ \text{for } t=1,2,\ldots. \tag{9.135}$$

■

Remark 340 *It is important at this stage to understand the implications of the assumption* (9.129)*. Let us choose any value of time* t *and let it be* t^**. Then for* t^* *the pool from which we will choose the matrices* $\mathbf{P}(t^*)$ *will be*

$$\mathfrak{P}_\nu(t^*)=\left\{\mathbf{P}_1(t^*),\mathbf{P}_2(t^*),\ldots,\mathbf{P}_\nu(t^*)\right\}.$$

According to assumption (9.129) *we will have that*

$$(\mathbf{I}-\mathbf{P}_1(t^*))\,\mathbf{1}^\top=\mathbf{p}_{k+1}^\top(t^*),\ (\mathbf{I}-\mathbf{P}_2(t^*))\,\mathbf{1}^\top=\mathbf{p}_{k+1}^\top(t^*),\ldots,$$
$$(\mathbf{I}-\mathbf{P}_\nu(t^*))\,\mathbf{1}^\top=\mathbf{p}_{k+1}^\top(t^*),$$

from which we get

$$\mathbf{p}_{k+1}^\top(t^*)=\mathbf{1}^\top-\mathbf{P}_1(t^*)\,\mathbf{1}^\top=\mathbf{1}^\top-\mathbf{P}_2(t^*)\,\mathbf{1}^\top=\ldots=\mathbf{1}^\top-\mathbf{P}_\nu(t^*)\,\mathbf{1}^\top,$$

and finally

$$\mathbf{P}_1(t^*)\,\mathbf{1}^\top=\mathbf{P}_2(t^*)\,\mathbf{1}^\top=\ldots=\mathbf{P}_\nu(t^*)\,\mathbf{1}^\top.$$

Hence, for any time t *all the matrices in the pool* $\mathfrak{P}_\nu(t)$ *have the same row sums. This is not restrictive since the values in any particular* $\mathbf{P}_i(t)$ *for* $i=1,2,\ldots,\nu$ *are free to fluctuate as much as it is needed for any particular application. In addition, since all* $\mathbf{P}_i(t)\in\mathfrak{P}_\nu(t)$ *for* $i=1,2,\ldots,\nu$ *and for every* $t=1,2,\ldots$ *are substochastic then also* $\mathbb{E}\left[\mathbf{P}(t)\right]$ *is a substochastic matrix.*

9.8.3 *Expected value of a random number of Bernoulli trials with probability of success a random variable*

Let us consider a number $\mathcal{B}(N, p)$ of Bernoulli trials where N and p are random variables. Denote by X the random variable being the number of successes in those N Bernoulli trials. Then it is known that

$$\mathbb{E}(X \mid N = n, p = \hat{p}) = n\hat{p},$$

and since this is true for any value of n and $\hat{p}$ we may write

$$\mathbb{E}(X \mid N, p).$$

Taking expected values on both sides of the above equation, we obtain

$$\mathbb{E}(X) = \mathbb{E}[\mathbb{E}(X \mid N, p)] = \mathbb{E}(N)\,\mathbb{E}(p). \tag{9.136}$$

9.8.4 *The expected population structure*

Let us refresh that we denote by $R(t)$ the random variable representing the number of memberships entering the population at time t, $N_i(t)$ the random variable representing the number of memberships who are in state i at the end of the interval $(t-1, t]$, $N_{ij}(t)$ the random variable representing the number of memberships that make a transition from state i to state j during the interval $(t-1, t]$ and finally denote by $F_i(t)$ the random variable representing the number of memberships who enter state i at the end of the interval $(t-1, t]$. It is easily seen that

$$N_i(t) = \sum_{j=1}^{k} N_{ji}(t-1) + F_i(t). \tag{9.137}$$

We now have that

$$\mathbb{E}(F_i(t)) = \mathbb{E}[\mathbb{E}(F_i(t) \mid R(t))] = \mathbb{E}[R(t)]\, p_{0i}(t), \tag{9.138}$$

and thinking similarly

$$\begin{aligned}\mathbb{E}[N_{ji}(t-1)] &= \mathbb{E}[\mathbb{E}(N_{ji}(t-1) \mid N_i(t-1), \mathfrak{P}_\nu(t))] \\ &= \mathbb{E}[N_i(t-1)]\,\mathbf{e}_j \mathbb{E}[\mathbf{P}(t)]\,\mathbf{e}_i\end{aligned}$$

$$= \mathbb{E}[N_j(t-1)]\,\mathbf{e}_j \left\{ \sum_{j=1}^{\nu} \mathbf{e}_j \left[\mathbf{e}_{i_0} \prod_{s=1}^{t} \mathbf{C}(s) \right]^\top \mathbf{P}_j(t) \right\} \mathbf{e}_i. \tag{9.139}$$

Now, taking expectations in (9.137) and using (9.138) and (9.139) we get that:

$$\mathbb{E}[N_i(t)] = \sum_{j=1}^{k} \mathbb{E}[N_j(t-1)]\,\mathbf{e}_j \left\{ \sum_{j=1}^{\nu} \mathbf{e}_j \left[\mathbf{e}_{i_0} \prod_{s=1}^{t} \mathbf{C}(s) \right]^\top \mathbf{P}_j(t) \right\} \mathbf{e}_i$$

$$+\mathbb{E}[R(t)]\, p_{0i}(t). \tag{9.140}$$

In an S-NHMS the expected number of memberships entering the system at the end of the interval $(t-1, t]$ are the expected number of memberships that left the system from the various states plus the number of memberships needed to meet the known total number of memberships needed to meet the known total number of memberships at time t. Hence, we have that:

$$\mathbb{E}[R(t)] = \Delta T(t) + \mathbb{E}[\mathbf{N}(t-1)]\,\mathbf{p}_{k+1}^\top(t), \tag{9.141}$$

Now substituting (9.141) into (9.140) we get that:

$$\mathbb{E}\left[N_i(t)\right]=\mathbb{E}\left[\mathbf{N}(t-1)\right]\left[\left\{\sum_{j=1}^{\nu}\mathbf{e}_j\left[\mathbf{e}_{i_0}\prod_{s=1}^{t}\mathbf{C}(s)\right]^{\top}\mathbf{P}_j(t)\right\}\mathbf{e}_i^{\top}+\right.$$

$$\left.\mathbf{p}_{k+1}^{\top}(t)\,\mathbf{p}_0(t)\,\mathbf{e}_i^{\top}\right]+\Delta T(t)\,p_{0i}(t). \tag{9.142}$$

If we write

$$\mathbb{E}\left[\mathbf{Q}(t)\right]=\sum_{j=1}^{\nu}\mathbf{e}_j\left[\mathbf{e}_{i_0}\prod_{s=1}^{t}\mathbf{C}(s)\right]^{\top}\mathbf{P}_j(t)+\mathbf{p}_{k+1}^{\top}(t)\,\mathbf{p}_0(t), \tag{9.143}$$

then (9.142) could be written in matrix notation as

$$\mathbb{E}\left[\mathbf{N}(t)\right]=\mathbb{E}\left[\mathbf{N}(t-1)\right]\mathbb{E}\left[\mathbf{Q}(t)\right]+\Delta T(t)\,\mathbf{p}_0(t). \tag{9.144}$$

Note that Equation (9.144) is illustratively similar in form to the respective equation of an NHMS. A first basic difference among the two cases is that in the place of $\mathbb{E}[\mathbf{Q}(t)]$ is $\mathbf{Q}(t)$, which is assumed to be known or being modeled from past data available to the practitioner. On the other hand $\mathbb{E}[\mathbf{Q}(t)]$ is a function of $\{\mathbf{C}(t)\}_{t=1}^{\infty}$ and $\mathfrak{P}_\nu(t)$. An important question which quickly arise is if the matrix $\mathbb{E}[\mathbf{Q}(t)]$ as defined in (9.143) is a stochastic matrix. The answer is provided in the following Proposition:

Proposition 341 *The matrix $\mathbb{E}[\mathbf{Q}(t)]$ is a stochastic matrix if and only if*

$$\sum_{j=1}^{\nu}\mathbf{e}_j\left[\mathbf{C}(0,s)\right]^{\top}\mathbf{e}_{i_0}^{\top}\mathbf{1}^{\top}+\mathbf{p}_{k+1}^{\top}(t)=\mathbf{1}^{\top}. \tag{9.145}$$

Proof. From equation (9.143) we get that

$$\begin{aligned}\mathbb{E}\left[\mathbf{Q}(t)\right]\mathbf{1}^{\top}&=\sum_{j=1}^{\nu}\mathbf{e}_j\left[\mathbf{e}_{i_0}\mathbf{C}(0,s)\right]^{\top}\mathbf{P}_j(t)\,\mathbf{1}^{\top}+\mathbf{p}_{k+1}^{\top}(t)\,\mathbf{p}_0(t)\,\mathbf{1}^{\top}\\&=\sum_{j=1}^{\nu}\mathbf{e}_j\left[\mathbf{e}_{i_0}\mathbf{C}(0,s)\right]^{\top}\mathbf{1}^{\top}+\mathbf{p}_{k+1}^{\top}(t)\\&=\sum_{j=1}^{\nu}\mathbf{e}_j\left[\mathbf{C}(0,s)\right]^{\top}\mathbf{e}_{i_0}^{\top}\mathbf{1}^{\top}+\mathbf{p}_{k+1}^{\top}(t),\end{aligned} \tag{9.146}$$

and consequently from (9.146) we get that $\mathbb{E}[\mathbf{Q}(t)]$ is a stochastic matrix if and only if the second part of (9.146) is equal with $\mathbf{1}^{\top}$. ■

Hence, from Proposition 341 we get that in general $\mathbb{E}[\mathbf{Q}(t)]$ is not a stochastic matrix. In the case that (9.145) is satisfied then $\mathbb{E}[\mathbf{Q}(t)]$ defines uniquely a non-homogeneous Markov chain which we call the *expected embedded non-homogeneous Markov chain* for the S-NHMS. Alternatively, we say that the S-NHMS has an *expected embedded non-homogeneous Markov chain.*

9.9 Maintainability in a stochastic environment

So far in the present chapter, we have studied the problem of maintainability of the expected relative population structure by input control. The study was deterministic in the sense that it assumes that the population structure will move along the expected path predicted by $\mathbb{E}[\mathbf{q}(t)]$ instead of using the respective random variables $\mathbf{q}(t)$. The first attempt to explore the stochastic nature of the process, by using the random population structure $\mathbf{q}(t)$ in the homogeneous case was done by Bartholomew (1975, 1977), and it was called *control in a stochastic environment.* We will now, following Tsantas and Vassiliou (1993), formulate the maintainability problem *in a stochastic environment for S-NHMS.* We will start by introducing the following definition for an S-NHMS:

Definition 342 *We say that a relative population structure $\mathbf{q}(t)$ of an S-NHMS is maintainable in a stochastic environment at time interval $[t-1,t)$ under input control, if and only if there exist an input policy $\mathbf{p}_O(t)$ such that $\mathbf{q}(t-1)=\mathbf{q}(t)$.*

From equation (9.137) it is evident that a relative population structure $\mathbf{q}(t)$ is maintainable in a stochastic environment if and only if there is room for entering memberships to meet the target, that is, for every state j

$$T(t)\,q_j(t) \geq \sum_{i=1}^{k} N_{ij}(t-1) \text{ for every } j \in S. \tag{9.147}$$

However, now $N_{ij}(t)$'s are random variables, hence, we can no longer guarantee that any structure $\mathbf{q}(t)$ can always be maintained. Instead we will determine the probability that a given structure $\mathbf{q}=\mathbf{q}(t-1)=\mathbf{q}(t)$ is maintainable. In this respect define by

$$P_m(t) = \mathbb{P}\left[T(t)\,q_j - \sum_{i=1}^{k} N_{ij}(t-1) \geq 0; j \in S \mid \mathbf{q}\right] \text{ for } t=1,2,..., \tag{9.148}$$

and let $\mathfrak{U}_t$ be the set of all maintainable relative population structures in the time interval $[t-1,t)$ with probability 1, i.e.

$$\mathfrak{U}_t = \{\mathbf{q} : P_m(t) = 1\}.$$

Define by $S_j(t)$ the space available for replacement and new memberships to enter state j. Then we have

$$S_j(t) = T(t)\,q_j - \sum_{i=1}^{k} N_{ij}(t-1) \text{ for every } j \in S, \tag{9.149}$$

then by taking expectations in (9.149) we get that

$$\mathbb{E}[S_j(t)] = T(t)\,q_j - T(t-1)\sum_{i=1}^{\nu} \mathbf{e}_i\left[\mathbf{e}_{i_0}\mathbf{C}(0,s)\right]^\top \mathbf{q}\mathbf{P}_i(t), \tag{9.150}$$

where $j \in \mathbb{S}$, and by writing the matrix

$$\mathbf{P}_i(t) = (\mathbf{p}_{i,1}, \mathbf{p}_{i,2}, ..., \mathbf{p}_{i,k}),$$

where, $\mathbf{p}_{i,j}(t) = (p_{i,1j}(t), p_{i,2j}(t), ..., p_{i,kj}(t))^\top$ is the j column of the matrix $\mathbf{P}_i(t)$ we have that:

$$Cov[S_j(t), S_m(t)] = T(t-1)\sum_{i=1}^{\nu} \mathbf{e}_i\left[\mathbf{e}_{i_0}\mathbf{C}(0,s)\right]^\top \delta_{jm}\mathbf{q}\mathbf{P}_i(t)\,\mathbf{e}_j^\top -$$

$$T(t-1)\sum_{i=1}^{\nu} \mathbf{e}_i\left[\mathbf{e}_{i_0}\mathbf{C}(0,s)\right]^\top \sum_{r=1}^{\nu} \mathbf{e}_r\left[\mathbf{e}_{i_0}\mathbf{C}(0,s)\right]^\top \mathbf{q}\left(\mathbf{p}_{i,j}(t) \vartriangle \mathbf{p}_{r,m}(t)\right), \tag{9.151}$$

where $j,m \in \mathbb{S}$ and δ_{jm} is the Kronecker's delta.

For every $j \in \mathbb{S}$ the empty space $S_j(t)$ is having a multinomial distribution with mean and variance given in (9.150) and (9.151). However, exact probabilities are almost prohibited to find when the multinomial distribution is involved. Nevertheless, there are useful approximating methods and the most commonly used is the χ^2 approximation (Johnson and Koltz (1969), p. 285). Moreover, Bartholomew (1977, p. 6), (1982, p. 174) used as approximation the multivariate normal distribution, which is known to be marginally identical to the χ^2 distribution with more than 30 degrees of freedom. That means that in all applications where the degrees of freedom are more than 30, the approximation would be very satisfactory. Tsantas and Vassiliou (1993) carried out a large number of simulations for manpower systems and observed that the approximation was satisfactorily for most practical purposes. Therefore

$$P_m(t) = \mathbb{P}\{S_j(t) \geq 0 \mid j \in \mathbb{S}\} = \left\{ Z_j \geq \frac{-\mathbb{E}[S_j(t)]}{(\mathbb{V}ar[S_j(t)])^{1/2}};\ j \in \mathbb{S} \right\}, \tag{9.152}$$

where $\mathbf{Z} = (Z_1, Z_2, ..., Z_k)$ has a standard normal distribution with the same correlation matrix as the $S_j(t)$'s. In order to find the normally distributed probabilities for the calculation of (9.152) the method of Schervish (1984) was suggested by Tsantas and Vassiliou (1993). Note also that, since the $S_j(t)$'s are discrete, the approximation of the probabilities $P_m(t)$ in (9.152) can be improved by using a continuity correction.

We will now prove the following basic result:

Theorem 343 *(Tsantas and Vassiliou (1993)). Let a complete probability space $(\Omega, \mathcal{F}, \mathbb{P})$ and an S-NHMS which is defined by the sequences $\{\mathfrak{P}_\nu(t)\}_{t=0}^\infty$, $\{\mathbf{C}(t)\}_{t=0}^\infty$, $\{\mathbf{p}_{k+1}(t)\}_{t=0}^\infty$, $\{T(t)\}_{t=0}^\infty$, and $\{\mathbf{p}_0(t)\}_{t=0}^\infty$. Then as $T(t) \to \infty$ the set $\mathfrak{U}_t$ contains all the relative population structures for which the following holds:*

$$q_j \geq \frac{T(t-1)}{T(t)} \sum_{i=1}^{\nu} \mathbf{e}_i \left[\mathbf{e}_{i_0}\mathbf{C}(0,s)\right]^\top \mathbf{q}\mathbf{P}_i(t)\,\mathbf{e}_j^\top. \tag{9.153}$$

Proof. It is known that if $\mathbf{Z}$ is a k-variate standard normal distribution then the multiple integral $\mathbb{P}(\mathbf{Z} \geq \mathbf{u})$ goes to 1 as the lower limits $\mathbf{u}$ tends to $-\infty$. Hence, it is sufficient to show that

$$\lim_{T(t)\to\infty} u_j = \lim_{T(t)\to\infty} \frac{-\mathbb{E}[S_j(t)]}{(\mathbb{V}ar[S_j(t)])^{1/2}} = -\infty. \tag{9.154}$$

In this respect, we have that

$$u_j = \frac{-\mathbb{E}[S_j(t)]}{(\mathbb{V}ar[S_j(t)])^{1/2}} =$$

$$-\left(T(t)\,q_j - T(t-1)\sum_{i=1}^{\nu} \mathbf{e}_i \left[\mathbf{e}_{i_0}\mathbf{C}(0,s)\right]^\top \mathbf{q}\mathbf{P}_i(t) \right) \times$$

$$(T(t-1)\sum_{i=1}^{\nu} \mathbf{e}_i \left[\mathbf{e}_{i_0}\mathbf{C}(0,s)\right]^\top \mathbf{q}\mathbf{P}_i(t)\,\mathbf{e}_j^\top -$$

$$T(t-1)\sum_{i=1}^{\nu} \mathbf{e}_i \left[\mathbf{e}_{i_0}\mathbf{C}(0,s)\right]^\top \sum_{r=1}^{\nu} \mathbf{e}_r \left[\mathbf{e}_{i_0}\mathbf{C}(0,s)\right]^\top \mathbf{q}\left(\mathbf{p}_{i,j}(t) \vartriangle \mathbf{p}_{r,m}(t)\right))^{-1/2}$$

$$= (T(t-1))^{1/2}\left(\frac{T(t)}{T(t-1)} q_j - \sum_{i=1}^{\nu} \mathbf{e}_i \left[\mathbf{e}_{i_0}\mathbf{C}(0,s)\right]^\top \mathbf{q}\mathbf{P}_i(t) \right) \times$$

$$(\sum_{i=1}^{\nu} \mathbf{e}_i \left[\mathbf{e}_{i_0}\mathbf{C}(0,s)\right]^\top \mathbf{q}\mathbf{P}_i(t)\,\mathbf{e}_j^\top - \tag{9.155}$$

$$\sum_{i=1}^{\nu} \mathbf{e}_i \left[\mathbf{e}_{i_0} \mathbf{C}(0,s)\right]^{\top} \sum_{r=1}^{\nu} \mathbf{e}_r \left[\mathbf{e}_{i_0} \mathbf{C}(0,s)\right]^{\top} \mathbf{q}\left(\mathbf{p}_{i,j}(t) \vartriangle \mathbf{p}_{r,m}(t)\right)\right)^{-1/2}$$

From (9.155) it is now apparent that (9.154) holds. ■

Remark 344 *When the pool of stochastic matrices $\{\mathfrak{P}_\nu(t)\}_{t=0}^{\infty}$ consists of one matrix namely the sequence $\{\mathbf{P}(t)\}_{t=0}^{\infty}$ then the matrix $\mathbf{C}(t)$ becomes a constant matrix of 1's and consequently the stochastic environment introduced in the present subsection vanishes. It is interesting in such a case to verify that the results of the present section coincide with the results of section 9.2.*

9.10 Strategies for attaining a structure in an S-NHMS

We so far found the probability of maintaining a certain structure in an S-NHMS. Logically the next reasonable question which arises is the design of strategies of attaining a desirable maintainable structure. Naturally, it is more realistic to attempt to design strategies to keep the population structure as near as possible to the desirable one. Such a strategy will include a proposition for a sequence of input policies to the population represented by the probability vectors $\mathbf{p}_0(t)$ as well as the time horizon in which these policies should be applied, in order to attain the population structure in an optimal way. In what follows, fixed time strategies are discussed, that is, the time required to attain a desirable structure is set in advance.

Let us now assume that a strategy is required to attain a desirable relative population structure $\mathbf{q}$ in t^* steps starting from the known present structure $\mathbf{q}(t-1)$. Apparently, we need some criteria of optimallity and in this respect we need a measure of distance between two relative population structures, that is, actually between two row vectors. Naturally, the norm in such a case is the appropriate measure of distance and in fact there is actually no difference which norm we will choose for that purpose. Also, there are various criteria of optimallity to be used depending on the actual application, and the characteristics of the population at hand. Without loss of generality a function to be minimized is proposed which could be used in many physical environments:

$$CF(t^*) = \sum_{\tau=1}^{t+t^*} \left\| c(\tau)\left[\mathbf{q} - \mathbf{q}(t)\right] \mathbf{W}_d(\tau) \right\|, \tag{9.156}$$

where the norm $\|.\|$ could be any norm, although the l_1-norm is suggested, $c(\tau)$ is the cost per one distance unit in step $(t-1, t]$ and $\mathbf{W}_d(\tau)$ is a $k \times k$ diagonal matrix with elements $w_j(\tau)$ $j \in \mathbb{S}$, a set of non-negative weights chosen to reflect the importance attached to the correct size of each state. The cost objective function (9.156) will be used as a model to present the methodology below for answering the problem imposed. Note however, that the methodology is irrespective of the chosen cost objective function.

In (9.156), the relative population structures $\mathbf{q}(t+1), ..., \mathbf{q}(t+t^*)$ are random variables and thus according to the principles of stochastic control (Dreyfus and Law (1977), Larson and Casti (1982) the best that could be done is to minimize the expectation $\mathbb{E}[CF(t^*)]$ which becomes:

$$\mathbb{E}[CF(t^*)] = \sum_{\tau=1}^{t+t^*} \left\| c(\tau)\left[\mathbf{q} - \mathbb{E}[\mathbf{q}(t)]\right] \mathbf{W}_d(\tau) \right\|. \tag{9.157}$$

There are two main control philosophies and which are the ones that we will use. In the first one, namely the *optimal feedback control,* the decision-maker has all the information needed about the present state $\mathbf{q}(t+\tau)$ of the population, at each point in time where the decisions for $\mathbf{q}(t+\tau)$ $(\tau = 1, ..., t^*-1)$ should be made. In the second one, called the *optimal open loop control,* all the decisions $\mathbf{p}_0(t+\tau)$ $(\tau = 1, ..., t^* - 1)$ should be made at time t, that is the beginning of the control horizon.

In the optimal open-loop scheme the solution is provided with the pair of sequences $\{\mathbf{p}_0(t+\tau), \mathbb{E}[\mathbf{q}(t+\tau)]\}_{\tau=1}^{t^*}$. Since we have no incoming information during our control exercise about the resulting actual relative population structures, as it is expected usually we end up with a larger expected cost than the optimal feedback scheme.

We will start with the optimal open-loop control strategy. We will formulate the problem as a mixture of stochastic dynamic and linear/non-linear programming. The basic feature of this strategy is that at each step on the sequential process which the principle of optimallity creates, the optimal decision is taken by solving a classical linear or non-linear programming problem. In this respect define the following optimal function:

$$F(\mathbf{q}(t+\tau), \mathbf{p}_0(t+\tau), t, \tau) = \{\text{the minimum expected operational}$$
$$\text{distance cost for the population structure from time } t \text{ to } t+\tau \text{ given}$$
$$\text{that in the time period } (t+\tau-1, t+\tau] \text{ the input policy is } \mathbf{p}_0(t+\tau)\}.$$

Then according to the principle of optimallity in dynamic programming, the recurrence relation needed to optimize the problem is the following:

$$\begin{aligned} F(\mathbf{q}(t+\tau), \mathbf{p}_0(t+\tau), t, \tau) &= \min\{c(t+\tau)\,\mathbf{q} - \mathbb{E}[\mathbf{q}(t+\tau)]\,\mathbf{W}_d(t+\tau)\} \\ &+F(\mathbf{q}(t+\tau-1), \mathbf{p}_0(t+\tau-1), t, \tau-1) = \\ &\min\{c(t+\tau)\,[\textstyle\sum^{+} w_j(t+\tau)(q_j - \mathbb{E}[q_j(t+\tau)]) \\ &+\textstyle\sum^{+} w_j(t+\tau)(\mathbb{E}[q_j(t+\tau)] - q_j)]\} \\ &+F(\mathbf{q}(t+\tau-1), \mathbf{p}_0(t+\tau-1), t, \tau-1), \end{aligned} \tag{9.158}$$

where by $\sum^{+}$ we denote the summation over $j = 1, 2, ..., k$ for which $q_j - \mathbb{E}[q_j(t+\tau)] \geq 0$ and $\sum^{-}$ the summation over the remaining j. The boundary condition of the problem is

$$F(\mathbf{q}(t), \mathbf{p}_0(t), t, 0) = 0 \text{ or a known constant.} \tag{9.159}$$

At every step the following conditions imposed by the S-NHMS must be satisfied:

$$\begin{aligned} \mathbb{E}[\mathbf{q}(t+\tau)] &= a(t+\tau-1)\,\mathbb{E}[\mathbf{q}(t+\tau-1)]\,\mathbb{E}[\mathbf{Q}(t+\tau)] \\ &+b(t+\tau-1)\,\mathbf{p}_0(t+\tau), \\ \text{with } a(t+\tau-1) &= \frac{T(t+\tau-1)}{T(t+\tau)}, b(t+\tau-1) = 1 - \frac{T(t+\tau-1)}{T(t+\tau)}, \\ &\mathbb{E}[\mathbf{q}(t+\tau)]\,\mathbf{1}^\top = 1, \\ \mathbf{p}_0(t+\tau) &\geq \mathbf{0} \text{ and } \mathbb{E}[\mathbf{q}(t+\tau)] \geq \mathbf{0} \end{aligned} \tag{9.160}$$

Now, the path to get a solution to the problem is quite evident. For each step of $\tau = 1, 2, ..., t^*$ solve the stochastic dynamic programming (9.158) of $2k$ variables with boundary condition (9.159) subject to the $k+1$ linear constraints (9.160). The linear programming problem which has to be solved at each step with $2k$ variables and $k+1$ equations could be solved computationally very satisfactorily by the method provided by Powell (1988).

A slight modification of the above procedure is sufficient to provide the steps needed to solve the same problem using the optimal feedback technique.

9.11 An illustrative application

In the present section, the previous results are illustrated through an example from manpower planning. The illustration is presented in the form of a nucleus and the reader is encouraged to complement it according to the instructions given. After completing the suggested exercise the reader may compare his results with the results in Tsantas and Vassiliou (1993) section 5 where a similar illustration is presented. Let that from the historical records of a firm with three grades it was possible to estimate the sequence of transition matrices $\{\mathbf{P}(t)\}_{t=1}^{\infty}$ and that it was found that a pool of three matrices consisted the pool from which transition matrices were drawn:

$$\mathbf{P}_1 = \begin{pmatrix} 0.81 & 0.17 & 0 \\ 0 & 0.8 & 0.1 \\ 0 & 0 & 0.9 \end{pmatrix}, \quad \mathbf{P}_2 = \begin{pmatrix} 0.75 & 0.2 & 0 \\ 0 & 0.7 & 0.1 \\ 0 & 0 & 0.81 \end{pmatrix},$$

$$\mathbf{P}_3 = \begin{pmatrix} 0.71 & 0.27 & 0 \\ 0 & 0.8 & 0.12 \\ 0 & 0 & 0.84 \end{pmatrix}. \tag{9.161}$$

Assume also that the historical records showed the following series of appearances of the above three transition matrices for the last 27 points in time:

$$\begin{gathered} \mathbf{P}_1 \to \mathbf{P}_3 \to \mathbf{P}_2 \to \mathbf{P}_1 \to \mathbf{P}_1 \to \mathbf{P}_2 \to \mathbf{P}_3 \to \mathbf{P}_3 \to \mathbf{P}_1 \to \\ \mathbf{P}_3 \to \mathbf{P}_2 \to \mathbf{P}_2 \to \mathbf{P}_2 \to \mathbf{P}_3 \to \mathbf{P}_3 \to \mathbf{P}_1 \to \mathbf{P}_1 \to \mathbf{P}_3 \to \\ \mathbf{P}_2 \to \mathbf{P}_2 \to \mathbf{P}_2 \to \mathbf{P}_3 \to \mathbf{P}_1 \to \mathbf{P}_3 \to \mathbf{P}_2 \to \mathbf{P}_1 \to \mathbf{P}_3. \end{gathered} \tag{9.162}$$

In (9.162) counting transitions the transition probabilities may be estimated for the compromise non-homogeneous Markov chain $\{\mathbf{C}(t)\}_{t=0}^{\infty}$. Use the maximum likelihood estimates for that purpose given in Forbes (1971), Vassiliou (1976) and Bartholomew and Forbes (1979) or in Chapter 5 for more difficult cases were certain values are missing. Divide the data (9.162) into three parts (9 points each) and test for time homogeneity of $\{\mathbf{C}(t)\}_{t=0}^{\infty}$. Use (9.132) to find $\mathbb{E}[\mathbf{P}(t)]$ and (9.143) to find $\mathbb{E}[\mathbf{Q}(t)]$. Assume or estimate for your population the sequence of total memberships $\{T(t)\}_{t=0}^{\infty}$ and the initial population structure $\mathbf{N}(0)$ and use (9.144) to find $\mathbb{E}[\mathbf{N}(t)]$ for various different input policies $\{\mathbf{p}_0(t)\}_{t=0}^{\infty}$.

The next important step is to find the probabilities $P_m(t)$ that particular relative structures can be maintained by input control. This could easily be done using the results in section 9 of the present Chapter. In addition find the area of the one step maintainable region $\mathfrak{B}_t$ based on the deterministic analysis, that is, using the expected values instead of the random variables. Then your results should show that outside the area $\mathfrak{B}_t$ the probabilities $P_m(t)$ of maintaining a relative population structure are small if you compare them with the relative population structures that lie within $\mathfrak{B}_t$. For the relative population structures that lie on the border of $\mathfrak{B}_t$ the probabilities of maintainability are between the former two cases. Choose various relative population structures and use the proposed strategies for attaining them in a number of time steps. Doing so it is possible to get considerable insight of how your stochastic population evolves under different conditions.

9.12 Research notes

One of the important areas of applications of the theory of NHMS's are in human populations. This area of applications was named *Manpower Planning* and has been studied by researchers from a vast diversity of research disciplines since the 60's. One of the important issues for apparent reasons has always been the costing of human populations. Georgiou and Vassiliou (1997) studied the problem of optimizing the cost of a manpower system to attain and maintain asymptotically a desired expected relative population structure under asymptotic control of the recruitment policy. They started by proposing a quite flexible class of models for costing the evolution of a manpower system through time. Then they distinguished three phases in the evolution of the expected relative population structure toward infinity: First, the phase which was called the *transient phase* which lasts until all the mobility parameters of the NHMS converge. Secondly, the *semi-transient phase* which starts from the end of the transient phase and ends when the sequences of the costing parameters converge and by now the homogeneous Markov system reaches its equilibrium state. Finally, the third phase called the *equilibrium phase* starts from the end of the semi-transient phase and last for ever. Optimization techniques are then provided for all three phases which minimize the aggregate cost and which are a combination of dynamic and linear programming techniques.

The interest in controlling a subset of the expected relative population structure $\mathbf{q}(t)$, using the input policy connected with a given subset of the states has its roots in Vajda (1975, 1978a, 1978b) for the case of homogeneous Markov population models. For NHMS, Georgiou (1992) studied control under input policies for partial maintainability and proved a basic theorem which characterizes the region of these expected relative population structures.

The techniques of goal programming as well as various cost considerations, as a general framework in manpower planning, can be found from the start of the large boom of research on the subject. Characteristic representatives of this effort was the important work of Charnes et al. (1971, 1975). Some other efforts are those by Clough et al. (1971), Price et al. (1972), Vajda (1978a), Morgan (1979) and Zanakis et al. (1981a, 1981b). The reader interested to learn the basics on goal programming and study some examples will find useful the books by Lee (1976) and Ignizio (1982). Georgiou (1999) used goal programming techniques to extend the work mentioned earlier by Georgiou and Vassiliou (1997). That is, he kept the three phase approach and using the advantages of goal programming gained considerable insight on the evolvement of an NHMS. That is, achievement functions, goals and rigid constraints are transformed to reflect at every step in time the aspirations and priorities imposed by the management. Algorithms of goal programming are provided for attainability, maintainability, partial maintainability of the expected population structure and relative expected structure.

The periodicity of the limiting behavior of an NHMS in a stochastic environment was studied in Georgiou and Tsantas (1994). They tackled the problem of asymptotic periodicity, that is, establishing the limiting of the system when the inherent non- homogeneous Markov chain converges asymptotically to a periodic homogeneous Markov chain. They proved that if d is the period of the homogeneous Markov chain then the expected population structure asymptotically splits into d converging subsequences, that depend on the limiting homogeneous Markov chain and independent of the initial population structure and other basic parameters of the NHMS. Also, the sets of all asymptotically attainable expected structures are characterized completely.

For closed homogeneous Markov systems (HMS), that is, for a population which has no inflow of new memberships and which is modeled by a homogeneous Markov chain, Tsaklidis (1994) studied the evolution of the attainable expected relative population structures in discrete case. We know from the study of the asymptotically attainable structures in the more general case of an NHMS that the volume of the attainable structures get smaller as $t \to \infty$. For a closed HMS with number of states k the set of attainable structures for each t, let say $A_k(t) \subset \mathbb{R}_k$. It is proved that if the transition probability matrix $\mathbf{Q}$ of the inherent Markov chain is non-singular then the volume of $A_k(t)$ depends geometrically on the absolute value of the determinant of $\mathbf{Q}$ and on the rank of $\mathbf{Q}$. The solution for $\mathbf{Q}$ singular is also established. It is also proved that the square of the distance of two arbitrary structures that are attainable at time t is a linear combination of $(\lambda_i \lambda_j)^t$ where $\lambda_i\,(i, j = 1, 2, ..., k)$ are the eigenvalues of $\mathbf{Q}$.

The problem of expectations, variances and covariances of the state sizes in an NHMS in a stochastic environment is studied in Tsantas (1995). The probabilities of attaining a particular expected population structure are found. Also optimal strategies to attain a desirable population structure are established and an algorithm is proposed which is a mixture of dynamic and quadratic programming techniques. In Tsantas and Georgiou (1998) the partiall maintainability theory for an NHMS is extended to an S-NHMS. The main result being the calculation of the probability that an expected population structure can be partially maintained in astochastic environment. In Tsantas (2001), the limiting behavior of an S-NHMS is studied, that is, the limiting behavior of the means, variances and covariances of the expected population structure. Also, conditions are found such that the convergence would be geometrically fast which apparently has great practical value. Finally, the sets of all asymptotically atainable structures under control of the input flow are characterized.

Evaluation of recruitment strategies for HMS in a stochastic environment were studied by Feyter and Guerry (2009). In this case, it is not assumred that the total number of members are known but the the total number of new recruits is another variable to be found in order that the strategy will satisfy some criteria. The proposed strategy is proposed for a step of one time unit attainability of a certain desired population structure. The optimal strategy is selected by evaluating possible recruitment strategies based on the concept of some degree of attainability, which is expressed by a membership function as it is too often done in fuzzy set theory. For a multivariate NHMS (see the section of Research Notes of Chapter 6) with human members Dimitriou et al. (2013) propose a fuzzy goal programming approach approach in order to determine optimal allocations of new entrants and personnel transfers among the various systems's departments.

The work by Georgiou et al. (2021) opens a new path of research by integrating Data Envelopment Analysis into Markov systems (and vice versa). It builds upon earlier studies that used Goal Programming to structure optimization problems in Markov systems and that is not peculiar at all since DEA models have similar goal programming structure in their core. In addition, the integration of DEA with NHMS provides the opportunity to assess potential feasible courses of action through convexity using a non-parametric methodology that enables handling of rather ill-defined problems regarding the degree of confidence on the components of inflow policy or the relative importance of the states of a target population structure. The study provides two hybrid models of Markov-DEA that are related to the celebrated attainability problem and also it introduces a scalar performance index based on the well-known slack based measures of DEA.

Chapter 10

LAWS OF LARGE NUMBERS FOR NON-HOMOGENEOUS MARKOV SYSTEMS

10.1 Introductory notes

One of the most celebrated theorems in probability theory is the Law of Large Numbers (Grimmett and Stirzaker (2001)). The Law of Large Numbers (LLN) were also studied for finite Markov chains (Kemeny and Snell (1976)). The Law of Large Numbers for a regular homogeneous Markov chain states, that if π_j is the limiting probability of being in state j independent of the initial state, then also π_j represents the fraction of time, that the process can be expected to be in state j for a large number of steps. The Law of Large Numbers for Markov chains is also linked with the Martingale Convergence Theorem (Kemeny, Snell and Knapp (1976)). These laws are of value for Markov chains exactly as they are for all stochastic processes: the LLN and CLT, in particular, provide the theoretical basis for many results in the statistical analysis of chains as they do in related fields. For this and other applications, the reader is referred to Hall and Heyde (1980).

LLN and CLT for homogeneous Markov chains were known from the time of Markov himself. In fact, as we have seen in Chapter 3 these problems have been the motive for Markov to define the class of stochastic processes known as the Markov chains. In the present chapter we will study LLN for non-homogeneous Markov systems and also for Cyclic NHMS an important category of NHMS presented in Chapter 8. We will start with some useful known results on the various modes of convergence of random variables in a probability space the proofs of which could be found in many known books in Probability theory. These results will be used repeatedly in the sections that follow. In section 10.3, we first prove a theorem which is a Law of Large Numbers for an NHMS. We prove, that for an NHMS under certain conditions the fraction of time that a membership is in a certain state, asymptotically converges in mean square to the limit of the expected relative population structure of memberships in that state. In a second theorem in the same section, we provide and prove under what conditions the mode of convergence in the previous basic result is almost surely. In section 10.5, we study the important category of Cyclic NHMS. We prove in two theorems, under what conditions the expected relative population structure of a Cycl-NHMS asymptotically converges in the strongly Cesaro sense. In section 10.5, we first prove a theorem which is a Law of Large Numbers for a Cycl-NHMS. We prove, that for a Cycl-NHMS under certain conditions the fraction of time that a membership is in a certain state, asymptotically converges in mean square to the limit of the expected relative population structure in the strongly Cesaro sense of membership in that state. We then proceed to establish a founding theorem, which provides the conditions under which the expected relative population structure asymptotically converges in the strongly Cesaro sense with geometrical rate. This theorem is the basic instrument missing to prove, under

DOI: 10.1201/b23204-10

what conditions the Law of Large Numbers for a Cycl-NHMS is with almost surely mode of convergence. In section 10.6, we provide applications of the present results in section 10.3 to geriatric and stroke patients. Also, we provide applications of the results in section 10.4 and 10.5 for the movements of students in a University system. In section 10.7, we study Laws of Large Numbers for NHMS the inherent non-homogeneous Markov chain of which the sequence of transition probabilities consist of arbitrary stochastic matrices.

10.2 Basic concepts and useful results

In the present section, we will firstly recall some basic concepts and useful results already introduced and discussed in the present book and also some well known concepts from basic probability theory. The criterion for the selection of these concepts and results is that they will play a part in proving Laws of Large Numbers for NHMS and Cy-NHNS.

Let an NHMS as it was introduced in Chapter 5 and with the same notation. Recall that $\mathbf{q}(t) = \mathbf{N}(t)/T(t)$ is denoted to be the relative population structure and

$$\mathbf{q}(s,t) = [q_1(s,t), q_2(s,t), ..., q_k(s,t)],$$

where

$$q_j(s,t) = \mathbb{P}[X_t = j \mid \mathbf{q}(s)], \quad \text{for} \quad s < t,$$

then from Chapter 9 or Georgiou and Vassiliou (1992) we get that

$$\mathbb{E}[\mathbf{q}(t-1,t)] = a(t-1)\mathbf{q}(t-1)\mathbf{Q}(t) + b(t-1)\mathbf{p}_0(t),$$

where

$$a(t-1) = \frac{T(t-1)}{T(t)} \quad \text{and} \quad b(t-1) = \frac{T(t) - T(t-1)}{T(t)},$$

also we get that

$$\mathbb{E}[\mathbf{q}(0,t)] = a(t-1)\mathbb{E}[\mathbf{q}(0,t-1)]\mathbf{Q}(t) + b(t-1)\mathbf{p}_0(t), \tag{10.1}$$

from which recursively we get that

$$\mathbb{E}[\mathbf{q}(0,t)] = \frac{T(0)}{T(t)}\mathbf{q}(0)\mathbf{Q}(0,t)$$

$$+\frac{1}{T(t)}\sum_{\tau=1}^{t}\Delta T(\tau)\mathbf{p}_0(\tau)\mathbf{Q}(\tau,t), \tag{10.2}$$

where $\mathbf{Q}(s,t) = \mathbf{Q}(s+1)\mathbf{Q}(s+2)...\mathbf{Q}(t)$ for $s \leq t$. We set $\mathbf{Q}(s,t) = \mathbf{I}$ the identity matrix for $s \geq t$. Note that we set $\mathbf{q}(s,t) = 0$ for $s > t$.

We denote by $\mathbb{E}\left[\mathbf{q}^{(i)}(s,t)\right]$ if it is known that

$$\mathbf{q}(s) = \left[0, 0, ..., \underbrace{1}_{i-th}, ..., 0\right],$$

and the vector $\mathbb{E}\left[\mathbf{q}^{(i)}(s,t)\right]$ is then given by

$$\mathbb{E}\left[\mathbf{q}^{(i)}(s,t)\right] = \left[\mathbb{E}\left[q_1^{(i)}(s,t)\right], \mathbb{E}\left[q_2^{(i)}(s,t)\right], ..., \mathbb{E}\left[q_k^{(i)}(s,t)\right]\right]. \tag{10.3}$$

In what follows in this chapter we will use the norm for a matrix $\|\mathbf{A}\| \in M_n(\mathbb{R}^n)$

$$\|\mathbf{A}\| = \sup_{i\in S}\sum_{j\in S}|a_{ij}|,$$

avoiding the subscript in the symbol of the norm to avoid complication in the notations since we will not use any other norm.

At this point it will be useful for the reader to refresh the section 1.12 of Chapter 1 entitled "Convergence of sequences of random variables". We also refresh from Grimmet and Stirzaker (2001) p. 310 the well known Chebychov inequality.

Theorem 345 *(Chebychov inequality). Let a probability space* $(\Omega, \mathcal{F}, \mathbb{P})$ *and a random variable* $X : \Omega \to \mathbb{R}$. *Then*

$$\mathbb{P}(|X| \geq a) \leq \frac{\mathbb{E}(X^2)}{a^2} \quad \text{if } a > 0.$$

From Grimmet and Stirzaker (2001) p. 311 we get the following theorem:

Theorem 346 *Let a probability space* $(\Omega, \mathcal{F}, \mathbb{P})$ *and a sequence of random variables* $\{X_n\}_{n=0}^{\infty}$ *with* $X_n : \Omega \to \mathbb{R}$ *and a random variable* $X : \Omega \to \mathbb{R}$. *Then (a) If* $r > s \geq 1$ *and* $X_n \overset{L^r}{\to} X$ *then* $X_n \overset{L^s}{\to} X$. *(b)* $X_n \overset{L^1}{\to} X$ *then* $X_n \overset{P}{\to} X$. *The converse assertions fail in general.*

Note that any sequence $\{X_n\}_{n=0}^{\infty}$ which satisfies $X_n \overset{P}{\to} X$ necessarily contains a subsequence $\{X_n : 1 \leq i < \infty\}$ which converge almost surely. From Grimmet and Stirzaker (2001) p. 311 we get the following theorem:

Theorem 347 *Let a probability space* $(\Omega, \mathcal{F}, \mathbb{P})$ *and a sequence of random variables* $\{X_n\}_{n=0}^{\infty}$ *with* $X_n : \Omega \to \mathbb{R}$ *and a random variable* $X : \Omega \to \mathbb{R}$. *If* $X_n \overset{P}{\to} X$ *there exists a non-random increasing sequence of integers* $n_1, n_2, ...$ *such that* $X_{n_i} \overset{a.s.}{\to} X$ *as* $i \to \infty$.

Also from Grimmet and Stirzaker (2001) p. 310 we get the following theorem:

Theorem 348 *Let a probability space* $(\Omega, \mathcal{F}, \mathbb{P})$ *and a sequence of random variables* $\{X_n\}_{n=0}^{\infty}$ *with* $X_n : \Omega \to \mathbb{R}$ *and a random variable* $X : \Omega \to \mathbb{R}$. *If* $P_n(\epsilon) = \mathbb{P}(|X_n - X| > \epsilon)$ *satisfies*

$$\sum_n P_n(\epsilon) < \infty \text{ for all } \epsilon > 0$$

then $X_n \overset{a.s.}{\to} X$.

10.3 Laws of large numbers for an NHMS

In the present section we will study Laws of Large Numbers for NHMS. Let X_t the random variable representing the state of a membership at time t. Define by

$$u_j(t) = \left\{ \begin{smallmatrix} 1 & if & X_t = j \\ 0 & if & X_t \neq j \end{smallmatrix} \right\}, \tag{10.4}$$

also let $y_j(t)$ be the random variable representing the number of times the membership is in state j up to time t, i.e., $X_s = j$, $1 \leq s \leq t$; $\nu_j(t)$ be the random variable representing the fraction of time the membership is in state j up to time t. We have that

$$y_j(t) = \sum_{s=1}^{t} u_j(s) \quad \text{and} \quad \nu_j(t) = \frac{y_j(t)}{t}. \tag{10.5}$$

Define by

$$\mathbf{u}(t) = [u_1(t), u_2(t), ..., u_k(t)], \quad \mathbf{y}(t) = [y_1(t), y_2(t), ..., y_k(t)],$$

$$\boldsymbol{\nu}(t) = [\nu_1(t), \nu_2(t), ..., \nu_k(t)].$$

We will start with a theorem about the mean square convergence and then we will proceed to prove almost sure convergence. Actually we will prove, that under certain conditions easily met in practice, the fraction of time the membership of an NHMS stays in a state after a large number of steps, converges in mean square to the limit of the expected relative population structure in that state. This result constitutes the Weak Law of Large Numbers for an NHMS. We will afterwards proceed and prove under which conditions the mode of convergence is almost surely.

Theorem 349 *(Vassiliou (2020)) Let a probability space $(\Omega, \mathcal{F}, \mathbb{P})$ and an NHMS as defined in Chapter 5. Assume that (a) $\lim_{t\to\infty} \|\mathbf{Q}(t) - \mathbf{Q}\| = 0$ and $\mathbf{Q}$ a regular stochastic matrix; (b) $\lim_{t\to\infty} \|\mathbf{p}_0(t) - \mathbf{p}_0\| = 0$; (c) $\lim_{t\to\infty} [\Delta T(t)/T(t)] = 0$. Then*

$$\boldsymbol{\nu}(t) \overset{L^2}{\to} \mathbf{q}(\infty), \tag{10.6}$$

where $\mathbf{q}(\infty) = \lim_{t\to\infty} \mathbb{E}[\mathbf{q}(s,t)]$ for every s.

Proof. In order to prove (10.6), it is equivalent to show that

$$\lim_{t\to\infty} \mathbb{E}\left[(\boldsymbol{\nu}(t) - \mathbf{q}(\infty))^2\right] = 0. \tag{10.7}$$

Since the dimensions of the vectors in (10.7) are finite it is equivalent to show that:

$$\lim_{t\to\infty} \mathbb{E}\left[(\nu_j(t) - q_j(\infty))^2\right] = 0 \text{ for every } j = 1, 2, ..., k. \tag{10.8}$$

We have that

$$\mathbb{E}\left[(\nu_j(t) - q_j(\infty))^2\right] = \mathbb{E}\left(\sum_{n=1}^{t} \frac{u_j(n)}{t} - q_j(\infty)\right)^2$$
$$= \frac{1}{t^2}\mathbb{E}\left(\sum_{n=1}^{t}(u_j(n) - q_j(\infty))\right)^2 \tag{10.9}$$
$$= \frac{1}{t^2}\mathbb{E}\left[\left(\sum_{n=1}^{t}(u_j(n) - q_j(\infty))\right)\left(\sum_{l=1}^{t}(u_j(l) - q_j(\infty))\right)\right].$$

Hence, we have that

$$\mathbb{E}\left[(\nu_j(t) - q_j(\infty))^2\right] = \frac{1}{t^2}\sum_{n=1}^{t}\sum_{l=1}^{t}\mathbb{E}[u_j(n)u_j(l)] \tag{10.10}$$
$$-\frac{1}{t^2}\sum_{n=1}^{t}\sum_{l=1}^{t} q_j(\infty)\mathbb{E}[u_j(n)] \tag{10.11}$$
$$-\frac{1}{t^2}\sum_{n=1}^{t}\sum_{l=1}^{t} q_j(\infty)\mathbb{E}[u_j(l)] \tag{10.12}$$
$$+\frac{1}{t^2}\sum_{n=1}^{t}\sum_{l=1}^{t} q_j^2(\infty). \tag{10.13}$$

Now, we have that

$$\lim_{t\to\infty}\left\{-\frac{1}{t^2}\sum_{n=1}^{t}\sum_{l=1}^{t} q_j(\infty)\mathbb{E}[u_j(n)]\right\}$$

$$= \lim_{t\to\infty} \left\{ -\frac{1}{t^2} \sum_{n=1}^{t} \sum_{l=1}^{t} q_j(\infty) \mathbb{E}\left[\mathbb{E}\left[u_j(n) \mid \mathbf{q}(0)\right]\right] \right\}$$

$$= \lim_{t\to\infty} \left\{ -\frac{1}{t} q_j(\infty) \sum_{n=1}^{t} \mathbb{E}\left[\mathbb{P}\left(X_n = j \mid \mathbf{q}(0)\right)\right] \right\}$$

$$\lim_{t\to\infty} \left\{ -q_j(\infty) \frac{1}{t} \sum_{n=1}^{t} \mathbb{E}\left[q_j(0,n)\right] \right\}$$

$$= (\text{by Theorem } 296) = -q_j^2(\infty). \tag{10.14}$$

Similarly we get that

$$\lim_{t\to\infty} \left\{ -\frac{1}{t^2} \sum_{n=1}^{t} \sum_{l=1}^{t} q_j(\infty) \mathbb{E}\left[u_j(l)\right] \right\} = -q_j^2(\infty). \tag{10.15}$$

Also it is easy to see that

$$\frac{1}{t^2} \sum_{n=1}^{t} \sum_{l=1}^{t} q_j^2(\infty) = q_j^2(\infty). \tag{10.16}$$

It remains to find the first term on the right hand side of (10.10), that is

$$\lim_{t\to\infty} \left\{ \frac{1}{t^2} \sum_{n=1}^{t} \sum_{l=1}^{t} \mathbb{E}\left[u_j(n) u_j(l)\right] \right\}$$

$$= \lim_{t\to\infty} \left\{ \frac{1}{t^2} \sum_{n=1}^{t} \sum_{l=1}^{t} \mathbb{E}\left[\mathbb{E}\left[u_j(n) u_j(l)\right] \mid \mathbf{q}(0)\right] \right\}$$

$$= \lim_{t\to\infty} \left\{ \frac{1}{t^2} \sum_{n=1}^{t} \sum_{l=1}^{t} \mathbb{E}\left[\mathbb{P}\left(X_n = j, X_l = j \mid \mathbf{q}(0)\right)\right] \right\} = A. \tag{10.17}$$

Define by $n \wedge l = \max\{n, l\}$ and $n \vee l = \min\{n, l\}$, then we have that

$$\begin{aligned} \mathbb{P}\left(X_{n\wedge l} = j, X_{n\vee l} = j \mid \mathbf{q}(0)\right) &= \mathbb{P}\left(X_{n\wedge l} = j \mid X_{n\vee l} = j, \mathbf{q}(0)\right) \\ &\quad \times \mathbb{P}\left(X_{n\vee l} = j \mid \mathbf{q}(0)\right) \end{aligned}$$

$$= \mathbb{P}\left(X_{n\wedge l} = j \mid q_j(0, n\vee l) = 1\right) \mathbb{P}\left(X_{n\vee l} = j \mid \mathbf{q}(0)\right). \tag{10.18}$$

Now from (10.17) and (10.18) we get that

$$A = \lim_{t\to\infty} \left\{ \frac{1}{t^2} \sum_{n=1}^{t} \sum_{l=1}^{t} \mathbb{E}\left[\mathbb{P}\left(X_n = j, X_l = j \mid \mathbf{q}(0)\right)\right] \right\}$$

$$\begin{aligned} = \lim_{t\to\infty} \{ \frac{1}{t^2} \sum_{n=1}^{t} \sum_{l=1}^{t} & \mathbb{E}\left[\mathbb{P}\left(X_{n\wedge l} = j \mid q_j(0, n\vee l) = 1\right)\right] \\ & \times \mathbb{E}\left[\mathbb{P}\left(X_{n\vee l} = j \mid \mathbf{q}(0)\right)\right] \}. \end{aligned}$$

$$= \lim_{t\to\infty} \left\{ \frac{1}{t^2} \sum_{n=1}^{t} \sum_{l=1}^{t} \mathbb{E}\left[q_j^{(j)}(n\vee l, n\wedge l)\right] \mathbb{E}\left[q_j(0, n\vee l)\right] \right\}$$

$$= \left(\text{since } q_j^{(j)}(s,t) = 0 \text{ for } s > t\right)$$

$$= \lim_{t\to\infty}\left\{\frac{1}{t^2}\sum_{n=1}^{t}\sum_{l=n}^{t}\mathbb{E}\left[q_j^{(j)}(n,l)\right]\mathbb{E}\left[q_j(0,n)\right]\right\}$$

$$= \lim_{t\to\infty}\left\{\frac{1}{t}\sum_{n=1}^{t}\mathbb{E}\left[q_j^{(j)}(0,n)\right]\frac{1}{t}\sum_{l=n}^{t}\mathbb{E}\left[q_j(n,l)\right]\right\} = q_j^2(\infty). \tag{10.19}$$

Therefore from Eqs $(10.10), (10.11), ..., (10.19)$ we complete the proof. ■

Now we will proceed in the following theorem to prove under which conditions the mode of convergence is almost surely.

Theorem 350 *(Vassiliou (2020)) Let a probability space* $(\Omega, \mathcal{F}, \mathbb{P})$ *and an NHMS as defined in Chapter 5. Assume that (a)* $\lim_{t\to\infty}\|\mathbf{Q}(t)-\mathbf{Q}\| = 0$ *geometrically fast and* $\mathbf{Q}$ *a regular stochastic matrix; (b)* $\lim_{t\to\infty}\|\mathbf{p}_0(t)-\mathbf{p}_0\| = 0$ *geometrically fast; (c)* $\lim_{t\to\infty}[\Delta T(t)/T(t)] = 0$ *geometrically fast. Then*

$$\boldsymbol{\nu}(t) \overset{a.s.}{\to} \mathbf{q}(\infty).$$

Proof. In the previous theorem, we have proved that under a subset of the present conditions

$$\nu_j(t) \overset{L^2}{\to} q_j(\infty) \text{ as } t\to\infty \text{ for every } j\in\mathbb{S}, \tag{10.20}$$

which is equivalent with

$$\frac{1}{t}\sum_{s=1}^{t}u_j(s) \overset{L^2}{\to} q_j(\infty) \text{ as } t\to\infty \text{ for every } j\in\mathbb{S}. \tag{10.21}$$

From Theorem 346 (a) and equation (10.21) we get that

$$\frac{1}{t}\sum_{s=1}^{t}u_j(s) \overset{L^1}{\to} q_j(\infty) \text{ as } t\to\infty \text{ for every } j\in\mathbb{S}. \tag{10.22}$$

From Theorem 346 (b) and equation (10.22) we get that

$$\frac{1}{t}\sum_{s=1}^{t}u_j(s) \overset{P}{\to} q_j(\infty) \text{ as } t\to\infty \text{ for every } j\in\mathbb{S}. \tag{10.23}$$

By Theorem 347 there exists a non-random increasing sequence of integers $t_1, t_2, ...$ such that

$$\frac{1}{t_i}\sum_{s=1}^{t_i}u_j(s) \overset{P}{\to} q_j(\infty) \text{ as } i\to\infty \text{ for every } j\in\mathbb{S}. \tag{10.24}$$

Will now show that such a choice of subsequence is $t_i = i^2$ for $i = 1, 2, ...$ It is sufficient according to Theorem 348 to show that

$$\text{for every } \epsilon>0 \sum_{i}^{\infty}\mathbb{P}\left(\left|\frac{1}{i^2}\sum_{s=1}^{i^2}u_j(s)-q_j(\infty)\right|>\epsilon\right)<\infty. \tag{10.25}$$

By Chebychov inequality, that is, Theorem 345 we get that

$$\mathbb{P}\left(\left|\frac{1}{i^2}\sum_{s=1}^{i^2}u_j(s)-q_j(\infty)\right|>\epsilon\right) \leq \frac{\mathbb{E}\left[\left(\frac{1}{i^2}\sum_{s=1}^{i^2}u_j(s)-q_j(\infty)\right)^2\right]}{\epsilon^2}. \tag{10.26}$$

Hence, what is remaining to prove is that

$$\text{For every } \epsilon > 0 \ \ B = \frac{1}{\epsilon^2} \sum_i \mathbb{E}\left[\left(\frac{1}{i^2} \sum_{s=1}^{i^2} u_j(s) - q_j(\infty) \right)^2 \right] < \infty. \tag{10.27}$$

From $(10.14), (10.15), (10.16)$ and (10.19) we get that

$$B = \frac{1}{\epsilon^2} \sum_i \frac{1}{i^4} \{ \sum_{n=1}^{i^2} \sum_{l=1}^{i^2} [- q_j(\infty) \mathbb{E}[q_j(0,n)] + q_j^2(\infty) - q_j(\infty) \mathbb{E}[q_j(0,l)]$$

$$+ \mathbb{E}\left[q_j^{(j)}(n \vee l, n \wedge l) \right] \mathbb{E}[q_j(0, n \vee l)]] \}$$

Assume that $n \vee l = l$. Then we have

$$B = \frac{1}{\epsilon^2} \sum_i \frac{1}{i^4} \{ \sum_{n=1}^{i^2} \sum_{l=1}^{i^2} \{ q_j(\infty) [q_j(\infty) - \mathbb{E}[q_j(0,n)]] +$$

$$\left[\mathbb{E}\left[q_j^{(j)}(n \vee l, n \wedge l) \right] - q_j(\infty) \right] \mathbb{E}[q_j(0,l)] \} \},$$

$$B \leq \frac{1}{\epsilon^2} \sum_i \frac{1}{i^4} \{ \sum_{n=1}^{i^2} \sum_{l=1}^{i^2} \{ |q_j(\infty) - \mathbb{E}[q_j(0,n)]| +$$

$$\left| \mathbb{E}\left[q_j^{(j)}(n \vee l, n \wedge l) \right] - q_j(\infty) \right| \} \} \tag{10.28}$$

We arrive similarly in the same result when $n \vee l = n$. From Theorem 295 we get that there exists constants $c > 0$ and $0 < b < 1$ such that

$$|q_j(\infty) - \mathbb{E}[q_j(0,n)]| \leq c b^n, \tag{10.29}$$

From Theorem 295 following a similar proof we get that there exists constants $c_1 > 0$ and $0 < b_1 < 1$ such that

$$\left| \mathbb{E}\left[q_j^{(j)}(n \vee l, n \wedge l) \right] - q_j(\infty) \right| \leq c_1 b_1^{n \wedge l - n \vee l} < \infty. \tag{10.30}$$

From $(10.28), (10.29)$ and (10.30) we get that

$$B \leq \frac{1}{\epsilon^2} \sum_i \frac{1}{i^4} \{ \sum_{n=1}^{i^2} \sum_{l=1}^{i^2} \left[c b^n + c_1 b_1^{n \wedge l - n \vee l} \right] < \infty.$$

Hence, we have proved that

$$\frac{1}{i^2} \sum_{s=1}^{i^2} u_j(s) \overset{a.s.}{\rightarrow} q_j(\infty). \tag{10.31}$$

Now we have that

$$\sum_{s=1}^{i^2} u_j(s) \text{ is monotonic non-decreasing,} \tag{10.32}$$

therefore

$$\sum_{s=1}^{i^2} u_j(s) \leq \sum_{s=1}^{t} u_j(s) \leq \sum_{s=1}^{(i+1)^2} u_j(s) \ \text{ if } \ i^2 \leq t \leq (i+1)^2, \tag{10.33}$$

from which we easily get

$$\frac{1}{(i+1)^2}\sum_{s=1}^{i^2} u_j(s) \leq \frac{1}{t}\sum_{s=1}^{t} u_j(s) \leq \frac{1}{i^2}\sum_{s=1}^{(i+1)^2} u_j(s) \quad \text{if} \quad i^2 \leq t \leq (i+1)^2, \tag{10.34}$$

if $i^2 \leq t \leq (i+1)^2$. In equation (10.34) let $t \to \infty$, use the fact that $\lim_{t\to\infty}\left(i^2/(i+1)^2\right) \to 1$, and relation (10.31) to get

$$\frac{1}{t}\sum_{s=1}^{t} u_j(s) \overset{a.s.}{\to} q_j(\infty).$$

■

10.4 An illustrative application

We will now discuss an illustrative application of the previous section to a hospital for geriatric patients. We will consider the hospital as an NHMS the population of which will be the geriatric patients. Its memberships in this case will be given to the patients at the reception and a membership is actually the assignment to a bed. The first basic assumption of an NHMS is that its total population $T(t)$ for $t = 1, 2, \ldots$ is known in advance. This applies naturally for a hospital system which is usually designed for a certain amount of beds well in advance and all plans for expansion must be done with a large lead time for apparent reasons.

There exists real applications of homogeneous Markov systems using transition probability matrices of a specific structure and which are called general Coxian phase type models, special forms of which has been used as stochastic models for geriatric patients and stroke patients by McClean and her co-authors McClean et al. (1998a,b), Taylor et al. (2000), Marshall et al. (2002), Marshall and McClean (2003, 2004), Garg et al. (2010), McClean et al. (2014a,b). In the present, we will use typical type of data from these applications but we will reduce the number of states for keeping the size of matrices presented to a reasonable size.

We distinguish three states in the NHMS for the hospital system which are called hospital pathways. In the case of geriatric patients the states are "Acute Care", the "Rehabilative" and the "Long Stay". From each state we have wastage from the hospital due to discharge or death for all three states. Also, geriatric patients may be thought as progressing through stages of acute care, rehabiliation and long-stay care, where most patients are eventually rehabilitated and discharged.

Geriatric medical services are an important asset in the care of the elderly, while at the same time they can be easy victims of the political pressure on savings in health expenditure. We have all been witnesses of what happened with the political options and strategies evaluated in different countries concerning the Cov-19 panademic. It is certainly an indication of a society's political civilization of how it treats its elderlies battered from the battle for survival even from real wars.

It is important to stress again that the number of pathways could be increased or decreased according to the researcher's needs to best fit the data. Also, it is of importance in the best management of hospital resources and certainly to the benefit of geriatric patients to know the tendencies of the system in the long run. That is, what proportion of the total population is going to be in each state. In the case that the population consists of other type of patients then the present model is easily adaptable. For example, in the case of stroke patients there are more types of transitions due to the nature of stroke, which allows for relapses and hence more transitions among the hospital pathways. In this case the only change in the model will be that the transition probability matrices will have more positive elements.

Consider a hospital which starts with $T(0) = 400$ patients with beds and in a rather short time reaches its full capacity of 435 patients. That is, $T(1) = 420$, $T(2) = 430$, $T(3) = 435$. Assume that the hospital system has the above mentioned three pathways and let that the initial relative population structure is

$$\mathbf{q}(0) = \begin{pmatrix} 0.5 & 0.25 & 0.25 \end{pmatrix}.$$

The vast majority of new patients enter the system in hospital pathway one, either by taking an empty place or as a virtual replacement of discharged patient, that is $q_{11}(t) = p_{11}(t) + p_{14}(t)\,p_{01}(t)$. In here state 4 expresses the external environment. The entrance probabilities are

$$\mathbf{p}_0(0) = \begin{pmatrix} 0.6 & 0.3 & 0.1 \end{pmatrix};\ \mathbf{p}_0(1) = \begin{pmatrix} 0.5 & 0.3 & 0.2 \end{pmatrix};$$

$$\mathbf{p}_0(2) = \begin{pmatrix} 0.75 & 0.25 & 0.1 \end{pmatrix} \text{ and } \mathbf{p}_0 = \mathbf{p}_0(t) = \begin{pmatrix} 0.7 & 0.2 & 0.1 \end{pmatrix},$$

for $t = 3, 4, \ldots$. The form of the transition probability matrices for the internal movements in the hospital system is

$$\mathbf{P}(t) = \begin{pmatrix} p_{11}(t) & p_{12}(t) & p_{13}(t) \\ 0 & p_{22}(t) & p_{23}(t) \\ 0 & 0 & p_{33}(t) \end{pmatrix},$$

also the inherent non-homogeneous Markov chain will evolve with the sequence of stochastic matrices $\mathbf{Q}(t) = \{q_{ij}(t)\}_{i,j\epsilon S}$ where $q_{ij}(t) = p_{ij}(t) + p_{14}(t)\,p_{0j}(t)$. We get the following typical set of $\mathbf{Q}(t)'$s which are easily estimated from the data available in all hospital systems by its maximum likelihood estimates:

$$\mathbf{Q}(0) = \begin{pmatrix} 0.7 & 0.2 & 0.1 \\ 0.2 & 0.6 & 0.2 \\ 0.6 & 0.1 & 0.3 \end{pmatrix};\quad \mathbf{Q}(1) = \begin{pmatrix} 0.65 & 0.25 & 0.1 \\ 0.1 & 0.7 & 0.2 \\ 0.5 & 0.1 & 0.4 \end{pmatrix};$$

$$\mathbf{Q}(2) = \begin{pmatrix} 0.5 & 0.3 & 0.2 \\ 0.2 & 0.5 & 0.3 \\ 0.4 & 0.2 & 0.4 \end{pmatrix};\quad \mathbf{Q}(t) = \mathbf{Q} = \begin{pmatrix} 0.6 & 0.3 & 0.1 \\ 0.14 & 0.6 & 0.26 \\ 0.49 & 0.14 & 0.37 \end{pmatrix},$$

for $t = 3, 4, \ldots$. The row of the stable matrix

$$\lim_{t\to\infty} \mathbf{Q}^t = \mathbf{Q}(\infty) = \begin{pmatrix} \mathbf{q}(\infty) & \mathbf{q}(\infty) & \mathbf{q}(\infty) \end{pmatrix}^{\top}$$

is

$$\mathbf{q}(\infty) = \begin{pmatrix} 0.41 & 0.37 & 0.22 \end{pmatrix},$$

where the convergence is geometrically fast, that is, for $t = 5$ it already converges. Now simulating Theorem 349 we find that for

$$t = 5 \text{ we get } \mathbb{E}\left[\left(\nu_1(t) - q_1(\infty)\right)^2\right] = 0.0138,$$

$$t = 6 \text{ we get } \mathbb{E}\left[\left(\nu_1(t) - q_1(\infty)\right)^2\right] = 0.0007.$$

From the above is apparent that

$$\nu_1(t) \xrightarrow{L^2} 0.41.$$

Analogous results are found also for the remaining of the hospital pathways hence

$$\boldsymbol{\nu}(t) \overset{L^2}{\to} \begin{pmatrix} 0.41 & 0.37 & 0.22 \end{pmatrix}.$$

In our application, we have chosen $\Delta T(t)/T(t) \to 0$ geometrically fast, that is for $t = 3$, and also we have seen that $\lim_{t\to\infty} \mathbf{Q}^t = \mathbf{Q}(\infty)$ geometrically fast, that is for $t = 5$ hence all the conditions of Theorem 350 hold and this is verified by the fact that $\boldsymbol{\nu}(t)$ converges geometrically fast hence

$$\boldsymbol{\nu}(t) \overset{a.s.}{\to} \begin{pmatrix} 0.41 & 0.37 & 0.22 \end{pmatrix}.$$

One useful conclusion for designing and planning this kind of hospital system is that a bed in the hospital system will be used for "Acute care" patients, in the long run, almost surely 0.41 of the time the hospital is in operation almost surely. Another useful practical physical meaning for hospital planning is that the relative population structure of memberships in the various hospital pathways tends asymptotically almost surely to $\begin{pmatrix} 0.41 & 0.37 & 0.22 \end{pmatrix}$.

10.5 Laws of large numbers for a Cy-NHMS

In Chapter 8 we introduced a Cy-NHMS and studied its asymptotic behavior and its asymptotic variability. We are now mature enough to study Laws of Large for such systems. Firstly, from Theorem 319 we may easily extract the result for $\mathbb{E}[\mathbf{q}_r(\infty)]$ from Equation 8.8. Also, it is not difficult to prove that under the same conditions as in Theorem 319 that

$$\lim_{t\to\infty} \left\| \frac{1}{t} \sum_{n=0}^{t} \mathbb{E}[\mathbf{q}(0,n)] - \mathbf{q}(\infty) \right\| = 0.$$

which will be used repeatedly in what follows. We will now state and prove the following basic theorem:

Theorem 351 *(Vassiliou (2018)). Define a probability space $(\Omega, \mathcal{F}, \mathbb{P})$ and a Cycl-NHMS. If $\mathbf{Q}_0 = \mathbf{Q}(0)\mathbf{Q}(1)...\mathbf{Q}(d-1)$ is a regular stochastic matrix and in addition $\lim_{t\to\infty} \Delta T(t)/T(t) = 0$ then*

$$\boldsymbol{\nu}(t) \overset{L^2}{\to} \frac{1}{d} \sum_{s=0}^{d-1} \mathbf{q}_s(\infty), \tag{10.35}$$

where $\mathbf{q}_s(\infty)$ could be extracted from Theorem 319 for both cases, $T(t) \to_{t\to\infty} \infty$ and $\lim_{t\to\infty} \Delta T(t)/T(t) = 0$ geometrically fast.

Proof. Relation (10.35) is equivalent with

$$\lim_{t\to\infty} \mathbb{E}\left[\left(\boldsymbol{\nu}(t) - \frac{1}{d}\sum_{s=0}^{d-1} \mathbf{q}_s(\infty)\right)^2\right] = 0. \tag{10.36}$$

Since the dimensions of the vectors in (10.36) are finite it is sufficient to show that

$$\lim_{t\to\infty} \mathbb{E}\left[\left(\nu_j(t) - \frac{1}{d}\sum_{s=0}^{d-1} q_{sj}(\infty)\right)^2\right] = 0 \quad \text{for } j = 1, 2, ..., k. \tag{10.37}$$

We have that

$$\mathbb{E}\left[\left(\nu_j(t) - \frac{1}{d}\sum_{s=0}^{d-1} q_{sj}(\infty)\right)^2\right] = \mathbb{E}\left[\left(\sum_{n=1}^{t} \frac{u_j(n)}{t} - \frac{1}{d}\sum_{s=0}^{d-1} q_{sj}(\infty)\right)^2\right]$$

$$= \frac{1}{t^2}\mathbb{E}\left[\left(\sum_{n=1}^{t}\left(u_j(n) - \frac{1}{d}\sum_{s=0}^{d-1} q_{sj}(\infty)\right)\right)^2\right] =$$

$$\frac{1}{t^2}\mathbb{E}\left[\left(\sum_{n=1}^{t}\left(u_j(n) - \frac{1}{d}\sum_{s=0}^{d-1} q_{sj}(\infty)\right)\right)\right.$$

$$\left.\times\left(\sum_{l=1}^{t}\left(u_j(l) - \frac{1}{d}\sum_{s=0}^{d-1} q_{sj}(\infty)\right)\right)\right]. \tag{10.38}$$

Hence we get that

$$\mathbb{E}\left[\left(\nu_j(t) - \frac{1}{d}\sum_{s=0}^{d-1} q_{sj}(\infty)\right)^2\right] = \frac{1}{t^2}\sum_{n=1}^{t}\sum_{l=1}^{t}\mathbb{E}\left[u_j(n)\,u_j(l)\right] \tag{10.39}$$

$$-\frac{1}{dt^2}\sum_{s=0}^{d-1}\sum_{n=1}^{l}\sum_{l=1}^{t} q_{sj}(\infty)\,\mathbb{E}\left[u_j(n)\right] \tag{10.40}$$

$$-\frac{1}{dt^2}\sum_{s=0}^{d-1}\sum_{n=1}^{l}\sum_{l=1}^{t} q_{sj}(\infty)\,\mathbb{E}\left[u_j(l)\right] \tag{10.41}$$

$$+\frac{1}{t^2}\sum_{n=1}^{t}\sum_{l=1}^{t}\left(\frac{1}{d}\sum_{s=0}^{d-1} q_{sj}(\infty)\right)^2. \tag{10.42}$$

Let us start with expression (10.40) which could be written as

$$\lim_{t\to\infty}\left\{-\frac{1}{dt^2}\sum_{s=0}^{d-1}\sum_{n=1}^{l}\sum_{l=1}^{t} q_{sj}(\infty)\,\mathbb{E}\left[u_j(n)\right]\right\}$$

$$= \lim_{t\to\infty}\left\{-\frac{1}{d}\sum_{s=0}^{d-1} q_{sj}(\infty)\,\frac{1}{t}\sum_{n=1}^{t}\mathbb{E}\left[u_j(n)\right]\right\}$$

$$= \lim_{t\to\infty}\left\{-\frac{1}{d}\sum_{s=0}^{d-1} q_{sj}(\infty)\,\frac{1}{t}\sum_{n=1}^{t}\mathbb{E}\left[\mathbb{E}\left[u_j(n) \mid \mathbf{q}(0)\right]\right]\right\}$$

$$= \lim_{t\to\infty}\left\{-\frac{1}{d}\sum_{s=0}^{d-1} q_{sj}(\infty)\,\frac{1}{t}\sum_{n=1}^{t}\mathbb{E}\left[\mathbb{P}\left(X_n = j \mid \mathbf{q}(0)\right)\right]\right\}$$

$$= \lim_{t\to\infty}\left\{-\frac{1}{d}\sum_{s=0}^{d-1} q_{sj}(\infty)\,\frac{1}{t}\sum_{n=1}^{t}\mathbb{E}\left[q_j(0,n)\right]\right\}$$

$$= -\left(\frac{1}{d}\sum_{s=0}^{d-1} q_{sj}(\infty)\right)^2. \tag{10.43}$$

Simiralry we have that

$$\lim_{t\to\infty}\left\{-\frac{1}{dt^2}\sum_{s=0}^{d-1}\sum_{n=1}^{l}\sum_{l=1}^{t} q_{sj}(\infty)\,\mathbb{E}\left[u_j(l)\right]\right\} = -\left(\frac{1}{d}\sum_{s=0}^{d-1} q_{sj}(\infty)\right)^2. \tag{10.44}$$

Also it is easy to see that

$$\frac{1}{t^2}\sum_{n=1}^{t}\sum_{l=1}^{t}\left(\frac{1}{d}\sum_{s=0}^{d-1} q_{sj}(\infty)\right)^2 = \left(\frac{1}{d}\sum_{s=0}^{d-1} q_{sj}(\infty)\right)^2. \tag{10.45}$$

From the relation (10.39), we have that

$$\lim_{t\to\infty}\left\{\frac{1}{t^2}\sum_{n=1}^{t}\sum_{l=1}^{t}\mathbb{E}\left[u_j(n)\,u_j(l)\right]\right\} =$$

$$\lim_{t\to\infty}\left\{\frac{1}{t^2}\sum_{n=1}^{t}\sum_{l=1}^{t}\mathbb{E}\left[\mathbb{E}\left[u_j(n)\,u_j(l)\mid \mathbf{q}(0)\right]\right]\right\} = B. \tag{10.46}$$

From relations (10.39) and (10.18), we have that

$$B = \lim_{t\to\infty}\left\{\frac{1}{t^2}\sum_{n=1}^{t}\sum_{l=1}^{t}\mathbb{E}\left[\mathbb{P}\left[u_j(n)\,u_j(l)\mid \mathbf{q}(0)\right]\right]\right\}$$

$$= (\text{From } (5.18)) =$$

$$\lim_{t\to\infty}\frac{1}{t^2}\{\sum_{n=1}^{t}\sum_{l=1}^{t}\mathbb{E}\left[\mathbb{P}\left(X_{n\wedge l}=j\mid q_j(0,n\vee l)=1\right)\right]$$

$$\times\mathbb{E}\left[X_{n\vee l}=j\mid \mathbf{q}(0)\right]\} =$$

$$\lim_{t\to\infty}\frac{1}{t^2}\left\{\sum_{n=1}^{t}\sum_{l=1}^{t}\mathbb{E}\left[q_j^{(j)}(n\vee l, n\wedge l)\right]\mathbb{E}\left[q_j(0,n\vee l)\right]\right\}$$

$$= \left(\text{ since } q_j^{(j)}(s,t)=0 \text{ for } s>t\right)$$

$$= \lim_{t\to\infty}\frac{1}{t^2}\left\{\sum_{n=1}^{t}\sum_{l=1}^{t}\mathbb{E}\left[q_j^{(j)}(n,l)\right]\mathbb{E}\left[q_j(0,n)\right]\right\}$$

$$= \lim_{t\to\infty}\left\{\frac{1}{t}\sum_{n=1}^{t}\mathbb{E}\left[q_j(0,n)\right]\sum_{l=n}^{t}\mathbb{E}\left[q_j^{(j)}(n,l)\right]\right\}$$

$$= \left(\frac{1}{d}\sum_{s=0}^{d-1} q_{sj}(\infty)\right)\left(\frac{1}{d}\sum_{s=0}^{d-1} q_{sj}(\infty)\right) = \left(\frac{1}{d}\sum_{s=0}^{d-1} q_{sj}(\infty)\right)^2. \tag{10.47}$$

From equations (10.38), (10.39) ,..., (10.47) we easily arrive at the desired conclusion. ■

In Chapter 8, we proved a basic theorem where we provided conditions under which in a Cycl-NHMS

$$\lim_{t\to\infty}\left\|\frac{1}{t}\sum_{n=0}^{t}\mathbb{E}\left[\mathbf{q}(0,t)\right]-\frac{1}{d}\sum_{s=0}^{d-1}\mathbf{q}_s(\infty)\right\| = 0,$$

is geometrically fast. Having proved this result, we are now in a position similarly to prove that:

Theorem 352 *(Vassiliou (2018). Define a probability space $(\Omega,\mathcal{F},\mathbb{P})$ and a Cy-NHMS. If $\mathbf{Q}_0 = \mathbf{Q}(0)\,\mathbf{Q}(1)\ldots\mathbf{Q}(d-1)$ is a regular stochastic matrix and in addition $\lim_{t\to\infty}\Delta T(t)/T(t)=0$ geometrically fast then*

$$\boldsymbol{\nu}(t) \overset{a.s}{\to} \frac{1}{d}\sum_{s=0}^{d-1}\mathbf{q}_s(\infty) \quad \text{as} \quad t\to\infty.$$

10.6 An illustrative application

In this section, we illustrate an application of a cyclic non-homogeneous Markov system in a University system. The introduction of cyclic Markov chains and their theoretical study first appeared in Bowerman et al. (1977). However the first real application appeared previously in Gani (1963) with the strong motive to study student enrolment at Michigan State University. The importance of cyclic behavior for homogeneous Markov systems was firstly stressed in Bartholomew (1982 p. 71), motivated by the application of Gani (1963).

In the present we consider the University system in Vassiliou and Tsantas (1984) with 3 years of study. In this system, there are 3 years of study with the rule that if a student fail their year repeat it in the following year. The estimates of transition probability matrices taken from Vassiliou and Tsantas (1984) assuming a cyclic behavior are the following:

$$\mathbf{P}(3t)=\begin{pmatrix}0.172 & 0.819 & 0\\ 0 & 0.299 & 0.686\\ 0 & 0 & 0.323\end{pmatrix},\quad \mathbf{P}(3t+1)=\begin{pmatrix}0.194 & 0.772 & 0\\ 0 & 0.335 & 0.655\\ 0 & 0 & 0.536\end{pmatrix},$$

$$\mathbf{P}(3t+2)=\begin{pmatrix}0.166 & 0.816 & 0\\ 0 & 0.343 & 0.646\\ 0 & 0 & 0.508\end{pmatrix},$$

$$\mathbf{p}_4(3t)=\begin{pmatrix}0.009 & 0.015 & 0.676\end{pmatrix},\quad \mathbf{p}_4(3t+1)=\begin{pmatrix}0.034 & 0.010 & 0.465\end{pmatrix},$$

$$\mathbf{p}_4(3t+2)=\begin{pmatrix}0.017 & 0.011 & 0.491\end{pmatrix},$$

and for $\mathbf{p}_0(t)=\begin{pmatrix}1 & 0 & 0\end{pmatrix}$ for every $t=1,2,\ldots$. The total population of students is $T(0)=8970$, $T(1)=9000$, $T(2)=9050$ and $T(t)=9050$ for $t=3,4,\ldots$. Then we get

$$\mathbf{Q}(3t)=\begin{pmatrix}0.181 & 0.819 & 0\\ 0.015 & 0.299 & 0.686\\ 0.677 & 0 & 0.323\end{pmatrix},\quad \mathbf{Q}(3t+1)=\begin{pmatrix}0.227 & 0.773 & 0\\ 0.010 & 0.335 & 0.655\\ 0.464 & 0 & 0.536\end{pmatrix},$$

$$\mathbf{Q}(3t+2)=\begin{pmatrix}0.184 & 0.816 & 0\\ 0.011 & 0.343 & 0.646\\ 0.491 & 0 & 0.509\end{pmatrix}\quad \text{for } t=0,1,2,\ldots.$$

From these we arrive easily at

$$\mathbf{Q}_0=\begin{pmatrix}0.277 & 0.182 & 0.541\\ 0.388 & 0.303 & 0.359\\ 0.147 & 0.427 & 0.426\end{pmatrix},\quad \mathbf{Q}_1=\begin{pmatrix}0.354 & 0.176 & 0.470\\ 0.433 & 0.305 & 0.262\\ 0.253 & 0.399 & 0.348\end{pmatrix},$$

$$\mathbf{Q}_2=\begin{pmatrix}0.274 & 0.167 & 0.559\\ 0.308 & 0.381 & 0.311\\ 0.179 & 0.470 & 0.351\end{pmatrix}.$$

Let us assume that the initial relative population structure was found to be

$$\mathbf{q}(0)=\begin{pmatrix}0.32 & 0.30 & 0.38\end{pmatrix}.$$

Now we have all the data needed to apply equation (5.2) recursively and we arrive at

$$\mathbb{E}[\mathbf{q}(0,3)]=\begin{pmatrix}0.242 & 0.315 & 0.443\end{pmatrix},\quad \mathbb{E}[\mathbf{q}(0,6)]=\begin{pmatrix}0.238 & 0.328 & 0.431\end{pmatrix},$$

$$\mathbb{E}[\mathbf{q}(0,9)]=\begin{pmatrix}0.241 & 0.328 & 0.431\end{pmatrix},\quad \mathbb{E}[\mathbf{q}(0,12)]=\begin{pmatrix}0.241 & 0.328 & 0.431\end{pmatrix}.$$

We observe that $\mathbb{E}[\mathbf{q}(0,3t)]$ converges as $t \to \infty$. Similar results were found for

$$\mathbb{E}[\mathbf{q}(0,3t+1)] \to_{t\to\infty} \begin{pmatrix} 0.340 & 0.295 & 0.365 \end{pmatrix},$$

and $\mathbb{E}[\mathbf{q}(0,3t+2)]$. Therefore

$$\frac{1}{t}\sum_{n=0}^{t-1}\mathbb{E}[0,n] \to_{t\to\infty} \begin{pmatrix} 0.278 & 0.331 & 0.391 \end{pmatrix}.$$

We choose above $\Delta T(t) = 0$ for $t = 3, 4, ...$ and also $\mathbf{Q}_0$ to be regular stochastic matrix so that the conditions of Theorems 351, 352 would be satisfied. Therefore, we would expect that by simulating the equations we would verify that

$$\lim_{t\to\infty}\mathbb{E}\left[\left(\boldsymbol{\nu}(t) - \frac{1}{d_r}\sum_{s=0}^{d-1}\mathbf{q}_s(\infty)\right)^2\right] = 0.$$

Actually this was found to be true with

$$\lim_{t\to\infty}\mathbb{E}\left[\left(\boldsymbol{\nu}(t) - \begin{pmatrix} 0.278 & 0.331 & 0.391 \end{pmatrix}\right)^2\right] = 0.$$

Hence,

$$\boldsymbol{\nu}(t) \xrightarrow{L^2} \begin{pmatrix} 0.278 & 0.331 & 0.391 \end{pmatrix}.$$

One of the physical meanings of the above result is for example that a membership of a student place remaining in the first year of study, in the long run, in mean square is 0.278 of the time the University is at work.

10.7 LLN for NHMS with arbitrary probability matrices

In the present section we will generalize our results for the Laws of Large numbers for an NHMS for the case that the sequence of transition probability matrices $\{\mathbf{P}(t)\}_{t=0}^{\infty}$ are given in chronological order but they are arbitrary stochastic matrices. The asymptotic behavior of such Markov chains was studied in Chapter 4 section 4.9. One basic assumption for the sequence $\{\mathbf{P}(t)\}_{t=0}^{\infty}$ was that all matrices $\mathbf{P}(t) = \bar{\mathbf{P}}(t)$ for every t, that is, they all have the same incidence matrix.

10.7.1 Cesaro convergence for Markov chains with arbitrary transition probability matrices

In Chapter 4 section 4.9 we studied the asymptotic behavior of a Markov chain with arbitrary stochastic matrices given in chronological order. Given the sequence $\{\mathbf{P}(t)\}_{t=0}^{\infty}$ of its transition probability matrices and assuming that $\mathbf{P}(t) = \bar{\mathbf{P}}(t)$ for every t in Theorme 276 we proved the case that under the assumption of asymptotic homegeneity the infinite product $\lim_{t\to\infty}\mathbf{P}(s,t)$ splits into d convergent subsequences for every s. We will now extend the results of Theorem 276 for the case of Cesaro convergence.

Theorem 353 *(Vassiliou, 2022a). Let $(\Omega, \mathcal{F}, \mathbb{P})$ be a complete probability space and $\{X_t\}_{t=0}^{\infty}$ a discrete time non-homogeneous Markov chain with $\{\mathbf{P}(t)\}_{t=0}^{\infty}$ the sequence of its transition probability matrices. Let also $\{\mathbf{P}(t)\}_{t=0}^{\infty}$ be a sequence of arbitrary stochastic matrices with the same incidence matrix and let with no loss of generality the canonical form for every t be*

$$\mathbf{P}(t) = \begin{pmatrix} \mathbf{P}_1(t) & \mathbf{0} & ... & \mathbf{0} & \mathbf{0} \\ \mathbf{0} & \mathbf{P}_2(t) & ... & \mathbf{0} & 0 \\ ... & ... & ... & ... & ... \\ ... & ... & ... & \mathbf{P}_r(t) & \mathbf{0} \\ \mathbf{T}_1(t) & \mathbf{T}_2(t) & ... & \mathbf{T}_r(t) & \mathbf{Q}(t) \end{pmatrix} \quad \text{for } t = 0, 1, ... \tag{10.48}$$

that is, the matrices $\mathbf{P}(t)$ *for* $t = 0, 1, ...$ *have* r *essential classes of states. Assume that each essential class is periodic with period* d_r*, respectively. Assume asymptotic homogeneity, that is,*

$$\lim_{t\to\infty} \|\mathbf{P}(t) - \mathbf{P}\| = 0, \tag{10.49}$$

with $\mathbf{P}$ *having the same incidence matrix as* $\mathbf{P}(t)$*. Now let*

$$s = \pi_{si} d_i + u_i, \qquad t = \pi_{ti} d_i + v_i \quad \text{and} \quad \nu_i = v_i - u_i + 1. \tag{10.50}$$

Then the limit $\lim_{t\to\infty} 1/(t-s) \sum_{k=0}^{t-s-1} \mathbf{P}(s, s+k)$ *of the Cesaro sum infinite product for every* s *exists and is given by*

$$\lim_{t\to\infty} \frac{1}{t-s} \sum_{k=0}^{t-s-1} \mathbf{P}(s, s+k) =$$

$$= \begin{pmatrix} \mathbf{\Pi}_1^* & \mathbf{0} & ... & \mathbf{0} & \mathbf{0} \\ \mathbf{0} & \mathbf{\Pi}_2^* & ... & \mathbf{0} & \mathbf{0} \\ ... & ... & ... & ... & ... \\ \mathbf{0} & 0 & ... & \mathbf{\Pi}_r^* & \mathbf{0} \\ \mathbf{T}_1^* & \mathbf{T}_2^* & ... & \mathbf{T}_r^* & \mathbf{0} \end{pmatrix}. \tag{10.51}$$

where

$$\mathbf{P}_i^* = \lim_{t\to\infty} \mathbf{P}_i^t \quad \text{and} \quad \mathbf{\Pi}_i^* = \frac{1}{d_i} \sum_{\nu_i=0}^{d_i-1} \mathbf{P}_i^* \mathbf{P}_i^{\nu_i}. \tag{10.52}$$

and

$$\mathbf{T}_i^* = \left(\mathbf{I} - \mathbf{Q}^{d_i}\right)^{-1} \sum_{l=0}^{d_i-1} \mathbf{Q}^l \mathbf{T}_i \mathbf{\Pi}_i^*. \tag{10.53}$$

Proof. Since $\mathbf{P}(t) = \bar{\mathbf{P}}(t) = \bar{\mathbf{P}}$ we get that also $\mathbf{P}_i(t) = \bar{\mathbf{P}}_i(t) = \bar{\mathbf{P}}_i$ and in addition

$$\lim_{t\to\infty} \|\mathbf{P}_i(t) - \mathbf{P}_i\| = 0.$$

Hence, by Theorem 253 we get that the following Cesaro convergent result holds

$$\lim_{t\to\infty} \left\| \frac{1}{t-s} \sum_{k=0}^{t-s-1} \mathbf{P}_i(s, s+k) - \mathbf{\Pi}_i^* \right\| = 0, \quad \text{for every } s \text{ and } i = 1, 2, ...r, \tag{10.54}$$

where

$$\mathbf{\Pi}_i^* = \frac{1}{d_i} \sum_{\nu_i=0}^{d_i-1} \mathbf{P}_i^* \mathbf{P}_i^{\nu_i} \quad \text{and} \quad \mathbf{P}_i^* = \lim_{t\to\infty} \mathbf{P}_i^t. \tag{10.55}$$

We are looking for the Cesaro convergence of

$$\lim_{t\to\infty} \frac{1}{t-s} \sum_{m=0}^{t-s-1} \mathbf{P}(s, s+m) = \tag{10.56}$$

$$\begin{pmatrix} \hat{\mathbf{\Sigma}}\mathbf{P}_1(s, s+m) & \mathbf{0} & ... & \mathbf{0} & \mathbf{0} \\ \mathbf{0} & \hat{\mathbf{\Sigma}}\mathbf{P}_2(s, s+m) & ... & \mathbf{0} & \mathbf{0} \\ ... & ... & ... & ... & ... \\ \mathbf{0} & \mathbf{0} & ... & \hat{\mathbf{\Sigma}}\mathbf{P}_r(s, s+m) & \mathbf{0} \\ \hat{\mathbf{\Sigma}}\mathbf{T}_1(s, s+m) & \hat{\mathbf{\Sigma}}\mathbf{T}_2(s, s+m) & ... & \hat{\mathbf{\Sigma}}\mathbf{T}_r(s, s+m) & \hat{\mathbf{\Sigma}}\mathbf{Q}(s, s+m) \end{pmatrix},$$

where

$$\hat{\Sigma}\mathbf{P}_i(s,s+m)=\lim_{t\to\infty}\frac{1}{t-s}\sum_{m=0}^{t-s-1}\mathbf{P}_i(s,s+m)\quad\text{for } i=1,2,...,r,\tag{10.57}$$

$$\hat{\Sigma}\mathbf{T}_i(s,s+m)=\lim_{t\to\infty}\frac{1}{t-s}\sum_{m=0}^{t-s-1}\mathbf{T}_i(s,s+m)\quad\text{for } i=1,2,...,r,\tag{10.58}$$

$$\hat{\Sigma}\mathbf{Q}(s,s+m)=\lim_{t\to\infty}\frac{1}{t-s}\sum_{k=0}^{t-s-1}\mathbf{Q}(s,s+m).\tag{10.59}$$

Note that as all over in the book, we define by $\mathbf{P}_i(s,s)=\mathbf{I}$ and simirarly $\mathbf{Q}(s,s)=\mathbf{I}$. Now in general $\{\mathbf{Q}(t)\}_{t=0}^{\infty}$ is a sequence of substochastic matrices hence there exists a $c_q>0$ and a $0<b_q<1$ such that

$$\|\mathbf{Q}(s,t)\|\leq c_q b_q^{t-s}.\tag{10.60}$$

Hence

$$\left\|\hat{\Sigma}\mathbf{Q}(s,s+m)\right\|\leq\lim_{t\to\infty}\frac{1}{t-s}\sum_{m=0}^{t-s-1}\|\mathbf{Q}(s,s+m)\|\leq\lim_{t\to\infty}\frac{c_q}{t-s}\sum_{m=0}^{t-s-1}b^m\leq 0.\tag{10.61}$$

From (10.61) it is apparent that

$$\hat{\Sigma}\mathbf{Q}(s,s+m)=0.\tag{10.62}$$

From the above it is obvious that it remains to find $\hat{\Sigma}\mathbf{T}_i(s,s+m)$. In this respect, we have that firstly it is easy to check that

$$\mathbf{T}_i(s,t)=\sum_{m=0}^{t-s-1}\mathbf{Q}(s-1,s+m-1)\,\mathbf{T}_i(s+m)\,\mathbf{P}_i(s+m,t).\tag{10.63}$$

Now, the reader could verify using (10.63) that

$$\frac{1}{t-s}\sum_{n=1}^{t-s}\mathbf{T}_i(s,s+n)=$$

$$\frac{1}{t-s}\sum_{n=1}^{t-s}\sum_{m=0}^{n-1}\mathbf{Q}(s-1,s+m-1)\,\mathbf{T}_i(s+m)\,\mathbf{P}_i(s+m,s+n),$$

and changing the summation order we get

$$\frac{1}{t-s}\sum_{n=1}^{t-s}\mathbf{T}_i(s,s+n)=$$

$$\frac{1}{t-s}\sum_{m=0}^{t-s-1}\mathbf{Q}(s-1,s+m-1)\,\mathbf{T}_i(s+m)\sum_{n=1}^{t-s-m}\mathbf{P}_i(s+m,s+m+n).\tag{10.64}$$

Now let that

$$\mathbf{M}_i(s+m,t)=\frac{1}{t-s}\sum_{n=0}^{t-s-m}\mathbf{P}_i(s+m,s+m+n)$$

$$-\frac{1}{d_i-1}\sum_{k=0}^{d_i-1}\mathbf{P}_i^*\mathbf{P}_i^{(t-s+k-m)\bmod d_i}.\tag{10.65}$$

Then equation (10.64) could be written in the form

$$\frac{1}{t-s}\sum_{n=1}^{t-s}\mathbf{T}_i(s,s+n)=$$

$$\sum_{m=0}^{t-s-1}\mathbf{Q}(s-1,s+m-1)\,\mathbf{T}_i(s+m)\,\mathbf{M}_i(s+m,t)$$

$$+\frac{1}{d_i-1}\sum_{m=0}^{t-s-1}\mathbf{Q}(s-1,s+m-1)\,\mathbf{T}_i(s+m)\sum_{k=0}^{d_i-1}\mathbf{P}^*\mathbf{P}^{(t-s+k-m)\bmod d_i}$$

$$=\langle\mathbf{A}\rangle+\langle\mathbf{B}\rangle. \tag{10.66}$$

We now have that

$$\|\langle\mathbf{A}\rangle\|=\left\|\sum_{m=0}^{t-s-1}\mathbf{Q}(s-1,s+m-1)\,\mathbf{T}_i(s+m)\,\mathbf{M}_i(s+m,t)\right\|$$

$$\le\sum_{m=0}^{t-s-1}\|\mathbf{Q}(s-1,s+m-1)\|\,\|\mathbf{T}_i(s+m)\|\,\|\mathbf{M}_i(s+m,t)\|$$

$$\le 2c_q\sum_{m=0}^{t-s-1}b_q^m.$$

Therefore

$$\lim_{t\to\infty}\|\langle\mathbf{A}\rangle\|=0\ \text{ for every } s\in\mathbb{Z}^+. \tag{10.67}$$

In (10.66) it remains to find the limit as $t\to\infty$ of $\langle\mathbf{B}\rangle$. With no loss of generality assume that (10.50) is true then it could be proved that

$$\langle\mathbf{B}\rangle=\sum_{m=0}^{t-s-1}\mathbf{Q}(s-1,s+m-1)\,\mathbf{T}_i(s+m)\,\frac{1}{d_i-1}\sum_{k=0}^{d_i-1}\mathbf{P}_i^*\mathbf{P}_i^{(t-s+k-m)\bmod d_i}$$

$$=\sum_{\tau=0}^{\pi_{ti}-\pi_{si}-2}\mathbf{Q}(s-1,s+\tau d_i-1)\times \tag{10.68}$$

$$[\sum_{l=0}^{v-u}\mathbf{Q}(s+\tau d_i-1,s+\tau d_i+l-1)\,\mathbf{T}_i(s+\tau d_i+l)$$

$$\times\frac{1}{d_i}\sum_{k=0}^{d_i-1}\mathbf{P}_i^*\mathbf{P}_i^{d_i+v_i-u_i+k-l}$$

$$+\sum_{l=v-u+1}^{d_i-1}\mathbf{Q}(s+\tau d_i-1,s+\tau d_i+l-1)\,\mathbf{T}_i(s+\tau d_i+l)$$

$$\times\frac{1}{d_i}\sum_{k=0}^{d_i-1}\mathbf{P}_i^*\mathbf{P}_i^{d_i+(d_i+v_i-u_i+k-l)}]+$$

$$\sum_{r=0}^{d_i+v-u-1}\mathbf{Q}(s-1,s+(\pi_{ti}-\pi_{si}-1)\,d_i+r-1)\,\mathbf{T}_i(s+(\pi_{ti}-\pi_{si}-1)\,d_i+r)$$

$$\times\frac{1}{d_i}\sum_{k=0}^{d_i-1}\mathbf{P}_i^*\mathbf{P}_i^{d_i+v_i-u_i+r+k}=\langle\mathbf{C}\rangle+\langle\mathbf{D}\rangle.$$

It is straightforward to see that

$$\lim_{t\to\infty}\langle\mathbf{D}\rangle=\mathbf{0}. \tag{10.69}$$

From (10.68) and (10.69) we get that

$$\lim_{t\to\infty}\langle\mathbf{B}\rangle=\lim_{t\to\infty}\sum_{\tau=0}^{\pi_{ti}-\pi_{si}-2}\mathbf{Q}\left(s-1,s+\tau d_i-1\right)\times$$

$$[\sum_{l=0}^{d_i-1}\mathbf{Q}\left(s+\tau d_i-1,s+\tau d_i+l-1\right)\mathbf{T}_i\left(s+\tau d+l\right)\mathbf{P}_i^{d_i-l}] \tag{10.70}$$

$$\times\frac{1}{d_i}\sum_{k=0}^{d_i-1}\mathbf{P}_i^*\mathbf{P}_i^{d_i+v_i-u_i+k}.$$

We now have

$$\mathbf{L}_P=\|\lim_{t\to\infty}\sum_{\tau=0}^{\pi_{ti}-\pi_{si}-2}\mathbf{Q}\left(s-1,s+\tau d_i-1\right)\times$$

$$[\sum_{l=0}^{d_i-1}\mathbf{Q}\left(s+\tau d_i-1,s+\tau d_i+l-1\right)\mathbf{T}_i\left(s+\tau d+l\right)\mathbf{P}_i^{d_i-l}]\times$$

$$\frac{1}{d_i}\sum_{k=0}^{d_i-1}\mathbf{P}_i^*\mathbf{P}_i^{d_i+v_i-u_i+k}-\left(\mathbf{I}-\mathbf{Q}^{d_i}\right)^{-1}[\sum_{l=0}^{d_i-1}\mathbf{Q}^l\mathbf{T}_i\mathbf{P}^{d_i-l}]\times$$

$$\frac{1}{d_i}\sum_{k=0}^{d_i-1}\mathbf{P}_i^*\mathbf{P}_i^{d_i+v_i-u_i+k}]\| \tag{10.71}$$

$$\leq\sum_{\tau=0}^{\infty}\|\mathbf{Q}\left(s-1,s+\tau d_i-1\right)[\sum_{l=0}^{d_i-1}\mathbf{Q}\left(s+\tau d_i-1,s+\tau d_i+l-1\right)\times$$

$$\mathbf{T}_i\left(s+\tau d+l\right)\mathbf{P}_i^{d_i-l}]-\left(\mathbf{Q}^{d_i}\right)^{\tau}[\sum_{l=0}^{d_i-1}\mathbf{Q}^l\mathbf{T}_i\mathbf{P}^{d_i-l}]\|.$$

Now define by

$$\mathbf{D}\left(s,\tau\right)=[\sum_{l=0}^{d_i-1}\mathbf{Q}\left(s+\tau d_i-1,s+\tau d_i+l-1\right)\mathbf{T}_i\left(s+\tau d_i+l\right)\mathbf{P}_i^{d_i-l}]$$

$$-[\sum_{l=0}^{d-1}\mathbf{Q}^l\mathbf{T}_i\mathbf{P}^{d_i-l}]. \tag{10.72}$$

Then we have that

$$\|\mathbf{D}\left(s,\tau\right)\|\leq\sum_{l=0}^{d_i-1}\left\|\mathbf{Q}\left(s+\tau d_i-1,s+\tau d_i+l-1\right)\mathbf{T}_i\left(s+\tau d_i+l\right)-\mathbf{Q}^l\mathbf{T}_i\right\|$$

$$\leq\sum_{l=0}^{d_i-1}\left\|\mathbf{Q}\left(s+\tau d_i-1,s+\tau d_i+l-1\right)-\mathbf{Q}^l\right\|+\sum_{l=0}^{d_i-1}\left\|\mathbf{T}_i\left(s+\tau d+l\right)-\mathbf{T}_i\right\|. \tag{10.73}$$

Since all $\mathbf{P}(t)'$s have the same incedence matrix, and since $\mathbf{Q}(t)$, $\mathbf{Q}$ are substochastic matrices we get that

$$\left\|\mathbf{Q}(s+\tau d_i-1, s+\tau d_i+l-1)-\mathbf{Q}^l\right\| \leq \epsilon/2d_i,$$

$$\|\mathbf{T}_i(s+\tau d+l)-\mathbf{T}_i\| \leq \epsilon/2d_i \quad \text{for} \quad \tau \geq \tau_0(s) \text{ and every } s. \tag{10.74}$$

Therefore,

$$\lim_{t\to\infty} \|\mathbf{D}(s,\tau)\| = 0. \tag{10.75}$$

From the above relations we arrive at the following

$$\begin{aligned}
\mathbf{L}_P \leq \sum_{\tau=0}^{\infty} &\|\mathbf{Q}(s-1, s+\tau d_i-1)\mathbf{D}(s,t)+\mathbf{Q}(s-1, s+\tau d_i-1) \\
&\times \sum_{l=0}^{d_i-1}\mathbf{Q}^l\mathbf{T}_i\mathbf{P}^{d_i-l} - \left(\mathbf{Q}^{d_i}\right)^{\tau}\sum_{l=0}^{d_i-1}\mathbf{Q}^l\mathbf{T}_i\mathbf{P}^{d_i-l}\| \\
&\leq \sum_{\tau=0}^{\infty}\|\mathbf{Q}(s-1, s+\tau d_i-1)\|\,\|\mathbf{D}(s,t)\| + \\
&\sum_{\tau=0}^{\infty}\left\|\mathbf{Q}(s-1, s+\tau d_i-1)-\left(\mathbf{Q}^{d_i}\right)^{\tau}\right\| d_i.
\end{aligned} \tag{10.76}$$

Now from (10.75) we get that $\|\mathbf{D}(s,t)\| \leq 4d_i$ for $\tau \geq \tau_1(s)$ for every s. Therefore

$$\begin{aligned}
\mathbf{L}_P \leq &\sum_{\tau=0}^{\infty}\|\mathbf{Q}(s-1, s+\tau d_i-1)\|\,4d_i + \sum_{\tau=0}^{\infty}\|\mathbf{Q}(s-1, s+\tau d_i-1)\|\,d_i \\
&+\sum_{\tau=0}^{\infty}\left\|\left(\mathbf{Q}^{d_i}\right)^{\tau}\right\| d_i \leq 5d_ic_qb_q^{d_i}\sum_{\tau=0}^{\infty}b_q^{\tau} + cd_ib^{d_i}\sum_{\tau=0}^{\infty}b^{\tau} \\
&\leq d_i(5c_q+c)\frac{\min(b_q,b_i)^{d_i}}{1-\min(b_q,b_i)} \quad \text{for } t \geq \max(\tau_0(s), \tau_1(s)).
\end{aligned}$$

■

10.7.2 Cesaro convergence for an NHMS with arbitrary transition probability matrices for the inherent Markov chain

Let an NHMS with the notation of section 5.2. Let also the expected relative population structure $\mathbb{E}[\mathbf{q}(0,t)]$ as is given in (10.1) and (10.2). The basic assumption in this subsection will be that the transition probability matrices of the inherent Markov chain $\mathbf{Q}(s,t)$ are arbitrary stochastic matrices and they all have the same incidence matrix. Under these basic general conditions we are interested to study the possible Cesaro convergence of the expected relative population structure $\mathbb{E}[\mathbf{q}(0,t)]$, that is we are interested in finding the set of conditions under which

$$\lim_{t\to\infty}\frac{1}{t}\sum_{n=1}^{t}\mathbb{E}[\mathbf{q}(0,t)] \quad \text{exists.}$$

The answer to this problem is given in a compact form in the following theorem:

Theorem 354 *(Vassiliou (2022a)). Let* $(\Omega, \mathcal{F}, \mathbb{P})$ *be a complete probability space and a discrete time non-homogeneous Markov system. Also let* $\{T(t)\}_{t=0}^{\infty}$ *the sequence of total memberships in the system,* $\{\mathbf{P}(t)\}_{t=0}^{\infty}$ *the sequence of its transition probability matrices among the set* $S = \{1, 2, ..., k\}$ *of the internal states,* $\{\mathbf{p}_0(t)\}_{t=0}^{\infty}$ *the sequences of input probabilities,* $\{\mathbf{p}_{k+1}(t)\}_{t=0}^{\infty}$ *the sequences of output probabilities. Let also* $\{\mathbf{Q}(t)\}_{t=0}^{\infty}$ *the transion probability matrices of the inherent Markov chain be a sequence of arbitrary stochastic matrices with the same incidence matrix and let with no loss of generality the canonical form for every* t *be*

$$\mathbf{Q}(t) = \begin{pmatrix} \mathbf{Q}_1(t) & \mathbf{0} & \dots & \mathbf{0} & \mathbf{0} \\ \mathbf{0} & \mathbf{Q}_2(t) & \dots & \mathbf{0} & 0 \\ \dots & \dots & \dots & \dots & \dots \\ \dots & \dots & \dots & \mathbf{Q}_r(t) & \mathbf{0} \\ \mathbf{T}_{Q1}(t) & \mathbf{T}_{Q2}(t) & \dots & \mathbf{T}_{Qr}(t) & \mathbf{R}(t) \end{pmatrix} \quad \text{for } t = 0, 1, ... \tag{10.77}$$

that is, the matrices $\mathbf{Q}(t)$ *for* $t = 0, 1, ...$ *have* r *essential classes of states. Assume that each essential class is periodic with period* d_i*, respectively. Assume asymptotic homogeneity, that is,*

$$\lim_{t\to\infty} \|\mathbf{Q}(t) - \mathbf{Q}\| = 0, \tag{10.78}$$

with $\mathbf{Q}$ *having the same incidence matrix as* $\mathbf{Q}(t)$*, that is, the canonical form of* $\mathbf{Q}$ *is*

$$\mathbf{Q} = \begin{pmatrix} \mathbf{Q}_1 & \mathbf{0} & \dots & \mathbf{0} & \mathbf{0} \\ \mathbf{0} & \mathbf{Q}_2 & \dots & \mathbf{0} & 0 \\ \dots & \dots & \dots & \dots & \dots \\ \dots & \dots & \dots & \mathbf{Q}_r & \mathbf{0} \\ \mathbf{T}_{Q1} & \mathbf{T}_{Q2} & \dots & \mathbf{T}_{Qr} & \mathbf{R} \end{pmatrix};$$

also let that $\lim_{t\to\infty} T(t) = T$ *and* $\Delta T(t) \geq 0$ *for every* t*. Then*

$$\lim_{t\to\infty} \frac{1}{t} \sum_{n=1}^{t} \mathbb{E}[\mathbf{q}(0,t)] = \frac{T(0)}{T} [\mathbf{q}(0) - \mathbf{p}_0] \mathbf{\Pi}_q^* + \mathbf{p}_0 \mathbf{\Pi}_q^*$$

with

$$\mathbf{\Pi}_q^* = \begin{pmatrix} \mathbf{\Pi}_{q1}^* & \mathbf{0} & \dots & \mathbf{0} & \mathbf{0} \\ \mathbf{0} & \mathbf{\Pi}_{q2}^* & \dots & \mathbf{0} & \mathbf{0} \\ \dots & \dots & \dots & \dots & \dots \\ \mathbf{0} & \mathbf{0} & \dots & \mathbf{\Pi}_{qr}^* & \mathbf{0} \\ \mathbf{T}_{Q1}^* & \mathbf{T}_{Q2}^* & \dots & \mathbf{T}_{Qr}^* & \mathbf{0} \end{pmatrix} \tag{10.79}$$

where

$$\mathbf{\Pi}_{qi}^* = \frac{1}{d_i} \sum_{\nu_i=0}^{d_i-1} \mathbf{Q}_i^* \mathbf{Q}_i^{\nu_i} \quad \text{with } \mathbf{Q}_i^* = \lim_{t\to\infty} \mathbf{Q}_i^t, \tag{10.80}$$

and

$$\mathbf{T}_{Qi}^* = \left[\mathbf{I} - \mathbf{R}^{d_i}\right]^{-1} \sum_{l=0}^{d_i-1} \mathbf{Q}_i^l \mathbf{T}_{Qi} \mathbf{\Pi}_{qi}^*. \tag{10.81}$$

Proof. From (10.2), we get that

$$\frac{1}{t} \sum_{n=1}^{t} \mathbb{E}[\mathbf{q}(0,n)] =$$

$$\frac{1}{t}\sum_{n=1}^{t}\left[\frac{T(0)}{T(n)}\mathbf{q}(0)\mathbf{Q}(0,n)+\frac{1}{T(n)}\sum_{\tau=1}^{n}\Delta T(\tau)\mathbf{p}_0(\tau)\mathbf{Q}(\tau,n)\right]=$$

$$\frac{1}{t}\sum_{n=1}^{t}\frac{T(0)}{T(n)}\mathbf{q}(0)\mathbf{Q}(0,n)+\frac{1}{t}\sum_{n=1}^{t}\frac{1}{T(n)}\sum_{\tau=1}^{n}\Delta T(\tau)\mathbf{p}_0(\tau)\mathbf{Q}(\tau,n)$$

$$=\mathbf{A}(t)+\mathbf{B}(t). \tag{10.82}$$

Now, we have

$$\|\mathbf{D}_1(t)\|=\left\|\mathbf{A}(t)-\frac{T(0)}{T}\mathbf{q}(0)\mathbf{\Pi}_q^*\right\|=$$

$$\left\|\frac{1}{t}\sum_{n=1}^{t}\frac{T(0)}{T(n)}\mathbf{q}(0)\mathbf{Q}(0,n)-\frac{1}{t}\sum_{n=1}^{t}\frac{T(0)}{T}\mathbf{q}(0)\mathbf{\Pi}_q^*\right\|\leq$$

$$\frac{1}{t}\sum_{n=1}^{t}\left|\frac{T(0)}{T(n)}-\frac{T(0)}{T}\right|\|\mathbf{q}(0)\|\|\mathbf{Q}(0,n)\|+\left\|\frac{T(0)}{T}\mathbf{q}(0)\left[\frac{1}{t}\sum_{n=1}^{t}\mathbf{Q}(0,n)-\mathbf{\Pi}_q^*\right]\right\|$$

$$\leq\frac{T(0)}{t}\sum_{n=1}^{t}\left(\frac{1}{T(n)}-\frac{1}{T}\right)+\left\|\frac{1}{t}\sum_{n=1}^{t}\mathbf{Q}(0,n)-\mathbf{\Pi}_q^*\right\|. \tag{10.83}$$

Now, it is not difficult to see that the first part on the right hand side of (10.83) goes to zero due to $\lim_{t\to\infty}T(t)=T$ and also the second part due to Theorem 286 also goes to zero as $t\to\infty$. Therefore, we get from (5.82) that

$$\lim_{t\to\infty}\frac{1}{t}\sum_{n=1}^{t}\frac{T(0)}{T(n)}\mathbf{q}(0)\mathbf{Q}(0,n)=\frac{T(0)}{T}\mathbf{q}(0)\mathbf{\Pi}_q^*. \tag{10.84}$$

From (10.82) we get that

$$\|\mathbf{D}_2(t)\|=\left\|\mathbf{B}(t)-\frac{T(t)-T(0)}{T}\mathbf{p}_0\mathbf{\Pi}_q^*\right\|=$$

$$\left\|\frac{1}{t}\sum_{n=1}^{t}\frac{1}{T(n)}\sum_{\tau=1}^{n}\Delta T(\tau)\mathbf{p}_0(\tau)\mathbf{Q}(\tau,n)-\frac{T(t)-T(0)}{T}\mathbf{p}_0\mathbf{\Pi}_q^*\right\|=$$

$$\left\|\sum_{\tau=1}^{t}\Delta T(\tau)\mathbf{p}_0(\tau)\frac{1}{t}\sum_{m=0}^{t-\tau}\frac{\mathbf{Q}(\tau,\tau+m)}{T(\tau+m)}-\sum_{\tau=1}^{t}\frac{\Delta T(\tau)}{T}\mathbf{p}_0\mathbf{\Pi}_q^*\right\|\leq$$

$$\sum_{\tau=1}^{t}\Delta T(\tau)\left\|\frac{\mathbf{p}_0(\tau)}{t-\tau+1}\sum_{m=0}^{t-\tau}\frac{\mathbf{Q}(\tau,\tau+m)}{T(\tau+m)}-\mathbf{p}_0\frac{\mathbf{\Pi}_q^*}{T}\right\|$$

$$\leq\sum_{\tau=1}^{t}\Delta T(\tau)\|\mathbf{p}_0(\tau)-\mathbf{p}_0\|\left\|\frac{1}{t-\tau+1}\sum_{m=0}^{t-\tau}\frac{\mathbf{Q}(\tau,\tau+m)}{T(\tau+m)}\right\|+$$

$$\sum_{\tau=1}^{t}\Delta T(\tau)\|\mathbf{p}_0\|\left\|\frac{1}{t-\tau+1}\sum_{m=0}^{t-\tau}\frac{\mathbf{Q}(\tau,\tau+m)}{T(\tau+m)}-\frac{\mathbf{\Pi}_q^*}{T}\right\|$$

$$=\Gamma(t)+\Theta(t). \tag{10.85}$$

From (10.85), we have that

$$\Gamma(t)=\sum_{\tau=1}^{t}\Delta T(\tau)\|\mathbf{p}_0(\tau)-\mathbf{p}_0\|\left\|\frac{1}{t-\tau+1}\sum_{m=0}^{t-\tau}\frac{\mathbf{Q}(\tau,\tau+m)}{T(\tau+m)}\right\|$$

$$\leq \sum_{\tau=1}^{t} \Delta T(\tau) \|\mathbf{p}_0(\tau) - \mathbf{p}_0\| \frac{1}{t-\tau+1} \sum_{m=0}^{t-\tau} \frac{1}{T(\tau+m)}$$

$$\leq \sum_{\tau=1}^{t} \Delta T(\tau) \frac{2}{t-\tau+1} \sum_{m=0}^{t-\tau} \frac{1}{T(\tau+m)}. \tag{10.86}$$

Now, since $\lim_{t\to\infty} T(t) = T$ by hypothesis then for every $\epsilon > 0$ there is a t_0 such that

$$\Delta T(t) < \epsilon / (t - t_0) \text{ for } t \geq t_0. \tag{10.87}$$

From (10.86) we get that

$$\Gamma(t) \leq \sum_{\tau=1}^{t_0} \Delta T(\tau) \frac{2\sum_{m=0}^{t-\tau} \frac{1}{T(\tau+m)}}{t-\tau+1} + \sum_{\tau=t_0}^{t} \Delta T(\tau) \frac{2\sum_{m=0}^{t-\tau} \frac{1}{T(\tau+m)}}{t-\tau+1}. \tag{10.88}$$

It is not difficult to see from (10.88) that

$$\lim_{t\to\infty} \Gamma(t) = 0. \tag{10.89}$$

Now going back to (10.85), we have

$$\Theta(t) = \sum_{\tau=1}^{t} \Delta T(\tau) \|\mathbf{p}_0\| \left\| \frac{1}{t-\tau+1} \sum_{m=0}^{t-\tau} \frac{\mathbf{Q}(\tau, \tau+m)}{T(\tau+m)} - \frac{\mathbf{\Pi}_q^*}{T} \right\|$$

$$\leq \sum_{\tau=1}^{t} \Delta T(\tau) \| \frac{1}{t-\tau+1} \sum_{m=0}^{t-\tau} \left[\frac{1}{T(\tau+m)} - \frac{1}{T} \right] \mathbf{Q}(\tau, \tau+m)$$

$$+ \frac{1}{T} \left[\frac{1}{t-\tau+1} \sum_{m=0}^{t-\tau} \mathbf{Q}(\tau, \tau+m) - \mathbf{\Pi}_q^* \right] \| \leq$$

$$\sum_{\tau=1}^{t} \Delta T(\tau) \frac{2}{t-\tau+1} \sum_{m=0}^{t-\tau} \left[\frac{1}{T(\tau+m)} - \frac{1}{T} \right]$$

$$+ \frac{1}{T} \left\| \frac{1}{t-\tau+1} \sum_{m=0}^{t-\tau} \mathbf{Q}(\tau, \tau+m) - \mathbf{\Pi}_q^* \right\| = \Phi(t) + \Psi(t). \tag{10.90}$$

From (10.90) since $\lim_{t\to\infty} T(t) = T$ and from Theorem 286 it is not difficult to see that

$$\lim_{t\to\infty} \Phi(t) = 0, \quad \lim_{t\to\infty} \Psi(t) = 0 \text{ and therefore } \lim_{t\to\infty} \Theta(t) = 0. \tag{10.91}$$

From (10.82), (10.84), (10.89), and (10.91) we get that

$$\lim_{t\to\infty} \frac{1}{t} \sum_{n=1}^{t} \mathbb{E}[\mathbf{q}(0,t)] = \frac{T(0)}{T} [\mathbf{q}(0) - \mathbf{p}_0] \mathbf{\Pi}_q^* + \mathbf{p}_0 \mathbf{\Pi}_q^* \tag{10.92}$$

It is important at this point to note that the right hand side of (5.92) is always positive since from the conditions of the theorem, we have that $\Delta T(t) \geq 0$ for every t, therefore, $T(0)/T \leq 1$. ■

10.7.3 Law of large numbers for Cesaro sums of expected population structures

We will now establish a LLN for the expected population structures in an NHMS. In this respect denote by $\delta = T(0)/T$ and we call it the *terminal population ratio.* Then $(1-\delta) = (T - T(0))/T$ which we will call the *relative terminal increment ratio.* Note that the terminal population ratio δ is characteristic of the design of an NHMS since it is a function of the initial number of memberships $T(0)$ and its terminal target T set by the targets of its management. Hence δ is a significant and important characteristic of an NHMS. Consider now the convex combination of the initial relative population structure $\mathbf{q}(0)$ and the final input probabilities vector $\mathbf{p}_0$ which is the limit of the input probability process $\{\mathbf{p}_0(t)\}_{t=0}^{\infty}$:

$$\delta \mathbf{q}(0) + (1-\delta)\mathbf{p}_0.$$

We call this the *convex limit population policy.* We baptize this expression in this way due to the fact that in most applications $\mathbf{p}_0$ is the most appropriate variable for control (see Bartholomew (1981), Vassiliou and Georgiou (1990) and Chapter 9). It is interesting to note that (10.92) could be written as

$$\lim_{t\to\infty} \frac{1}{t} \sum_{n=1}^{t} \mathbb{E}[\mathbf{q}(0,t)] = [\delta \mathbf{q}(0) + (1-\delta)\mathbf{p}_0]\, \mathbf{\Pi}_q^*. \tag{10.93}$$

We are now in a position to establish the following Law of Large numbers for Cesaro sums of expected population structures for an NHMS with arbitrary transition probability matrices for the inherent Markov chain.

Theorem 355 *(Vassiliou (2022a)). Let a complete probability space $(\Omega, \mathcal{F}, \mathbb{P})$ and an NHMS as defined in chapter 5 with arbitrary transition probability matrices for the inherent Markov chain. Let also $\{\mathbf{Q}(t)\}_{t=0}^{\infty}$ the transition probability sequence of the inherent Markov chain is such that all matrices have the same incidence matrix which with no loss of generality we may assume that they have the canonical form*

$$\mathbf{Q}(t) = \begin{pmatrix} \mathbf{Q}_1(t) & \mathbf{0} & \dots & \mathbf{0} & \mathbf{0} \\ \mathbf{0} & \mathbf{Q}_2(t) & \dots & \mathbf{0} & 0 \\ \dots & \dots & \dots & \dots & \dots \\ \dots & \dots & \dots & \mathbf{Q}_r(t) & \mathbf{0} \\ \mathbf{T}_{Q1}(t) & \mathbf{T}_{Q2}(t) & \dots & \mathbf{T}_{Qr}(t) & \mathbf{R} \end{pmatrix} \quad \text{for } t = 0, 1, ... \tag{10.94}$$

that is, the matrices $\mathbf{Q}(t)$ for $t = 0, 1, ...$ have r essential classes which in general are periodic with periods respectively d_i. Assume asymptotic homogeneity, that is,

$$\lim_{t\to\infty} \|\mathbf{Q}(t) - \mathbf{Q}\| = 0 \text{ and } \|\mathbf{p}_0(t) - \mathbf{p}_0\| = 0. \tag{10.95}$$

with $\mathbf{Q}$ having the same incidence matrix as $\mathbf{Q}(t)$, that is, the canonical form of $\mathbf{Q}$ is

$$\mathbf{Q} = \begin{pmatrix} \mathbf{Q}_1 & \mathbf{0} & \dots & \mathbf{0} & \mathbf{0} \\ \mathbf{0} & \mathbf{Q}_2 & \dots & \mathbf{0} & 0 \\ \dots & \dots & \dots & \dots & \dots \\ \dots & \dots & \dots & \mathbf{Q}_r & \mathbf{0} \\ \mathbf{T}_{Q1} & \mathbf{T}_{Q2} & \dots & \mathbf{T}_{Qr} & \mathbf{R} \end{pmatrix};$$

also let that $\lim_{t\to\infty} T(t) = T$ and $\Delta T(t) \geq 0$ for every t. Then

$$\boldsymbol{\nu}(t) \overset{L^2}{\to} [\delta \mathbf{q}(0) + (1-\delta)\mathbf{p}_0]\, \mathbf{\Pi}_q^*, \tag{10.96}$$

with

$$\mathbf{\Pi}_q^* = \begin{pmatrix} \mathbf{\Pi}_{q1}^* & \mathbf{0} & \dots & \mathbf{0} & \mathbf{0} \\ \mathbf{0} & \mathbf{\Pi}_{q2}^* & \dots & \mathbf{0} & \mathbf{0} \\ \dots & \dots & \dots & \dots & \dots \\ \mathbf{0} & \mathbf{0} & \dots & \mathbf{\Pi}_{qr}^* & \mathbf{0} \\ \mathbf{T}_{Q1}^* & \mathbf{T}_{Q2}^* & \dots & \mathbf{T}_{Qr}^* & \mathbf{0} \end{pmatrix} \tag{10.97}$$

where

$$\mathbf{\Pi}_{qi}^* = \frac{1}{d_i} \sum_{\nu_i=0}^{d_i-1} \mathbf{Q}_i^* \mathbf{Q}_i^{\nu_i} \quad \textit{with} \;\; \mathbf{Q}_i^* = \lim_{t\to\infty} \mathbf{Q}_i^t, \tag{10.98}$$

and

$$\mathbf{T}_{Qi}^* = \left[\mathbf{I} - \mathbf{R}^{d_i}\right]^{-1} \sum_{l=0}^{d_i-1} \mathbf{Q}_i^l \mathbf{T}_{Qi} \mathbf{\Pi}_{qi}^*. \tag{10.99}$$

Proof. In order to show (10.96), it suffices to show that

$$\lim_{t\to\infty} \mathbb{E}\left[\left(\boldsymbol{\nu}(t) - [\delta \mathbf{q}(0) + (1-\delta)\mathbf{p}_0]\mathbf{\Pi}_q^*\right)^2\right] = 0, \tag{10.100}$$

or equivalently

$$\lim_{t\to\infty} \mathbb{E}\left[\left(\boldsymbol{\nu}_j(t) - [\delta \mathbf{q}(0) + (1-\delta)\mathbf{p}_0]\mathbf{\Pi}_q^* \mathbf{e}_j^\top\right)^2\right] = 0, \text{ for } j = 1, 2, ..., k. \tag{10.101}$$

Now we have that

$$\mathbb{E}[y_j(t)] = \sum_{n=1}^t \mathbb{E}[u_j(t)] = \sum_{n=1}^t \mathbb{E}[\mathbb{E}[u_j(t) \mid \mathbf{q}(0)]] \tag{10.102}$$

$$= \sum_{n=1}^t \mathbb{E}[\mathbb{P}(X_n = j \mid \mathbf{q}(0))].$$

Also

$$\mathbb{E}[y_j(t)\, y_l(t)] = \sum_{n=1}^t \sum_{l=1}^t \mathbb{E}[u_j(t)\, u_l(t)] = \sum_{n=1}^t \sum_{l=1}^t \mathbb{E}[\mathbb{E}[u_j(t)\, u_l(t) \mid \mathbf{q}(0)]]$$

$$= \sum_{n=1}^t \sum_{l=1}^t \mathbb{E}[\mathbb{P}[X_n = j, X_l = j \mid \mathbf{q}(0)]]. \tag{10.103}$$

We now have that

$$\mathbb{P}[X_n = j, X_l = j \mid \mathbf{q}(0)] = \mathbb{P}[X_{n\wedge l} = j, X_{n\vee l} = j, \mathbf{q}(0)]$$

$$\times \mathbb{P}[X_{n\vee l} = j \mid \mathbf{q}(0)]. \tag{10.104}$$

From (10.101) we get that

$$\lim_{t\to\infty} \mathbb{E}\left[\left(\boldsymbol{\nu}_j(t) - [\delta \mathbf{q}(0) + (1-\delta)\mathbf{p}_0]\mathbf{\Pi}_q^* \mathbf{e}_j^\top\right)^2\right] =$$

$$= \lim_{t\to\infty} \mathbb{E}\left[\frac{y_j(t)}{t} - [\delta \mathbf{q}(0) + (1-\delta)\mathbf{p}_0]\mathbf{\Pi}_q^* \mathbf{e}_j^\top\right]^2$$

$$\lim_{t\to\infty} \frac{1}{t^2} \mathbb{E}\left[\sum_{n=1}^t \left(u_j(n) - [\delta \mathbf{q}(0) + (1-\delta)\mathbf{p}_0]\mathbf{\Pi}_q^* \mathbf{e}_j^\top\right)\right]^2 = \mathbf{A}. \tag{10.105}$$

From which we have that

$$\mathbf{A} = \lim_{t\to\infty} \frac{1}{t^2} \sum_{n=1}^{t} \sum_{l=1}^{t} \left[\left[\delta \mathbf{q}(0) + (1-\delta) \mathbf{p}_0 \right] \mathbf{\Pi}_q^* \mathbf{e}_j^\top \right]^2 + \tag{10.106}$$

$$\lim_{t\to\infty} -\frac{1}{t^2} \sum_{n=1}^{t} \sum_{l=1}^{t} \mathbb{E}\left[u_j(n)\right] \left[\delta \mathbf{q}(0) + (1-\delta) \mathbf{p}_0 \right] \mathbf{\Pi}_q^* \mathbf{e}_j^\top + \tag{10.107}$$

$$\lim_{t\to\infty} -\frac{1}{t^2} \sum_{n=1}^{t} \sum_{l=1}^{t} \mathbb{E}\left[u_j(l)\right] \left[\delta \mathbf{q}(0) + (1-\delta) \mathbf{p}_0 \right] \mathbf{\Pi}_q^* \mathbf{e}_j^\top + \tag{10.108}$$

$$\lim_{t\to\infty} \frac{1}{t^2} \sum_{n=1}^{t} \sum_{l=1}^{t} \mathbb{E}\left[u_j(n)\, u_j(l)\right] = \mathbf{A}_1 + \mathbf{A}_2 + \mathbf{A}_3 + \mathbf{A}_4. \tag{10.109}$$

Relation (10.106) using becomes

$$\mathbf{A}_1 = \left[\left[\delta \mathbf{q}(0) + (1-\delta) \mathbf{p}_0 \right] \mathbf{\Pi}_q^* \mathbf{e}_j^\top \right]^2. \tag{10.110}$$

Now, from relation (10.107) using relation (10.102) we get

$$\mathbf{A}_2 = \lim_{t\to\infty} -\frac{1}{t^2} \sum_{n=1}^{t} \mathbb{E}\left[\mathbb{P}\left(X_n = j \mid \mathbf{q}(0)\right)\right] \left[\delta \mathbf{q}(0) + (1-\delta) \mathbf{p}_0 \right] \mathbf{\Pi}_q^* \mathbf{e}_j^\top$$

$$= \lim_{t\to\infty} - \left[\delta \mathbf{q}(0) + (1-\delta) \mathbf{p}_0 \right] \mathbf{\Pi}_q^* \mathbf{e}_j^\top \frac{1}{t} \sum_{n=1}^{t} \mathbb{E}\left[q_j(0,n)\right]$$

$$= -\left[\left[\delta \mathbf{q}(0) + (1-\delta) \mathbf{p}_0 \right] \mathbf{\Pi}_q^* \mathbf{e}_j^\top \right]^2. \tag{10.111}$$

Similarly from (10.108) we get

$$\mathbf{A}_3 = -\left[\left[\delta \mathbf{q}(0) + (1-\delta) \mathbf{p}_0 \right] \mathbf{\Pi}_q^* \mathbf{e}_j^\top \right]^2. \tag{10.112}$$

Finally, from (10.109) using (10.103) and (10.104), we have

$$\mathbf{A}_4 = \lim_{t\to\infty} \frac{1}{t^2} \sum_{n=1}^{t} \sum_{l=1}^{t} \mathbb{E}\left[\mathbb{P}\left[X_n = j, X_l = j \mid \mathbf{q}(0)\right]\right] =$$

$$\lim_{t\to\infty} \frac{1}{t^2} \sum_{n=1}^{t} \sum_{l=1}^{t} \mathbb{E}\left[\mathbb{P}\left[X_{n\wedge l} = j \mid q_j(0, n \vee l) = 1\right]\right]$$

$$\times \mathbb{E}\left[\mathbb{P}\left[X_{n\vee l} = j \mid \mathbf{q}(0)\right]\right] =$$

$$= \left(\text{since } q^j(s,t) = 0 \text{ for } s > t\right)$$

$$= \lim_{t\to\infty} \frac{1}{t} \sum_{n=1}^{t} \mathbb{E}\left[q^{(j)}(n,l)\right] \frac{1}{t} \sum_{l=n}^{t} \mathbb{E}\left[q(n,l)\right]$$

$$= (\text{by Theorem 253}) = \left[\left[\delta \mathbf{q}(0) + (1-\delta) \mathbf{p}_0 \right] \mathbf{\Pi}_q^* \mathbf{e}_j^\top \right]^2. \tag{10.113}$$

Hence, from equations $(10.100), (10.101), \ldots, (10.113)$ the proof is competed. ■

10.8 An illustrative example

We will use a hospital system to apply and illustrate the previous results. Homogeneous Markov systems have played an important role in the planning and performance modeling of hospital systems and also for the better use of resources, and improved adherence to targets (see in this respect Jones et al. (2019), Fady and McClean (2000, 2005), Marshall and McClean (2003), McClean et al. (2007), McClean et al. (2011), McClean et al. (2014)). The Hospital system has three large sections. One section contains the stroke patients and the debilitating nature of a stroke and complex makeup of the disease is demanding stochastic models for bed occupancy analysis, capacity planning which also can be seen in previous studies referred above. Another section is the one which contains the geriatric patients. Geriatric patients who occupy a bed in the geriatric section if they get a stroke they remain and treated in this section without changing environment. The third section contains the reception, external emergency patients, and patients with injuries from accidents that need immediate care and diagnosis before getting discharged or transferred to other specialized hospitals or to different sections of the same hospitals.

The patients in the geriatric section are categorized into four states which are called hospital pathways. These four pathways are "Surgical care", "Acute care", the "Rehabilitative", the "Long Stay". These will be numbered as states from 1 to 4. Also, geriatric patients may be thought as progressing through the various stages and eventually rehabilitated and discharged.

The patients in the stroke section are categorized into four states which are also called hospital pathways. The four states are "Hemographic" for those patients with bleed in the brain, "Cerebral Infraction" for those patients with clot in the brain, "Transient Ischemic attack" (TIA) for those patients with minor stroke, and finally "Unspecified" for those with undetermined type of stroke. These will be numbered as states from 5 to 8. All patients are eventually being discharged from the Hospital system to a nursing home or to their unusual residence or else.

The reception and external emergency patients are categorized into the "Emergency state" and the "Diagnostic state". These will be numbered as states 9 and 10. From both states they progress to each other, to the other sections of the hospital and are being transferred to more specialized hospitals.

The hospital system has $T(0) = 500$ beds, and is planning to have $T(1) = 580$, $T(2) = 660$, $T(3) = 725 = T(t)$ for $t = 3, 4, \ldots$. We have already distinguished 10 pathways in the Hospital system and let that the first 4 are in the geriatric section, the other 4 to the stroke section and finally the last two to the reception and emergency section. The data needed to estimate the transition probability matrices of the inherent Markov chain are easily accessible from administration in the Hospital system. The estimation method is the well known maximum likelihood estimators. Now assume that the initial relative population structure of the beds in the various states are:

$$\mathbf{q}(0) = \begin{pmatrix} 0.08 & 0.16 & 0.2 & 0.06 & 0.04 & 0.12 & 0.17 & 0.03 & 0.06 & 0.08 \end{pmatrix}$$

Let a typical set of transition probability matrices for the inherent non-homogeneous Markov system, which they satisfy due to the nature of the transitions the assumption that they all have the same incidence matrix. Let them be the following:

$$\mathbf{Q}(1) = \begin{pmatrix} 0 & 0 & 0.35 & 0.65 & 0 & 0 & 0 & 0 & 0 & 0 \\ 0 & 0 & 0.84 & 0.16 & 0 & 0 & 0 & 0 & 0 & 0 \\ 0.33 & 0.67 & 0 & 0 & 0 & 0 & 0 & 0 & 0 & 0 \\ 0.32 & 0.68 & 0 & 0 & 0 & 0 & 0 & 0 & 0 & 0 \\ 0 & 0 & 0 & 0 & 0 & 0 & 0.74 & 0.26 & 0 & 0 \\ 0 & 0 & 0 & 0 & 0 & 0 & 0.43 & 0.67 & 0 & 0 \\ 0 & 0 & 0 & 0 & 0.18 & 0.82 & 0 & 0 & 0 & 0 \\ 0 & 0 & 0 & 0 & 0.66 & 0.34 & 0 & 0 & 0 & 0 \\ 0.14 & 0.06 & 0.08 & 0 & 0.12 & 0.14 & 0.16 & 0 & 0.18 & 0.12 \\ 0.12 & 0.07 & 0.06 & 0 & 0.18 & 0.16 & 0.11 & 0.09 & 0 & 0.21 \end{pmatrix},$$

$$\mathbf{Q}(2) = \begin{pmatrix} 0 & 0 & 0.3 & 0.7 & 0 & 0 & 0 & 0 & 0 & 0 \\ 0 & 0 & 0.8 & 0.2 & 0 & 0 & 0 & 0 & 0 & 0 \\ 0.25 & 0.75 & 0 & 0 & 0 & 0 & 0 & 0 & 0 & 0 \\ 0.4 & 0.6 & 0 & 0 & 0 & 0 & 0 & 0 & 0 & 0 \\ 0 & 0 & 0 & 0 & 0 & 0 & 0.8 & 0.2 & 0 & 0 \\ 0 & 0 & 0 & 0 & 0 & 0 & 0.4 & 0.6 & 0 & 0 \\ 0 & 0 & 0 & 0 & 0.1 & 0.9 & 0 & 0 & 0 & 0 \\ 0 & 0 & 0 & 0 & 0.6 & 0.4 & 0 & 0 & 0 & 0 \\ 0.15 & 0.05 & 0.03 & 0 & 0.17 & 0.1 & 0.2 & 0 & 0.2 & 0.1 \\ 0.10 & 0.07 & 0.03 & 0 & 0.20 & 0.2 & 0.1 & 0.1 & 0 & 0.2 \end{pmatrix},$$

$$\mathbf{Q}(3) = \begin{pmatrix} 0 & 0 & 0.4 & 0.6 & 0 & 0 & 0 & 0 & 0 & 0 \\ 0 & 0 & 0.9 & 0.1 & 0 & 0 & 0 & 0 & 0 & 0 \\ 0.35 & 0.65 & 0 & 0 & 0 & 0 & 0 & 0 & 0 & 0 \\ 0.45 & 0.55 & 0 & 0 & 0 & 0 & 0 & 0 & 0 & 0 \\ 0 & 0 & 0 & 0 & 0 & 0 & 0.7 & 0.3 & 0 & 0 \\ 0 & 0 & 0 & 0 & 0 & 0 & 0.48 & 0.52 & 0 & 0 \\ 0 & 0 & 0 & 0 & 0.19 & 0.81 & 0 & 0 & 0 & 0 \\ 0 & 0 & 0 & 0 & 0.68 & 0.32 & 0 & 0 & 0 & 0 \\ 0.15 & 0.05 & 0.03 & 0 & 0.17 & 0.1 & 0.2 & 0 & 0.2 & 0.1 \\ 0.06 & 0.09 & 0.01 & 0 & 0.24 & 0.16 & 0.1 & 0.1 & 0 & 0.24 \end{pmatrix}$$

$$\mathbf{Q}(t) = \mathbf{Q} = \begin{pmatrix} 0 & 0 & 0.33 & 0.67 & 0 & 0 & 0 & 0 & 0 & 0 \\ 0 & 0 & 0.69 & 0.31 & 0 & 0 & 0 & 0 & 0 & 0 \\ 0.32 & 0.68 & 0 & 0 & 0 & 0 & 0 & 0 & 0 & 0 \\ 0.49 & 0.51 & 0 & 0 & 0 & 0 & 0 & 0 & 0 & 0 \\ 0 & 0 & 0 & 0 & 0 & 0 & 0.74 & 0.26 & 0 & 0 \\ 0 & 0 & 0 & 0 & 0 & 0 & 0.47 & 0.53 & 0 & 0 \\ 0 & 0 & 0 & 0 & 0.14 & 0.86 & 0 & 0 & 0 & 0 \\ 0 & 0 & 0 & 0 & 0.48 & 0.52 & 0 & 0 & 0 & 0 \\ 0.16 & 0.04 & 0.07 & 0 & 0.13 & 0.14 & 0.16 & 0 & 0.22 & 0.08 \\ 0.11 & 0.08 & 0.02 & 0 & 0.18 & 0.22 & 0.12 & 0.08 & 0 & 0.19 \end{pmatrix},$$

for $t = 3, 4, \ldots$. Finally, the input in the system is coming into the reception department and let that the input probability process converge after a long time into

$$\mathbf{p}_0 = \begin{pmatrix} 0 & 0 & 0 & 0 & 0 & 0 & 0 & 0 & 0.4 & 0.6 \end{pmatrix}.$$

We now have that the terminal population ratio is equal with $\delta = T(0)/T = 500/725 = 0.69$. The relative terminal increment ratio is $1 - \delta = (T - T(0))/T = 0.31$. Therefore, the convex population policy will be

$$\delta \mathbf{q}(0) + (1 - \delta)\, \mathbf{p}_0 =$$

$$\begin{pmatrix} 0.06 & 0.11 & 0.14 & 0.04 & 0.028 & 0.08 & 0.12 & 0.02 & 0.17 & 0.24 \end{pmatrix}.$$

Now from Theorem 355 the condition that all matrices $\mathbf{Q}(t)$ have the same incidence matrix for all t is satisfied, also condition (10.95) holds and $\lim_{t\to\infty} T(t) = T = 725$ with $\Delta T(t) \geq 0$ for every t. The matrix $\mathbf{Q}_1$ a submatrix of $\mathbf{Q}$ with elements the probabilities of transition between states $1, 2, 3$ and 4 is periodic with period 2 and from (10.98) we get that

$$\mathbf{\Pi}_{q1}^{*} = \begin{pmatrix} 0.2 & 0.3 & 0.27 & 0.23 \\ 0.2 & 0.3 & 0.27 & 0.23 \\ 0.2 & 0.3 & 0.27 & 0.23 \\ 0.2 & 0.3 & 0.27 & 0.23 \end{pmatrix}.$$

Also the matrix $\mathbf{Q}_2$ a submatrix of $\mathbf{Q}$ with elements the probabilities of transition between the states $5, 6, 7$ and 8 is periodic with period 2 and from (10.98) we get that

$$\mathbf{\Pi}_{q2}^{*} = \begin{pmatrix} 0.15 & 0.35 & 0.28 & 0.22 \\ 0.15 & 0.35 & 0.28 & 0.22 \\ 0.15 & 0.35 & 0.28 & 0.22 \\ 0.15 & 0.35 & 0.28 & 0.22 \end{pmatrix}.$$

Now we are in a position to find the remaining submatrices of $\mathbf{\Pi}_q^*$ in (10.97) given in (10.99) and, therefore, we get

$$\mathbf{T}_{Q1}^{*} = \begin{pmatrix} 0.1 & 0.22 & 0.18 & 0.13 \\ 0.11 & 0.26 & 0.21 & 0.16 \end{pmatrix}, \ \mathbf{T}_{Q1}^{*} = \begin{pmatrix} 0.06 & 0.13 & 0.1 & 0.08 \\ 0.05 & 0.08 & 0.07 & 0.06 \end{pmatrix}.$$

Finally from (10.96), we get that

$$\boldsymbol{\nu}(t) \xrightarrow{L^2} \begin{pmatrix} 0.11 & 0.20 & 0.17 & 0.13 & 0.06 & 0.12 & 0.12 & 0.09 & 0 & 0 \end{pmatrix}.$$

The physical meaning of the above is that a membership of a patient remains in hospital pathway 1, in the long run in mean square, 0.11 of the time the hospital is in operation.

Chapter 11

THE S-NHMS IN CONTINUOUS TIME

11.1 Introductory notes

We now turn to the development of continuous time version of an NHMS as introduced in Chapter 5. The choice, in practice, between the discrete and continuous versions as a model of a stochastic process is partly a matter of realism and partly one of convenience. On grounds of realism, for example, one would usually want to model movements in a population between the various states in continuous time, but in practice the computational advantages of treating time as discrete as well as the mental process of the modeler have to all to often led to the choice of a discrete time process. Continuous time processes are often more amenable to mathematical analysis and this may count in their favor when realism calls for a discrete time process. In any case by presenting in the present chapter, the continuous time version of an NHMS and S-NHMS we will have both versions on our disposal and hence have a more complete version of the entire theory. The parts of the discrete time theory which we will not present their continuous time versions could easily be done using the present Chapter as a nucleus theory to build upon.

We will start by studying non-homogeneous Markov chains in continuous time and some useful results on their ergodicity. We will then proceed to study S-NHMS in continuous time and solving problems of asymptotic behavior of expected population structures, the set of all asymptotically attainable expected relative population structures when we control the limiting input probabilities. Apparently, the study of NHMS in continuous time is a special case of the above.

11.2 Asymptotic behavior of NHMP in continuous time

The theory of homogeneous Markov chains was developed almost in parallel with the relative discrete time one. There is a rich and interesting theory on the subject and a classic account of the theory of homogeneous Markov processes is the book Chung (1960). However, from the point of view of applications, it is rare that the random factors driving the system are constant in time. An example that almost all intuitively understands is a queueing system. If $X(t)$ describes the state at time t of some complex queueing system the evolution will depend on such random factors as the arrival rate of customers, which will usually vary with time (see Kingman (2009) for more details). In most early accounts of homogeneous Markov process comes the assumption, sometimes explicit but sometimes only implicit, that the conditional distribution of $X(t)$ given $X(s)$ depends only on $X(s)$ and the time difference $t-s$. Some authors use the terminology of *autonomous* in this case (see Kingman (2012) for details).

A first concise presentation of the Theory of non-homogeneous Markov process exists in Iosifescu (1980, section 8.9). In there apart from a rigorous foundation of the subject in the respective references one could find also the initial founders of the subject. The theory of non-homogeneous Markov processes has realized an interesting growth during the 70's and through the 80's starting with Goodman (1970). Strong ergodicity for continuous time

DOI: 10.1201/b23204-11

non-homogeneous Markov processes using mean visit times was studied by Scott et al. (1982). Important results on the strong ergodicity for continuous time non-homogeneous Markov processes using criteria on the functions of intensity transition matrices were provided by Johnson (1984), Johnson et al. (1988) and Yong (1976). We will make extensive use of these results in the present. The estimates of rate of convergence for non-homogeneous Markov processes were studied in a series of papers by Zeifman (1989), Zeifman et al. (1994), (2015), (2018) and Mitrophanov (2003). For Markov systems in continuous time results could be found in McClean (1976, 1978), Bartholomew (1982), Gerontidis (1990). Estimation of the transition intensities in NHMS in continuous time were provided by McClean et al. (1998) for various cases of missing data. More recent results on NHMS in continuous time could be found Dimitriou and Georgiou (2020) while the more recent review on the subject was given by Esquivel et al. (2021). In the section Research Notes of the present chapter more related work on Markov systems in continuous time are cited and discussed.

For a non-homogeneous Markov chain $\{X_t\}_{t=0}^{\infty}$ one of the basic important parameters as we have seen is the sequence of transition probability matrices $\{\mathbf{P}(t)\}_{t=0}^{\infty}$, where $\mathbf{P}(t)$ contains the transition probabilities at the unit interval $(t-1,t]$. Also we denoted by

$$p_{ij}(s,t) = \mathbb{P}\{X_t = j \mid X_s = i\},$$

and $\mathbf{P}(s,t) = \{p_{ij}(s,t)\}_{i,j\in\mathbb{S}}$. In the case of a non-homogeneous Markov process, where time is continuous, $\mathbf{P}(s,t)$ will still have the same meaning but now the time scale is not a unit and in order that the time be continuous we need the transition probabilities at least from the physical point of view to be able to represent changes, that is, transitions in infinitesimal time intervals. This fact led to the following formulation (Iosifescu (1980)) of transition probabilities in an infinitesimal interval of time:

$$p_{ij}(t,t+\delta t) = \mathbb{P}\{X_{t+\delta t} = j \mid X_t = i\} = q_{ij}(t)\,\delta t + o(\delta t) \text{ for } i \neq j, \tag{11.1}$$

where $o(\delta t)$ goes faster to zero than δt when $\delta t \to 0$. Also $q_{ij}(t)$ is a measurable function which represents the "*intensity*" of an "instantaneous" transition out of state i into state j at time t. In the case that $i = j$ then we define

$$p_{ii}(t,t+\delta t) = \mathbb{P}\{X_{t+\delta t} \neq i \mid X_t = i\} = -q_{ii}(t)\,\delta t + o(\delta t), \tag{11.2}$$

that is, $-q_{ii}(t)$ represents a measure of the "*intensity*" with which the process exits state i. We collect the intensities into an intensity matrix $\mathbf{Q}(t)$ which has the form

$$\mathbf{Q}(t) = \begin{pmatrix} -q_{11}(t) & q_{12}(t) & \dots & q_{1k}(t) \\ q_{21}(t) & -q_{22}(t) & \dots & q_{1k}(t) \\ \dots & \dots & \dots & \dots \\ q_{k1}(t) & q_{k2}(t) & \dots & -q_{kk}(t) \end{pmatrix}, \tag{11.3}$$

where

$$q_{ij}(t) \geq 0 \text{ and } \mathbf{Q}(t)\,\mathbf{1}^{\top} = 0, \tag{11.4}$$

with the assumption that

$$\int_0^t q_{ii}(s)\,ds < \infty, \text{ for all } t \geq 0 \text{ and } i \in \mathbb{S}. \tag{11.5}$$

Now, (11.1) and (11.2) could be equivalently be defined as:

$$q_{ii}(t) = \lim_{\delta t \to 0+} \frac{p_{ii}(t,t+\delta t)}{\delta t} \text{ and } q_{ij}(t) = \lim_{\delta t \to 0+} \frac{p_{ij}(t+\delta t)}{\delta t}, \tag{11.6}$$

for every $i,j \in \mathbb{S}$ and all $t \notin \mathbb{E}$, where $\mathbb{E} \subset [0,\infty)$ is a set of Lebesque measure zero. If $\mathbf{Q}(t)$ is continuous then $\mathbb{E}$ is the empty set. We also assume that the limits in (11.6) are uniform in i. Under these assumptions the following forward and backward Kolmogorov equations could be derived (see Feller (1968)):

$$\frac{\partial}{\partial t} p_{ij}(s,t) = \sum_{k \in \mathbb{S}} p_{ik}(s,t) q_{kj}(t), \quad \frac{\partial}{\partial s} p_{ij}(s,t) = -\sum_{k \in \mathbb{S}} q_{ik}(t) p_{kj}(s,t), \tag{11.7}$$

for any $i,j \in \mathbb{S}$. Equations (11.7) in matrix notation could be written as

$$\frac{\partial}{\partial t}\mathbf{P}(s,t) = \mathbf{P}(s,t)\mathbf{Q}(t) \text{ and } \frac{\partial}{\partial s}\mathbf{P}(s,t) = -\mathbf{Q}(s)\mathbf{P}(s,t). \tag{11.8}$$

Equivalently (11.8) could be written as the integral equations

$$\mathbf{P}(s,t) = \mathbf{I} + \int_s^t \mathbf{P}(s,x)\mathbf{Q}(x)\,dx \text{ and } \mathbf{P}(s,t) = \mathbf{I} + \int_s^t \mathbf{Q}(x)\mathbf{P}(x,t)\,dx. \tag{11.9}$$

Hence, $\mathbf{P}(s,.)$ is absolutely continuous on $[s,\infty)$ for all $s \geq 0$ and $\mathbf{P}(.,t)$ is absolutely continuous on $[0,t]$ for all $t \geq 0$. The transition matrix $\mathbf{P}(s,t)$ could be expressed in terms of $\mathbf{Q}(t)$ (see Iosifescu (1980, p. 264)) as follows for all $0 \leq s \leq t$:

$$\begin{aligned}\mathbf{P}(s,t) &= \mathbf{I} + \int_s^t \mathbf{Q}(x)\,dx \\ &+ \sum_{n\geq 2}\int_s^t ds_1 \int_{s_1}^t ds_2 ... \int_{s_{n-1}}^t \mathbf{Q}(s_1)\mathbf{Q}(s_2)...\mathbf{Q}(s_n)\,ds_n \\ &= \mathbf{I} + \int_s^t \mathbf{Q}(x)\,dx \\ &+ \sum_{n\geq 2}\int_s^t dt_1 \int_s^{t_1} dt_2 ... \int_s^{t_{n-1}} \mathbf{Q}(t_n)\mathbf{Q}(t_{n-1})...\mathbf{Q}(t_1)\,dt_n\end{aligned} \tag{11.10}$$

As it is intuitively expected we may use the transition probability matrices $\mathbf{P}(s,t)$ to define irreducibility, weak ergodicity, and strong ergodicity. Periodicity has no meaning for Markov processes. We now provide these definitions in terms of the transition probabilities $\mathbf{P}(s,t)$:

Definition 356 *A Markov process* $\{X_t\}_{t=0}^{\infty}$ *is irreducible if for every* $i,j \in \mathbb{S}$ *and* $s \geq 0$ *there exists* $t_1 > s$ *and* $t_2 > s$ *such that* $p_{ij}(s,t_1) > 0$ *and* $p_{ji}(s,t_2) > 0$.

Definition 357 *A Markov process* $\{X_t\}_{t=0}^{\infty}$ *is weakly ergodic if for every* $s \geq 0$, $\lim_{t\to\infty} \delta(\mathbf{P}(s,t)) = 0$.

Definition 358 *A Markov process* $\{X_t\}_{t=0}^{\infty}$ *is strongly ergodic if there exists a row-constant matrix* $\mathbf{\Pi}$ *such that, for all* $s \geq 0$

$$\lim_{t\to\infty} \|\mathbf{P}(s,t) - \mathbf{\Pi}\| = 0. \tag{11.11}$$

As the reader have by now may recognized the generator of a non-homogeneous Markov process is the sequence of intensity matrices $\{\mathbf{Q}(t)\}_{t=0}^{\infty}$. This is so in the sense that the transition probability matrix $\mathbf{P}$ could be seen in a sense as the generator of a homogeneous Markov chain, and the sequence of transition probability matrices $\{\mathbf{P}(t)\}_{t=1}^{\infty}$ as the generator of a non-homogeneous Markov chain. Hence, our goal now will be to find conditions for strong ergodicity for a non-homogeneous Markov process based on the convergence of the sequence of intensity matrices $\{\mathbf{Q}(t)\}_{t=1}^{\infty}$. We start with the following Lemma from Johnson and Isaacson(1988) or Johnson (1984):

Lemma 359 *Let a complete probability space $(\Omega, \mathcal{F}, \mathbb{P})$ and a non-homogeneous Markov process $\{X_t\}_{t=0}^{\infty}$ with sequence of intensity matrices $\{\mathbf{Q}(t)\}_{t=0}^{\infty}$ which is such that $\sup_{t\geq 0}\|\mathbf{Q}(t)\| \leq c$. Let also a homogeneous Markov process $\left\{\hat{X}_t\right\}_{t=0}^{\infty}$ with intensity matrix $\mathbf{Q}$ such that $\|\mathbf{Q}\| \leq c$. Then*

$$\|\mathbf{Q}(s_1)\mathbf{Q}(s_2)...\mathbf{Q}(s_n) - \mathbf{Q}^n\| \leq c^{n-1}\sum_{i=1}^{n}\|\mathbf{Q}(s_i) - \mathbf{Q}\|. \tag{11.12}$$

Let $\mathbf{P}(s,t)$ be the transition probability matrices for the Markov process $\{X_t\}_{t=0}^{\infty}$ and $\mathbf{P}(t)$ the transition probability matrices for $\left\{\hat{X}_t\right\}_{t=0}^{\infty}$ then

$$\|\mathbf{P}(t,t+m) - \mathbf{P}(m)\| \leq \int_t^{t+m}\|\mathbf{Q}(s_1) - \mathbf{Q}\|\,ds_1 +$$

$$\sum_{n\geq 2}\int_t^{t+m}ds_1\int_{s_1}^{t+m}ds_2...\int_{s_{n-1}}^{t+m}c^{n-1}\sum_{i=1}^{n}\|\mathbf{Q}(s_i) - \mathbf{Q}\|\,ds_n. \tag{11.13}$$

Proof. We have that

$$\|\mathbf{Q}(s_1)\mathbf{Q}(s_2)...\mathbf{Q}(s_n) - \mathbf{Q}^n\| \leq$$
$$\|\mathbf{Q}(s_1)\mathbf{Q}(s_2)...\mathbf{Q}(s_n) - \mathbf{Q}(s_1)\mathbf{Q}(s_2)...\mathbf{Q}(s_{n-1})\mathbf{Q}\| +$$
$$\left\|\mathbf{Q}(s_1)\mathbf{Q}(s_2)...\mathbf{Q}(s_{n-1})\mathbf{Q} - \mathbf{Q}(s_1)\mathbf{Q}(s_2)...\mathbf{Q}(s_{n-2})\mathbf{Q}^2\right\|$$
$$+... + \left\|\mathbf{Q}(s_1)\mathbf{Q}^{n-1} - \mathbf{Q}^n\right\|$$
$$\leq \|\mathbf{Q}(s_1)\|\,\|\mathbf{Q}(s_2)\|...\|\mathbf{Q}(s_{n-1})\|\,\|\mathbf{Q}(s_n) - \mathbf{Q}\| +$$
$$\|\mathbf{Q}(s_1)\|\,\|\mathbf{Q}(s_2)\|...\|\mathbf{Q}(s_{n-2})\|\,\left\|\mathbf{Q}(s_{n-1})\mathbf{Q} - \mathbf{Q}^2\right\|$$
$$+... + \|\mathbf{Q}(s_1) - \mathbf{Q}\|\,\left\|\mathbf{Q}^{n-1}\right\| \leq c^{n-1}\sum_{i=1}^{n}\|\mathbf{Q}(s_i) - \mathbf{Q}\|.$$

Now, we proceed to prove (11.13). Using (11.10) we get that

$$\|\mathbf{P}(t,t+m) - \mathbf{P}(m)\| = \|\mathbf{I} + \int_t^{t+m}\mathbf{Q}(x)\,dx$$

$$+\sum_{n\geq 2}\int_t^{t+m}ds_1\int_{s_1}^{t+m}ds_2...\int_{s_{n-1}}^{t+m}\mathbf{Q}(s_1)\mathbf{Q}(s_2)...\mathbf{Q}(s_n)\,ds_n$$

$$-\mathbf{I} - \int_t^{t+m}\mathbf{Q}dx - \sum_{n\geq 2}\int_t^{t+m}ds_1\int_{s_1}^{t+m}ds_2...\int_{s_{n-1}}^{t+m}\mathbf{Q}^n ds_n\|$$

$$\leq \int_t^{t+m}\|\mathbf{Q}(x) - \mathbf{Q}\|\,dx + \sum_{n\geq 2}\int_t^{t+m}ds_1\int_{s_1}^{t+m}ds_2$$

$$...\int_{s_{n-1}}^{t+m}c^{n-1}\sum_{i=1}^{n}\|\mathbf{Q}(s_1) - \mathbf{Q}\|\,ds_n.$$

■

We will now in the following Lemma improve on the bound (11.13) for the case when the non-homogeneous Markov chain $\{X_t\}_{t=0}^{\infty}$ converges to the homogeneous Markov chain $\left\{\hat{X}_t\right\}_{t=0}^{\infty}$, in the sense that the sequence of intensity matrices $\{\mathbf{Q}(t)\}_{t=0}^{\infty}$ converge to $\mathbf{Q}$.

Lemma 360 *(Johnson and Isaacson (1988)). Let a complete probability space* $(\Omega, \mathcal{F}, \mathbb{P})$ *and a non-homogeneous Markov process* $\{X_t\}_{t=0}^{\infty}$ *with sequence of intensity matrices* $\{\mathbf{Q}(t)\}_{t=0}^{\infty}$ *which is such that* $\sup_{t\geq 0}\|\mathbf{Q}(t)\| \leq c$. *Let also a homogeneous Markov process* $\left\{\hat{X}_t\right\}_{t=0}^{\infty}$ *with intensity matrix* $\mathbf{Q}$ *such that* $\|\mathbf{Q}\| \leq c$. *If* $\lim_{t\to\infty}\|\mathbf{Q}(t)-\mathbf{Q}\| = 0$, *then for any* $m \geq 0$ *and* $\delta > 0$ *there exists a* t_0 *such that for* $t \geq t_0$ *the difference in the transition matrices satisfies*

$$\|\mathbf{P}(t, t+m) - \mathbf{P}(m)\| \leq \delta m \exp(cm). \tag{11.14}$$

Proof. From the conditions of the Theorem, we get that there exists a $\delta > 0$ such that $\|\mathbf{Q}(t) - \mathbf{Q}\| < \delta$ for $t \geq t_0$. From (11.13) we get

$$\|\mathbf{P}(t, t+m) - \mathbf{P}(m)\| \leq \sum_{n=1}^{\infty} \int_t^{t+m} ds_1 \int_{s_1}^{t+m} ds_2 ...$$

$$\int_{s_{n-1}}^{t+m} c^{n-1} \sum_{i=1}^{n} \|\mathbf{Q}(s_1) - \mathbf{Q}\| \, ds_n \leq \sum_{n=1}^{\infty} c^{n-1} n\delta \frac{\left(\int_t^{t+m} ds\right)^n}{n!}$$

$$= \delta m \sum_{n=1}^{\infty} \frac{c^{n-1} m^{n-1}}{(n-1)!} = \delta m \exp(cm).$$

■

We are in a position to prove a basic theorem concerning strong ergodicity for a non-homogeneous Markov chain based on its sequence of intensity matrices.

Theorem 361 *(Johnson (1984), Johnson and Isaacson (1988)). Let a complete probability space* $(\Omega, \mathcal{F}, \mathbb{P})$ *and a non-homogeneous Markov process* $\{X_t\}_{t=0}^{\infty}$ *with sequence of intensity matrices* $\{\mathbf{Q}(t)\}_{t=0}^{\infty}$ *which is such that* $\sup_{t\geq 0}\|\mathbf{Q}(t)\| \leq c$. *Let also a homogeneous Markov process* $\left\{\hat{X}_t\right\}_{t=0}^{\infty}$ *with intensity matrix* $\mathbf{Q}$ *such that* $\|\mathbf{Q}\| \leq c$ *and which is strongly ergodic. If* $\lim_{t\to\infty}\|\mathbf{Q}(t)-\mathbf{Q}\| = 0$ *then if* $\mathbf{\Pi}$ *is the stable stochastic matrix, the limit of* $\left\{\hat{X}_t\right\}_{t=0}^{\infty}$ *then* $\{X_t\}_{t=0}^{\infty}$ *is also strongly ergodic with limit* $\mathbf{\Pi}$.

Proof. We have that

$$\|\mathbf{P}(s,t) - \mathbf{\Pi}\| \leq \|\mathbf{P}(s, t-r)\mathbf{P}(t-r, t) - \mathbf{P}(s, t-r)\mathbf{P}(r)\|$$
$$+ \|\mathbf{P}(s, t-r)\mathbf{P}(r) - \mathbf{P}(s, t-r)\mathbf{\Pi}\|$$
$$\leq \|\mathbf{P}(s, t-r)\| \, \|\mathbf{P}(t-r, t) - \mathbf{P}(r)\| + \|\mathbf{P}(s, t-r)\| \, \|\mathbf{P}(r) - \mathbf{\Pi}\|. \tag{11.15}$$

Since $\left\{\hat{X}_t\right\}_{t=0}^{\infty}$ is strongly ergodic there exists an m such that

$$\|\mathbf{P}(t) - \mathbf{\Pi}\| \leq \epsilon/2 \ \text{ for } t \geq m. \tag{11.16}$$

Now, in Lemma 360 we can always choose t large enough such that $\delta = (\epsilon/2m)\exp(-mc)$ thus from (11.15) and (11.16) we get that

$$\|\mathbf{P}(s,t) - \mathbf{\Pi}\| \leq (\epsilon/2m)\exp(-mc)\, m\exp(mc) + \epsilon/2 = \epsilon.$$

■

It is well known that for a homogeneous Markov process, with intensity matrix $\mathbf{Q}$ and transition matrix $\mathbf{P}(t)$, which is strongly ergotic, then the rate of convergence with which $\mathbf{P}(t)$ converges to a stable stochastic matrix is exponential. Logically, this fact creates the intuition that possibly for a non-homogeneous Markov process with sequence of intensity matrices $\mathbf{Q}(t)$ the rate at which the transition probability matrices converge to a stable stochastic matrix is also exponential. The answer to this is negative, since we need one more condition for this to be true, and that is $\lim_{t\to\infty}\|\mathbf{Q}(t)-\mathbf{Q}\|=0$ with an exponential rate of convergence. This result is stated formally in the following theorem which could be proved following the analogous steps as in the discrete time case where we proved geometric rate of convergence. For more details one could look at Johnson (1984).

Theorem 362 *Let a complete probability space $(\Omega,\mathcal{F},\mathbb{P})$ and a non-homogeneous Markov process $\{X_t\}_{t=0}^{\infty}$ with sequence of intensity matrices $\{\mathbf{Q}(t)\}_{t=0}^{\infty}$ which is strongly ergodic. Let also a homogeneous Markov process $\left\{\hat{X}_t\right\}_{t=0}^{\infty}$ with intensity matrix $\mathbf{Q}$ which is strongly ergodic. Let $g:\mathbb{R}^+\to\mathbb{R}^+$ be a monotonically increasing function. If $\lim_{t\to\infty}g(2t)\|\mathbf{Q}(t)-\mathbf{Q}\|=0$ then*

$$\lim_{t\to\infty}\sup_{s\geq 0}\{\min(\exp(\lambda t)), g(t)\|\mathbf{P}(s,t)-\mathbf{\Pi}\|\}=0, \tag{11.17}$$

where $0<\lambda<\beta/2$ and $\beta>0$ is the constant parameter of the exponential rate of convergence at which $\left\{\hat{X}_t\right\}_{t=0}^{\infty}$ converges.

Remark 363 *At this point let us refresh the fact that for finite homogeneous, discrete or continuous Markov chains the concept of ergodicity, strong ergodicity and weak ergodicity coincide. For infinite chain the notions of ergodicity and strong ergodicity are separated.*

We will present an important result from Yong (1976). Let that $\mathbf{Q}$ is the intensity matrix of a homogeneous Markov process $\{X_t\}_{t=0}^{\infty}$ and let that $\sup_{i\in\mathbb{S}}\{|q_{ii}|\}<c<\infty$ and $b>c$ then define

$$\hat{\mathbf{P}}=\mathbf{I}+\frac{\mathbf{Q}}{b},$$

then $\hat{\mathbf{P}}$ generates a discrete Markov chain $\left\{\hat{X}_t\right\}_{t=0}^{\infty}$.

Theorem 364 *(Yong (1976)). Let a complete probability space $(\Omega,\mathcal{F},\mathbb{P})$ and a finite homogeneous Markov process $\{X_t\}_{t=0}^{\infty}$, then it is ergodic if and only if the Markov chain $\left\{\hat{X}_t\right\}_{t=0}^{\infty}$ generated by $\hat{\mathbf{P}}=\mathbf{I}+\mathbf{Q}/b$ is ergodic.*

11.3 The S-NHMS in continuous time

We will start by presenting the concept of a non-homogeneous Markov system in stochastic environment in continuous time (S-NHMSC). Let $\{T(t), t\geq 0\}$ a known continuous function of time or a realization of a known stochastic process denoting the total number of members in the system. Also let that among the states of the system at the infinitesimal time interval $[t,t+\delta t)$ the probabilities of members of the system to move from state i to state j are generated by the transition intensities $r_{ij}(t)$, i.e.

$$p_{ij}(t,t+\delta t)=r_{ij}(t)\,\delta t+o(\delta t),\ \text{for } i\neq j\in\mathbb{S}. \tag{11.18}$$

Also let that state $k+1$ represents members leaving the population and assume that $r_{i,k+1}(t)$ is the transition intensity for a member of the population in state i to leave in the time interval $[t,t+\delta t)$, i.e.

$$p_{ik+1}(t,t+\delta t)=r_{ik+1}(t)\,\delta t+o(\delta t), \text{for } i\neq j\in\mathbb{S} \text{ for } i\in\mathbb{S}. \tag{11.19}$$

The transition intensities $r_{jj}(t)$ are defined by

$$r_{jj}(t)=-\sum_{\substack{i=1\\ i\neq j}}^{k+1}r_{ji}(t)\quad\text{for } j\in\mathbb{S}. \tag{11.20}$$

Let $\mathbf{R}(t) = \{r_{ij}(t)\}_{i,j\in S}$ be the matrix of transition intensities at time t and $\mathbf{r}_{k+1}^{\top}(t) = [r_{1,k+1}(t), r_{2,k+1}(t), ..., r_{k,k+1}(t)]^{\top}$ be the vector of leaving intensities at time t. Now, let $p_{0i}(t, t+\delta t)$ be the probability of a new member to enter the population in state i given that it will enter the population in the time interval $[t, t+\delta t)$ and let $\mathbf{p}_0(t, t+\delta t) = [p_{01}(t, t+\delta t), p_{02}(t, t+\delta t), ..., p_{0k}(t, t+\delta t)]$. Define the following probabilities

$$\begin{aligned}\hat{p}_{ij}(t, t+\delta t) &= p_{ij}(t, t+\delta t) + p_{ik+1}(t, t+\delta t)\, p_{0j}(t, t+\delta t)\\ &= r_{ij}(t)\,\delta t + r_{i,k+1}(t)\,\delta t p_{0j}(t, t+\delta t) + o(\delta t). \qquad (11.21)\end{aligned}$$

Now, let

$$q_{ij}(t) = \lim_{\delta t\to 0}\frac{\hat{p}_{ij}(t, t+\delta t)}{\delta t} = r_{ij}(t) + r_{i,k+1}(t)\, p_0(t) \ \text{ for } i\neq j, \qquad (11.22)$$

to be the transition intensity of a membership to move to state j in the time interval $[t, t+\delta t)$ given that it was in state i at time t. Also let $\mathbf{Q}(t) = \{q_{ij}(t)\}_{i,j\in\mathbb{S}}$ the matrix of transition intensities of the memberships. Assume that $\mathbf{Q}(t)$ is measurable and that $\sup_{ij\in\mathbb{S}}\{|q_{ij}(t)|\}$ is integrable on every finite interval of t. We call the Markov process defined by the matrix of intensities $\{\mathbf{Q}(t), t\geq 0\}$ the *imbedded or inherent Markov process* of the S-NHMSC.

Assume now, that in the infinitesimal time interval $[t, t+\delta t)$ the system has the choice of selecting a transition intensity matrix from the pool

$$\mathcal{R}_I(t) = \{\mathbf{R}_1(t), \mathbf{R}_2(t), ..., \mathbf{R}_\nu(t)\}. \qquad (11.23)$$

Further assume that it makes its choice in a stochastic way and more specifically let that in the infinitesimal time interval $[t, t+\delta t)$ the probability of selecting an intensity matrix from the set $\mathcal{R}_I(t)$ is given by

$$\begin{aligned}c_{ij}(t, t+\delta t) &= \mathbb{P}\{\mathbf{R}(t+\delta t) = \mathbf{R}_j(t+\delta t) \mid \mathbf{R}(t) = \mathbf{R}_j(t)\}\\ &= z_{ij}(t)\,\delta t + o(\delta t) \quad \text{for} \quad i\neq j, t\geq 0, \qquad (11.24)\end{aligned}$$

and $z_{ii}(t)$ is defined to be

$$z_{ii}(t) = -\sum_{j\neq i} z_{ij}(t)\,,\ i,j\in I, t\geq 0,$$

and let $c_i(0)$ for $i = 1, 2, ..., k$ be the probabilities of the initial states.

Let $\mathbf{Z}(t) = \{z_{ij}(t)\}_{i,j\in I}$ be the above intensity matrix and assume that $\mathbf{Z}(t)$ is measurable for every $t\geq 0$ and that $\sup_{i\in I}\{|z_{ii}(t)|\}$ is integrable on every finite interval of time. Then the intensity matrices $\{\mathbf{Z}(t)\}_{t\geq 0}$ define a non-homogeneous Markov process which we call the *compromise non-homogeneous Markov process* of the S-NHMSC. The name compromise is selected in the sense that is the outcome of the choice of strategy under the various pressures in the environment. We call a process like the one described above a *non-homogeneous Markov system in a stochastic environment in continuous time (S-NHMSC)*.

11.4 The expected population structure of the S-NHMSC

We will now study the problem of finding the expected population structure $\mathbb{E}[\mathbf{N}(t)]$ in terms of the basic functions of parameters of the system. We call basic functions of parameters the least number of parameters that uniquely determine an S-NHMSC. These are the functions $\{\mathcal{R}_I(t)\}_{t\geq 0}$, $\{\mathbf{Z}(t)\}_{t\geq 0}$, $\{T(t)\}_{t\geq 0}$, $\{\mathbf{r}_{k+1}(t)\}_{t\geq 0}$, $\{\mathbf{p}_0(t)\}_{t\geq 0}$, the initial population structure $\mathbf{N}(0)$ and the initial probabilities $c_j(0)$. Define by

$$D(t) = dT(t)/dt \quad \text{or} \quad T(t+\delta t) - T(t) = D(t)\,\delta t + o(\delta t). \tag{11.25}$$

Let $N_0(t, t+\delta t)$ be the random variable which represents the number of new members entering the population in the infinitesimal time interval $[t, t+\delta t)$. Then since the number of losses from the population is a random variable with distribution for each state $i \in \mathbb{S}$ the binomial $\mathcal{B}(N_i(t), r_{i,k+1}(t)\,\delta t)$ conditional on $N_i(t)$ we have

$$\mathbb{E}[N_0(t, t+\delta t)] = \sum_{i=1}^{k} \mathbb{E}[N_i(t)]\, r_{i,k+1}(t)\,\delta t + D(t)\,\delta t. \tag{11.26}$$

Also let $N_{ij}(t, t+\delta t)$ be the random variable representing the number of members of the system moving from state i to state j in the time interval $[t, t+\delta t)$. Then these flows from i to $j \in \mathbb{S}$ are multinomial random variables in the sense that

$$\begin{aligned}\mathbb{E}[N_{ij}(t, t+\delta t)] &= \mathbb{E}[\mathbb{E}[N_{ij}(t, t+\delta t) \mid N_i(t), \mathcal{R}_I(t)]] \\ &= \mathbb{E}[N_i(t)]\,\mathbb{E}[p_{ij}(t, t+\delta t)] \\ &= \mathbb{E}[N_i(t)]\,\mathbb{E}[r_{ij}(t)]\,\delta t + o(\delta t) \ \text{ for } i \neq j \in \mathbb{S},\end{aligned} \tag{11.27}$$

and

$$\begin{aligned}\mathbb{E}[N_{ii}(t, t+\delta t)] &= \mathbb{E}[\mathbb{E}[N_{ii}(t, t+\delta t) \mid N_i(t), \mathcal{R}_I(t)]] \\ &= \mathbb{E}[N_i(t)]\,\mathbb{E}[p_{ii}(t, t+\delta t)] \\ &= \mathbb{E}[N_i(t)] + \mathbb{E}[N_i(t)]\,\mathbb{E}[r_{ii}(t)]\,\delta t + o(\delta t) \ \text{ for } i \neq j \in \mathbb{S}.\end{aligned} \tag{11.28}$$

Consequently we have

$$\begin{aligned}\mathbb{E}[N_j(t+\delta t)] = &\sum_{i\neq j} \mathbb{E}[N_i(t)]\,[\mathbb{E}[r_{ij}(t)]\,\delta t + r_{i,k+1}(t)\,p_{0j}(t, t+\delta t)] \\ &+\mathbb{E}[N_j(t)]\,[1 + \mathbb{E}[r_{jj}(t)]\,\delta t + r_{jk+1}(t)\,\delta t p_{0j}(t, t+\delta t)] \\ &+D(t)\,\delta t p_{0j}(t, t+\delta t) + o(\delta t).\end{aligned} \tag{11.29}$$

Equation (11.29) for all $j \in \mathbb{S}$ could be written in matrix notation

$$\frac{d\mathbb{E}[\mathbf{N}(t)]}{dt} = \mathbb{E}[\mathbf{N}(t)]\,\mathbb{E}[\mathbf{Q}(t)] + D(t)\,\mathbf{p}_0(t), \tag{11.30}$$

where

$$\mathbb{E}[\mathbf{Q}(t)] = \mathbb{E}[\mathbf{R}(t)] + \mathbf{r}_{k+1}^{\top}(t)\,\mathbf{p}_0(t). \tag{11.31}$$

We will now prove that the sum of the rows of the matrix $\mathbb{E}[\mathbf{Q}(t)]$ is equal to zero. We have

$$\begin{aligned}\mathbb{E}[\mathbf{Q}(t)]\,\mathbf{1}^{\top} &= \mathbb{E}[\mathbf{R}(t)]\,\mathbf{1}^{\top} + \mathbf{r}_{k+1}^{\top}(t)\,\mathbf{p}_0(t)\,\mathbf{1}^{\top} \\ &= \sum_{j=1}^{\nu} \mathbb{P}[\mathbf{R}(t) = \mathbf{R}_j(t)]\,\mathbf{R}_j(t)\,\mathbf{1}^{\top} + \mathbf{r}_{k+1}^{\top}(t) \\ &= -\sum_{j=1}^{\nu} \mathbb{P}[\mathbf{R}(t) = \mathbf{R}_j(t)]\,\mathbf{r}_{k+1}^{\top}(t) + \mathbf{r}_{k+1}^{\top}(t) \\ &= -\mathbf{r}_{k+1}^{\top}(t) + \mathbf{r}_{k+1}^{\top}(t) = 0.\end{aligned} \tag{11.32}$$

Hence, the matrix $\mathbb{E}[\mathbf{Q}(t)]$ is an intensity matrix and defines a non-homogeneous Markov process which by analogy with the NHMS we call *the expected embedded non-homogeneous Markov process* for the S-NHMSC. The matrix $\mathbb{E}[\mathbf{Q}(t)]$ is a function of $\{\mathbf{Z}(t)\}_{t\geq 0}$ and $\{\mathcal{R}_I(t)\}_{t\geq 0}$ as we show in what follows:

$$\mathbb{E}[\mathbf{R}(t)] = \sum_{j=1}^{\nu} \mathbb{P}[\mathbf{R}(t) = \mathbf{R}_j(t)]\, \mathbf{R}_j(t)$$

$$= \sum_{i=1}^{\nu}\sum_{j=1}^{\nu} \mathbb{P}[\mathbf{R}(t) = \mathbf{R}_j(t) \mid \mathbf{R}(0) = \mathbf{R}_i(0)]\, \mathbb{P}[\mathbf{R}(0) = i]\, \mathbf{R}_j(t). \tag{11.33}$$

Now the choice of the intensity matrix $\mathbf{R}_j(t)$ is stochastically described by the compromise non-homogeneous Markov process which is defined by the intensity matrix $\{\mathbf{Z}(t)\}_{t\geq 0}$ which is measurable and with $\sup_{i\in I}\{|z_{ii}(t)|\}$ integrable on every finite interval of time and thus according to Iosifescu (1980) we get that

$$\mathbb{P}[\mathbf{R}(t) = \mathbf{R}_j(t) \mid \mathbf{R}(0) = \mathbf{R}_i(0)] = \mathbf{e}_i[\mathbf{I} + \int_0^t \mathbf{Z}(u)\, du$$

$$+\sum_{n\geq 2}\int_0^t dt_1 \int_{t_1}^t dt_2 ... \int_{t_{n-1}}^t \mathbf{Z}(t_1)\mathbf{Z}(t_2)...\mathbf{Z}(t_n)\, dt_n]\mathbf{e}_j^\top, \tag{11.34}$$

and thus from $(11.31), (11.33)$ and (11.34) we get

$$\mathbb{E}[\mathbf{Q}(t)] = \sum_{i=1}^{\nu}\sum_{j=1}^{\nu} \mathbf{e}_i[\mathbf{I} + \int_0^t \mathbf{Z}(u)\, du$$

$$+\sum_{n\geq 2}\int_0^t dt_1 \int_{t_1}^t dt_2 ... \int_{t_{n-1}}^t \mathbf{Z}(t_1)\mathbf{Z}(t_2)...\mathbf{Z}(t_n)\, dt_n]\mathbf{e}_j^\top c_i(0)\, \mathbf{R}_j(t)$$

$$+\mathbf{r}_{k+1}(t)\, \mathbf{p}_0(t). \tag{11.35}$$

In what follows we will use a probabilistic argument in order to find $\mathbb{E}[\mathbf{N}(t)]$ which will also be the solution of the differential equation (11.30). The initial number of memberships $T(0) = \mathbf{N}(0)\mathbf{1}^\top$ at time t will be distributed to the various states with probabilities $\mathbb{E}[\mathbf{P}_q(0,t)]$ which are the probabilities of transitions of the expected embedded non-homogeneous Markov process generated by the intensity matrix $\mathbb{E}[\mathbf{Q}(t)]$. Thus

$$\mathbb{E}[\mathbf{P}_q(0,t)] = \mathbf{I} + \int_0^t \mathbb{E}[\mathbf{Q}(u)]\, du +$$

$$\sum_{n\geq 2}\int_0^t dt_1 \int_{t_1}^t dt_2 ... \int_{t_{n-1}}^t \mathbb{E}[\mathbf{Q}(t_1)]\,\mathbb{E}[\mathbf{Q}(t_2)]...\mathbb{E}[\mathbf{Q}(t_n)]\, dt_n, \tag{11.36}$$

and the expected distribution across the states of the initial memberships will be

$$\mathbf{N}(0)\,\mathbb{E}[\mathbf{P}_q(0,t)]. \tag{11.37}$$

Now, let the time interval $[x, x+\delta x)$ then the new memberships entering in that time interval are $D(x)\,\delta x$ and their expected values in the various states at the end of the interval are given by $\mathbf{p}_0(x, x+\delta x)\, D(x)\,\delta x$. After time $t-x$ the expected number of new memberships will be distributed to the various states of the population and their expected values will be $\mathbf{p}_0(x, x+\delta x) D(x) \delta x \mathbb{E}[\mathbf{P}_q(x,t)]$ therefore integrating x from 0 to t we get

$$\mathbb{E}[\mathbf{N}(t)] = \mathbf{N}(0)\mathbb{E}[\mathbf{P}_q(0,t)] + \int_0^t \mathbf{p}_0(x) D(x) \mathbb{E}[\mathbf{P}_q(x,t)]\, dx, \tag{11.38}$$

where

$$\mathbb{E}[\mathbf{P}_q(x,t)] = \mathbf{I} + \int_x^t \mathbb{E}[\mathbf{Q}(u)]\, du +$$

$$\sum_{n\geq 2} \int_x^t dx_1 \int_{x_1}^t dx_2 ... \int_{x_{n-1}}^t \mathbb{E}[\mathbf{Q}(x_1)]\mathbb{E}[\mathbf{Q}(x_2)]...\mathbb{E}[\mathbf{Q}(x_n)]\, dx_n. \tag{11.39}$$

11.5 The asymptotic behavior of the S-NHMSC

As we have seen in the previous chapters the central problems in the theory of NHMS which in the present sections we will study for S-NHMSC are basically of two natures. The first classical problem is that of finding the conditions under which the asymptotic behavior of the expected population structure $\mathbb{E}[\mathbf{N}(t)]$ as $t \to \infty$ exists and finding its limit in closed analytic form as a function of the limits of the basic parameters of the system. The second classical problem is finding which expected relative population structures are possible as limiting ones, provided that we control the limiting vector of input probabilities into the population. In what follows we will use as a norm of a matrix $\mathbf{A} \in M_{k\times k}(\mathbb{R})$ the following

$$\|\mathbf{A}\| = \sup_i \sum_i |a_{ij}|,$$

and since is the only type of norm used in the present chapter we will keep the natation without the subscript. We will now prove the following basic theorem:

Theorem 365 *(Vassiliou (2022b)). Let a complete probability space $(\Omega, \mathcal{F}, \mathbb{P})$ and a finite S-NHMSC as defined in section 11.3. Assume that the following conditions hold:*

$$(1) \lim_{t\to\infty} \|\mathbf{R}_j(t) - \mathbf{R}_j\| = 0, (2) \lim_{t\to\infty} \left\|\mathbf{r}_{k+1}^\top(t) - \mathbf{r}_{k+1}^\top\right\| = 0,$$

$$(3) \lim_{t\to\infty} \|\mathbf{p}_0(t) - \mathbf{p}_0\|, (4) \lim_{t\to\infty} \|\mathbf{Z}(t) - \mathbf{Z}\| = 0, \textit{with}$$

$$\sup_{t\geq 0} \|\mathbf{Z}(t)\| < \infty, \ \sup_{i\in I}\{|z_{ii}|\} < z < \infty \textit{ and let } \mathbf{P}_Z = \mathbf{I} + \frac{\mathbf{Z}}{c_1}$$

$$\textit{with } c_1 > z, \mathbf{P}_Z \textit{ an irreducible, aperiodic matrix}$$

$$(5) \ \sup_{i\in\mathbb{S}}\{r_{j,ii}\} < a < \infty, \ \sup\{r_{i,k+1}\} < b < \infty$$

$$\textit{where } r_{j,ii} \textit{ the } (i,j) \textit{ element of } \mathbf{R}_j,$$

then as $t \to \infty$ $\mathbb{E}[\mathbf{Q}(t)]$ converges in norm to the intensity matrix

$$\mathbb{E}[\mathbf{Q}] = \sum_{j=1}^{\nu} \pi_{z_j}\mathbf{R}_j + \mathbf{r}_{k+1}^\top \mathbf{p}_0,$$

where $\boldsymbol{\pi}_Z = (\pi_{z_1}, \pi_{z_2}, ..., \pi_{z_k})$ is the left eigenvector of the eigenvalue 1 of the matrix $\mathbf{P}_Z$.

Proof. From condition (4) since $\mathbf{P}_Z$ is an irreducible, aperiodic stochastic matrix then there exists a stable stochastic matrix $\mathbf{\Pi}_Z$ with common row $\boldsymbol{\pi}_Z = (\pi_{z_1}, \pi_{z_2}, ..., \pi_{z_k})$ which is the left eigenvector of the eigenvalue 1 of the matrix $\mathbf{P}_Z$, that is,

$$\lim_{t\to\infty} \left\|\mathbf{P}_Z^t - \mathbf{\Pi}_Z\right\| = 0. \tag{11.40}$$

Also from condition (4), we have that the intensity matrices $\{\mathbf{Z}(t)\}_{t\geq 0}$ converge to the intensity matrix $\mathbf{Z}$ which from (11.40) we know that it generates an ergodic Markov process. Therefore, $\{\mathbf{Z}(t)\}_{t\geq 0}$ due to Theorem 364 generates an ergodic non-homogeneous Markov process and we have that

$$\lim_{t\to\infty} \|\mathbf{C}(s,t) - \mathbf{\Pi}_Z\| = 0. \tag{11.41}$$

From (11.33) we get that

$$\mathbb{E}[\mathbf{R}(t)] = \sum_{i=1}^{\nu}\sum_{j=1}^{\nu} c_{ij}(0,t)\, c_i(0)\, \mathbf{R}_j(t). \tag{11.42}$$

Now, consider

$$\left\|\mathbb{E}[\mathbf{Q}(t)] - \sum_{j=1}^{\nu}\pi_{z_j}\mathbf{R}_j - \mathbf{r}_{k+1}^{\top}\mathbf{p}_0\right\| \leq$$

$$\left\|\mathbb{E}[\mathbf{R}(t)] - \sum_{j=1}^{\nu}\pi_{z_j}\mathbf{R}_j\right\| + \left\|\mathbf{r}_{k+1}^{\top}(t)\,\mathbf{p}_0(t) - \mathbf{r}_{k+1}^{\top}\mathbf{p}_0\right\| \leq$$

$$\left\|\sum_{i=1}^{\nu}\sum_{j=1}^{\nu} c_{ij}(0,t)\, c_i(0)\, \mathbf{R}_j(t) - \sum_{i=1}^{\nu}\sum_{j=1}^{\nu}\pi_{z_j} c_i(0)\,\mathbf{R}_j\right\| +$$

$$\left\|\mathbf{r}_{k+1}^{\top}(t) - \mathbf{r}_{k+1}^{\top}\right\| + \left\|\mathbf{r}_{k+1}^{\top}\right\| \left\|\mathbf{p}_0(t) - \mathbf{p}_0\right\| \leq$$

$$\sum_{i=1}^{\nu}\sum_{j=1}^{\nu}\left\|c_{ij}(0,t)\,\mathbf{R}_j(t) - \pi_{z_j}\mathbf{R}_j\right\| c_i(0) +$$

$$\left\|\mathbf{r}_{k+1}^{\top}(t) - \mathbf{r}_{k+1}^{\top}\right\| + \left\|\mathbf{r}_{k+1}^{\top}\right\| \left\|\mathbf{p}_0(t) - \mathbf{p}_0\right\| \leq$$

$$\sum_{i=1}^{\nu}\sum_{j=1}^{\nu}\left[\left\|c_{ij}(0,t) - \pi_{z_j}\right\| \left\|\mathbf{R}_j(t)\right\| + \left\|\mathbf{R}_j(t) - \mathbf{R}_j\right\|\right]$$

$$+ \left\|\mathbf{r}_{k+1}^{\top}(t) - \mathbf{r}_{k+1}^{\top}\right\| + \left\|\mathbf{r}_{k+1}^{\top}\right\| \left\|\mathbf{p}_0(t) - \mathbf{p}_0\right\|. \tag{11.43}$$

We have that

$$\left\|\mathbf{R}_j(t)\right\| \leq \left\|\mathbf{R}_j(t) - \mathbf{R}_j\right\| + \left\|\mathbf{R}_j\right\|,$$

and since $|r_{j,ii}| = \sum_{l\neq i}|r_{j,il}| \ + |r_{j,i\ k+1}|$ and by condition (5) we have $\sup_{i\in\mathbb{S}}\{r_{j,ii}\} < a < \infty$ we could easily prove that

$$\left\|\mathbf{R}_j\right\| \leq 2\left\{\sum_{l\neq i}|r_{j,il}|\right\} + \sup_{i\in\mathbb{S}}|r_{j,i\ k+1}| < b < \infty. \tag{11.44}$$

By condition (1) one can choose t_0^* such that for $t > t^*$, $\left\|\mathbf{R}_j(t) - \mathbf{R}_j\right\| < 1$. Let $M^* = \sup_{0\leq t\leq t^*}\{\left\|\mathbf{R}_j(t) - \mathbf{R}_j\right\|\}$ and denote by $M = M^* + 1 + b$. Then

$$\left\|\mathbf{R}_j(t)\right\| < M < \infty. \tag{11.45}$$

From (11.43) , (11.45) and the conditions of the theorem we get that for $t > t_0$:

$$\left\| \mathbb{E}\left[\mathbf{Q}(t)\right] - \sum_{j=1}^{\nu} \pi_{z_j} \mathbf{R}_j + \mathbf{r}_{k+1}^{\top} \mathbf{p}_0 \right\| \leq \epsilon.$$

Also, it is not difficult using the conditions of the theorem to see that

$$\mathbb{E}\left[\mathbf{Q}\right] \mathbf{1}^{\top} = \sum_{j=1}^{\nu} \pi_{z_j} \mathbf{R}_j \mathbf{1}^{\top} + \mathbf{r}_{k+1}^{\top} \mathbf{p}_0 \mathbf{1}^{\top} = -\sum_{j=1}^{\nu} \pi_{z_j} \mathbf{r}_{k+1}^{\top} + \mathbf{r}_{k+1}^{\top} = 0.$$

■

In analogy with the discrete case for an S-NHMS we provide the following definition:

Definition 366 *We say that an S-NHMSC has an asymptotically attainable expected relative population structure* $\mathbb{E}\left[\mathbf{q}(\infty)\right]$ *under asymptotic input control, if there exists a* $\mathbf{p}_0 = \lim_{t\to\infty} \mathbf{p}_0(t)$ *such that* $\lim_{t\to\infty} \mathbb{E}\left[\mathbf{q}(t)\right] = \mathbb{E}\left[\mathbf{q}(\infty)\right]$. *We denote by* $\mathcal{A}^{\infty}$ *the set of asymptotically expected relative population structures under asymptotic input control of the S-NHMSC.*

We now provide the following basic theorem concerning the asymptotic behavior of the S-NHMSC.

Theorem 367 *(Vassiliou (2022b). Let a complete probability space* $(\Omega, \mathcal{F}, \mathbb{P})$ *and a finite S-NHMSC as defined in section 11.3. Assume that the conditions* (1) $\to$ (5) *of Theorem 365 hold and in addition that the following conditions are true:* (6)

$$\lim_{t\to\infty} T(t) = T,$$

where $T(t)$ *is a non-dicresing continuous function.* (7) *The matrix*

$$\mathbf{P}_q = \mathbf{I} + \frac{\mathbb{E}\left[\mathbf{Q}\right]}{c_2},$$

with $c_2 > \sup_{i\in\mathbb{S}} \{|\mathbb{E}(q_{ii})|\}$ *is an irreducible, aperiodic stochastic matrix. Then (i) As* $t \to \infty$, $\mathbb{E}\left[\mathbf{q}(t)\right]$ *converges to* $\boldsymbol{\pi}_q = (\pi_{q1}, \pi_{q2}, ..., \pi_{qk})$ *which is the left eigenvector of the eigenvalue* 1 *of the matrix* $\mathbf{P}_q$. *(ii) The set* $\mathcal{A}^{\infty}$ *is the convex hull of the points*

$$\mu_i \left[\mathbf{e}_i \left(\sum_{j=1}^{\nu} \pi_{z_j} \mathbf{R}_j \right)^{-1} \right], \text{ where } \mu_i = \mathbf{e}_i \left(\sum_{j=1}^{\nu} \pi_{z_j} \mathbf{R}_j \right)^{-1} \mathbf{1}^{\top}.$$

Proof. Since $\boldsymbol{\pi}_z$ is the left eigenvector of the eigenvalue 1 of the irreducible, aperiodic matrix $\mathbf{P}_Z$, we have that $0 \leq \pi_{z_j} \leq 1$ for $j = 1, 2, ..., \nu$. Also, condition (5) of Theorem 365 is also valid for the present, hence $\sup_{i\in\mathbb{S}} \{r_{j,ii}\} < \infty$ and $\sup\{r_{i,k+1}\} < b < \infty$ and consequently from the expression of $\mathbb{E}\left[\mathbf{Q}\right]$ in Theorem 365 we get that

$$c_2 > \sup_{i\in\mathbb{S}} \{\mathbb{E}\left[q_{ii}\right]\} < \infty. \tag{11.46}$$

Now, since $\mathbf{P}_q$ is an irreducible, aperiodic stochastic matrix, we have that

$$\lim_{t\to\infty} \left\|\mathbf{P}_q^t - \mathbf{\Pi}_q\right\| = 0,$$

where $\mathbf{\Pi}_q$ is a stable stochastic matrix with row $\boldsymbol{\pi}_q = (\pi_{q1}, \pi_{q2}, ..., \pi_{qk})$ which is the left eigenvector of the eigenvalue 1 of the matrix $\mathbf{P}_q$. From (12.46), Theorem 364 and Theorem

361 we have that if we denote by $\mathbb{E}\left[\mathbf{P}_q(s,t)\right]$ the probability transition matrix of the non-homogeneous Markov process defined by the intensities $\{\mathbb{E}\left[\mathbf{Q}(t)\right]\}_{t\geq 0}$ then

$$\lim_{t\to\infty} \left\|\mathbb{E}\left[\mathbf{P}_q(s,t)\right] - \mathbf{\Pi}_q\right\| = 0 \text{ for every } s \in \mathbb{N}. \tag{11.47}$$

Therefore, as $t \to \infty$ the first part of the right hand side of the equation

$$\lim_{t\to\infty} \mathbf{N}(0)\,\mathbb{E}\left[\mathbf{P}_q(s,t)\right] = \mathbf{N}(0)\,\mathbf{\Pi}_q = T(0)\,\boldsymbol{\pi}_q. \tag{11.48}$$

Now, consider

$$\mathbf{U}(t) = \left\|\int_0^t \mathbf{p}_0(x)\,D(x)\,\mathbb{E}\left[\mathbf{P}_q(x,t)\right]dx - \int_0^t \mathbf{p}_0 D(x)\,\mathbf{\Pi}_q dx\right\|$$

$$\leq \int_0^t \left\|\mathbf{p}_0(x)\,\mathbb{E}\left[\mathbf{P}_q(x,t)\right] - \mathbf{p}_0\mathbf{\Pi}_q\right\| D(x)\,dx$$

$$\leq \int_0^t \left\|\mathbf{p}_0(x)\right\| \left\|\mathbb{E}\left[\mathbf{P}_q(x,t)\right] - \mathbf{\Pi}_q\right\| D(x)\,dx + \int_0^t \left\|\mathbf{p}_0(x) - \mathbf{p}_0\right\| D(x)\,dx$$

$$= \int_0^t \left\|\mathbb{E}\left[\mathbf{P}_q(x,t)\right] - \mathbf{\Pi}_q\right\| D(x)\,dx + \int_0^t \left\|\mathbf{p}_0(x) - \mathbf{p}_0\right\| D(x)\,dx$$

$$= \mathbf{A}(t) + \mathbf{B}(t)$$

From (11.47), we have that there exists a $t_0 > 0$ such that for $t - x > t_0$, $\left\|\mathbb{E}\left[\mathbf{P}_q(x,t)\right] - \mathbf{\Pi}_q\right\| < \epsilon$. Thus

$$\mathbf{A}(t) \leq \int_0^{t-t_0} \left\|\mathbb{E}\left[\mathbf{P}_q(x,t)\right] - \mathbf{\Pi}_q\right\| D(x)\,dx +$$

$$\int_{t-t_0}^t \left\|\mathbb{E}\left[\mathbf{P}_q(x,t)\right] - \mathbf{\Pi}_q\right\| D(x)\,dx$$

$$\leq \epsilon\left\{T(t-t_0) - T(0)\right\} + 2\left\{T(t) - T(t-t_0)\right\}. \tag{11.49}$$

Now, from condition (3), we have that there exists a $t_1 > 0$ such that for $t > t_1$, $\left\|\mathbf{p}_0(t) - \mathbf{p}_0\right\| < \epsilon$. Thus

$$\mathbf{B}(t) = \int_0^t \left\|\mathbf{p}_0(x) - \mathbf{p}_0\right\| D(x)\,dx \leq$$

$$\int_0^{t_1} \left\|\mathbf{p}_0(x) - \mathbf{p}_0\right\| D(x)\,dx + \int_{t_1}^t \left\|\mathbf{p}_0(x) - \mathbf{p}_0\right\| D(x)\,dx$$

$$\leq 2\left(T(t_1) - T(0)\right) + \epsilon\left(T(t) - T(t_1)\right).$$

Hence, we have that $\lim_{t\to\infty} \mathbf{A}(t) = 0$ and $\lim_{t\to\infty} \mathbf{B}(t) < \infty$. Thus, $\mathbf{U}(t)$ is an increasing function bounded from above and $\lim_{t\to\infty} \mathbf{U}(t) = 0$. Therefore, from (11.49) we have that

$$\lim_{t\to\infty} \int_0^t \mathbf{p}_0(x)\,D(x)\,\mathbb{E}\left[\mathbf{P}_q(x,t)\right]dx = \int_0^t \mathbf{p}_0 D(x)\,\mathbf{\Pi}_q dx$$

$$= \mathbf{p}_0\mathbf{\Pi}_q\left[T - T(0)\right] = \boldsymbol{\pi}_q\left[T - T(0)\right]. \tag{11.50}$$

Hence, from $(11.38), (11.48)$, and (11.50) we get that

$$\lim_{t\to\infty} \mathbb{E}\left[\mathbf{N}(t)\right] = \mathbb{E}\left[\mathbf{N}(t)\right] = T\boldsymbol{\pi}_q. \tag{11.51}$$

(*ii*). Since $\|\mathbb{E}[\mathbf{Q}]\|$ is finitely bounded and defines an ergodic Markov process it is known that

$$\boldsymbol{\pi}_q \mathbb{E}[\mathbf{Q}] = 0. \tag{11.52}$$

From Theorem 365 and equation (11.52) we get that

$$\boldsymbol{\pi}_q \sum_{j=1}^{\nu} \pi_{z_j} \mathbf{R}_j = -\boldsymbol{\pi}_q \mathbf{r}_{k+1} \mathbf{p}_0 = \boldsymbol{\pi}_q \mathbf{R}_j \mathbf{1}^\top \mathbf{p}_0. \tag{11.53}$$

The matrix $\sum_{j=1}^{\nu} \pi_{z_j} \mathbf{R}_j$ due to condition (8) is irreducible and aperiodic and is part of the intensity matrix $\mathbb{E}[\mathbf{Q}]$. Hence, (Darroch and Seneta (1967)) $\left(\sum_{j=1}^{\nu} \pi_{z_j} \mathbf{R}_j\right)^{-1}$ exists and is nonnegative. Therefore

$$\boldsymbol{\pi}_q = \boldsymbol{\pi}_q \mathbf{R}_j \mathbf{1}^\top \mathbf{p}_0 \left(\sum_{j=1}^{\nu} \pi_{z_j} \mathbf{R}_j\right)^{-1}, \tag{11.54}$$

and

$$\boldsymbol{\pi}_q = \boldsymbol{\pi}_q \mathbf{R}_j \mathbf{1}^\top \sum_{\mathbf{i}=\mathbf{1}}^{\mathbf{k}} p_{0i} \mathbf{e}_i \left(\sum_{j=1}^{\nu} \pi_{z_j} \mathbf{R}_j\right)^{-1}. \tag{11.55}$$

Multiplying both sides of (11.55) by $\mathbf{1}^\top$ we obtain

$$1 = \boldsymbol{\pi}_q \mathbf{R}_j \mathbf{1}^\top \sum_{\mathbf{i}=\mathbf{1}}^{\mathbf{k}} p_{0i} \mathbf{e}_i \left(\sum_{j=1}^{\nu} \pi_{z_j} \mathbf{R}_j\right)^{-1} \mathbf{1}^\top. \tag{11.56}$$

Let

$$\mu_i = \mathbf{e}_i \left(\sum_{j=1}^{\nu} \pi_{z_j} \mathbf{R}_j\right)^{-1} \mathbf{1}^\top. \tag{11.57}$$

Then

$$1 = \boldsymbol{\pi}_q \mathbf{R}_j \mathbf{1}^\top \sum_{\mathbf{i}=\mathbf{1}}^{\mathbf{k}} p_{0i} \mu_i. \tag{11.58}$$

Therefore from (11.51) and the above we get that

$$\lim_{t\to\infty} \mathbb{E}[\mathbf{q}(t)] = \boldsymbol{\pi}_q = \sum_{\mathbf{i}=\mathbf{1}}^{\mathbf{k}} \frac{p_{0i}\mu_i}{\sum_{\mathbf{j}=\mathbf{1}}^{\mathbf{k}} p_{0j}\mu_j} \mu_i^{-1} \left[\mathbf{e}_i \left(\sum_{j=1}^{\nu} \pi_{z_j} \mathbf{R}_j\right)^{-1}\right]. \tag{11.59}$$

Hence, $\mathbb{E}[\mathbf{q}(\infty)]$ is a convex combination of the vertices

$$\mu_i^{-1} \left[\mathbf{e}_i \left(\sum_{j=1}^{\nu} \pi_{z_j} \mathbf{R}_j\right)^{-1}\right].$$

■

An important question which logically arises is what is the rate of convergence to asymptotically attainable structures in an S-NHMSC. In fact we are interested in finding conditions under which the rate is exponential, because then the practical value of the asymptotic result is greater. The following theorem answers the question of the rate of convergence of the expected structure of an S-MHMSC.

Theorem 368 *(Vassiliou (2022b)). Let a complete probability space* $(\Omega, \mathcal{F}, \mathbb{P})$ *and a finite S-NHMSC as defined in section 11.3. Also, let that the conditions* $(1) \rightarrow (7)$ *of Theorem 367 hold and in addition assume that the convergences in conditions* $(1) \rightarrow (4)$ *and* (6) *are exponentially fast. Then the convergence of* $\mathbb{E}[\mathbf{N}(t)]$ *as* $t \rightarrow \infty$ *is exponentially fast.*

Proof. Since $\lim_{t\rightarrow\infty} \|\mathbf{Z}(t) - \mathbf{Z}\| = 0$ exponentially fast and in addition $\mathbf{Z}$ is strongly ergodic then due to Lemma 360 and Theorem 362 there are constants c_3 and $\lambda_1 > 0$ such that

$$\|\mathbf{C}(s, s+t) - \mathbf{\Pi}_z\| \leq c_3 e^{-\lambda_1 t} \text{ for every } s, t > 0. \tag{11.60}$$

Since the convergences in conditions $(1), (2)$, and (3) are exponentially fast we have that

$$\exists\ c_0 > 0,\ a_0 > 0 \text{ such that } \|\mathbf{p}_0(t) - \mathbf{p}_0\| \leq c_0 e^{-a_0 t} \text{ for every } t. \tag{11.61}$$

$$\exists\ c_1 > 0,\ a_1 > 0 \text{ such that } \left\|\mathbf{r}_{k+1}^{\top}(t) - \mathbf{r}_{k+1}^{\top}\right\| \leq c_1 e^{-a_1 t} \text{ for every } t. \tag{11.62}$$

$$\exists\ c_2 > 0,\ a_2 > 0 \text{ such that } \|\mathbf{R}_j(t) - \mathbf{R}_j\| \leq c_2 e^{-a_2 t} \text{ for every } t. \tag{11.63}$$

From $(11.43), (11.60) \rightarrow (11.63)$ we arrive at

$$\|\mathbb{E}[\mathbf{Q}(t)] - \mathbb{E}[\mathbf{Q}]\| \leq c e^{-at} \text{ with } c > 0, a > 0. \tag{11.64}$$

Now, from (11.64) condition (7) of Theorem 367 and Theorem 362 we get that

$$\|\mathbb{E}[\mathbf{P}_q(s, s+t) - \mathbf{\Pi}_q]\| \leq c_q e^{-\lambda_2 t}, c_q, \lambda_2, t > 0 \text{ for every } t. \tag{11.65}$$

We now have the following

$$\|\mathbb{E}[\mathbf{N}(t)] - T\boldsymbol{\pi}_q\| = \|\mathbf{N}(0)\mathbb{E}[\mathbf{P}_q(0,t)] \tag{11.66}$$

$$+ \int_0^t \mathbf{p}_0(x) D(x) \mathbb{E}[\mathbf{P}_q(x,t)]\, dx - T\boldsymbol{\pi}_q\| \leq$$

$$\|\mathbf{N}(0)\| \, \|\mathbb{E}[\mathbf{P}_q(0,t)] - \mathbf{\Pi}_q\| +$$

$$\left\|\int_0^t \mathbf{p}_0(x) D(x) \mathbb{E}[\mathbf{P}_q(x,t)]\, dx - (T - T(0))\boldsymbol{\pi}_q\right\|$$

$$\leq \|\mathbf{N}(0)\| \, \|\mathbb{E}[\mathbf{P}_q(0,t)] - \mathbf{\Pi}_q\| + \|(T(t) - T)\boldsymbol{\pi}_q\| +$$

$$\left\|\int_0^t \mathbf{p}_0(x) D(x) \mathbb{E}[\mathbf{P}_q(x,t)]\, dx - (T(t) - T(0))\boldsymbol{\pi}_q\right\| \leq$$

$$\|\mathbf{N}(0)\| \, \|\mathbb{E}[\mathbf{P}_q(0,t)] - \mathbf{\Pi}_q\| + |(T(t) - T)| +$$

$$\int_0^t \|\mathbb{E}[\mathbf{P}_q(x,t)] - \mathbf{\Pi}_q\| D(x)\, dx + \int_0^t \|\mathbf{p}_0(x) - \mathbf{p}_0\| D(x)\, dx \tag{11.67}$$

From (11.66), condition (3), the fact that the convergence as $t \rightarrow \infty$ of $T(t)$ is exponentially fast and (11.67) we arrive at the following relation

$$\|\mathbb{E}[\mathbf{N}(t)] - T\boldsymbol{\pi}_q\| \leq c e^{-\lambda t} \text{ with } c, \lambda, t > 0 \text{ and for every } t > 0,$$

which proves the theorem. ∎

11.6 An illustrative example from manpower planning

In the present section, the previous results are illustrated through an example from manpower planning. Interesting examples on such systems can be found in Bartholomew, Forbes and McClean (1991). Let that from the historical records of a firm with three grades the transition intensities were estimated and they found that three were repeatedly exercised and so the pool $\mathcal{R}_I(t)$ has the elements

$$\mathbf{R}_1(t) = \begin{pmatrix} -4-2e^{-3t} & 3+e^{-3t} & 0 \\ 0 & -5-3e^{-t} & 3+2e^{t} \\ 0 & 0 & -7-e^{-5t} \end{pmatrix},$$

$$\mathbf{R}_2(t) = \begin{pmatrix} -5-10e^{-3t} & 4+9e^{-3t} & 0 \\ 0 & -6-9e^{-t} & 4+7e^{t} \\ 0 & 0 & -7-e^{-5t} \end{pmatrix},$$

$$\mathbf{R}_1(t) = \begin{pmatrix} -3-4e^{-3t} & 2+3e^{-3t} & 0 \\ 0 & -7-3e^{-t} & 5+e^{t} \\ 0 & 0 & -7-e^{-5t} \end{pmatrix}.$$

Let also

$$\mathbf{r}_{k+1}(t) = \left[1+e^{-3t}, 2+e^{-7}, 7+e^{-5t}\right], \quad \mathbf{p}_0(t) = \begin{pmatrix} 0.2 & 0.3 & 0.5 \end{pmatrix}.$$

In addition assume that utilizing the well known maximum likelihood estimates of transition intensities (Bartholomew et al (1991)) the transition intensities of the compromise non-homogeneous Markov process $\{\mathbf{Z}(t)\}_{t\geq 0}$ under the assumption that they are time independent was found to be

$$\hat{\mathbf{Z}} = \begin{pmatrix} -5 & 3 & 2 \\ 4 & -9 & 5 \\ 2 & 5 & -8 \end{pmatrix}.$$

Applying Theorem 365 to the above data, we have that conditions $(1),(2),(3)$ are satsfied with

$$\mathbf{R}_1 = \begin{pmatrix} -4 & 3 & 0 \\ 0 & -5 & 3 \\ 0 & 0 & 7 \end{pmatrix}, \mathbf{R}_2 = \begin{pmatrix} -5 & 4 & 0 \\ 0 & -6 & 4 \\ 0 & 0 & 7 \end{pmatrix},$$

$$\mathbf{R}_3 = \begin{pmatrix} -3 & 2 & 0 \\ 0 & -7 & 5 \\ 0 & 0 & -7 \end{pmatrix} \text{ and } \mathbf{r}_{k+1} = \begin{pmatrix} 1 & 2 & 7 \end{pmatrix}.$$

Obviously, $\sup_{t\geq 0}\|\mathbf{Z}(t)\| < \infty$ and with $c_1 = 10$ we get

$$\mathbf{P}_Z = \begin{pmatrix} 0.5 & 0.3 & 0.2 \\ 0.4 & 0.1 & 0.5 \\ 0.3 & 0.5 & 0.2 \end{pmatrix},$$

which is obviously an irreducible regular stochastic matrix and thus condition (4) of Theorem 365 is satisfied and also condition (5). Now, the asymptotic expected intensity matrix is found to be

$$\mathbb{E}[\mathbf{Q}] = \begin{pmatrix} -3.8 & 3.3 & 0.5 \\ 0.4 & -5.3 & 4.9 \\ 1.4 & 2.1 & -3.5 \end{pmatrix},$$

which, apparently is a matrix of transition intensities.

11.7 Research notes

In the work Tsaklidis (1996), the continuous time version of the problem in discrete time (Tsaklidis (1994)) presented on the Research notes of Chapter 9 was studied. It was established that if the matrix of intensities $\mathbf{Q}$ of the inherent Markov process is irreducible then the volume of $A_k(t)$ depends geometrically on the trace of $\mathbf{Q}$ and its rank. Also, it is established that the distance of two arbitrary expected relative population structures which are attainable is expressed as a linear combination of the functions $\exp(\lambda_i + \lambda_j)\, t$, where λ_i, λ_j are eigenvalues of $\mathbf{Q}$.

The continuous time version of the problem of the size order studied in Kipouridis and Tsaklidis (2001) for the discrete time case is presented in Kipouridis and Tsaklidis (2001b). It is established that if the intensity matrix of the inherent homogeneous Markov process is irreducible and the left eigenvector of associated with the eigenvalue 0 has unequal elements, then a number t_0 is evaluated such that the size order of the elements become stable for $t \geq t_0$.

The continuous time version of the problem of the state sizes studied in Vasiliadis and Tsaklidis (2008) for the discrete time case is presented in Vasiliadis and Tsaklidis (2009). They evaluate the distributions of expected population structure using equations in analytic forms as functions of the intensity matrix of the inherent homogeneous Markov chain. Finally, they study the form of asymptotic distributions.

A related study is the work Vasiliadis (2012) were the author discusses the case of a closed HMS with finite state capacities. The author mentions the closed relation of this process with queueing systems with finite source as far as closeness is concerned. Note that the author means closed in term of the number of members and not memberships. At the same time finite capacity corresponds in queueing systems with capacity constraints or buffer capacities. A useful remark might be that these processes are closely related with multiple renewal processes introduced by Bartholomew (1982). The author studies the expected population structure of the system, the transition probabilities for a finite interval and also the joint p.m.f. of the expected population structure. In Vasiliadis (2014) the close relation between the M/M/k/T/T and a closed continuous HMS is presented. Then using the known results for closed continuous HMS the expected population structure and the joint p.m.f. of the queueing system are evaluated for its transient behavior.

The study of discete-time queueing systems have lately become important due to their applications in computer and telecommunication systems. Transient behavior in queueing systems are scarcest due to the fact that they are thought to be hard problems and also with not so much interest as the steady state studies. The article by Vasileiadis (2016) is devoted to the study of the transient behavior of a finite source discete time queueing system. This is done by modeling the queueing system as a closed HMS with finite capacities for its states and with priorities on its transitions. Using the known results for the homogeneous Markov system the expected population structure and the joint probability mass function of the expected population structure are evaluated for the queueing system.

Chapter 12

THE PERTURBED NON-HOMOGENEOUS MARKOV SYSTEM

12.1 Introductory notes

As we have seen so far in the previews Chapters, one of the basic assumptions in the theory on inhomogeneous Markov chains and non-homogeneous Markov systems was that the imbedded (inherent) non-homogeneous Markov chain of the NHMS or the non-homogeneous Markov chain itself converged as $t \to \infty$ to a homogeneous Markov chain.

Real applications of NHMS in many cases indicated strongly the need to replace this assumption with one that will answer the problem which rather informally could be described as "what happens when the sequence of transition probability matrices $\{\mathbf{Q}(t)\}_{t=0}^{\infty}$ of the imbedded non-homogeneous Markov chain does not converge but oscillates to a large number of matrices with no pattern but which are rather close to a stochastic matrix $\mathbf{Q}$". The above question was posed to the author of the present book in approximately the above words for the first time by Prof. D.J. Bartholomew in 1981. The need to provide an answer to this realistic problem is included among others in the theme of the present chapter that follows.

12.2 The group inverse and the asymptotic behavior

In the present section we will continue the results presented, in Chapter 2 sections 2.9 to 2.13, on the relation between the asymptotic behavior of a homogeneous Markov chain and the group inverse of the matrix $\mathbf{A} = \mathbf{I} - \mathbf{Q}$; where $\mathbf{Q}$ is the transition probability matrix for the Markov chain. Recall that $\mathbf{A}$ belongs to some multiplicative matrix group G and possess an inverse in G. This matrix is called the *group inverse* of $\mathbf{A}$ and is denoted by $\mathbf{A}^{\sharp}$. Moreover, Meyer (1975) pointed out that the computation of $\mathbf{A}^{\sharp}$ is quite simple. Further properties of $\mathbf{A}^{\sharp}$ are presented in Campbell and Meyer (1979).

Suppose now that $\tilde{\mathbf{Q}} = \mathbf{Q} - \mathcal{E}$ where $\mathcal{E}$ is a perturbation (see Chapter 2) for the matrix $\mathbf{Q}$ a regular stochastic matrix. That is, $\tilde{\mathbf{Q}}$ uniquely defines an ergodic homogeneous Markov chain. We have that $\tilde{\mathbf{A}} = \mathbf{I} - \tilde{\mathbf{Q}} = \mathbf{I} - \mathbf{Q} + \mathcal{E} = \mathbf{A} + \mathcal{E}$. In the theorem that follows we will find $(\mathbf{A}+\mathcal{E})^{\sharp}$ in terms of $\mathbf{A}$ and $\tilde{\mathbf{A}}$ which will be proved to be a basic result that we will use to find answers to important problems.

Theorem 369 *(Meyer (1980)). Let a complete probability space $(\Omega, \mathcal{F}, \mathbb{P})$ and a finite homogeneous Markov chain $\{X_n\}_{n=0}^{\infty}$ which is uniquely determined by the regular stochastic matrix $\mathbf{Q}$. Let also $\tilde{\mathbf{Q}}$ be a perturbation of $\mathbf{Q}$, that is, $\tilde{\mathbf{Q}}$ is a regular stochastic matrix such that $\tilde{\mathbf{Q}} = \mathbf{Q} - \mathcal{E}$. If $\mathbf{A} = \mathbf{I} - \mathbf{Q}$ then $\mathbf{I} + \mathcal{E}\mathbf{A}^{\sharp}$ is non-singular and*

$$(\mathbf{A}+\mathcal{E})^{\sharp} = \mathbf{A}^{\sharp} - \mathbf{A}^{\sharp}\mathcal{E}\mathbf{A}^{\sharp}\left(\mathbf{I} + \mathcal{E}\mathbf{A}^{\sharp}\right)^{-1} - \mathbf{Q}^{*}\left(\mathbf{I} + \mathcal{E}\mathbf{A}^{\sharp}\right)^{-1}\mathbf{A}^{\sharp}\left(\mathbf{I} + \mathcal{E}\mathbf{A}^{\sharp}\right)^{-1}$$

Proof. From Theorem 164 we know that

$$\lim_{t \to \infty} \mathbf{Q}^{t} = \mathbf{Q}^{*} = \mathbf{I} - \mathbf{A}\mathbf{A}^{\sharp}, \tag{12.1}$$

DOI: 10.1201/b23204-12

a stable stochastic matrix. It is easy to see that

$$\mathbf{A}\mathbf{1}^{\top} = (\mathbf{I}-\mathbf{Q})\,\mathbf{1}^{\top} = \mathbf{0} \;\text{ and }\; (\mathbf{A}+\mathcal{E})\,\mathbf{1}^{\top} = \mathbf{0}. \tag{12.2}$$

Since $\mathbf{I}-\mathbf{A}\mathbf{A}^{\sharp}$ is a stable stochastic matrix we get that

$$\mathcal{E}\left(\mathbf{I}-\mathbf{A}\mathbf{A}^{\sharp}\right) = \mathcal{E}\left(\mathbf{1}^{\top}\mathbf{q}^{*}\right) = \mathbf{0}, \tag{12.3}$$

where $\mathbf{q}^{*}$ is the row of the matrix $\mathbf{Q}^{*}$. Hence from (12.3) we straightforwardly get

$$\mathcal{E}\mathbf{A}\mathbf{A}^{\sharp} = \mathcal{E}. \tag{12.4}$$

Since $\mathbf{A} \in G$ a multiplicative matrix group and assuming that the state space of $\{X_n\}_{n=0}^{\infty}$ is $\mathbb{S} = \{1, 2, ..., k\}$ we get that $\mathbf{A} \in M_n(\mathbb{R})$ then there exists nonsingular matrices (see Meyer (2000)) $\mathbf{L}$ and $\mathbf{B} \in M_{n-1}(\mathbb{R})$ such that

$$\mathbf{A} = \mathbf{L}\begin{pmatrix}\mathbf{B} & \mathbf{0}\\ \mathbf{0} & 0\end{pmatrix}\mathbf{L}^{-1}\;,\quad \mathbf{A}^{\sharp} = \mathbf{L}\begin{pmatrix}\mathbf{B}^{-1} & \mathbf{0}\\ \mathbf{0} & 0\end{pmatrix}\mathbf{L}^{-1}\;,$$

$$\text{and}\quad \left(\mathbf{I}-\mathbf{A}\mathbf{A}^{\sharp}\right) = \mathbf{L}\begin{pmatrix}\mathbf{0} & \mathbf{0}\\ \mathbf{0} & 1\end{pmatrix}\mathbf{L}^{-1}\;. \tag{12.5}$$

It is helpful if we split $\mathcal{E}$ in a 2×2 blog matrix in the following way

$$\mathcal{E} = \mathbf{L}\begin{pmatrix}\mathcal{E}_1 & \mathcal{E}_3\\ \mathcal{E}_2 & \mathcal{E}_4\end{pmatrix}\mathbf{L}^{-1}, \tag{12.6}$$

where $\mathcal{E}_1 \in M_{n-1}(\mathbb{R})$. We have that

$$\mathcal{E}\mathbf{A}\mathbf{A}^{\sharp} = \mathbf{L}\begin{pmatrix}\mathcal{E}_1 & \mathbf{0}\\ \mathcal{E}_2 & 0\end{pmatrix}\mathbf{L}^{-1}, \tag{12.7}$$

and from (12.4) we get that $\mathcal{E}_3 = \mathbf{0}$ and $\mathcal{E}_4 = 0$. Hence, we have that

$$\tilde{\mathbf{A}} = \mathbf{A}+\mathcal{E} = \mathbf{L}\begin{pmatrix}\mathbf{B}+\mathcal{E}_1 & \mathbf{0}\\ \mathcal{E}_2 & 0\end{pmatrix}\mathbf{L}^{-1}. \tag{12.8}$$

From Campbell and Meyer (1979) or Meyer and Rose (1977), we get that for a blog matrix of the type in (12.8) the group inverse is given by

$$\begin{pmatrix}\mathbf{C} & \mathbf{0}\\ \mathbf{D} & \mathbf{0}\end{pmatrix}^{\sharp} = \begin{pmatrix}\mathbf{C}^{\sharp} & \mathbf{0}\\ \mathbf{D} & \mathbf{C}^{2\sharp}\end{pmatrix}. \tag{12.9}$$

Therefore from (12.8) using formula (12.9) we get that

$$\tilde{\mathbf{A}}^{\sharp} = \mathbf{L}\begin{pmatrix}(\mathbf{B}+\mathcal{E}_1)^{\sharp} & \mathbf{0}\\ \mathcal{E}_2\,(\mathbf{B}+\mathcal{E}_1)^{2\sharp} & 0\end{pmatrix}\mathbf{L}^{-1}. \tag{12.10}$$

From (12.8) and (12.10) after some algebra we get that

$$\mathbf{I}-\tilde{\mathbf{A}}\tilde{\mathbf{A}}^{\sharp} = \mathbf{L}\begin{pmatrix}\mathbf{I}-(\mathbf{B}+\mathcal{E}_1)\,(\mathbf{B}+\mathcal{E}_1)^{\sharp} & \mathbf{0}\\ -\mathcal{E}_2\,(\mathbf{B}+\mathcal{E}_1) & 1\end{pmatrix}\mathbf{L}^{-1}. \tag{12.11}$$

Since $\tilde{\mathbf{Q}}$ is a perturbation for $\mathbf{Q}$ it is a regular stochastic matrix and therefore due to Theorem 164 the matrix $\mathbf{I}-\tilde{\mathbf{A}}\tilde{\mathbf{A}}^{\sharp}$ is a stable stochastic matrix which implies that $\mathbf{I}-$

$(\mathbf{B}+\mathcal{E}_1)(\mathbf{B}+\mathcal{E}_1)^\sharp = \mathbf{0}$. Therefore $(\mathbf{B}+\mathcal{E}_1)$ is a nonsingular matrix. Since $(\mathbf{B}+\mathcal{E}_1) = \left(\mathbf{I}+\mathcal{E}_1\mathbf{B}^{-1}\right)\mathbf{B}$ it implies that the matrix $\left(\mathbf{I}+\mathcal{E}_1\mathbf{B}^{-1}\right)$ is also nonsingular. From (12.5) and (12.6) it is easy to calculate

$$\mathbf{I}+\mathcal{E}\mathbf{A}^\sharp = \mathbf{L}\begin{pmatrix} \left(\mathbf{I}+\mathcal{E}_1\mathbf{B}^{-1}\right) & \mathbf{0} \\ \mathcal{E}_2\mathbf{B}^{-1} & 1 \end{pmatrix}\mathbf{L}^{-1}, \tag{12.12}$$

which is also nonsingular and apparently we get the existence of

$$\left(\mathbf{I}+\mathcal{E}\mathbf{A}^\sharp\right)^{-1} = \mathbf{L}\begin{pmatrix} \left(\mathbf{I}+\mathcal{E}_1\mathbf{B}^{-1}\right)^{-1} & \mathbf{0} \\ -\mathcal{E}_2\mathbf{B}^{-1}\left(\mathbf{I}+\mathcal{E}_1\mathbf{B}^{-1}\right)^{-1} & 1 \end{pmatrix}\mathbf{L}^{-1}. \tag{12.13}$$

Now using (12.9) for (12.8) and taking into account that $\left(\mathbf{I}+\mathcal{E}_1\mathbf{B}^{-1}\right)$ is nonsingular we get for $(\mathbf{A}+\mathcal{E})^\sharp$

$$(\mathbf{A}+\mathcal{E})^\sharp = \mathbf{L}\begin{pmatrix} \mathbf{B}^{-1}\left(\mathbf{I}+\mathcal{E}_1\mathbf{B}^{-1}\right)^{-1} & \mathbf{0} \\ \mathcal{E}_2\mathbf{B}^{-1}\left(\mathbf{I}+\mathcal{E}_1\mathbf{B}^{-1}\right)^{-1}\mathbf{B}^{-1}\left(\mathbf{I}+\mathcal{E}_1\mathbf{B}^{-1}\right)^{-1} & 0 \end{pmatrix}\mathbf{L}^{-1}. \tag{12.14}$$

From (12.6) and (12.13) it is straightforward to see that

$$\mathbf{A}^\sharp\left(\mathbf{I}+\mathcal{E}\mathbf{A}^\sharp\right)^{-1} = \mathbf{L}\begin{pmatrix} \mathbf{B}^{-1}\left(\mathbf{I}+\mathcal{E}_1\mathbf{B}^{-1}\right)^{-1} & \mathbf{0} \\ \mathbf{0} & 0 \end{pmatrix}\mathbf{L}^{-1}, \tag{12.15}$$

$$\left(\mathbf{I}-\mathbf{A}\mathbf{A}^\sharp\right)\left(\mathbf{I}+\mathcal{E}\mathbf{A}^\sharp\right)^{-1} = \mathbf{L}\begin{pmatrix} \mathbf{0} & \mathbf{0} \\ -\mathcal{E}_2\mathbf{B}^{-1}\left(\mathbf{I}+\mathcal{E}_1\mathbf{B}^{-1}\right)^{-1} & 0 \end{pmatrix}\mathbf{L}^{-1}, \tag{12.16}$$

and

$$\left(\mathbf{I}-\mathbf{A}\mathbf{A}^\sharp\right)\left(\mathbf{I}+\mathcal{E}\mathbf{A}^\sharp\right)^{-1}\mathbf{A}^\sharp\left(\mathbf{I}+\mathcal{E}\mathbf{A}^\sharp\right)^{-1} =$$

$$\mathbf{L}\begin{pmatrix} \mathbf{0} & \mathbf{0} \\ -\mathcal{E}_2\mathbf{B}^{-1}\left(\mathbf{I}+\mathcal{E}_1\mathbf{B}^{-1}\right)^{-1}\mathbf{B}^{-1}\left(\mathbf{I}+\mathcal{E}_1\mathbf{B}^{-1}\right)^{-1} & 0 \end{pmatrix}\mathbf{L}^{-1}. \tag{12.17}$$

Therefore (12.14) becomes

$$(\mathbf{A}+\mathcal{E})^\sharp = \mathbf{A}^\sharp - \mathbf{A}^\sharp\mathcal{E}\mathbf{A}^\sharp\left(\mathbf{I}+\mathcal{E}\mathbf{A}^\sharp\right)^{-1} - \mathbf{Q}^*\left(\mathbf{I}+\mathcal{E}\mathbf{A}^\sharp\right)^{-1}\mathbf{A}^\sharp\left(\mathbf{I}+\mathcal{E}\mathbf{A}^\sharp\right)^{-1},$$

where $\mathbf{Q}^* = \lim_{t\to\infty}\mathbf{Q}^t = \left(\mathbf{I}-\mathbf{A}\mathbf{A}^\sharp\right)$. ■

We are now in a position quite easily to answer the main problem of the present section. That is, to find the asymptotic distribution of a perturbed Markov chain as a function of the asymptotic distribution of the original Markov chain.

Theorem 370 *(Meyer (1980)). Let a complete probability space $(\Omega,\mathcal{F},\mathbb{P})$ and a finite homogeneous Markov chain $\{X_n\}_{n=0}^{\infty}$ which is uniquely determined by the regular stochastic matrix $\mathbf{Q}$. Let also $\tilde{\mathbf{Q}}$ be a perturbation of $\mathbf{Q}$, that is, $\tilde{\mathbf{Q}}$ is a regular stochastic matrix such that $\tilde{\mathbf{Q}} = \mathbf{Q}-\mathcal{E}$. Let also $\left\{\tilde{X}_n\right\}_{n=0}^{\infty}$ the perturbed Markov chain with limiting distribution $\tilde{\mathbf{Q}}^* = \lim_{t\to\infty}\tilde{\mathbf{Q}}^t$ and also $\mathbf{Q}^* = \lim_{t\to\infty}\mathbf{Q}^t$, then*

$$\tilde{\mathbf{Q}}^* = \mathbf{Q}^*\left(\mathbf{I}+\mathcal{E}\mathbf{A}^\sharp\right)^{-1}.$$

Proof. From Theorem 164 we know that $\tilde{\mathbf{Q}}^* = \mathbf{I}-\tilde{\mathbf{A}}\tilde{\mathbf{A}}^\sharp = \mathbf{I}-(\mathbf{A}+\mathcal{E})(\mathbf{A}+\mathcal{E})^\sharp$ and from the fact that $\left(\mathbf{I}+\mathcal{E}\mathbf{A}^\sharp\right)$ is nonsingular and Theorem 369 we arrive at (12.18). ■

12.3 The perturbed non-homogeneous Markov system

In Chapter 2, including sections 2.9–2.13, we introduced the concept of generalized inverses, and we studied the close connection of the group inverse of a stochastic matrix and the theory of homogeneous Markov chains. We also introduced the concept of the perturbation matrix of a stochastic regular matrix as well as a perturbation vector of a stochastic matrix. In what follows we will use the above concepts and results in order to progress the theory and define the *Perturbed non-homogeneous Markov system* which was first defined in Vassiliou and Symeonaki (1999).

Following the notation introduced in Chapter 5 for an NHMS assume that for each $t \in \mathbb{N}$ the matrix $\mathbf{P}(t)$ is of the form

$$\mathbf{P}(t) = \mathbf{P} - \mathcal{E}_p(t) \quad \text{for } t = 0, 1, 2, ..., \tag{12.19}$$

with

$$\mathbf{P}(t)\mathbf{1}^\top \leq \mathbf{1}^\top, \ \mathbf{P}\mathbf{1}^\top \leq \mathbf{1}^\top \ \text{ and } \mathbf{P} \geq 0, \tag{12.20}$$

where the matrix $\mathcal{E}_p(t)$ is randomly selected from the finite set

$$\mathcal{E}_I = \{\mathcal{E}_p(1), \mathcal{E}_p(2), ..., \mathcal{E}_p(v)\}, \tag{12.21}$$

according to the probabilities

$$\mathbb{P}\{\mathcal{E}_p(t) = \mathcal{E}_p(i)\} = c_i > 0 \ \text{ for } i = 1, 2, ..., v. \tag{12.22}$$

Assume also that

$$(\mathbf{I} - \mathbf{P} + \mathcal{E}_p(t))\mathbf{1}^\top = \mathbf{p}_{k+1}^\top(t) \ \text{ for every } t = 0, 1, 2, ... \tag{12.23}$$

Assume moreover that for each t the sequence of input probabilities is of the form

$$\mathbf{p}_0(t) = \mathbf{p}_0 - \varepsilon_0(t), \tag{12.24}$$

with

$$\mathbf{p}_0\mathbf{1}^\top = 1 \ \text{ and } \mathbf{p}_0(t)\mathbf{1}^\top = 1 \ \text{ for every } t = 0, 1, 2, ..., \tag{12.25}$$

where the vectors $\varepsilon_0(t)$ are perturbations vectors for the vector $\mathbf{p}_0$ since from (12.25) $\mathbf{p}_0(t)$ for $t = 0, 1, 2, ...$ are stochastic vectors. Moreover, assume that the vectors $\varepsilon_0(t)$ are randomly selected from the finite set

$$\mathcal{E}_0 = \{\varepsilon_0(1), \varepsilon_0(2), ..., \varepsilon_0(m)\}, \tag{12.26}$$

according to the probabilities

$$\mathbb{P}\{\varepsilon_0(t) = \varepsilon_0(i)\} = c_{0i} > 0 \quad \text{for } i = 1, 2, ..., m. \tag{12.27}$$

From (12.19) to (12.27) we get that

$$\mathbf{p}_{k+1}^\top(t) = \mathbf{p}_{k+1}^\top - \varepsilon_{k+1}^\top(t) \ \text{ for every } t = 0, 1, 2, ..., \tag{12.28}$$

where

$$\mathbf{p}_{k+1}^\top = (\mathbf{I} - \mathbf{P})\mathbf{1}^\top \ \text{ and } \ \varepsilon_{k+1}^\top(t) = -\mathcal{E}_p(t)\mathbf{1}^\top, \tag{12.29}$$

with

$$\mathbb{P}\{\varepsilon_{k+1}(t) = \varepsilon_{k+1}(i)\} = c_i > 0,$$

for every t due to (12.22) and (12.27).

Now, consider the imbedded non-homogeneous Markov chain $\{\mathbf{Q}(t)\}_{t=0}^{\infty}$ then for each t, $\mathbf{Q}(t)$ is of the form:

$$\begin{aligned}\mathbf{Q}(t) &= (\mathbf{P} - \mathcal{E}_p(t)) + \left(\mathbf{p}_{k+1}^{\top} - \varepsilon_{k+1}^{\top}(t)\right)(\mathbf{p}_0 - \varepsilon_0(t)) \\ &= \mathbf{P} + \mathbf{p}_{k+1}^{\top}\mathbf{p}_0 - \mathcal{E}_q(t) = \mathbf{Q} - \mathcal{E}_q(t),\end{aligned} \tag{12.30}$$

where

$$\mathcal{E}_q(t) = \mathcal{E}_p(t) + \mathbf{p}_{k+1}^{\top}\varepsilon_0(t) + \varepsilon_{k+1}^{\top}(t)\,\mathbf{p}_0 - \varepsilon_{k+1}^{\top}(t)\,\varepsilon_0(t) \text{ for every } t, \tag{12.31}$$

and let that

$$\mathbb{P}\{\mathcal{E}_q(t) = \mathcal{E}_q(i)\} = c_{qi} > 0, \quad \text{for } i = 1, 2, ..., mv. \tag{12.32}$$

Assume in addition that the matrices $\mathcal{E}_q(t)$ are perturbation matrices for the matrix $\mathbf{Q}$. That is $\mathbf{Q}$ an irreducible non-periodic matrix (regular) so are all $\mathbf{Q} - \mathcal{E}_q(t)$ for all t. We call such an NHMS a *perturbed non-homogeneous Markov system (P-NHMS).*

12.4 The oscillation of the matrix Q in manpower systems

In the previous section, we defined the process P-NHMS in discrete time and space in such a way as to be directly applicable as a model to a large variety of applications. In the present section, we will discuss how the asymptotic theory of P-NHMS, which will follow later, is closely connected with the answer to the question posed in the Introductory notes of the present chapter which is important for manpower systems.

In many cases in manpower systems, the tail of the sequence of transition probability matrices $\{\mathbf{Q}(t)\}_{t=0}^{\infty}$ of the imbedded non-homogeneous Markov chain has elements of the form $q_{ij} \pm \varepsilon_{ij}(t)$, where $\varepsilon_{ij}(t)$ are small positive numbers compared with q_{ij}. Consider the matrix $\mathcal{E}_q(t) = \{\varepsilon_{ij}(t)\}_{i,j\in\mathbb{S}}$ then the sequence $\{\mathbf{Q}(t)\}_{t=0}^{\infty}$ would look like the following:

$$\mathbf{Q} - \mathcal{E}_q(1), \quad \mathbf{Q} - \mathcal{E}_q(2), \quad \mathbf{Q} - \mathcal{E}_q(3), \quad \mathbf{Q} - \mathcal{E}_q(3), \quad \mathbf{Q} - \mathcal{E}_q(1),$$

$$\mathbf{Q} - \mathcal{E}_q(2), \quad \mathbf{Q} - \mathcal{E}_q(3), \quad \mathbf{Q} - \mathcal{E}_q(2), \quad \mathbf{Q} - \mathcal{E}_q(3), \quad \mathbf{Q} - \mathcal{E}_q(1), ...$$

Given enough data we could estimate the probability distributions of the occurrences of $\mathcal{E}_q(t)$ for $mv = 3$. Now, in the theory of P-NHMS it is assumed that the $\mathcal{E}_q(t)$'s are perturbation matrices for the matrix $\mathbf{Q}$. This assumption is equivalent with assuming that all matrices $\mathbf{Q} - \mathcal{E}_q(t)$ for all t are irreducible regular stochastic matrices. A necessary condition for this is that the $\mathcal{E}_q(t)$'s are such that they do not change the position of the zero's in the matrices $\mathbf{Q} - \mathcal{E}_q(t)$ which is a realistic assumption since the $\varepsilon_{ij}(t)$'s due to their nature are small positive numbers compared with the q_{ij}'s.

12.5 Asymptotic behavior of the P-NHMS

In this section, we will study as we state in the Introductory notes the central problem for P-NHMS, that is, the asymptotic behavior of the population structure in the sense of finding the $\lim_{t\to\infty} \mathbb{E}[\mathbf{N}(t)]$. Retaining the notation introduced in Chapter 5 and by denoting by $F_j(t)$ the random variable which is the number of new entrants to state i at the end of the interval $(t-1, t]$ we get that

$$N_j(t) = \sum_{i=1}^{k} N_{ij}(t) + F_j(t). \tag{12.33}$$

In what follows we will use repeatedly the results in subsection 9.8.3 for Bernoulli trials with a random number of trials and probability which is a random variable. In this respect we get that

$$\mathbb{E}\left[N_{ij}(t)\right] = \mathbb{E}\left[\mathbb{E}\left[N_{ij}(t)\right] \mid N_i(t), \mathcal{E}_I\right] = \mathbb{E}\left[N_i(t)\right]\mathbb{E}\left[p_{ij}(t)\right], \text{ for } i,j \in \mathbb{S}, \tag{12.34}$$

where from (12.19) we get that

$$\mathbb{E}\left[\mathbf{P}(t)\right] = \mathbf{P} - \mathbb{E}\left[\mathcal{E}_p(t)\right]. \tag{12.35}$$

Now, in a P-NHMS the flows $F_j(t)$ for each $j \in \mathbb{S}$ are multinomial random variables with probabilities $\mathbb{E}\left[\mathbf{p}_0(t)\right]$ conditional on $R(t)$. Hence

$$\mathbb{E}\left[F_i(t)\right] = \mathbb{E}\left[\mathbb{E}\left[F_i(t) \mid R(t), \mathcal{E}_0\right]\right] = \mathbb{E}\left[R(t)\right]\mathbb{E}\left[p_{0i}(t)\right], \tag{12.36}$$

where

$$\mathbb{E}\left[\mathbf{p}_0(t)\right] = \mathbf{p}_0 - \mathbb{E}\left[\varepsilon_0(t)\right]. \tag{12.37}$$

We take expectations in (12.33) and with the use of (12.34) $\rightarrow$ (12.37) we get that

$$\mathbb{E}\left[N_j(t)\right] = \sum_{i=1}^{k}\mathbb{E}\left[N_i(t)\right]\mathbb{E}\left[p_{ij}(t)\right] + \mathbb{E}\left[R(t)\right]\mathbb{E}\left[p_{0j}(t)\right], \tag{12.38}$$

where the expected number of recruits $\mathbb{E}\left[R(t)\right]$ in the time interval $(t-1, t]$ is given by

$$\mathbb{E}\left[R(t)\right] = \Delta T(t) + \sum_{i=1}^{k}\mathbb{E}\left[N_i(t)\right]\mathbb{E}\left[p_{i,k+1}(t)\right], \tag{12.39}$$

with

$$\mathbb{E}\left[\mathbf{p}_{k+1}^{\top}(t)\right] = \mathbf{p}_{k+1}^{\top} - \mathbb{E}\left[\varepsilon_{k+1}^{\top}(t)\right]. \tag{12.40}$$

We now have that

$$\mathbb{E}\left[\mathbf{Q}(t)\right] = \mathbb{E}\left[\mathbf{P}(t)\right] + \mathbb{E}\left[\mathbf{p}_{k+1}(t)\right]^{\top}\mathbb{E}\left[\mathbf{p}_0(t)\right], \tag{12.41}$$

and by substituting $\mathbb{E}\left[R(t)\right]$ from (2.39) , (2.40) and by (2.41) we get that

$$\mathbb{E}\left[\mathbf{N}(t)\right] = \mathbb{E}\left[\mathbf{N}(t-1)\right]\mathbb{E}\left[\mathbf{Q}(t)\right] + \Delta T(t)\,\mathbb{E}\left[\mathbf{p}_0(t)\right]. \tag{12.42}$$

From the Definition of the P-NHMS it is easy to see that

$$\mathbb{E}\left[\mathbf{Q}(t)\right] = \sum_{i=1}^{mv}\left[\mathbf{Q} - \mathcal{E}_q(i)\right]c_{qi} \tag{12.43}$$

In the next theorem we will prove an important property for $\mathbb{E}\left[\mathbf{Q}(t)\right]$:

Theorem 371 *(Vassiliou and Symeonaki (1999)). Let a complete probability space* $(\Omega, \mathcal{F}, \mathbb{P})$ *and a finite P-NHMS as defined in section 12.3. The matrix* $\mathbb{E}\left[\mathbf{Q}(t)\right]$ *is an irreducible aperiodic matrix.*

Proof. We will start by proving that $\mathbb{E}\left[\mathbf{Q}(t)\right]$ is a stochastic matrix. From (12.41) we get that

$$\mathbb{E}\left[\mathbf{Q}(t)\right]\mathbf{1}^{\top} = \mathbb{E}\left[\mathbf{P}(t)\right]\mathbf{1}^{\top} + \mathbb{E}\left[\mathbf{p}_{k+1}(t)\right]^{\top}\mathbb{E}\left[\mathbf{p}_0(t)\right]\mathbf{1}^{\top}. \tag{12.44}$$

From equation (12.37) we get that

$$\mathbb{E}\left[\mathbf{p}_0(t)\right]\mathbf{1}^{\top} = \mathbf{p}_0\mathbf{1}^{\top} - \mathbb{E}\left[\varepsilon_0(t)\right]\mathbf{1}^{\top} = 1 - \mathbb{E}\left[\varepsilon_0(t)\right]\mathbf{1}^{\top} = 1, \tag{12.45}$$

since from equations (12.24) and (12.25) we easily get that $\mathbb{E}[\varepsilon_0(t)]\mathbf{1}^\top = 0$ for every t. Hence from equations (12.18), (12.23), (12.44), and (12.45), we get that $\mathbb{E}[\mathbf{Q}(t)]\mathbf{1}^\top = 1$, that is, $\mathbb{E}[\mathbf{Q}(t)]$ is a stochastic matrix.

We will now prove by induction that $\mathbb{E}[\mathbf{Q}(t)]$ as given in (12.43) is irreducible and aperiodic, that is,a regular stochastic matrix. We start by proving that this is true for $mv = 2$. For finite $[\mathbf{Q} - \mathcal{E}_p(1)]$ is a regular matrix (Gantmacher (1959)) if and only if there exists $n_1 \in \mathbb{N}$ such that $[\mathbf{Q} - \mathcal{E}_p(1)]^{n_1} > 0$. Similarly for $[\mathbf{Q} - \mathcal{E}_p(2)]$ there exists a $n_2 \in \mathbb{N}$ such that $[\mathbf{Q} - \mathcal{E}_p(1)]^{n_2} > 0$. Now let $n = \min(n_1, n_2)$. Then we have that

$$\{[\mathbf{Q} - \mathcal{E}_p(1)]c_{q1} + [\mathbf{Q} - \mathcal{E}_p(2)]c_{q2}\}^n =$$

$$[\mathbf{Q} - \mathcal{E}_p(1)]^n c_{q1}^n + [\mathbf{Q} - \mathcal{E}_p(2)]^n c_{q2}^n +$$
$$f([\mathbf{Q} - \mathcal{E}_p(1)]c_{q1}, [\mathbf{Q} - \mathcal{E}_p(2)]c_{q2}) > 0,$$

since $[\mathbf{Q} - \mathcal{E}_p(1)]^n > 0$ and $[\mathbf{Q} - \mathcal{E}_p(2)]^n > 0$. Hence, $[\mathbf{Q} - \mathcal{E}_p(1)]c_{q1} + [\mathbf{Q} - \mathcal{E}_p(2)]c_{q2}$ is a regular matrix. Similarly assuming that this is true for $mv = k$, we can prove that it is true for $mv = k + 1$. Therefore, $\mathbb{E}[\mathbf{Q}(t)]$ is a regular matrix. ∎

Apparently, $\mathbb{E}[\mathbf{Q}(t)]$ defines uniquely a homogeneous Markov chain which we will call the *expected perturbed homogeneous Markov chain.*

We will now prove the basic theorem for the asymptotic behavior of the expected population structure of an P-NHMS.

Theorem 372 *(Vassiliou and Symeonaki (1999)). Let a complete probability space $(\Omega, \mathcal{F}, \mathbb{P})$ and a finite P-NHMS as defined in section 12.3. In addition, let that $\lim_{t\to\infty} T(t) = T$. Then*

$$\lim_{t\to\infty} \mathbb{E}[\mathbf{N}(t)] = \mathbb{E}[\mathbf{N}(\infty)] = T\mathbf{e}_i\mathbf{Q}^*\left(\mathbf{I} + \mathcal{E}_q^*\mathbf{A}^\sharp\right)^{-1},$$

where $\mathbf{Q}^* = \lim_{t\to\infty}\mathbf{Q}^t$, $\mathbf{A}^\sharp$ is the group inverse of $\mathbf{A} = \mathbf{I} - \mathbf{Q}$, and

$$\mathcal{E}_q^* = \sum_{i=1}^{v}\mathcal{E}_p(i)c_i + \mathbf{p}_{k+1}^\top\sum_{i=1}^{m}\varepsilon_0(i)c_{0i} + \left\{\sum_{i=1}^{v}\varepsilon_{k+1}^\top(i)c_i\right\}\mathbf{p}_0$$
$$-\sum_{i=1}^{v}\sum_{j=1}^{m}\varepsilon_{k+1}^\top(i)\varepsilon_0(j)c_ic_{0j}.$$

Proof. From (12.43) we get that

$$\mathbb{E}[\mathbf{Q}(t)] = \mathbf{Q} - \mathbb{E}[\mathcal{E}_q(t)], \tag{12.46}$$

and from equation (12.31) by taking expectations we get that

$$\mathbb{E}[\mathcal{E}_q(t)] = \sum_{i=1}^{v}\mathcal{E}_p(i)c_i + \mathbf{p}_{k+1}^\top\sum_{i=1}^{m}\varepsilon_0(i)c_{0i} + \left\{\sum_{i=1}^{v}\varepsilon_{k+1}^\top(i)c_i\right\}\mathbf{p}_0$$
$$-\sum_{i=1}^{v}\sum_{j=1}^{m}\varepsilon_{k+1}^\top(i)\varepsilon_0(j)c_ic_{0j} = \mathcal{E}_q^*. \tag{12.47}$$

Also from (12.37) we get that

$$\mathbb{E}[\mathbf{p}_0(t)] = \mathbf{p}_0 - \mathbb{E}[\varepsilon_0(t)] = \mathbf{p}_0 - \sum_{j=1}^{m}\varepsilon_0(j)c_{0j}. \tag{12.48}$$

Hence, $\mathbb{E}[\mathcal{E}_q(t)]$ and $\mathbb{E}[\mathbf{p}_0(t)]$ are independent of t and it is easy to apply (12.42) recursively to get that

$$\mathbb{E}[\mathbf{N}(t)] = \mathbb{E}[\mathbf{N}(0)]\left[\mathbf{Q}-\mathcal{E}_q^*\right]^t + [\mathbf{p}_0 - \mathbb{E}[\varepsilon_0(t)]] \times$$

$$\sum_{\tau=1}^{t} \Delta T(\tau)\left[\mathbf{Q}-\mathcal{E}_q^*\right]^{t-\tau}. \tag{12.49}$$

Define by

$$\mathbf{U}(t) = \{\mathbf{p}_0 - \mathbb{E}[\varepsilon_0(t)]\}\sum_{\tau=1}^{t} \Delta T(\tau)\left[\mathbf{Q}-\mathcal{E}_q^*\right]^{t-\tau}. \tag{12.50}$$

In what follows in the present chapter we will use the norm $\|.\|_r$ of a matrix defined in section 2.5. However, since there will be no other norm used we will omit the subscript r. Let us define by $\mathbf{A} = \mathbf{I} - \mathbf{Q}$ then we have that:

$$\left\|\mathbf{U}(t) - [T(t) - T(0)]\{\mathbf{p}_0 - \mathbb{E}[\varepsilon(t)]\}\mathbf{Q}^*\left(\mathbf{I}+\mathcal{E}_q^*\mathbf{A}^\natural\right)^{-1}\right\|$$

$$= \| \{\mathbf{p}_0 - \mathbb{E}[\varepsilon_0(t)]\}\sum_{\tau=1}^{t} \Delta T(\tau)\left(\mathbf{Q}-\mathcal{E}_q^*\right)^{t-\tau}$$

$$- \{\mathbf{p}_0 - \mathbb{E}[\varepsilon(t)]\}\sum_{\tau=1}^{t} \Delta T(\tau)\mathbf{Q}^*\left(\mathbf{I}+\mathcal{E}_q^*\mathbf{A}^\natural\right)^{-1} \|$$

$$\leq \sum_{\tau=1}^{t} \Delta T(\tau)\left\|\left(\mathbf{Q}-\mathcal{E}_q^*\right)^{t-\tau} - \mathbf{Q}^*\left(\mathbf{I}+\mathcal{E}_q^*\mathbf{A}^\natural\right)^{-1}\right\|. \tag{12.51}$$

From Theorem 371 and the definition of perturbation it is apparent that $\mathcal{E}_q^*$ is a perturbation for the matrix $\mathbf{Q}$ and hence from section 12.2 we get that $\lim_{t\to\infty}\left(\mathbf{Q}-\mathcal{E}_q^*\right)^t = \mathbf{Q}^*\left(\mathbf{I}+\mathcal{E}_q^*\mathbf{A}^\natural\right)^{-1}$. Therefore, choose m such that for $\epsilon > 0$ and $t \geq m$

$$\left\|\left(\mathbf{Q}-\mathcal{E}_q^*\right)^{t-m} - \mathbf{Q}^*\left(\mathbf{I}+\mathcal{E}_q^*\mathbf{A}^\natural\right)^{-1}\right\| < \epsilon. \tag{12.52}$$

Therefore, for t large, we have

$$\left\|\mathbf{U}(t) - [T(t) - T(0)]\{\mathbf{p}_0 - \mathbb{E}[\varepsilon_0(t)]\}\mathbf{Q}^*\left(\mathbf{I}+\mathcal{E}_q^*\mathbf{A}^\natural\right)^{-1}\right\|$$

$$\leq \sum_{\tau=1}^{t-m} \Delta T(\tau)\left\|\left\{\left(\mathbf{Q}-\mathcal{E}_q^*\right)^{t-\tau}\right\} - \mathbf{Q}^*\left(\mathbf{I}+\mathcal{E}_q^*\mathbf{A}^\natural\right)^{-1}\right\|$$

$$+ \sum_{\tau=t-m+1}^{t} \Delta T(\tau)\left\|\left\{\left(\mathbf{Q}-\mathcal{E}_q^*\right)^{t-\tau}\right\} - \mathbf{Q}^*\left(\mathbf{I}+\mathcal{E}_q^*\mathbf{A}^\natural\right)^{-1}\right\|$$

$$\leq [T(t-m) - T(0)]\epsilon + 2[T(t) - T(t-m)]. \tag{12.53}$$

We have that $\lim_{t\to\infty} T(t) = T$ therefore from (12.53) we get that

$$\lim_{t\to\infty}\mathbf{U}(t) = [T - T(0)]\{\mathbf{p}_0 - \mathbb{E}[\varepsilon_0(t)]\}\mathbf{Q}^*\left(\mathbf{I}+\mathcal{E}_q^*\mathbf{A}^\natural\right)^{-1}. \tag{12.54}$$

Now applying (12.42) recursively and using (12.37), (12.43), and (12.54) we rather easily arrive at

$$\lim_{t\to\infty}\mathbb{E}[\mathbf{N}(t)] = \mathbb{E}[\mathbf{N}(\infty)] = \mathbb{E}[\mathbf{N}(0)]\mathbf{Q}^*\left(\mathbf{I}+\mathcal{E}_q^*\mathbf{A}^\natural\right)^{-1}$$

$$+T\left\{\mathbf{p}_0 - \mathbb{E}\left[\varepsilon_0(t)\right]\right\}\mathbf{Q}^*\left(\mathbf{I}+\mathcal{E}_q^*\mathbf{A}^\natural\right)^{-1}$$
$$-T(0)\left\{\mathbf{p}_0 - \mathbb{E}\left[\varepsilon_0(t)\right]\right\}\mathbf{Q}^*\left(\mathbf{I}+\mathcal{E}_q^*\mathbf{A}^\natural\right)^{-1}. \tag{12.55}$$

Using the fact that $\mathbf{Q}^*\left(\mathbf{I}+\mathcal{E}_q^*\mathbf{A}^\natural\right)^{-1}$ is a stable stochastic matrix and $\left\{\mathbf{p}_0 - \mathbb{E}\left[\varepsilon_0(t)\right]\right\}$ is a stochastic vector, we have that

$$\begin{aligned}
\mathbb{E}\left[\mathbf{N}(\infty)\right] &= \mathbb{E}\left[\mathbf{N}(0)\right]\mathbf{Q}^*\left(\mathbf{I}+\mathcal{E}_q^*\mathbf{A}^\natural\right)^{-1} + T\mathbf{e}_i\mathbf{Q}^*\left(\mathbf{I}+\mathcal{E}_q^*\mathbf{A}^\natural\right)^{-1} \\
&\quad -T(0)\,\mathbf{e}_i\mathbf{Q}^*\left(\mathbf{I}+\mathcal{E}_q^*\mathbf{A}^\natural\right)^{-1} \\
&= \mathbb{E}\left[\mathbf{N}(0)\right]\mathbf{Q}^*\left(\mathbf{I}+\mathcal{E}_q^*\mathbf{A}^\natural\right)^{-1} + T\mathbf{e}_i\mathbf{Q}^*\left(\mathbf{I}+\mathcal{E}_q^*\mathbf{A}^\natural\right)^{-1} \\
&\quad -\mathbb{E}\left[\mathbf{N}(0)\right]\mathbf{1}^\top\mathbf{e}_i\mathbf{Q}^*\left(\mathbf{I}+\mathcal{E}_q^*\mathbf{A}^\natural\right)^{-1} \\
&= \mathbb{E}\left[\mathbf{N}(0)\right]\mathbf{Q}^*\left(\mathbf{I}+\mathcal{E}_q^*\mathbf{A}^\natural\right)^{-1} + T\mathbf{e}_i\mathbf{Q}^*\left(\mathbf{I}+\mathcal{E}_q^*\mathbf{A}^\natural\right)^{-1} \\
&\quad -\mathbb{E}\left[\mathbf{N}(0)\right]\mathbf{Q}^*\left(\mathbf{I}+\mathcal{E}_q^*\mathbf{A}^\natural\right)^{-1} \\
&= T\mathbf{e}_i\mathbf{Q}^*\left(\mathbf{I}+\mathcal{E}_q^*\mathbf{A}^\natural\right)^{-1}.
\end{aligned}$$

■

12.6 Sensitivity of the limiting distributions

It is quite evident from the previous chapters that the asymptotic behavior of the expected relative population structure is one of the central problems for NHMS's. Also, we have extensively studied the control of the asymptotic relative expected population structure by controlling the limiting input probabilities which is the best instrument for control for NHMS. We did so by finding the convex set of all possible asymptotic relative population structures that are possible when the limiting input probabilities take all possible values.

In the present section, we study the following problem: Let $\mathbf{p}_0$ be the vector of limiting input probabilities for an NHMS and let that we perturb the vector $\mathbf{p}_0$ and $\mathbf{p}_0 - \varepsilon_0$ is the new limiting input probability vector. The important question is what are the consequences on the limiting relative expected population structure?

From Chapter 6 and in later chapters, we studied the asymptotic behavior of the relative population structure under some conditions easily met in practice and we have found that a central result was that the limiting population structure was

$$\mathbf{q}^* \text{ the row of the stable matrix } \mathbf{Q}^* = \lim_{t\to\infty}\mathbf{Q}^t \text{ when } \lim_{t\to\infty}\left\|\mathbf{Q}(t)-\mathbf{Q}\right\| = 0,$$

where $\left\{\mathbf{Q}(t)\right\}_{t=0}^{\infty}$ is the non-homogeneous inherent Markov chain of the NHMS. It is important to note at this point that

$$\mathbf{Q}(t) = \mathbf{P}(t) + \mathbf{p}_{k+1}^\top(t)\,\mathbf{p}_0(t) \quad \text{and } \mathbf{Q} = \mathbf{P} + \mathbf{p}_{k+1}^\top\mathbf{p}_0.$$

Hence, it is evident that the key to the answer of our question is the study of the sensitivity of the inherent (imbedded) Markov chain of the NHMS. In this respect we are led to the following theorem which is a synthesis of theorem 2.1 and and Corollary 2.2 from Meyer and Shoaf (1980) and whose proofs are quite accessible mathematically.

Theorem 373 *Let $\mathbf{A} \in M_n(\mathbb{R})$ be such that that $\mathbf{A}^\sharp$ exist and let*

$$\mathbf{B} = \mathbf{A} + \mathbf{c}\mathbf{d}^\top,$$

where $\mathbf{c} \in M_{n,1}(\mathbb{R})$, $\mathbf{d} \in M_{n,1}(\mathbb{R})$, and $\mathbf{c}, \mathbf{d} \neq \mathbf{0}$. Denote by

$$\mathbf{v}^\top = \mathbf{d}^\top \left(\mathbf{I} - \mathbf{A}\mathbf{A}^\sharp\right), \ \mathbf{u} = \left(\mathbf{I} - \mathbf{A}\mathbf{A}^\sharp\right)\mathbf{c}, \ a = \mathbf{d}^\top \left(\mathbf{I} - \mathbf{A}\mathbf{A}^\sharp\right)\mathbf{c},$$

$$\beta = 1 + \mathbf{d}^\top \mathbf{A}^\sharp \mathbf{c}, \ \mathbf{h}^\top = \mathbf{d}^\top \mathbf{A}^\sharp$$

Then if $\mathbf{B}^\sharp$ exists and

$$a = 0, \ \beta = 0, \ \mathbf{u} \neq \mathbf{0}, \ \text{and } \mathbf{v}^\top = \mathbf{0}$$

then

$$\mathbf{I} - \mathbf{B}\mathbf{B}^\sharp = \left(\mathbf{I} - \mathbf{A}\mathbf{A}^\sharp\right)\left(\mathbf{I} - \frac{1}{\beta}\mathbf{c}\mathbf{h}^\top\right).$$

We will now answer the basic question posed at the beginning of the section in the form of the following theorem:

Theorem 374 *(Vassiliou and Symeonaki (1999)). Let a complete probability space $(\Omega, \mathcal{F}, \mathbb{P})$ and a finite NHMS defined uniquely by the sequences*

$$\{\mathbf{P}(t)\}_{t=0}^{\infty}, \ \{\mathbf{p}_{k+1}(t)\}_{t=0}^{\infty}, \ \{\mathbf{p}_0(t)\}_{t=0}^{\infty}, \ \{T(t)\}_{t=0}^{\infty} \ \text{and } \mathbf{N}(0).$$

Assume also that the four sequences satisfy the conditions:

$$(i) \ \lim_{t\to\infty} \|\mathbf{P}(t) - \mathbf{P}\| = 0, \ \lim_{t\to\infty} \|\mathbf{p}_{k+1}(t) - \mathbf{p}_{k+1}\| = 0,$$

$$\lim_{t\to\infty} \|\mathbf{p}_0(t) - \mathbf{p}_0\| = 0, \ \text{and } \lim_{t\to\infty} \Delta T(t)/T(t) = 0.$$

Assume also that the matrix $\mathbf{Q} = \mathbf{P} + \mathbf{p}_{k+1}^\top \mathbf{p}_0$ is a regular stochastic matrix and the system is expanding ($T(t) \geq T(t-1)$). Let also that we perturb $\mathbf{p}_0$ and $\hat{\mathbf{p}} = \mathbf{p}_0 - \varepsilon_0$ is its perturbation and assume that $\hat{\mathbf{Q}} = \mathbf{Q} - \mathbf{p}_{k+1}^\top \varepsilon_0$ is a regular stochastic matrix. If we denote by $\mathbb{E}[\hat{\mathbf{q}}(\infty)]$ the limiting expected relative population structure of the NHMS with inherent non-homogeneous Markov chain uniquely determined by $\hat{\mathbf{Q}}$ and $\mathbb{E}[\mathbf{q}(\infty)]$ the respective limiting relative expected population structure determined by $\mathbf{Q}$, then we have

$$\mathbb{E}[\hat{\mathbf{q}}(\infty)] - \mathbb{E}[\mathbf{q}(\infty)] = -\mathbb{E}[\mathbf{q}(\infty)]\left(\mathbf{I} + \varepsilon_0 \mathbf{A}^\sharp\right)^{-1} \mathbf{p}_{k+1}^\top \varepsilon_0 \mathbf{A}^\sharp,$$

where $\mathbf{A}^\sharp$ is the group inverse of the matrix $\mathbf{A} = \mathbf{I} - \mathbf{Q}$.

Proof. Now since $\hat{\mathbf{p}}_0 = \mathbf{p}_0 - \varepsilon_0$, we have that

$$\hat{\mathbf{Q}} = \mathbf{P} + \mathbf{p}_{k+1}^\top \hat{\mathbf{p}}_0 = \mathbf{P} + \mathbf{p}_{k+1}^\top (\mathbf{p}_0 - \varepsilon_0) = \mathbf{Q} - \mathbf{p}_{k+1}^\top \varepsilon_0. \tag{12.56}$$

Let us define by

$$\hat{\mathbf{A}} = \mathbf{I} - \hat{\mathbf{Q}} = \mathbf{A} + \mathbf{p}_{k+1}^\top \varepsilon_0 \ \text{ where } \ \mathbf{A} = \mathbf{I} - \mathbf{Q}. \tag{12.57}$$

Since $\mathbf{Q}$ is a stochastic matrix then it is well known (Meyer and Shoaf (1980)) that $\mathbf{A}^\sharp$ exists and for the same reason also $\hat{\mathbf{A}}$ exists. Now, since $\mathbf{A} = \mathbf{I} - \mathbf{Q}$ with $\mathbf{Q}$ a regular stochastic matrix then we know from Theorem 164 that

$$\mathbf{Q}^* = \lim_{t\to\infty} \mathbf{Q}^t = \mathbf{I} - \mathbf{A}\mathbf{A}^\sharp \ \text{a stable stochastic matrix with row } \mathbf{q}^*. \tag{12.58}$$

From Theorem 326, we have that the present theorem satisfies it's conditions therefore we get that

$$\mathbb{E}[\mathbf{q}(\infty)] = \ \mathbf{q}^* \ \text{the row of the stable matrix } \mathbf{I} - \mathbf{A}\mathbf{A}^\sharp. \tag{12.59}$$

Also, since $\hat{\mathbf{Q}}$ is a regular stochastic matrix with the same reasoning we get that $\hat{\mathbf{A}}^{\sharp}$ exists and

$$\hat{\mathbf{Q}}^{*} = \lim_{t\to\infty} \hat{\mathbf{Q}}^{t} = \mathbf{I} - \hat{\mathbf{A}}\hat{\mathbf{A}}^{\sharp} \text{ a stable stochastic matrix with row } \hat{\mathbf{q}}^{*}. \tag{12.60}$$

$$\mathbb{E}\left[\hat{\mathbf{q}}\left(\infty\right)\right] = \hat{\mathbf{q}}^{*} \text{ the row of the stable matrix } \mathbf{I} - \hat{\mathbf{A}}\hat{\mathbf{A}}^{\sharp}. \tag{12.61}$$

We will now examine if the conditions of Theorem 373 are valid in our case. We first observe that $\mathbf{B} = \hat{\mathbf{A}}$, $\mathbf{c} = \mathbf{p}_{k+1}^{\top}$ and $\mathbf{d}^{\top} = \boldsymbol{\varepsilon}_0$. The third condition for our case is

$$\left(\mathbf{I} - \mathbf{A}\mathbf{A}^{\sharp}\right)\mathbf{p}_{k+1}^{\top} = \mathbf{Q}^{*}\mathbf{p}_{k+1}^{\top} = \mathbf{1}^{\top}\mathbf{q}^{*}\mathbf{p}_{k+1}^{\top} \neq \mathbf{0}, \tag{12.62}$$

since the matrix $\mathbf{Q}$ is a regular stochastic matrix and thus $\mathbf{q}^{*} > 0$.

Now consider the fourth condition which in our case is

$$\boldsymbol{\varepsilon}_0\left(\mathbf{I} - \mathbf{A}\mathbf{A}^{\sharp}\right) = \boldsymbol{\varepsilon}_0\mathbf{Q}^{*} = \boldsymbol{\varepsilon}_0\mathbf{1}^{\top}\mathbf{q}^{*} = \mathbf{0} \text{ since } \boldsymbol{\varepsilon}_0\mathbf{1}^{\top} = 0. \tag{12.63}$$

Consider now the first condition

$$\boldsymbol{\varepsilon}_0\left(\mathbf{I} - \mathbf{A}\mathbf{A}^{\sharp}\right)\mathbf{p}_{k+1}^{\top} = \boldsymbol{\varepsilon}_0\mathbf{Q}^{*}\mathbf{p}_{k+1}^{\top} = \boldsymbol{\varepsilon}_0\mathbf{1}^{\top}\mathbf{q}^{*}\mathbf{p}_{k+1}^{\top} = 0. \tag{12.64}$$

From (12.62), (12.63), and (12.64) we get that the three conditions out of the four in Theorem 373 are satisfied therefore the fourth condition must also be true since it is the only possibility. Thus equation (2.7) in Corollary 2.2 of Meyer and Shoaf (1980) holds and thus we have that

$$\left(\mathbf{I} - \hat{\mathbf{A}}\hat{\mathbf{A}}^{\sharp}\right) = \left(\mathbf{I} - \mathbf{A}\mathbf{A}^{\sharp}\right)\left[\mathbf{I} - \frac{1}{1+\boldsymbol{\varepsilon}_0\mathbf{A}^{\sharp}}\mathbf{p}_{k+1}^{\top}\boldsymbol{\varepsilon}_0\mathbf{A}^{\sharp}\right],$$

or

$$\hat{\mathbf{Q}}^{*} = \mathbf{Q}^{*}\left[\mathbf{I} - \frac{1}{1+\boldsymbol{\varepsilon}_0\mathbf{A}^{\sharp}}\mathbf{p}_{k+1}^{\top}\boldsymbol{\varepsilon}_0\mathbf{A}^{\sharp}\right],$$

from which we easily get

$$\hat{\mathbf{q}}^{*} - \mathbf{q}^{*} = -\mathbf{q}^{*}\left(1+\boldsymbol{\varepsilon}_0\mathbf{A}^{\sharp}\right)^{-1}\mathbf{p}_{k+1}^{\top}\boldsymbol{\varepsilon}_0\mathbf{A}^{\sharp}$$

and finally from (12.59) and (12.61) we get

$$\mathbb{E}\left[\hat{\mathbf{q}}\left(\infty\right)\right] - \mathbb{E}\left[\mathbf{q}\left(\infty\right)\right] = -\mathbb{E}\left[\mathbf{q}\left(\infty\right)\right]\left(\mathbf{I}+\boldsymbol{\varepsilon}_0\mathbf{A}^{\sharp}\right)^{-1}\mathbf{p}_{k+1}^{\top}\boldsymbol{\varepsilon}_0\mathbf{A}^{\sharp}.$$

■

12.7 Asymptotic variability of the P-NHMS

In the present section we will study firstly the problem of evaluating the variances and covariances of the state sizes in P-NHMS. Secondly, we will study their asymptotic behavior. We will follow the notation of Chapter 5 and we also assume that the systems starts at time $t = 0$. Following similar arguments as these in section 12.5, we rather easily arrive at the following relations:

$$\begin{aligned}\mathbb{E}\left[N_{ij}(t)\,N_{lr}(t)\right] &= \mathbb{E}\left[\mathbb{E}\left[N_{ij}(t)\,N_{lr}(t)\right] \mid N_i(t), N_l(t), \mathcal{E}_q(t)\right] \\ &= \mathbb{E}\left[N_i(t)\right]\mathbb{E}\left[N_l(t)\right]\mathbb{E}\left[q_{ij}(t)\right]\mathbb{E}\left[q_{lr}(t)\right] \quad \text{for } i \neq l,\ j \neq r.\end{aligned} \tag{12.65}$$

and

$$\mathbb{E}\left[N_{ij}(t)\,N_{ir}(t)\right] = \mathbb{E}\left[N_i(t)\right]^2\mathbb{E}\left[q_{ij}(t)\right]\mathbb{E}\left[q_{ir}(t)\right]$$

$$-\mathbb{E}\left[N_i(t)\right]\mathbb{E}\left[q_{ij}(t)\right]\mathbb{E}\left[q_{ir}(t)\right]. \tag{12.66}$$

Now, when the system is expanding, i.e. $\Delta T(t) > 0$ then we assume that the memberships of the expansion will be allocated to the states independent of the replacement memberships; then these flows will be independent of all other flows, so their contribution to the variances and covariances can simply be added. Following the arguments and notation of section 5.3 we arrive at

$$\boldsymbol{\mu}(t) = \boldsymbol{\mu}(t-1)\,\mathbb{E}\left[\mathbf{V}_q(t)\right] + \Delta T(t)\,\mathbb{E}\left[\boldsymbol{\mu}_0(t)\right], \tag{12.67}$$

where

$$\mathbb{E}\left[\mathbf{V}_q(t)\right] = \begin{pmatrix} \mathbb{E}\left[\mathbf{Q}(t)\right] & \mathbb{E}\left[\mathbf{Q}(t)\right]\mathbf{J} - \mathbf{J}\left(\mathbb{E}\left[\mathbf{Q}(t)\right]\otimes\mathbb{E}\left[\mathbf{Q}(t)\right]\right) \\ \mathbf{0} & \mathbb{E}\left[\mathbf{Q}(t)\right]\otimes\mathbb{E}\left[\mathbf{Q}(t)\right] \end{pmatrix} =$$

$$\begin{pmatrix} \mathbf{Q}-\mathbb{E}\left[\mathcal{E}_q(t)\right] & \left(\mathbf{Q}-\mathbb{E}\left[\mathcal{E}_q(t)\right]\right)\mathbf{J} - \mathbf{J}[\left(\mathbf{Q}-\mathbb{E}\left[\mathcal{E}_q(t)\right]\right)\otimes\left(\mathbf{Q}-\mathbb{E}\left[\mathcal{E}_q(t)\right]\right)] \\ \mathbf{0} & \left(\mathbf{Q}-\mathbb{E}\left[\mathcal{E}_q(t)\right]\right)\otimes\left(\mathbf{Q}-\mathbb{E}\left[\mathcal{E}_q(t)\right]\right) \end{pmatrix},$$

$$\mathbb{E}\left[\boldsymbol{\mu}_0(t)\right] = [\mathbb{E}\left[p_{01}(t)\right], \mathbb{E}\left[p_{02}(t)\right], ..., \mathbb{E}\left[p_{0k}(t)\right], \mathbb{E}\left[p_{01}(t)\right]\left(1-\mathbb{E}\left[p_{01}(t)\right]\right),$$

$$\mathbb{E}\left[p_{01}(t)\right]\mathbb{E}\left[p_{02}(t)\right], ..., \mathbb{E}\left[p_{0k}(t)\right]\left(1-\mathbb{E}\left[p_{0k}(t)\right]\right)], \tag{12.68}$$

with

$$\mathbb{E}\left[p_{0i}(t)\right] = p_{0i} - \mathbb{E}\left[\varepsilon_{0i}(t)\right]. \tag{12.69}$$

Now, applying recursively equation (12.67) we get that

$$\boldsymbol{\mu}(t) = \boldsymbol{\mu}(0)\left\{\mathbb{E}\left[\mathbf{V}_q(t)\right]\right\}^t + \sum_{\tau=1}^{t}\Delta T(\tau)\,\mathbb{E}\left[\boldsymbol{\mu}_0(t)\right]\left\{\mathbb{E}\left[\mathbf{V}_q(t)\right]\right\}^{t-\tau}. \tag{12.70}$$

From (12.68) and Proposition 309 we get that

$$\left\{\mathbb{E}\left[\mathbf{V}_q(t)\right]\right\}^t =$$

$$\begin{pmatrix} \left\{\mathbf{Q}-\mathbb{E}\left[\mathcal{E}_q(t)\right]\right\}^t & \begin{matrix}\left(\mathbf{Q}-\mathbb{E}\left[\mathcal{E}_q(t)\right]\right)^t\mathbf{J} - \\ \mathbf{J}[\left(\mathbf{Q}-\mathbb{E}\left[\mathcal{E}_q(t)\right]\right)\otimes\left(\mathbf{Q}-\mathbb{E}\left[\mathcal{E}_q(t)\right]\right)]\end{matrix} \\ \mathbf{0} & \left(\mathbf{Q}-\mathbb{E}\left[\mathcal{E}_q(t)\right]\right)^t\otimes\left(\mathbf{Q}-\mathbb{E}\left[\mathcal{E}_q(t)\right]\right)^t \end{pmatrix},$$

and from Theorems 370 and 372 if we denote by

$$\mathbb{E}\left[\mathbf{Q}(\infty)\right] = \mathbf{Q}^*\left(\mathbf{I}+\mathcal{E}_q^*\mathbf{A}^\sharp\right)^{-1}, \tag{12.71}$$

we get that

$$\lim_{t\to\infty}\left\{\mathbb{E}\left[\mathbf{V}_q(t)\right]\right\}^t = \mathbb{E}\left[\mathbf{V}_q(\infty)\right] = \tag{12.72}$$

$$\begin{pmatrix} \mathbb{E}\left[\mathbf{Q}(\infty)\right] & \mathbb{E}\left[\mathbf{Q}(\infty)\right]\mathbf{J} - \mathbf{J}\left(\mathbb{E}\left[\mathbf{Q}(\infty)\right]\otimes\mathbb{E}\left[\mathbf{Q}(\infty)\right]\right) \\ \mathbf{0} & \mathbb{E}\left[\mathbf{Q}(\infty)\right]\otimes\mathbb{E}\left[\mathbf{Q}(\infty)\right] \end{pmatrix}.$$

From (12.68) we get that

$$\mathbb{E}\left[\boldsymbol{\mu}_0(t)\right] = \begin{pmatrix}\mathbb{E}\left[\mathbf{p}_0(t)\right] & \mathbb{E}\left[\mathbf{p}_0(t)\right]\otimes\mathbb{E}\left[\mathbf{p}_0(t)\right]\end{pmatrix}. \tag{12.73}$$

It is important to note that $\mathbb{E}\left[\mathbf{Q}(\infty)\right] = \mathbf{Q}^*\left(\mathbf{I}+\mathcal{E}_q^*\mathbf{A}^\sharp\right)^{-1}$ is a stable stochastic matrix and denote by $\mathbb{E}\left[\mathbf{q}(\infty)\right]$ its row. Then we have that

$$\mathbb{E}\left[\mathbf{q}(\infty)\right] = \mathbf{e}_i\mathbf{Q}^*\left(\mathbf{I}+\mathcal{E}_q^*\mathbf{A}^\sharp\right)^{-1}. \tag{12.74}$$

It is not difficult to see that due to Theorem 131, we get that

$$\boldsymbol{\mu}(0)\,\mathbb{E}\left[\mathbf{V}_q(\infty)\right] =$$

$$\boldsymbol{\mu}(0)\begin{pmatrix}\mathbb{E}[\mathbf{Q}(\infty)] & \mathbb{E}[\mathbf{Q}(\infty)]\,\mathbf{J}-\mathbf{J}\left(\mathbb{E}[\mathbf{Q}(\infty)]\otimes\mathbb{E}[\mathbf{Q}(\infty)]\right)\\ \mathbf{0} & \mathbb{E}[\mathbf{Q}(\infty)]\otimes\mathbb{E}[\mathbf{Q}(\infty)]\end{pmatrix}$$
$$= T(0)\left(\mathbb{E}[\mathbf{q}(\infty)] \quad \mathbb{E}[\mathbf{q}(\infty)]\,\mathbf{J}-\mathbf{J}\left(\mathbb{E}[\mathbf{q}(\infty)]\otimes\mathbb{E}[\mathbf{q}(\infty)]\right)\right) \tag{12.75}$$
$$= T(0)\,\mathbb{E}[\mathbf{v}_q(\infty)].$$

From (12.68) and (12.69), we note that $\mathbb{E}[\boldsymbol{\mu}_0(t)]$ is independent of t, hence we have that

$$(T-T(0))\,\mathbb{E}[\boldsymbol{\mu}_0(t)]\,\mathbb{E}[\mathbf{V}_q(\infty)] =$$
$$(T-T(0))\left(\mathbb{E}[\mathbf{p}_0(t)] \quad \mathbb{E}[\mathbf{p}_0(t)]\otimes\mathbb{E}[\mathbf{p}_0(t)]\right)\times$$
$$\begin{pmatrix}\mathbb{E}[\mathbf{Q}(\infty)] & \mathbb{E}[\mathbf{Q}(\infty)]\,\mathbf{J}-\mathbf{J}\left(\mathbb{E}[\mathbf{Q}(\infty)]\otimes\mathbb{E}[\mathbf{Q}(\infty)]\right)\\ \mathbf{0} & \mathbb{E}[\mathbf{Q}(\infty)]\otimes\mathbb{E}[\mathbf{Q}(\infty)]\end{pmatrix}$$
$$= (T-T(0))\,\mathbb{E}[\mathbf{v}_q(\infty)]. \tag{12.76}$$

From (12.75) and (12.76) we get that

$$\boldsymbol{\mu}(0)\,\mathbb{E}[\mathbf{V}_q(\infty)] + (T-T(0))\,\mathbb{E}[\boldsymbol{\mu}_0(t)]\,\mathbb{E}[\mathbf{V}_q(\infty)] =$$
$$T\mathbb{E}[\mathbf{v}_q(\infty)]. \tag{12.77}$$

Now we have that

$$\sum_{\tau=1}^{t}\Delta T(\tau)\,\mathbb{E}[\boldsymbol{\mu}_0(t)]\left\{\mathbb{E}[\mathbf{V}_q(t)]\right\}^{t-\tau} =$$
$$\mathbb{E}[\boldsymbol{\mu}_0(t)]\sum_{\tau=1}^{t}\Delta T(\tau)\left\{\mathbb{E}[\mathbf{V}_q(t)]\right\}^{t-\tau},$$

and by using the techniques used in previous chapters the reader should be in a position to show as an exercise that

$$\lim_{t\to\infty}\sum_{\tau=1}^{t}\Delta T(\tau)\,\mathbb{E}[\boldsymbol{\mu}_0(t)]\left\{\mathbb{E}[\mathbf{V}_q(t)]\right\}^{t-\tau} =$$
$$(T-T(0))\,\mathbb{E}[\boldsymbol{\mu}_0(t)]\,\mathbb{E}[\mathbf{V}_q(\infty)]. \tag{12.78}$$

From (12.70) $\to$ (12.78) we get that

$$\boldsymbol{\mu}(\infty) = T\mathbb{E}[\mathbf{v}_q(\infty)]. \tag{12.79}$$

Thus we have proved the following theorem:

Theorem 375 *Let a complete probability space $(\Omega, \mathcal{F}, \mathbb{P})$ and a finite P-NHMS as defined in section 12.3 In addition let that $\lim_{t\to\infty} T(t) = T$. Then*

$$\lim_{t\to\infty}\boldsymbol{\mu}(t) = \boldsymbol{\mu}(\infty) = T\mathbb{E}[\mathbf{v}_q(\infty)],$$

where

$$\mathbb{E}[\mathbf{v}_q(\infty)] = \left(\mathbb{E}[\mathbf{q}(\infty)] \quad \mathbb{E}[\mathbf{q}(\infty)]\,\mathbf{J}-\mathbf{J}\left(\mathbb{E}[\mathbf{q}(\infty)]\otimes\mathbb{E}[\mathbf{q}(\infty)]\right)\right),$$

with

$$\mathbb{E}[\mathbf{q}(\infty)] = \mathbf{e}_i\mathbf{Q}^*\left(\mathbf{I}+\mathcal{E}_q^*\mathbf{A}^\sharp\right)^{-1},$$

where $\mathbf{Q}^* = \lim_{t\to\infty}\mathbf{Q}^t$, $\mathbf{A}^\sharp$ is the group inverse of $\mathbf{A} = \mathbf{I}-\mathbf{Q}$, and

$$\mathcal{E}_q^* = \sum_{i=1}^{v}\mathcal{E}_p(i)\,c_i + \mathbf{p}_{k+1}^\top\sum_{i=1}^{m}\varepsilon_0(i)\,c_{0i} + \left\{\sum_{i=1}^{v}\varepsilon_{k+1}^\top(i)\,c_i\right\}\mathbf{p}_0$$
$$-\sum_{i=1}^{v}\sum_{j=1}^{m}\varepsilon_{k+1}^\top(i)\,\varepsilon_0(j)\,c_ic_{0j}.$$

12.8 Research notes

In several papers, the school of research by C.D. Meyer established the usefulness of the group generalized inverse in the study and computations of various aspects of Markov chains. We have already referred to some seminal of these papers and we will just mention in here for the reader the book Campbell and Meyer (2009). On the sensitivity of the stationary distribution to perturbation of finite homogeneous Markov chains important studies are those of Seneta (1988a, 1988b, 1993). While for the bounds on the condition number of a finite homogeneous Markov chain and on the error in the computation of the stationary distribution vector Kirkland et al (1998) used Paz's inequality and improved by a factor 2, some of the constants in previous studies.

For P-NHMS the study of the rate of convergence in the limiting expected population structure could be found in Symeonaki (1998). Also in the same reference there exist the study of the set of all possible limiting expected population structures when we control the input distribution and we allow it to span all possible values. The P-NHMS in continuous time is studied in Vassiliou and Symeonaki (1998). It is showed that if we perturb the intensity matrix $\mathbf{Q}$ of the imbedded Markov chain then $\mathbb{E}\left[\mathbf{Q}\left(t\right)\right]$ is an ergodic intensity matrix. Also the asymptotic behavior of the expected population structure and its asymptotic variability are found in elegant closed analytic forms. The Cyclic P-NHMS is defined in Symeonaki (1998) and its asymptotic behavior is studied. The theory of Perturbed Markov systems is extended to perturbed non-homogeneous semi-Markov systems (P-NHSMS) in Vassiliou and Tsakiridou (2004). The development of multidimensional sequence of matrices and multiple iterated limits for their convergence is the key theory for studying the limiting behavior of the expected population structure. Also, Pauwelyn and Guerry (2020, 2021) studied some other interesting kind of perturbations for stochastic matrices of non-diagonalizable stochastic matrices with preservation of spectral properties.

Chapter 13

NON-HOMOGENEOUS MARKOV SET SYSTEM

13.1 Introduction

All too often in the theory and applications of stochastic processes the values of the various parameters are assumed to be exact while in practice these are estimated from the data. Naturally the same applied in the present book for the theory and applications of non-homogeneous Markov chains and NHMS. However, the values of the parameters is more realistic to be viewed as being contained in intervals with the desired probability confidence. This is a classical approach for systems of linear equations and in this case the solutions are given as the set of all possible solutions. Two books have been written on this topic by Neumaier (1990) and Alefield and Herzberger (1983). For the analogous problem for differential equations a book was written by Ben-Haim and Elishakoff (1990).

In the present chapter in this respect the non-homogeneous Markov set system (NHMSS) is defined. An NHMSS was first defined and studied in Vassiliou (2021*b*) and the present Chapter is based to these findings. The essence of the definition is the extension of the definition of the NHMS given in Chapter 5 by allowing its basic parameters to lie in an interval. We will start with the definition of an interval for a stochastic vector following Hartfiel (1998) who was the first to allow the parameters of a homogeneous Markov chain to be defined in an interval.

Definition 376 *Let SM_{1n} the set of all $1 \times n$ stochastic vectors. Also let $\boldsymbol{\lambda}$ and $\boldsymbol{\mu}$ be nonnegative $1 \times n$ vectors with $\boldsymbol{\lambda} \leq \boldsymbol{\mu}$ componentwise. Then define the corresponding interval in SM_{1n} by*

$$[\boldsymbol{\lambda}, \boldsymbol{\mu}] = \{\mathbf{p} : \mathbf{p} \in SM_{1,n} \text{ with } \boldsymbol{\lambda} \leq \mathbf{p} \leq \boldsymbol{\mu}\},$$

where $\boldsymbol{\lambda}$,$\boldsymbol{\mu}$ are chosen such that $[\boldsymbol{\lambda}, \boldsymbol{\mu}] \neq \varnothing$.

Now, a geometrical example will be provided for the reader which will be helpful in the understanding of some concepts:

Example 377 *Let $SM_{1,3}$ the set of all 1×3 stochastic vectors. It is well known that, this is the convex hull of the vectors $\mathbf{e}_1 = \begin{pmatrix}1 & 0 & 0\end{pmatrix}$, $\mathbf{e}_2 = \begin{pmatrix}0 & 1 & 0\end{pmatrix}$ and $\mathbf{e}_3 = \begin{pmatrix}0 & 0 & 1\end{pmatrix}$ in $\mathbb{R}^3$. Then it is easy to see that, if $\mathbf{x}$ is a non-negative vector, such that $\boldsymbol{\lambda} \leq \mathbf{x} \leq \boldsymbol{\mu}$ then all possible values of $\mathbf{x}$ is the set which is included within and the surface of a rectangle the coordinates of which are determined by $\boldsymbol{\lambda}, \boldsymbol{\mu}$. The interval $[\boldsymbol{\lambda}, \boldsymbol{\mu}]$ will be the intersection of the two above described spaces. We can visualize this more easily if we consider the triangle $\mathbf{e}_1\mathbf{e}_2\mathbf{e}_3$ in $\mathbb{R}^2$. Let $\boldsymbol{\lambda} = \begin{pmatrix}0.1 & 0.2 & 0.3\end{pmatrix}$ and $\boldsymbol{\mu} = \begin{pmatrix}0.5 & 0.7 & 0.8\end{pmatrix}$ then the interval $[\boldsymbol{\lambda}, \boldsymbol{\mu}]$ is given in Fig 13.1.*

Looking at the figure it is self explanatory how the shadowed part which is the interval $[\boldsymbol{\lambda}, \boldsymbol{\mu}]$ is found geometrically. The doted lines represent the intersection of the planes in $\mathbb{R}^3$ and $SM_{1,3}$.

DOI: 10.1201/b23204-13

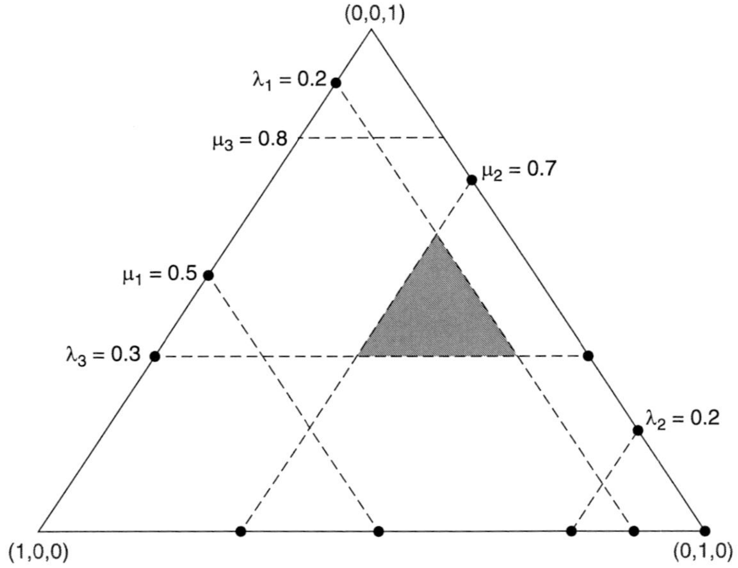

Figure 13.1 *Interval* $[\boldsymbol{\lambda}, \boldsymbol{\mu}]$

The definition of a *tight interval* will be proved useful in what follows:

Definition 378 *Let* $[\boldsymbol{\lambda}, \boldsymbol{\mu}]$ *be an interval, then if*

$$\lambda_i = \min_{\mathbf{x} \in [\boldsymbol{\lambda}, \boldsymbol{\mu}]} x_i \quad \text{and} \quad \mu_i = \max_{\mathbf{x} \in [\boldsymbol{\lambda}, \boldsymbol{\mu}]} x_i,$$

then λ_i, μ_i *are called tight respectively. If* λ_i, μ_i *are tight for all* i*, then the interval* $[\boldsymbol{\lambda}, \boldsymbol{\mu}]$ *is called tight.*

The following Lemma can be used to test intervals for tightness and also to tight up an interval which is not tight.

Lemma 379 *(Hartfiel (1998)). Let* $[\boldsymbol{\lambda}, \boldsymbol{\mu}]$ *be an interval. Then for each coordinate* i

$$(i) \ \lambda_i \text{ is tight if and only if } \lambda_i + \sum_{k \neq i} \mu_k \geq 1.$$

$$(ii) \ \mu_i \text{ is tight if and only if } \mu_i + \sum_{k \neq i} \lambda_k \geq 1.$$

The following definition of a *free* vector that belongs to a tight interval is useful.

Definition 380 *Let* $[\boldsymbol{\lambda}, \boldsymbol{\mu}]$ *be a tight interval and* $\mathbf{p} \in [\boldsymbol{\lambda}, \boldsymbol{\mu}]$*. Then if* $\lambda_i < p_i < \mu_i$ *for some coordinate* i*, then the coordinate* p_i *in* $\mathbf{p}$ *is called free.*

The link between convex sets and tight intervals is provided in the following useful Lemma:

Lemma 381 *(Hartfiel 1998). Let* $[\boldsymbol{\lambda}, \boldsymbol{\mu}]$ *be a tight an interval . Then* $[\boldsymbol{\lambda}, \boldsymbol{\mu}]$ *is a convex polytope. A vector* $\mathbf{p} \in [\boldsymbol{\lambda}, \boldsymbol{\mu}]$ *is a vertex of* $[\boldsymbol{\lambda}, \boldsymbol{\mu}]$ *if and only if* $\mathbf{p}$ *has at most one free component.*

The definition of an interval of a matrix and of a tight interval of matrices extending that of a vector are now provided from Harfiel (1988):

Definition 382 *Let SM_n the set of all $1 \times n$ stochastic matrices. Also let $\mathbf{\Lambda}$ and $\mathbf{M}$ be nonnegative $n \times n$ matrices with $\mathbf{\Lambda} \leq \mathbf{M}$ componentwise. Then define the corresponding interval in SM_n by*

$$[\mathbf{\Lambda}, \mathbf{M}] = \{\mathbf{P} : \mathbf{P} \in SM_n \text{ with } \mathbf{\Lambda} \leq \mathbf{P} \leq \mathbf{M}\},$$

where $\mathbf{\Lambda}, \mathbf{M}$ are chosen such that $[\mathbf{\Lambda}, \mathbf{M}] \neq \varnothing$.

We now proceed to define a tight interval of matrices:

Definition 383 *Let $[\mathbf{\Lambda}, \mathbf{M}]$ be an interval of matrices. If*

$$\lambda_{ij} = \min_{\mathbf{P} \in [\mathbf{\Lambda}, \mathbf{M}]} p_{ij} \text{ and } \mu_{ij} = \max_{\mathbf{P} \in [\mathbf{\Lambda}, \mathbf{M}]} p_{ij},$$

for all i and j, then $[\mathbf{\Lambda}, \mathbf{M}]$ is called tight.

The interval $[\mathbf{\Lambda}, \mathbf{M}]$ can be constructed also by rows. That is,

$$[\mathbf{\Lambda}, \mathbf{M}] = \left\{ \begin{array}{c} \mathbf{P} : \ \mathbf{p}_i \in [\boldsymbol{\lambda}_i, \boldsymbol{\mu}_i] \text{ for all } i, \text{ with } \mathbf{p}_i, \boldsymbol{\lambda}_i, \boldsymbol{\mu}_i \\ \text{being the rows of the respective matrices } \mathbf{P}, \mathbf{\Lambda}, \mathbf{M} \end{array} \right\}.$$

13.2 Non-homogeneous Markov set system

The entire notation introduced in Chapter 5 for an NHMS will be kept the same and we will build on that to define a non-homogeneous Markov set system.

Let $[\mathbb{M}] = \left[\check{\mathbf{Q}}, \hat{\mathbf{Q}}\right]$ be an interval of $k \times k$ stochastic matrices with $\check{\mathbf{Q}} \leq \mathbf{Q}(t) \leq \hat{\mathbf{Q}}$ for every $t \in \mathbb{N}$ which is tight. That is,

$$[\mathbb{M}] = \left\{\mathbf{Q}(t) : \text{is an } k \times k \text{ stochastic matrix with } \check{\mathbf{Q}} \leq \mathbf{Q}(t) \leq \hat{\mathbf{Q}}\right\}$$

with

$$\check{q}_{ij} = \min_{\mathbf{Q}(t) \in \left[\check{\mathbf{Q}}, \hat{\mathbf{Q}}\right]} q_{ij}(t) \text{ for every } t \in \mathbb{N},$$

$$\hat{q}_{ij} = \max_{\mathbf{Q}(t) \in \left[\check{\mathbf{Q}}, \hat{\mathbf{Q}}\right]} q_{ij}(t) \text{ for every } t \in \mathbb{N},$$

and the notation $\left[\check{\mathbf{Q}}, \hat{\mathbf{Q}}\right]$ will be taken to imply that $\left[\check{\mathbf{Q}}, \hat{\mathbf{Q}}\right] \neq \varnothing$.

The following basic assumptions will be made:

Assumption $\mathbf{A}_1$. *Let that the imbedded non-homogeneous Markov chain of the NHMS has all its probability matrices in $[\mathbb{M}]$.*

We call $[\mathbb{M}]$ the *probability transition matrix set (PTMS) of the imbedded non-homogeneous Markov chain.*

Let that

$$\left[\mathbb{M}^2\right] = \{\mathbf{Q}(0)\,\mathbf{Q}(1) : \mathbf{Q}(0), \mathbf{Q}(1) \in [\mathbb{M}]\},$$

$$\ldots\ldots\ldots\ldots\ldots\ldots\ldots\ldots,$$

$$[\mathbb{M}^n] = \{\mathbf{Q}(0)\,\mathbf{Q}(1) \ldots \mathbf{Q}(n-1) : \mathbf{Q}(0), \ldots, \mathbf{Q}(n-1) \in [\mathbb{M}]\}.$$

We call the sequence $\{[\mathbb{M}^n]\}_{n=1}^{\infty}$ the *inherent or imbedded non-homogeneous Markov set chain.*

Assumption $\mathbf{A}_2$. Let $[\mathbb{S}_0]$ be the set of $1 \times k$ stochastic vectors from which the initial distribution $\mathbf{q}(0)$ is chosen.

$$[\mathbb{S}_0] = [\check{\mathbf{q}}_0, \hat{\mathbf{q}}_0] = \{\mathbf{q}(0) : \text{is a stochastic vector with } \mathbf{q}_0(t) \in [\mathbb{S}_0]\}.$$

Assumption $\mathbf{A}_3$. Let $[\mathbb{R}_0]$ be the set of $1\times k$ stochastic vectors from which the allocation probabilities $\mathbf{p}_0(t)$ are being selected. That is

$$[\mathbb{R}_0] = [\check{\mathbf{p}}_0, \hat{\mathbf{p}}_0] = \{\mathbf{p}_0(t) : \text{ is a stochastic vector with } \mathbf{p}_0(t) \in [\mathbb{R}_0] \text{ for every } t\}.$$

An NHMS whose parameters are assumed to be in intervals as in Assumptions $\mathbf{A}_1$, $\mathbf{A}_2$, and $\mathbf{A}_3$ is called a *Non-homogeneous Markov Set System (NHMSS)*.

It is easy to see that, Markov set chains that were initiated by Hartfiel and Seneta (1994), Hartfiel (1991, 1994) and Hartfiel (1998) are special cases of an NHMSS.

13.3 The set of the expected relative population structures

The well known in geometry Minkowski sum (also known as dilation) of two sets of position vectors $\mathbb{A}$ and $\mathbb{B}$ in Euclidean space is formed by adding each vector in $\mathbb{A}$ to each vector in $\mathbb{B}$. That is

$$\mathbb{A} + \mathbb{B} = \{a + b : a \in \mathbb{A}, b \in \mathbb{B}\}.$$

Example 384 *Let two sets $\mathbb{A}$ and $\mathbb{B}$ consisting of three position vectors (informally, three points) representing the vertices of two triangles in $\mathbb{R}^2$ with coordinates*

$$\mathbb{A} = \{(1,0), (0,1), (0,-1)\} \text{ and } \mathbb{B} = \{(0,0), (1,1), (1,-1)\},$$

then their Minkowski sum is

$$\mathbb{A} + \mathbb{B} = \{(1,0), (2,1), (2,-1), (0,1), (1,2), (0,-1), (1,-2)\},$$

which comprises the vertices of a hexagon.

The zero set containing only the zero vector $\mathbf{0}$, in the Minkowski addition is the identity element. That is, $\mathbb{V}+\{0\} = \mathbb{V}$ in the vector space. Also it is very important to note that $\mathbb{V}+\varnothing = \varnothing$.

The following Lemma is borrowed from Eggleston (2008):

Lemma 385 *If $\mathbb{V}$ is a convex set then $\mu\mathbb{V} + \lambda\mathbb{V}$ is also a convex set and further more $\mu\mathbb{V} + \lambda\mathbb{V} = (\mu + \lambda)\mathbb{V}$ for every $\lambda, \mu > 0$. Conversely, if this "distributive property" holds for all non-negative real numbers $\lambda, \mu > 0$ then the set is convex.*

Remark 386 *For two convex polygons $\mathbb{V}_1$ and $\mathbb{V}_2$ in the plane with m and n vertices, their Minkowski sum is a convex polygon with at most $m + n$ vertices and may be computed in time $O(m + n)$ by a very simple procedure.*

Let the following sets which will be used in what follows

$$Rng(\mathbf{q}) = \{\mathbf{y} : \mathbf{y} = \mathbf{q}\mathbf{Q} \text{ for some } \mathbf{Q} \in [\mathbb{M}] \text{ and any } \mathbf{q} \in SM_{1,n}\}.$$

Also

$$Rng(\mathbb{S}) = \cup_{\mathbf{q}\in\mathbb{S}} Rng(\mathbf{q}) = \{\mathbf{y} : \mathbf{y} = \mathbf{q}\mathbf{Q} \text{ for some } \mathbf{Q} \in [\mathbb{M}] \text{ and some } \mathbf{q} \in \mathbb{S}\}.$$

Lemma 387 *(Vassiliou (2021b). Let an NHMSS with $T(t) \geq 0$ and finite, which is expanding $(\Delta T(t) \geq 0)$. Also let $[\mathbb{S}_0]$ the set from which the initial distribution of memberships is drawn, $[\mathbb{R}_0]$ the set from which the allocation probabilities in the various states are chosen at every time step and finally let $[\mathbb{M}]$ the set to which all the transition probability matrices of memberships belong. Then the set $Rng_t(\mathbb{S}_0, \mathbb{R}_0)$ of all possible expected relative population structures at time t is given by*

$$Rng_t(\mathbb{S}_0, \mathbb{R}_0) = \{\mathbb{E}[\mathbf{q}(0,t)] : \mathbb{E}[\mathbf{q}(0,t)] \in \frac{T(0)}{T(t)} \cup_{\mathbf{q}(0)\in[\mathbb{S}_0]} \mathbf{q}(0) [\mathbb{M}^t] \qquad (13.1)$$

$$+\frac{1}{T(t)}\sum_{\tau=1}^{t}\Delta T(\tau)\cup_{\mathbf{p}_0(\tau)\in[\mathbb{R}_0]}\mathbf{p}_0(\tau)\left[\mathbb{M}^{t-\tau}\right]\},$$

with $\left[\mathbb{M}^0\right]=[\mathbb{M}]$ *and* $\left[\mathbb{M}^{-1}\right]=\{\mathbf{I}\}$.

Proof. Following Vassiliou (1981), Vassiliou (1982) and the relevant proofs that lead to equations (2.1a), (2.1b) and (3.4) in Georgiou and Vassiliou (1992) we could prove (13.1). ■

The following Lemma from Hartfiel (1998) p. 39 will be borrowed:

Lemma 388 *In a non-homogeneous Markov set chain let the set of initial distributions* $[\mathbb{S}_0]$ *be convex and let* $[\mathbb{M}]$ *be a tight interval from which the transition probability sequence of matrices is being selected. Then the set* $[\mathbb{S}_t]$ *of all possible probability distributions at the various states at time t is a convex set.*

The next theorem from Vassiliou (2021b) shows under what conditions the set of all possible expected relative population structures in an NHMSS is a convex set.

Theorem 389 *(Vassiliou (2021b)). Let an NHMSS with* $T(t)$ *finite and which is expanding* $(\Delta T(t)\geq 0)$. *If* $[\mathbb{S}_0],[\mathbb{R}_0]$ *are convex sets and* $[\mathbb{M}]$ *a tight interval then the set of all possible expected relative population structures is a convex set.*

Proof. Define $\{\mathbb{E}[\mathring{\mathbf{q}}(0,t)]\}$ be the set of all possible expected relative structures of the initial memberships then from Lemma 387, we have

$$\{\mathbb{E}[\mathring{\mathbf{q}}(0,t)]\}=\{\mathbb{E}[\mathring{\mathbf{q}}(0,t)]:\mathbb{E}[\mathring{\mathbf{q}}(0,t)]\in\frac{T(0)}{T(t)}\cup_{\mathbf{q}(0)\in[\mathbb{S}_0]}\mathbf{q}(0)\left[\mathbb{M}^t\right]\},$$

since $\frac{T(0)}{T(t)}\geq 0$, $[\mathbb{S}_0]$ is a convex set and $[\mathbb{M}]$ a tight interval, from Lemmas 383 and 385 we get that $\{\mathbb{E}[\mathring{q}(0,t)]\}$ is a convex set. Also, the set

$$\{\mathbb{E}[\mathbf{r}_\tau]\}=\{\mathbb{E}[\mathbf{r}_\tau]:\mathbb{E}[\mathbf{r}_\tau]\in\Delta T(\tau)\cup_{\mathbf{p}(\tau)\in[\mathbb{R}_0]}\mathbf{p}_0(\tau)\left[\mathbb{M}^{t-\tau}\right]\},$$

the set of all possible relative expected structures of new memberships at time t which entered in the system at time τ. The system is expanding, that is $\Delta T(\tau)\geq 0$, $[\mathbb{R}_0]$ is a convex set and $[\mathbb{M}]$ a tight interval, from Lemma 381, we get that $\{\mathbb{E}[\mathbf{r}_\tau]\}$ is a convex set. Also, from Remark 386 we get that the Minkowski sum of sets

$$\frac{1}{T(t)}\sum_{\tau=1}^{t}\{\mathbb{E}[\mathbf{r}_\tau]\},$$

is a convex set. Hence, since the two sets in the right-hand side of equation (13.1) are convex and according to Lemma 385 their Minkowski sum $Rng_t(\mathbb{S}_0,\mathbb{R}_0)$ is a convex set. ■

The following theorem from Hartfiel (1998 p. 40) is borrowed:

Theorem 390 *In a non-homogeneous Markov set chain let the set of initial distributions* $[\mathbb{S}_0]$ *be a convex polytope and let* $[\mathbb{M}]$ *be a tight interval from which the transition probability sequence of matrices is being selected. Then the set* $[\mathbb{S}_t]$ *of all possible probability distributions at the various states at time t is a convex polytope with vertices of the form* $\mathcal{E}_iE_{i_1}...E_{i_k}$ *for some vertices* $\mathcal{E}_i$ *of* $[\mathbb{S}_0]$ *and some vertices* E_{i_j} *of* $[\mathbb{M}]$.

The next theorem from Vassiliou (2021b) shows under which conditions, the set of all possible expected relative population structures in an NHMSS is a convex polytope.

Theorem 391 *Let an NHMSS with* $T(t)$ *finite and which is expanding* $(\Delta T(t)\geq 0)$. *If* $[\mathbb{S}_0],[\mathbb{R}_0]$ *are convex polytopes and* $[\mathbb{M}]$ *a tight interval then the set of all possible expected relative population structures is a convex polytope.*

Proof. The proof follows the steps of the proof of Theorem 389. ■

The following Theorem from Vassiliou (2021b) concerns the evolution of the set $Rng_t(\mathbb{S}_0, \mathbb{R}_0)$ as $t \to \infty$.

Theorem 392 *Let an NHMSS with $T(t)$ finite and which is expanding ($\Delta T(t) \geq 0$) also let that $\mathbb{S}_0 = \mathbb{R}_0 = SM_{1,n}$. Then it is true that*

$$Rng_{t+1}(\mathbb{S}_0, \mathbb{R}_0) \subseteq Rng_t(\mathbb{S}_0, \mathbb{R}_0).$$

Also, define by

$$Rng_\infty(\mathbb{S}_0, \mathbb{R}_0) = \lim_{t\to\infty} Rng_t(\mathbb{S}_0, \mathbb{R}_0) = \cap_{t=1}^{\infty} Rng_t(\mathbb{S}_0, \mathbb{R}_0)$$

then $Rng_\infty(\mathbb{S}_0, \mathbb{R}_0)$ is not empty and is a convex set if in addition $[\mathbb{S}_0], [\mathbb{R}_0]$ are convex sets and $[\mathbb{M}]$ a tight interval.

Proof. Let $\mathbb{E}[\mathbf{q}(0,t)] \in Rng_t(\mathbb{S}_0, \mathbb{R}_0)$ be an element of the set. Then there exist $\mathbf{q}(0) \in [\mathbb{S}_0]$, $\mathbf{Q}(1), ..., \mathbf{Q}(t) \in [\mathbb{M}]$ and $\mathbf{p}_0(1), ..., \mathbf{p}_0(t) \in [\mathbb{R}_0]$ such that

$$\mathbb{E}[\mathbf{q}(0,t)] = \frac{T(0)}{T(t)}\mathbf{q}(0)\mathbf{Q}(0,t) + \frac{1}{T(t)}\sum_{\tau=1}^{t}\Delta T(\tau)\mathbf{p}_0(\tau)\mathbf{Q}(\tau,t), \tag{13.2}$$

denote by

$$\mathbf{q}(0)\mathbf{Q}(1) = \mathbf{q}^*(1) \in \cup_{\mathbf{q}(0)\in[\mathbb{S}_0]}\mathbf{q}(0)[\mathbb{M}] \subseteq [\mathbb{S}_0], \tag{13.3}$$

now, $[\mathbb{S}_0]$ and $[\mathbb{M}]$ uniquely defines a non-homogeneous Markov set chain and $T(t) \geq T(t-1)$ therefore we get that

$$\frac{T(0)}{T(t)}\mathbf{q}(0)\mathbf{Q}(0,t) = \frac{T(0)}{T(t)}\mathbf{q}^*(1)\mathbf{Q}(1,t) \leq \frac{T(0)}{T(t-1)}\mathbf{q}^*(1)\mathbf{Q}(1,t) \tag{13.4}$$

$$\in \{\mathbb{E}[\mathring{\mathbf{q}}(0,t)] : \mathbb{E}[\mathring{\mathbf{q}}(0,t)] \in \frac{T(0)}{T(t-1)} \cup_{\mathbf{q}^*(0)\in[\mathbb{S}_0]} \mathbf{q}^*(1)\left[\mathbb{M}^t\right]\}.$$

The following is true

$$\begin{aligned}\mathbf{p}_0(\tau)\mathbf{Q}(\tau) &= \mathbf{p}_0^*(\tau+1) \in \cup_{\mathbf{p}_0^*(\tau+1)\in[\mathbb{R}_0]}\mathbf{p}_0^*(\tau+1)\left[\mathbb{M}^{t-\tau}\right] \subseteq [\mathbb{R}_0] \\ \text{for every } \tau &= 1,2,...,t,\end{aligned} \tag{13.5}$$

from the fact that $\mathbf{p}_0(\tau) \in [\mathbb{R}_0]$ is a stochastic vector and $\mathbf{Q}(\tau) \in [\mathbb{M}]$ uniquely defines a non-homogeneous Markov set chain and $T(t) \geq T(t-1)$ we arrive at

$$\frac{\Delta T(\tau)}{T(t)}\mathbf{p}_0(\tau)\mathbf{Q}(\tau,t) = \frac{\Delta T(\tau)}{T(t)}\mathbf{p}_0^*(\tau+1)\mathbf{Q}(\tau+1,t)$$

$$\leq \frac{\Delta T(\tau)}{T(t)}\mathbf{p}_0^*(\tau+1)\mathbf{Q}(\tau+1,t)$$

$$\in \{\frac{1}{T(t)}\Delta T(\tau) \cup_{\mathbf{p}^*(\tau+1)\in[\mathbb{R}_0]} \mathbf{p}_0^*(\tau)\left[\mathbb{M}^{t-\tau}\right]\} \text{ for every } \tau.$$

Therefore,

$$\frac{1}{T(t)}\sum_{\tau=1}^{t}\Delta T(\tau)\mathbf{p}_0(\tau)\mathbf{Q}(\tau,t) \tag{13.6}$$

$$\in \{\frac{1}{T(t-1)}\sum_{\tau=1}^{t-1}\Delta T(\tau) \cup_{\mathbf{p}_0(\tau+1)\in[\mathbb{R}_0]} \mathbf{p}_0(\tau+1)\left[\mathbb{M}^{t-\tau}\right]\}.$$

From (13.2), (13.3), (13.4), (13.5) and (13.6) we get that

$$\mathbb{E}\left[\mathbf{q}\left(0,t\right)\right] \in Rng_{t-1}\left(\mathbb{S}_0, \mathbb{R}_0\right),$$

and thus

$$Rng_t\left(\mathbb{S}_0, \mathbb{R}_0\right) \subseteq Rng_{t-1}\left(\mathbb{S}_0, \mathbb{R}_0\right).$$

From Vassiliou (1981a, 1982, 2015) and Georgiou and Vassiliou (1992) we get various conditions under which $\lim_{t\to\infty} \mathbb{E}\left[\mathbf{q}\left(0,t\right)\right]$ exists, thus $Rng_\infty\left(\mathbb{S}_0, \mathbb{R}_0\right) \neq \varnothing$. Also, by Theorem 389 $Rng_t\left(\mathbb{S}_0, \mathbb{R}_0\right)$ is convex for every t, hence $Rng_\infty\left(\mathbb{S}_0, \mathbb{R}_0\right)$ is a convex set. ■

13.4 Asymptotic behavior of NHMSS

As we learned so far in this book the problem of asymptotic behavior has been one of central importance for, homogeneous Markov chains, non-homogeneous Markov chains, NHMS, and non-homogeneous Markov set chains. In the present section we will establish a series of theorems concerning the asymptotic behavior of an NHMSS.

The coefficient of ergodicity $\mathcal{T}\left(\mathbf{Q}\right)$ of a $k \times k$ stochastic matrix $\mathbf{Q} = \{q_{ij}\}_{i,j=1}^{k}$, as we have seen so far, has been a fundamental tool in the study of Markov chains. As is also shown in Seneta (1981)

$$\mathcal{T}\left(\mathbf{Q}\right) = \frac{1}{2}\max_{ij}\sum_{l=1}^{k}\left|q_{il} - q_{jl}\right|,$$

thus $0 \leq \mathcal{T}\left(\mathbf{Q}\right) \leq 1$. We repeat in here that if $\mathcal{T}\left(\mathbf{Q}\right) < 1$ the stochastic matrix $\mathbf{Q}$ is called *scrambling.* Scrambling matrices are regular, but not all regular matrices are scrambling. Now, if $\mathbf{Q}$ is a regular stochastic matrix then some power of $\mathbf{Q}$, say $\mathbf{Q}^n$ is scrambling. We denote by

$$\bar{\mathcal{T}}\left(\left[\mathbb{M}\right]\right) = \max_{\mathbf{Q}\in\left[\check{\mathbf{Q}},\hat{\mathbf{Q}}\right]} \mathcal{T}\left(\mathbf{Q}\right),$$

if $\bar{\mathcal{T}}\left(\left[\mathbb{M}\right]\right) < 1$ we say that $\left[\mathbb{M}\right]$ is *uniformly scrambling.* The interpretation of the coefficient of ergodicity of a matrix $\mathbf{A} \in \mathbb{R}^{n\times n}$ as a matrix norm, also in Hartfiel and Rothblum (1998). For explicit forms for ergodicity coefficients and properties see also Rhodius (1993) and Seneta (1993).

The concept of $\bar{\mathcal{T}}\left(\left[\mathbb{M}\right]\right) < 1$ is useful in the study of asymptotic behavior in an NHMSS. Consider as $\left(\left[\mathbb{M}^C\right], \|.\|_1\right)$ the space of non-empty compact subsets of $\left(\left[\mathbb{M}\right], \|.\|_1\right)$ then $\left(\left[\mathbb{M}^C\right], \|.\|_1\right)$ is a metric space (Berge (1963)). This space can be topologized using the Hausdorff metric $d\left(.,.\right)$ defined by

$$d\left(\mathbb{S}_1, \mathbb{S}_2\right) = \max\left\{\delta\left(\mathbb{S}_1, \mathbb{S}_2\right), \delta\left(\mathbb{S}_2, \mathbb{S}_1\right)\right\},$$

where

$$\delta\left(\mathbb{S}_1, \mathbb{S}_2\right) = \max_{\mathbf{Q}_1\in S_1}\min_{\mathbf{Q}_2\in S_2}\left\|\mathbf{Q}_1 - \mathbf{Q}_2\right\|_1 \text{ and } \mathbb{S}_1, \mathbb{S}_2 \in \left[\mathbb{M}^C\right],$$

From Berge (1963), we get also that $\left(\left[\mathbb{M}^C\right], d\left(.,.\right)\right)$ is a metric space.

When the initial structures for two different sets of expected relative population structures are different as well as their allocation probabilities then the following theorem answers the problem of what happens asymptotically.

Theorem 393 *(Vassiliou (2021b)). Let an NHMSS for which* $\left[\mathbb{M}\right]$ *is a tight interval which is uniformly scrambling. Assume that*

$$\left\{\frac{\Delta T\left(t\right)}{T\left(t\right)}\right\}_{t=0}^{\infty} \textit{ converges to zero geometrically fast.}$$

Also, assume that $[\mathbb{S}], [\mathbb{R}_0]$ *are compact and convex. Let* $\mathbb{S}_1, \mathbb{S}_2 \subseteq [\mathbb{S}^C]$ *and* $\mathbb{R}_1, \mathbb{R}_2 \subseteq [\mathbb{R}_0^C]$ *then*

$$d\left(Rng_t\left(\mathbb{S}_1, \mathbb{R}_1\right), Rng_t\left(\mathbb{S}_2, \mathbb{R}_2\right)\right) \to_{t\to\infty} 0 \;\; \textit{geometrically fast.}$$

Proof. From (13.1) we get that

$$\begin{aligned} Rng_t\left(\mathbb{S}_1, \mathbb{R}_1\right) &= \{\mathbb{E}\left[\mathbf{q}\left(0,t\right)\right] : \mathbb{E}\left[\mathbf{q}\left(0,t\right)\right] \in \frac{T\left(0\right)}{T\left(t\right)} \cup_{\mathbf{q}(0)\in[\mathbb{S}_1]} \mathbf{q}\left(0\right)\left[\mathbb{M}^t\right] + \\ &\quad \frac{1}{T\left(t\right)} \sum_{\tau=1}^{t} \Delta T\left(\tau\right) \cup_{\mathbf{p}_0(\tau)\in[\mathbb{R}_1]} \mathbf{p}_0\left(\tau\right)\left[\mathbb{M}^{t-\tau}\right]\}, \end{aligned} \tag{13.7}$$

and by replacing in (13.7), $\mathbf{q}(0)$ with $\bar{\mathbf{q}}(0)$, $[\mathbb{S}_1]$ with $[\mathbb{S}_2]$; $\mathbf{p}_0(\tau)$ with $\bar{\mathbf{p}}_0(\tau)$; $[\mathbb{R}_1]$ with $[\mathbb{R}_2]$ and $\mathbb{E}[\mathbf{q}(0,t)]$ with $\mathbb{E}[\bar{\mathbf{q}}(0,t)]$ we get $Rng_t(\mathbb{S}_2, \mathbb{R}_2)$. From the fact that $[\mathbb{S}_1], [\mathbb{R}_1]$ are compact and $[\mathbb{M}]$ is a tight interval, it is straightforward to prove that $Rng_t(\mathbb{S}_1, \mathbb{R}_1)$ is compact. The same is true for $Rng_t(\mathbb{S}_2, \mathbb{R}_2)$ since $[\mathbb{S}_2], [\mathbb{R}_2]$ are also compact. Therefore, there is a meaning in taking their Hausdorf metric. We get that

$$\begin{aligned} &\delta\left(Rng_t\left(\mathbb{S}_1, \mathbb{R}_1\right), Rng_t\left(\mathbb{S}_2, \mathbb{R}_2\right)\right) \\ =& \max_{\substack{\mathbf{q}(0)\in[\mathbb{S}_1], \mathbf{p}_0(\tau)\in[\mathbb{R}_1], \\ \mathbf{Q}(t)\in[\mathbb{M}] \text{ for all } \tau,t}} \; \min_{\substack{\bar{\mathbf{q}}(0)\in[\mathbb{S}_2], \bar{\mathbf{p}}_0(\tau)\in[\mathbb{R}_2] \\ \bar{\mathbf{Q}}(t)\in[\mathbb{M}] \text{ for all } \tau,t}} \left\|\mathbb{E}\left[\mathbf{q}\left(0,t\right)\right] - \mathbb{E}\left[\bar{\mathbf{q}}\left(0,t\right)\right]\right\|_1 . \end{aligned} \tag{13.8}$$

by continuity of

$$f\left(\mathbb{E}\left[\bar{q}\left(0,t\right)\right]\right) = \min \left\|\mathbb{E}\left[q\left(0,t\right)\right] - \mathbb{E}\left[\bar{q}\left(0,t\right)\right]\right\|_1$$

then for some $\mathbf{q}^*(0) \in [\mathbb{S}_1]$, $\mathbf{p}_0^*(\tau) \in [\mathbb{R}_1]$ for every $\tau \in [0,t]$, $\mathbf{Q}^*(t) \in [\mathbb{M}]$ for every $t \in [0,t]$ and by setting with

$$\mathbb{E}\left[\mathbf{q}^*\left(0,t\right)\right] = \frac{T\left(0\right)}{T\left(t\right)}\mathbf{q}^*\left(0\right)\mathbf{Q}^*\left(0,t-1\right) + \frac{1}{T\left(t\right)}\sum_{\tau=1}^{t}\Delta T\left(\tau\right)\mathbf{p}_0^*\left(\tau\right)\mathbf{Q}^*\left(\tau,t\right), \tag{13.9}$$

we have that

$$\begin{aligned} &\delta\left(Rng_t\left(\mathbb{S}_1, \mathbb{R}_1\right), Rng_t\left(\mathbb{S}_2, \mathbb{R}_2\right)\right) = \\ &\min_{\substack{\bar{\mathbf{q}}(0)\in[\mathbb{S}_2], \bar{\mathbf{p}}_0(\tau)\in[\mathbb{R}_2] \\ \bar{\mathbf{Q}}(t)\in[\mathbb{M}] \text{ for all } \tau,t}} \left\|\mathbb{E}\left[\mathbf{q}^*\left(0,t\right)\right] - \mathbb{E}\left[\bar{\mathbf{q}}\left(0,t\right)\right]\right\|_1 \\ &\leq \left\|\mathbb{E}\left[\mathbf{q}^*\left(0,t\right)\right] - \mathbb{E}\left[\bar{\mathbf{q}}^*\left(0,t\right)\right]\right\|_1 , \end{aligned} \tag{13.10}$$

where $\mathbb{E}[\bar{\mathbf{q}}^*(0,t)]$ is driven from (13.9) by replacing $\mathbf{q}^*(0)$ with any $\bar{\mathbf{q}}^*(0) \in [\mathbb{S}_2]$, $\mathbf{p}_0^*(\tau)$ with any $\bar{\mathbf{p}}_0^*(\tau) \in [\mathbb{R}_2]$ for every $\tau \in [0, t-1]$. Also, from (13.9) and (13.10) we get that

$$\begin{aligned} \delta\left(Rng_t\left(\mathbb{S}_1, \mathbb{R}_1\right), Rng_t\left(\mathbb{S}_2, \mathbb{R}_2\right)\right) &\leq \frac{T\left(0\right)}{T\left(t\right)}\left\|\left[\mathbf{q}^*\left(0\right) - \bar{\mathbf{q}}^*\left(0\right)\right]\mathbf{Q}^*\left(0,t\right)\right\|_1 \\ &+\frac{1}{T\left(t\right)}\sum_{\tau=1}^{t}\Delta T\left(\tau\right)\left\|\mathbf{p}_0^*\left(\tau\right)\mathbf{Q}^*\left(\tau,t\right) - \bar{\mathbf{p}}_0^*\left(\tau\right)\mathbf{Q}^*\left(\tau,t\right)\right\|_1 \\ &\leq \frac{T\left(0\right)}{T\left(t\right)}\mathcal{T}\left(\mathbf{Q}^*\left(0,t\right)\right)\left\|\left[\mathbf{q}^*\left(0\right) - \bar{\mathbf{q}}^*\left(0\right)\right]\right\|_1 \\ &+\frac{1}{T\left(t\right)}\sum_{\tau=1}^{t}\Delta T\left(\tau\right)\mathcal{T}\left(\mathbf{Q}^*\left(\tau,t\right)\right)\left\|\mathbf{p}_0^*\left(\tau\right) - \bar{\mathbf{p}}_0^*\left(\tau\right)\right\|_1 \end{aligned}$$

(since $[\mathbb{M}]$ is uniformly scrambling $\tilde{\mathcal{T}} = \max_{\mathbf{Q}^*(t)\in [\mathbb{M}]} \mathcal{T}\left(\mathbf{Q}^*\left(t\right)\right) < 1$)

$$\leq \frac{T(0)}{T(t)}\tilde{\mathcal{T}}^{t}\left\|[\mathbf{q}^{*}(0)-\bar{\mathbf{q}}^{*}(0)]\right\|_{1}$$

$$+\frac{1}{T(t)}\sum_{\tau=1}^{t}\Delta T(\tau)\tilde{\mathcal{T}}^{t-\tau}\left\|\mathbf{p}_{0}^{*}(\tau)-\bar{\mathbf{p}}_{0}^{*}(\tau)\right\|_{1}$$

$$\leq \frac{T(0)}{T(t)}\tilde{\mathcal{T}}^{t}\delta(\mathbb{S}_1,\mathbb{S}_2)+\frac{1}{T(t)}\sum_{\tau=1}^{t}\Delta T(\tau)\tilde{\mathcal{T}}^{t-\tau}\boldsymbol{\delta}(\mathbb{R}_1,\mathbb{R}_2). \tag{13.11}$$

From the fact that $\left\{\frac{\Delta T(t)}{T(t)}\right\}_{t=0}^{\infty}$ converges to zero geometrically fast, from Lemma 387 and 388 we conclude that there are c_1, $c_2>0$ and $0<b_1,b_2<1$ such that

$$\left|\frac{1}{T(t)}-\frac{1}{T}\right|\leq c_1 b_1^t \quad \text{and} \quad \left|\frac{\Delta T(t)}{T(t)}\right|\leq c_2 b_2^t. \tag{13.12}$$

From $(13.11),(13.12)$ we get that

$$\delta(Rng_t(\mathbb{S}_1,\mathbb{R}_1),Rng_t(\mathbb{S}_2,\mathbb{R}_2))\leq T(0)\left(\frac{1}{T}+c_1b_1^t\right)\tilde{\mathcal{T}}^{t}\delta(\mathbb{S}_1,\mathbb{S}_2)$$

$$+\boldsymbol{\delta}(\mathbb{R}_1,\mathbb{R}_2)\left(\frac{1}{T}+c_1b_1^t\right)\sum_{\tau=1}^{t}c_2b_2^{\tau}\tilde{\mathcal{T}}^{t-\tau},$$

(assuming $\tilde{\mathcal{T}}<b_2$)

$$\leq T(0)\left(\frac{1}{T}+c_1b_1^t\right)\tilde{\mathcal{T}}^{t}\delta(\mathbb{S}_1,\mathbb{S}_2)+\boldsymbol{\delta}(\mathbb{R}_1,\mathbb{R}_2)\left(\frac{1}{T}+c_1b_1^t\right)c_2b_2^{t-1}\sum_{\tau=1}^{t}\left(\frac{\tilde{\mathcal{T}}}{b_2}\right)^{t-\tau}$$

$$\leq T(0)\left(\frac{1}{T}+c_1b_1^t\right)\tilde{\mathcal{T}}^{t}\delta(\mathbb{S}_1,\mathbb{S}_2)+\boldsymbol{\delta}(\mathbb{R}_1,\mathbb{R}_2)\left(\frac{1}{T}+c_1b_1^t\right)c_2b_2^{t-1}\left(1-\frac{\tilde{\mathcal{T}}}{b_2}\right)^{-1}. \tag{13.13}$$

Therefore, from (13.13), we have that $\delta(Rng_t(\mathbb{S}_1,\mathbb{R}_1),Rng_t(\mathbb{S}_2,\mathbb{R}_2))$ converges to zero geometrically fast. Analogously, we arrive at the relative conclusion if $\tilde{\mathcal{T}}>b_2$.

Similarly, it is possible to prove that $\delta(Rng_t(\mathbb{S}_2,\mathbb{R}_2),Rng_t(\mathbb{S}_1,\mathbb{R}_1))$ converges to zero geometrically fast. ■

In the following theorem the conditions under which there is geometrically fast convergence is studied.

Theorem 394 *(Vassiliou (2021b). Let an NHMSS for which* $[\mathbb{M}]$ *is a tight interval which is uniformly scrambling. Assume that*

$$\left\{\frac{\Delta T(t)}{T(t)}\right\}_{t=0}^{\infty} \quad \textit{converges to zero geometrically fast.}$$

Also, let that $[\mathbb{S}],[\mathbb{R}_0]$ *are compact and convex. Then we have that*

$$d(Rng_t(\mathbb{S},\mathbb{R}_0),Rng_{\infty}(\mathbb{S},\mathbb{R}_0)\to 0, \quad \textit{geometrically fast.} \tag{13.13}$$

Proof. The first step is to prove that

$$Rng(Rng_t(\mathbb{S},\mathbb{R}_0))=Rng_{t+1}(\mathbb{S},\mathbb{R}_0). \tag{13.14}$$

By the evaluation of the function $Rng(.,.)$ given in Lemma 381 we get that

$$Rng(Rng_t(\mathbb{S},\mathbb{R}_0))=\frac{T(t)}{T(t+1)}\cup_{\mathbb{E}[\mathbf{q}(0,t)]\in Rng_t(\mathbb{S},\mathbb{R}_0)}\mathbb{E}[\mathbf{q}(0,t)][\mathbb{M}]$$

$$+\frac{1}{T(t+1)}\Delta T(t+1)\cup_{\mathbf{p}_0(t)\in[\mathbb{R}_0]}\mathbf{p}_0(t)=$$

$$\frac{T(t)}{T(t+1)}\frac{T(0)}{T(t)}\cup_{\mathbf{q}(0)\in[\mathbb{S}]}\mathbf{q}(0)\left[\mathbb{M}^t\right][\mathbb{M}]+$$

$$\frac{T(t)}{T(t+1)}\frac{1}{T(t)}\sum_{\tau=1}^{t}\Delta T(\tau)\cup_{\mathbf{p}_0(\tau)\in[\mathbb{R}_0]}\mathbf{p}_0(\tau)\left[\mathbb{M}^{t-\tau}\right][\mathbb{M}]$$

$$+\frac{1}{T(t+1)}\Delta T(t+1)\cup_{\mathbf{p}_0(t)\in[\mathbb{R}_0]}\mathbf{p}_0(t)$$

$$=\frac{T(0)}{T(t+1)}\cup_{\mathbf{q}(0)\in[\mathbb{S}]}\mathbf{q}(0)\left[\mathbb{M}^{t+1}\right]+$$

$$\frac{1}{T(t+1)}\sum_{\tau=1}^{t+1}\Delta T(\tau)\cup_{\mathbf{p}_0(\tau)\in[\mathbb{R}_0]}\mathbf{p}_0(\tau)\left[\mathbb{M}^{t-\tau+1}\right]=Rng_{t+1}(\mathbb{S},\mathbb{R}_0).$$

Secondly it will proved the following

$$Rng_t\left(Rng_\infty(\mathbb{S},\mathbb{R}_0)\right)=Rng_\infty(\mathbb{S},\mathbb{R}_0). \tag{13.15}$$

Let that $\mathbb{E}[\mathbf{q}(\infty)]$ is an element of $Rng_\infty(\mathbb{S},\mathbb{R}_0)$ then from Lemma 381, we have that

$$Rng\left(Rng_\infty(\mathbb{S},\mathbb{R}_0)\right)=\frac{T}{T}\cup_{\mathbb{E}[\mathbf{q}(\infty)]\in Rng_\infty(\mathbb{S},\mathbb{R}_0)}\mathbb{E}[\mathbf{q}(\infty)][\mathbb{M}]+$$

$$\frac{1}{T}(T-T)\cup_{\mathbf{p}_0(\infty)\in[\mathbb{R}_0]}\mathbf{p}_0(\infty)=Rng_\infty(\mathbb{S},\mathbb{R}_0)[\mathbb{M}]=Rng_\infty(\mathbb{S},\mathbb{R}_0), \tag{13.16}$$

where the last equality will be proved in Theorem 398. From (13.16) we recursively get (13.15).

From Theorem 393 is also valid in the present theorem and hence we have:

$$d\left(Rng_t(\mathbb{S}_1,\mathbb{R}_1),Rng_t(\mathbb{S}_2,\mathbb{R}_2)\right)\to 0 \text{ geometrically fast.} \tag{13.17}$$

Now, in (13.17) replace $Rng_t(\mathbb{S}_1,\mathbb{R}_1)$ with $Rng_t\left(Rng_\infty(\mathbb{S},\mathbb{R}_0)\right)$ and $Rng_t(\mathbb{S}_2,\mathbb{R}_2)$ with $Rng_t(\mathbb{S},\mathbb{R}_0)$,

$$d\left(Rng_t\left(Rng_\infty(\mathbb{S},\mathbb{R}_0)\right),Rng_t(\mathbb{S},\mathbb{R}_0)\right)\to 0,\ \text{ geometrically fast.} \tag{13.18}$$

From $(13.15),(13.18)$ we arrive at

$$d(Rng_t(\mathbb{S},\mathbb{R}_0),Rng_\infty(\mathbb{S},\mathbb{R}_0)\to 0,\ \text{ geometrically fast.}$$

■

The following theorem could be proved by just following the analogous steps as in the proofs of Theorems 393 and 394 and Theorem 3.3 in Vassiliou (1981a):

Theorem 395 *Consider an NHMS and assume that*

$$\left\{\frac{\Delta T(t)}{T(t)}\right\}_{t=0}^{\infty} \text{ converges geometrically fast.}$$

Let $\{\mathbf{Q}(t)\}_{t=0}^{\infty}$ *be the sequence of transition matrices and* $\{\mathbf{p}_0(t)\}_{t=0}^{\infty}$ *be the sequence of allocation probabilities. If* $\sup_t \tilde{\mathcal{T}}(\mathbf{Q}(t))<1$ *then*

$$\lim_{t\to\infty}\mathbb{E}(\mathbf{q}(0,t))=\mathbf{q}(\infty) \text{ geometrically fast,}$$

where $\mathbf{q}(\infty)$ *is the row of the stable matrix* $\mathbf{Q}(\infty)=\lim_{t\to\infty}\mathbf{Q}(0,t)$.

We now simply state the following result which could be proved analogously.

Theorem 396 *Let two NHMS* **a** *and* **b** *which have the same number of states, a common sequence of transition probability matrices of memberships* $\{\mathbf{Q}(t)\}_{t=0}^{\infty}$ *and the same total population of memberships. Assume that*

$$\left\{\frac{\Delta T(t)}{T(t)}\right\}_{t=0}^{\infty} \text{ converges geometrically fast.}$$

Let $\mathbf{q_a}(0)$ *and* $\mathbf{q_b}(0)$ *be their initial relative population structures respectively, and* $\{\mathbf{p_a}(t)\}_{t=0}^{\infty}$ *and* $\{\mathbf{p_b}(t)\}_{t=0}^{\infty}$ *their sequences of allocation probabilities. If* $\sup_t \tilde{\mathcal{T}}(\mathbf{Q}(t)) < 1$ *then*

$$\lim_{t\to\infty} \mathbb{E}(\mathbf{q_a}(0,t)) = \lim_{t\to\infty} \mathbb{E}(\mathbf{q_b}(0,t)) = \mathbf{q}(\infty) \quad \text{geometrically fast,}$$

where $\mathbf{q}(\infty)$ *is the row of the stable matrix* $\mathbf{Q}(\infty) = \lim_{t\to\infty} \mathbf{Q}(0,t)$.

13.5 Properties of the limiting set $Rng_\infty(\mathbb{S}, \mathbb{R}_0)$

Important properties of the limit set $Rng_\infty(\mathbb{S}, \mathbb{R}_0)$ will be studied in the present section. Let the following definition:

Definition 397 *Let* n *be an integer such that* $\mathcal{T}(\mathbf{Q}_1\mathbf{Q}_2...\mathbf{Q}_n) < 1$ *for all* $\mathbf{Q}_1, \mathbf{Q}_2, ..., \mathbf{Q}_n \in [\mathbb{M}]$. *Then* $[\mathbb{M}]$ *is said to be product scrambling and* n *its scrambling integer.*

The following Theorem 3.3 in Hartfiel (1998) is transferred

Theorem 398 *(Hartfiel (1998)). Let* $\mathbf{x}$, $\mathbf{y}$ *be non-compact subsets of* $SM_{1,n}$. *Then using the Hausdorff metric, we have*

$$d(\mathbf{x}\mathbb{M}, \mathbf{y}\mathbb{M}) \le \mathcal{T}(\mathbb{M})\, d(\mathbf{x}, \mathbf{y}).$$

A theorem with some important properties for the limiting set $Rng_\infty(\mathbb{S}, \mathbb{R}_0)$ will be established.

Theorem 399 *(Vassiliou (2021b)). Consider an NHMSS for which* $[\mathbb{M}]$ *is an interval which is uniformly scrambling. Assume that* $T(t) \ge T(t-1) > 0$ *for all* $t = 1, 2, ...$ *and*

$$\left\{\frac{\Delta T(t)}{T(t)}\right\}_{t=0}^{\infty} \text{ converges to zero geometrically fast.}$$

Also, assume that $[\mathbb{S}_0], [\mathbb{R}_0]$ *are compact and convex. Then if we define by*

$$Rng_\infty(\mathbb{S}_0, \mathbb{R}_0) = \lim_{t\to\infty} Rng_t(\mathbb{S}_0, \mathbb{R}_0) = \cap_{t=1}^{\infty} Rng_t(\mathbb{S}_0, \mathbb{R}_0),$$

the limit set $Rng_\infty(\mathbb{S}_0, \mathbb{R}_0)$ *satisfies*

$$Rng_\infty(\mathbb{S}_0, \mathbb{R}_0) = Rng_\infty(\mathbb{S}_0, \mathbb{R}_0)[\mathbb{M}].$$

If in addition $[\mathbb{M}]$ *is product scrambling with integer* n *then it is the unique set that has this property.*

Proof. Since $Rng_\infty(\mathbb{S}_0, \mathbb{R}_0)$ is compact, it is sufficient to show that

$$d(Rng_\infty(\mathbb{S}_0, \mathbb{R}_0), Rng_\infty(\mathbb{S}_0, \mathbb{R}_0)[\mathbb{M}]) = 0.$$

We have that

$$\delta(Rng_\infty(\mathbb{S}_0, \mathbb{R}_0), Rng_\infty(\mathbb{S}_0, \mathbb{R}_0)[\mathbb{M}]) =$$

$$\max_{\substack{\mathbf{q}(0)\in[\mathbb{S}_0],\\ \mathbf{p}_0(\tau)\in[\mathbb{R}_0]\\ \mathbf{Q}(t)\in[\mathbb{M}] \text{ for all } \tau,t}} \min_{\substack{\mathbf{q}(0)\in[\mathbb{S}_0],\\ \mathbf{p}_0(\tau)\in[\mathbb{R}_0]\\ \mathbf{Q}(t)\in[\mathbb{M}] \text{ for all } \tau,t}} \left\| \lim_{t\to\infty} \{\mathbb{E}[\mathbf{q}(0,t+1)]\} - \lim_{t\to\infty} \{\mathbb{E}[\mathbf{q}(0,t)]\} [\mathbb{M}] \right\|,$$

where

$$\{\mathbb{E}[\mathbf{q}(0,t+1)]\} = \cup_{\mathbf{q}(0)\in[\mathbb{S}_0]} \frac{T(0)}{T(t+1)} \mathbf{q}(0) \left[\mathbb{M}^{t+1}\right]$$

$$+\frac{1}{T(t+1)} \sum_{\tau=1}^{t+1} \Delta T(\tau) \cup_{\mathbf{p}_0(\tau)\in[\mathbb{R}_0]} \mathbf{p}_0^*(\tau) \left[\mathbb{M}^{t-\tau+1}\right].$$

There exists a $\mathbf{q}^*(0) \in [\mathbb{S}_0]$, $\mathbf{Q}^*(t) \in [\mathbb{M}]$ for $t \in [0,t]$, $\mathbf{p}_0^*(\tau) \in [\mathbb{R}_0]$ for $\tau \in [1,t]$, that is,

$$\mathbb{E}[\mathbf{q}^*(0,t+1)] = \frac{T(0)}{T(t+1)} \mathbf{q}^*(0) \mathbf{Q}^*(0,t) +$$

$$\frac{1}{T(t+1)} \sum_{\tau=1}^{t+1} \Delta T(\tau) \mathbf{p}_0^*(\tau) \mathbf{Q}^*(\tau,t),$$

such that

$$\delta(Rng_\infty(\mathbb{S}_0,\mathbb{R}_0), Rng_\infty(\mathbb{S}_0,\mathbb{R}_0)[\mathbb{M}]) =$$

$$\min_{\substack{\mathbf{q}(0)\in[\mathbb{S}_0],\mathbf{p}_0(\tau)\in[\mathbb{R}_0],\\ \mathbf{Q}(t)\in[\mathbb{M}] \text{ for all } \tau,t}} \left\| \lim_{t\to\infty} \mathbb{E}[\mathbf{q}^*(0,t+1)] - \lim_{t\to\infty} \{\mathbb{E}[\mathbf{q}(0,t)]\} [\mathbb{M}] \right\|_1. \tag{13.19}$$

The values of the parameters of $\mathbb{E}[\mathbf{q}(0,t)]$ that does not maximize it, the difference above is still greater or equal than the ones which minimize $\mathbb{E}[\mathbf{q}(0,t)]$. Thus, we are free to choose the parameters $\mathbf{q}^*(0) \in [\mathbb{S}_0]$, $\mathbf{Q}^*(0,t) \in [\mathbb{M}^t]$ and $\mathbf{p}_0^*(\tau) \in [\mathbb{R}_0]$ for every $\tau \in [0,t]$. With the same reasoning we could choose $\mathbf{Q}^*(t)$ in the place of $[\mathbb{M}]$. Thus, we get that

$$\delta(Rng_\infty(\mathbb{S}_0,\mathbb{R}_0), Rng_\infty(\mathbb{S}_0,\mathbb{R}_0)[\mathbb{M}]) =$$

$$\leq \lim_{t\to\infty} \left\| \frac{T(0)}{T(t+1)} \mathbf{q}^*(0) \mathbf{Q}^*(0,t) - \frac{T(0)}{T(t)} \mathbf{q}^*(0) \mathbf{Q}^*(0,t) \right\|_1$$

$$+ \lim_{t\to\infty} \| \frac{1}{T(t+1)} \sum_{\tau=1}^{t+1} \Delta T(\tau) \mathbf{p}_0^*(\tau) \mathbf{Q}^*(\tau,t) -$$

$$\frac{1}{T(t)} \sum_{\tau=1}^{t} \Delta T(\tau) \mathbf{p}_0^*(\tau) \mathbf{Q}^*(\tau,t) \|_1$$

$$\leq \lim_{t\to\infty} \left| \frac{T(0)}{T(t+1)} - \frac{T(0)}{T(t)} \right| \|\mathbf{q}^*(0)\|_1 \|\mathbf{Q}^*(0,t)\|_1 +$$

$$\lim_{t\to\infty} \left| \frac{1}{T(t+1)} - \frac{1}{T(t)} \right| \left\| \sum_{\tau=1}^{t} \Delta T(\tau) \mathbf{p}_0^*(\tau) \mathbf{Q}^*(\tau,t) \right\|_1$$

$$+ \lim_{t\to\infty} \left| \frac{\Delta T(t)}{T(t+1)} \right| \|\mathbf{p}_0^*(t)\|_1. \tag{13.20}$$

From the fact that $[\mathbb{M}]$ is uniformly scrambling we get that $\bar{\mathcal{T}} = max_{t\in\mathbb{N}} \mathcal{T}(\mathbf{Q}^*(t)) < 1$ and thus $\|\mathbf{Q}^*(\tau,t)\| < \bar{\mathcal{T}}^{t-\tau}$ and

$$\left\| \sum_{\tau=1}^{t} \Delta T(\tau) \mathbf{p}_0^*(\tau) \mathbf{Q}^*(\tau,t) \right\|_1 \leq \sum_{\tau=1}^{t} \Delta T(\tau) \bar{\mathcal{T}}^{t-\tau},$$

which goes to zero as $t \to \infty$. Therefore

$$\delta\left(Rng_{\infty}(\mathbb{S}_0, \mathbb{R}_0), Rng_{\infty}(\mathbb{S}_0, \mathbb{R}_0)[\mathbb{M}]\right) = 0.$$

Similarly we prove that

$$\delta\left(Rng_{\infty}(\mathbb{S}_0, \mathbb{R}_0)[\mathbb{M}], Rng_{\infty}(\mathbb{S}_0, \mathbb{R}_0)\right) = 0,$$

and therefore

$$d\left(Rng_{\infty}(\mathbb{S}_0, \mathbb{R}_0)[\mathbb{M}], Rng_{\infty}(\mathbb{S}_0, \mathbb{R}_0)\right) = 0.$$

When in addition $[\mathbb{M}]$ is a product scrambling with index n then assume there is a second $Rng^*_{\infty}(\mathbb{S}_0, \mathbb{R}_0)$ which is compact for which

$$Rng^*_{\infty}(\mathbb{S}_0, \mathbb{R}_0)[\mathbb{M}] = Rng^*_{\infty}(\mathbb{S}_0, \mathbb{R}_0).$$

In such a case, we have

$$\begin{aligned} d\left(Rng^*_{\infty}(\mathbb{S}_0, \mathbb{R}_0), Rng_{\infty}(\mathbb{S}_0, \mathbb{R}_0)\right) &= d\left(Rng^*_{\infty}(\mathbb{S}_0, \mathbb{R}_0)[\mathbb{M}^n], Rng_{\infty}(\mathbb{S}_0, \mathbb{R}_0)[\mathbb{M}^n]\right) \\ &\leq \mathcal{T}[\mathbb{M}^n]\, d\left(Rng^*_{\infty}(\mathbb{S}_0, \mathbb{R}_0), Rng_{\infty}(\mathbb{S}_0, \mathbb{R}_0)\right), \end{aligned}$$

and from $\mathcal{T}[\mathbb{M}^n] < 1$ we have

$$d\left(Rng^*_{\infty}(\mathbb{S}_0, \mathbb{R}_0), Rng_{\infty}(\mathbb{S}_0, \mathbb{R}_0)\right) = 0,$$

which lead to the conclusion that $Rng^*_{\infty}(\mathbb{S}_0, \mathbb{R}_0)$ and $Rng_{\infty}(\mathbb{S}_0, \mathbb{R}_0)$ are the same set. ■

We borrow the following Lemma from Hartfiel (1998) p. 70 or Hartfiel (1993).

Lemma 400 *Let* $\mathbf{q}(0)$, $\tilde{\mathbf{q}}(0)$ *be stochastic vectors and* $\mathbf{Q}(t) \in [\mathbb{M}]; \tilde{\mathbf{Q}}(t) \in \left[\tilde{\mathbb{M}}\right]$ *for* $t = 1, 2, \ldots$ *for* $\mathcal{T}(\mathbb{M}) \leq \mathcal{T} < 1$ *and* $\mathcal{T}\left(\tilde{\mathbb{M}}\right) \leq \mathcal{T} < 1$. *Then*

$$\left\|\mathbf{q}(0)\mathbf{Q}(1,t) - \tilde{\mathbf{q}}(0)\tilde{\mathbf{Q}}(1,t)\right\| \leq \mathcal{T}^t \left\|\mathbf{q}(0) - \tilde{\mathbf{q}}(0)\right\|_1 + \left(\mathcal{T}^{t-1} + \ldots + 1\right)\mathbf{D}$$

where $\mathbf{D} = \max_n \left\|\mathbf{Q}(n) - \tilde{\mathbf{Q}}(n)\right\|$.

Now the following is an interesting result:

Theorem 401 *(Vassiliou (2021b)). Let two NHMSS with inherent Markov set chains with two different tight intervals* $[\mathbb{M}]$ *and* $\left[\tilde{\mathbb{M}}\right]$ *which have a common bound, that is,* $\mathcal{T}(\mathbb{M}) \leq \mathcal{T} < 1$ *and* $\mathcal{T}\left(\tilde{\mathbb{M}}\right) \leq \mathcal{T} < 1$. *Let that for the first NHMSS, we have* $\mathbf{q}(0) \in [\mathbb{S}_1]$ *and* $\mathbf{p}_0(\tau) \in [\mathbb{R}_1]$ *for* $\tau = 0, 1, \ldots, t$ *where* $[\mathbb{S}_1]$, $[\mathbb{R}_1]$ *are convex and compact. For the second NHMSS we assume that* $\tilde{\mathbf{q}}(0) \in [\mathbb{S}_2]$ *and* $\tilde{\mathbf{p}}_0(\tau) \in [\mathbb{R}_2]$ *for* $\tau = 0, 1, \ldots, t$ *where* $[\mathbb{S}_2]$, $[\mathbb{R}_2]$ *are convex and compact. Let that the two NHMSS have common total population of memberships* $\{T(t)\}_{t=0}^{\infty}$ *which is known. Assume that* $T(t) \geq T(t-1) > 0$ *and*

$$\left\{\frac{\Delta T(t)}{T(t)}\right\}_{t=0}^{\infty} \quad \text{converges to zero geometrically fast.}$$

Denote by $Rng_t(\mathbb{S}_1, \mathbb{R}_1, \mathbb{M})$ *the set of all possible expected relative population structures for the first NHMSS and* $Rng_t\left(\mathbb{S}_2, \mathbb{R}_2, \tilde{\mathbb{M}}\right)$ *for the second one respectively. Then the Hausdorff metric of the two limit sets of expected relative population structures is bounded by*

$$d\left(Rng_{\infty}(\mathbb{S}_1, \mathbb{R}_1, \mathbb{M}), Rng_{\infty}\left(\mathbb{S}_2, \mathbb{R}_2, \tilde{\mathbb{M}}\right)\right) \leq (1 - \mathcal{T})^{-1} d\left(\mathbb{M}, \tilde{\mathbb{M}}\right).$$

Proof. From Lemma 387, we have

$$Rng_t(\mathbb{S}_1,\mathbb{R}_1,\mathbb{M}) = \{\mathbb{E}[\mathbf{q}(0,t)] : \mathbb{E}[\mathbf{q}(0,t)] \in \frac{T(0)}{T(t)} \cup_{\mathbf{q}(0)\in[\mathbb{S}_1]} \mathbf{q}(0)\left[\mathbb{M}^t\right] +$$

$$+\frac{1}{T(t)}\sum_{\tau=1}^{t}\Delta T(\tau)\cup_{\mathbf{p}_0(\tau)\in[\mathbb{R}_0]}\mathbf{p}_0(\tau)\left[\mathbb{M}^{t-\tau}\right]\},$$

also the set $Rng_t\left(\mathbb{S}_2,\mathbb{R}_2,\tilde{\mathbb{M}}\right)$ could be expressed analogously. From the fact that the intervals $[\mathbb{S}_1]$, $[\mathbb{R}_1]$, $[\mathbb{S}_2]$, $[\mathbb{R}_2]$ are convex and compact and $[\mathbb{M}]$ and $\left[\tilde{\mathbb{M}}\right]$ are tight intervals we get that the sets $Rng_t(\mathbb{S}_1,\mathbb{R}_1,\mathbb{M})$ and $Rng_t\left(\mathbb{S}_2,\mathbb{R}_2,\tilde{\mathbb{M}}\right)$ are convex and compact, therefore we may get their Hausdorff metric

$$\begin{aligned} & d\left(Rng_t(\mathbb{S}_1,\mathbb{R}_1,\mathbb{M}), Rng_t\left(\mathbb{S}_2,\mathbb{R}_2,\tilde{\mathbb{M}}\right)\right) = \\ & \max\{\delta\left(Rng_t(\mathbb{S}_1,\mathbb{R}_1,\mathbb{M}), Rng_t\left(\mathbb{S}_2,\mathbb{R}_2,\tilde{\mathbb{M}}\right)\right), \\ & \delta\left(Rng_t\left(\mathbb{S}_2,\mathbb{R}_2,\tilde{\mathbb{M}}\right), Rng_t(\mathbb{S}_1,\mathbb{R}_1,\mathbb{M})\right)\}. \end{aligned} \tag{13.22}$$

The following is true

$$\begin{aligned} & \delta\left(Rng_t(\mathbb{S}_1,\mathbb{R}_1,\mathbb{M}), Rng_t\left(\mathbb{S}_2,\mathbb{R}_2,\tilde{\mathbb{M}}\right)\right) = \\ & \max_{\substack{\mathbf{q}(0)\in[\mathbb{S}_1],\mathbf{p}_0(\tau)\in[\mathbb{R}_1],\\ \mathbf{Q}(t)\in[\mathbb{M}] \text{ for all } \tau,t}} \ \min_{\substack{\tilde{\mathbf{q}}(0)\in[\mathbb{S}_2],\tilde{\mathbf{p}}_0(\tau)\in[\mathbb{R}_2],\\ \tilde{\mathbf{Q}}(t)\in[\tilde{\mathbb{M}}] \text{ for all } \tau,t}} \left\|\mathbb{E}[\mathbf{q}(0,t)] - \mathbb{E}[\tilde{\mathbf{q}}(0,t)]\right\|_1. \end{aligned} \tag{13.23}$$

There exists some $\mathbf{q}^*(0) \in [\mathbb{S}_1]$, $\mathbf{p}_0^*(\tau)$ for every $\tau \in [0,t]$, $\mathbf{Q}^*(t) \in [\mathbb{M}]$ for every $t \in [0,t]$ which determines a $\mathbb{E}[\mathbf{q}^*(0,t)]$; also for any $\tilde{\mathbf{q}}^*(0) \in [\mathbb{S}_2]$, any $\tilde{\mathbf{p}}_0^*(\tau)$ for every $\tau \in [0,t]$, and finally any $\tilde{\mathbf{Q}}^*(t) \in \left[\tilde{\mathbb{M}}\right]$ for every $t \in [0,t]$ which determines a $\mathbb{E}[\tilde{\mathbf{q}}^*(0,t)]$ for which we have

$$\begin{aligned} \delta\left(Rng_t(\mathbb{S}_1,\mathbb{R}_1,\mathbb{M}), Rng_t\left(\mathbb{S}_2,\mathbb{R}_2,\tilde{\mathbb{M}}\right)\right) & \leq \left\|\mathbb{E}[\mathbf{q}^*(0,t)] - \mathbb{E}[\tilde{\mathbf{q}}^*(0,t)]\right\|_1 \\ & \leq \frac{T(0)}{T(t)}\left\|\mathbf{q}^*(0)\mathbf{Q}^*(0,t) - \tilde{\mathbf{q}}^*(0)\bar{\mathbf{Q}}^*(0,t)\right\|_1 \\ & +\frac{1}{T(t)}\sum_{\tau=1}^{t}\Delta T(\tau)\left\|\mathbf{p}_0^*(\tau)\mathbf{Q}^*(\tau,t) - \tilde{\mathbf{p}}_0^*(\tau)\tilde{\mathbf{Q}}^*(\tau,t)\right\|_1. \end{aligned} \tag{13.24}$$

From Lemma 400, we have

$$\begin{aligned} & \frac{T(0)}{T(t)}\left\|\mathbf{q}^*(0)\mathbf{Q}^*(0,t) - \tilde{\mathbf{q}}^*(0)\tilde{\mathbf{Q}}^*(0,t)\right\|_1 \leq \\ & \frac{T(0)}{T(t)}\mathcal{T}^t\left\|\mathbf{q}^*(0) - \tilde{\mathbf{q}}^*(0)\right\|_1 + \frac{T(0)}{T(t)}\left\|\mathcal{T}^{t-1} + \ldots + 1\right\|_1 \max_t \left\|\mathbf{Q}^*(t) - \tilde{\mathbf{Q}}^*(t)\right\|_1 \\ & \leq \frac{T(0)}{T(t)}\mathcal{T}^t\delta(\mathbb{S}_1,\mathbb{S}_2) + \frac{T(0)}{T(t)}\left\|\mathcal{T}^{t-1} + \ldots + 1\right\|_1\delta\left(\mathbb{M},\tilde{\mathbb{M}}\right). \end{aligned} \tag{13.25}$$

Following similar steps, we get that

$$\frac{1}{T(t)}\sum_{\tau=1}^{t}\Delta T(\tau)\left\|\mathbf{p}_0^*(\tau)\mathbf{Q}^*(\tau,t) - \tilde{\mathbf{p}}_0^*(\tau)\tilde{\mathbf{Q}}^*(\tau,t)\right\|_1$$

$$\leq \frac{1}{T(t)}\sum_{\tau=1}^{t}\Delta T(\tau)\,\delta(\mathbb{R}_1,\mathbb{R}_2)\,\mathcal{T}^{t-\tau}+$$

$$\frac{1}{T(t)}\sum_{\tau=1}^{t}\Delta T(\tau)\left(\mathcal{T}^{t-1}+...+1\right)\delta\left(\mathbb{M},\tilde{\mathbb{M}}\right). \tag{13.26}$$

From $(13.24), (13.25)$ and (13.26) as $t \to \infty$, we have

$$\delta\left(Rng_{\infty}(\mathbb{S}_1,\mathbb{R}_1,\mathbb{M}), Rng_{\infty}\left(\mathbb{S}_2,\mathbb{R}_2,\tilde{\mathbb{M}}\right)\right) \leq (1-\mathcal{T})^{-1}\,\delta\left(\mathbb{M},\tilde{\mathbb{M}}\right), \tag{13.27}$$

from the following

$$\lim_{t\to\infty}\frac{1}{T(t)}\sum_{\tau=1}^{t}\Delta T(\tau)\left(\mathcal{T}^{t-1}+...+1\right)\delta\left(\mathbb{M},\tilde{\mathbb{M}}\right)$$

$$=\lim_{t\to\infty}\frac{1}{T(t)}\delta\left(\mathbb{M},\tilde{\mathbb{M}}\right)\sum_{\tau=1}^{t}\Delta T(\tau)\sum_{k=\tau}^{t-1}\mathcal{T}^{t-k}\leq$$

$$\lim_{t\to\infty}\frac{1}{T(t)}(1-\mathcal{T})^{-1}\delta\left(\mathbb{M},\tilde{\mathbb{M}}\right)\sum_{\tau=1}^{t}\Delta T(\tau)=\left[1-\frac{T(0)}{T}\right](1-\mathcal{T})^{-1}\delta\left(\mathbb{M},\tilde{\mathbb{M}}\right),$$

also

$$\lim_{t\to\infty}\frac{T(0)}{T(t)}\mathcal{T}^{t}\delta(\mathbb{S}_1,\mathbb{S}_2)+\lim_{t\to\infty}\frac{T(0)}{T(t)}\left\|\mathcal{T}^{t-1}+...+1\right\|_1\delta\left(\mathbb{M},\tilde{\mathbb{M}}\right)$$

$$=\frac{T(0)}{T}(1-\mathcal{T})^{-1}\delta\left(\mathbb{M},\tilde{\mathbb{M}}\right)$$

and finally

$$\lim_{t\to\infty}\frac{1}{T(t)}\sum_{\tau=1}^{t}\Delta T(\tau)\,\delta(\mathbb{R}_1,\mathbb{R}_2)\,\mathcal{T}^{t-\tau+1}=0,$$

as we have seen in the proof of Theorem 396. In a similar way, as it is proved in (13.25) the following could be proved

$$\delta\left(Rng_{\infty}\left(\mathbb{S}_2,\mathbb{R}_2,\tilde{\mathbb{M}}\right), Rng_{\infty}(\mathbb{S}_1,\mathbb{R}_1,\mathbb{M})\right) \leq (1-\mathcal{T})^{-1}\,\delta\left(\tilde{\mathbb{M}},\mathbb{M}\right). \tag{13.28}$$

Relations(13.27) and (13.28) prove the theorem. ∎

13.6 An illustrative representative example

An illustrative example will be illustrated for a Geriatric and Stroke Patients system. Also it will present the methodology in terms of computational geometry algorithms needed. A general Coxian phase type model is used, special forms of which could be found in the studies by the school of research by McClean and her co-authors McClean (1998a,b), Taylor et al. (2000), Marshall et al. (2002), Marshall and McClean (2003, 2004), McClean et al. (2014a,b). For the present example three states which are called hospital pathways are distinguished. For the system of Geriatric and Stroke Patients these stages are labeled as "Acute Care", the "Rehabilative" and the "Long Stay". There are transitions outside the hospital due to discharge or death. Other transitions are those of geriatric patients progressing through stages of acute care, rehabilitation and long-stay care, where most patients are eventually rehabilitated and discharged. It is certainly an indication of the level of civilization in a society the geriatric medical services the availability of which is an asset for the elderly. At the same time, their funding could be easily reduced due to political pressure on savings in healthcare expenditure.

There is no need to consider in here a larger number of states due to the restriction of space.

Consider a hospital which starts with $T(0) = 400$ patients and in a very short time reaches its full capacity of 435 patients. That is, $T(1) = 420$, $T(2) = 430$, $T(3) = 435$. Assume three hospital pathways and let that the initial relative population structure be any stochastic vector which lies in the set

$$\mathbb{S}_0 = \{[0 \;\; 0 \;\; 0], \quad [1 \;\; 1 \;\; 1]\}.$$

The vast majority of new patients enter the system in hospital pathway one, either by taking an empty place or as a virtual replacement of a discharged patient. Therefore, let $\mathbb{R}_0$ be the convex set from which allocation probabilities are drown and let it be the interval

$$\mathbb{R}_0 = \{[0.5 \;\; 0.2 \;\; 0.1], \quad [0.7 \;\; 0.4 \;\; 0.1]\}$$

The set of stochastic vectors of allocation probabilities that are in the above interval is a convex set with vertices $r_1 = [0.7 \;\; 0.2 \;\; 0.1]$ and $r_2 = [0.5 \;\; 0.4 \;\; 0.1]$. Assume from past data that the interval of transition matrices are $[\mathbb{M}] = \left[\check{\mathbf{Q}}, \hat{\mathbf{Q}}\right]$. We need that $[\mathbb{M}] = \left[\check{\mathbf{Q}}, \hat{\mathbf{Q}}\right]$ should be tight. In order to test that interval $[\mathbb{M}]$ is tight we do the following.

We choose $[\mathbb{M}]$ to be

$$\check{\mathbf{Q}} = \begin{pmatrix} 0.5 & 0.2 & 0.1 \\ 0.2 & 0.6 & 0.2 \\ 0.3 & 0.2 & 0.3 \end{pmatrix} \quad \text{and} \quad \hat{\mathbf{Q}} = \begin{pmatrix} 0.7 & 0.4 & 0.1 \\ 0.2 & 0.6 & 0.2 \\ 0.5 & 0.5 & 0.3 \end{pmatrix},$$

in order that $[\mathbb{M}]$ is tight every row should be tight. Therefore, the stochastic matrices that belong to $[\mathbb{M}]$ is a convex polytope and we need to find its vertices. The same applies for $\mathbb{S}_0$ and $\mathbb{R}_0$ which are tight and we need to find the vertices of the convex polytope on which all stochastic vectors of the interval lie. The following Algorithm from Vassiliou (2021b) describes these steps

Algorithm A1: Finding the vertices of a tight interval $[\mathbf{p}, \mathbf{q}]$.

For each $i = 1, 2, 3$ construct the vectors:

For $i = 1$ and the vector $\mathbf{p}$

$$[p_1 \; p_2 \;\; p_3], \; [p_1 \;\; p_2 \;\; q_3], \; [p_1 \; q_2 \;\; p_3], \; [p_1 \; q_2 \;\; q_3],$$

replace p_1 with $\tilde{p}_1$ such that the above vectors will become stochastic.

If any of the four resulting stochastic vectors

$$[\tilde{p}_1 \; p_2 \;\; p_3], \; [\tilde{p}_1 \;\; p_2 \;\; q_3], \; [\tilde{p}_1 \; q_2 \;\; p_3], \; [\tilde{p}_1 \; q_2 \;\; q_3],$$

belongs in the interval $[\mathbf{p}, \mathbf{q}]$ then it is a vertex.

Do the same for the vector $\mathbf{q}$.

END.

Test that the set $\mathbb{S}_0$ is tight and apply Algorithm A1 to find that $\mathbb{S}_0$ is convex with vertices $\boldsymbol{v}_1 = (1 \;\; 0 \;\; 0)$, $\boldsymbol{v}_2 = (0 \;\; 1 \;\; 0)$ and $\boldsymbol{v}_3 = (0 \;\; 0 \;\; 1)$. The physical meaning of the set $\mathbb{S}_0$ in here is that at the start of our study at time 0 we allow that the hospital could have any relative population structure. Find that the set $\mathbb{R}_0$ is tight and apply Algorithm A1 to get that $\mathbb{R}_0$ is convex with vertices $\mathbf{r}_1 = (0.7 \;\; 0.2 \;\; 0.1)$, $\mathbf{r}_2 = (0.5 \;\; 0.4 \;\; 0.1)$.

In order to find the vertices of the convex set of stochastic matrices that belong to the tight interval $[\mathbb{M}]$ apply Algorithm A1 for each row vector in $\left[\check{\mathbf{Q}}, \hat{\mathbf{Q}}\right]$ and find that the vertices are

$$\mathbf{V}_1 = \begin{pmatrix} 0.7 & 0.2 & 0.1 \\ 0.2 & 0.6 & 0.2 \\ 0.5 & 0.2 & 0.3 \end{pmatrix}, \quad \mathbf{V}_2 = \begin{pmatrix} 0.7 & 0.2 & 0.1 \\ 0.2 & 0.6 & 0.2 \\ 0.3 & 0.4 & 0.3 \end{pmatrix},$$

$$\mathbf{V}_3 = \begin{pmatrix} 0.5 & 0.4 & 0.1 \\ 0.2 & 0.6 & 0.2 \\ 0.5 & 0.2 & 0.3 \end{pmatrix}, \quad \mathbf{V}_4 = \begin{pmatrix} 0.5 & 0.4 & 0.1 \\ 0.2 & 0.6 & 0.2 \\ 0.3 & 0.4 & 0.3 \end{pmatrix}.$$

Now compute all the row vectors $\boldsymbol{v}_i \mathbf{V}_j$ for $i = 1, 2, 3$ and $j = 1, 2, 3, 4$:

$$\begin{pmatrix} 0.7 & 0.2 & 0.1 \end{pmatrix}, \begin{pmatrix} 0.7 & 0.2 & 0.1 \end{pmatrix}, \begin{pmatrix} 0.5 & 0.4 & 0.1 \end{pmatrix}, \begin{pmatrix} 0.5 & 0.4 & 0.1 \end{pmatrix},$$

$$\begin{pmatrix} 0.2 & 0.6 & 0.2 \end{pmatrix}, \begin{pmatrix} 0.2 & 0.6 & 0.2 \end{pmatrix}, \begin{pmatrix} 0.2 & 0.6 & 0.2 \end{pmatrix}, \begin{pmatrix} 0.2 & 0.6 & 0.2 \end{pmatrix}$$

$$\begin{pmatrix} 0.5 & 0.2 & 0.3 \end{pmatrix}, \begin{pmatrix} 0.3 & 0.4 & 0.5 \end{pmatrix}, \begin{pmatrix} 0.5 & 0.2 & 0.3 \end{pmatrix}, \begin{pmatrix} 0.3 & 0.4 & 0.3 \end{pmatrix}.$$

These 12 stochastic vectors belong to a convex set, therefore using any of the computational geometry methods in Preparata and Shamos (1985) find that the vertices of the convex hull of these vectors are

$$\boldsymbol{\omega}_1 = \begin{pmatrix} 0.7 & 0.2 & 0.1 \end{pmatrix}, \boldsymbol{\omega}_2 = \begin{pmatrix} 0.5 & 0.4 & 0.1 \end{pmatrix}, \boldsymbol{\omega}_1 = \begin{pmatrix} 0.2 & 0.6 & 0.2 \end{pmatrix},$$

$$\boldsymbol{\omega}_4 = \begin{pmatrix} 0.5 & 0.2 & 0.3 \end{pmatrix}, \boldsymbol{\omega}_5 = \begin{pmatrix} 0.3 & 0.4 & 0.3 \end{pmatrix}.$$

Thus

$$Rng_1 (\mathbb{S}_0, \mathbb{R}_0) = \frac{T(0)}{T(1)} conv \{\boldsymbol{\omega}_1, \boldsymbol{\omega}_2, \boldsymbol{\omega}_3, \boldsymbol{\omega}_4, \boldsymbol{\omega}_5\} +$$

$$\frac{1}{T(1)} [T(1) - T(0)] \, conv \{r_1, r_2\},$$

from which you get that $Rng_1 (\mathbb{S}_0, \mathbb{R}_0)$ is the convex set with vertices given by the above Minkowski sum

$$\begin{pmatrix} 0.7 & 0.2 & 0.1 \end{pmatrix}, \begin{pmatrix} 0.51 & 0.39 & 0.1 \end{pmatrix}, \begin{pmatrix} 0.22 & 0.58 & 0.2 \end{pmatrix}, \begin{pmatrix} 0.51 & 0.2 & 0.29 \end{pmatrix},$$

$$\begin{pmatrix} 0.32 & 0.39 & 0.29 \end{pmatrix}, \begin{pmatrix} 0.69 & 0.21 & 0.1 \end{pmatrix}, \begin{pmatrix} 0.5 & 0.4 & 0.1 \end{pmatrix},$$

$$\begin{pmatrix} 0.21 & 0.59 & 0.2 \end{pmatrix}, \begin{pmatrix} 0.5 & 0.21 & 0.29 \end{pmatrix}, \begin{pmatrix} 0.31 & 0.4 & 0.29 \end{pmatrix},$$

$$\begin{pmatrix} 0.51 & 0.2 & 0.29 \end{pmatrix}.$$

Taking into account the rounding errors done with all the multiplications and additions compute the vertices with one decimal point of accuracy and get that $Rng_1 (\mathbb{S}_0, \mathbb{R}_0)$ is the convex hull of the vertices

$$\begin{pmatrix} 0.7 & 0.2 & 0.1 \end{pmatrix}, \begin{pmatrix} 0.5 & 0.4 & 0.1 \end{pmatrix}, \begin{pmatrix} 0.2 & 0.6 & 0.2 \end{pmatrix}, \begin{pmatrix} 0.5 & 0.2 & 0.3 \end{pmatrix},$$

$$\text{and} \quad \begin{pmatrix} 0.3 & 0.4 & 0.3 \end{pmatrix}.$$

Having now the vertices of the convex set $Rng_1 (\mathbb{S}_0, \mathbb{R}_0)$ we repeat the previous process to find $Rng_2 (\mathbb{S}_0, \mathbb{R}_0)$. It is apparent that $Rng_1 (\mathbb{S}_0, \mathbb{R}_0) \subset Rng_0 (\mathbb{S}_0, \mathbb{R}_0) = \mathbb{S}_0$.

We need a sufficient condition as a criterion to verify that a tight interval of transition probability matrices is uniformly scrambling. Such a criterion is provided in the following easily proved Lemma.

Lemma 402 *Let* $[\mathbb{M}] = \left[\check{\mathbf{Q}}, \hat{\mathbf{Q}}\right]$ *then* $[\mathbb{M}]$ *is uniformly scrambling if the following holds* $\mathcal{T}\left(\hat{\mathbf{Q}} - \check{\mathbf{Q}}\right) < 1$.

For our application it is easy to check that $\mathcal{T}\left(\hat{\mathbf{Q}} - \check{\mathbf{Q}}\right) < 1$ and hence any stochastic matrix selected from $[\mathbb{M}]$ will be scrambling. Let $\mathbf{Q}(1)$, $\mathbf{Q}(2)$, $\mathbf{Q}(3)$, $\mathbf{Q}(4)$ be the four vertices of the convex set of stochastic matrices in $[\mathbb{M}]$ and as $\mathbf{Q}(6)$, $\mathbf{Q}(7)$ any convex combination of them. Let $\mathbf{q}(0) = \begin{pmatrix} 0.5 & 0.25 & 0.25 \end{pmatrix}$ and let vectors of allocation probabilities be the following $\mathbf{p}_0(1) = \begin{pmatrix} 0.6 & 0.3 & 0.1 \end{pmatrix}$ and $\mathbf{p}_0(2) = \begin{pmatrix} 0.5 & 0.4 & 0.1 \end{pmatrix}$ which both belong to $\mathbb{R}_0$. We get

$$\mathbb{E}[\mathbf{q}(0,1)] = \begin{pmatrix} 0.5 & 0.3 & 0.2 \end{pmatrix}$$

$$\mathbb{E}[\mathbf{q}(0,2)] = \begin{pmatrix} 0.48 & 0.36 & 0.16 \end{pmatrix}$$

$$\mathbb{E}[\mathbf{q}(0,3)] = \begin{pmatrix} 0.4 & 0.43 & 0.17 \end{pmatrix}$$

$$\mathbb{E}[\mathbf{q}(0,4)] = \begin{pmatrix} 0.34 & 0.48 & 0.18 \end{pmatrix}$$

$$\mathbb{E}[\mathbf{q}(0,5)] = \begin{pmatrix} 0.35 & 0.46 & 0.19 \end{pmatrix}$$

$$\mathbb{E}[\mathbf{q}(\infty)] = \begin{pmatrix} 0.4 & 0.4 & 0.2 \end{pmatrix}.$$

Therefore the expected relative population structure converges geometrically fast. It is easy to verify that

$$\mathbb{E}[\mathbf{q}(\infty)]\,\mathbf{V}_1 = \begin{pmatrix} 0.4 & 0.4 & 0.2 \end{pmatrix} \begin{pmatrix} 0.7 & 0.2 & 0.1 \\ 0.2 & 0.6 & 0.2 \\ 0.5 & 0.2 & 0.3 \end{pmatrix} = \begin{pmatrix} 0.4 & 0.4 & 0.2 \end{pmatrix},$$

and the same happens with all the vertices of $[\mathbb{M}]$.

Let now another NHMS with initial population structure $\mathbf{q}(0) = \begin{pmatrix} 0.6 & 0.2 & 0.2 \end{pmatrix}$; allocation probabilities $\mathbf{p}_0(1) = \begin{pmatrix} 0.5 & 0.4 & 0.1 \end{pmatrix}$ and $\mathbf{p}_0(2) = \begin{pmatrix} 0.7 & 0.2 & 0.1 \end{pmatrix}$ and the remaining parameters the same. Then the asymptotically relative population structure is again $\mathbb{E}[\mathbf{q}(\infty)] = \begin{pmatrix} 0.4 & 0.4 & 0.2 \end{pmatrix}$ which verifies our proven theorem. The physical meaning of the previous result is that when the hospital is in full capacity for a sometime then the relative population structure remains unchanged under the condition that the maximum ergodicity coefficient of the transition probability matrices is less than one.

In order to use the last theorem we need a numerical value that will replace $d\left(\mathbb{M}, \tilde{\mathbb{M}}\right)$. The following Lemma, provides an answer to that demand.

Lemma 403 *(Vassiliou(2021b). Let* $\mathbb{M}, \tilde{\mathbb{M}}$ *be two tight intervals of transition probability matrices for the memberships. Let* $\mathbf{V}_1, \mathbf{V}_2, ..., \mathbf{V}_m$ *be the vertices of the convex set* $\mathbb{M}$*, and* $\mathbf{U}_1, \mathbf{U}_2, ..., \mathbf{U}_n$ *be the vertices of the convex set* $\tilde{\mathbb{M}}$*. Then*

$$\delta\left(\mathbb{M}, \tilde{\mathbb{M}}\right) = \left\| a_1^*\mathbf{V}_1 + a_2^*\mathbf{V}_2 + ... + a_m^*\mathbf{V}_m - \left[b_1^*\mathbf{U}_1 + b_2^*\mathbf{U}_2 + ... + b_n^*\mathbf{U}_n\right]\right\|,$$

where $a_1^*, a_2^*, ..., a_m^*$ *is the solution of the optimization problem*

$$\max\left[a_1\mathbf{V}_1 + a_2\mathbf{V}_2 + ... + a_m\mathbf{V}_m\right] \quad with$$

$$a_1 + a_2 + ... + a_m = 1, a_1 \geq 0, a_2 \geq 0, ..., a_m \geq 0.$$

and b_1^*,b_2^*,..., b_n^* *is the solution of the optimization problem*

$$\min\left[b_1\mathbf{U}_1 + b_2\mathbf{U}_2 + ... + b_n\mathbf{U}_n\right] \quad with$$

$$b_1 + b_2 + ... + b_n = 1, b_1 \geq 0, b_2 \geq 0, ..., b_m \geq 0.$$

In what follows we summarize in an Algorithm the previous steps which are necessary for using the present results in any population system.

Algorithm A2. (Vassiliou (2021b)).

Use Lemma 379 to check that $[\mathbb{S}_0]$ is tight.

Apply **Algorithm A1** to find the vertices of the convex set $[\mathbb{S}_0]$:

$$\boldsymbol{v}_1, \boldsymbol{v}_2, ..., \boldsymbol{v}_s.$$

Use Lemma 379 to check that $[\mathbb{R}_0]$ is tight.

Apply **Algorithm A1** to find the vertices of the convex set $[\mathbb{R}_0]$:

$$\mathbf{r}_1, \mathbf{r}_2, ..., \mathbf{r}_\pi.$$

Use Lemma 379 to check that each row in $\left[\check{\mathbf{Q}}, \hat{\mathbf{Q}}\right]$ is tight. If yes then $[\mathbb{M}]$ is tight.

Apply **Algorithm A1** for each row vector in $\left[\check{\mathbf{Q}}, \hat{\mathbf{Q}}\right]$ to find the vertices of $[\mathbb{M}]$:

$$\mathbf{V}_1, \mathbf{V}_2, ..., \mathbf{V}_m.$$

Compute all the raw vectors

$$\boldsymbol{v}_i \mathbf{V}_j \text{ for all } i = 1, 2, ..., s; j = 1, 2, ..., m.$$

Use any of the computational geometry methods in Preparata and Shamos (1985) to find the vertices of the convex hull of the vectors $\boldsymbol{v}_i \mathbf{V}_j$ for all $i = 1, 2, ..., s; j = 1, 2, ..., m$.

Let that

$$\boldsymbol{\omega}_1, \boldsymbol{\omega}_2, ..., \boldsymbol{\omega}_\nu,$$

the vertices found. Compute using properties of the Minkowski sum of vectors

$$Rng_1(\mathbb{S}_0, \mathbb{R}_0) = \frac{T(0)}{T(1)} conv\{\boldsymbol{\omega}_1, \boldsymbol{\omega}_2, \boldsymbol{\omega}_3, \boldsymbol{\omega}_4, \boldsymbol{\omega}_5\} +$$

$$\frac{1}{T(1)} [T(1) - T(0)]\, conv\{r_1, r_2\}.$$

Repeat the process until $Rng_\infty(\mathbb{S}_0, \mathbb{R}_0)$ geometrically fast, that is in 6 to 8 steps.

Use Lemma 402 to verify that $[\mathbb{M}] = \left[\check{\mathbf{Q}}, \hat{\mathbf{Q}}\right]$ is uniformly scrambling, that is $\mathcal{T}\left(\hat{\mathbf{Q}} - \check{\mathbf{Q}}\right) < 1$.

Use Lemma 403 to find bounds for

$$\delta\left(\mathbb{M}, \tilde{\mathbb{M}}\right) < \mu \text{ and } \delta\left(\tilde{\mathbb{M}}, \mathbb{M}\right) < \tilde{\mu}.$$

Set

$$d\left(\mathbb{M}, \tilde{\mathbb{M}}\right) < \max(\mu, \tilde{\mu}).$$

Chapter 14

MARKOV SYSTEMS ON A GENERAL STATE SPACE

14.1 Introductory notes

This chapter is about Markov systems on general state spaces (MSGS). Let X be an abstract set of points and $\mathcal{B}(X)$ a countably generated $\sigma-$field on X. In the present, we will study a Markov system, that is a population which its present state contains all the information needed for its future evolution that lives on $(X, \mathcal{B}(X))$. A $\sigma-$field $\mathcal{B}$ is generated by a collection of sets $\mathcal{A}$ in $\mathcal{B}$ if $\mathcal{B}$ is the smallest $\sigma-$field containing the sets $\mathcal{A}$, and then we write $\mathcal{B} = \sigma(\mathcal{A})$; a $\sigma-$field $\mathcal{B}$ is *countably generated* if it is generated by a countable collection $\mathcal{A}$ of sets in $\mathcal{B}$. When X is topological, then X is equipped with a locally compact, separable, metrizable topology with $\mathcal{B}(X)$ as the Borel $\sigma-$field, but otherwise it may be arbitrary. The theory of Markov chains on general state spaces have matured considerably over the past thirty years in many ways which makes it very much more accessible. The bible on Markov chains in general state spaces is the book by Meyn and Tweedie (2009) and we will lean heavily on that since it has been brought up to date to reflect all developments in the field. Our approach will be to use definitions and theorems from Meyn and Tweedie (2009) on Markov chains on general state space that will be used in developing the theory of MSGS and refer for proofs in there. Occasionally depending on the needs at hand we will provide possibly useful intuitions.

A Markov chain on general state space is a discrete time stochastic process $\mathbf{\Phi} = \{\Phi_0, \Phi_1, ...\}$ or $\mathbf{\Phi} = \{\Phi_n\}_{n \in \mathbb{N}^+}$ which takes values on $(X, \mathcal{B}(X))$ which has the Markov property. These random variables $\Phi_0, \Phi_1, ...$ with each Φ_n taking values in X are assumed measurable individually with respect to $\mathcal{B}(X)$. We shall in general denote elements of X by letters $x, y, z, ...$ and elements of $\mathcal{B}(X)$ by A, B, C. For $\mathbf{\Phi}$ to be defined as a random variable in its own right, Ω will be equipped with a $\sigma-$field $\mathcal{F}$, and for each state $x \in X$, thought as an initial condition in the sample path, there will be a probability measure P_x such that the probability of the event $\{\mathbf{\Phi} \in A\}$ is well defined for any set $A \in \mathcal{F}$; the initial condition requires, of course, that $P_x(\Phi_0 = x) = 1$.

If $P = \{P(x, A), x \in X, A \in \mathcal{B}(X)\}$ is such that

(i) for each $A \in \mathcal{B}(X), P(., A)$ is a non-negative measurable function on X;

(ii) for each $x \in X, P(x, .)$ is a probability measure on $\mathcal{B}(X)$,

then we call P a *transition probability kernel or Markov transition function.*

Apparently, $P_x(\Phi_1 = A) = P(x, A)$.

The analogue Chapman-Kolmogorov equations for a Markov chain in general state space is given (Meyn and Tweedie (2009), p. 61):

Theorem 404 *Let a Markov chain living in the general state space $(X, \mathcal{B}(X))$ with transition probability kernel $P(x, A), x \in X, A \in \mathcal{B}(X)$. Then for any $0 \leq m \leq n$,*

$$P^n(x, A) = \int_X P^m(y, dy) P^n(y, A).$$

DOI: 10.1201/b23204-14

14.2 A mental image for MSGS

In the present section, it is provided a mental image useful in understanding the many concepts and results presented in the present chapter. In this respect will be used the traditional mental image created by Howard (1971) for a Markov chain which includes a frog and a lily pond as was presented for the first time in Vassiliou (2014b). To do so the mental image will be expanded in order to accommodate the more general concept of a Markov system. Imagine – although a real situation is described- a lily pond which is covered by the leaves of the lilies. In such a case the leaves are of different sizes and given that all the lilies are not of the same type, the leaves have different characteristics. Imagine that the pond has the shape of a large bean, let say $5 \times 3\ m^2$ and that in the edges there exists a necklace like set of large pebbles which surround the edges. Our general set X consists of the surface of water of the pond and the set of pebbles that surround the edges. The natural partition of the set X consists of the various leaves, let say sets A_i, and the set of pebbles, let say set W, that is,

$$\{A_i\}_{i=1}^{k+1}, \text{ with } A_i \in \mathcal{B}(X) \text{ and } A_i \cap A_j = \varnothing \text{ for } i \neq j \text{ and such that}$$

$$\cup_{i=1}^{k+1} A_i = X \cup W \text{ and with no loss of generality } A_{k+1} = W.$$

Smelling the water a population $T(0)$ of frogs of the kind *bufo viridis* arrive in the pond. The characteristic of the kind *bufo viridis* is that the males of the species in order to attract the females they make sounds that in the human ear are like the songs of paradise birds. This adds immensely in the beauty of the environment. Living and reproducing in the pond is antagonistic for the frogs, hence we have wastage from the pond through the gate and also new arrivals again through the gate. The total population of the frogs at time t, denoted as $T(t)$ is assumed to be the realization of a known stochastic process. The first problem answered among others in Vassiliou (2014*b*) was to find the expected number of frogs in each leaf. The second and important problem answered was proving, under certain conditions, the existence of an asymptotic expected population of frogs in each leaf. It was also proved in that theorem that the asymptotic expected number of frogs in each leaf is independent of the initial time at the first leaf that they went at their entrance in the pond from gate W. Every mid November the frogs go on underground hiding places and their population at the pond becomes 0. This is a regenerative point for the Markov system since every spring an initial population of frogs reappears.

14.3 The foundation of a MSGS

Let X be a general set and $\mathcal{B}(X)$ denote a countably generated σ-field on X (see also Vassiliou(2014b). We assume that a population of total number $T(t)$ at time t lives in $(X, \mathcal{B}(X))$ where $\{T(t)\}_{t=0}^{\infty}$ is a known stochastic process in discrete time and state space $\mathbb{N}^+$.

We assume the existence of a positive σ-finite measure $\lambda : X \to \mathbb{R}^+$, that is,

$$\lambda(A) \geq 0,\ \lambda(A) < \infty \text{ for every } A \in \mathcal{B}(X),$$

and which is normalized, that is, $\lambda(X) = 1$. The measure $\lambda(.)$ represents the initial probabilities of a member of the population in the space $(X, \mathcal{B}(X))$. Assume now that, there is a set $W \in B(X)$ for which $\lambda(W) = 0$, which we will call the "gate" of the space $(X, \mathcal{B}(X))$, for reasons which will be apparent in what follows.

Assume also that, each member of the population holds a membership, which when it leaves the space $(X, \mathcal{B}(X))$ is left at the "gate". Therefore, the number of memberships at the "gate" W is the sum of memberships left by leavers, plus the number $\Delta T(t) = T(t) - T(t-1)$ of memberships to reach the desired target $T(t)$, assuming that $\Delta T(t) \geq 0$. New members collect at the "gate" a membership and then move to one of the sets $A \in B(X)$. Denote by

$$\{P(x,A),\ x \in X,\ A \in \mathcal{B}(X)\}, \tag{14.1}$$

the transition probability kernel (see Meyn and Tweedie (2009) of a member of the population in $x \in X$ to make a transition to the set $A \in \mathcal{B}(X)$ in one time step. We denote by $\mathbf{\Phi}_p = \left\{\Phi_p^{(n)}\right\}_{n\in\mathbb{N}^+}$ the Markov chain in $(X, \mathcal{B}(X))$ that is being uniquely defined by the probability kernel (14.1). We now make the following assumption:

Assumption A.1.

The set W is an atom for the Markov chain $\mathbf{\Phi}_p = \left\{\Phi_p^{(n)}\right\}_{n\in\mathbb{N}^+}$.

Which means that, there exist a measure $\mu(.)$ on $\mathcal{B}(X)$ such that

$$P(x,A) = \mu(A) \text{ for every } x \in W \text{ and } A \in \mathcal{B}(X), \tag{14.2}$$

with $\mu(X) = 1$. Also, it is vital that $\mu(W) = 0$.

Denote by

$$Q(x,A) = P(x,A) + P(x,W)\mu(A), \text{ for every } x \in X,\ A \in \mathcal{B}(X),\ x \notin W, \tag{14.3}$$

$$Q(x,A) = P(x,A) = \mu(A), \text{ for every } x \in W,\ A \in \mathcal{B}(X). \tag{14.4}$$

The transition probability kernel $Q(x,A)$ in (14.3) is the probability of a membership to make a transition from point x into the set A in one time step, either by direct transition of the member holding the membership with probability $P(x,A)$, or by the member leaving the system through the gate W with probability $P(x,W)$ and the entrance of a new member, who gets the membership of the member from gate W, into set A with probability $\mu(A)$. The transition probability kernel $Q(x,A)$ in (14.4) is the probability of new memberships entering the system at each time interval $(t-1,t]$ entering the space $(X,\mathcal{B}(X))$ through the gate of the atom W and being distributed according to the σ-finite probability measure $\mu(.) : \mathcal{B}(X) \to [0,1]$.

Denote by $\mathbf{\Phi}_Q = \left\{\Phi_Q^{(n)}\right\}_{n\in\mathbb{N}^+}$ the Markov chain in the general state space $(X,\mathcal{B}(X))$ that is being uniquely defined by the transition probability kernel $Q(x,A)$ in (14.3) and (14.4). Denote also by $Q^n(x,A)$ the probability that a membership in x will move in n time steps to the set $A \in \mathcal{B}(X)$.

Denote by:

$N(x,t,A)$: the number of memberships of the population that are in set $A \in \mathcal{B}(X)$ at time t given that were initially at $x \in X$.

In many applications there will be a natural partition of the space $(X,\mathcal{B}(X))$ that is,

$$\{A_i\}_{i=1}^{k+1},\ \text{with } A_i \in \mathcal{B}(X) \text{ and } A_i \cap A_j = \varnothing \text{ for } i \neq j \text{ and such that} \tag{14.5}$$

$$\cup_{i=1}^{k+1} A_i = X \cup W \text{ and with no loss of generality } A_{k+1} = W. \tag{14.6}$$

It is of interest the *expected population structure* defined as:

$$\mathbb{E}_{\mathbb{P}}[\mathbf{N}(t,X)] = [\mathbb{E}_{\mathbb{P}}[N(x,t,A_1)], \mathbb{E}_{\mathbb{P}}[N(x,t,A_2)], ..., \mathbb{E}_{\mathbb{P}}[N(x,t,A_k)]], \tag{14.7}$$

where $\mathbf{N}(t,X)$ is a non-negative measurable function on X.

Also the *relative expected population structure* is defined by

$$\mathbb{E}_{\mathbb{P}}[\mathbf{q}(t,X)] = \mathbb{E}_{\mathbb{P}}[\mathbf{N}(t,X)/T(t)]. \tag{14.8}$$

Then the measure $q : X \to \mathbb{R}^+$ is a positive σ-finite probability measure due to the fact that

$$q(x,t,A_i) = \frac{N(x,t,A_i)}{T(t)} \geq 0;\ q(x,t,A_i) < \infty\ \forall i \in \mathbb{N}^+ \text{ and } q(x,t,X) = 1. \tag{14.9}$$

The evolution of the expected population structure $\mathbb{E}_{\mathbb{P}}[\mathbf{N}(t,X)]$ or the relative expected population structure $\mathbb{E}_{\mathbb{P}}[\mathbf{q}(t,X)]$ is of central importance in the study of Markov systems in general state space. It is not difficult to prove that

$$\mathbb{E}_{\mathbb{P}}[N(x,t,A)] = T(0)\int_X \lambda(dx)\,Q^t(x,A) + \sum_{n=1}^{t-1} \Delta T(n) \int_X \mu(dx)\,Q^{t-n}(x,A). \tag{14.10}$$

Note that

$$T(0)\int_X \lambda(dx)\,Q^t(x,A),$$

is the expected number of memberships from the initial population $T(0)$ that survived up to time t and are in set A. The part

$$\sum_{n=1}^{t-1} \Delta T(n) \int_X \mu(dx)\,Q^{t-n}(x,A),$$

is the expected number of memberships that entered the system in the interval $[0,t]$ and survived up to time t and are in set A.

A population that lives in a general state space $(X, \mathcal{B}(X))$ and may be modeled by the random process described above will be called a *Markov system in a general state space.* In addition, we will call the Markov chain $\mathbf{\Phi}_Q = \left\{\Phi_Q^{(n)}\right\}_{n\in\mathbb{N}^+}$ the *inherent Markov chain* of the Markov system.

We introduced the concept of the Markov system in a general state space. In what follows we will exploit the Markov chain $\mathbf{\Phi}_Q = \left\{\Phi_Q^{(n)}\right\}_{n\in\mathbb{N}^+}$ in $(X, \mathcal{B}(X))$. We start with some known results from the theory of Markov chains in general state spaces.

For any set $A \in \mathcal{B}(X)$ we denote by $oc(A)$ the number of visits by $\mathbf{\Phi}_Q$ to A after time zero and is given by

$$oc(A) = \sum_{n=1}^{\infty} \mathbf{1}\left\{\Phi_Q^{(n)} \in A\right\}.$$

Define the kernels U and L as follows

$$U(x,A) := \sum_{n=1}^{\infty} Q^n(x,A) = \mathbb{E}_{\mathbb{P}}[oc_x(A)] \text{ with } U : X \times \mathcal{B}(X) \to \mathbb{R}\cup\{\infty\},$$

and for any set $A \in \mathcal{B}(X)$ we define

$$\tau(A) := \min\left\{n \geq 1\colon \Phi_Q^{(n)} \in A\right\},\ \textit{i.e.}, \text{ the first return time},$$

$$\sigma(A) := \min\left\{n \geq 0\colon \Phi_Q^{(n)} \in A\right\},\ \textit{i.e.}, \text{ the first hitting time},$$

and the return time probabilities

$$L(x,A) := \mathbb{P}(\tau_x(A) < \infty) = \mathbb{P}(\mathbf{\Phi}_Q \text{ ever enters } A \text{ given that initially is at } x).$$

Let that $\mathbf{\Phi}_Q$ is ϕ-irreducible, that is, there exists a measure ϕ on $\mathcal{B}(X)$ such that, whenever $\phi(A) > 0$, we have $L(x, A) > 0$ for all $x \in X$. Equivalently, $\mathbf{\Phi}_Q$ is ϕ-irreducible if for all $x \in X$, whenever $\phi(A) > 0$, $U(x, A) > 0$ and also alternatively for all $x \in X$, whenever $\phi(A) > 0$, there exist some $n > 0$, possibly depending both on A and x, such that $Q^n(x, A) > 0$.

Since $\mathbf{\Phi}_Q$ is ϕ-irreducible then (see Meyn and Tweedie (2009 p. 83)) there exists a "maximal" irreducibility probability measure ψ on $\mathcal{B}(X)$ such that

(i) $\mathbf{\Phi}_Q$ is ψ-irreducible;

(ii) for any other measure ϕ', the chain $\mathbf{\Phi}_Q$ is ϕ'-irreducible if and only if $\psi \succ \phi'$ ($\succ$: equivalent);

(iii) if $\psi(A) = 0$, then $\psi\{y : L(y, A) = 0\} = 0$;

(iv) the probability measure ψ is equivalent to

$$\psi'(A) := \int_X \phi'(dy) K_{a_{\frac{1}{2}}}(y, A),$$

for any finite irreducibility measure ϕ', where

$$K_{a_{\frac{1}{2}}}(y, A) := \sum_{n=0}^{\infty} Q^n(y, A) 2^{-(n+1)}, \; y \in X, A \in \mathcal{B}(X).$$

The ψ-irreducibility concept for Markov chains on general state spaces was essentially developed by Doeblin (1937) and (1940), and followed by many authors, including Doob (1990), Harris (1956), Chung (1967), Orey (1959), Tweedie (1974a, 1974b, 1975) and Numellin (1984).

Denote by

$$\mathcal{B}^+(X) := \{A \in \mathcal{B}(X) : \psi(A) > 0\},$$

for the sets of positive ψ-probability measure; the equivalence of maximal irreducibility measures means that $\mathcal{B}^+(X)$ is uniquely defined.

A set $C \in \mathcal{B}(X)$ is called a *small* set if there exists an $m > 0$, and a non-trivial measure ν_m on $\mathcal{B}(X)$, such that for all $x \in C$, $A \in \mathcal{B}(X)$, such that for all $x \in C, A \in \mathcal{B}(X)$

$$Q^m(x, A) \geq \nu_m(A),$$

and when this holds we say C is ν_m-small. Small sets themselves behave, in many ways, analogously to atoms.

The following is a useful theorem and it's proof could be found in Meyn and Tweedie (2009) p 105.

Theorem 405 *If $\mathbf{\Phi}_Q$ is ψ-irreducible, then for every $A \in \mathcal{B}^+(X)$, there exists $m \geq 1$ and a ν_m-small set $C \subseteq A$ such that $C \in \mathcal{B}^+(X)$ and $\nu_m(C) > 0$.*

The set $C \in \mathcal{B}^+(X)$ described in the previous theorem could be used to define a cycle for $\mathbf{\Phi}_Q$. Since, C is ν_m-small and $\nu_m(C) > 0$ then when the Markov chain starts in C, there is a positive probability that the chain will return in C at time m. Let

$$R_C = \{n \geq 1\text{: the set } C \text{ is } \nu_n\text{-small, with } \nu_n = \delta_n \nu_m \text{ for some } \delta_n > 0\},$$

be the set of time points for which C is a small set with minorizing measure proportional to ν_m. It is known that R_C is closed under addition. Therefore, there is a natural "period" for the set C given the greatest common divisor of R_C. It is known from Number theory that there exists $n_0 < \infty$ such that $nd \in R_C$ for all $n \geq n_0$.In the theorem that follows it is established that this property holds for the whole chain $\mathbf{\Phi}_Q$ and is independent of the chosen small set chosen (the proof could be found in Meyn and Tweedie (2009) p. 113).

Theorem 406 *Suppose that $\mathbf{\Phi}_Q$ is a ψ-irreducible Markov chain on X. Let $C \in \mathcal{B}^+(X)$ be a ν_m-small set and let d be the greatest common divisor of the set R_C. Then there exists disjoint sets $D_1, ..., D_d \in \mathcal{B}(X)$ a "$d-$cycle" such that*

(i) for $x \in D_i$, $Q(x, D_{i+1}) = 1$, $i = 0, 1, ..., d-1$.

(ii) the set $N = \left[\cup_{i=1}^d D_i\right]^c$ is ψ-null.

The d-cycle $\{D_i\}$ is maximal in the sense that for any other collection $\left\{d', D'_i, k = 0, 1, ..., d'\right\}$ satisfying $(i) - (ii)$, we have d' dividing d; if $d = d'$, then by re-ordering the indices if necessary $D'_i = D_i$ a.e.-ψ.

In such a case d is called the *period* of the Markov chain $\mathbf{\Phi}_Q$. If $d = 1$ the chain is called *aperiodic.* When there exists a ν_1-small set C with $\nu_1(C) > 0$ then the chain is called *strongly aperiodic.*

Definition 407 *The chain $\mathbf{\Phi}_Q$ is called recurrent if it is ψ-irreducible and $U(x, A) = \infty$ for every $x \in X$ and every $A \in \mathcal{B}^+(X)$.*

The kernel $U(x, A)$ and the concept of recurrence have equivalent properties that were studied by Tuominen (1976). In what follows the concept of Harris recurrent chains will be introduced and used in what follows. Harris chains are named after T.E. Harris who introduced many of the essential ideas in Harris (1956). Many important results were also introduced by Orey (1959, 1971).

Let the event $\mathbf{\Phi}_Q \in A$ infinitely often or $oc(A) = \infty$, that is

$$\{\mathbf{\Phi}_Q \in A,\ i.o\} := \cup_{n=1}^{\infty} \cap_{k=n}^{\infty} \left\{\Phi_Q^{(k)} \in A\right\},$$

which is well defined as an $\mathcal{F}$-measurable event on Ω.

For $x \in X$, $A \in \mathcal{B}(X)$ we write

$$Q_{i.o}(x, A) := \mathbb{P}\left(\mathbf{\Phi}_Q \in A, i.o \mid \Phi_Q^{(0)} = x\right).$$

Obviously, for any $x \in X$, $A \in \mathcal{B}(X)$, we have $Q_{i.o}(x, A) \leq L(x, A)$.

Definition 408 *A set $A \in \mathcal{B}^+(X)$ is called Harris recurrent if*

$$Q_{i.o}(x, A) = \mathbb{P}(oc_x(A) = \infty) = 1, \text{ for any } x \in A.$$

The chain $\mathbf{\Phi}_Q$ is called Harris recurrent if it is ψ-irreducible and every set in $\mathcal{B}^+(X)$ is Harris recurrent.

Let $\mathbf{a} = \{a(n)\}$ be a probability measure, on $\mathbb{N}^+$, and consider the Markov chain ${}^{(\mathbf{a})}\mathbf{\Phi}_Q$ with probability transition kernel

$$K_{\mathbf{a}}(x, A) = \sum_{n=0}^{\infty} Q^n(x, A)\, a(n) \text{ with } x \in X, A \in \mathcal{B}(X).$$

We call the Markov chain ${}^{(\mathbf{a})}\mathbf{\Phi}_Q$ with transition probability kernel $K_{\mathbf{a}}(x, A)$ the *sampled chain* of the Markov chain $\mathbf{\Phi}_Q$. In the ψ-irreducible case Small sets always exist and due to their properties they are very useful. A generalization of small sets, *petite sets,* have even more tractable properties, especially in topological analysis.

Definition 409 *We will call a set $C \in \mathcal{B}(X)$ ν_a-petite if the sampled chain satisfies the inequality*

$$K_{\mathbf{a}}(x, A) \geq \nu_a(A),$$

for all $x \in C, A \in \mathcal{B}(X)$, where ν_a is a non-trivial measure on $\mathcal{B}(X)$.

Petite sets as defined here were introduced by Meyn and Tweedie (1992). The following theorem is important and its proof exists in Meyn and Tweedie (2009).

Theorem 410 *If $\mathbf{\Phi}_Q$ is irreducible and aperiodic, then every petite set is small.*

The concept of *invariant measures* will be introduced which is useful in finding conditions under which the distribution of $\Phi_Q^{(n)}$ does not change as n takes various values.

Definition 411 *Let a Markov system that lives on $(X, \mathcal{B}(X))$. A σ-finite measure π with the property*

$$\pi(A) = \int_X \pi(dx) Q(x, A), \ A \in \mathcal{B}(X),$$

will be called an invariant measure.

The existence of invariant probability measures has been a central topic of Markov chain theory since the inception of the subject. Doob (1990) and Orey (1971) provide a good background. Two important theorems now will be stated and be used latter.

The existence of π.

Theorem 412 *(Meyn and Tweedie (2009)). If the chain $\mathbf{\Phi}_Q$ is recurrent it admits a unique (up to constant multiples) invariant measure π, and the measure π has the representation, for any $A \in \mathcal{B}^+(X)$*

$$\pi(B) = \int_A \pi(dx) \mathbb{E}_{\mathbb{P}} \left[\sum_{n=1}^{\tau_A} \mathbf{1}\left\{ \Phi_Q^{(n)} \in A, \Phi_Q^{(0)} = x \right\} \right] \ \textit{with } B \in \mathcal{B}(X).$$

The invariant measure π is finite (rather than merely σ-finite) if there exists a petite set C such that

$$\sup_{x \in C} \mathbb{E}_{\mathbb{P}} \left[\tau_C \mid \Phi_Q^{(0)} = x \right] < \infty.$$

We state the following definition:

Definition 413 *Let a Markov system that lives on $(X, \mathcal{B}(X))$. If the inherent Markov chain $\mathbf{\Phi}_Q$ is ψ-irreducible, and admits an invariant measure π, then $\mathbf{\Phi}_Q$ is called a positive chain and the respective Markov system a positive Markov system. If the inherent Markov chain $\mathbf{\Phi}_Q$ is Harris recurrent and positive, then $\mathbf{\Phi}_Q$ is called a positive Harris chain and the respective Markov system a positive Harris Markov system.*

Since in a Markov system we have an atom W then Kac's theorem has an immediate application:

Theorem 414 *(Kac (1947)). If $\mathbf{\Phi}_Q$ is ψ-irreducible and admits an atom $W \in \mathcal{B}^+(X)$, then $\mathbf{\Phi}_Q$ is positive recurrent if and only if*

$$\mathbb{E}_{\mathbb{P}} \left[\tau_W \mid \Phi_Q^{(0)} \in W \right] < \infty;$$

and if π is an invariant probability measure for $\mathbf{\Phi}_Q$, then

$$\pi(W) = \left(\mathbb{E}_{\mathbb{P}} \left[\tau_W \mid \Phi_Q^{(0)} \in W \right] \right)^{-1}.$$

14.4 Asymptotic behavior or ergodicity of MSGS

The problem of asymptotic behavior of Markov chains has been central for finite or countable chains as we have seen in Chapters 3 and 4. Also for NHMS's the problem of asymptotic behavior of the expected population structure is of central importance both from a theoretical and a practical point of view. We studied these problems analytically in Chapters 7,8 and 9. In the present section we will study the important problem of the asymptotic behavior of the expected population structure given in (14.7) and (14.10) as well as the asymptotic behavior of the relative population structure given in (14.8) and (14.9).

We will start with finding the limit as $t \to \infty$ of

$$\mathbb{E}_{\mathbb{P}}\left[\mathbf{N}\left(t,X\right)\right] = \left[\mathbb{E}_{\mathbb{P}}\left[N\left(x,t,A_1\right)\right], \mathbb{E}_{\mathbb{P}}\left[N\left(x,t,A_2\right)\right], ..., \mathbb{E}_{\mathbb{P}}\left[N\left(x,t,A_k\right)\right]\right]$$

Where as in many applications we assume that there exists a natural partition of the space $(X, \mathcal{B}(X))$ that is,

$$\{A_i\}_{i=1}^{k+1}, \text{ with } A_i \in \mathcal{B}(X) \text{ and } A_i \cap A_j = \varnothing \text{ for } i \neq j \text{ and such that}$$

$$\cup_{i=1}^{k+1} A_i = X \cup W \text{ and with no loss of generality } A_{k+1} = W.$$

We will start by studying the limit for any $A \in \mathcal{B}(X)$ of

$$\mathbb{E}_{\mathbb{P}}\left[N\left(x,t,A\right)\right] = T\left(0\right)\int_X \lambda\left(dx\right) Q^t\left(x,A\right) + \sum_{n=1}^{t-1} \Delta T\left(n\right)\int_X \mu\left(dx\right) Q^{t-n}\left(x,A\right).$$

In this respect we will use the total variation norm.

If μ is a signed measure on $(X, \mathcal{B}(X))$, then the total variation norm $\|\mu\|$ is defined as

$$\|\mu\| := \sup_{f:|f|\leq 1} |\mu(f)| = \sup_{A\in\mathcal{B}(X)} \mu(A) - \inf_{A\in\mathcal{B}(X)} \mu(A). \tag{14.11}$$

Our interest in this section will be limits of the form

$$\lim_{n\to\infty} \|Q^n(x,.) - \pi(.)\| = 2 \lim_{n\to\infty} \sup_A |Q^n(x,.) - \pi(.)|. \tag{14.12}$$

The total variation norm is the traditional approach in contrast to the countable state space case. We will find conditions under which there exist pointwise limits of the form

$$\lim_{n\to\infty} |P^n(x,y) - \pi(y)| = 0;$$

but the results achieved are related to the signed measure $(P^n - \pi)$, and so concern not merely such pointwise or even setwise convergence, but a more global convergence in terms of the total variation norm. When (14.12) is valid in a countable space, then the previous limit also holds and indeed holds uniformly in the end point y. The transition to the total variation norm, is being dictated, as has been revealed in the case of Markov chains in general spaces, by the typical lack of structure of pointwise transitions in the general state space. As a functional, total variational finds applications in several branches of Mathematics like optimal control, numerical analysis and calculus of variations. Boundedness of hitting times from arbitrary initial measures will become important. From (Meyn and Tweedie (2009) p. 115) we need the following Proposition:

Proposition 415 *Suppose Φ is a ψ-irreducible chain with period d and a D-cycle $\{D_i, i = 0, ..., d-1\}$. Then each of the sets D_i is an absorbing ψ-irreducible set for the chain Φ_d corresponding to the transition kernel P^d, and Φ_d on each D_i is aperiodic.*

The chain Φ is called recurrent (Meyn and Tweedie (2009) if it is ψ-irreducible and $U(x, A) = \infty$ for every $x \in X$ and every $A \in \mathcal{B}^+(X)$. If the chain Φ is irreducible and admits an atom $a \in \mathcal{B}^+(X)$ then if a is recurrent, then every set in $\mathcal{B}^+(X)$ is recurrent.

Definition 416 *The set $A \in \mathcal{B}^+(X)$ is called Harris recurrent if*

$$P[oc_x(A) = \infty] = 1.$$

A chain Φ is called Harris recurrent if it is ψ-irreducible and every set in $\mathcal{B}^+(X)$ is Harris recurrent.

We will be needing the following four theorems in order to prove the basic theorems of the present section. Their proofs could be found in Meyn and Tweedie (2009).

Theorem 417 *If $\mathbf{\Phi}$ in $(X, \mathcal{B}(X))$ is a positive Harris recurrent and strongly aperiodic chain, then for any initial probability measure λ*

$$\left\| \int_x \lambda(dx) P^n(x, .) - \pi(.) \right\| \to 0 \text{ as } n \to \infty. \tag{14.13}$$

Theorem 418 *If $\mathbf{\Phi}$ in $(X, \mathcal{B}(X))$ is a positive Harris and aperiodic chain, then for any initial probability measure λ*

$$\left\| \int_x \lambda(dx) P^n(x, .) - \pi(.) \right\| \to 0 \text{ as } n \to \infty. \tag{14.14}$$

Theorem 419 *Let $\mathbf{\Phi}$ in $(X, \mathcal{B}(X))$ be ψ-irreducible and aperiodic, and suppose that there exists some ν-small set $C \in \mathcal{B}^+(X)$ and some $P^\infty(C) > 0$ such that as $n \to \infty$*

$$\int_C \nu_C(dx) [P^n(x, C) - P^\infty(C)] \to 0. \tag{14.15}$$

where $\nu_C(.) = \nu(.)/\nu(C)$ is normalized to a probability in C. Then the chain is positive and there exists a ψ-null set N, such that for every initial distribution λ with $\lambda(N) = 0$

$$\left\| \int_x \lambda(dx) P^n(x, .) - \pi(.) \right\| \to 0 \text{ as } n \to \infty. \tag{14.16}$$

Theorem 420 *If Φ is recurrent, then we can write*

$$X = H \cup N$$

where H is a non-empty maximal Harris set and N is transient.

The following is one of the basic theorems of this section

Theorem 421 *(Vassiliou (2014b)). Let a Markov system that lives in $(X, \mathcal{B}(X))$ as described in the previous section and let that*

$$\lim_{t \to \infty} T(t) = T, \ \Delta T(t) \geq 0 \text{ for every } t \in \mathbb{Z}^+. \tag{14.17}$$

If the inherent Markov chain $\mathbf{\Phi}_Q$ that lives in $(X, \mathcal{B}(X))$ is such that

a) If $\mathbf{\Phi}_Q$ in $(X, \mathcal{B}(X))$ is a positive Harris recurrent and strongly aperiodic chain, then for any initial probability measures λ, μ

or

b) If $\mathbf{\Phi}_Q$ *in* $(X, \mathcal{B}(X))$ *is a positive Harris aperiodic chain, then for any initial probability measures* λ, μ

or

c)Let $\mathbf{\Phi}_Q$ *in* $(X, \mathcal{B}(X))$ *be* ψ*-irreducible and aperiodic, and suppose that there exists some* ν*-small set* $C \in \mathcal{B}^+(X)$ *and some* $P^\infty(C) > 0$ *such that as* $n \to \infty$

$$\int_C \nu_C(dx)\left[P^n(x, C) - P^\infty(C)\right] \to 0. \tag{14.18}$$

where $\nu_C(.) = \nu(.)/\nu(C)$ *is normalized to a probability in* C*. Then for any initial probability measures* λ, μ

$$\|\mathbb{E}_\mathbb{P}[N(x,t,A)] - T\pi(A)\| \to 0 \text{ as } t \to \infty.$$

$$\|\mathbb{E}_\mathbb{P}[q(x,t,A)] - \pi(A)\| \to 0 \text{ as } t \to \infty.$$

Proof. From (14.10) using the total variation norm (14.11) and (14.12), we have that

$$\begin{aligned}
&\|\mathbb{E}_\mathbb{P}[N(x,t,A)] - T\pi(A)\| = \\
&\|\mathbb{E}_\mathbb{P}[N(x,t,A)] - T(0)\pi(A) - (T - T(0))\pi(A)\| = \\
&\|T(0)\left[\int_X \lambda(dx)\, Q^t(x,A) - \pi(A)\right] + \\
&\sum_{n=1}^{t-1} \Delta T(n)\left[\int_X \mu(dx)\, Q^{t-n}(x,A) - \pi(A)\right]\| \\
&\le T(0)\left\|\int_X \lambda(dx)\, Q^t(x,A) - \pi(A)\right\| + \\
&\sum_{n=1}^{t-1} \Delta T(n)\left\|\int_X \mu(dx)\, Q^{t-n}(x,A) - \pi(A)\right\|.
\end{aligned} \tag{14.19}$$

As $t \to \infty$ we have that if *a*), or *b*) or *c*) holds then by Theorems 418, 419, and 420 we get that

$$\lim_{t\to\infty}\left\|\int_X \lambda(dx)\, Q^t(x,A) - \pi(A)\right\| = 0. \tag{14.20}$$

Now, for t large if *a*), or *b*) or *c*) holds, then by Theorems 418, 419, and 420 we may always choose a t_0 such that for $t - n \ge t_0$ or equivalently $n \le t - t_0$

$$\left\|\int_X \mu(dx)\, Q^{t-n}(x,A) - \pi(A)\right\| < \varepsilon \text{ for } n \le t - t_0. \tag{14.21}$$

Consequently, we have for t large

$$\sum_{n=1}^{t-1} \Delta T(n)\left\|\int_X \mu(dx)\, Q^{t-n}(x,A) - \pi(A)\right\| =$$

$$\sum_{n=1}^{t-t_0} \Delta T(n) \left\| \int_X \mu(dx) Q^{t-n}(x,A) - \pi(A) \right\| + \tag{14.22}$$

$$\sum_{n=t-t_0+1}^{t-1} \Delta T(n) \left\| \int_X \mu(dx) Q^{t-n}(x,A) - \pi(A) \right\|$$

Now for $n \geq t - t_0$

$$\left\| \int_X \mu(dx) Q^{t-n}(x,A) - \pi(A) \right\| = \left\| \int_X \mu(dx) Q^{t-n}(x,A) - \int_X \mu(dx) \pi(A) \right\|$$

$$\leq \int_X \mu(dx) \left\| Q^{t-n}(x,A) - \pi(A) \right\|$$

$$\leq \int_X \mu(dx)\, 2 \sup_A \left| Q^{t-n}(x,A) - \pi(A) \right| \leq 2 \int_X \mu(dx) = 2. \tag{14.23}$$

From (14.21) (14.22) and (14.23) for t large we get that

$$\sum_{n=1}^{t-1} \Delta T(n) \left\| \int_X \mu(dx) Q^{t-n}(x,A) - \pi(A) \right\| \leq \tag{14.24}$$

$$[T(t-t_0) - T(0)]\varepsilon + 2[T(t) - T(t-t_0)].$$

Therefore from (14.24), we have that

$$\lim_{t\to\infty} \sum_{n=1}^{t-1} \Delta T(n) \left\| \int_X \mu(dx) Q^{t-n}(x,A) - \pi(A) \right\| = 0. \tag{14.25}$$

From (14.20) and (14.25) we easily arrive at

$$\left\| \mathbb{E}_{\mathbb{P}}[N(x,t,A)] - T\pi(A) \right\| \to 0 \text{ as } t \to \infty,$$

therefore

$$\left\| \mathbb{E}_{\mathbb{P}}[q(x,t,A)] - \pi(A) \right\| \to 0 \text{ as } t \to \infty.$$

■

With similar techniques we could show that

Theorem 422 *(Vassiliou (2014b)) Let a Markov system that lives in $(X, \mathcal{B}(X))$ as described in the previous section and let that*

$$\lim_{t\to\infty} [\Delta T(t)/T(t)] = 0, \ \Delta T(t) \geq 0 \textit{ for every } t \in \mathbb{Z}^+.$$

If the inherent Markov chain $\mathbf{\Phi}_Q$ that lives in $(X, \mathcal{B}(X))$ is such that

a) If $\mathbf{\Phi}_Q$ in $(X, \mathcal{B}(X))$ is a positive Harris recurrent and strongly aperiodic chain, then for any initial probability measures λ, μ

or

b) If $\mathbf{\Phi}_Q$ in $(X, \mathcal{B}(X))$ is a positive Harris aperiodic chain, then for any initial probability measures λ, μ

or

c)Let $\mathbf{\Phi}_Q$ *in* $(X, \mathcal{B}(X))$ *be* ψ*-irreducible and aperiodic, and suppose that there exists some* ν*-small set* $C \in \mathcal{B}^+(X)$ *and some* $P^\infty(C) > 0$ *such that as* $n \to \infty$

$$\int_C \nu_C(dx)\left[P^n(x, C) - P^\infty(C)\right] \to 0.$$

where $\nu_C(.) = \nu(.)/\nu(C)$ *is normalized to a probability in* C*. Then for any initial probability measures* λ, μ

$$\left\|\mathbb{E}_{\mathbb{P}}\left[q(x, t, A)\right] - \pi(A)\right\| \to 0 \text{ as } t \to \infty.$$

14.5 Total variability from the invariant measure; a coupling theorem

Let us assume that the conditions for a limit asymptotically to exist for the expected population structure hold for a MSGS. In this section for such a case study of the total variability from the invariant measure of the Markov system will be the main course. It will be shown that the total variation is finite. This will be called the coupling problem for a MSGS. On the coupling problem for Markov chains that live in a general space see also the book by Lindvall (1992). The following theorem from Meyn and Tweedie (2009, p. 332) is borrowed:

Theorem 423 *Suppose* $\mathbf{\Phi}$ *is aperiodic positive Harris chain and suppose that the chain admits an atom* $a \in \mathcal{B}^+(X)$*. Then for any regular distributions* λ, μ

$$\sum_{n=1}^{\infty} \int_X \int_X \lambda(dx)\,\mu(dy)\left\|P^n(x, .) - P^n(y, .)\right\| < \infty, \tag{14.26}$$

and in particular if $\mathbf{\Phi}$ *is regular, then for every* $x, y \in X$

$$\sum_{n=1}^{\infty} \left\|P^n(x, .) - P^n(y, .)\right\| < \infty.$$

The coupling theorem for Markov systems that live in a general space is established below.

Theorem 424 *(Vassiliou (2014b)). Let a Markov system that lives in* $(X, \mathcal{B}(X))$ *as described in a previous section. Let that*

$$\lim_{t \to \infty} T(t) = T,\ \ \Delta T(t) \geq 0 \text{ for every } t \in \mathbb{Z}^+. \tag{14.27}$$

Then for any regular initial distributions λ_1, μ_1 *and* λ_2, μ_2*, we have*

$$\sum_{n=1}^{\infty} \left\|\mathbb{E}_{\mathbb{P}}\left[N(x, t, A)\right] - \mathbb{E}_{\mathbb{P}}\left[N(y, t, A)\right]\right\| < \infty. \tag{14.28}$$

Proof. Apparently we get that

$$\mathbb{E}_{\mathbb{P}}\left[N(x, t, A)\right] = T(0)\int_X \lambda_1(dx)\,Q^t(x, A) + \sum_{n=1}^{t-1} \Delta T(n)\int_X \mu_1(dx)\,Q^{t-n}(x, A), \tag{14.29}$$

and

$$\mathbb{E}_{\mathbb{P}}\left[N(y, t, A)\right] = T(0)\int_X \lambda_2(dy)\,Q^t(y, A) + \sum_{n=1}^{t-1} \Delta T(n)\int_X \mu_2(dy)\,Q^{t-n}(y, A). \tag{14.30}$$

Let us consider the following

$$\int_X\int_X \lambda_1(dx)\,\lambda_2(dy)\left[Q^t(x,A)-Q^t(y,A)\right] = \int_X\int_X \lambda_1(dx)\,\lambda_2(dy)\,Q^t(x,A) - \int_X\int_X \lambda_1(dx)\,\lambda_2(dy)\,Q^t(y,A)$$

$$= \int_X \lambda_1(dx)\; Q^t(x,A) - \int_X \lambda_2(dy)\,Q^t(y,A). \tag{14.31}$$

From(14.29), (14.30) and (14.31) we get that

$$\mathbb{E}_{\mathbb{P}}\left[N(x,t,A)\right]-\mathbb{E}_{\mathbb{P}}\left[N(y,t,A)\right] = T(0)\int_X\int_X \lambda_1(dx)\,\lambda_2(dy)\left[Q^t(x,A)-Q^t(y,A)\right]$$

$$+\sum_{n=0}^{t-1}\Delta T(n)\int_X\int_X \mu_1(dx)\,\mu_2(dy)\left[Q^t(x,A)-Q^t(y,A)\right]. \tag{14.32}$$

In (14.32), we take the total variation norm on both sides

$$\sum_{t=1}^{\infty}\left\|\mathbb{E}_{\mathbb{P}}\left[N(x,t,A)\right]-\mathbb{E}_{\mathbb{P}}\left[N(y,t,A)\right]\right\| \leq$$

$$T(0)\sum_{t=1}^{\infty}\int_X\int_X \lambda_1(dx)\,\lambda_2(dy)\left\|Q^t(x,A)-Q^t(y,A)\right\| +$$

$$\sum_{t=1}^{\infty}\sum_{n=0}^{t-1}\Delta T(n)\int_X\int_X \mu_1(dx)\,\mu_2(dy)\left\|Q^{t-n}(x,A)-Q^{t-n}(y,A)\right\|$$

$$\leq T(0)\,K_1+\sum_{n=0}^{\infty}\Delta T(n)\sum_{t-n=1}^{\infty}\int_X\int_X \mu_1(dx)\,\mu_2(dy)\left\|Q^{t-n}(x,A)-Q^{t-n}(y,A)\right\|$$

$$\leq T(0)\,K_1+\left(T-T(0)\right)K_2<\infty.$$

■

14.6 Rate of convergence of MSGS

There is an increasing interest in spectral theory and rates of convergence for Markov chains in general spaces (see Meyn and Tweedie (2009), Kontoyannis and Meyn (2012, 2003, 2005) and Mitrophanov (2005). Research is motivated by elegant mathematics as well as a range of applications. The word spectrum invites techniques in a Hilbert space frame-work. The majority of the quantitative results on rates of convergence are obtained using such methods, within the Hilbert space $L_2 = L_2(\pi)$, where π denotes the stationary distribution for the Markov chain in question. Note that, for countable inhomogeneous Markov systems it was proved that the space of all possible expected population structures is a Hilbert space (Vassiliou (2013) and Chapter 5).

Let Q_1 and Q_2 be Markov transition kernels and for a positive function $\infty > V \geq 1$, define the V-*norm distance* between Q_1 and Q_2 as

$$|||Q_1 - Q_2|||_V := \sup_{x \in X} \frac{\|Q_1(x,.) - Q_2(x,.)\|_V}{V(x)}. \tag{14.33}$$

We define the *outer product* of the function 1 and the invariant measure π the kernel

$$[1 \otimes \pi](x, A) = \pi(A), \ x \in X, \ A \in \mathcal{B}(X). \tag{14.34}$$

In many applications we consider the distance $|||Q^n - 1 \otimes \pi|||_V$ for large n. We provide the following definition from Meyn and Tweedie (2009):

Definition 425 ***(V−uniform ergodicity).*** *An ergodic chain Φ with transition kernel Q is called V-uniformly ergodic if*

$$|||Q^n - 1 \otimes \pi|||_V \to 0, \text{ as } n \to \infty. \tag{14.35}$$

Of great interest is the special case when $V = 1$. In this case, we provide the following definition:

Definition 426 ***(Uniform ergodicity).*** *A chain Φ with transition kernel Q is called uniformly ergodic if it is V-uniformly ergodic with $V = 1$, that is, if*

$$\sup_{x \in X} \|Q^n(x,.) - \pi\| \to 0, \text{ as } n \to \infty. \tag{14.36}$$

In what follows the conditions under which the rate of convergence of the expected population structure of an ergodic Markov system is uniformly ergodic and V-uniformly ergodic will be established. From Meyn and Tweedie (2009) p. 393 we get the following two theorems:

Theorem 427 *Suppose that $\mathbf{\Phi}$ is ψ-irreducible and aperiodic with transition kernel Q. Then the following are equivalent for $V \geq 1$:*
(i) $\mathbf{\Phi}$ is V-uniformly ergodic.
(ii) There exist $r > 1$ and $R < \infty$ such that for all $n \in \mathbb{Z}_+$

$$|||Q^n - 1 \otimes \pi|||_V \leq Rr^{-n}.$$

(iii) There exists some $n > 0$ such that $\left|\left|\left|Q^i - 1 \otimes \pi\right|\right|\right|_V < \infty$ for $i \leq n$ and

$$\left|\left|\left|Q^i - 1 \otimes \pi\right|\right|\right|_V < 1.$$

Theorem 428 *For any Markov chain $\mathbf{\Phi}$ with transition kernel Q the following are equivalent:*
(i) $\mathbf{\Phi}$ is uniformly ergodic.
(ii) There exist $r > 1$ and $R < \infty$ such that for all x

$$\|Q^n(x,.) - \pi(.)\| \leq Rr^{-n}; \tag{14.37}$$

that is, the convergence takes place at a uniform geometric rate.
(iii) For some $n \in \mathbb{Z}_+$

$$\sup_{x \in X} \|Q^n(x,.) - \pi(.)\| < 1.$$

(iv) The chain is aperiodic and Doeblin condition holds : that is, there is a probability measure ϕ on $\mathcal{B}(X)$ and $\varepsilon < 1$, $\delta > 0$, $m \in \mathbb{Z}_+$ such that whenever $\phi(A) > \varepsilon$

$$\inf_{x \in X} Q^m(x, A) > \delta.$$

(*v*) *The state space X is ν_m-small for some m.*

(*vi*) *The chain is aperiodic and there is a petit set C with*

$$\sup_{x\in X}\mathbb{E}_x\left[\tau_C\right]<\infty,$$

in which case for every set $A\in\mathcal{B}^+(X)$, $\sup_{x\in X}\mathbb{E}_x\left[\tau_A\right]<\infty$.

(*vii*) *The chain is aperiodic and there is a petite set C and $\kappa>1$ with*

$$\sup_{x\in X}\mathbb{E}_x\left[\kappa^{\tau_C}\right]<\infty,$$

in which case for every $A\in\mathcal{B}^+(X)$ we have for some $\kappa_A>1$,

$$\sup_{x\in X}\mathbb{E}_x\left[\kappa_A^{\tau_A}\right]<\infty,$$

(*viii*) *The chain is aperiodic and there is a bounded solution $V\geq 1$ to*

$$\Delta V(x)\leq-\beta V(x)+b\mathbb{I}_C(x),\ x\in X.$$

for some $\beta>0, b<\infty$, and some petit set C,

Under (v), we have in particular that for any x,

$$\left\|Q^n(x,.)-\pi(.)\right\|\leq 2\rho^{[n/m]},$$

where $\rho=1-\nu_m(X)$.

The theorem that follows establish Uniform ergodicity for a MSGS firstly, for simplicity reasons and then proceed to study V-uniform ergodicity.

Definition 429 ***(Uniform ergodicity for a MSGS).*** *Let a Markov system in a general state space $(X,\mathcal{B}(X))$. Also let $\{T(t)\}_{t=0}^{\infty}$ be the total population of memberships with $\lim_{t\to\infty}T(t)=T$; $Q(x,A)$ with $x\in X$ and $A\in\mathcal{B}(X)$ the transition kernel of the inherent Markov chain of memberships and $\mathbb{E}_Q[\mathbf{N}(x,A)]$ the expected population structure. We say that the MSGS is uniformly ergodic if and only if there exists a $C<\infty$ and an $0<a<1$ such that*

$$\left\|\mathbb{E}_Q\left[N(x,t,A)\right]-T\pi(A)\right]\|\leq Ca^t \text{ for every } x\in X,\ A\in\mathcal{B}(X). \tag{14.38}$$

Theorem 430 *(Vassiliou (2020b) Let a Markov system that lives in $(X,\mathcal{B}(X))$. Assume in addition that*

$$\lim_{t\to\infty}T(t)=T \text{ at a geometric rate, } \Delta T(t)\geq 0 \text{ for every } t\in\mathbb{N}_+,$$

and that the inherent Markov chain with transition kernel $Q(x,A)$ is uniformly ergodic. Then the MSGS is uniformly ergodic.

Proof. We have that

$$\left\|\mathbb{E}_Q\left[N(x,t,A)\right]-T\pi(A)\right]\|=\left\|\mathbb{E}_Q\left[N(x,t,A)\right]-T(0)\pi(A)-(T-T(0)\pi(A))\right]\|$$

$$=(\text{From equation } (14.10)\ )$$

$$=\|\ T(0)\left[\int_X\lambda(dx)\,Q^t(x,A)-\pi(A)\right]+$$

$$\sum_{n=1}^{t-1} \Delta T(n) \int_X \mu(dx) Q^{t-n}(x,A) - \sum_{n=1}^{\infty} \Delta T(n) \pi(A) \parallel$$

$$\leq \parallel T(0) \left[\int_X \lambda(dx) Q^t(x,A) - \pi(A) \right] +$$

$$\sum_{n=1}^{t-1} \Delta T(n) \left[\int_X \mu(dx) Q^{t-n}(x,A) - \pi(A) \right] - \sum_{n=t}^{\infty} \Delta T(n) \pi(A) \parallel$$

$$\leq T(0) \parallel \int_X \lambda(dx) Q^t(x,A) - \pi(A) \parallel + \sum_{n=1}^{t-1} \Delta T(n) \left[\int_X \mu(dx) Q^{t-n}(x,A) - \pi(A) \right]$$

$$\sum_{n=1}^{t-1} \Delta T(n) \left[\int_X \mu(dx) Q^{t-n}(x,A) - \pi(A) \right] + \parallel \sum_{n=t}^{\infty} \Delta T(n) \pi(A) \parallel . \qquad (14.39)$$

From (14.39) and since $\lim_{t\to\infty} T(t) = T$ geometrically fast and $\Delta T(t) \geq 0$ for every $t \in \mathbb{N}_+$ from Theorem 428 we get that there exist a $c > 0$ and a $1 > b > 0$ such that

$$\parallel \sum_{n=t}^{\infty} \Delta T(n) \pi(A) \parallel \leq \|\pi(A)\| \sum_{n=t}^{\infty} \Delta T(n) \leq c(1-b)^{-1} b^t. \qquad (4.40)$$

Also from (14.39), since $\int_X \lambda(dx) = 1$ and since the inherent Markov chain with kernel $Q(x,A)$ is uniformly ergodic, from Theorem 428 (ii) there exists $R < \infty$ and $r > 1$ such that

$$T(0) \parallel \int_X \lambda(dx) Q^t(x,A) - \pi(A) \parallel = T(0) \parallel \int_X \lambda(dx) \left[Q^t(x,A) - \pi(A) \right] \parallel$$

$$\leq T(0) \int_X \lambda(dx) \left\| Q^t(x,A) - \pi(A) \right\| \leq T(0) \int_X \lambda(dx) Ra^t \leq T(0) Ra^t, \qquad (14.41)$$

with $a = r^{-1}$, $0 < a < 1$.

Now, $\lim_{t\to\infty} T(t) = T$ geometrically fast and $\Delta T(t) \geq 0$ for every $t \in \mathbb{N}_+$ then from Theorem 428 we get that there exist a $c > 0$ and a $1 > b > 0$ such that $\Delta T(t) \leq cb^t$; in addition since the inherent Markov chain with kernel $Q(x,A)$ is uniformly ergodic, from Theorem 428(ii) there exists $R < \infty$ and $r > 1$, that is, $0 < a = r^{-1} < 1$ such that

$$\parallel \sum_{n=1}^{t-1} \Delta T(n) \left[\int_X \mu(dx) Q^{t-n}(x,A) - \pi(A) \right] \parallel$$

$$\leq \sum_{n=1}^{t-1} \Delta T(n) \left\| \int_X \mu(dx) Q^{t-n}(x,A) - \pi(A) \right\|$$

$$\leq \sum_{n=1}^{t-1} \Delta T(n) \int_X \mu(dx) \left\| Q^{t-n}(x,A) - \pi(A) \right\| \leq \sum_{n=1}^{t-1} \Delta T(n) \int_X \mu(dx) Ra^{t-n}$$

$$\leq \sum_{n=1}^{t-1} cb^t Ra^{t-n} \leq cR(1-b/a)^{-1} \left(1 - (b/a)^t\right) a^t = R^* a^t. \qquad (14.42)$$

Now, from (14.39), (14.40), (14.41) and (14.42), we have that

$$\|\mathbb{E}_Q[N(x,t,A)] - T\pi(A)]\| \leq T(0)Ra^t + R^*a^t + c(1-b)^{-1}b^t. \tag{14.43}$$

Let $a_1 = \max\{a,b\}$ then $0 < a_1 < 1$ and

$$\begin{aligned} \|\mathbb{E}_Q[N(x,t,A)] - T\pi(A)]\| &\leq T(0)Ra_1^t + R^*a_1^t + c(1-b)^{-1}a_1^t \\ &\leq \left[T(0)R + R^* + c(1-b)^{-1}\right]a_1^t \\ &\leq c^*a^t, \end{aligned}$$

which concludes the proof that MSGS is uniformly ergodic. ■

Extending the above results to *V−iniform ergodicity* is now straight forward. Let us provide the following essential definition

Definition 431 ***(V-Uniform ergodicity for a MSGS).*** *Let a Markov system in a general state space $(X, \mathcal{B}(X))$. Also let $\{T(t)\}_{t=0}^{\infty}$ be the total population of memberships with $\lim_{t\to\infty} T(t) = T$; $Q(x,A)$ with $x \in X$ and $A \in \mathcal{B}(X)$ the transition kernel of the inherent Markov chain of memberships and $\mathbb{E}_Q[\mathbf{N}(x,A)]$ the expected population structure. We say that the MSGS is V-uniformly ergodic if and only if there exists a $C < \infty$ and an $0 < a < 1$ such that*

$$|||\ \mathbb{E}_Q[N(x,t,A)] - T\pi(A)]\ |||_V \leq Ca^t \text{ for every } x \in X,\ A \in \mathcal{B}(X). \tag{14.44}$$

Following exactly the proof of Theorem 430 we arrive at the following result:

Theorem 432 *Let a Markov system that leaves in $(X, \mathcal{B}(X))$. Assume in addition that*

$$\lim_{t\to\infty} T(t) = T \text{ at a geometric rate, } \Delta T(t) \geq 0 \text{ for every } t \in \mathbb{N}_+,$$

and that the inherent Markov chain with transition kernel $Q(x,A)$ is V-uniformly ergodic. Then the MSGS is V-uniformly ergodic.

14.7 Asymptotic periodicity of a MSGS

The study the asymptotic periodicity of the expected population structure, which apparently is an important problem for a population that lives in $(X, \mathcal{B}(X))$ will be the main theme of the present section (see also Vassiliou (2021b)). Let that the inherent Markov chain of the MSGS with kernel Q is positive Harris with period $d \geq 1$, and let that the d-cycle, as described before, be the set $\{D_0, D_1, ..., D_{d-1}\}$, where $D_i \in \mathcal{B}(X)$ for $i = 0, 1, ..., d-1$. Refresh that $\lambda(x)$ is the initial distribution on X. Then the distribution on the set D_i is

$$\lambda_i(x) = \frac{\lambda(x)}{\lambda(D_i)} \text{ where } x \in D_i. \tag{14.45}$$

From Theorem 407, we have that if $x \in D_0$ then $Q(x, D_1) = 1$ and consequently $Q^2(x, D_2) = \int_{D_1} Q(x,dy)Q(y,D_2)$. However, for $y \in D_1$ we have $Q(y, D_2) = 1$, hence $Q^2(x, D_2) = \int_{D_1} Q(x,dy) = Q(x,D_1) = 1$. Similarly, since the period is d we have

$$\text{If } x \in D_i \text{ then } Q^{nd}(x, D_i) = 1,\ Q^{nd}(x, D_j) = 0 \text{ for } i \neq j. \tag{14.46}$$

In general for $x \in D_i$ and for $d > i + v$, we have

$$\begin{aligned} Q^{nd+v}(x, D_{i+v}) &= \int_{D_i} Q^{nd}(x,dy)Q^v(y, D_{i+v}) \\ &= (\text{for } y \in D_i \text{ and } d > i+v,\ Q^v(y, D_{i+v}) = 1) \\ &= \int_{D_i} Q^{nd}(x,dy) = 1. \end{aligned} \tag{14.47}$$

It is apparent from (14.47) that

$$\text{If } x \in D_i \text{ and for } d > i + v, \text{ we have } Q^{nd+v}(x, D_j) = 0 \text{ for } j \neq i + v. \tag{14.48}$$

If $x \in D_i$ and for $d \leq i + v$, we have

$$\begin{aligned} Q^{nd+v}(x, D_{i+v-d}) &= \int_{D_i} Q^{nd}(x, dy)\, Q^{v}(y, D_{i+v-d}) \\ &= (\text{for } y \in D_i \text{ and } d \leq i + v,\ Q^{v}(y, D_{i+v-d}) = 1) \\ &= \int_{D_i} Q^{nd}(x, dy) = 1 \end{aligned}$$

From the fact that the inherent Markov chain with kernel Q is assumed to be positive Harris with period d by Proposition 416 when restricted to each cyclic set is aperiodic and positive Harris on the d-skeleton, and by Theorem 418 it is straightforward to check the following theorem:

Theorem 433 *(Vassiliou 2020b). If the inherent Markov chain of the MSGS is positive Harris and periodic with period d, then for the sets $\{D_0, D_1, ..., D_{d-1}\}$ with $D_i \in \mathcal{B}(X)$ we have that*

$$\lim_{n\to\infty} \left\| \int_{D_i} \lambda_i(dx)\, P^{nd+v}(x, A) - \pi_{i+v}(A) \right\| = 0 \text{ if } d > i + v \text{ for } i = 0, 1, ..., d-1,$$

for any set $A \subseteq D_{i+v}$, and

$$\lim_{n\to\infty} \left\| \int_{D_i} \lambda_i(x)\, P^{nd+v}(x, A) - \pi_{i+v-d}(A) \right\| = 0 \text{ if } d \leq i + v \text{ for } i = 0, 1, ..., d-1,$$

for any set $A \subseteq D_{i+v-d}$.

Now, from the above Theorem and the decomposition theorem we immediately get the following:

Theorem 434 *Vassiliou (2020b). If the inherent Markov chain of the MSGS is positive recurrent and periodic with period d, then for the sets $\{D_0, D_1, ..., D_{d-1}\}$ with $D_i \in \mathcal{B}(X)$, then for each i there is a π_i-null set N_i such that for every initial distribution λ_i with $\lambda_i(N_i) = 0$*

$$\lim_{n\to\infty} \left\| \int_{D_i} \lambda_i(dx)\, P^{nd+v}(x, A) - \pi_{i+v}(A) \right\| = 0 \text{ if } d > i + v \text{ for } i = 0, 1, ..., d-1,$$

for any set $A \subseteq D_{i+v}$, and

$$\lim_{n\to\infty} \left\| \int_{D_i} \lambda_i(dx)\, P^{nd+v}(x, A) - \pi_{i+v-d}(A) \right\| = 0 \text{ if } d \leq i + v \text{ for } i = 0, 1, ..., d-1,$$

for any set $A \subseteq D_{i+v-d}$.

Let us now start building the proof for the periodicity theorem for a MSGS and conclude by stating the theorem in the end. The important question is the periodicity of the expected population structure

$$\mathbb{E}_{\mathbb{P}}\left[\mathbf{N}\left(t,X\right)\right]=\left[\mathbb{E}_{\mathbb{P}}\left[N\left(x,t,A_1\right)\right],\mathbb{E}_{\mathbb{P}}\left[N\left(x,t,A_2\right)\right],...,\mathbb{E}_{\mathbb{P}}\left[N\left(x,t,A_k\right)\right]\right],$$

when the inherent Markov chain with kernel Q is positive Harris and periodic with period d or is positive recurrent and periodic with period d. With no loss of generality and for simplicity reasons in what follows the set A will be used instead of any of the sets $A_i, (i=1,2,...,k)$. From equation (14.10), we have that

$$\mathbb{E}_{\mathbb{P}}\left[N\left(x,t,A\right)\right]=T\left(0\right)\int\limits_X\lambda\left(dx\right)Q^t\left(x,A\right)+\sum_{n=1}^{t-1}\Delta T\left(n\right)\int\limits_X\mu\left(dx\right)Q^{t-n}\left(x,A\right). \tag{14.49}$$

Let us start with the first part of the right hand side of equation (14.49). We have that

$$\begin{aligned}\int\limits_X\lambda\left(dx\right)Q^t\left(x,A\right) &= \lambda\left(D_0\right)\int\limits_{D_0}\lambda_0\left(dx\right)Q^t\left(x,A\right)+\lambda\left(D_1\right)\int\limits_{D_1}\lambda_1\left(dx\right)Q^t\left(x,A\right)\\ &\quad+...+\lambda\left(D_{d-1}\right)\int\limits_{D_{d-1}}\lambda_{d-1}\left(dx\right)Q^t\left(x,A\right)\end{aligned} \tag{14.50}$$

The set A in general could be written as follows

$$A=A\cap D_0\cup A\cap D_1\cup...\cup A\cap D_{d-1},\text{ with }\left(A\cap D_i\right)\cap\left(A\cap D_j\right)=\varnothing,\text{ for }i\neq j. \tag{14.51}$$

Let that $t=\pi d+v$ and consider in what follows that for all $i=0,1,...,d-1,\ d>i+v$. Then we have that

$$\begin{aligned}Q^t\left(x,A\right) &= Q^{\pi d+v}\left(x,A\right)=Q^{\pi d+v}\left(x,A\cap D_0\right)+Q^{\pi d+v}\left(x,A\cap D_1\right)+...\\ &\quad+Q^{\pi d+v}\left(x,A\cap D_{d-1}\right).\end{aligned} \tag{14.52}$$

We have that

$$\lim_{t\to\infty}\left\Vert\int\limits_X\lambda\left(dx\right)Q^t\left(x,A\right)-\sum_{i=0}^{d-1}\lambda\left(D_i\right)\pi_{i+v}\left(A\cap D_{i+v}\right)\right\Vert=$$

$$=\lim_{\pi\to\infty}\left\Vert\sum_{i=0}^{d-1}\lambda\left(D_i\right)\int\limits_{D_i}\lambda_i\left(dx\right)Q^{\pi d+v}\left(x,A\right)-\sum_{i=0}^{d-1}\lambda\left(D_i\right)\pi_{i+v}\left(A\cap D_{i+v}\right)\right\Vert$$

$$\leq\lim_{\pi\to\infty}\sum_{i=0}^{d-1}\lambda\left(D_i\right)\left\Vert\int\limits_{D_i}\lambda_i\left(dx\right)Q^{\pi d+v}\left(x,A\right)-\pi_{i+v}\left(A\cap D_{i+v}\right)\right\Vert. \tag{14.53}$$

Now, we get that

$$\lim_{t\to\infty}\left\Vert\int\limits_X\lambda\left(dx\right)Q^t\left(x,A\right)-\sum_{i=0}^{d-1}\lambda\left(D_i\right)\pi_{i+v}\left(A\cap D_{i+v}\right)\right\Vert$$

$$\leq\lim_{\pi\to\infty}\sum_{i=0}^{d-1}\lambda\left(D_i\right)\left\Vert\sum_{j=0}^{d-1}\int\limits_{D_i}\lambda_i\left(dx\right)Q^{\pi d+v}\left(x,A\cap D_j\right)-\pi_{i+v}\left(A\cap D_{i+v}\right)\right\Vert$$

$$\leq \lim_{\pi\to\infty} \sum_{i=0}^{d-1} \lambda(D_i) \left\| \int_{D_i} \lambda_i(dx)\, Q^{\pi d+v}(x, A\cap D_{i+v}) - \pi_{i+v}(A\cap D_{i+v}) \right\|$$

$$+ \lim_{\pi\to\infty} \sum_{i=0}^{d-1} \lambda(D_i) \sum_{j=0, j\neq i+v}^{d-1} \left\| \int_{D_i} \lambda_i(dx)\, Q^{\pi d+v}(x, A\cap D_j) \right\| \leq \in .$$

Therefore, it was proved that when $t = \pi d + v$ and $d > i + v$ then the limit of $\int_X \lambda(dx)\, Q^t(x, A)$ as $t \to \infty$ is

$$\sum_{i=0}^{d-1} \lambda(D_i)\, \pi_{i+v}(A\cap D_{i+v}). \tag{14.54}$$

Analogously, we get that for $t = \pi d + v$ and $d \leq i + v$ then

$$\lim_{t\to\infty} \left\| \int_X \lambda(dx)\, Q^t(x, A) - \sum_{i=0}^{d-1} \lambda(D_i)\, \pi_{i+v-d}(A\cap D_{i+v-d}) \right\| = 0. \tag{14.55}$$

It is not difficult to check that

$$\lim_{t\to\infty} \| \int_X \lambda(dx)\, Q^t(x, A) - \sum_{i=0}^{d-v-1} \lambda(D_i)\, \pi_{i+v}(A\cap D_{i+v})$$

$$- \sum_{i=d-v}^{d-1} \lambda(D_i)\, \pi_{i+v-d}(A\cap D_{i+v-d}) \| = 0. \tag{14.56}$$

Assume now that $\lim_{t\to\infty} T(t) = T$ and $\Delta T(t) \geq 0$ for every t. Let again that $t = \pi d + v$ then

$$U(t) = \sum_{n=1}^{t-1} \Delta T(n) \int_X \mu(dx)\, Q^{t-n}(x, A) \tag{14.57}$$

$$= \sum_{n=1}^{\pi d+v-1} \Delta T(n) \int_X \mu(dx)\, Q^{\pi d+v-n}(x, A)$$

$$= \sum_{m=0}^{d-1} \sum_{\tau=0}^{\pi-1} \Delta T(\tau d + m) \int_X \mu(dx)\, Q^{\pi d+v-(\tau d+m)}(x, A)$$

$$+ \sum_{m=0}^{v-1} \Delta T(\pi d + m) \int_X \mu(dx)\, Q^{\pi d+v-(\pi d+m)}(x, A).$$

Since $\lim_{t\to\infty} \Delta T(t) = 0$, we have

$$\lim_{\pi\to\infty} \left\| \sum_{m=0}^{v-1} \Delta T(\pi d + m) \int_X \mu(dx)\, Q^{\pi d+v-(\pi d+m)}(x, A) \right\| \tag{14.58}$$

$$\leq \sum_{m=0}^{v-1} \lim_{\pi\to\infty} \Delta T(\pi d + m) \lim_{\pi\to\infty} \left\| \int_X \mu(dx)\, Q^{\pi d+v-(\pi d+m)}(x, A) \right\| = 0.$$

Therefore, we get that

$$\lim_{t\to\infty} U(t) = \lim_{\pi\to\infty} \sum_{m=0}^{d-1}\sum_{\tau=0}^{\pi-1} \Delta T(\tau d+m) \int_X \mu(dx)\, Q^{\pi d+\upsilon-(\tau d+m)}(x,A) \tag{14.59}$$

$$= \lim_{\pi\to\infty} \sum_{m=0}^{\upsilon}\sum_{\tau=0}^{\pi-1} \Delta T(\tau d+m) \int_X \mu(dx)\, Q^{(\pi-\tau)d+\upsilon-m}(x,A)$$

$$+ \lim_{\pi\to\infty} \sum_{m=\upsilon+1}^{d-1}\sum_{\tau=0}^{\pi-1} \Delta T(\tau d+m) \int_X \mu(dx)\, Q^{(\pi-\tau-1)d+(d+\upsilon-m)}(x,A).$$

Since $\Delta T(t) \geq 0$

$$\sum_{\tau=0}^{\pi-1} \Delta T(\tau d+m) \leq \sum_{t=0}^{\pi d+r} \Delta T(t) = T(\pi d+r) - T(0),$$

and

$$\sum_{\tau=0}^{\pi-1} \Delta T(\tau d+m), \tag{14.60}$$

is a series of non-negative elements and it is bounded by $T - T(0)$ and therefore it converges; so let

$$w_m = \frac{\lim_{\pi\to\infty} \sum_{\tau=0}^{\pi-1} \Delta T(\tau d+m)}{\lim_{t\to\infty} \sum_{\tau=0}^{t} \Delta T(t)}. \tag{14.61}$$

Define by

$$\lim_{\pi\to\infty} U_1(t) = \sum_{m=0}^{\upsilon} \lim_{\pi\to\infty} \sum_{\tau=0}^{\pi-1} \Delta T(\tau d+m) \int_X \mu(dx)\, Q^{(\pi-\tau)d+\upsilon-m}(x,A). \tag{14.62}$$

Now we get that

$$\lim_{\pi\to\infty} \sum_{\tau=0}^{\pi-1} \Delta T(\tau d+m) = w_m \lim_{t\to\infty} \sum_{\tau=0}^{t} \Delta T(t) = w_m (T - T(0)). \tag{14.63}$$

Thus

$$\lim_{\pi\to\infty} U_1(t) = (T - T(0)) \sum_{m=0}^{\upsilon} w_m \times \tag{14.64}$$

$$\lim_{\pi\to\infty} \sum_{\tau=0}^{\pi-1} \frac{\Delta T(\tau d+m)}{\sum_{\tau=0}^{\pi-1} \Delta T(\tau d+m)} \int_X \mu(dx)\, Q^{(\pi-\tau)d+\upsilon-m}(x,A).$$

The following is a useful Lemma:

Lemma 435 *If $\{a_n\}$ is a sequence of non-negative numbers such that $a_n / \sum_{k=0}^{n} a_k \to 0$ as $n \to \infty$, and $\{b_n\} \to b$ then*

$$\lim_{n\to\infty} \frac{b_0 a_n + b_1 a_{n-1} + \ldots + b_n a_0}{\sum_{k=0}^{n} a_k} = b. \tag{14.65}$$

From the first Lemma of the present Chapter, (14.64) and the result in (14.56) where we may replace $\lambda(dx)$ by $\mu(dx)$ it is not difficult to check that the following holds

$$\lim_{\pi\to\infty} \| U_1(t) - (T - T(0)) \times \sum_{m=0}^{v} \sum_{i=0}^{d-(v-m)-1} w_m \mu(D_i) \pi_{i+(v-m)} \left(A \cap D_{i+(v-m)}\right) - (T - T(0)) \sum_{m=0}^{v} \sum_{\substack{i=d-(v-m) \\ i\le d-1}}^{d-1} w_m \mu(D_i) \pi_{i-d+(v-m)} \left(A \cap D_{i-d+(v-m)}\right) \| = 0. \tag{14.66}$$

Define by

$$\lim_{t\to\infty} U_2(t) = \lim_{\pi\to\infty} \sum_{m=v+1}^{d-1} \sum_{\tau=0}^{\pi-1} \Delta T(\tau d + m) \times \int_X \mu(dx) Q^{(\pi-\tau-1)d+(d+v-m)}(x, A). \tag{14.67}$$

Analogously it could be prove that

$$\lim_{\pi\to\infty} \| U_2(t) - (T - T(0)) \sum_{m=v+1}^{d-1} \sum_{i=0}^{d-(d+v-m)-1} w_m \mu(D_i) \pi_{i+d+(v-m)} \left(A \cap D_{i+d+(v-m)}\right) -(T - T(0)) \sum_{m=v+1}^{d-1} \sum_{\substack{i=d-(d+m-v) \\ i\le d-1}}^{d-1} w_m \mu(D_i) \times \pi_{i-d+(d+v-m)} \left(A \cap D_{i-d+(d+v-m)}\right) \| = 0. \tag{14.68}$$

Thus from $(14.49), (14.56), (14.59), (14.65)$ and (14.67) for $t = \pi d + v$ we get that

$$\begin{aligned} \lim_{t\to\infty} \| \mathbb{E}_{\mathbb{P}}[N(x,t,A)] &- T(0) \sum_{i=0}^{d-v-1} \lambda(D_i) \pi_{i+v} (A \cap D_{i+v}) \\ &-T(0) \sum_{i=d-v}^{d-1} \lambda(D_i) \pi_{i+v-d} (A \cap D_{i+v-d}) \\ &- (T - T(0)) \sum_{m=0}^{v} \sum_{i=0}^{d-(v-m)-1} w_m \mu(D_i) \pi_{i+(v-m)} \left(A \cap D_{i+(v-m)}\right) \\ &- (T - T(0)) \sum_{m=0}^{v} \sum_{\substack{i=d-(v-m) \\ i\le d-1}}^{d-1} w_m \mu(D_i) \pi_{i-d+(v-m)} \left(A \cap D_{i-d+(v-m)}\right) \\ &-(T - T(0)) \sum_{m=v+1}^{d-1} \sum_{i=0}^{(m-v)-1} w_m \mu(D_i) \pi_{i+d+(v-m)} \left(A \cap D_{i+d+(v-m)}\right) \\ &-(T - T(0)) \sum_{m=v+1}^{d-1} \sum_{\substack{i=(m-v) \\ i\le d-1}}^{d-1} w_m \mu(D_i) \pi_{i+(v-m)} \left(A \cap D_{i+(v-m)}\right) \| = 0, \end{aligned} \tag{14.69}$$

for $v = 0, 1, ..., d-1$. Therefore, we have proved the following theorem:

Theorem 436 *(Vassiliou (2020b)). Let a MSGS and let that the inherent Markov chain of the MSGS be positive Harris and periodic with period d,*

OR let that the inherent Markov chain of the MSGS is positive recurrent and periodic with period d, and let the sets $\{D_0, D_1, ..., D_{d-1}\}$ *with* $D_i \in \mathcal{B}(X)$*. Then the expected population structure splits into d converging subsequences, that is,*

$$\lim_{t\to\infty} \parallel \mathbb{E}_{\mathbb{P}}\left[N(x,t,A)\right] - T(0)\sum_{i=0}^{d-v-1}\lambda(D_i)\,\pi_{i+v}(A\cap D_{i+v})$$

$$-T(0)\sum_{i=d-v}^{d-1}\lambda(D_i)\,\pi_{i+v-d}(A\cap D_{i+v-d})$$

$$-(T-T(0))\sum_{m=0}^{v}\sum_{i=0}^{d-(v-m)-1} w_m\mu(D_i)\,\pi_{i+(v-m)}\left(A\cap D_{i+(v-m)}\right)$$

$$-(T-T(0))\sum_{m=0}^{v}\sum_{\substack{i=d-(v-m)\\ i\le d-1}}^{d-1} w_m\mu(D_i)\,\pi_{i-d+(v-m)}\left(A\cap D_{i-d+(v-m)}\right)$$

$$-(T-T(0))\sum_{m=v+1}^{d-1}\sum_{i=0}^{(m-v)-1} w_m\mu(D_i)\,\pi_{i+d+(v-m)}\left(A\cap D_{i+d+(v-m)}\right)$$

$$-(T-T(0))\sum_{m=v+1}^{d-1}\sum_{\substack{i=(m-v)\\ i\le d-1}}^{d-1} w_m\mu(D_i)\,\pi_{i+(v-m)}\left(A\cap D_{i+(v-m)}\right)\parallel = 0,$$

for $v = 0, 1, 2, ..., d-1$.

14.8 Total variability from the invariant measures

The total variability from the invariant measures in the periodic case for a MSGS is studied in the present section. It is shown that, the total variation in the periodic case is finite, which is known as the coupling problem. On the coupling problem see also Lindvall (1992). We will make use of the following theorems from Meyn and Tweedie (2009 p. 332).

Theorem 437 *Suppose that* Φ *is an aperiodic positive Harris chain and suppose that the chain admits an atom* $a \in \mathcal{B}^+(X)$*. Then for any regular initial distributions* λ, μ

$$\sum_{n=1}^{\infty}\int_X\int_X \lambda(dx)\,\mu(dy)\,\|P^n(x,.) - P^n(y,.)\| < \infty,$$

and in particular, if Φ *is regular, then for every* $x, y \in X$

$$\sum_{n=1}^{\infty}\|P^n(x,.) - P^n(y,.)\| < \infty.$$

Theorem 438 *Suppose* Φ *is an aperiodic positive Harris chain. For any regular initial distributions* λ, μ

$$\sum_{n=1}^{\infty}\int_X\int_X \lambda(dx)\,\mu(dy)\,\|P^n(x,.) - P^n(y,.)\| < \infty.$$

Theorem 439 *Suppose that Φ is an aperiodic positive Harris chain and that α is an accessible atom. If $\mathbb{E}_{\alpha}\left[\tau_{\alpha}^{2}\right]<\infty$, then for any regular initial distribution*

$$\sum_{n=1}^{\infty}\|\lambda P^{n}-\pi\|<\infty.$$

We assume that the inherent Markov chain Φ_Q with kernel Q of the MSGS is positive Harris with period d.Then each of the cyclic sets as described in Proposition 416 is aperiodic and positive Harris on the d-skeleton. From the three previous Theorems collected from Meyn and Tweedie (2009) an immediate result is the following theorem:

Theorem 440 *(Vassiliou (2020b)). If the inherent Markov chain Φ_Q with kernel Q of the MSGS is positive Harris with period d, then for the sets $\{D_0, D_1, ..., D_{d-1}\}$ with $D_i \in \mathcal{B}(X)$ we have that*

a) For any regular initial distributions λ, μ and for $d>i+v$,

$$\begin{aligned}\sum_{n=1}^{\infty}\int_{D_i}\int_{D_i}&\lambda(dx)\,\mu(dy)\left\|Q^{nd+v}(x,A\cap D_{i+v})-Q^{nd+v}(y,A\cap D_{i+v})\right\|\\ &<\infty,\ x,y\in D_i,\end{aligned}$$

for $i,v=0,1,...,d-1$. Also, for $d\leq i+v$

$$\begin{aligned}\sum_{n=1}^{\infty}\int_{D_i}\int_{D_i}&\lambda(dx)\,\mu(dy)\left\|Q^{nd+v}(x,A\cap D_{i+v-d})-Q^{nd+v}(y,A\cap D_{i+v-d})\right\|\\ &<\infty,\ x,y\in D_i,\end{aligned}$$

for $i,v=0,1,...,d-1$.

b) If for the accessible atom W, we have that $\mathbb{E}_w\left[\tau_w^2\right]<\infty$, then for any regular initial distribution λ we have for $d>i+v$, $x\in D_i$ and $i,v=0,1,...,d-1$

$$\sum_{n=1}^{\infty}\left\|\lambda(dx)\,Q^{nd+v}(x,A\cap D_{i+v})-\pi_{i+v}(A\cap D_{i+v})\right\|<\infty,$$

and for $d\leq i+v$, $x\in D_i$ and $i,v=0,1,...,d-1$

$$\sum_{n=1}^{\infty}\left\|\lambda(dx)\,Q^{nd+v}(x,A\cap D_{i+v-d})-\pi_{i+v-d}(A\cap D_{i+v-d})\right\|<\infty.$$

The following coupling theorem for a MSGS with inherent Markov chain Φ_Q which is assumed to be positive Harris and periodic will be established.

Theorem 441 *(Vassiliou (2020b)). Let a Markov system that lives in $(X,\mathcal{B}(X))$ and let that the inherent Markov chain Φ_Q with kernel $Q(x,.)$ is positive Harris with period d. Let also that*

$$\lim_{t\to\infty}T(t)=T,\ \Delta T(t)\geq 0 \text{ for every } t\in\mathbb{Z}^{+}.$$

Then for any regular distributions $\lambda_1,\mu_1,\lambda_2,\mu_2$, we have

$$\sum_{t=1}^{\infty}\|\mathbb{E}_P N(x,t,A)-\mathbb{E}_P[N(y,t,A)]\|<\infty.$$

Proof. Let us consider the following

$$\sum_{t=1}^{\infty} \|\mathbb{E}_P N(x,t,A) - \mathbb{E}_P [N(y,t,A)]\| = \sum_{t=1}^{\infty} \| T(0) \int_X \lambda_1(dx) Q^t(x,A) +$$

$$\sum_{n=1}^{t-1} \Delta T(n) \int_X \mu_1(dx) Q^{t-n}(x,A) - T(0) \int_X \lambda_2(dy) Q^t(y,A) -$$

$$-\sum_{n=1}^{t-1} \Delta T(n) \int_X \mu_2(dy) Q^{t-n}(y,A) \| \tag{14.70}$$

$$\leq \sum_{t=1}^{\infty} T(0) \| \int_X \lambda_1(dx) Q^t(x,A) - \int_X \lambda_2(dy) Q^t(y,A)] \| +$$

$$\sum_{t=1}^{\infty}\sum_{n=1}^{t-1} \Delta T(n) \quad \| \quad \int_X \mu_1(dx) Q^{t-n}(x,A) - \int_X \mu_2(dy) Q^{t-n}(y,A) \|$$

$$= \quad U_\lambda(t) + U_\mu(t)$$

Assume that $t = \pi d + v$. Denote by

$$\lambda_{1i}(dx) = \frac{\lambda_1(dx)}{\lambda_1(D_i)} \text{ and } \lambda_{2i}(dx) = \frac{\lambda_2(dx)}{\lambda_2(D_i)}. \tag{14.71}$$

From (14.70) and (14.71), we have that

$$U_\lambda(t) \leq \sum_{\pi=0}^{\infty}\sum_{i=0}^{d-1} T(0) \| \lambda_1(D_i) \int_{D_i} \lambda_{1i}(dx) Q^{\pi d+v}(x,A) - \tag{14.72}$$

$$\lambda_2(D_i) \int_{D_i} \lambda_{2i}(dy) Q^{\pi d+v}(y,A) \| .$$

with

$$\sum_{i=0}^{d-1} \lambda_1(D_i) = 1 \text{ and } \sum_{i=0}^{d-1} \lambda_2(D_i) = 1. \tag{14.73}$$

From (14.73) it is apparent that we may assume that

$$\begin{aligned} \lambda_1(D_i) &\leq \lambda_2(D_i) \text{ for } i = 0,1,...j \text{ and } \lambda_1(D_i) \geq \lambda_2(D_i) \\ \text{for } j &= j+1,...,d-1, \end{aligned} \tag{14.74}$$

and $d > j + v$. Now from (14.72) and (14.74) we get

$$U_\lambda(t) \leq \sum_{\pi=0}^{\infty}\sum_{i=0}^{j} T(0) \lambda_2(D_i) \| \int_{D_i} \lambda_{1i}(dx) Q^{\pi d+v}(x, A \cap D_{i+v})$$

$$-\int_{D_i} \lambda_{2i}(dy) Q^{\pi d+v}(y, A \cap D_{i+v}) \|$$

$$+\sum_{\pi=0}^{\infty}\sum_{i=j+1}^{d-v-1} T(0)\,\lambda_1(D_i)\parallel \int_{D_i}\lambda_{1i}(dx)\,Q^{\pi d+v}(x,A\cap D_{i+v}) \tag{14.75}$$

$$-\int_{D_i}\lambda_{2i}(dy)\,Q^{\pi d+v}(y,A\cap D_{i+v})\parallel$$

$$+\sum_{\pi=0}^{\infty}\sum_{i=d-v}^{d-1} T(0)\,\lambda_1(D_i)\parallel \int_{D_i}\lambda_{1i}(dx)\,Q^{\pi d+v}(x,A\cap D_{i+v-d})$$

$$-\int_{D_i}\lambda_{2i}(dy)\,Q^{\pi d+v}(x,A\cap D_{i+v-d})\parallel = U_\lambda^{(1)}(t)+U_\lambda^{(2)}(t)+U_\lambda^{(3)}(t).$$

Now, it is easy to see that

$$\int_{D_i}\lambda_{1i}(dx)\,Q^{\pi d+v}(x,A\cap D_{i+v}) = \int_{D_i}\int_{D_i}\lambda_{1i}(dx)\,\lambda_{2i}(dy)\times Q^{\pi d+v}(x,A\cap D_{i+v}), \tag{14.76}$$

and

$$\int_{D_i}\lambda_{2i}(dy)\,Q^{\pi d+v}(y,A\cap D_{i+v}) = \int_{D_i}\int_{D_i}\lambda_{1i}(dx)\,\lambda_{2i}(dy)\times Q^{\pi d+v}(y,A\cap D_{i+v}). \tag{14.77}$$

Therefore, from (14.75), (14.76) and (14.77), we have that

$$U_\lambda^{(1)}(t) \le \sum_{i=0}^{j} T(0)\,\lambda_2(D_i)\sum_{\pi=0}^{\infty}\int_{D_i}\int_{D_i}\lambda_{1i}(dx)\,\lambda_{2i}(dy) \times \parallel Q^{\pi d+v}(x,A\cap D_{i+v}) - Q^{\pi d+v}(y,A\cap D_{i+v})\parallel. \tag{14.78}$$

From Theorem 440 and (14.78) we get that

$$U_\lambda^{(1)}(t) \le \sum_{i=0}^{j} T(0)\,\lambda_2(D_i)\,K_1 < \infty, \tag{14.79}$$

analogously, we have that $U_\lambda^{(2)}(t) < \infty$ and $U_\lambda^{(3)}(t) < \infty$ and therefore from (14.75), we get that

$$U_\lambda(t) < \infty. \tag{14.80}$$

Similarly we prove the case for $d \le j+v$.

We have that

$$U_\mu(t) = \sum_{t=1}^{\infty}\Big[\sum_{n=0}^{t-1}\Delta T(n)\int_X \mu_1(dx)\,Q^{t-n}(x,A) - \sum_{n=0}^{t-1}\Delta T(n)\int_X \mu_2(dx)\,Q^{t-n}(x,A)\Big]$$

$$= \sum_{\pi=0}^{\infty}\sum_{\upsilon=0}^{d-1}[\sum_{n=0}^{\pi d+\upsilon-1} \Delta T(n) \times$$

$$\left\{\int_X \mu_1(dx)\, Q^{t-n}(x,A) - \int_X \mu_2(dy)\, Q^{t-n}(y,A)\right\}]$$

$$= \sum_{\pi=0}^{\infty}\sum_{\upsilon=0}^{d-1}[\sum_{m=0}^{d-1}\sum_{\tau=0}^{\pi-1} \Delta T(\tau d+m) \times$$

$$\{\int_X \mu_1(dx)\, Q^{(\pi d+\upsilon)-(\tau d+m)}(x,A) - \int_X \mu_2(dy)\, Q^{(\pi d+\upsilon)-(\tau d+m)}(y,A)\}]$$

$$+\sum_{\pi=0}^{\infty}[\sum_{m=0}^{\upsilon-1} \Delta T(\tau d+m) \times$$

$$\{\int_X \mu_1(dx)\, Q^{(\pi d+\upsilon)-(\pi d+m)}(x,A) - \int_X \mu_2(dy)\, Q^{(\pi d+\upsilon)-(\pi d+m)}(y,A)\}]$$

$$= U_\mu^{(\pi)}(t) + U_\mu^{(\upsilon)}(t). \tag{14.81}$$

The function $U_\mu^{(\upsilon)}(t)$ will be studied first

$$\left\|U_\mu^{(\upsilon)}(t)\right\| \leq \sum_{\pi=0}^{\infty}[\sum_{m=0}^{\upsilon-1} \Delta T(\tau d+m)$$

$$\times \left\|\int_X \mu_1(dx)\, Q^{\upsilon-m}(x,A) - \int_X \mu_2(dy)\, Q^{\upsilon-m}(y,A)\}]\right\|$$

$$\leq \sum_{\pi=0}^{\infty}[\sum_{m=0}^{\upsilon-1} \Delta T(\tau d+m) \sum_{i=0}^{d-1} \| \int_{D_i} \mu_1(dx)\, Q^{\upsilon-m}(x, A\cap D_{i+\upsilon-m})$$

$$- \int_{D_i} \mu_2(dy)\, Q^{\upsilon-m}(y, A\cap D_{i+\upsilon-m})\}] \| . \tag{14.82}$$

Define by

$$\mu_{1i}(dx) = \frac{\mu_1(dx)}{\mu_1(D_i)} \text{ and } \mu_{2i}(dx) = \frac{\mu_2(dx)}{\mu_2(D_i)}. \tag{14.83}$$

We assume that, there is an r such that

$$\mu_1(D_i) \leq \mu_2(D_i) \text{ for } i=0,1,...,r \text{ and } \mu_1(D_i) > \mu_2(D_i) \text{ for } i=r+1,...,d-1. \tag{14.84}$$

This assumption retains the generality of the proof. Relations (14.82), (14.83), (14.84) and Theorem 440 lead us to

$$\left\|U_\mu^{(\upsilon)}(t)\right\| \leq \sum_{\pi=0}^{\infty}[\sum_{m=0}^{\upsilon-1} \Delta T(\tau d+m) \sum_{i=0}^{r} \mu_2(D_i) \times$$

$$\| \int_{D_i}\int_{D_i} \mu_{1i}(dx)\, \mu_{2i}(dy)\, Q^{\upsilon-m}(x, A\cap D_{i+\upsilon-m})$$

$$-\int_{D_i}\int_{D_i} \mu_{1i}(dx)\,\mu_{2i}(dy)\,Q^{\upsilon-m}(y, A\cap D_{i+\upsilon-m})\}]\;\|$$

$$+\sum_{\pi=0}^{\infty}[\sum_{m=0}^{\upsilon-1}\Delta T(\tau d+m)\sum_{j=r+1}^{d-1}\mu_1(D_i)\times$$

$$\|\int_{D_i}\int_{D_i}\mu_{1i}(dx)\,\mu_{2i}(dy)\,Q^{\upsilon-m}(x, A\cap D_{i+\upsilon-m})$$

$$-\int_{D_i}\int_{D_i} \mu_{1i}(dx)\,\mu_{2i}(dy)\,Q^{\upsilon-m}(y, A\cap D_{i+\upsilon-m})\}]\;\|$$

$$\leq\sum_{\pi=0}^{\infty}\sum_{m=0}^{\upsilon-1}\Delta T(\tau d+m)\,[\sum_{j=0}^{r}\mu_2(D_i)\,K_1(\upsilon-m)$$

$$+\sum_{j=r+1}^{d-1}\mu_1(D_i)\,K_2(\upsilon-m)]<\infty. \tag{14.85}$$

Now, for the function $U_\mu^{(\pi)}(t)$, we have that

$$\left\|U_\mu^{(\pi)}(t)\right\|\leq\sum_{\pi=0}^{\infty}\sum_{\upsilon=0}^{d-1}[\sum_{m=0}^{\upsilon}\sum_{\tau=0}^{\pi-1}\Delta T(\tau d+m)\;\|\int_X\mu_1(dx)\,Q^{(\pi-\tau)d+\upsilon-m}(x,A)$$

$$-\int_X\mu_2(dy)\,Q^{(\pi-\tau)d+\upsilon-m}(y,A)\;\|$$

$$\sum_{\pi=0}^{\infty}\sum_{\upsilon=0}^{d-1}[\sum_{m=\upsilon+1}^{d-1}\sum_{\tau=0}^{\pi-1}\Delta T(\tau d+m)\;\|\int_X\mu_1(dx)\,Q^{(\pi-\tau-1)d+d+\upsilon-m}(x,A)$$

$$-\int_X\mu_2(dy)\,Q^{(\pi-\tau-1)d+d+\upsilon-m}(y,A)\;\|=\left\|U_\mu^{(\pi,1)}(t)\right\|+\left\|U_\mu^{(\pi,2)}(t)\right\|. \tag{14.86}$$

For $U_\mu^{(\pi,1)}(t)$ and we have that

$$\left\|U_\mu^{(\pi,1)}(t)\right\|\leq\sum_{\tau=0}^{\infty}\sum_{\upsilon=0}^{d-1}\sum_{m=0}^{\upsilon}\Delta T(\tau d+m)\sum_{\pi-\tau=1}^{\infty}\sum_{i=0}^{d-1}$$

$$\|\int_{D_i}\mu_1(dx)\,Q^{(\pi-\tau)d+\upsilon-m}(x,A)-\int_{D_i}\mu_2(dy)\,Q^{(\pi-\tau)d+\upsilon-m}(y,A)\;\|$$

$$\leq\sum_{\tau=0}^{\infty}\sum_{\upsilon=0}^{d-1}\sum_{m=0}^{\upsilon}\Delta T(\tau d+m)\sum_{\pi-\tau=1}^{\infty}\sum_{i=0}^{r}\mu_2(D_i)\times$$

$$\|\int_{D_i}\int_{D_i}\mu_{1i}(dx)\,\mu_{2i}(dy)\,Q^{(\pi-\tau)d+\upsilon-m}(x,A)$$

$$-\int_{D_i}\int_{D_i}\mu_{1i}(dx)\,\mu_{2i}(dy)\,Q^{(\pi-\tau)d+\upsilon-m}(x,A)\;\|$$

$$+\sum_{\tau=0}^{\infty}\sum_{v=0}^{d-1}\sum_{m=0}^{v}\Delta T(\tau d+m)\sum_{\pi-\tau=1}^{\infty}\sum_{i=r+1}^{d-1}\mu_1(D_i)$$

$$\| \int_{D_i}\int_{D_i}\mu_{1i}(dx)\,\mu_{2i}(dy)\,Q^{(\pi-\tau)d+v-m}(x,A)$$

$$-\int_{D_i}\int_{D_i}\mu_{1i}(dx)\,\mu_{2i}(dy)\,Q^{(\pi-\tau)d+v-m}(x,A)\,\| \tag{14.87}$$

$$=\| U_{\mu}^{(\pi,1,1)}(t)\| + \| U_{\mu}^{(\pi,1,2)}(t)\| .$$

From (14.87) and Theorem 441 we get that

$$\| U_{\mu}^{(\pi,1,1)}(t)\| \le \sum_{\tau=0}^{\infty}\sum_{v=0}^{d-1}\sum_{m=0}^{v}\Delta T(\tau d+m)\times \tag{14.88}$$

$$\sum_{\pi-\tau=1}^{\infty}\sum_{i=0}^{r}\mu_2(D_i)\,C(v-m)<\infty.$$

In an analogous way we prove that

$$\| U_{\mu}^{(\pi,1,2)}(t)\| <\infty \text{ and } \| U_{\mu}^{(\pi,2)}(t)\| <\infty. \tag{14.89}$$

From (14.80), (14.81), (14.88), and (14.89) we conclude the proof of the theorem. ■

With the use of Theorem 440.b the following theorem could similarly be proved:

Theorem 442 *(Vassiliou (2020b)). Let a Markov system that lives in $(X,\mathcal{B}(X))$ and let that the inherent Markov chain is positive Harris with period d. Assume in addition that for the accessible atom W we have $\mathbb{E}_{\omega}\left[\tau_{\omega}^2\right]<\infty$ and that*

$$\lim_{t\to\infty}T(t)=T,\ \Delta T(t)\ge 0 \ \ \textit{for every } t\in\mathbb{Z}^{+}.$$

Then for any regular distribution λ, we have

$$\sum_{t=1}^{\infty}\| \mathbb{E}[N(x,t,A)]-T(0)\sum_{v=0}^{d-1}\sum_{i=0}^{d-v-1}\lambda(D_i)\,\pi_{i+v}(A\cap D_{i+v})$$

$$-T(0)\sum_{v=0}^{d-1}\sum_{i=d-v}^{d-1}\lambda(D_i)\,\pi_{i+v-d}(A\cap D_{i+v-d})$$

$$-(T-T(0))\sum_{v=0}^{d-1}\sum_{m=0}^{v}\sum_{i=0}^{d-(v-m)-1}w_m\mu(D_i)\,\pi_{i+(v-m)}\left(A\cap D_{i+(v-m)}\right)$$

$$-(T-T(0))\sum_{v=0}^{d-1}\sum_{m=0}^{v}\sum_{\substack{i=d-(v-m)\\ i\le d-1}}^{d-1}w_m\mu(D_i)\,\pi_{i-d+(v-m)}\left(A\cap D_{i-d+(v-m)}\right)$$

$$-(T-T(0))\sum_{v=0}^{d-1}\sum_{m=v+1}^{d-1}\sum_{i=0}^{(m-v)-1}w_m\mu(D_i)\,\pi_{i+d+(v-m)}\left(A\cap D_{i+d+(v-m)}\right)$$

$$-(T-T(0))\sum_{v=0}^{d-1}\sum_{m=v+1}^{d-1}\sum_{\substack{i=(m-v)\\ i\le d-1}}^{d-1}w_m\mu(D_i)\,\pi_{i+(v-m)}\left(A\cap D_{i+(v-m)}\right)\|=0.$$

Chapter 15

THE $\mathcal{G}$–NON-HOMOGENEOUS MARKOV SYSTEM OF HIGH ORDER

15.1 Introductory notes

In section 5.5, we introduced the concept of $\mathcal{G}$-non-homogeneous Markov system. In the present chapter, we will expand this class of stochastic process by considering $\mathcal{G}$-NHMS's of high order, that is of order $m > 1$. We will start the rigorous definition of a $\mathcal{G}$-NHMS of high order using as a basis the definition of the first order given in section 5.5. The need for the study of Markov process of high order in manpower systems existed since the work in Vassiliou (1978), were in an actual application with data from two British firms, it was demonstrated, that a high order non-homogeneous Markov model provided impressively better fit than any other Markov model of first order. More recently an application of non-homogeneous semi Markov system of high order in the game of basket ball appeared (Stavropoulos et al. (2021)) extending the concept into the semi-Markov process. Note that the main difference between the $\mathcal{G}$-NHMS and the $\mathcal{G}$-NHMS of high order is that in the latter all the non-homogeneous Markov chains that defines it are assumed to be of order $m > 1$. Naturally, as we will find out, this creates some statistical problems such as for example the estimation of a large number of parameters. In what follows we will find out the proposed methods to overcome such problems.

15.2 The $\mathcal{G}$-non-homogeneous Markov system of high order

Consider a $\mathcal{G}$-NHMS as defined in Chapter 5. We will use this definition as a shell to build on the process $\mathcal{G}$-NHMS of high order m with $m > 1$ which was first defined in Vassiliou et al. (2010). Recall that we defined $X_t^{(i)}$ be the stochastic process which describes the state the i-member is in at time t. We assume that $X_t^{(i)}$ for a particular i follows a discrete time non-homogeneous Markov chain under the real probability measure $\mathbb{P}$ with respect to $\mathcal{G}_t$ of order m. Let $\mathbb{S}^* = \{1, 2, ..., k, k+1\}$ then for any function $h : \mathbb{S}^* \to \mathbb{R}$, we have:

$$\mathbb{E}_{\mathbb{P}}\left[h\left(X_{t+1}^{(i)}\right) \mid \mathcal{G}_t\right] = [\mathbb{E}_{\mathbb{P}}h\left(X_{t+1}^{(i)}\right) \mid \sigma\left(X_t^{(i)}\right) \vee \sigma\left(X_{t-1}^{(i)}\right) \vee \\ ... \vee \sigma\left(X_{t-m+1}^{(i)}\right)]. \tag{15.1}$$

Note that the process $X_t^{(i)}$ is a $\mathcal{G}$-inhomogeneous Markov chain of m order, thus it is also an $\mathcal{F}_t^{X_t^{(i)}}$-non homogeneous Markov chain under $\mathbb{P}$, although the converse is not true in general. The state space $\mathbb{S}^*$is finite, thus condition (15.1) is equivalent to the following one

$$\mathbb{P}\left(X_{t+1}^{(i)} = j_{m+1} \mid \mathcal{G}_t\right) = \\ \mathbb{P}\left(X_{t+1}^{(i)} = j_{m+1} \mid X_t^{(i)} = j_m, X_{t-1}^{(i)} = j_{m-1}, ..., X_{t-m+1}^{(i)} = j_1\right). \tag{15.2}$$

DOI: 10.1201/b23204-15

Assume that $X_t^{(i)}$ for $i = 1, 2, ..., T(t)$ are independent for every t. We have that, by defining for any member i of the system (population)

$$\mathbb{P}\left(X_{t+1}^{(i)} = j_{m+1} \mid X_t^{(i)} = j_m, X_{t-1}^{(i)} = j_{m-1}, ..., X_{t-m+1}^{(i)} = j_1\right) =$$

$$p(j_1, j_2, ..., j_m, j_{m+1}, t), \tag{15.3}$$

for $j_1, j_2, ..., j_m \in \mathbb{S}, j_{m+1} \in \mathbb{S}^*$.

Denote by $R_{t+1}^{(i)}$ the random variable representing the state in which the replacement of an individual member i leaving the system in the time interval $[t, t+1)$, enters the system in that interval. Assume that $R_{t+1}^{(i)}$ for $i = 1, 2, ..., T(t)$ are independent for every t and

$$\mathbb{P}\left(R_{t+1}^{(i)} = j_{m+1} \mid \mathcal{G}_t\right) =$$

$$\mathbb{P}\left(R_{t+1}^{(i)} = j_{m+1} \mid X_t^{(i)} = j_m, X_{t-1}^{(i)} = j_{m-1}, ..., X_{t-m+1}^{(i)} = j_1\right) =$$

$$p_0(j_1, j_2, ..., j_m, j_{m+1}, t). \tag{15.4}$$

Now, denote by Y_t the random variable representing the state of the i-th membership at time t. Then, we have that

$$\mathbb{P}\left(Y_{t+1}^{(i)} = j_{m+1} \mid \mathcal{G}_t\right) =$$

$$\mathbb{P}\left(X_{t+1}^{(i)} = j_{m+1} \mid \mathcal{G}_t\right) + \mathbb{P}\left(X_{t+1}^{(i)} = k+1 \mid \mathcal{G}_t\right) \mathbb{P}\left(R_{t+1}^{(i)} = j_{m+1} \mid \mathcal{G}_t\right). \tag{15.5}$$

Using (15.3) and (15.4) we get that

$$\mathbb{P}\left(Y_{t+1}^{(i)} = j_{m+1} \mid \mathcal{G}_t\right) =$$

$$\mathbb{P}\left(X_{t+1}^{(i)} = j_{m+1} \mid X_t^{(i)} = j_m, X_{t-1}^{(i)} = j_{m-1}, ..., X_{t-m+1}^{(i)} = j_1\right)$$

$$+\mathbb{P}\left(X_{t+1}^{(i)} = k+1 \mid X_t^{(i)} = j_m, X_{t-1}^{(i)} = j_{m-1}, ..., X_{t-m+1}^{(i)} = j_1\right)$$

$$\times\mathbb{P}\left(R_{t+1}^{(i)} = j_{m+1} \mid X_t^{(i)} = j_m, X_{t-1}^{(i)} = j_{m-1}, ..., X_{t-m+1}^{(i)} = j_1\right)$$

$$= \mathbb{P}\left(Y_{t+1}^{(i)} = j_{m+1} \mid X_t^{(i)} = j_m, X_{t-1}^{(i)} = j_{m-1}, ..., X_{t-m+1}^{(i)} = j_1\right)$$

$$= \mathbb{P}\left(Y_{t+1}^{(i)} = j_{m+1} \mid Y_t^{(i)} = j_m, Y_{t-1}^{(i)} = j_{m-1}, ..., Y_{t-m+1}^{(i)} = j_1\right)$$

$$:= q(j_1, j_2, ..., j_{m+1}, t), \tag{15.6}$$

and finally

$$q(j_1, j_2, ..., j_{m+1}, t) = p(j_1, j_2, ..., j_m, j_{m+1}, t) +$$

$$p(j_1, j_2, ..., j_m, k+1, t)\, p_0(j_1, j_2, ..., j_m, j_{m+1}, t). \tag{15.7}$$

It could easily proved that

$$\sum_{j_{m+1}} q(j_1, j_2, ..., j_{m+1}, t) = 1. \tag{15.8}$$

Now, define by

$$\mathbf{Q}(j_1, j_2, ..., j_m, j_{m+1}, t) = \{q(j_1, j_2, ..., j_m, j_{m+1}, t)\}_{j_1, j_2, ..., j_{m+1} \in \mathbb{S}},$$

which is a stochastic matrix. Hence, $\{Y_t\}_{t=0}^{\infty}$ is a $\mathcal{G}$-inhomogeneous Markov chain under $\mathbb{P}$ of m order, which we call the *inherent $\mathcal{G}$-inhomogeneous Markov chain under $\mathbb{P}$ of order m* and $\mathbf{Q}(j_1, j_2, ..., j_m, j_{m+1}, t)$ is its transition probability matrix.

It is apparent that the inherent $\mathcal{G}$-inhomogeneous Markov chain under $\mathbb{P}$ of m order moves among the states, which are all possible combinations of m states $(j_1, j_2, ..., j_m)$. Hence, we could structure the inherent $\mathcal{G}$-inhomogeneous Markov chain with all possible states $(j_1, j_2, ..., j_m)$ and treat it like this set of states $\mathbb{S}_{j_1, j_2, ..., j_m}$ is its state space. We will call now $\{Y_t\}_{t=0}^{\infty}$ structured with the state space $\mathbb{S}_{j_1, j_2, ..., j_m}$ the *structured inherent $\mathcal{G}$-inhomogeneous Markov chain under $\mathbb{P}$ of order m*. Note that the matrix $\mathbf{Q}(j_1, j_2, ..., j_m, j_{m+1}, t)$ will have many zeros since in order that a transition is possible between two states they must have $m-2$ common elements. It will be more clear at this point for the reader if he tries to construct an example with $k = 3$ and $m = 2$ and construct in this case $\mathbf{Q}(j_1, j_2, j_3, t)$. To check if the construction is correct Example 2.1 in Vassiliou et al (2010) could be consulted.

Remark 443 *From Raftery (1985) and Bertchtold et. al (2002) one could verify that the matrix $\mathbf{Q}(j_1, j_2, ..., j_m, j_{m+1}, t)$ of the $\mathcal{G}$-non-homogeneous Markov system of high order m under $\mathbb{P}$ will consist of blocks of matrices of sizes $k \times k^{m-1}$ and its total size will be $(k.k^{m-1}) \times (k.k^{m-1})$. Also, within each block of size $k \times k^{m-1}$ only one column is possible to be positive. This column is the i-th column of the $k \times k^{m-1}$ block if that block is in the i-th row of the blocks of size $k \times k^{m-1}$ which construct the transition matrix. A drawback are the large number of parameters needed to be estimated when the number of the states increase and the order increase. For example, if the initial states are 3 and the order is 3 which are thought to be moderate numbers in practice, then the number of parameters to be estimated for a time period $[0, 1]$ are 54. While if the number of states are 5 and the order is 3 for the same period the number of states increases to 500.*

15.3 The population structure of the $\mathcal{G}$-NHMS

Following Vassiliou et.al. (2010) we denote by $N(j_1, j_2, ..., j_m, t+1)$ the random variable representing at time $t+1$ the number of memberships who are in state j_m at time $t+1$, while they were in state j_{m-1} at time t, in state j_{m-2} at time $t-1$ and so on.

Also, denote by $\mathbf{N}(j_1, j_2, ..., j_m, t+1)$ the random vector, which is the row vector with elements the random variables $N(j_1, j_2, ..., j_m, t+1)$ for all $(j_1, j_2, ..., j_m) \in \mathbb{S}_{j_1, j_2, ..., j_m}$ written in an order, which relates it with the physical meaning of the phenomenon being modeled and with the construction of the transition matrix of the structured inherent $\mathcal{G}$-inhomogeneous Markov chain under $\mathbb{P}$ of order m.

We call the random vector $\mathbf{N}(j_1, j_2, ..., j_m, t+1)$ the *random population structure of the structured* $\mathcal{G}$-inhomogeneous Markov chain under $\mathbb{P}$ of order m.

Define by

$$\sigma(\mathbf{N}_t) = \sigma(\mathbf{N}(j_1, j_2, ..., j_m, t)) = . \tag{15.9}$$

$$\bigvee_{(j_1, j_2, ..., j_m) \in \mathbb{S}_{j_1, j_2, ..., j_m}} \sigma(N(j_1, j_2, ..., j_m, t)).$$

Moreover, assume that $\sigma(\mathbf{N}_t) \subset \mathcal{G}_t$. Now, we have for an arbitrary chosen t:

$$\mathbb{E}_{\mathbb{P}}[\mathbf{N}(j_1, j_2, ..., j_m, t+1)] = \mathbb{E}_{\mathbb{P}}[\mathbb{E}_{\mathbb{P}}[\mathbf{N}(j_1, j_2, ..., j_m, t+1)] \mid \mathcal{G}_t]. \tag{15.10}$$

Using the relation (15.10) and the results in section 2 of the present chapter it is not difficult using the techniques of Chapter 5 to show that

$$\mathbb{E}_{\mathbb{P}}[\mathbf{N}(j_1, j_2, ..., j_m, t+1)] = \mathbb{E}_{\mathbb{P}}[\mathbf{N}(j_0, j_1, ..., j_{m-1}, t)] \times$$

$$\mathbf{Q}(j_0, j_1, ..., j_{m-1}, j_m, t) + \Delta T(t) p_0(j_0, j_1, ..., j_{m-1}, j_m, t). \tag{15.11}$$

Relation (15.11) provides the expected population structure of the *structured $\mathcal{G}$ non-homogeneous Markov system.*

The maximum likelihood estimates of the transition probabilities in the matrix $\mathbf{Q}(j_0, j_1, ..., j_{m-1}, j_m, t)$ are simple extensions of the estimates of the transition probabilities $\mathbf{Q}(t)$ for an NHMS provided in Chapter 5.

15.4 The inhomogeneous mixture transition distribution model

In the Remark 443 of the present chapter, we identified the problem of the large number of parameters that, have to be estimated when the number of states and the time interval increases, a problem known even for homogeneous Markov chains of high order (Bertchtold and Raftery (2002)). The solution of the problem contains the introduction of the mixture transition distribution by Raftery (1985). The mixture transition distribution model approximates satisfactorily homogeneous Markov chains of high order. However, the most notably property of the mixture transition distribution model, as it was proved by Raftery (1985) and Aake and Deshmukh (1988), is that it has the same equilibrium distribution as the first order homogeneous Markov chain. This is an appealing and desirable for many reasons result. Among the reasons being the fact that the parameters defining the equilibrium distribution, are specified separately from those defining the lag distribution of the model.

The mixture transition distribution model has been very succesful in approximating high order Markov chains in many applications. Such applications could be found in Bertchtold and Raftery (2002) and Bertchtold (1988) in a large range of subjects. Wind modeling applications could be found in Raftery (1985), Craig (1989), Raftery and Tavare (1994) and MacDonald et al. (1977). Social behavior applications are exibited in the book by Bertchtold (1988). DNA sequences applications are provided in Raftery and Tavare (1994). Financial and economic time series applications are studied in Bollerslev et al. (1992) and Hamilton (1994). Biological applications are presented in Raftery and Tavare (1994), Bertchtold (1998, 2001).

We will define the non-homogeneous mixture transition model for a general $\mathcal{G}$-non-homogeneous high-order Markov chain through the inherent $\{Y_t\}_{t=0}^{\infty}$ $\mathcal{G}$-non-homogeneous high order Markov chain under $\mathbb{P}$ defined in section 2 of the present Chapter. Apparently, there is no loss of generality in this way.

We introduce what we call *the superficial razor cut mixture transition distribution model* for the inherent $\{Y_t\}_{t=0}^{\infty}$ $\mathcal{G}$-non-homogeneous high order Markov chain under $\mathbb{P}$ with values on $\mathbb{S}$ of m order as follows:

$$q(j_1, j_2, ..., j_m, j_{m+1}, t) = \sum_{i=1}^{m} \lambda_i(t)\, q(j_{m+1-i}, j_{m+1}, t) \text{ with}$$

$$\sum_{i=1}^{m} \lambda_i(t) = 1 \text{ and } 0 \leq \lambda_i(t) \leq 1, \tag{15.12}$$

that is, the conditional probabilities of the m order non-homogeneous Markov chain at time t are convex combinations of the first order Markov chain transition probabilities also at time t.

We call the $\mathcal{G}$-non-homogeneous Markov system under $\mathbb{P}$ of order m with its inherent $\mathcal{G}$-non-homogeneous high order Markov chain under $\mathbb{P}$ approximated by the model (15.12) the *$\mathcal{G}$-inhomogeneous Markov mixture transition distribution system under* $\mathbb{P}$.

Now, denote the n-step transition probabilities of the m-order $\mathcal{G}$-non-homogeneous Markov chain under $\mathbb{P}$ by

$$q(j_1, j_2, ..., j_m, j, t, t+n) = \mathbb{P}(Y_{t+n} = j \mid Y_{t-m+1} = j_1, ..., Y_t = j_m), \tag{15.13}$$

and the corresponding n-step transition probability matrices by

$$\mathbf{Q}(j_1, j_2, ..., j_m, j, t, t+n) = \{q(j_1, j_2, ..., j_m, j, t, t+n)\}_{j_1, j_2, ..., j_m, j \in \mathbb{S}}. \tag{15.14}$$

Also, the inherent first order $\mathcal{G}$-non-homogeneous Markov chain n-step transition probabilities

$$q(i, j, t, t+n) = \mathbb{P}(Y_{t+n} = j \mid Y_t = i),$$
$$\mathbf{Q}(i, j, t, t+n) = \mathbf{Q}(t, t+n) = \{q(i, j, t, t+n)\}_{i,j \in \mathbb{S}} = \{q(t, t+n)\}_{i,j \in \mathbb{S}}, \tag{15.15}$$

in the usual notation of first order Markov chains.

Now define by $\mathbf{J}(t)$ the probability distribution of $\{Y_t\}_{t=0}^{\infty}$, that is

$$\mathbf{J}(t) = \mathbb{P}(Y_t = j_m, Y_{t-1} = j_{m-1}, ..., Y_{t-m+1} = j_1), \tag{15.16}$$

which is a k^m-element row vector. Then it is not difficult to show that

$$\mathbf{J}(t+1) = \mathbf{J}(t)\,\mathbf{Q}(j_1, j_2, ..., j_m, j, t). \tag{15.17}$$

Now, exercising the above equation recursively we get

$$\mathbf{J}(t+1) = \mathbf{J}(0)\,\mathbf{Q}(j_{01}, j_{02}, ..., j_{0m}, j, 0, t), \tag{15.18}$$

where it is not difficult, in analogy with the first order non-homogeneous Markov chains, to prove that

$$\begin{aligned} \mathbf{Q}(j_{01}, j_{02}, ..., j_{0m}, j, 0, t) &= \mathbf{Q}(j_{01}, j_{02}, ..., j_{0m}, j_{1,m+1}, 0) \\ &\times \mathbf{Q}(j_{02}, ..., j_{0m}, j_{1,m+1}, j_{2,m+2}, 1) \\ &\times ... \times \mathbf{Q}(j_1, j_2, ..., j_m, j, t). \end{aligned} \tag{15.19}$$

Define the row vector

$$\boldsymbol{\pi}(t) = \{\mathbb{P}[Y_t = j_m]\},\ j_m = 1, 2, ..., k, \tag{15.20}$$

then the elements of $\boldsymbol{\pi}(t)$ provide the marginal probability mass function of $\{Y_t\}_{t=0}^{\infty}$ and are of the form

$$\pi(j_m, t) = \sum_{j_1}\sum_{j_2} ... \sum_{j_{m-1}} \mathbb{P}(Y_t = j_m, Y_{t-1} = j_{m-1}, ..., Y_{t-m+1} = j_1). \tag{15.21}$$

We now provide the following definition of the *collapsed matrix*.

Definition 444 *For an $\mathcal{G}$-non-homogeneous Markov system under $\mathbb{P}$ of order m and with the relative first order having state space $\mathbb{S} = \{1, 2, ..., k\}$ we define the collapsed matrix $\mathbf{C}(t)$ to be the $k^m \times k$ matrix, which is constructed from the matrix $\mathbf{Q}(j_1, j_2, ..., j_m, j_{m+1}, t)$ by extracting the elements which are by definition 0 and pressing into their position the nearest from the right elements.*

Example 445 *Let $k = 3$ and $m = 2$. Then the state space is*

$$\mathbb{S}_{j_1, j_2} = \{(1,1), (2,1), (3,1), (1,2), (2,2), (3,2), (1,3), (2,3), (3,3)\},$$

and it is not difficult for the reader to construct the matrix $\mathbf{Q}(i, j, l, t)$ for all possible values of (i, j) and for $l = 1, 2, 3$. Then the collapsed matrix in this case is

$$\mathbf{C}(t) = \begin{pmatrix} q_{111}(t) & q_{112}(t) & q_{113}(t) \\ q_{211}(t) & q_{212,t} & q_{213,t} \\ q_{311}(t) & q_{312,t} & q_{313,t} \\ q_{121}(t) & q_{122}(t) & q_{123}(t) \\ q_{221}(t) & q_{222}(t) & q_{223}(t) \\ q_{321}(t) & q_{322}(t) & q_{323}(t) \\ q_{131}(t) & q_{132}(t) & q_{133}(t) \\ q_{231}(t) & q_{232}(t) & q_{233}(t) \\ q_{331}(t) & q_{332}(t) & q_{333}(t) \end{pmatrix}.$$

It is not difficult to show that

$$\boldsymbol{\pi}(t+1) = \mathbf{J}(t)\,\mathbf{C}(t). \tag{15.22}$$

15.5 The asymptotic behavior of the inherent Markov chain

In the present section, we will prove a basic desired result concerning the success of the approximation by the superficial razor cut model of mixture transition distribution on the inherent $\mathcal{G}$-non-homogeneous Markov chain of high order. Naturally, this result applies for any $\mathcal{G}$-non-homogeneous Markov chain of high order establishing in this way a quite general result.

Theorem 446 *(Vassiliou et.al (2010)). Let a $\mathcal{G}$-non-homogeneous Markov chain of high order m under $\mathbb{P}$, modeled by the superficial razor cut model of mixture transition distribution with matrix transition probability sequence of first order $\{\mathbf{Q}(t)\}_{t=0}^{\infty}$. In addition let that*

$$\lim_{t\to\infty} \|\mathbf{Q}(t) - \mathbf{Q}\| = 0 \text{ and } \lim_{t\to\infty} \lambda_i(t) = \lambda_i(\infty), \tag{15.23}$$

with

$$\sum_{i=1}^{m} \lambda_i(\infty) = 1,\ 0 \le \lambda_i(\infty) \le 1. \tag{15.24}$$

Assume that the matrix $\mathbf{Q} = \{q_{ij}\}_{i,j\in\mathbb{S}}$ and denote by $\mathbf{Q}(\infty)$ the limit of $\mathbf{Q}^t$ as $t \to \infty$. Then

$$\lim_{t\to\infty} \|\boldsymbol{\pi}(t) - \boldsymbol{\pi}\| = 0, \tag{15.25}$$

where $\boldsymbol{\pi}$ is the asymptotic distribution of the Markov chain uniquely defined by $\mathbf{Q} = \{q_{ij}\}_{i,j\in\mathbb{S}}$ and in fact is the row of the stable matrix $\mathbf{Q}(\infty)$. If in addition the convergence as $t \to \infty$ of the sequences $\mathbf{Q}(t)$ and $\lambda_i(t)$ are geometrically fast then the convergence in (15.25) is geometrically fast.

Proof. It is not difficult to check that the collapsed matrix $\mathbf{C}(t)$ for the superficial razor cut model is given in the following form

$$\mathbf{C}(t) = \sum_{i=1}^{m} \lambda_i(t)\,\mathbf{U}_i \mathbf{Q}(t), \tag{15.26}$$

where

$$\mathbf{U}_1 = \mathbf{I} \otimes \mathbf{1}^{\top} \otimes ... \otimes \mathbf{1}^{\top},$$
$$\mathbf{U}_2 = \mathbf{1}^{\top} \otimes \mathbf{I} \otimes \mathbf{1}^{\top} \otimes ... \otimes \mathbf{1}^{\top},$$
$$\dots\dots\dots\dots\dots\dots\dots\dots$$
$$\mathbf{U}_m = \mathbf{1}^{\top} \otimes ... \otimes \mathbf{1}^{\top} \otimes \mathbf{I},$$

where the number of $\mathbf{1}^{\top}$ for each $\mathbf{U}_i, i = 1, 2, ...m$ is exactly $m-1$. We will prove now that $\mathbf{C}(t)$ is stochastic.

$$\mathbf{C}(t)\,\mathbf{1}^{\top} = \sum_{i=1}^{m} \lambda_i(t)\,\mathbf{U}_i \mathbf{Q}(t)\,\mathbf{1}^{\top} = \sum_{i=1}^{m} \lambda_i(t)\,\mathbf{U}_i \mathbf{1}^{\top} = \sum_{i=1}^{m} \lambda_i(t)\,\mathbf{1}^{\top} = \mathbf{1}^{\top}. \tag{15.27}$$

Also, if we denote by

$$\mathbf{C} = \sum_{i=1}^{m} \lambda_i(\infty)\,\mathbf{U}_i \mathbf{Q}, \tag{15.28}$$

then

$$\|\mathbf{C}(t)-\mathbf{C}\| \leq \sum_{i=1}^{m} \|\lambda_i(t)\,\mathbf{U}_i\mathbf{Q}(t)-\lambda_i(\infty)\,\mathbf{U}_i\mathbf{Q}\|$$

$$\leq \sum_{i=1}^{m} \lambda_i(t)\,\|\mathbf{Q}(t)-\mathbf{Q}\| + \sum_{i=1}^{m} |\lambda_i(t)-\lambda_i(\infty)|\,. \tag{15.29}$$

From conditions (15.23) and (15.29), we easily get that $\lim_{t\to\infty}\|\mathbf{C}(t)-\mathbf{C}\| = 0$. The superficial razor cut model of mixture transition distribution, use the transition probability sequence $\{\mathbf{Q}(t)\}_{t=0}^{\infty}$ to approximate the transition probabilities matrices $\{\mathbf{Q}(j_1,j_2,...,j_m,j_{m+1},t)\}_{t=0}^{\infty}$. Since $\mathbf{Q}(t)$ converges to $\mathbf{Q}$ then we have that there exists a matrix $\mathbf{Q}(j_1,j_2,...,j_m,j_{m+1})$ related with $\mathbf{Q}$ by the superficial razor cut model and for which

$$\lim_{t\to\infty}\|\mathbf{Q}(j_1,j_2,...,j_m,j_{m+1},t)-\mathbf{Q}(j_1,j_2,...,j_m,j_{m+1})\| = 0. \tag{15.30}$$

In addition, since $\mathbf{Q}$ is regular we could easily show that $\mathbf{Q}(j_1,j_2,...,j_m,j_{m+1})$ is regular. Therefore there exists a stochastic vector $\mathbf{J}$ such that

$$\lim_{t\to\infty}\|\mathbf{J}(t)-\mathbf{J}\| = 0, \tag{15.31}$$

and for which we have

$$\mathbf{J} = \mathbf{J}\mathbf{Q}(j_1,j_2,...,j_m,j_{m+1}) \quad \text{and} \quad \mathbf{J}\mathbf{1}^\top = 1.$$

From (15.22) we get that

$$\mathbf{J}\mathbf{C} = \boldsymbol{\pi}. \tag{15.32}$$

Now, from the conditions of the theorem we get that

$$\|\boldsymbol{\pi}(t+1)-\boldsymbol{\pi}\| = \|\mathbf{J}(t)\,\mathbf{C}(t)-\mathbf{J}\mathbf{C}\| \leq \|\mathbf{J}(t)-\mathbf{J}\|\,\|\mathbf{C}(t)\|$$

$$+\|\mathbf{J}(t)\|\,\|\mathbf{C}(t)-\mathbf{C}\| \leq \|\mathbf{J}(t)-\mathbf{J}\| + \|\mathbf{C}(t)-\mathbf{C}\| < \epsilon. \tag{15.33}$$

Now, since the convergence of $\lambda_i(t)$ to λ_i and $\mathbf{Q}(t)$ to $\mathbf{Q}$ is geometrically fast it is not difficult to show that there exist $c>0$ and $0<b<1$ such that $\|\boldsymbol{\pi}(t+1)-\boldsymbol{\pi}\| < cb^{t+1}$. ■

15.6 The asymptotic expected population structure

In the present section, we will study the classical problem of finding the asymptotic expected population structure for the $\mathcal{G}$-non-homogeneous Markov system of high order m under $\mathbb{P}$. We will establish for this purpose a basic theorem the proof of which perchaps will be different from what the reader reaching the end of the book might expect.

Theorem 447 *(Vassiliou et al. (2010)). Let a $\mathcal{G}$-non-homogeneous Markov system of high order m under $\mathbb{P}$ and let the sequence of transition probability matrices be $\mathbf{Q}(j_1,j_2,...,j_m,j_{m+1},t)$. Let $\{\mathbf{Q}(t)\}_{t=0}^{\infty}$ be the first order $\mathcal{G}$-non-homogeneous Markov chain used to model the related superficial razor cut model. Assume that*

$$\lim_{t\to\infty}\|\mathbf{Q}(t)-\mathbf{Q}\| = 0,$$

$\mathbf{Q}$ *regular,* $\lim_{t\to\infty} T(t) = T$ *with* $\Delta T(t) \geq 0$ *for every* $t\in\mathbb{N}$. *Then we have that*

$$\lim_{t\to\infty}\mathbb{E}_{\mathbb{P}}[\mathbf{N}(t)] = \mathbb{E}_{\mathbb{P}}[\mathbf{N}(\infty)] = T\mathbf{q}(\infty)\,. \tag{15.34}$$

If in addition

$$\lim_{t\to\infty} T(t) = T \text{ and } \lim_{t\to\infty}\|\mathbf{Q}(t)-\mathbf{Q}\| = 0 \text{ geometrically fast,}$$

then

$$\mathbb{E}_{\mathbb{P}}\left[\mathbf{N}(\infty)\right] = T\mathbf{q}(\infty) \quad \textit{geometrically fast.}$$

Proof. *From Theorem 418, we get that* $q(j_1, j_2, ..., j_m, x, t, \infty) = q(x, \infty)$. *Therefore, the probability of a membership which at time 0 is in state* $(j_1, j_2, ..., j_m)$ *to be asymptotically in* x *is given by* $q(x, \infty)$. *Hence, of the initial memberships* $N(j_1, j_2, ..., j_m, 0)$ *it is expected that* $N(j_1, j_2, ..., j_m, 0)\, q(x, \infty)$ *will be in state* x *asymptotically. Denote by* $\mathbb{E}_{\mathbb{P}}\left[\hat{N}_x(\infty)\right]$ *the expected number of initial memberships which will be in state* x *asymptotically then:*

$$\mathbb{E}_{\mathbb{P}}\left[\hat{N}_x(\infty)\right] = \sum_{j_1, j_2, \cdots, j_m} N(j_1, j_2, ..., j_m, 0)\, q(x, \infty) = T(0)\, q(x, \infty). \tag{15.35}$$

Denote the vector

$$\mathbb{E}_{\mathbb{P}}\left[\hat{\mathbf{N}}(\infty)\right] = \left[\mathbb{E}_{\mathbb{P}}\left[\hat{N}_1(\infty)\right], \mathbb{E}_{\mathbb{P}}\left[\hat{N}_2(\infty)\right], ..., \mathbb{E}_{\mathbb{P}}\left[\hat{N}_k(\infty)\right]\right],$$

then it is not difficult to see that

$$\mathbb{E}_{\mathbb{P}}\left[\hat{\mathbf{N}}(\infty)\right] = T(0)\, \mathbf{q}(\infty), \tag{15.36}$$

where $\mathbf{q}(\infty)$ *is the row of the stable matrix* $\mathbf{Q}(\infty)$. *It is known that* $\Delta T(t)$ *is the number of memberships entering the system in the end of the interval* $(t-1, t]$ *and they will be distributed to the various states with probabilities* $p_0(j_1, j_2, ..., j_m, y, t)$ *with* $y \in \mathbb{S}$. *Therefore, the expected number of new memberships which asymptotically will be in state* x *at time* t *are*

$$\begin{aligned} &\Delta T(t)\, p_0(j_1, j_2, ..., j_m, y, t)\, q(j_1, j_2, ..., j_m, y, x, t, \infty) \\ &= \Delta T(t)\, p_0(j_1, j_2, ..., j_m, y, t)\, q(x, \infty). \end{aligned} \tag{15.37}$$

The expected number of new memberships which will be comming from all possible experiences into state $(j_1, j_2, ..., j_m, y)$ *at time* t *and will be asymptotically in state* x *will be*

$$\begin{aligned} \mathbb{E}_{\mathbb{P}}\left[\hat{N}_{0x}(t, \infty)\right] &= \sum_{j_1, j_2, \cdots, j_m \in \mathbb{S}} \Delta T(t)\, p_0(j_1, j_2, ..., j_m, y, t)\, q(x, \infty) \\ &= \Delta T(t) \left[\sum_{j_1, j_2, \cdots, j_m \in \mathbb{S}} p_0(j_1, j_2, ..., j_m, y, t)\right] q(x, \infty) \\ &= \Delta T(t)\, \hat{p}_0(y, t)\, q(x, \infty). \end{aligned} \tag{15.38}$$

Denote the vectors

$$\mathbb{E}_{\mathbb{P}}\left[\hat{\mathbf{N}}_0(t, \infty)\right] = \left[\mathbb{E}_{\mathbb{P}}\left[\hat{N}_{01}(t, \infty)\right], \mathbb{E}_{\mathbb{P}}\left[\hat{N}_{02}(t, \infty)\right], ..., \mathbb{E}_{\mathbb{P}}\left[\hat{N}_{0k}(t, \infty)\right]\right], \tag{15.39}$$

$$\tilde{\mathbf{p}}_0(t) = \left[\tilde{p}_0(1, t), \tilde{p}_0(2, t), ..., \tilde{p}_0(k, t)\right], \tag{15.40}$$

then

$$\mathbb{E}_{\mathbb{P}}\left[\hat{\mathbf{N}}_0(t, \infty)\right] = \Delta T(t)\, \tilde{\mathbf{p}}_0(t)\, \mathbf{Q}(\infty) = \Delta T(t)\, \mathbf{q}(\infty). \tag{15.41}$$

Now the expected population structure of the new memberships which will enter in the time interval $[0, \infty)$ *will be*

$$\sum_{t=1}^{\infty} \mathbb{E}_{\mathbb{P}}\left[\hat{\mathbf{N}}_0(t, \infty)\right] = \sum_{t=1}^{\infty} \Delta T(t)\, \mathbf{q}(\infty) = \left[T - T(0)\right] \mathbf{q}(\infty). \tag{15.42}$$

Finally from equations (15.36) *and* (15.42), *we get that*

$$\mathbb{E}_{\mathbb{P}}\left[\mathbf{N}(\infty)\right] = \mathbb{E}_{\mathbb{P}}\left[\hat{\mathbf{N}}(\infty)\right] + \sum_{t=1}^{\infty} \mathbb{E}_{\mathbb{P}}\left[\hat{\mathbf{N}}_0(t,\infty)\right] = T\mathbf{q}(\infty).$$

The rate of convergegence under the conditions given in the Theorem it is not difficult to prove that it will be geometric. ■

15.7 An illustrative example from manpower systems

In the present section, we will provide an illustrative example of applying the results of the present Chapter to an University system. The University system is an appropriate example due to the small number of grades which allow the necessary matrices to be presented in a book. Therefore, let state 1 representing the grade of Professors, state 2 the grade of Associate Professors, and state3 the grade of Assistant Professors. Typically $\mathbf{P}(t)$ for every t the transition probability matrix of order 1 of a member is upper diagonal and diagonally dominant. The elements of the recruitment probability vector $\mathbf{p}_0(t)$ are of increasing size as we go down the hierarchy, a universal phenomenon. The transition probability matrix of a membership as we have seen is given by $\mathbf{Q}(t) = \mathbf{P}(t) + \mathbf{p}_{k+1}^{\top}\mathbf{p}_0(t)$. From the available records of staff it is easy to estimate these probabilities, with the maximum likelihood estimates given in Chapter 5. Assume that it has been found that:

$$\mathbf{Q}(1) = \begin{pmatrix} 0.8 & 0.1 & 0.1 \\ 0.3 & 0.6 & 0.1 \\ 0.1 & 0.4 & 0.5 \end{pmatrix}, \mathbf{Q}(2) = \begin{pmatrix} 0.9 & 0.1 & 0 \\ 0.2 & 0.7 & 0.1 \\ 0 & 0.4 & 0.6 \end{pmatrix},$$

$$\mathbf{Q}(3) = \begin{pmatrix} 0.75 & 0.15 & 0.1 \\ 0.25 & 0.6 & 0.15 \\ 0.15 & 0.35 & 0.5 \end{pmatrix}, \mathbf{Q}(t) = \begin{pmatrix} 0.69 & 0.16 & 0.15 \\ 0.23 & 0.62 & 0.15 \\ 0.12 & 0.38 & 0.5 \end{pmatrix},$$

for $t = 4, 5, ...$ Now, we start applying a second order Markov chain estimating from the data the collapsed matrix of transition probability matrices and assume that we found

$$\mathbf{C}(1) = \begin{pmatrix} 0.8 & 0.1 & 0.1 \\ 0.6 & 0.3 & 0.1 \\ 0.5 & 0.2 & 0.3 \\ 0.4 & 0.4 & 0.2 \\ 0.3 & 0.4 & 0.3 \\ 0.2 & 0.3 & 0.5 \\ 0.35 & 0.2 & 0.45 \\ 0.2 & 0.3 & 0.5 \\ 0.15 & 0.25 & 0.6 \end{pmatrix}, \quad \mathbf{C}(2) = \begin{pmatrix} 0.7 & 0.2 & 0.1 \\ 0.55 & 0.35 & 0.1 \\ 0.4 & 0.25 & 0.35 \\ 0.44 & 0.36 & 0.2 \\ 0.25 & 0.45 & 0.3 \\ 0.18 & 0.32 & 0.5 \\ 0.4 & 0.25 & 0.35 \\ 0.2 & 0.3 & 0.5 \\ 0.12 & 0.28 & 0.6 \end{pmatrix},$$

$$\mathbf{C}(3) = \begin{pmatrix} 0.6 & 0.2 & 0.2 \\ 0.5 & 0.3 & 0.2 \\ 0.4 & 0.25 & 0.35 \\ 0.34 & 0.46 & 0.2 \\ 0.24 & 0.46 & 0.3 \\ 0.25 & 0.20 & 0.55 \\ 0.45 & 0.05 & 0.50 \\ 0.15 & 0.35 & 0.5 \\ 0.2 & 0.25 & 0.55 \end{pmatrix} \quad \text{and} \quad \mathbf{C}(t) = \begin{pmatrix} 0.54 & 0.26 & 0.2 \\ 0.47 & 0.33 & 0.2 \\ 0.4 & 0.27 & 0.33 \\ 0.3 & 0.4 & 0.3 \\ 0.22 & 0.48 & 0.3 \\ 0.25 & 0.25 & 0.5 \\ 0.40 & 0.14 & 0.46 \\ 0.17 & 0.39 & 0.44 \\ 0.17 & 0.29 & 0.54 \end{pmatrix},$$

for $t = 4, 5, ...$ Model the superficial cut mixture transition model to the collapsed matrices and estimate the $\lambda_i(t)$'s by any regression method. Use the initial value of $\mathbf{J}(0)$ and use recurssively equation (15.17) until it converges to $\mathbf{J}$. This value should coincide with the value of $\mathbf{J}$ given in Theorem 446. Using equation (15.17) find $\boldsymbol{\pi}(t)$ recursively until it converges to $\boldsymbol{\pi}$ and it should coincide with $\mathbf{JC}$ given in Theorem 446. Find $\mathbf{N}(\infty)$ using Theorem 447.

15.8 Research notes

The variability in $\mathcal{G}$-inhomogeneous Markov systems of high order, that is finding the variances and covariances of the state sizes were studied by Vassiliou et al. (2010). Also in the same work the imporant problem of the asymptotic variability of the state sizes was solved and the solution was provided in elegant closed analytic form. Also, further related results on $\mathcal{G}$-inhomogeneous Markov systems of high order could be found in Moysiadis (2010).

Research progress in extending the present ideas to semi-Markov processes was done by the research team headed by Prof A. Papadopoulou in the present days that the book is being published. A first publication in this respect for semi-Markov processes of order two applied in a novel way to the game of basketball appeared in Stavropoulos et al. (2021).

REFERENCES

Adke, S.R. and Deshmurkh, S.R. (1988). Limit distribution of high order Markov chain. *J. R. Stat. Soc. B* **50**, 105–108.

Afonso, L.B., Cardoso, R.M., Egidio dos Reis, A.D., Guerreiro, G.R. (2020). Ruin probabilities and capital recruitment for open automobile portfolios with a bonus-malus system based on claim counts. *J. Risk Insur.* **87**, 501–522.

Alefeld, G., Herzberger, J. (1983). *Introduction to Interval Computations.* Academic Press, New York.

Amblard, C. (1988). Seasonal succession and thestrategies of phytoplakton on the development in two lakes of different tropic states. *J. Plankton Res.,* **10**, 1189–1208.

Anderson, T.W. and Goodman, L.A. (1957). Stastical inference about Markov chains. *Ann. Math. Statist.* **28**, 89–110.

Baez, J.C., Fong, B., Pollard, B.S. (2016). A compositional framework for Markov processes. *J. Math. Phys.* **57**, 033301.

Barbour, A.D. and Pugliese, A. (2005). Asymptotic behavior of a metapopulation. *Ann. Appl. Prob.* **15**, 1306–1338.

Barbu, V.S, D'Amico, G., Manca, R., Petroni, F. (2016). Step semi-Markov models and application to manpower management. *ESAIM:PS* **20**, 555–571.

Bachelier, L. (1900). *Théorie de la Spé*culation. *Annales Scientifiques de l' Ecole Normal Superier,* 21–86.

Bartholomew, D.J. (1963). A multistage renewal process. *J. R. Statist. Soc.* **B**, 25, 150–168.

Bartholomew, D.J. (1967). *Stochastic Models for Social Processes.* 1st edition. New York: Wiley.

Bartholomew, D.J. (1973). *Stochastic Models for Social Processes.* 2nd edition. New York: Wiley.

Bartholomew, D.J. (1975). A stochastic control problem in the social sciences. *Bull. Int. Statist. Inst.* **46**, 670–680.

Bartholomew, D.J. (1977). Maintaining a grade or age structure in a stochastic enviroment. *Adv. Appl. Prob.* **9**, 1–17.

Bartholomew, D.J. (1979). The control of a grade structure in a stochastic enviroment using promotion control. *Adv. Appl. Prob.* **11**, 603–615.

Bartholomew, D.J. and Forbes, A.C. (1979). *Statistical Techniques for Manpower Planning.* Wiley, Chichester.

Bartholomew, D.J. (1982). *Stochastic Models for Social Processes.* 3rd edition. New York: Wiley.

Bartholomew, D.J., Forbes, A. and McClean, S. I. (1991). *Statistical Techniques in Manpower Planning.* Wiley, Chichester, UK.

Barnett, S. (1979). *Matrix methods for Engineers and Scientists.* Mc Graw Hill, London.

Barnett, S. (1994). *Matrices: Methods and Applications.* Oxford, Applied Mathematics and Computing Science Series.

Basharin, G.P., Langville, A.N. and Naumov, V.A. (2004). The life and work of A.A. Markov, *Linear Algebra and its Applications.* **386**, 3–26.

Beers, J.C., Reed, M.M., and Steward, G.L. (1982). Seasonal abundance of the mikroplankton population in the North Pacific central gyre. *Deep Sea Res.* **29**, 227–245.

Ben-Haim, Y., Elishakoff, I. (1990). *Convex Models of Uncertainty in Applied Mathematics.* Elsevier. New York.

Berge, C. (1963). *Topological Spaces.* Oliver Boyd. Edinburgh.

Berman, A. and Plemmons, R.J. (1979). *Non-negative Matrices in the Mathematical Sciences.* Academic Press. London.

Bernstein, S.N. (1946). *Teoriya Veroiatnostei.* 4th edition. Gostehizdat, Moscow-Leningrad.

Bertchtold, A. (1988). Chaines de Markov et modeles de transition: applications aux sciences sociales. Hermes, Paris.

Bertchtold, A. (2001). Estimation in the mixture transition distribution model. *J. Time Ser. Anal.* **22**, 379–397.

Bertchtold, A. and Raftery, A.E. (2002). The mixture transition distribution model to high-order Markov chains and non-Gaussian time series. *Stat. Sci.* 17, 3, 328–356.

Bhat, U.N. (1984). *Elements of Applied Stochastic Processes.* John Wiley. New York.

Bielecki, T.R., Rutkowski, M. (2004). *Credit Risk: Modeling, Valuation and Hedging.* Berlin: Springer.

Billingsley, P. (1965). *Ergodic Theory and Information,* John Wiley, New York.

Billingsley, P. (1968). *Convergence of Probability Measures,* John Wiley, New York.

Bingham, N.H. (2005). Doob: A half century on. *Journal of Applied Probability,* 42, 257–266.

Bolloerslev, T, Chou, R.Y. and Kroner, K.F. (1992). ARCH modelingin finance: a review of the theory and empirical evidence. *J. Econom* **52**, 5–59.

Bowerman, B. (1974). Nonstationary Markov decision processes and related topics in nonstationary Markov chains. Ph.D. thesis, Iowa State University, Ames, Iowa.

Bowerman, B., David, H.T. and Isaacson, D. (1977). The convergence of Cesaro averages for certain nonstationary Markov chains. *Stochastic Processes and their Applications* **5**, 221–230.

Bremaud, P. (1981). *Point Process and Queues. Martingale Dynamics.* Springer.

Bremaud, P. (2020). *Markov Chains. Gibbs Fields, Monte Carlo Simulation and Queues.* 2nd edition. Springer.

Bru, B. and Yor, M. (2002). Comments on the life and mathematical legacy of Wolfgang Doeblin. *Finance and Stochast.* **6**, 3–47.

Brudel, J., Preisendorfer, P., Ziegler, R. (1993). Upward mobility in organizations-the effects of hierarchy and opportunity structure. *European Sociological Review,* **9**, 2, 173-188.

Buckley, F.M. and Pollet, P.K. (2010). Limit theorems for discrete-time metapopulation models. *Probab Surv.* **7**, 53-83.

Campbell, S.L. and Meyer, C.D. (1979). *Generalized Inverses of Linear Transformations.* SIAM, Philadelfia.

Campbell, S.L. and Meyer, C.D. (2009). *Generalized Inverses of Linear Transformations.* Surveys and Reference Works in Mathematics, Pitman Pub. Co. London.

Carreras-Carbonell, J., Macpherson, E. and Pascual, M. (2007). High self-recruitment levels in a Mediterrranean littoral fish population revealed by microsatellite markers. *Mar Biol.* **151**, 719-727.

Charns, A., Cooper, W.W., Niehaus, R.J., Shultz, D. (1971). A model for civilian manpower management and planning in US Navy. In Smith, A.R. (Ed.) *Models of Manpower Systems.* Elsevier, Amsterdam.

Charns, A., Cooper, W.W., Niehaus, R.J. (1975). Dynamic multiattribute models for fixed manpower systems. *Naval Research Logistics Quarterly* **22**, 205-220.

Chung, K.L. (1960). *Markov chains with stationary transition probabilities.* Springer, Berlin.

Chatterjee, S. and Seneta, E. (1977). Towards consensus: some convergence theorems repeated averaging. *Journal of Applied Probability,* **14,** 89-97.

Chen, G., Sun, J., Xia, J. (2018). Robust sampled data for Itô stochastic Markovian jump systems with state delay. *International Journal of Robust Nonlinear Control,* **28**, 4345-4366.

Cho, G.E. and Meyer, C.D. (2000). Markov chain sensitivity measured by mean first passage times. *Linear Algebra and its Applications.* 316, 21-28.

Cho, G.E. and Meyer, C.D. (2001). Comparison of perturbation bounds for the stationary distribution of a Markov chain. *Linear Algebra and its Applications,* 335, 137-150.

Chow, YS. and Teicher, H. (1997). *Probability Theory, Independence, Interchangeability, Martingales.* Springer-Verlag, Berlin.

Chung, K.L. (1964). The general theory of Markov processes according to Doeblin. *Zet. fur Wahr.* **2,** 230-254.

Chung, K.L. (1967). *Markov Chains with stationary transition probabilities.* 2nd edition. Springer-Velag, Berlin.

Chung, K.L. (1992). Doeblin's Big Limit Theorem. *Theo. Prob.* **6**, 417-426.

Clark, C.W. (2010). Mathematical Bioeconomics: *The Mathematics of Conservation.* Vol 91. John Wiley and Sons, New Jersey.

Clough, D.J., Dudding, R.C., Price, W.L. (1971). Mathematical programming models of quasi-independent subsystem. In Smith, A.R. (Ed.) *Models of Manpower Systems.* Elsevier, Amsterdam.

Cramer, H. (1976). Half a century with probability theory: Some personal recollections. *The Annals of Probability,* 4, 4, 509-546.

Craig, P. (1989). Time series analysis of directional data. Ph.D. thesis, Dept. Statistics, Trinity College, Dublin.

Cohen, J.E. (1976). Ergodicity of age structure in populations with Markovian vital rates I: countable states. *J. Amer. Statist. Assoc.* **71**, 335-339.

Cohen, J.E. (1977). Ergodicity of age structure in populations with Markovian vital rates II: general states. *Adv. Appl. Prob.* **9**, 18-37.

Cohen, J.E., Hajnal, J., Newman, C.M. (1986). Approaching consensus can be delicate when positions harden. *Stochastic processes and their Applications* **22,** 315-322.

Cohn, H. (ed). (1993). Doeblin and modern probability, Blaubeuren 1991 (Contemporary Mathematics 149) Providence : AMS.

Colebrook, J.M. (1979). Continuous plankton records. Seasonal cycles of phytoplankton and corepods in the North Atlantic Ocean and North Sea. *Mar. Biol.* **51**, 23–32.

Conlisk, J. (1978). A stability theorem for an interactive Markov chain. *J. Math. Sociol.* **6**, 163–168.

Cox, D.R. (1955). A use of complex probabilities in the theory of stochastic processes. *Proceedings of the Cambridge Philosophical Society* **51**: 313–319.

Courtault, J. (2000). Louis Bachelier. On the centenary of *Théorie de la Spé*culation. *Mathematical Finance,* **10**, 3, 341–353.

Costa, O.L.V., Fragoso, M.D., Marques, R.P. (2005). *Discrete-time Markovian Jump Linear Systems,* London: Springer-Verlag.

Cox, D.R. and Miller, H.D. (1968). *The Theory of Stochastic Processes.* Methuen and Co. Ltd.

Cox, D.R. and Oakes, D. (1984). *Analysis of Survival Data.* Chapman and Hall. London.

D' Amico, G., Janssen, J., Manca, R. (2007). Valuing credit default swap in a non-homogeneous semi-Markov rating based model. *Computational Economics,* **29**, (2), 119–138.

D' Amico, G., Janssen, J., Manca, R. (2010). Initial and final backward and forward discrete time non-homogeneous semi-Markov credit risk models. *Methodology and Computing in Applied Probability,* **12**, 2, 215–225.

Darroch, J.N. and Seneta (1988). On quasi-stationary distributions in absorbing continuous-time finite Markov chains. *J. Appl. Prob.* **25**, 34–42.

Davies, G.S. (1973). Structural control in a graded manpower system. *Man. Sci.* **20**, 76–84.

Davies, G.S. (1975). Maintainability of structures in Markov chains under recruitment control. *J. Appl. Prob.* **12**, 376–382.

Davies, G.S. (1981). Maintainable regions in a Markov manpower model. *J. Appl. Prob.* **18**, 738–742.

Dans, S.L., Degrati, M., and Pedraza, S.N. (2012). Effects of tour boats on dolphin activity examined with sensitivity analysis of Markov chains. *Conservation Biology* **26**, 4, 708–716.

Daubechies, I. and Lagarias, J.C. (1992). Sets of matrices all infinite products of which converge. *Linear Alg. Appl.* **161**:227–263.

De Groot, M.H. (1974). Reaching a consensus. *Journal of the American Statistical Association,* **69**, 345, 118–121.

De Farias, D.P., Genomel, J.C., Do Val, J.B.R., Costa, O.L.V. (2000). Output feedback control of Markov jump linear systems in continuous time. *IEEE Transactions of Automatic Control,* **45**, 5, 944–949.

De Feyter, T. (2006). Modelling heterogeneity in manpower planning: dividing the personnel system into more homogeneous subgroups. *Applied Stchastic Models in Business and Industry.* **22**, 4, 321–334.

De Lourdes Centeno, M., Manuel Andrade e Silva, J. (2001). Bonus systems in an open portfolio. *Insur. Math. Econ.* 2001, 28, 341–350.

Dimitriou, V.A. and Tsantas, N. (2010). Evolution of a time dependent Markov model for training and recruitment decisions in manpower planning. *Linear Algebra and its Applications,* **433**, 1950–1972.

Dimitriou, V.A., Georgiou, A.C., and Tsantas, N. (2013). The multivariate non-homogeneous Markov manpower system in a depatmental mobility framework. *European Journal of Operational Research,* **228**, 1, 112–121.

Dimitriou, V.A., Georgiou, A.C., and Tsantas, N. (2015). On the equilibrium personnel structure in the presence of vertical and horizontal mobility via multivariate Markov chains. *Journal of the Operational Research Society,* **66**, 6, 993–1006.

Dobrushin, R.L. (1956). Central limit theorem for non-stationary Markov chains, I,II. *Theory Prob. Appl.* **1,** 65–80, 329–383. (English transilation).

Doeblin, W. (1937). Le cas discontinu des probabilités en chaîne. *Publ. Fac. Sci. Univ. Masaryk* **236**, 3–13.

Doeblin, W. (1938). Theorie des chaines simples constantes de Markoff a un nombre fini d'etats. *Revue mathematique de l Union Interbolkanique* **2**, 77–105.

Doeblin, W. (1940). Elements d'une theorie general des chaines simple constantes de Markoff. *Ann. Sci. Ec. Norm. Sup.,* 57, 61–111.

Dong, H., Wang, Z., Ho, D.W.C., Gao, H. (2011). Robust H_∞ filtering for Markovian jump systems with randomly occuring nonlinearities and sensor saturation: The finite horizon case. *IEEE Transactions on Signal Processing,* **59,** 7, 3048–3057.

Doob, J.L. (1990). *Stochastic Processes.* Wiley Classics Library, John Wiley, New York.

Doob, J.L. (1981). *Classical Potential Theory and its Probabilistic Counterpart.* Springer, New York.

Drazin, M.P. (1958). Pseudoinverse in associative rings and semigroups. *Amer. Math. Monthly,* **65,** 506–514.

Dreyfus, S.E. and Law, A.M. (1977). *The Art and Theory of Dynamic Programming.* Academic Press, New York.

Dunkin, E. (1965). *Markov processes* (two volumes). Springer-Verlag, Berlin.

Durret, R. (1996). *Probability: Theory and Examples,* second edition, Duxbury Press, New York.

Einstein, A. (1905). On the movement of small particles suspended in stationary liquid demanded by the molecular-kinetic theory of heat. *Ann. d. Physic.* **17**.

Eggleston, H.G. (2008). *Convexity.* Cambridge University Press.

Elandt-Johnson, R.C. and Johnson, N.L. (1980). *Survival Models and Data Analysis.* John Wiley. New York.

Ellner, S. (1984). Asymptotic behavior of some stochastic difference equation population models. *J. Math. Biology* **19**, 169–200.

Erdèrley, I. (1967). On the matrix equation $\mathbf{Ax} = \lambda \mathbf{Bx}$. *J. Math. Anal. Appl.* **17**, 119–132.

Erlang, A.K. (1909). The theory of probabilities and telephone conversations. *Nyt. Tidskr. Mat. B,* **20**, 33–41.

Esquivel, M.L., Fernandes, J.M., Guerreiro, G.R. (2014). On the evolution and asymptotic analysis of open Markov populations: Application to consumption credit. *Stoch. Model.* **30**, 365–389.

Esquivel, M.L., Guerreiro, G.R., Fernandes, J.M. (2017). Open Markov chain sheme models fed by second order stationary and non stationary processes. *REVSTAT-Statistical Journal,* **15**, 277.

Esquivel, M.L., Guerreiro, G.R., Oliveira, M.C., Real, P.C. (2021a). Calibration of Transition Intensities for aMultistate Model: Application to long-term care. *Risks* **9**, 2.

Esquivel, M.L., Krasil, N.P., Guerreiro, G.R.(2021b). Open type population models: From discrete to continuous time. *Mathematics,* **9**, 13.

Evans, S.N., Ralph, P.L. and Sen, A. (2013). Stochastic population growth in spatially heterogeneous enviroments. *J. Math. Biology* **66**, 3, 423–476.

Ewens, W.J. (1979). *Mathematical Population Genetics.* Springer-Verlag. Berlin.

Faddy, M. and McClean, S. (1999). Analysing data on lengths of stay of hospital patients using phase-type distributions. *Appl Stoch Models and Data Anal.* **15**, 311–317.

Faddy, M. and McClean, S. (2005). Markov chain modeling for geriatric patient care. *Methods Archiv* **44**, 3, 369–373.

Feller, W. (1968). *An introduction to Probability and its Applications,* Vol. 1. New York: Wiley.

Feller, W. (1971). *An introduction to Probability and its Applications,* Vol. 2. New York: Wiley.

Feichtinger, G. (1976). On the generalization of stable age distribution to Gani-type person flow models. *Adv. Appl. Prob.* **8**, 433–445.

Filinkov, A., Richmond, M., Nicholson, R. (2011). Modelling personnel sustainability: a tool for military force structure analysis. *Journal of the Operational Research Society* **62**, 8, 1485–1497.

Floriani, E., Lima, R., Ourrad, O., Spinelli, L. (2016). Flux through a Markov chain. *Chaos Solitons Fractals* **93**, 136–146.

Forbes, A.F. (1971). Markov chain models for manpower systems. In *Manpower and Management Science,* ed. D.J. Bartholomew and A.R. Smith, English University Press, London, pp. 93–113.

Funderlic, R. and Meyer, C.D. (1986). Sensitivity of the stationary distribution vector for an egodic Markov chain. *Linear Algebra and its Applications. II,* **76**, 1–17.

Gani, J. (1963). Formula for projecting enrollments and degrees awarded in Universities. *J. R. Stat. Soc. A* **126**, 400–409.

Gantmacher, F.R. (1959a). *Applications of the Theory of Matrices.* Interscience. New York.

Gantmacher, F.R. (1959b). *The Theory of Matrices. Vol 1.* Chelsea. New York.

Gantmacher, F.R. (1959c). *The Theory of Matrices. Vol 2.* Chelsea. New York.

Garg, L., McClean, S.I., Meenan, B., Millard, P. (2010). A non-homogeneous discrete time Markov model for admission scheduling and resource planning in a cost capacity constraint healthcare system. *Health care Management Science* **13,** 2, 155-169.

Garg, L., McClean, S.I., Meenan, B., Millard, P. (2011). Phase-Type survival trees and mixed distribution survival trees for clustering patient's hospital length of stay. *Informatica* **22**, 57–72.

Garg, L., McClean, S.I., Barton, M., Meenan, B., Fullerton, K. (2011). Intelligent patient management and resource planning for complex, heterogeneous and stochastic healthcare systems. *IEEE Transactions on systems man, and cybernetics-Part A: Systems and Humans* **42**, 6.

Garg, L., McClean, S.I., Meenan, B., Millard, P. (2009). Non-homogeneousMarkov models for sequential pattern mining of healthcare data. *IMA Journal of Management Mathematics,* **20**, 327–344.

Gass, S.I. (1991). Military manpower planning models. *Computer and Operations Research,* **18**, 1, 65–73.

Georgiou, A.C. (1989). *Control of the asymptotic behavior of non-homogeneous Markov systems.* Ph.D thesis, Mathematics Department, Aristotle University of Thessaloniki.

Georgiou, A.C. and Vassiliou, P.-C.G. (1992). Periodicity of asymptotically attainable structures in Non-homogeneous Markov systems. *Linear Algebra and its applications* **176,** 137–174.

Georgiou, A.C. (1992). Partial maintainability and control in non-hmogeneous Markov manpower systems. *European Journal of Operational Research,* **62**, 241–251.

Georgiou, A.C. and Tsantas, N. (1996). Nonstationary Cyclic behavior in Markov systems. *Linear Algebra and its Applications* **237/238:** 549–578.

Georgiou, A.C. and Vassiliou, P.-C.G. (1997). Cost models in nonhomogeneous Markov systems. *European Journal of Operational Research,* **100**, 81–96.

Georgiou, A.C. (1999). Aspirations and priorities in a three phase approach of a nonhomogeneous Markov system. *European Journal of Operational Research,* **116**, 3, 565–583.

Georgiou, A.C. and Tsantas, N. (2002). Modelling recruitment training in mathematical human resource planning. *Applied Stochastic Models in Business and Industry* **18**, 1, 53–74.

Georgiou, A.C., Thanassoulis, E., and Papadopoulou, A. (2022). Using data envelopment analysis in markovian decision making. *European Journal of Operational Research,* **298**, 276–292.

Gerontidis, I. (1990). On certain aspects of non-homogeneous Markov systems in continuous time. *J. Appl. Prob.* **27**, 530–544.

Giplin, M.E. and Hanski, I. (1991). *Metapopulation Dynamics.* Academic Press, New York.

Grimmett, G. and Stirzaker, D. (2001). *Probability and Random Processes,* third edition, Cambridge University Press, Cambridge, UK.

Grabill, F.A. (1983). *Matrices with Applications to Statistics.* 2d ed. Wadsworth, Belmont, California.

Grinold, R.D. and Stanford, R.E. (1974). Optimal control od a graded manpower system. *Man. Sc.* **20**, 1201–1216.

Goodman, G.S. (1970). An intrinsic time for non-stationary Markov chains. *Z. Wahrscheinlichkeitsth,* **16**, 165–180.

Guerreiro, G., Mexia, J. (2004). An alternative approach to bonus malus. *Discassiones Mathematicae Probability and Statistics,* **24**, 197–213.

Guerreiro, G., Mexia, J., Miguens, M. (2010) A model for open populations subject to periodical re-classifications. *J. Stat. Theory Pract.* **4**, 303–321.

Guerreiro, G., Mexia, J., Miguens, M. (2013). Statistica approach for bonus malus. *ASTIN Bulletin,* 1–21.

Guerry, M.A. (1997). Properties of calculated predictions of grade sizes and the associated integer valued vectors. *J. Appl. Prob.* **34,** 94–100.

Guerry, M.A. (2008). On the evolution of stock vectors in a deterministic integer-valued Markov system. *Linear Algebra and its Applications.* **429**, 1944–1953.

Guerry, M.A. (2014). Some results on the embeddable problem for discrete time Markov models in manpower planning. *Communications in Statistics-Theory and Methods* **43**: 1575–1584.

Gyllenberg, M. and Silvestrov, D.S. (1994). Quasi-stationary distributions of a stochastic metapopulation model. *J. Math. Biol.* **33**, 35–70.

Gyllenberg, M., Hognas, H. and Koski, T. (1994). Population models with enviromental stochasticity. *J. Math. Biol.* **32**, 93-108.

Hajnal, J. (1956). The ergotic properties of non-homogeneous finite Markov chains. *Proc. Cambridge Philosoph. Soc.* **52**, 67–77.

Hajnal, J. (1958). Weak ergodicity in nonhomogeneous Markov chains. *Proc. Cambridge Philos. Soc.* **54**, 233–246.

Hajnal, J. (1976). On products of non-negative matrices. *Math. Proc. Cambridge Phil. Soc.* ***79,*** 521–30.

Hajnal, J. (1993). Shuffling with two matrices. *Contempor. Math.* **149**, 271–287.

Hall, P. and Heyde, C.C. (1980). *Martingale limit theory and its applications.* Academic Press. New York.

Hamilton, J.D. (1994). *Time series analysis.* Prinston University Press.

Harris, T.E. (1956). The existence of stationary measures for certain Markov processes. In *Proceedings of the 3^d Berkley Simposium on Mathematical Statistics and Probability, V2, pages 113–124*. University of California Press, Berkley.

Hartfiel, D.J. (1974). On infinite products of non-negative matrices. *SIAM J. Appl. Math.* **26**: 297–301.

Hartfiel, D.J. (1991). Sequential limits in Markov set chains. *Journal of Applied Probability* **28**, (4) 910–913.

Hartfiel, D.J. (1993). Results on limiting sets of Markov set chains. *Linear Algebra and its Applications* **105**, 155–163

Hartfiel, D.J. (1994). Homogeneous Markov chains with bounded transition matrix. *Stoch. Proc. Appl.* **50**, 275–279.

Hartfiel, D.J., Seneta, E. (1994). On the theory of Markov set-chains. *Advances in Applied Probability* **26**, 947–964.

Hartfiel, D.J. (1995). Intermediate Markov systems. *Applied Mathematics and Computation* **72**, 51–59.

Hartfiel, D.J. and Rothblum, U. (1998). Convergence of inhomogeneous products of matrices and coefficients of ergodicity. *Linear Algebra and its Applications.* **277**, 1–9.

Hartfiel, D.J. (1998). *Markov Set-Chains.* Lecture Notes in Mathematics 1695. Springer.

Hartfiel, D.J. (2002). *Nonhomogeneous Matrix Products.* World Scientific. London.

Haviv, M. and Van der Heyden, L. (1984). Perturbation bounds for the stationary probabilities of a finite Markov chain. *Adv. Appl. Prob.* 16, 804-818.

Hening, A. and Tran, K.Q. (2020). Harvesting and seeding of stochastic populations: analysis and numerical approximation. *J. Math. Biol.* **81**, 65–112.

Hening, A. and Nguyen, D. (2018). Coexistence and exctinction for Kolmogorov systems. Ann Appl Prob **28**, 3, 1893–1942.

Hening, A. (2021). Coexistence, Extinction and Optimal Haevesting in discrete-time stochastic population models. *Journal of Nonlinear Science,* **31**, 1.

Henry, N.W., McGinnis, R., and Tegtmeger, H.W. (1971). A finite model for mobility. *J. Math. Sociology,* **1**, 107–118.

Hilker, F.M. and Liz, E. (2019). Proportional threshold harvesting in discrete-time population models. *J. Math. Biol.* **79**, 5, 1927–1951.

Hodge, R.W. (1966). Occupational mobility as a probability process. *Demography,* **3**, 19–34.

Horn, R.A. and Johnson, C.J. (1985). *Matrix Analysis.* Cambridge University Press.

Horn, R.A. and Johnson, C.J. (1991). *Topics in Matrix Analysis.* Cambridge University Press.

Horn, R.A. and Johnson,C.J. (2013). *Matrix Analysis.*Second Ed. Cambridge University Press.

Howard, R.A. (1971). *Dynamic Probabilistic Systems. Vol.1. Markov Models.* John Wiley.

Howard, R.A. (1971). *Dynamic Probabilistic Systems.* Vol. 2. John Wiley.

Huang, C., Isaacson, D. and Vinograde, B. (1976) The rate of convergence of certain nonhomogeneous Markov chains. *Zeitsh. Wahrsch. and Gebiete,* **35**, 141–46.

Huang, C. and Isaacson, D. (1976). Ergodicity using mean visit times. *J. London Math. Soc.* (2), **14** 570–576.

Hyslop, J.M. (1965). *Infinite series.* Oliver and Boyd. Edinburgh.

Ignizio, J.P. (1982). Linear programming in single and multiple objective systems. Prentice-Hall, Englewood Cliffs, NJ.

Iosifescu, M. (1980). *Finite Markov Processes and Applications.* John Wiley.

Isaacson, D.L. and Madsen, R.W. (1976). *Markov Chains Theory and Applications.* John Wiley.

Isaacson, D.L. (1979). A characterization of geometric ergodicity. *Zeitsh. Wahrsch. and Gebiete* **49**, 267–273.

Isaacson, D.L. and Seneta, E. (1982). Ergodicity for countable inhomogeneous Markov chains. *Linear Algebra and its Applications.* **48**, 37–44.

Jacod, J. and Protter, P. (2004). *Probability Essentials,* Springer-Verlag, Berlin.

Johnson, N.I. and Kotz, S. (1969). *Distributions in Statistics. Discrete Distributions.* Wiley, New York.

Johnson, J.T. (1984). Ergodic properties of nonhomogeneous, continuous-time Markov chains. Ph.d thesis. Iowa State University.

Johnson, J.T. and Isaacson, D. (1988). Conditions for strong ergodicity using intensity matrices. *J. Appl. Prob.* **25**, 34–42.

Jones, B., McClean, S. and Stanford, D. (2019). Modelling mortality and discharge of hospiltalized stroke patients using a phase- type recovery model. *Health Care Management Science* **22**, 570–588.

Kac, M. (1947). On the notion of recurrence in discrete stochastic processes. *Bull. Amer. Math. Soc.,* 53, 1002–1010.

Kalamatianou, A.G. (1987). Attainable and Maintainable structures in Markov manpower systems with pressure in the grades. *Journal of the Operational Research Society,* **38**, 183–190.

Kalamatianou, A.G. (1988). A model for responding to promotion blockages. *Journal of Appl.Prob.* **25**, 268–278.

Kalbfleisch, J.D. and Prentice, R.L. (2002). *The statistical analysis of failure time data.* 2nd edn. John Wiley, New York.

Kallenberg, O. (2002). *Foundations of Modern Probability,* second edition, Springer-Verlag, Berlin.

Kang, Y., Li, Z, Dong, Y. and Xi, H. (2012). Markovian based fault-tolerant control for wheeled mobile manipulators. *IEEE Trans on Control System Technology,* **20**, 266–276.

Karlin, S. and Taylor, S.J. (1975). *A First Course in Stochastic Process* (2nd edn.). Academic Press, New York.

Kasymov, A.G. (1988). Abundance of zooplankton and zoobenthos in Baku Bay, Caspian Sea. *Oceanology* **28**, 524–526.

Keenay, G.A, Morgan, R.W. and Ray, K.H. (1975). Convergence properties of age distributions. *J. Appl. Prob.* **12**, 684–691.

Kemeny, J.G. and Snell, J.L. (1976). *Finite Markov Chains.* Springer.

Kemeny, J.G. and Snell, J.L. (1976). *Finite Markov Chains.* Springer.

Kemeny, J.G., Snell, J.L. and Knapp A.W. (1976). *Denumerable Markov Chains.* Springer.

Kijima, M. (1997). *Markov Process for Stochastic Modeling.* Chapman and Hall.

Kirkland, S., Neumann, M. and Shader, B. (1998). Applications of Paz's inequality to pertubation bounds for Markov chains. *Linear Algebra and ita Applications.* 268, 183–196.

Kingman, J.F.C. and Williams, D (1973). The combinatorial structure of nonhomogeneous Markov chains. *Z. Wahrscheinlichkeitsth,* **26**, 77–86.

Kingman, J.F.C. (2009). The first Erlang century-and the next. *Queueing Syst.* **63**, 3–12.

Kipouridis, I. and Tsaklidis, G. (2001). The size order of the state vector of discrete-time homogeneous Markov systems. *J. Appl. Prob.* **38**, 357–368.

Kipouridis, I. and Tsaklidis, G. (2001b). The size order of the state vector of continuous-time homogeneous Markov system with fixed size. *J. Appl. Prob.* **38**, 635–646.

Kolmogorov, A.N. (1931). Über die analytichen Methoden in der Wahrscheinlichkeitscrechnung. *Math. Ann.* **104,** 415–458.

Konta, S.L. and Stewman, S. (1980). An opportunity labor demand model and Markovian labor supply models-Comparative tests in an organization. *American Sociological Review,* **45**, 2, 276–301.

Kontoyannis, I. and Meyn, S. (2003). Spectral theory and limit theorems for geometrically ergodic Markov processes. *Ann. Appl. Probab.,* 13, 304–362.

Kontoyannis, I. and Meyn, S. (2005). Large deviation asymptotics and the spectral theory of multiplicative regular Markov processes. *Electron. J. Probab.,* **10,** 3, 61–123.

Kontoyannis, I. and Meyn, S. (2012). Geometric ergodicity and the spectral gap of non-reversible Markov chains. *Probab. Theory Relat. Fields.* **154,** 327–339.

Kopp, P.E. (1984). *Martingales and Stochastic Intergals.* Cambridge University Press, Cambridge.

Kühne, R. and Rhodius, A. (1990). A characterization of Dobrushin's coefficient of ergodicity. *Z. Anal. Anwendungen* **9**, 187–188.

Lancaster, P. (1969). Theory of Matrices. Academic Press. London.

Lancaster, P. and Tismentesky M. (1985). *The Theory of Matrices with applications.* 2d ed. Academic Press. New York.

Larson, R.E. and Casti, J.L. (1982). *Principles of Dynamic Programming,* Part II. Marcel Dekker. New York.

Lee, S.M. (1976). *Goal Programming for Decision Analysis.* Auerbach, Philadelphia.

Lee, E.T. (1992). *Statistical Methods for Survival Data Analysis.* 2nd edn. John Wiley. New York.

Leeson, G.W. (1981). Markov models of 2-hierarchically graded manpower systems. *European Journal of Operations Research* **31**, 3, 247–256.

Leizarowitz, A. (1992). On infinite products of stochastic matrices. *Linear Algebra and its Applications.* **168**, 189–219.

Leslie, P.H. (1945). On the use of matrices in certain populations mathematics. *Biometrika* **33**, 183–212.

Leslie, P.H. (1948). Some further notes on the use of matrices in certain populations mathematics. *Biometrika* **35**, 213–245.

Lešanovský, A. (1990). Coefficients of ergodicity generated by non-symetrical vector norms. *Czechoslovak Mathematical Journal,* 40, 115.

Levy, P. (1955). Wolfgang Doeblin (V. Doblin) Rev. Histoire Sci., pp. 107–115.

Lewy, P. and Nielsen, A. (2003). Moddelling stochastic fish stock dynamics using Markov Chain Monte Carlo. *ICES Journal of Marine Science,* **60**, 743-752.

Lindvall, T. (1991). W. Doeblin 1915-1940. *Ann. Prob.* **19,** 929–934.

Lindley, J.A. (1987). Continuous plankton records, the geographicl distribution and seasonal cycles of decapod crustacean larrae and pelagic post larrae in the north eastern Atlandic Ocean and the North Sea. *J. Mar. Biol. Assoc.* **67**, 145–147.

Liu, Y., Yin.,Y., Teo, K.L., Wang, S., Liu, F. (2019). Probabilistic control of Markov Jump Systems by scenario optimization approach. *Journal of Industrial and Management Optimization* **15**, 3, 1447–1453.

Loumponias, K. and Tsaklidis, G. (2017). Interpretation of the evolution of the homogeneous Markov system (or equivalently, of the embedded Markov chain) as a deformation of a viscoelastic medium. The 3-D case. *Methodology and Computing in Applied Probability* **19**, 4, 1227–1239.

Lundberg, F. (1903). Approximerad framstalling avsannolikhetsfunktionen. Aterfor saking av Kollektivrisker. Thesis Uppsala.

Luan, X., Liu, F., Shi, P., (2011). Observer based finite time stabilization for extended Markov jump systems. *Asian Journal of Control,* **13**, 925–935.

Luan, X., Zhao, S., Shi, P., Liu, F. (2014). H_∞ filtering for discrete time Markov jump systems with unknown transition probabilities. *International Journal of Adaptive Control and Signal Processing,* **28**, 138–148.

MacDonald, I.L. and Zucchini, W. (1997). *Hidden Markov and other models for discrete-valued time series.* Chapman and Hall, London.

Markov, A.A. (1907). Investigation of an important case of dependent trials. (In Russian). *Isvestya Akad Nauk.* S.P.B. **6**, 61–80.

Marshall, A.H., McClean, S.I., Shapcott, C.M., Millard, P. (2002). Modelling patient duration of stay to facilitate resource management of geriatric hospitals. *Health Care Management Science,* **5,** 3, 313–319.

Marshall, A.H. and McClean, S.I. (2003). Conditional phase type distributionsfor modelling patient length of stay in hospital. *Int Trans Op Res* **10**, 565–576.

Marshall, A.H. and McClean, S.I. (2004). Using Coxian Phase type distributions to identify patient characteristics for duration of stay in hospital. *Health Care Management Science* **7**, 285–289.

Marshall, A.H., Shaw, B. and McClean, S.I. (2004). Estimating the costs for a group of geriatric patients using the Coxian phase-type distribution. *Statistics in Medicine* **26**, 2716–2729.

Mayer, T. (1972). *Models of intragenerational mobility.* Houghton Mifflin, Boston.

McClean, S.I. (1976). A continuous-time population model with Poisson recruitment. *J. Appl. Prob.* **13**, 348–354.

McClean, S.I. (1978). Continuous-time stochastic models for multigrade population. *J. Appl. Prob.* **15**, 26-37.

McClean, S.I. (1980). A semi-Markov model for multigrade population with Poisson recruitment. *J. Appl. Prob.* **17**, 846–852.

McClean, S.I. and Gribbin, J.O. (1987). Estimation for incomplete manpower data. *Appl. Stoch. Models Data Anal.* **3**, 13–25.

McClean, S.I. and Millard, P. (1998a). A three compartment model of the patient flows in a geriatric department: a decision support approach. *Health care Management Science* **7**, 285–289.

McClean, S.I., McAlea,B. and Millard, P. (1998b). Using a Markov reward model to estimate spend-down costs for geriatric department. *J Oper Res Soc* **49**, 1021–1025.

McClean, S.I. (1991). Manpower planning models and their estimation. *European Journal of Operational Research* **51**, 179–187.

McClean, S.I. and Gribbin, J.O. (1991). A non-parametric competing risks model for manpower planning. *Appl. Stoch. Models Data Anal.* **7**, 327–341.

McClean, S.I., Papadopoulou, A.A., Tsaklidis, G. (2004). Discrete time reward models for homogeneous semi-Markov systems. *Communication in statistics Theory and Methods* **33**, 3, 623–638.

McClean, S.I. and Millard, P. (2007). Where to treat the older patient? Can Markov models help us better understand the relationship between hospital and community care? *J. Oper. Res. Soc.* **58**, 2, 255–261.

McClean, S.I., Barton, M., Garg, L., Fullerton, K. (2011). A modeling framework that combines Markov models and discrete simulation for stroke patient care. *ACM Transactions on Modelling and Computer Simulations.* **21,** 4, 25.

McClean, S.I., Gillespie, J., Scotney, B., Fullerton, K. (2012). Using phase type models to cost a cohort of stroke patients. 25th *IEEE International Symposium on Computer Based Medical Systems, IEEE Xplore.*

McClean, S.I., Gillespie, J., Garg, L., Barton, M., Scotney, B., Fullerton, K. (2014a). Using phase-type models to cost stroke patient care across health, social and community services. *Eur J Oper Res* **236,** 1, 190–199.

McClean, S.I., Garg, L., Fullerton, K. (2014b). Costing mixed Coxian phase type systems with Poisson arrivals. *Communication in statistics Theory and Methods* **43**, 7, 1437–1452.

McClean, S.I. (2021). Using Markov models to characterize and predict orocess target compliance. *Mathematics,* **9**, 1187.

McKean, H.P.Jr. (1969). Stochastic integrals. New York. Academic Press.

McVinish, R. and Pollet, P.K. (2012). The limiting behavior of a mainland-island metapopulation. *J. Math. Biol.* **64**, 775–801.

Mehlmann, A. (1977). A note on the limiting behavior of discrete-time Markovian manpower modela with inhomogeneous independent poisson input. *J. Appl. Prob.* **14**, 611–613.

Moran, P.J. and Grant, T.R. (1989). The effects of industrial pollution on the development and the succession of marine fouling communities. *Mar. Ecol.* **10**, 231–246.

Morgan, R.W. (1979). Some models for hierarchical manpower systems. *Journal of the Operational Research Society* **10**, 260–269.

Myers, G.C., McGinnis, R., and Masnick (1967). The duration of residence approach to a dynamic stochastic model of internal migration: a test of the axiom of the cummulative inertia. *Eugenics Quartely,* **14**, 122–126.

Meyer, C.D. (1975). The role of the group generalized inverse in the theory of finite Markov chains. *SIAM Review,* **17,** 3, 443–464.

Meyer, C.D. (1980). The condition of a finite Markov chain and perturbation bounds for the limiting probabilities. *SIAM J. Algebraic Discrete Methods* **1,** 273–283.

Meyer, C.D. and Rose, N.J. (1977). The index and the Drazin inverse of block triangular matrices. *SIAM J. Appl. Math.* **33,** 1–7.

Meyer, C.D. and Shoaf, J. (1980). Updating finite Markov chains by using techniques for group matrix inversion. *J. Statist. Comput. Simulation* **11**, 163–181.

Meyer, C.D. (2000). *Matrix Analysis and Applied Linear Algebra.* SIAM.

Meyn, S. and Tweedie, R.L. (1992). Stability of Markovian processes I: discrete time chains. *Adv. Appl. Prob.* 24, 542–574.

Meyn, S. and Tweedie, R.L. (2009). *Markov Chains and Stochastic Stability.* Cambridge University Press.

Minc, H. (1988). *Non-negative matrices.* J. Wiley. New York.

Mitrophanov, A.Y. (2003). Stability and exponential convergence of continuous time Markov chains. *J. Appl. Prob.* **40**, 970–979.

Mitrophanov, A.Y. (2005) Sensitivity and convergence of uniformly ergodic Markov chains. *J. Appl. Prob.* **42,** 1003–1014.

Moore, E.H. (1920). On the reciprocal of the general algebraic matrix. *Bull. Am. Math. Soc.,* **26**, 394–395.

Mott, J.L. (1957). Conditions for the ergodicity of non-homogeneous finite Markov chains. *Proc. Roy. Soc. of Edinburgh. Sec. A Mathemat. Phys. Sci.* **64**, 369–380.

Mott, J.L., Schneider, H. (1957). Matrix norms applied to weakly ergodic Markov chains. *Archiv. der Mathematik.* **8**, 331–333.

Musiella, M. and Rutkwoski, M. (2005). *Martingale Methods in Financial Modelling,* second edition, Springer-Verlag, Berlin.

Neumaier, A. (1990). *Interval methods for systems of equations.* Cambridge University Press. Cambridge.

Neumann, M. and Schneider, H. (1999). The convergence of general products of matrices and the weak ergodicity of Markov chains. *Linear Algebra and its Applications.* **287**, 307–314.

Nilakantan, K. and Raghavendra, B.G. (2005). Control aspects in proportionality Markov manpower systems. Applied Mathematical Modelling **29**, 1, 85–116.

Nikaido, H. (1968). *Convex structures and Economic Theory.* Academic Press, New York.

Noutsos, D. (2006). On Perron-Frobenius property of matrices having some negative entries. *Linear Algebra Appl.* **412**, 132–153.

Nummelin, E. (1984). *General Irreducible Markov chains and Nonnegative Operators.* Cambridge University Press, Cambridge

O'Cinneide (1993). Entrywise perturbation theory and error analysis for Markov chains. *Numer. Math.* **65**, 109–120.

O'Connor, J.J. and Robertson, E.F. Wolfgang Doeblin, *Mac Tutor History of Mathematics archive,* University of St Andrews.

Ondar, K.O. (1981). *The correspondence between A.A. Markov and A.A. Chuprov on the Theory of Probability and Mathematical Statistics.* Springer-Verlag, New York.

Orey, S. (1959). Recurrent Markov chains. *Pacific J. Math.,* **9**, 805–827.

Orey, S. (1971). *Limit Theorems for Markov chain transition probabilities.* Van Nostrand Reinhold, London.

Ousama, M.A., Van Utterbeeck, F., Van Acker, G., Guerry, M.A. (2020). Multidimensional military manpower planning based on a career path approach. *Operations Management Research* **13**, 3–4, 249–263.

Papadopoulou, A.A. and Vassiliou, P.-C.G. (1994). Asymptotic behavior of non-homogeneous Markov systems. *Linear Algebra and its Applications* **210**, 153–198.

Papadopoulou, A.A. (1997). Counting transitions entrance probabilities in non-homogemeous semi-Markov systems. *Applied Stochastic Models and Data Analysis* **13**, 3–4, 199–206.

Papadopoulou, A.A. and Vassiliou, P.-C.G. (1999). Continuous time non-homogeneous semi-Markov systems. *Semi-Markov models and Applications.* Springer Verlag. pp. 241–251.

Papadopoulou, A.A. (2004). Economic rewards in non-homogeneous Markov systems. *Communications in Statistics-Theory and Methods* **33**, 3, 681–696.

Papadopoulou, A.A., Tsaklidis, G.(2005). Some reward paths in semi Markov models with stochastic selection of the transition probabilities. *Methodology and Computing in Applied Probability* **9**, 3, 399–411.

Papadopoulou, A.A and Tsaklidis, G. (2007). Some reward paths in semi-Markov models with stochastic selection of the transition probabilities. *Methodology and Computing in Applied Probability,* **9**, (3), 399–411.

Papadopoulou, A.A., Tsaklidis, G., McClean, S, Garg, L. (2012). On the moments and the distribution of the cost of a semi-Markov model for Hralthcare systems. *Methodology and Computing in Applied Probability* **14**, 3, 717–737.

Papadopoulou, A.A. (2013) Some results on modelling biological sequences and web navigation with semi-Markov chain. *Communications in Statistics-Theory and Methods* **42**, 16, 2853–2871.

Papadopoulou, A.A. and Vassiliou, P.-C.G. (2014). On the variances and covariances of the duration state sizes of semi-Markov systems. *Communications in Statistics-Theory and Methods* **43**, 7, 1470–1483.

Parthasarathy, K.R. (1988). Obituary: Andrei Nikolaevich Kolmogorov. *J. Appl. Prob.* **25**, 445–450.

Patoucheas, P.D. and Stamou, (1993). Non-homogeneous Markovian models in ecological modeling a study of the zoobenthos dynamics in Thermaikos Gulf, Greece. *Ecological Modelling,* **66**, 197–215.

Pauwelyn, P.J. and Guerry, M.A. (2020). On perturbations of non-diagonizable stochastic matrices of order 3. *Statistics and Probability Letters,* **157**, 108633.

Pauwelyn, P.J. and Guerry, M.A. (2020). Perturbations of non-diagonizable stochastic matrices with preservation of spectral properties. *Linear and Multilinear Algebra.* DOI:10.1080/03081087.2021.1904813.

Paz (1971). *Introduction to Probabilistic Automata.* Academic Press.

Paz, A. and Reichaw, M. (1967). Ergodic theorems for sequences of infinite stochastic matrices. *Camb. Phil. Soc.* **63**, 777–784.

Pease, M.C. (1965). *Methods of Matrix Algebra.* Academic Press. New York and London.

Penrose, R.A. (1955). A generalized inverse of matrices. *Proc. Cambridge Philos. Soc.,* **51**, 406–413.

Pitman, J.W. (1974). Uniform rates of convergence for Markov chains transition probabilities. *Z. Warhrscleinlichkeitstheorie und Verw. Geb.* **29**, 193–227.

Pollard, J.H. (1966). On the use of the direct matrix product in analysing certain stochastic population models. *Biometrika,* **53**, 397–415.

Pollard, J.H. (1967). Hierarchical population models with Poisson recruitment. *J. Appl. Prob.* **4**, 209–213.

Pollard, J.H. (1973). *Stochastic processes and population growth.* Cambridge University Press.

Pollard, B.S. (2016). Open Markov processes: A compositional perspective on a non-equilibrium steady states in biology. *Entropy,* **18**, 140.

Pollard, B.S. (2017). Open Markov processes and reaction networks. Ph.D. Thesis, University of California, Riverside, CA, USA.

Prabbu, N.U. (2007). *Stochastic Processes, Basic Theory and its Applications.* World Scientific.

Prais, S.J. (1955). Measuring social mobility. *J. R. Statist. Soc.* A, **118**, 56–66.

Preparata, F.P. and Shamos, M.I. (1985). *Computational Geometry. An Introduction.* Springer

Price, W.L., Priskor, W.G. (1972). The application of goal programming to manpower planning. INFOR **10**, 221-231.

Powell, M.J.D. (1988). A tolerant algorithm for linearly constrained optimization calculations. Report DAMPT/1988/NA17. University of Cambridge.

Raftery, A.E. (1985). A model for high order Markov chains. *J. R. Stat. Soc. B* **47**, 528–539.

Raftery, A.E. and Tavare, S. (1994). Estimation and modelling repeated patterns in high order Markov chains with the mixture transition distribution model. *Appl. Stat. Sci.* **1**, 403-423.

Rao, C.R. and Mitra, S.K. (1971). *Generalized Inverse Matrices and its Applications,* John Wiley. New York.

Rao, C.R. (1973). *Linear Statistical Inference and Its Applications,* second edition, John Wiley. New York.

Reeves, G.R. and Reid, R.C. (1999). A military reserve manpower planning model. *Computers and Operations Research* **26**, 12, 1231-1242.

Rhodius, A. (1988). The maximal value for coefficients of ergodicity. *Stochastic Processes Appl.* **29**, 141-143.

Rhodius, A. (1993). On explicit forms for ergodicity coefficients. *Linear Algebra and its Applications* **194**, 71-83.

Rhodius, A. (1997). On the maximum of ergodicity coefficients, the Dobrusin ergodicity coefficient, and products of stochastic matrices. *Linear Algebra and its Applications* **253**, 141-154.

Roff, J.C., Middlebrook, K., and Evans, F. (1988). Long term variability in North Sea, zooplankton of the Northumber land coast: productivity of small corepods an analysis of trofic interactions. *J. Mar. Biol. Assoc.* **68**, 143-164.

Rogers, L.C.G. and Williams, D. (2000). *Diffusions Markov processes and martingales: Vol. 1, Foundations.* Cambridge University Press.

Rogers, L.C.G. and Williams, D. (2000). *Diffusions Markov processes and martingales: Vol. 2, Ito Calculus.* Cambridge University Press.

Ross, S.M. (1996). *Stochastic Process.* Second Edition. John Wiley. New York.

Rolski, T., Schmidilli, H., Schmidt, V., and Teugels, J. (1999). *Stochastic Processes for Insurance and Finance.* J. Wiley.

Rothblum, U.G. and Schneider, H. (1980). Chatacterizations of optimal scalings of matrices. *Mathematical Programming* 19, 121-136.

Rothblum, U.G. and Tan, C.P. (1985). Upper bounds on the maximum modulus of subdominant eigenvalues of nonnegative matrices. *Linear Algebra and its Applications* **66:** 45-86.

Rucker, R. (2005). *Infinite and the Mind.* Princeton University Press.

Rudin, W. (1987). *Real and Complex Analysis.* Academic Press.

Rutherford, E. and Geiger, H. (1908). An electrical method of counting the number of particles for radiactive substances. *Proc. Roy. Soc. A,* **81**, 141-161.

Salgado-Garcia, R. (2021). Open Markov chains: Cumulant dynamics fluctuations and correlations. *Entropy,* **23**, 256.

Sarymsakov, T.A. (1953). On the ergodic principle for inhomogeneous Markov chains [in Russian] *Doklady A.N. S.S.S.R.* **90**, 25–8.

Sarymsakov, T.A. (1956). On the theory of inhomogeneous Markov chains [In Russian]. *Doklady A.N. Uzbeck. S.S.R.,* **8,** 3–7.

Sarymsakov, T.A. (1958). On inhomogeneous Markov chains. *Dokl. Akad. Nauk. S.S.S.R.* **120**, 465–467.

Sarymsakov, T.A. (1961). Inhomogeneous Markov chains (in Russian). *Teo. Veroiat. Primenen.* **6**, 194-201.

Savage, L.J. (1954). *The foundations of Statistics.* Wiley, New York.

Schott, J.R. (2005). *Matrix Analysis for Statistics.* Second edition. John Wiley. New York.

Schervish, M.J. (1984). Algorithm AS 195. Multivariate normal probabilitieswith error bound. *Appl. Statist.* **33**, 81–94.

Schreiber, S.I., Benaim, M. and Atchade, K.A.S. (2011). Persistence in fluctuating enviroments. *J. Math. Biology* **62**, 5, 655–683.

Schweitzer, P. (1968). Perturbation theory and finite Markov chains. *J. Appl. Prob.* **5**, 410–413.

Scott, M., Arnold, B.C. and Isaacson, D.L. (1982). Strong ergodicity for continuous time non-homogeneous Markov chains. *J. Appl. Prob.* **19**, 692–694.

Shannon, G.P. (1995). Andrew Young. *Q. J. R. astr. Soc.* **36**, 465–466.

Searle, S.R. (1982). *Matrix Algebra Useful for Statistics.* John Wiley. New York.

Sefrozo, R. (2009). *Basics of Applied Stochastic Processes.* Springer.

Seneta, E. (1973). On the historical development of the theory of finite inhomogeneous Markov chains. *Proc. Camb. Phil. Soc.* **74,** 507–513.

Seneta, E. (1979). Coefficients of Ergodicity: Structure and Applications. *Adv. Appl. Prob.* **11**, 576–590.

Seneta, E. (1982). Review of the correspondence between A.A. Markov and A.A. Chuprov on the theory of probability and mathematical statistics. *Ann. Sci.* **39**, 614–617.

Seneta, E. (1984). Explicit forms for ergodicity coefficients and spectrum localization. *Linear Algebra and its Applications.* **268**, 183–196.

Seneta, E. (1988a). Sensitivity of the stationary distribution measured by ergodicity coefficient. Adv. Appl. Prob. **20**, 228–230.

Seneta, E. (1988b). Sensitivity to perturbation of the stationary distribution. Somerefinements. *Linear Algebra and its Applications,* **108**, 121–126.

Seneta, E. (1993). Sensitivity of finite Markov chains under perturbation. *Statistics and Probability Letters*, **17**, 163–168.

Seneta, E. (1996). Markov and the birth of chain dependence theory. *Int. Statist. Rev.* 64, 3, 255–263.

Seneta, E. (2014). Inhomogeneous Markov chains and Ergodicity Coefficients: John Hajnal (1924-2008).

Seneta, E. (1981). *Non-negative Matrices and Markov Chains.* Springer-Verlag, Berlin.

Shen, J. (2000a). A geometric approach to ergodic non-homogeneous Markov chains. *Notes in Pure and Applied Mathematics*, 344–366.

Shen, J. (2000b). Compatification of a set of matrices with convergent infinite products. *Linear Algebra and its Applications.* **311**, 177–186.

Sheynin, O.B. (1988). A.A. Markov's work on probability. *Arch. His. Exact Sci.* **39**, 337–377.

Shreve, S.E. (2004). *Stochastic Calculus for Finance II, Continuous-Time Models,* Springer-Verlag, Berlin.

Sierksma, G. (1999). Limiting polytopes and periodic Markov chains. *Linear Multilinear Algebra* 46, 281–298.

Simmons, G.F. (1963). *Topology and Modern Analysis.* McGraw-Hill. New York.

Skulj, D., Vehovar, V. and Stamfelj, D. (2008). The modelling of manpower planning by Markov chains-A case study of the Slovenian armed forces. *Informatica* **32**, 289–291.

Smith, D. and Keyfitz, N. (1977). *Mathematical Demography. Selected Papers.* Springer-Verlag. Berlin.

Smith, A.R. and Bartholomew, D.J. (1988). Manpower planning in the United Kingdom: An Historical review. *J. Opl Res. Soc.* **39**, 3, 235–248.

Snell, J.L. (1997). A conversation with Joe Doob. *Statist. Sci.* **12**, 301–311.

Snell, J.L. (2005). Obituary: Joseph Leonard Doob. *Journal of Applied Probability,* **42**, 1, 247–256.

Stadje, W. (1999). Stationarity of a stochastic population model. *J. Appl. Probab.* **36**, 291–294.

Stavropoulos, N., Papadopoulou, A. and Kolias, P. (2021). Evaluating the efficiency of off-ball screens in elite basketball teams via second order Markov modelling. *Mathematics,* **9**, 1991.

Stewman, S. (1975). Application of job vacancy chain model to a civil service internal labor market. *Mathematical Sociology,* **4**, 37–59.

Stewman, S. (1976a). Markov models of occupational mobility. Theoretical development and empirical support. Part 1: Careers. *Journal of Mathematical Sociology,* **4**, 201–245.

Stewman, S. (1976b). Markov models of occupational mobility. Theoretical development and empirical support. Part 2: Contiuously operative job systems. *Journal of Mathematical Sociology,* **4**, 247–278.

Stewman, S. and Konda, S.L. (1983). Careers and organizational labour-markets-demographic-models of organizational-behavior. *American Journal of Sociology,* **88**, 4, 637–685.

Stewman, S. (1986). Demographic models of internal labor markets. *Administrative Science Quarterly,* **31**, 2, 212–247.

Stewman, S. (1988). Organizational Demography. *Annual Review of Sociology,* **14**, 173–202.

Symeonaki, M., Stamou G.B. and Tzafestas, S.G. (2002). Fuzzy Non-Homogeneous Markov Systems. *Applied Intelligence* **17**, 203–214.

Symeonaki, M., Stamou G.B. (2004). Theory of Markov systems with fuzzy states. *Fuzzy sets and systems,* **143**, 427–445.

Symeonaki, M., Stamou G.B. (2006). Rate of convergence, asymptotically attainable structures and sensitivity in non-homogeneous Markov systems with fuzzy states. *Fuzzy sets and systems,* **157**, 143–159.

Symeonaki, M. and Vassiliou, P.-C.G (2007). Periodicity of the perturbed non-homogeneous Markov system. In *Recent advances in stochastic modelling and data analysis.* C Skiadas (Ed.) 182–189. World Sci. Publ. Hackensack, N.J.

Symeonaki, M. and Stamatopoulou, G. (2014). Exploring the transitions to higher education in Greece: Issues of intergenerational educational mobility. *Policy features in education* **12**, 5, 681–694.

Symeonaki, M. (2015). Theory of fuzzy non homogeneous Markov systems with fuzzy states. *Qual. Quant.,* **49**, 2369–2385.

Symeonaki, M. and Filopoulou, C. (2017). Quantifying gender distances in education, occupation and employment. *Equality, Diversity and Inclusion: An International Journal* **36**, 4, 340–361.

Symeonaki, M., Parsanoglou, D. and Stamatopoulou, G. (2019). The evolution of Early Job Insecurity in Europe. *Sage Open* **9**, 2, 215.

Tan, C.P. (1982). A functional form of a particular coefficient of ergodicity. *J. Appl. Prob.* 19, 858–863.

Taqqu, M.S. (2001). Bachelier and his times: a conversation with Bernard Bru. *Finance and Stochastics,* **5**, 1, 3–32.

Tarazaga, P., Raydan, M., Hurman, A. (2001). Perron-Frobenius theorem for matrices with some negative entries, *Linear Algebra Appl.* **328**, 57–68.

Taylor, S.J. and Karlin, S. (1998). *An Introduction to Stochastic Modelling.* Academic Press.

Taylor, G.J., McClean, S.I., Millard, P. (2000). Stochastic models of geriatric patient bed occupancy behavior. *J. R. Stat Soc A 163 Part 1*, 39–48.

Tsaklidis, G. and Vassiliou, P.-C.G. (1988). Asymptotic periodicity of the variances and covariances of the state sizes in non-homogeneous Markov systems. *J. Appl. Prob.* **25**, 21–33.

Tsaklidis, G. and Vassiliou, P.-C.G. (1989). Periodicity of infinite products of matrices with some negative elements and row sums equal to one. *Linear Algebra Appl.* **176**, 175–196.

Tsaklidis, G. and Vassiliou, P.-C.G. (1992). V-matrices as covariance matrices in homogeneous Markov systems, in M.J.C. Gover, S. Barnett (Eds). *Applications of Matrix Theory,* Oxford Science Publications.

Tsaklidis, G. and Vassiliou, P.-C.G. (1992b). Periodicity of infinite products of matrices with some negative elements and row sums equal to one. *Linear Algebra Appl.* **176**, 175–196.

Tsaklidis, G. (1994). The evolution of the attainable structures of a homogeneous Markov system with fixed size. *J. Appl. Prob.* **31**, 348–361.

Tsaklidis, G. (1996). The evolution of the attainable structures of a continuous time homogeneous Markov system with fixed size. *J. Appl. Prob.* **33**, 34–47.

Tsaklidis, G.M. (1997). Continuous-time homogeneous Markov system with fixed size as a Newtonian fluid. *Applied Stochastic models and Data Analysis* **13**, 3–4, 177–182.

Tsaklidis, G.M. (1999). The stress tensor and the energy of a continuous time homogeneous Markov system with fixed size. *J. Appl. Probab.* **36**, 21–29.

Tsaklidis, G.M. and Soldatos, K.P. (1999). Modelling of continuous time homogenuous Markov system with fixed size as elastic solid: the two dimensional case. *Applied mathematical modelling* **27**, 11, 877–887.

Tsantas, N. and Vassiliou, P.-C.G. (1993). The non-homogeneous Markov system in a stochastic environment. *J. Appl. Prob.* **30**, 285–301.

Tsantas, N. and Georgiou, A.C. (1994). Periodicity of equilibrium structures in a time dependent Markov model under stochastic environment. *Applied Stochastic Models and Data Analysis,* **10**, 269–277.

Tsantas, N. (1995). Stochastic analysis of a non-homogeneous Markov system. *European Journal of Operational Research,* **85**, 670–685.

Tsantas, N. and Georgiou, A.C. (1998). Partial maintainability of a population model in a stochastic environment. *Applied Stochastic Models and Data Analysis,***13**, 183–189.

Tsantas, N. (2001). Ergodic behavior of a Markov chain model in a stochastic environment. *Mathematical Methods of Operations Research,* **54**, 101–117.

Tuominen. P. (1976). Notes on l-recurrent Markov chains. *Z. Warhrscleinlichkeitstheorie und Verw. Geb.*, 36, 111–118.

Tweedie, R.L. (1974a). R-theory for Markov chains on a general state space I: solidarity properties and R-recurrent chains. *Adv. Appl. Probab.* **2**, 840–864.

Tweedie, R.L. (1974b). R-theory for Markov chains on a general state space II: R-subinvariant measures for R-transient chains. *Adv. Appl. Probab.* **2**, 865–878.

Tweedie, R.L. (1975). Sufficient conditions for ergodicity and recurrence of Markov chains on a general state space. *Stoch. Proc. Applns.* **3**, 385–403.

Ugwuogo, F.I. and McClean, S.I.(2000). Modelling heterogeneity in manpower systems: a review. *Appl. Stoch. Models Bus. Ind.* **2**, 99–110.

Vajda, S. (1947). The stratified semi-stationary population. *Biometrika 34*, 3/4, 243–254.

Vajda, S. (1975). Mathematical aspects of manpower planning. *Operational Research Quarterly* **26**, 3, 527–542.

Vajda, S. (1978a). *Mathematical of manpower planning.* Wiley, Chichester.

Vajda, S. (1978b). Maintainability and preservation of a graded population structure. *Studies in Management Sciences,* **8**, 219–230.

Vasiliadis, G. and Tsaklidis, G. (2008). On the distribution of the state sizes of discrete time homogeneous Markov systems. *Method. Comput. Appl. Probab.* **10**, 55–71.

Vasiliadis, G. and Tsaklidis, G. (2009). On the distribution of the state sizes of closed continuous time homogeneous Markov systems. *Method. Comput. Appl. Probab.* **11**, 561–582.

Vasiliadis, G. (2012). On the distributions of the state sizes of the continuous time homogeneous Markov system with finite state capacities. *Methodol. Comput. Appl. Probab.* **14**, 863–882.

Vasiliadis, G. (2014). Transient analysis of the M/M/k/N/N queue using acontinuous time homogeneous Markov system with finite state size capacity. *Communications in Statistics, Theory and Methods* **43**, 1548–1562.

Vasiliadis, G. (2016). Transient analysis of a finite source discrete-time queueing system using homogeneous Markov system with state size capacities. *Communications in Statistics-Theory and Methods,* **45**, 1403–1423.

Vassiliou, P.-C.G. (1976). A Markov model for wastage in manpower systems. *Oper. Res. Quart.* **27**, 57–70.

Vassiliou, P.-C.G. (1978). A high order non-linear Markovian model for promotion in manpower systems. *J. R. Stat. Soc. A,* **141**, 86–94.

Vassiliou, P.-C.G. (1981a). On the limiting behavior of a non-homogeneous Markov model in manpower systems. *Biometrika* **68**, 557–561.

Vassiliou, P.-C.G. (1981b). Stability in a non-homogeneous Markov chain model in manpower systems. *J. Appl. Prob.* **18**,

Vassiliou, P.-C.G. (1981c). On the asymptotic behavior of age distributions in manpower systems. *J. Opl. Res. Soc.* **32**, 503–506.

Vassiliou, P.-C.G. (1982). Asymptotic behavior of Markov systems., *J. Appl. Prob.* **19**, 851–857.

Vassiliou, P.-C.G. (1982b). On the limiting behavior of a non-homogeneous Markovian manpower model with Poisson input. *J. Appl. Probab.* **19**, 433–438.

Vassiliou, P.-C.G. (1984). Cyclic behavior and asymptotic stability of non-homogeneous Markov systems. *J. Appl. Prob.* **21**, 315–325.

Vassiliou, P.-C.G. and Tsantas, N. (1984a). Stochastic Control in Non-Homogeneous Markov Systems. *Intern. J. Computer. Math.* **16**, 139–155.

Vassiliou, P.-C.G. and Tsantas, N. (1984b). Maintainability of structures in non-homogeneous Markov systems under cyclic behavior and input control. *SIAM J. Appl. Math.* **44**, 5, 1014–1022.

Vassiliou, P.-C.G. and Gerontidis, I. (1985). Variances and covariances of the grade sizes in manpower systems. *J. Appl. Prob.* **22**, 583–597.

Vassiliou, P.-C.G. (1986). Asymptotic variability of non-homogeneous Markov systems under cyclic behavior. *European J. Oper. R*es. **27**, 215–228.

Vassiliou, P.-C.G. and Tsaklidis, G. (1989). The rate of convergence of the vector of variances and covariances in Non-Homogeneous Markov Systems. *J. Appl. Prob.* **27**, 776–783.

Vassiliou, P.-C.G. and Georgiou, A.C. (1990). Asymptotically attainable structures in non-homogeneous Markov systems. *Operations Research* **38**, 5, 537–545.

Vassiliou, P.-C.G., Georgiou, A.C. and Tsantas, N. (1990). Control of asymptotic variability in non-homogeneous Markov systems. *J. Appl. Prob.* **28**, 756–766.

Vassiliou, P.-C.G. and Papadopoulou, A.A. (1992). Non-homogeneous semi-Markov systems and maintainability of the state sizes. *J. Appl. Probab.* **29**, 3, 519–534.

Vassiliou, P.-C.G. (1997). The evolution of the theory of non-homogeneous Markov systems. *Appl. Stoch. Models Data Anal.* **13**, 159–176.

Vassiliou, P.-C.G. and Symeonaki, M.A. (1998). The perturbed nonhomogeneous Markov system in continuous time. *Appl. Stochastic Models and Data Anal.* **13**, 207–216.

Vassiliou, P.-C.G. and Symeonaki, M.A. (1999). The perturbed nonhomogeneous Markov system. *Linear Algebra Its Appl.* **289**, 319–332.

Vassiliou, P.-C.G. and Tsaklidis, G. (2001). *Applied Matrix Theory.* Ziti Press. Thessaloniki.

Vassiliou, P.-C.G. and Tsakiridou, H.L. (2004). Asymptotic behavior of first passage probabilities in the perturbed non-homogeneous semi-Markov systems. *Communications in Statistics, Theory and Methods* **33**, 3, 651–679.

Vasileiou, A. and Vassiliou, P.-C.G. (2006). An inhomogeneous semi-Markov model for the term structure of credit risk spreads. *Adv. Appl. Probab.* **38**, 171–198.

Vassiliou, P.-C.G. and Moysiadis, T.P. (2010). Variability in $\mathcal{G}-$Inhomogeneous Markov systems of high order. International Workshop Applied Probability. Ed. $\mathcal{G}-$Inhomogeneous Markov systems of high order. Chiquet, J., Glaz, J., Limnios, N. and Moyal, P. Universite de Technologie de Compiegne. France.

Vassiliou, P.-C.G. and Moysiadis, T.P. (2010). $\mathcal{G}-$Inhomogeneous Markov systems of high order. *Methodology and Computing in Applied Probability,* **12**, 2, 271–292.

Vassiliou, P.-C.G. and Vasileiou, A. (2013a). Asymptotic behavior of the survival probabilities in an inhomogeneous semi-Markov model for the migration process in credit risk. *Linear Algebra Its Appl.* **438**, 2880–2903.

Vassiliou, P.-C.G. (2013b). Fuzzy semi-Markov migration process in credit risk. *Fuzzy Sets Syst.* **223**, 39–58.

Vassiliou, P.-C.G. (2013c). Exotic properties of non-homogeneous Markov and semi-Markov systems. *Communication in Statistics-Theory and Methods* **42**, 2971–2990.

Vassiliou, P.-C.G. (2014b). Semi-Markov migration process in a stochastic market in credit risk. *Linear Algebra Its Appl.* **450**, 13–43.

Vassiliou, P.-C.G. (2014). Markov systems in general state space. *Communications in Statistics: Theory and Methods,* **43**, 7, 1322-1339.

Vassiliou, P.-C.G. (2015). On the periodicity of non-homogeneous Markov chains and systems. *Linear Algebra and its Applications.* **471**, 654–684.

Vassiliou, P.-C.G. (2010). *Discrete-time Asset Pricing Models in Applied Stochastic Finance,* John Wiley, New York.

Vassiliou, P.-C.G. (2020). Laws of Large numbers for Non-Homogeneous Markov Systems. *Methodology and Computing in Applied Probability,* **22**, 4, 1631–1658.

Vassiliou, P.-C.G. (2020b) Rate of convergence and periodicity of the expected population structure of Markov systems that live in a general state space. *Mathematics* **8**, (6) 1021.

Vassiliou, P.-C.G. and Georgiou, A.C. (2021). *Markov and Semi-Markov Chains, Processes, Systems and Emerging Related Fields.* MDPI, Basel.

Vassiliou, P.-C.G. (2021a). Non-homogeneous semi-Markov and Markov renewal processes and change of measure in credit risk. *Mathematics* **9**, 55.

Vassiliou, P.-C.G. (2021b). Non-Homogeneous Markov set systems. *Mathematics* **9**, 471.

Vassiliou, P.-C.G. (2022a). Laws of Large Numbers for Non-Homogeneous Markov Systems with arbitrary transition probability matrices. *Journal of Statistical Theory and Practice,* **16**, 2, 1–26.

Vassiliou, P.-C.G. (2022b). Limiting distributions of a Non-Homogeneous Markov System in a stochastic environment in continuous time. *Mathematics,* 10 (8) 1214.

Ville, J. (1939). Etude critique de la notion de collectif. *Bull. Amer. Math. Soc.* 45, 824.

Xia, H., Wang, H., Xuan, Z. (2011). Opinion dynamics: A multidisciplinary review and perspective on future research. *International Journal of Knowledge and Systems Science,* **2,** 4, 72–91.

Wagon, S. (1994). *The Banach-Tarski Paradox,* Cambridge University Press.

Whitlle, P. (2000). *Probability via Expectation (4th edn). Springer, New York.*

Willard, S. (1970). *General Topology. Addison-Wesley.*

Williams, D. (2004). *Weighing the Odds,* Cambridge University Press, Cambridge, UK.

Williams, D. (2000). *Probability with Martingales.* Cambridge University Press, Cambridge, UK.

Yadavalli, V.S.S., Natarajan, R., Udayabhaskaran, S. (2002). Time dependent behavior of stochastic models of manpower system-Impact of pressure on promotion. *Stochastic Analysis and Applications,* **20**, 4, 863–882.

Yadavalli,V.S.S. Setlhare, K., Forcheh, N. (2006). Stochastic structures of graded size in manpower planning systems. Proceedings of the sixth IASTED International Conference on modelling, simulation, and optimization: Science and Technology for development in the 21st Century. 128–133.

Yin,Y. and Lin, Z. (2017). Constrained control of uncertain nonhomogeneous Markovian jump systems. *International Journal of Robust and Nonlinear Control,* **27**, 3937–3950.

Yin,Y., Lin, Z., Liu, Y., Teo K.L., (2018a). Event-triggered constrained control of positive systems with input saturation. *International Journal of Robust and Nonlinear Control,* **28**, 3532–3542.

Yin,Y., Liu, Y., Teo, K.L., and Wang, S. (2018b). Event triggered probabilistic robust control of linear systems with input constrainrs: By scenario optimization approach. *International Journal of Robust and Nonlinear Control,* **28**, 144–153.

Young, A. and Almond, G. (1961). Predicting distibutions of staff. *Comput. J.,* **3**, 246–250.

Young, A. (1965). Models for planning recruitment and promotion of staff. *Brit. J. Indust. Rel.,* **3**, 301–310.

Young, A. (1971). Demographic and ecological models for manpower planning. In *Aspects of Manpower Planning* (D.J. Bartholomew and B.R. Morris, eds). London: English University Press.

Young, A. and Vassiliou, P.-C.G. (1974). A non-linear model on the promotion of staff. *J. R. Stat. Soc. A,* **138**, 584–595.

Yang, W. (2002). Convergence in the Cesaro sence and strong law of large numbers for non-homogeneous Markov chains. *Linear Algebra and its Applications,* **354**, 275–288.

Yong, P.L. (1976). Some results related to q-bounded Markov processes. *Nanta Mathematica,* **8**, 34–41.

Wang, J. (2005). A review of Operations Research applications in workforce planning and potential modelling of military training. Australian Goverment, Department of Defence, *DSTO-TR-1688.*

Zais, M. and Zhang, D. (2016). A Markov chain model of military personnel dynamics. *International Journal of Production Research,* **54**, 6, 1863–1885.

Zanakis, S.H. and Maret, MW. (1981a). A Markovian goal programming approach to aggregate manpower planning. *Journal of the Operational Research society* **32**, 1143–1144.

Zanakis, S.H. and Maret, MW. (1981b). A note on aggregate manpower planning using a Markovian goal programming approach. *Journal of the Operational Research society* **32**, 55–63.

Zeifman, A.I. (1989). Quasi-ergodicity for non-homogeneous continuous time Markov chains. *J. Appl. Prob.* **26**, 643–648.

Zeifman, A.I. and Isaacson, D.L. (1994). On strong ergodicity for non-homogeneous continuous-time Markov chains. *Stochastic Processes and their Applications,* **50**, 263–273.

Zeifman, A.I. and Korolev, V.Yu. (2015). Two sided bounds on the rate of convergence for continuous-time finite inhomogeneous Markov chains. *Statistics and Probability Letters* **137**, 84–90.

Zeifman, A.I., Korolev, V.Yu., Satin, Ya.A., Kiseleva, K.M. (2018). Lower bounds for the rate of convergence for continuous-time inhomogeneous Markov chains with a finite state space. *Statistics and Probability Letters* **103**, 30–36.

Zhang, R., Tang, C., Ma, S., Yuan, H., Gao, L., Fan, W. (2011). Using Markov chains to analyze changes in wetland trends in arid Yinchuan Plain, China.

Zhao, J., Shen, H, Li, B., Wang, J. (2013). Finite time H_∞ control for a class of Markovian jump delayed systems with input saturation. *Non-Linear Dynamics,* **73**, 1099–1110.

INDEX

D